ANIMAL PHYSIOLOGY:

Adaptation and environment

FOURTH EDITION

ANIMAL PHYSIOLOGY:
Adaptation and environment

KNUT SCHMIDT-NIELSEN

James B. Duke Professor of Physiology, Department of Zoology, Duke University

CAMBRIDGE
UNIVERSITY PRESS

Published by the Press Syndicate of the University of Cambridge
The Pitt Building, Trumpington Street, Cambridge CB2 1RP
40 West 20th Street, New York, NY 10011-4211, USA
10 Stamford Road, Oakleigh, Melbourne 3166, Australia

First published 1975
Second edition 1979
Third edition 1983
Fourth edition 1990
Reprinted 1991, 1993, 1994

Printed in the United States of America

Library of Congress Cataloging-in-Publication Data is available

British Library Cataloguing in Publication applied for

ISBN 0-521-38196-7 hardback

Contents

CONTENTS

About this book

This book is about animals and how they function in their world. First of all it is about problems and their solutions. It is also about aspects of physiology that I myself happen to find particularly interesting. The book is written for students who want to know how things work, who want to know what animals do and how they do it.

The book deals with the familiar subjects of physiology: respiration, circulation, digestion, and so on. These subjects are arranged according to major environmental features: oxygen, food and energy, temperature, and water. This arrangement is important, for there is no way to be a good physiologist, or a good biologist for that matter, without understanding how living organisms function in their environment.

The book is elementary and the needed background is minimal. There are few demands on previous knowledge, but I have assumed that the student is familiar with a few simple concepts, no more than provided by a good high school text. I have, however, included in the text sufficient background information to make physiological principles understandable in terms of simple physics and chemistry.

The quantity and complexity of scientific information today are steadily increasing, and students are already overburdened with material to remember. However, the mere recital of more facts does not signify understanding; the student needs a framework of principles on which to hang the facts. This book should help the student discover that many problems can be understood, once a few fundamental principles are familiar.

I also feel that clear concepts are more important than the learning of technical terms. However, because concepts cannot be conveyed without words, terminology is necessary. But terms are of no use unless they are clearly and accurately defined.

Much of this book explores how animals can live in environments that seem to place insurmountable obstacles in their way. The book discusses possible approaches and solutions. Animals with anatomical and physiological specializations often contribute much to our understanding of general principles. However, unless we look for these general principles, comparative physiology is apt to become a description of functions peculiar to uncommon animals – uncommon not because they are rare, but because they are outside our daily experience with other humans and with well known pets and laboratory animals such as dogs, cats, rats, and frogs. What we want is to place information into general concepts that help us understand how all animals function.

The text contains references to the scientific literature. These are arranged at the end of each chapter to help interested students satisfy their curiosity without having to search for information that is often hard to come by. The vast amount of scientific information made it necessary to be highly selective, and opinions about the proper selection will differ.

To bring more advanced information within the reach of the reader, I have arranged a list of Additional Readings for each chapter, placed after the list of text references. To spare the student from a feeling of helplessness, I have made these lists short. The titles vary from brief and simple essays to large, comprehensive treatises. Except for a few classical works, these lists contain reasonably recent and up-to-date material.

Like most authors I hope that friendly and perhaps less friendly readers will let me know about errors and what I could have done better.

Preface to the
fourth edition

When I started revising this book, a friend warned me against the dinosaur syndrome – when they got too big, they died out. I have therefore tried to make the book better without substantially increasing its size. I have revised existing material and have made additions where there is essential new information. However, there is no change in the general outline or the nature of the book.

A novel feature of this edition is that each chapter now begins with a brief synopsis, outlining its essential features and the fundamental principles to be discussed.

The separate Chapters 1 and 2 in the third edition, which both dealt with respiration, have been combined in one chapter. The disadvantage is that the chapter now is much longer; the advantage is more unified treatment of the principles.

New figures have been added where recent information is helpful, and some old figures have been deleted. Nearly 200 new references have been added, both as text references and for additional reading. These are chosen with emphasis on recent developments, judiciously selected from the burgeoning scientific literature but of necessity leaving much important new information unlisted.

For those who teach physiology I will mention some of the subjects where new information has been added. These include: Gas exchange in eggs, including eggs of mound-building birds and dinosaur eggs; regulation of blood flow in open circulation systems; composition of mammalian milk; recent information about deep sea rift animals and their metabolism of hydrogen sulfide from hot springs; digestion of waxes in marine communities; energy cost and economy of different gaits in horses; how biological antifreezes act; nucleating agents and freeze tolerance in vertebrates; survival of small Arctic birds in winter cold and darkness; heater organs that keep the brain and eye of swordfish warm; how the perturbing effect of urea on enzyme function in the kidney is counteracted; new ideas about Henle's loop and achievement of high urine concentrations; revised treatment of endocrine function with information about natriuretic factor, cell surface receptors, diuretic hormone in blood-sucking insects, and formation of endogenous morphine; perception of surface waves and absolute pitch in *Xenopus;* hearing in fish; electrosensitive organs in mammals; perception of infrasound for communication among elephants.

These and many other new concepts have been incorporated without making the book appreciably longer.

What is physiology?

Physiology is about the functions of living organisms – how they eat, breathe, and move about, and what they do just to keep alive. To use more technical words, physiology is about food and feeding, digestion, respiration, transport of gases in the blood, circulation and function of the heart, excretion and kidney function, muscle and movements, and so on. The dead animal has the structures that carry out these functions; in the living animal the structures work.

Physiology is also about how the living organism adjusts to the adversities of the environment – obtains enough water to live or avoids too much water, escapes freezing to death or dying from excessive heat, moves about to find suitable surroundings, food, and mates – and how it obtains information about the environment through its senses. Finally, physiology is about the regulation of all these functions – how they are correlated and integrated into a smooth-functioning organism.

Physiology is not only a description of function; it also asks why and how. To understand how an animal functions, it is necessary to be familiar both with its structure and with some elementary physics and chemistry. For example, we cannot understand respiration unless we know about oxygen. Since ancient times breathing movements have been known as a sign of life or death, but the true meaning of respiration could not be understood until chemists had discovered oxygen.

The understanding of how living organisms function is helped enormously by using a comparative approach. By comparing different animals and examining how each has solved its problem of living within the constraints of the available environment, we gain insight into general principles that otherwise might remain obscure. No animal exists, or can exist, independently of an environment, and the animal that utilizes the resources of the environment must also be able to cope with the difficulties it presents. Thus, a comparative and environmental approach provides deeper insight into physiology.

Examining how an animal copes with its environment often tends to show what is good for the animal. This may bring us uncomfortably close to explanations that suggest evidence of purpose, or teleology, and many biologists consider this scientifically improper. However, we all do tend to ask why? or what good is it for the animal? Anyway, the animal has to survive, and there is nothing improper or unscientific in finding out how and why it succeeds. If it did not arrive at solutions to the problem of survival, it would no longer be around to be studied. And the study of the living organism is what physiology is all about.

PART ONE

OXYGEN

CHAPTER ONE

Respiration

Why is oxygen important? It is because most animals satisfy their energy requirement by oxidation of food materials, forming carbon dioxide and water in the process.

The process of oxygen uptake and release of carbon dioxide is called *respiration*. Aquatic animals take up oxygen from the small amount of this gas dissolved in the water, terrestrial animals from the abundant oxygen in the air.

Many small animals can take up sufficient oxygen through the general body surface, but most animals need special respiratory organs for oxygen uptake. Carbon dioxide follows the opposite path, being released from the general body surface or from the respiratory organs. The water formed in the oxidation processes merely enters the general pool of water in the body and presents no special problems.

The most important, and sometimes the only physical process in the movement of oxygen from the external medium to the cells is *diffusion*, a process in which a substance moves from a higher to a lower concentration. The movement of carbon dioxide in the opposite direction also follows the concentration gradients.

Diffusion may be aided by *bulk movement,* such as the movement of air in and out of the lungs, but concentration gradients remain as the fundamental driving force for moving the respiratory gases. To understand respiration it is therefore necessary to know about the respiratory gases, their solubility, and the physics of diffusion processes.

Life presumably originated in the sea, and most animals (except insects) are marine. Large-scale evolutionary adaptation to air breathing has occurred only among arthropods and vertebrates. Some snails are well adapted to terrestrial life, and a small number of other invertebrates live in various terrestrial microhabitats.

Easy access to oxygen in the atmosphere permits a high rate of metabolism and a high degree of organizational development. The greatest drawback to breathing in air is the evaporation of water.

THE ATMOSPHERE

Composition of dry atmospheric air

The physiologically most important gases are oxygen, carbon dioxide, and nitrogen. They are present in atmospheric air in the proportions shown in Table 1.1. In addition, the atmosphere contains water vapor in highly variable amounts.

What physiologists usually call nitrogen is actually a mixture of nitrogen with about 1% of the noble gases, and for accuracy these should be listed as well. However, in physiology it is customary to lump these gases with nitrogen, the main reason being that in most physiological processes, nitrogen and the noble gases are equally inert to the organism. Another reason is that the analysis of respiratory gases is usually carried out by determining oxygen and carbon dioxide values and calling the remainder ''nitrogen.'' To the physiologist the amount of ''nitrogen'' in air is, therefore, 78.09 + 0.93%, or 79.02%.

The nearly 1% argon is of physiological interest only in some quite special circumstances: for example, in connection with the secretion of gases into the swimbladder of a fish. The complete analysis of all the gases in an air sample can be carried out with the aid of a mass spectrometer, an expensive and rather elaborate instrument that is unavailable to most physiologists.

The composition of the atmosphere remains extremely constant. Convection currents cause extensive mixing to a height of at least 100 km, and no discernible changes in the percentage composition have been demonstrated, although the pressure of the air is greatly reduced at high altitudes. The statement that the lighter gases, notably hydrogen and helium, are enriched in the outer reaches of the atmosphere applies to the very outermost layers, which are of no physiological interest whatsoever. For our purposes, the open atmosphere has a constant gas composition, except for its water vapor.

The composition of the air is maintained as a balance between the use of oxygen in oxidation

TABLE 1.1 Composition of dry atmospheric air. All atmospheric air contains water vapor in highly variable amounts. The less common noble gases (helium, neon, krypton, and xenon) together make up only 0.002% of the total.

Component	%
Oxygen	20.95
Carbon dioxide	0.03
Nitrogen	78.09
Argon	0.93
Total	100.00

processes (primarily oxidation of organic compounds to carbon dioxide) and the assimilation of carbon dioxide by plants, which in the process release oxygen.

The fear that our use of fossil fuels – oil, coal, and natural gas – may deplete the atmosphere of oxygen and add large amounts of carbon dioxide is probably unfounded. In 1910 an extremely accurate oxygen analysis showed the value of 20.948%, and during 1967 to 1970 repeated measurements gave a value of 20.946% ± 0.006. The investigators who made these very accurate analyses then calculated that if all known recoverable fossil fuel reserves were depleted, there would still be 20.8% oxygen left in the atmosphere (Machia and Hughes 1970). Physiologically this change would be of no consequence.

The greenhouse effect

The slight increase in carbon dioxide caused by the combustion of all the fuel would likewise have negligible physiological effects, but this is not to say it would be harmless. Even a slight change in carbon dioxide alters the absorption of solar radiation in the atmosphere and may have an unpredictable *greenhouse effect* that over the years may drastically change climatic conditions on the earth's surface.

The atmosphere is more transparent to incoming short-wave radiation than to the long-wave radiation emitted by the earth. The outgoing long-wave radiation is absorbed in the atmosphere mainly by carbon dioxide and water vapor. It is estimated that a doubling of the atmospheric carbon dioxide content would increase world temperature by 1.3° C if atmospheric water remained constant.

This would cause extensive melting of the polar ice covers and a rise in the ocean level of 20 to 30 m or more, enough to submerge many of the world's major cities. In addition, at higher temperature the atmosphere can hold more water vapor, which enhances the blanketing effect and causes further temperature rise. However, increased water vapor in the atmosphere may augment formation of clouds, which in turn reflect more of the incoming solar radiation, thus having the opposite effect. The complexity of these relationships makes predictions about the greenhouse effect of increased carbon dioxide highly uncertain (Hansen et al. 1981; Bolin et al. 1986; Dickinson and Cicerone 1986; Ramanathan 1988).

Having stressed the constancy of the atmospheric composition, we must add a few words about special cases. For example, microenvironments, such as burrows occupied by animals, have more variable air composition, with the oxygen as low as 15% or even less (Darden 1972). The carbon dioxide content is increased, but not necessarily to the same extent. However, carbon dioxide may rise to above 5%, an amount that has considerable physiological effects (Kuhnen 1986).

The air contained in soil – in open spaces between the soil particles – is often low in oxygen. The reason is that the soil may contain oxidizable material that can severely deplete the oxygen. Not only organic matter, but also substances such as iron sulfide, can consume oxygen until practically all free oxygen has been removed. These oxidation processes depend on temperature, humidity, and other factors, as well as on the amount of exchange with the atmosphere. Rain, for example, may block the surface porosity of the soil and at the same time provide humidity for increased oxidation, and the microatmosphere may then change drastically.

Water vapor in air

The preceding information about the percentage composition of the atmosphere referred to dry air, and we must now turn to the water content of air. The pressure of water vapor over a free water surface changes with temperature (Table 1.2). At the freezing point the vapor pressure is 4.6 mm Hg (0.61 kPa).* It increases with increasing temperature, and reaches 760 mm Hg (101.3 kPa) at 100°C. For this reason water boils at 100°C if the atmospheric pressure is 760 mm Hg. If the atmospheric pressure is lower, water boils at a lower temperature; for example, if the pressure is reduced to 17.5 mm Hg (2.34 kPa) water boils at 20°.

Any mixture of gases, such as atmospheric air, that is in equilibrium with free water contains water vapor at a pressure corresponding to the temperature, and the fraction of the air sample that is made up of water vapor therefore increases with the temperature (column 4 of Table 1.2). At 37°C, the usual body temperature of mammals, the water vapor pressure is about 47 mm Hg (6.28 kPa), and water vapor then makes up 6.2% of the air volume (see also Figure 8.10, page 323).

The lung air of humans and other air-breathing vertebrates is always saturated with water vapor at body temperature, but the outside atmospheric air usually is not. When air is saturated with water vapor, we say that the *relative humidity* (r.h.) is 100%. If the air contains less water vapor, the humidity can be expressed as a percent of the amount required for saturation at that temperature; for example, 50% relative humidity means that the air contains half the water it would contain if saturated with water vapor at that temperature.

For some purposes relative humidity is a con-

*The unit millimeters of mercury (mm Hg) is traditionally used in physiology. It is derived from the use of mercury manometers, and 1 mm Hg at 0°C is also known as 1 torr. In the International System of Units (the SI System) the pressure unit is the pascal (Pa), defined as 1 newton per square meter (N m^{-2}). Thus, 1 mm Hg = 133.3 Pa or 0.133 kPa; 1 atm or 760 mm Hg = 101.3 kPa. Further details on the use of the SI System are given in Appendix A.

TABLE 1.2 Water vapor over a free water surface at various temperatures.

Temperature (°C)	Water vapor			
	mm Hg	kPa	% of 1 atm	mg H$_2$O per liter air
0	4.6	0.61	0.6	4.8
10	9.2	1.23	1.2	9.4
20	17.5	2.34	2.3	17.3
30	31.7	4.24	4.2	30.3
40	55.1	7.38	7.3	51.1
50	92.3	12.33	12.2	83.2
100	760.0	101.33	100.0	598.0
37	46.9	6.28	6.2	43.9

venient expression, but when we want to know the total *amount* of water vapor in the air, this can be expressed as milligrams of water per liter of air. Because cold air has a very low water vapor content, even at 100% r.h., the absolute amount of water in cold air is small. Therefore, if saturated outside air in winter enters our houses and is heated, the indoor relative humidity will be extremely low (although the absolute humidity of the air is unchanged), and we say that "the air is very dry." This dryness causes moist surfaces and mucous membranes to dry out, often to great discomfort of sensitive persons.

Altitude and atmospheric pressure

Climbing in high mountains or ascent to high altitude in nonpressurized aircraft has serious physiological effects. At an altitude of 3000 m humans begin to feel the effects of altitude as a reduction in physical performance, and at 6000 m (about 20 000 ft) most humans can just barely survive. This is because of lack of oxygen, although the air still contains the usual 20.95% oxygen.

At sea level, where atmospheric pressure is 760 mm Hg and 20.95% of this is oxygen, the *partial pressure* of oxygen in dry air is 159 mm Hg (21.2

kPa).* At 6000 m the atmospheric pressure is half that at sea level, or about 380 mm Hg (50.7 kPa). The partial pressure of oxygen is also half that at sea level, or about 80 mm Hg or 10.6 kPa (20.95% of 380 mm Hg or 50.7 kPa). It is this decrease in partial pressure of oxygen that produces such severe effects.

The relation between altitude and atmospheric pressure is shown in Figure 1.1. The lower scale of the abscissa shows the partial pressure of oxygen in inhaled air. The zero of this scale does not coincide with zero atmospheric pressure, but instead falls where atmospheric pressure is slightly

---

*In a mixture of gases, the total pressure is the sum of the pressure each gas would exert if it were present alone. In dry atmospheric air at standard barometric pressure (760 mm Hg), the partial pressure of oxygen (P_{O_2}) is 159.2 mm Hg (20.95% of 760 mm Hg), of nitrogen (P_{N_2}) 600.6 mm Hg (79.02% of 760 mm Hg), and of carbon dioxide (P_{CO_2}) 0.2 mm Hg (0.03% of 760 mm Hg).

Atmospheric air is never completely dry, and its water vapor exerts a partial pressure (P_{H_2O}) corresponding to the water vapor content in the air. The partial pressure of the other gases is then reduced in exact proportion. If the air at 760 mm Hg contains 5% water vapor ($P_{H_2O} = 38$ mm Hg), the total pressure of the remaining gases is 722 mm Hg, and their individual partial pressures are in the proportion of their relative concentrations to make up the total of 722 mm Hg.

FIGURE 1.1 The relationship between altitude and atmospheric pressure. 1. Altitude where most un-acclimatized persons will lose consciousness because of oxygen lack. 2. Highest permanent human habitation. 3. Highest altitude where acclimatized humans can survive for a few hours when breathing air. 4. Highest altitude possible for humans breathing pure oxygen. [Dejours 1966]

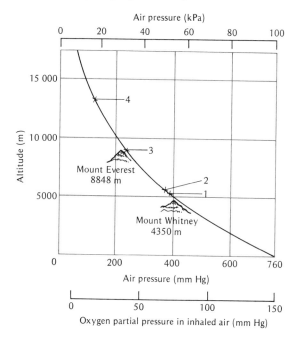

TABLE 1.3 Solubilities of gases in water at 15 °C when the gas is at 1 atm pressure.

Oxygen	34.1 ml O_2 per liter water
Nitrogen	16.9 ml N_2 per liter water
Carbon dioxide	1019.0 ml CO_2 per liter water

molecules enter the water and go into solution. This continues until an equilibrium has been established and an equal number of gas molecules enters and escapes from the water per unit time. The amount of gas that is then dissolved in the water depends on (1) the nature of the gas, for the solubility is not the same for all gases; (2) the pressure of the gas in the gas phase; (3) the temperature; and (4) the presence of other solutes. For an understanding of physiology it is necessary to be familiar with the fundamentals of these relationships.

The solubilities of different gases in water are very different; those of the physiologically most important gases are listed in Table 1.3. The *solubility coefficient*, α, is defined as the volume of gas (in milliliters STPD) dissolved in 1 liter of water, when the pressure of the gas itself, without the water vapor, is 1 atm. The amount of gas is expressed as the volume this gas would occupy if the dry gas were at 0°C and 1 atm pressure; this is designated *standard temperature and pressure dry* (STPD).

We immediately see that nitrogen is only about half as soluble as oxygen, but carbon dioxide is roughly 30 times as soluble as oxygen or 60 times as soluble as nitrogen. This high solubility of carbon dioxide in water makes possible soda water or champagne. In an unopened bottle of soda water, which usually is at between 2 and 3 atm pressure, the small amount of gas at the top of the neck is under the pressure, and the remainder of the carbon dioxide is dissolved in the liquid. As we know, this must be a substantial amount, for large numbers of bubbles continue to rise to the surface for a long time. If nitrogen were dissolved at 3 atm pressure, only a few bubbles would be formed, for the

less than 50 mm Hg (6.67 kPa). The reason is simple. At the body temperature of humans (37°C) the water vapor pressure is 47 mm Hg (6.28 kPa). Therefore, if a person were placed at an atmospheric pressure of 47 mm Hg (19 000 m or 63 000 ft altitude), his lungs would be filled with water vapor and no air or oxygen could enter his lungs.

We shall later, at the end of Chapter 5, return to some of the effects of low oxygen partial pressure on animals at high altitude.

SOLUBILITY OF GASES

Gases are soluble in water. If a sample of pure water is brought into contact with a gas, some gas amount of dissolved nitrogen would be only about $\frac{1}{60}$ of the carbon dioxide.

Effects of pressure and temperature

The amount of gas dissolved in a given volume of water depends on the pressure of the gas. If the gas pressure is doubled, twice as much gas will be dissolved. The proportionality between gas pressure and the amount dissolved is known as *Henry's law* and can be expressed as follows:

$$V_g = \alpha \frac{P_g}{760} \cdot V_{H_2O}$$

α is the solubility coefficient (see Table 1.3), and the equation tells us the number of milliliters of the gas (V_g, at STPD) that is dissolved in the water at the pressure (P_g) given in millimeters of mercury.

The solubility of a mixture of gases depends on the partial pressure of each gas present in the gas phase. Any one of the gases is dissolved according to its own partial pressure in the gas phase, independently of the presence of other gases.

The solubility of gases decreases with increasing temperature. Most of us know this from our own experience. When we open a bottle of warm soda water or beer (Heaven forbid!), the drink has a much greater tendency to foam and overflow than if it is cold. Also, if we watch a pot of water being heated on a stove, small bubbles begin to form on its walls long before boiling begins; these bubbles consist of gas driven out of solution as temperature rises. The solubility of gases in water is thus exactly the reverse of the solubility of solids, which for the most part are more soluble in hot water than in cold (sugar, for example).

We can now examine the solubility of a gas in greater detail. As an example, the solubility of oxygen is given in Table 1.4. First, we must note that this table refers to the amount of oxygen dissolved in water in equilibrium with atmospheric air (not with 1 atm pure oxygen as in Table 1.3). We can see that the solubility of oxygen decreases to about half as the temperature is raised from the

TABLE 1.4 The temperature effect on the amount of oxygen dissolved in fresh water and in sea water in equilibrium with atmospheric air. [Krogh 1941]

Temperature (°C)	Fresh water (ml O_2 liter^{-1} water)	Sea water (ml O_2 liter^{-1} water)
0	10.29	7.97
10	8.02	6.35
15	7.22	5.79
20	6.57	5.31
30	5.57	4.46

freezing point to 30°C. This decreased solubility is quite important for many aquatic animals.

Table 1.4 also gives the solubility of atmospheric oxygen in sea water, which on the whole is some 20% lower than in fresh water. This is because salts reduce the solubility of gases. This reduction is a characteristic effect of dissolved solids, but does not apply to dissolved gases; gases do not affect the solubility of other gases under conditions with which we deal in physiology.

Partial pressure and tension

We have discussed the solubility of gases in terms of the amount of a gas that enters into solution when the gas has a certain pressure. Let us look at it the other way. Take a sample of water that has a certain gas dissolved in it; the amount of gas in the water sample must correspond to one specific gas pressure in the gas phase. This pressure is called the *tension* of this gas in the water sample.* If a water sample has several gases dissolved in it, the tension of each gas corresponds to the partial pressure of that particular gas in the atmosphere with

---

*Physiologists often refer to the *tension* of a gas in a liquid, rather than its partial pressure, and define tension of a gas in solution as the pressure of this gas in an atmosphere with which that particular liquid sample is in equilibrium. The major reason for using the word *tension* is that a dissolved gas as such exerts no measurable pressure, and the term *partial pressure* is therefore conceptually somewhat misleading. However, we frequently find the two terms used interchangeably.

which the water is equilibrated. The tension of a gas in solution is thus defined as the partial pressure of that gas in an atmosphere in equilibrium with the solution.

When the gas pressure over a water sample is reduced, gases tend to leave the solution. If we reduce the gas pressure to about half the original value, gas leaves the solution until equilibrium is reached when the amount of dissolved gas has reached half its original value.

If the gas pressure is reduced to zero, which is the same as exposing the water sample to a vacuum, all the gas leaves or is extracted from the water. Such vacuum extraction is one way of removing all dissolved gas from a liquid; in fact, it is a commonly used method in the analysis of the gas content of blood samples.

Because the gas dissolved in a liquid is in equilibrium with a given partial pressure in the gas phase, we can say that the gas in the liquid is under that particular partial pressure. If we have a container of water that has been equilibrated with atmospheric air, and we introduce a tiny bubble of, for example, pure nitrogen into the water, oxygen (as well as carbon dioxide) will diffuse from the water into the bubble, equilibrium being reached when the bubble contains 20.95% oxygen. Some nitrogen will initially dissolve into the water, for the initial nitrogen pressure in the bubble is 1 atm and the tension in the water only 0.79 atm. However, as oxygen enters, the loss of nitrogen subsides, and the final concentrations within the bubble are those of the initial equilibration atmosphere. (This argument depends on the bubble's volume being so small, relative to the volume of water, that it does not materially influence the gas concentration in the water.)

Solubility of carbon dioxide and its rate of diffusion

The solubility of carbon dioxide in water is some 30 times as high as that of oxygen (see Table 1.3). However, because the amount of carbon dioxide in the atmosphere is very small (0.03%), the total quantity dissolved in water is very small. The amount can be calculated from the solubility coefficient and the fractional concentration of carbon dioxide in the atmosphere as follows:

$$\text{volume dissolved } CO_2 = \frac{1019 \times 0.03}{100}$$
$$= 0.3 \text{ ml } CO_2 \text{ per liter water}$$

In addition to the dissolved carbon dioxide, natural water contains a variable amount of carbon dioxide found in bicarbonates and carbonates. The total amount of carbon dioxide present in water in nature therefore can be quite high and varies with the cations present. Hard water, for example, contains large amounts of dissolved calcium bicarbonate, $Ca(HCO_3)_2$, which adds to the total amount of carbon dioxide.

In sea water, which is slightly alkaline (pH = ca 8.2), the total amount of carbon dioxide may range between 34 and 56 ml CO_2 per liter sea water (Nicol 1960), yet the amount of carbon dioxide present as dissolved gas is still that which is in equilibrium with the atmosphere, about 0.3 ml CO_2 per liter; the remainder is primarily in the form of bicarbonate ion. Thus, in spite of the carbon dioxide *content* of sea water (and many other natural waters), oceanic sea water still has a carbon dioxide tension close to that in the atmosphere, 0.23 mm Hg (or 30.7 Pa).

In an isolated body of water the concentrations of oxygen and carbon dioxide can change appreciably. The two major processes are (1) that oxygen is formed in photosynthesis by plants and (2) that plants as well as animals use oxygen and produce carbon dioxide. In the day-time photosynthesis will usually dominate so that the oxygen content increases substantially in the water, which becomes supersaturated. At night, on the other hand, both animals and plants use oxygen and produce carbon dioxide, almost totally depleting all available oxygen.

These two opposing biological processes can lead to extreme conditions. Measurements made in an

FIGURE 1.2 The water in a tidal rock pool undergoes great variations in gas tensions when it is isolated from the sea at low tide. These records were made in a rock pool on the coast of Brittany (France).

Gas tensions, P_{O_2} and P_{CO_2}, are in millimeters of mercury. The normal value for each variable in sea water is indicated by a broken line. [Dejours 1981]

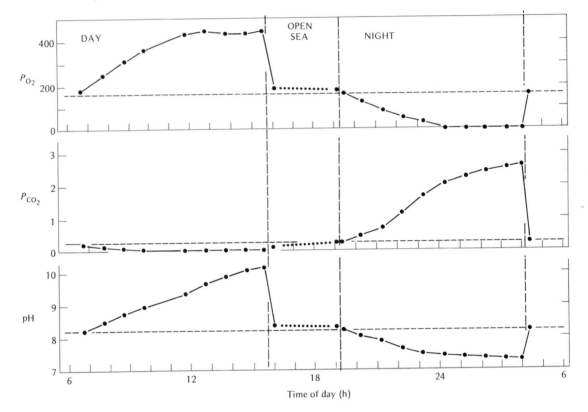

isolated rock tidal pool are recorded in Figure 1.2. When the pool was isolated from the sea during the day, the P_{O_2} increased from the normal level of sea water to as much as *555* mm Hg by midafternoon. The carbon dioxide used for photosynthesis was taken from the sea water, but as the amount dissolved is small, most of the CO_2 was removed from the bicarbonate in the sea water. This resulted in a tremendous shift in the pH of the sea water, which became highly alkaline, reaching a pH of 10.1, while the P_{CO_2} decreased to 10^{-4} mm Hg.

When, at high tide in the late afternoon, the rock

pool was connected to the open sea, the values immediately approached those in normal sea water. At night, when the pool again was isolated, the respiring organisms, both animals and plants, rapidly removed oxygen, which fell to values as low as 1.6 mm Hg. The pH then fell to 7.3, a whole pH unit below that in normal sea water, and the P_{CO_2} increased to 2.7 mm Hg. The reason that the P_{CO_2} increased by only a few millimeters of mercury is that most of the CO_2 again entered into the bicarbonate.

Similar variations in the gas tensions have been reported for fresh water. The magnitude of the

FIGURE 1.3 Diffusion between air and water of oxygen and carbon dioxide. The amount of carbon dioxide dissolved at the surface of water (at 15 °C) is 29.8 times as high as the amount of oxygen when the two gases are at the same partial pressure in the gas phase. Carbon dioxide molecules diffuse more slowly (at a rate 0.86 times the rate for oxygen), but because of their higher concentration in the surface layer, the amount of carbon dioxide diffusing from the same pressure in the gas phase into the water is 25.6 times the amount of oxygen ($29.8 \times 0.86 = 25.6$).

changes depends on local conditions, on the composition of the water, and on the amounts of plants and animals in the water. The general rule is that the oxygen tension tends to increase in the daytime, in particular in strong sunlight, and may fall substantially and approach zero at night.

Gas diffusion

In the respiratory organs of animals (e.g., gills, lungs) gases diffuse between the environment and the organism; oxygen enters and carbon dioxide leaves the animal. It is therefore of interest to know how fast the gases diffuse, their *rate of diffusion*. This subject needs special attention, for many biologists have been led to believe that carbon dioxide diffuses much faster than oxygen. This is not so.

The rate of diffusion of a gas is inversely proportional to the square root of its molecular weight; carbon dioxide is heavier than oxygen and there-

fore diffuses more slowly. The molecular weight of carbon dioxide is 44 and that of oxygen 32; the square roots of these numbers are 6.6 and 5.7. Because the diffusion rates are inversely proportional to the square roots, the diffusion rates of carbon dioxide and oxygen are in the proportion of 5.7/6.6, or 0.86. In other words, carbon dioxide diffuses at a rate which is 0.86 that of oxygen.

When carbon dioxide diffuses between air and water, its high solubility in water makes carbon dioxide appear to diffuse faster. The situation can be explained by reference to Figure 1.3. Let us assume we have an atmosphere that contains oxygen at 100 mm Hg and carbon dioxide at 100 mm Hg pressure (i.e., the partial pressures, or concentrations, of the two gases are equal). Each gas dissolves in the surface water independently of the other, in proportion to its concentration in the gas phase and the solubility coefficient. The amount of the two gases dissolved at the surface is therefore 4.5 ml O_2 per liter and 134 ml CO_2 per liter (at 15°C). In other words, at the surface the concentration of carbon dioxide in solution is 30 times as high as that of oxygen (29.8 times, to be exact), although in the gas phase their concentrations are equal.

Carbon dioxide and oxygen now diffuse from the surface into the body of water. The amount of carbon dioxide dissolved at the surface is much higher, and more carbon dioxide therefore diffuses into the water. The carbon dioxide molecules as such diffuse at a rate 0.86 times that of the oxygen molecules, and the amount of carbon dioxide diffusing, relative to oxygen, is $29.8 \times 0.86 = 25.6$.

We now understand that, when carbon dioxide and oxygen diffuse from equal concentrations in air into water, the total amount of carbon dioxide diffusing is about 25 times higher. Likewise, if the carbon dioxide and oxygen tensions in the water are equal (in our case, 100 mm Hg at the surface of the water), diffusion into the atmosphere is also higher for carbon dioxide. It is important to note that this apparently faster diffusion of carbon dioxide is applicable only when diffusion takes place

TABLE 1.5 Comparison of air and water as respiratory medium.

	Water	Air	Ratio: water/air
O_2 concentration (liter/liter)	0.007	0.209	~1:30
Density, ρ (kg/liter)	1.000	0.0013	~800:1
Dynamic viscosity, η (cP)	1	0.02	~50:1
Heat capacity (cal/liter °C)	1000	0.31	~3000:1
Heat conductivity (cal/s cm °C)	0.0014	0.000 057	~25:1
Diffusion coefficient, D_{O_2} (cm²/s)	0.000 025	0.198	~1:8000
D_{CO_2} (cm²/s)	0.000 018	0.155	~1:9000
Diffusion constant, K_{O_2} (cm²/atm min)	34×10^{-6}	11	~1:300 000
K_{CO_2} (cm² atm min)	850×10^{-6}	9.4	~1:11 000
Liters of medium per liter O_2	143	4.8	~30:1
Kilograms of medium per liter O_2	143	0.0062	~23 000:1

in water or between gas and water, and we refer to *the tension or partial pressure of the gases rather than to their molar concentrations*. This fast diffusion is attributable to the higher solubility of carbon dioxide; as a molecular species carbon dioxide still diffuses according to the laws of physics (i.e., somewhat slower than oxygen).*

COMPARING WATER AND AIR

The differences between water and air are extremely important for animals and their supply of oxygen. We can illustrate these differences by comparing some of the physical properties of water and air, as listed in Table 1.5.

For example, let us compare how much oxygen is available in water and in air. Water in equilibrium with atmospheric air at 15°C contains 7 ml O_2 per liter (1000 ml). This 7 ml O_2 weighs 0.01 g, and this amount is found in a weight of water 100 000 times as great. To obtain a given amount of oxygen, we must therefore move 100 000 times its weight† of water over the respiratory organs.

*In a gas the molar concentrations are directly proportional to the pressures and no confusion can arise as to the rate of diffusion.

†More correctly, mass. Because in daily speech we say that we weigh an object when we determine its mass, it is very common to say weight when we should say mass. See footnote on page 30.

In contrast, 1 liter of air contains 209 ml O_2, which weighs 280 mg. The remainder of the air, 791 ml N_2, weighs 910 mg. To obtain the oxygen, we must move only 3.5 times its mass of inert gas. This difference is illustrated in Figure 1.4.

This tremendous difference in the *mass* of the inert medium has one important consequence: In aquatic respiratory systems the movement of water is almost always unidirectional. If the flow of water were in and out or back and forth, a large mass of water would have to be accelerated, then stopped, and again accelerated in the opposite direction. A high expenditure of energy would be required for the continuous changes in kinetic energy of the water. For respiratory organs that use air, such as the lung, an in-and-out flow is not very expensive, for the inert mass will be only a few times more than the mass of oxygen used. However, air-breathing animals may use a unidirectional flow of air, as we shall see later in this chapter when we discuss bird and insect respiration.

In addition to the mass of the respiratory medium, another factor contributes to the amount of work required to move it. Water has a higher viscosity than air, and this increases the work required to pump the fluid, for the driving pressure must be increased in proportion to the higher viscosity. Because the energy required for moving the fluid increases in direct proportion to the pressure,

FIGURE 1.4 Air has a high oxygen content (21%) and contains inert gas (nitrogen) in an amount weighing only 3.5 times as much as the weight of the oxygen. Water, in contrast, contains only 0.7% of its volume as dissolved oxygen (at 15 °C, P_{O_2} 159 mm Hg). In this case the weight of inert medium is 100 000 times as great as the weight of oxygen.

Air (1 liter)

O₂ (209 ml)

N₂ (791 ml)

(280 mg)

(910 mg)

Water (1 liter)

H₂O (1000 ml)

(1000 000 mg)

O₂ (7 ml; 10 mg)

it also increases in direct proportion to the viscosity.

The *viscosity* of water at 20°C is 1 cP (centipoise), and the viscosity of air is 0.02 cP (i.e., water is about 50 times as viscous as air). The work required for pumping increases accordingly.

Another advantage of air respiration is the high rate of *diffusion* of oxygen in air, which, at the same partial pressure or tension, is some 10 000 times as rapid as in water (see Appendix B). This fast diffusion in air permits very different dimensions in the respiratory organs. The distances over which a gas can diffuse, for example in a lung, may be several millimeters, but the diffusion distance in the fish gill is a minute fraction of a millimeter.

To prevent undue *evaporation* from the respiratory surfaces, these should not be freely accessible to the outside air. The gas-exchange surfaces are usually located in specialized respiratory cavities (lungs), and this greatly limits the access of air.

The renewal of air in the cavity is often very carefully regulated, being no greater than dictated by the requirement for oxygen.

Gas exchange across the general body surface is usually possible only in a moist habitat. Earthworms, for example, in which the entire respiratory gas exchange takes place through the body surface, are very susceptible to water loss. They live in moist habitats, and if they remain exposed on the surface of the earth they rapidly dry out and die.

With regard to water and the need for gas exchange, plants are really in a much worse situation than animals. Plants need carbon dioxide for photosynthesis, and they must obtain this gas from the air, which, as we have seen, contains only 0.03% CO_2 (0.23 mm Hg). The diffusion gradient for carbon dioxide into the plant is therefore extremely small. Even if the plant can maintain a zero concentration of carbon dioxide inside its tissues, the driving force for carbon dioxide diffusion is no more than 0.23 mm Hg. The water vapor pressure of plant tissues varies with temperature; at 25°C it is 24 mm Hg. The driving force for diffusion of water to dry air is, therefore, 24 mm Hg. In moist air it is less; at 50% relative humidity, for example, it is half as much. In any event, the magnitude of the vapor pressure that drives the outward diffusion of water is some 100 times as high as the inward driving pressure for carbon dioxide diffusion. As a consequence of this unfavorable situation, plants have a very high requirement for water.

The situation for plants is compared with that for animals in Figure 1.5. In animals the inward driving force for oxygen is the 21% O_2 in the air, or 159 mm Hg. At 25°C the outward driving force for loss of water to dry air is 24 mm Hg, or only a small fraction of the inward driving pressure for oxygen. Even at high temperatures, say 37°C, the water vapor pressure is small compared with the pressure for inward movement of oxygen. There is thus a great contrast between the unfavorable water balance plants must cope with and the favorably high oxygen concentration animals enjoy (at least, hopefully so).

FIGURE 1.5 For plants it is difficult to obtain by diffusion the necessary carbon dioxide from the atmosphere without incurring a relatively tremendous water loss. For animals to obtain oxygen, the water loss presents less of a problem because of the high partial pressure of oxygen in air (ca. 700 times as high as the carbon dioxide pressure).

One way for plants to reduce the water loss associated with photosynthesis and uptake of CO_2 is used by many drought-tolerant plants, one of special interest being the pineapple. During the night when the air humidity is high, these plants open their stomata and CO_2 can diffuse in with a lower loss of water than would occur in the daytime. Of course, the CO_2 cannot be photosynthesized in the dark, but it is stored in the plant, bound in the form of malic acid. In the daytime the stomata are kept closed, thus reducing water loss, and the now available sunlight is used for photosynthesis of the stored CO_2 (Joshi et al. 1965; Szarek et al. 1973).

This process is known as CAM respiration. The name is derived from the term crassulacean acid metabolism, but the phenomenon is not restricted to the plant family Crassulaceae. Many other plants show CO_2 uptake at night with accumulation of malic acid, followed by depletion of the malic acid

and photosynthesis during the day (Kluge and Ting 1978).

With the knowledge we now have of the physical properties of gases in both air and water, we can move on to the question of how animals manage to live and function within these bounds.

RESPIRATION IN WATER

Some of the simplest arrangements for respiratory gas exchange are found in aquatic animals. Many small organisms obtain oxygen by diffusion through their body surfaces, without having any special respiratory organs and without circulating blood. Larger and more complex animals often have specialized surfaces for gas exchange and also a blood system to transport oxygen more rapidly than diffusion alone can provide.

Animals without specialized respiratory organs

The simplest geometrical shape of an organism is a sphere. For our considerations it is important that a sphere has the smallest possible surface corresponding to a given volume; any deviation from the spherical shape gives a relative enlargement of the surface area. If we assume that a spherical organism is to be supplied with oxygen by diffusion through the surface and into every part of the body, the longest diffusion distance is from that surface to the center.

For oxygen to reach the center, the oxygen concentration at the surface must be of a certain magnitude, for as oxygen diffuses inward it is consumed by the metabolism of the organism. The necessary oxygen tension at the surface, sufficient to supply the entire organism with oxygen by diffusion, can be calculated from an equation developed by E. Newton Harvey (1928), who was well known for his studies of luminescent organisms in the sea:

$$F_{O_2} = \frac{\dot{V}_{O_2} r^2}{6\,K}$$

In this equation F_{O_2} is the concentration of oxygen at the surface expressed in fractions of an atmosphere, V_{O_2} is the rate of oxygen consumption as cm^3 of oxygen per cm^3 tissue per minute, r is the radius of the sphere in centimeters, and K is the diffusion constant in square centimeters per atmosphere per minute. K signifies the cubic centimeters of oxygen that will diffuse per minute through an area of 1 cm^2 when the gradient is 1 atm cm^{-1}. For a further discussion of diffusion, see Appendix B.

Taking as a hypothetical example a spherical organism with a radius of 1 cm, an oxygen consumption of 0.001 ml O$_2$ g^{-1} min^{-1}, and a diffusion constant of 11×10^{-6} cm^2 atm^{-1} min^{-1} (the same as for connective tissue and many other animal tissues), we find that the required oxygen concentration at the surface, necessary to supply the entire organism to the center by diffusion, is 15 atm. This shows clearly that the organism cannot be supplied by diffusion alone if it has the postulated metabolic rate, which, incidentally, is rather low even for an invertebrate.

The conclusion is that the organism, to be supplied with oxygen, must either be much smaller or have a much lower metabolic rate. Let us consider a smaller organism, choosing a radius of 1 mm. We then find that the required oxygen concentration at the surface is 0.15 atm. Because well-aerated water is in equilibrium with the atmosphere, which contains 0.21 atm oxygen, such an organism could obtain enough oxygen by diffusion only, and would be quite feasible.

If we consider real animals we find that these calculations have given quite reasonable orders of magnitude. Organisms that are supplied with oxygen by diffusion only (e.g., protozoans, flatworms) are mostly quite small, less than 1 mm or so, or have very low metabolic rates, as jellyfish do. Although a jellyfish can be very large, it may contain less than 1% organic matter; the rest is water and salts. It has a very low average rate of oxygen consumption, and the actively metabolizing cells are located along the surfaces, where the diffusion distances are relatively short.

An organism that deviates from the spherical shape has a larger relative surface and shorter diffusion distance than a sphere. This holds for a variety of relatively simple organisms. They are flattened or threadlike, have pseudopodia, or have very large and complex surfaces (such as corals and sponges) and can thus obtain enough oxygen by diffusion, although some may be much larger than the sizes used in the preceding calculations.

Animals with respiratory organs

Although a small organism can get enough oxygen by diffusion through the surface, this is usually not true for larger organisms. Of course, any shape deviating from the sphere has a larger surface, and the diffusion distances are also reduced. However, in most cases this does not suffice, and we find specialized respiratory organs with greatly enlarged surfaces. Often these organs also have a thinner cuticle than other parts of the body, thus facilitating gas exchange.

If the respiratory surface is turned out, forming an evagination, the resulting organ is usually called a *gill*. Secondarily, the gill may be enclosed in a cavity, such as in a fish, but this does not change the fact that gills fundamentally are evaginations.

If the general body surface is turned in, or invaginated, the resulting hollow is called a *lung*. Our own lungs are a good example, although secondarily they are finely subdivided and have a quite complex structure. Simpler lungs exist; a pulmonate land snail, for example, has a lung that is little more than a simple saclike invagination in which gas exchange takes place. The term *lung* is used whether the respiratory medium is water or air.

Insects have a special form of respiratory system. Small openings on an insect's body surface connect to a system of tubes (*tracheae*) that branch and lead to all parts of the body. In this case the respiratory organ combines a distribution system (the tubes) with the gas-exchange system, for most of the gas passes through the walls of the finest branches of this system and diffuses directly to the cells.

In general, gills mostly serve for aquatic breath-

Mackerel — 2551

Butterfish — 1725

Menhaden — 1685

Rudderfish — 1532

Scup — 1323

Sea trout — 1253

Sea bass — 1111

Eel — 902

Remora — 892

Sea robin, red — 805

Sea robin, brown — 712

Puffer — 505

Window pane —462

Tautog — 442

Summer flounder — 268

Toadfish — 137

Goosefish — 51

FIGURE 1.6 Highly active, fast-swimming fish have larger gill areas than sluggish bottom-living fish. The number after each name indicates the total surface area of the gills expressed in arbitrary units per gram body weight of the fish. [I. E. Gray, based on Gray 1954]

ing and lungs for breathing in air. There are exceptions: Sea cucumbers have water lungs in which most of the gas exchange seems to take place. Gills may also be modified for use in air, but on the whole they are rather unsuited for atmospheric respiration. For example, most fish when taken out of water rapidly become asphyxiated, although there is far more oxygen in air than in water. The reason is that in water the weight of the gills is well supported, but in air the gills cannot support their own weight.

This is because an effective respiratory organ requires (1) a large surface and (2) a thin cuticle. Both these demands make it difficult to provide the mechanical rigidity for support of a gill in air. Furthermore, in air the surfaces of the fish gill tend to stick together because of surface adhesion. Therefore, the surface area exposed to air is reduced to a minute fraction of what is exposed to water, severely impeding oxygen uptake.

Ventilation of gills

If a gill removes oxygen from completely still water, the immediately adjacent boundary layer of water will soon be depleted of oxygen. Renewal of this water is therefore important in supplying oxygen, and various mechanical devices serve to increase the flow of water over the gill surface. Increased flow can be achieved in two ways: by moving the gill through the water or by moving water over the gill.

Moving the gill through the water is practical only for small organisms. Some aquatic insect larvae – mayfly larvae (Ephemeridae), for example – ventilate their gills in this way. The difficulty is that, if a gill is moved through the water with its

base serving as a pivot point, the force needed to overcome the resistance to the movement is too great. This resistance increases with the square of the linear velocity of the organ, and therefore the energy needed to move the gill increases in the same proportion. The mechanical strength of the gill would also need to be increased, again with the square of the linear velocity, as would the force applied to the base of the gill to make it move through the water. The large aquatic salamander known as the mudpuppy (*Necturus*) does move its gills, but the movements are very slow.

Moving water over the respiratory surface is a much more feasible solution. The movement may be achieved by ciliary action, as in protozoans and in the gills of mussels and clams. Sponges move water through their ostia by the action of flagella.

Moving the water with a mechanical pumplike device is more common. Fish and crabs, for example, move water over their gills in this way. As a matter of principle, it is less expensive to move water slowly over a large surface than to move water fast over a smaller surface.

For some animals their own locomotion contributes to the movement of water. This is true of many pelagic fish; the large, fast-swimming tunas have practically immobile gill covers and obtain the required high water flow over the gills by swimming rapidly through the water. They probably cannot survive if kept from swimming forward, and when these fish are maintained in captivity it is common to keep them in large circular tanks so that they can keep moving without meeting obstacles.

Also in squid and octopus there is a close correlation between locomotion and water flow over the gills, but in these animals the relationship is in a way reversed. A squid or octopus ventilates its gills by taking water into the mantle cavity, and by ejecting the water through the siphon, it propels itself through the water by jet propulsion. In this case the ventilatory system has been modified for locomotion, but as is true of fish, as increased swimming increases the call for oxygen, oxygen is automatically provided in greater amounts.

Other functions of gills

Gills may have functions other than the respiratory gas exchange, and sometimes it is difficult to decide what is the primary or only function of a gill. Some so-called anal gills of mosquito larvae function in osmotic regulation: They absorb ions from the surrounding water, and it is doubtful that they play any major role in respiration. The gills of both fish and crabs serve in osmotic regulation, but in these cases it is quite clear that the gills also serve a primary function in respiration.

To evaluate the role in gas exchange of a gill or other suspected respiratory organ, we must have information about the amount of oxygen taken up, and preferably also about the carbon dioxide given off, through this organ. By comparing this with the oxygen consumption of the entire organism, we can see whether the organ is responsible for virtually all oxygen uptake, a large part, a small part, or just a trivial amount.

If an organism has definite and well-developed respiratory organs, there is usually a need for a circulatory system as well to carry oxygen to the various parts of the body. (This does not hold for the tracheal system of insects, which in principle is independent of a circulatory system.)

The gills of mussels and clams were mentioned above as establishing water currents by ciliary action. In the filter-feeding bivalves the gills are arranged so that they act as a sieve, retaining particles suspended in the water, which afterward are carried to the mouth and ingested. The gills therefore have a primary function in food uptake; whether they also are of importance in gas exchange is less certain. Bivalves on the whole have quite low metabolic rates, and it is possible that the surface of the mantle is sufficient to provide the required gas exchange.

Gas exchange and water flow

The fact that gill surface area must be large enough to provide adequate gas exchange is well expressed in fish. Highly active fish have the largest relative gill areas (Figure 1.6). The fast-swimming mackerel's gill surface area, expressed per unit body weight, is some 50 times as high as the sluggish, bottom-living goosefish's.

For the gas exchange to be adequate, a high rate of water flow and a close contact between the water and the gill are necessary. This is achieved by the anatomical structure of the gill apparatus. The gills are enclosed in a gill cavity, which provides protection for these rather fragile organs and, equally important, permits water to be perfused over the gills in a most effective way. A special advantage is gained by an arrangement that makes the stream of water flowing over the gill and the stream of blood running within the gill flow in opposite directions to each other. This we call a *countercurrent flow*. To understand the importance of this arrangement, it is necessary to know the structure of the gill (Figure 1.7).

Countercurrent flow

Fish gills consist of several major *gill arches* on each side. From each gill arch extend two rows of *gill filaments*. The tips of these filaments from adjoining arches meet, forcing water to flow between the filaments. Each filament carries densely packed, flat *lamellae* in rows. Gas exchange takes place in these lamellae as water flows between them in one direction and blood within them in the opposite direction.

This countercurrent type of flow has an important consequence. Just as the blood is about to leave the gill lamella, it encounters water whose oxygen has not yet been removed. Thus, this blood takes up oxygen from water which still has the full oxygen content of inhaled water, and this permits the oxygen content of the blood to reach the highest possible level. As the water runs further between the lamellae, it meets blood with a lower and lower oxygen content, and it therefore continues to give up more oxygen. Thus, the lamella, along its entire length, serves in taking up oxygen from the water, and the water may leave the gill having lost as much as 80 or 90% of its initial oxygen content (Hazelhoff and Evenhuis 1952). This is considered a very

FIGURE 1.7 The gills of a fish consist of several gill arches on each side. Each arch carries two rows of gill filaments. Each filament carries thin parallel, platelike lamellae. In these lamellae the blood flows, in a direction opposite to that of the water which flows between the lamellae. [Randall 1968]

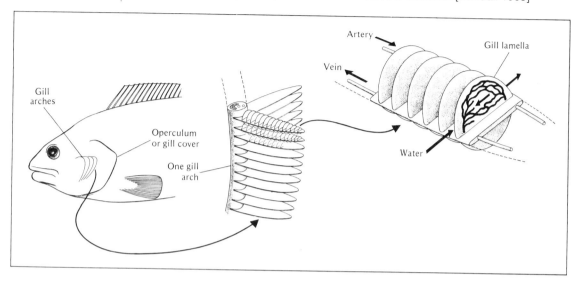

high oxygen extraction; for example, mammals remove only about one-quarter of the oxygen present in the lung air before it is exhaled.

We can express the effect of the countercurrent type of gas exchange in a diagram (Figure 1.8). This figure shows how blood, as it flows through the gill lamellae, takes up more and more oxygen and approaches the oxygen tension of the incoming water. The outflowing water has had most of its oxygen removed and has a tension far lower than that of the blood leaving the gill. If the flows of water and blood were in the same direction, this would be impossible, for the blood could at best reach the oxygen tension of the outflowing water. Because the pumping of water over the gills requires energy, the countercurrent flow through the increased oxygen extraction also reduces the energetic cost of pumping.

It was previously believed that sharks have no countercurrent flow in their gills, although they are excellent swimmers and might be expected to have superbly functioning respiratory organs. However,

FIGURE 1.8 Diagram of the countercurrent flow in fish gills. The numbers indicate partial pressures of oxygen (P_{O_2}) in water and blood. The blood enters the gill at a low P_{O_2} (in this example 20 mm Hg), and oxygen diffuses from the water into the blood. As the blood runs along the lamella, it takes up more and more oxygen from the water, and as it leaves the lamella, it has nearly reached the P_{O_2} of the incoming oxygen-rich water. The water, flowing in the opposite direction, gradually loses more and more of its oxygen and leaves the gill after having lost most of its oxygen content.

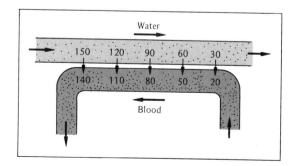

FISH GILL Scanning electron micrograph of gill filaments of a sturgeon (*Acipenser transmontanus*). White arrows show the direction of water flow and black arrows the direction of blood flow. The bar in the upper left corner represents 0.05 mm. [Burggren et al. 1979; courtesy of Warren W. Burggren, University of Massachusetts]

sharks do indeed have countercurrent flow in their gills (Grigg 1970).

Countercurrent flow is found also in the gills of some crabs, but in these the efficiency of oxygen removal from the water often is far less than in fish. In part this may be because the water flow is less effective, but probably it is mainly because the gill–blood diffusion barrier is greater. The latter explanation seems to be the case in the European shore crab (*Carcinus*), in which the oxygen extraction, in spite of countercurrent flow, may range between no more than 7 and 23% (Hughes et al. 1969). In other crabs, for example the Dungeness crab (*Cancer magister*), the oxygen extraction from the respired water often exceeds 50% (McDonald et al. 1980).

Gill resistance. Because the lamellae in the fish gill are very close, it has been suggested that there is a high resistance and virtually no flow of water between them. There can be as many as 30 or 40 lamellae per millimeter with a space between them of less than 0.02 mm. The pressure that drives the water through the gills is often less than 10 mm H_2O, and such a low pressure appears insufficient to drive water through the narrow space.

A careful analysis of this problem was carried out by Hughes (1966), based on the width of the passageway between the lamellae, its height, and its length. Using a modified form of Poiseuille's equation (see Chapter 3), Hughes calculated that the flow through the gills of a 150-g tench (*Tinca*), for a pressure of 5 mm H_2O, would be 10.1 ml

FIGURE 1.9 Water is pumped over the gills of a fish by a dual pumping system. With the aid of suitable valves, the pumps provide a unidirectional flow of water over the gill surface. [Hughes 1960]

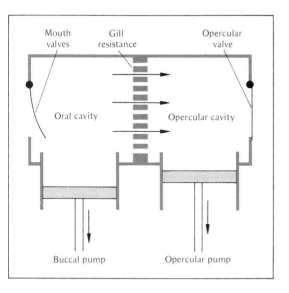

FIGURE 1.10 Record of pressure changes in the respiratory pump of a carplike fish, the roach (*Rutilus*). The lower curve shows the difference between the pressures in the oral and the opercular cavities. A pressure of 1 cm H_2O ≈ 0.1 kPa. [Hughes and Shelton 1958]

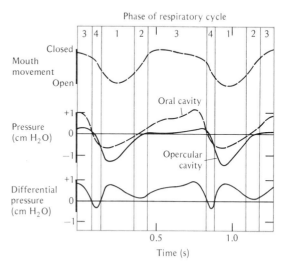

s^{-1}. The normal volume of water pumped through the gills of this fish is about 1 to 2 ml s^{-1}; the obvious conclusion is that the gill lamellae do not offer much resistance to flow, as they would allow a greater flow than actually occurs in the living fish. Similar calculations for more than a dozen other species of fish showed that in all cases the calculated ventilation volumes could easily be greater than those measured in the living fish.

It is interesting that the highest water flows are found in some Antarctic fish, the icefish, which are very unusual because they have no hemoglobin in their blood. We shall return later to these peculiar fish and how they can live without the normal oxygen-carrying mechanism of the blood.

Water pumping

To move water over the gills, teleost fish use a combined pumping action of the mouth and the opercular covers, aided by suitable valves to control the flow. A model of this pump is shown in Figure

1.9. The system actually consists of a double set of pumps. The volume of the first, the oral cavity, can be enlarged by lowering the jaw and especially the floor of the mouth. The volume of the second, the opercular cavity, can be increased by movements of the opercular covers while backflow of water around the edges is prevented by a skin flap acting as a passive valve. The diagram shows only one opercular pump; in reality there are two opercular chambers, one on each side.

The action of the two pumps is such that a flow of water through the gills is maintained throughout nearly the entire respiratory cycle. This flow continues although the pressure in the mouth during part of the cycle may be less than that in the surrounding water; the reason is simply that the pressure in the opercular cavities is maintained even lower than in the mouth.

This is evident from pressure recordings made in the mouth and the opercular cavities during the respiratory cycle (Figure 1.10). The graph shows

FIGURE 1.11 When the swimming speed of a mackerel increases to between 0.5 and 0.8 m s^{-1}, opercular pumping ceases and the fish breathes entirely by ram ventilation of the gills. The records in this graph were obtained from five individuals weighing about 70 g each. [Roberts 1975]

that the pressure changes are synchronized with the movements of the mouth and the opercula. Furthermore, the difference between the pressures in the mouth and in the opercular cavities (as shown by the curve at the bottom of the graph) remains positive almost throughout the cycle and provides the pressure that drives water through the gills. Only during a brief moment is there a slight pressure reversal. The details of the pressure curves differ from fish to fish, but all those studied so far are, in principle, the same: During almost the entire respiratory cycle the pressure in the oral cavity remains higher than in the opercular cavities, providing for a virtually continuous flow of water over the gills.

Ram ventilation

Some fish are unable to breathe this way. Fish biologists have long known that the large tunas cannot be kept alive in captivity unless they can swim continuously; this can be arranged by keeping them in large ring-shaped tanks where they can cruise without stopping. The fish swims with its mouth partly open, there are no visible breathing movements, and water flows continuously over the gills; this is called *ram ventilation*.

Ram ventilation is not restricted to large, fast-swimming pelagic fish. Many fish breathe by pumping at low speed and change to ram ventilation at higher speeds. The transition takes place around speeds of 0.5 to 1 m s^{-1}, and above this speed active breathing movements cease (Figure 1.11).

The change to ram ventilation does not mean that the gills are ventilated for free; it means only that the work of breathing is transferred from the muscles of the opercular pumps to the swimming muscles of the body and tail. The open mouth causes increased drag, and this has to be paid for in increased muscular work. However, the continuous flow during ram ventilation is more economical in energy than opercular pumping at the high rates required for fast swimming.

The reduced cost in energy as a fish changes from opercular pumping to ram ventilation is shown in Figure 1.12. The rate of oxygen consumption, which expresses the energy cost of swimming, increases with increasing speed. This is expected because the drag on the fish increases with its speed, and the cost of respiratory pumping also increases as more water must be pumped over the gills to supply the increased need for oxygen. The surprise comes when the fish changes to ram ventilation. In spite of the higher speed, the oxygen requirement now decreases. It then increases again as the speed of the fish increases further. We can surmise that, had opercular pumping been maintained at the higher swimming speeds, the cost of pumping would by far exceed what was measured when the fish used ram ventilation.

The degree of opening of the mouth during ram ventilation seems to be adjusted to give just the right flow of water over the gills, keeping the drag

FIGURE 1.12 The rate of oxygen consumption of a swimming fish, indicating its total energy expenditure, increases with increasing swimming speed. However, when the fish changes from opercular pumping to ram ventilation, there is a drop in oxygen consumption in spite of the higher swimimng speed. The record is from a striped bass (*Morone saxatilis*) of 25 cm length and 214 g body mass. [Data from Freadman 1981]

to the lowest possible level compatible with the ventilation requirement. This conclusion is based on experiments in which mackerel were kept swimming at constant speeds, but subjected to a progressive lowering of the oxygen content in the water. This resulted in a graded increase in mouth gape so that the lowered oxygen supply was compensated for by increased water flow (Brown and Muir 1970; Steffenson 1985).

As water flows over the gills, solid particles suspended in the water tend to get caught in the gills. Such material can be dislodged by a sudden reversal of the flow, which is brought about by enlarging the oral cavity with closed lips, causing a sudden lowering of the pressure in the mouth. This

maneuver is analogous to coughing, which removes material from the respiratory passageways of mammals.

Crabs keep their gills clean in a somewhat similar way. The flow of water over the gills is unidirectional practically all the time, but at intervals the pumping is stopped and there is a sudden reversal of the flow for a few seconds. The frequency of the abrupt reversal varies greatly. It may happen once a minute or once in 10 minutes, or even less frequently (Hughes et al. 1969). We assume that such abrupt reversals of water flow serve to remove particles that have become lodged in the gills.

The physiological basis for respiratory gas exchange as outlined here depends on some very simple physical principles. We have seen how these principles are used, and how animal function is adapted to the requirements of the physical environment. We shall now examine how equally simple physical principles apply to the physiology of respiratory gas exchange in animals that breathe air.

RESPIRATORY ORGANS

Air-breathing animals have three major types of respiratory organs: gills, lungs, and tracheae.

Gills. On the whole, gills are rather poorly suited to respiration in air and are used by only a few animals – mostly some that relatively recently have invaded the terrestrial habitat, taking with them the remnants of their previous mode of aquatic respiration. Land crabs are good examples. The coconut crab (*Birgus latro*), which has adopted an almost completely terrestrial existence (including climbing coconut palms), has gills that are sufficiently rigid to remain useful for respiration in air.

Another land crab, *Cardiosoma,* is particularly interesting because it can survive indefinitely either in air *or* in water, whereas the coconut crab will drown if it is kept submerged (Cameron and Mecklenburg 1973). Another crustacean group with air-breathing gills is the terrestrial isopods (commonly known as sow bugs, pill bugs, or wood lice); these animals usually prefer to live in moist surroundings, and those that are most successful in the terrestrial habitat have their gills within cavities that can be regarded as functional lungs.

Among fish that can breathe air, functional gills have been maintained in some, but not in all. The common eel (*Anguilla vulgaris*) survives quite well in air if it is kept reasonably cool and moist. Much of the oxygen is then taken up through the skin and less through the gills, for the filaments tend to stick together and expose only a small surface to the air in the gill chamber. As a consequence, the eel does not obtain its normal oxygen requirement, and the oxygen uptake in air is reduced to only about one-half of that in water (Berg and Steen 1965).

Lungs. We can distinguish two types of lungs: *diffusion lungs* and *ventilation lungs*. Diffusion lungs are characterized by the fact that air exchange with the surrounding atmosphere takes place by diffusion only. Such lungs are found in relatively small animals such as pulmonate snails, scorpions, and some isopods.

Ventilation lungs are typical of vertebrates. Substantial and regular renewal of the air in the lung is necessary for a large body size combined with a high metabolic rate. Vertebrate respiratory systems are ventilated by an in-and-out, or tidal, flow of air. The respiratory system of birds, however, which is far more complex than that of mammals, is arranged so that the air can flow unidirectionally through the lung during both inspiration and expiration (see under Bird Respiration, later in this chapter).

Tracheae. This type of respiratory organ is characteristic of insects. It consists of a system of tubes that supply oxygen directly to the tissues, thus obviating the need for circulation of blood for the purpose of gas transport. Exchange of gas in the tracheal system may take place by diffusion only, but in many, especially in highly active insects, there is active unidirectional pumping of air through parts of the tracheal system. The advantage of uni-directional flow is that it permits a far better gas exchange than is obtained by pumping air in and out.

Respiratory movements

Vertebrate lungs are ventilated by active pumping of air. (Tracheal ventilation will be discussed later, for insects differ from other animals in so many respects that it is convenient to treat them separately.) Ventilation of the vertebrate lung can be achieved in two different ways. Filling of the lung can take place with the use of a pressure pump, as in amphibians, or a suction pump, as in most reptiles, birds, and mammals.

A frog fills its lungs by taking air into its mouth cavity, closing its mouth and nostrils, and pressing air into its lungs by elevating the floor of its mouth. As a result of this filling mechanism, a frog can continue to take in repeated volumes of air several times in sequence without letting air out, and thus can blow itself up to a considerable size.

Nevertheless, Bentley and Shield (1973) have shown that, contrary to commonly accepted opinion, at least some amphibians can also breathe with the aid of suction-type pumping. Whether this is universal for amphibians is not known.

Positive pumping, similar to the amphibian mechanism, is found in some reptiles. The chuckawalla (*Sauromalus*), a desert lizard from southwestern North America, often hides in a rock crevice where, by inflating its lungs, it lodges itself so firmly that it cannot be pulled out. One way to dislodge the animal, used by Indians who want it for food, is to puncture it with a pointed stick.

The normal mechanism for filling the lungs in reptiles is the same as in birds and mammals: The lungs are filled by suction. Exhalation can be passive, following inhalation by elastic recoil, or it can be actively aided by muscular contraction. A suction-type pump requires a closed thoracic cavity where the pressure during inhalation is less than the surrounding atmosphere. In mammals inhalation is aided by contraction of the muscular diaphragm. Birds have a membranous diaphragm at-

FIGURE 1.13 Oxygen uptake through the skin of frogs is nearly constant throughout the year. The increased oxygen consumption during summer is covered by a greatly increased oxygen uptake through the lung. [Dolk and Postma 1927]

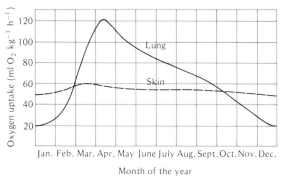

FIGURE 1.14 Pulmonary and cutaneous gas exchange in the toad *Bufo americanus* at different temperatures. [Hutchison et al. 1968]

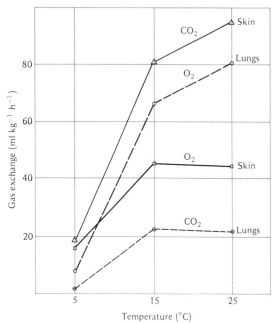

tached to the body wall by muscles, but its function differs from the mammalian diaphragm's. The common statement that birds have no diaphragm is incorrect.

Role of the skin in respiration

Gas exchange through the skin is normal and important for amphibians, which have moist and well-vascularized skin. In fact, some small salamanders (plethodont salamanders) have no lungs, and all gas exchange takes place through the skin surface, except for a small contribution by the oral mucosa (Feder and Burggren 1985).

In frogs the relative roles of skin and lungs change through the year (Figure 1.13). In winter, when the oxygen uptake is quite low, the skin takes up more oxygen than the lungs. In summer, when oxygen consumption is high, the uptake through the lungs increases several-fold and far exceeds the cutaneous uptake. The fact that oxygen uptake through the skin remains nearly constant throughout the year is related to the constant oxygen concentration in the atmosphere, which provides a constant diffusion head. If the oxygen concentration in the blood remains uniformly low through the year, the diffusion through the skin should not change much,

for diffusion rates change very little with temperature. Because the need for oxygen increases greatly in summer, the increase could not be handled by the skin and must be covered by additional uptake in the lung, as is indeed the case.

Is the change with the seasons a result of the changing temperature? This question has been studied with toads (*Bufo americanus*), which were kept at three different temperatures: 5, 15, and 25°C (Figure 1.14). At the two higher temperatures, the pulmonary oxygen uptake exceeds that through the skin, but at the lowest temperature the cutaneous oxygen uptake is greater. This is similar to the situation in the frog. For carbon dioxide exchange the skin is more important at all temperatures. At the lowest temperature the skin is therefore more important than the lungs, both for carbon dioxide and for oxygen.

Salamanders of the family Plethodontidae are

A FROG THAT BREATHES THROUGH ITS SKIN
The Titicaca frog (*Telmatobius culeus*) lives in the depths of Lake Titicaca at 3812 m altitude. This animal does not surface to breathe and obtains oxygen entirely by diffusion through the skin surface, which is highly vascularized and enlarged by loose folds. [Courtesy of Victor H. Hutchison, University of Oklahoma]

unusual because they have neither lungs nor gills. These salamanders are by no means uncommon – they constitute about 70% of existing salamander species. They are quite small and occur both in terrestrial and aquatic habitats. Gas exchange and the role of the blood in gas transport have been studied in *Desmognathus fuscus,* which as adult weighs about 5 to 7 g.

This animal is essentially a skin breather, although about 15% of the total gas exchange takes place across the mucosa of the mouth and pharynx. The blood has no exceptional characteristics: Both its hemoglobin content and its oxygen affinity are in the same range as in other salamanders, both aquatic and terrestrial. There is no special blood vessel carrying the oxygenated blood from the skin to the heart, and oxygen-rich blood leaving the skin is mixed with the general venous blood.

In other amphibians there is a partial separation of oxygenated and venous blood in the heart, but such separation appears to be absent in the plethodont salamanders, and as a result their arterial blood is never fully saturated with oxygen. Nevertheless, their common occurrence indicates that the plethodont salamanders are quite successful with what from our viewpoint appears a rather inadequate respiratory apparatus (Gatz et al. 1974).

Reptiles, in contrast to amphibians, are often believed to have nearly impermeable skin. They breathe with lungs, and most of them are terrestrial. However, the true sea snakes are marine and some even bear living young at sea. They are excellent swimmers and are able to dive to a depth of at least 20 m. When the sea snake *Pelamis platurus* is submerged, it can take up oxygen through the skin at rates up to 33% of its total standard oxygen uptake and excrete carbon dioxide at rates up to 94% of the total. Although the lung is the primary organ of gas exchange, the skin obviously is of considerable aid in gas exchange as this snake pursues juvenile fish on which it preys (Graham 1974).

In mammals gas exchange through the skin is trivial. There is an oft-repeated legend about some children who for a religious procession in Italy were painted with gold paint; the story goes that they all died of asphyxiation because the skin could not "breathe." Death from asphyxiation is out of the question, for oxygen uptake through the skin is barely measurable and carbon dioxide loss from the skin is less than 1% of that from the lung (Alkalay et al. 1971). The gold-painted children must have died from other causes. A plausible explanation is that the gold paint was made by amalgamating gold and mercury and suspending the amalgam in oil, a common paint base. Mercury emulsifies readily in oil and is then rapidly absorbed through the skin, and the children may well have died from acute mercury poisoning.

Bats have a relatively much larger skin surface than other mammals; the large, thin, hairless wing membranes are highly vascularized and may contribute to gas exchange. There is, in fact, some carbon dioxide loss from the wing membrane. In the bat *Eptesicus fuscus* 0.4% of the total carbon dioxide production is lost from the wing skin at 18°C. The amount increases with temperature, and at an air temperature of 27.5°C as much as 11.5% of the total carbon dioxide is lost this way (Herreid et al. 1968). The uptake of oxygen through the wing membranes, however, is not sufficiently great to be of any significance; as discussed earlier, the diffusion between water and air is some 25-fold slower for oxygen than for carbon dioxide.

MAMMALIAN LUNGS

As we move up through the vertebrate classes, the lungs become increasingly complex. In amphibians the lung is a single sac, subdivided by a few ridges that give an increased surface. The mammalian lung is much more finely divided into small sacs, the *alveoli,* which vastly increase the surface area available for gas exchange. Measurements of the surface of the frog lung indicate that 1 cm^3 lung tissue has a total gas-exchange surface of 20 cm^2; the corresponding figure for normal mice is over 800 cm^2 surface per 1 cm^3 lung tissue (Weibel 1979). The large surface area is essential for the high rate of oxygen uptake required for the high metabolic rate of warm-blooded animals.

The membrane that separates the air in the lungs from the blood must be thin so that oxygen can diffuse readily into the blood while carbon dioxide moves in the opposite direction. In the human lung much of the alveolar membrane where gas exchange takes place is no more than 0.2 μm thick. What this means is difficult to imagine. The thickness of a single page in this book is about 50 μm; if one page could be cut into 250 parallel layers, each layer would be the thickness of the alveolar membrane. Consider now that the total surface area of the human lung is about 100 m^2, or nearly the size of a tennis court (Weibel 1984). This large area in combination with the minute thickness of the membrane provides a physical barrier between

FIGURE 1.15 The lung volumes of mammals, when plotted against body size, fall close to a straight line with a slope of 1.06 [Gehr et al. 1981]

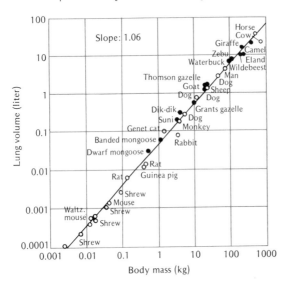

expect that these animals, in order to stay under water longer, would have very large lungs that they could fill with air before a dive. This is not the case; as we shall see later, diving animals do not depend for diving on the oxygen reserves in their lungs.

The regression line in Figure 1.15 can be described by the equation:

$$V_L = 0.046 \cdot M_b^{1.06}$$

where V_L is the lung volume in liters and M_b the body mass in kilograms.* For a mammal of 1 kg body mass the expected lung volume would be 0.046 liter, or 46 ml. Assuming that the body volume in liters equals the body mass in kilograms (this is for practical purposes correct, for we know that a person is close to neutral buoyancy in water and therefore has a density near 1.0), the lung volume is 4.6% of the body volume. (For those unfamiliar with the arithmetic manipulation of exponential equations, a brief discussion is included in Appendix C.)

air and blood that permits very high rates of gas exchange.

Lung volume

The lung volume of a mammal constitutes about 5% of the body volume, irrespective of the body weight (Figure 1.15). If the lung volume were to remain the exact same proportion of body size, the slope of the regression line in Figure 1.15 would be exactly 1.0. The best fitting regression line has a slope of 1.06 (i.e., there is a slight deviation from strict proportionality: Larger animals tend to have proportionately larger lung volumes).

As expected, the individual points do not fall exactly on the line, but there are no large characteristic deviations. This means that small mammals, which have high specific metabolic rates, obtain sufficient oxygen with lungs of the same relative size as in large animals.

It is worth noting that diving animals, such as porpoise, manatee, and whale, follow the common mammalian pattern in regard to lung size. One might

Inhalation and exhalation

Gas exchange in the lung takes place in the alveoli; the trachea, the bronchi, and their branches are only connecting tubes. At the end of an exhalation these tubes are filled with ''used'' air from the lung, and when inhalation follows, this air is pulled back into the lungs before fresh outside air enters.

---

*Mass is a fundamental property of matter, and weight refers to the force exerted on a given mass by a specified gravitational field. In the gravitational field of the earth mass and weight, if both are expressed in kilograms, numerically have the same value. In a different gravitational field the mass is unchanged but the weight is different. For example, on the moon a 70-kg person still has a mass of 70 kg, but his weight is only one-sixth of that on the earth, or about 12 kg (more correctly, kilogram force or kgf). Because of the similarity between mass and weight on earth, it is very common not to distinguish between them. However, the correct statement is that the mass of our person is 70 kg, whether he is on the earth, on the moon, or in space; but his weight will be 686 N on earth, about 110 N on the moon, and zero in space.

MAMMALIAN LUNG Scanning electron micrograph of the lung structure of a wildebeest (*Connochaetes taurinus*). The capillaries, filled with red blood cells, are visible as a bulging network in the alveolar walls. The distance across the photograph corresponds to about 0.5 mm in the lung. [Courtesy of Ewald Weibel, University of Berne, Switzerland]

The volume of air in the passageways reduces the amount of fresh air that enters the lung and is called the *dead space*. The volume of air inhaled in a single breath is the *tidal volume*. A normal person at rest has a tidal volume of about 500 cm^3. Because the dead space is about 150 cm^3, only 350 cm^3 of fresh air will reach the lungs. The dead space thus constitutes about one-third of the tidal volume at rest. In exercise, the relative role of the dead space is less. For example, if a person who breathes heavily inhales 3000 cm^3 air in a single breath, a dead space of 150 cm^3 is now only about one-twentieth of the tidal volume. Therefore, the dead space is a substantial fraction of the tidal volume at rest, but in exercise it is relatively insignificant.

An important aspect of respiration is that the lungs are never completely emptied of air. Even if a person exhales as much as possible, there is still about 1000 cm^3 of air left in his lungs. It is therefore impossible for a person to fill his lungs completely with ''fresh'' air, for the inhaled air is always mixed with air that remained in the lungs and the dead space.

In respiration at rest a person may have about 1650 cm^3 of air in the lungs when inhalation begins. During inhalation, 350 cm^3 fresh air reach the lungs and are mixed with the 1650 cm^3 already there. The renewal of air is therefore only about one part in five. The result is that the composition of the alveolar gas remains quite constant at about 15% oxygen and 5% carbon dioxide. This composition of alveolar air remains the same during exercise; in other words, the increased ventilation during exercise is adjusted to match accurately the increased use of oxygen.

Surface tension

Every person who has blown soap bubbles knows that, when the connection to the atmosphere is open, the bubble tends to contract and expel air until it collapses. The vertebrate lung is somewhat similar; the bubble-like shape and high curvature of the alveoli mean that the surface tension of the moist inner surface tends to make the ''bubbles'' contract and disappear. The surface tension should make the lung collapse, but this tendency is minimized

ALVEOLAR STRUCTURE Cross section of the alveolar structure of a dog's lung. The irregular dark spots are red blood cells, which become distorted as they flow through the capillaries. The thickness of the alveolar membrane, separating blood from air, is only a fraction of 1 μm. The distance across the photograph corresponds to about 0.15 mm in the lung. [Courtesy of Ewald Weibel, University of Berne, Switzerland]

FIGURE 1.16 The amount of surfactant that can be extracted from the lungs of various vertebrates plotted against the surface area of the lung. The extracted amount is uniformly greater than the theoretical minimum amount needed to cover the lung surface with a monomolecular layer. [Clements et al. 1970]

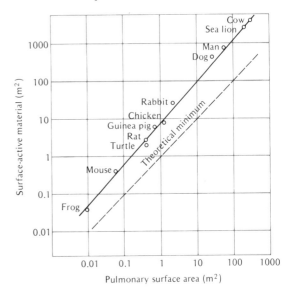

tebrate lung seems always to be above the minimum required to cover the pulmonary surface with a monomolecular layer (Figure 1.16).

Mechanical work of breathing

The movement of air in and out of the lung requires work, and it is of interest to compare the cost of pumping with the amount of oxygen provided. The simplest way is to compare the amount of oxygen it takes to run the pump with the amount of oxygen the organism obtains in the same period of time.

Determinations of the work of breathing are rather difficult and have a substantial margin of error. Most determinations indicate that the cost of breathing in a person at rest is 1.2% of the total resting oxygen consumption. At rest the lung ventilation is about 5 liters of air per minute and the cost of respiration about 0.5 cm^3 O$_2$ per liter ventilated air.

by the presence on the inner surface of the alveoli of substances that greatly reduce the surface tension (Goerke and Clements 1986).

These substances are phospholipids, and their effect on surface tension has given them the name *surfactants* (Bourbon and Rieutort 1987). Surfactants are found in the lungs of all vertebrates — mammals, birds, reptiles, and amphibians. Even the lung surface of lungfish is covered with surfactants. The amount of surfactant present in the ver-

With increasing ventilation, however, the cost of breathing increases. If the ventilation increases to 10 liters per minute, the cost increases to 1 cm³ O_2 per liter air, and in very heavy respiration (50 liters per minute) the cost increases to 2 cm³O_2 per liter ventilated air (Otis 1954). Other determinations indicate that the maximum work of breathing during exercise is no more than 3% of the total oxygen consumed, or slightly less than the figure just mentioned (Margaria et al. 1960). Cost of respiration has also been determined with some success in dogs; the figures are of the same magnitude as those for humans.

What is the cost of pumping for fish, which must move a much heavier and more viscous medium? Accurate determinations are difficult to make, and the question remains somewhat controversial. It has been suggested that the cost of breathing in fish may be as high as 30% or even 50% of the oxygen obtained (Schumann and Piiper 1966).

These figures seem unrealistic, and recent determinations indicate that the cost of breathing is much lower – a few percent of the total oxygen uptake. If the mechanical work of breathing is calculated from the pressure drop across the gill and the volume of water flow, the calculated work comes out to less than 1%, a much more reasonable figure. Without a unidirectional flow of water, such a low cost of ventilation would probably be impossible.

Regulation of respiration

If the need for oxygen increases, the ventilation of the respiratory organs must be increased accordingly. Likewise, if the oxygen concentration in the medium falls, there must be compensation by increasing the ventilation, or by increasing the amount of oxygen removed from the respired air, or by both means.

In warm-blooded vertebrates – mammals and birds – the ventilation of the lungs is very precisely adjusted to the need for oxygen, but interestingly, the primary agent responsible for the regulation is the carbon dioxide concentration in the lung air. This is readily demonstrated by adding carbon dioxide to the inhaled air. This causes a rapid in-

FIGURE 1.17 The effect on pulmonary ventilation volume of increased carbon dioxide content of the inhaled air in chickens. [Johnston and Jukes 1966]

crease in pulmonary ventilation (Figure 1.17). Normal atmospheric air contains virtually no carbon dioxide (0.03%), and if 2.5% CO_2 is added to the inhaled air, the ventilation volume is approximately doubled. The effect is similar in mammals and birds.

This 2.5% CO_2 is really not much, for the lung air of mammals already contains about 5% CO_2. If the carbon dioxide concentration in inhaled air is increased to what is normally found in the lung, the increase in respiratory ventilation volume is several-fold. In much higher concentrations carbon dioxide becomes narcotic and therefore induces abnormal responses.

Oxygen has a much smaller effect on ventilation. If we reduce the oxygen concentration in inhaled air by 2.5%, from 21% to 18.5% O_2, there is virtually no change in respiration.

The respiratory center

Breathing in mammals is brought about by the muscles of the diaphragm and the intercostal muscles. The rhythmic movements of these muscles are controlled by a respiratory center, which is located in the area of the medulla oblongata and the pons of the brain.

The respiratory center contains a number of separate neurons for inspiration and expiration that alternate in their activity. When the inspiratory neurons send stimulatory impulses to the respiratory muscles, inhibitory impulses to the expiratory neurons keep these from activity. When the activity of the inspiratory neurons ceases, the expiratory neurons become active, and in this way the alternating action of the two sets continues indefinitely.

The respiratory center is sensitive to the increased carbon dioxide or increased acidity in the cerebrospinal fluid, and an increased P_{CO_2} in the blood acts because carbon dioxide rapidly diffuses from the blood to the cerebrospinal fluid. Acidity in the blood, without an increase in carbon dioxide, is less effective because the hydrogen ion does not readily diffuse into the cerebrospinal fluid.

As an adequate supply of oxygen is essential, it may seem peculiar that the respiratory center is very insensitive to a decrease in the oxygen in the blood. However, there are peripheral chemoreceptors that are sensitive to a decrease in oxygen. These are located partly in a carotid body and partly in the aortic arch near the heart. A decrease in the P_{O_2} in the arterial blood activates these chemoreceptors, which cause an increase in both the depth and the frequency of respiration. However, the effect will not be very profound because an increased ventilation volume leads to a loss of carbon dioxide from the blood, which in turn reduces the ventilation volume. As a result, a moderate decrease in oxygen has very minor effects on the respiration, which under normal circumstances is regulated primarily to maintain a normal P_{CO_2} and acid–base balance in the blood.

In exercise respiration increases, and this increase occurs before the increased oxygen consumption and carbon dioxide production of the muscles have altered the oxygen and carbon dioxide content of the blood. What happens is that impulses from the motor cortex of the brain, as well as from proprioceptors in the muscles, affect the

respiration so that in the end the P_{CO_2} of the blood during exercise is maintained at very nearly the normal resting level and oxygen is supplied at the rate with which it is consumed by the muscles.

The sensitivity of respiration to the carbon dioxide concentration is often used by swimmers who wish to stay under water for long periods. By breathing deeply for some time, a person can increase the loss of carbon dioxide from the lungs and from the blood. This removes the usual stimulus to respiration, and as a consequence, a person who has hyperventilated can remain under water longer before he is forced to the surface by the urge to breathe.

This practice is extremely dangerous. As a person swims under water, the oxygen in his blood is gradually depleted, but in the absence of the usual concentration of carbon dioxide the urge to breathe is not very strong. He therefore remains submerged, and as the blood oxygen falls, he may lose consciousness without even being aware of the danger. If in this state he is not immediately discovered and rescued, he will drown. This sequence of events has in fact been the cause of many drowning accidents, particularly in swimming pools, when good swimmers competitively attempt long underwater swimming feats (Craig 1961a,b).

Diving animals

It has been stated that seals and whales are less sensitive to carbon dioxide than other animals and therefore are able to stay under water longer. However, the duration of a dive is probably limited by the amount of oxygen available, and it is unlikely that a dive could be prolonged merely by a decreased sensitivity to carbon dioxide.

If we want to examine the response to inhaled carbon dioxide in various animals, it is not enough to measure the respiration frequency. In some animals the respiration frequency increases considerably in response to carbon dioxide, but in others there may be little or no change in frequency and yet a considerable increase in tidal volume. This

FIGURE 1.18 An increase in the carbon dioxide content of inhaled air causes a several-fold increase in respiration volume. Note that the seal is more sensitive to carbon dioxide than the dog. [Bentley et al. 1967]

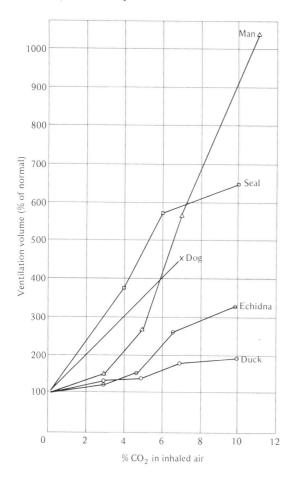

has, for example, been found in the spiny anteater (the echidna, *Tachyglossus*) (Bentley et al. 1967). The information we need is the *ventilation volume* (i.e., the product of respiration frequency and tidal volume).

If we determine the increase in ventilation volume in response to carbon dioxide, it turns out that the seal, at carbon dioxide concentrations up to 6%, is even more sensitive to carbon dioxide than non-diving animals such as human and dog (Figure 1.18). This contradicts the statements about diving seals being insensitive to carbon dioxide.

Comparison of air and aquatic respiration

In many aquatic invertebrates the regulation of respiration is poor or even absent. This is true especially in marine species that normally live in well-aerated water with a relatively constant oxygen supply. Some lower animals are quite tolerant of lack of oxygen; bivalve molluscs can keep the shells closed for long periods and, in the absence of ventilation, utilize anaerobic metabolic processes (see Chapter 5).

For most aquatic animals the primary stimulus to respiration is lack of oxygen. This is usual in crustaceans, octopus, fish, and so on. The effect of carbon dioxide on aquatic invertebrates is never very pronounced and may be absent. The carbon dioxide tension in natural water is almost always low, and, as we shall see in Chapter 2, because of the high solubility of carbon dioxide in water, aquatic animals cannot build up a high tension of this gas. If they were to depend on an increase in carbon dioxide tension for stimulation of respiration, an adequate supply of oxygen could not be assured.

If we compare two closely related forms, the marine lobster (*Homarus*) and the fresh-water crayfish (*Astacus*), we notice a characteristic difference. The lobster, a marine animal, does not show much change in ventilation with a decrease in oxygen in the water (Thomas 1954). The fresh-water crayfish, in contrast, responds to a decrease in oxygen with an increase in ventilation. The difference is readily understood. The lobster lives in cold waters where the oxygen is always high, and an elaborate mechanism for regulation of ventilation would be superfluous. The crayfish, on the other hand, may readily encounter fresh-water environments where oxygen is depleted, and a ventilatory response is needed.

Fish in general respond to a decrease in oxygen, and their response to changes in carbon dioxide is minimal. In this regard they resemble other aquatic animals rather than the air-breathing vertebrates. Insects are for the most part highly sensitive to carbon dioxide.

Can we generalize these differences? We have already mentioned that aquatic animals cannot readily base their respiratory regulation on carbon dioxide, not only because the carbon dioxide tension in natural waters usually is low, but also because it is an unreliable measure of the oxygen content. Sea water is highly buffered, so that the carbon dioxide tension never builds up to any appreciable extent. Stagnant fresh water, on the other hand, may have high carbon dioxide concentrations, usually associated with low oxygen. Aquatic animals could not possibly depend on something as unreliable as the carbon dioxide concentration; when their respiration is regulated, it is in response to oxygen concentration.

The question now becomes: Why have terrestrial air-breathing animals abandoned regulation on oxygen and gone over to carbon dioxide as the primary stimulus? The answer is probably that, as oxygen becomes easily available in the atmosphere, the carbon dioxide concentration tends to build up in the respiratory organs. For example, if a mammal reduces the oxygen in the respired air from 21% to 16%, this entails a simultaneous buildup of carbon dioxide to about 5%, an amount that greatly affects the acid–base balance of the organism. A decrease in oxygen to 16% does not have any profound physiological effect, but a change in 1% of the carbon dioxide concentration, say from 4% to 5%, constitutes a 25% increase in carbonic acid.

Also, it may be easier, physiologically speaking, to design a precise control system based on carbon dioxide and the detection of small changes in hydrogen ion concentration than to devise a system sensitive to small changes in oxygen concentration. Be this as it may, we universally find that air-breathing animals are far more sensitive to changes in carbon dioxide than to changes in oxygen.

AIR-BREATHING FISH

Many fish will respond to a decrease in the oxygen content in the water by swimming to the surface to take a bubble of air into the mouth, a response that greatly improves the oxygen supply. Goldfish that were kept in water with an oxygen tension of only 20 mm Hg would gulp air at the surface as often as 180 times per hour. This nearly doubled the oxygen content of the arterial blood of the oxygen-starved fish (Burggren 1982).

A number of other fish are specifically adapted to air breathing to a much greater extent than the goldfish. Everybody has heard about lungfish, but these are not the only fish that depend on air. Many fish will resort to breathing air only when the oxygen content in the water is low; relatively few depend on air to such an extent that they drown if kept submerged.

There are two main ecological reasons for using accessory or exclusive air breathing: (1) depletion of oxygen in the water and (2) the occurrence of periodic droughts. A lungfish, for example, during dry periods burrows deep into the mud, encases itself in a cocoon, and remains inactive until the next flood.

Most of the air-breathing fish are tropical freshwater or estuarine species; few are truly marine. Oxygen-deficient fresh water is much more common in the tropics than in temperate climates. This is because there is much decaying organic matter in the water, the temperature is high and speeds up bacterial action, small bodies of water are often heavily shaded by overhanging jungle (which reduces photosynthesis and oxygen production in the water), and there is little temperature change between day and night and therefore minimal thermal convection to bring oxygen-rich surface water to deeper layers. Not all air-breathing fish belong in the tropics, however. The well-known bowfin (*Amia calva*) is found in the northern United States even where the lakes are frozen over in winter. During such periods they manage well without air breathing because the low temperature reduces their oxygen consumption.

It was explained that fish gills are not well suited for respiration in air. They lack the necessary rigidity and tend to stick together, but even so, some oxygen can be taken up through gills in air. Any other moist surface will supplement the gas exchange, provided it is a surface that has access to air and is supplied with blood. Some gas exchange can always take place through the skin and the surface of the mouth cavity, but in addition there may be other anatomically more specialized organs that aid in gas exchange. Organs commonly utilized in air breathing are gills, skin, mouth, opercular cavities, stomach, intestine, swimbladder, and lung.

Some air-breathing fish are listed in Table 1.6, with a few comments to place them in a familiar frame of reference. It is worth noting that all except the last five are ordinary higher bony fishes (Actinopterygii). The bowfin (*Amia*) is a representative of the primitive group Holostei, and the bichir (*Polypterus*) is peculiar because it has a lung, but both are considered primitive representatives of the Actinopterygii. Only the last three fish listed are the true lungfish (Dipnoi), which presumably are closely related to the Crossopterygii (the subclass that contains the famous coelacanth, *Latimeria*). There is one genus of true lungfish in each of three continents: Australia, Africa, and South America.

The need for air breathing depends on the amount of oxygen in the water and the temperature, for the rate of oxygen consumption increases with temperature. For this reason it is not always possible to say whether a given fish does or does not depend on air breathing. However, some fish are so dependent on air breathing that they cannot survive even in well-aerated water. They are obligatory air breathers and will ''drown'' if deprived of access to air. The fish listed in Table 1.7 are such obligatory air breathers; interestingly, the Australian lungfish, *Neoceratodus,* is not included.

We shall now discuss how some of the various mechanisms function and how effective they are by describing some fish that have been studied by physiologists.

Common eel

The common European and North American eel (*Anguilla vulgaris*) supposedly is able to crawl considerable distances over land and cross from one watercourse to another, especially at night and through moist grass. Fishermen often keep live eels in a box for days, merely covered by a wet sack to keep them from drying out.

When an eel is out of water, it keeps its gill cavities inflated with air. About once a minute it renews the air in the gill cavity. In contrast, when in water at 20°C an eel respires at a frequency of about 20 per minute.

The oxygen consumption of an eel in air is about half of what it is in water at the same temperature. When an eel is moved to air, oxygen taken from the air is initially supplemented with oxygen removed from the swimbladder; this amount may, during the first hour, nearly equal that taken from the air. However, when an eel is kept in air longer, it gradually accumulates a substantial oxygen debt, and the lactic acid concentration in the blood increases. When returned to water, recovery is rapid, and in 2 hours the blood lactic acid has returned to the normal level.

At a lower temperature, 7°C, the rate of oxygen consumption is much lower and there is no rise in lactic acid in the blood and apparently no oxygen debt. The permeability of the skin to oxygen should be nearly independent of temperature, and at the low temperature the combined oxygen uptake through skin and gills seems sufficient to cover the metabolic rate (Figure 1.19).

The relative importance to the eel of gills and skin differs in water and in air. In water (at 20°C) about 90% of the total oxygen uptake is via the gills; the remainder through the skin. In air, only about one-third of the total oxygen uptake is via the gills; the remaining two-thirds via the skin. Although the gills are still of some use in air, they are far from adequate, and even with the aid of the skin the total available oxygen is insufficient to avoid an oxygen debt and formation of lactic acid, except at the lowest temperatures.

TABLE 1.6 Fish that can use accessory air breathing or depend completely on air. Most are modern teleosts; only the last three are true lungfish.

Organ used for respiration in air	Fish	Habitat	Comment
Gills	*Synbranchus*	South America, fresh water	An eel-shaped fish without any common English name
Skin	*Anguilla*	North America, Europe	The common eel; breeds in the sea; larva migrates to fresh water
Skin	*Periophthalmus*	Tropical estuarine beaches	A common fish, often called mud skipper
Mouth and opercular cavities	*Electrophorus*	South America, fresh water	The electric eel
Mouth and opercular cavities	*Anabas*	Southeast Asia, fresh water	Called climbing perch, but not really a perch; related to betta, the Siamese fighting fish
Mouth and opercular cavities	*Clarias*	Southeast Asia, (Florida, introduced), fresh water	A catfish, known also as the walking catfish
Mouth and opercular cavities	*Gillichthys*	Pacific Coast of North America	Also called the mudsucker
Stomach	*Plecostomus*	South America, fresh water	A small catfish common in home aquaria
Stomach	*Anicistrus*	South America, fresh water	An armored catfish, protected by heavy spines and bony plates
Intestine	*Hoplosternum*	South America, fresh water	An armored catfish
Swimbladder	*Arapaima*	South America, rivers	The world's largest freshwater fish
Swimbladder	*Amia*	North America, fresh water	The bowfin; range extends north to areas where lakes remain ice-covered through winter; belongs to primitive group Holostei
Swimbladder	*Lepisosteus*	North America, fresh water	The garpike; belongs to the primitive group Holostei
Lung	*Polypterus*	Africa, fresh water	The bichir; has a lung, but is not a true lungfish (see text)
Lung	*Lepidosiren*	South America, fresh water	A true lungfish
Lung	*Protopterus*	Africa, fresh water	A true lungfish
Lung	*Neoceratodus*	Australia, fresh water, rivers	A true lungfish

TABLE 1.7 Some fish that are obligatory air breathers.

Fish	Respiration organ	Habitat
Protopterus	Lung	Africa
Lepidosiren	Lung	South America
Arapaima	Swimbladder	South America
Hoplosternum	Intestine	South America
Ophiocephalus	Pharyngeal cavities	South Asia and Africa
Electrophorus	Mouth	South America

FIGURE 1.19 When an eel is transferred from water to air (at 7 °C), its rate of oxygen consumption gradually decreases and stabilizes at a lower level (top curve). Initially, oxygen is taken from the swimbladder, but when this is used up, the combined oxygen uptake from skin and gills suffices to sustain the lowered rate of oxygen consumption. [Berg and Steen 1965]

Mud skipper

Another fish that depends primarily on cutaneous respiration is the well-known mud skipper (*Periophthalmus*), which is common on tropical beaches, especially in muddy estuarine areas. These fish are up to about 20 cm long; they prefer to remain in air for extended periods and move about supported on their pectoral fins. When disturbed, a mud skipper rapidly disappears into a mud hole that usually has some water at the bottom. When kept submerged in water, however, the fish displays circulatory reactions that indicate it suffers from partial asphyxia.

Synbranchus

This South American swamp fish has no common English name. It is notable because it seems able to breathe equally well in water and air. It uses primarily the gills for respiration, but the walls of the gill chamber are highly vascularized and contribute to gas exchange. When the fish is in air, the gill chamber is periodically inflated, and the air is retained for some 10 to 15 minutes while about half its oxygen is extracted.

When the fish is in water, the arterial oxygen saturation is always rather low, never more than 50 or 60%. Inflation of the gill chamber with air is invariably followed by a substantial increase in the arterial oxygen, which then approaches saturation (Johansen 1966).

When *Synbranchus* breathes air, the arterial carbon dioxide concentration increases, a change characteristic of all air breathers, and when the fish is returned to water, there is a prompt fall in the carbon dioxide again. If the carbon dioxide concentration in the water is increased, the respiration rate decreases; thus carbon dioxide inhibits rather than stimulates respiration, in contrast to the case in other air breathers. The decreased respiration caused by a high carbon dioxide level in the water leads to hypoxia, and this stimulates air breathing. The primary stimulus in the respiratory regulation of *Synbranchus* is therefore a decrease in oxygen, which makes the fish turn to air respiration.

Electric eel

The South American electric eel (*Electrophorus electricus*) is known for its powerful electric discharges, which have been measured at 550 V, enough to stun or kill other fish and perhaps even a human. The electric eel is a mouth breather and

FIGURE 1.20 Air kept in the oral cavity of the electric eel gradually loses oxygen as it is used by the fish. The carbon dioxide in the mouth air gradually increases, but not to the same degree. When about one-third of the oxygen has been removed, the fish seeks to surface for renewal of the air. [Johansen et al. 1968]

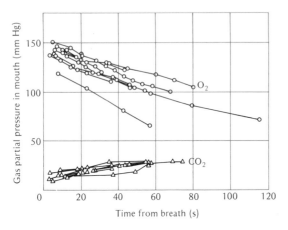

drowns if denied access to air. The oral cavity, which is immensely vascularized, has multiple foldings and papillations that greatly enlarge its surface area.

The electric eel takes air into the mouth at intervals from a few seconds to several minutes. The oxygen in this air gradually decreases while the carbon dioxide increases, although not in the same proportion (Figure 1.20). When the oxygen has fallen by about one-third, the fish slowly moves to the surface for renewal of the air.

In an air breather in which the entire gas exchange takes place between the blood and a body of air, the buildup of carbon dioxide should correspond roughly to the oxygen depletion. As this is not the case for the electric eel, carbon dioxide must be lost elsewhere, perhaps through the rather rudimentary gills or, more likely, through the skin.

The regulation of respiration in the electric eel can be examined by changing the carbon dioxide and oxygen concentrations in the atmosphere it is allowed to breathe. The result is that a substantial

increase in the carbon dioxide content in the air gives only a moderate increase in the rate of breathing, but a drop in the oxygen content causes a several-fold increase. Likewise, increasing the oxygen pressure above that in normal atmospheric air has a somewhat depressing action on the respiration (Johansen 1968).

It is interesting that this fish is an obligatory air breather, but has retained the aquatic mode of regulating respiration on oxygen. Evidently, this is associated with the fact that carbon dioxide escapes through the relatively permeable skin, thus not building up to sufficient concentrations to be physiologically important.

Garpike

The garpike (*Lepisosteus osseus*) is a relic of the ancient group Holostei, which presently is confined in North America. It is a long-nosed fish with hard scales and not popular as a game fish. Often it can be seen to break through the water surface as it comes up to breathe air. It is a large fish that may exceed 10 kg in weight and is widely distributed and occurs in the northern United States even where the lakes are frozen over for several months in winter.

The garpike has gills and also uses its modified swimbladder for respiration. The frequency with which it comes to the surface to take fresh air into the swimbladder is a function of water temperature. At 22°C the interval between breaths averages 8 minutes in an undisturbed fish, but in winter, when the water is covered by ice, the fish obviously cannot breathe air and depends entirely on its gills. At 20 to 25°C, air breathing meets about 70 to 80% of the total oxygen requirement; the remainder is met by the gills. Carbon dioxide elimination from the swimbladder is, of course, zero in winter when the fish depends entirely on the gills. In summer, carbon dioxide eliminated from the swimbladder approaches 10%, and the bulk of the carbon dioxide is eliminated via the gills and skin (Rahn et al. 1971). This is characteristic of many air-breathing fish: Although air breathing is impor-

AIR-BREATHING FISH A garpike (*Lepisosteus osseus*) breathing air at the water surface. A bubble of air is visible at the posterior end of the head; it has just been expelled from the gill slit, and the fish is ready to take in air by lowering the floor of the mouth. The whole sequence of breathing requires only slightly more than 0.5 second before the fish returns below the surface. [Courtesy of Katharine Rahn, Buffalo, New York]

tant for the uptake of oxygen, carbon dioxide, because of its high solubility, is eliminated mostly to the water.

Lungfish

The African and the South American lungfish (*Protopterus* and *Lepidosiren*) were listed in Table 1.7 as obligatory air breathers. They live in stagnant bodies of water and in lakes where long droughts may cause complete drying out of their habitat. They estivate until the next wet period, when they come out from their cocoons in the mud and resume normal life.

The Australian lungfish (*Neoceratodus*) lives in rivers and slow streams. It also estivates in dry periods, but it depends much less on the lung and is primarily a gill breather. If it is kept in a tank from which the water is drained slowly, it becomes frantic and seeks to keep itself submerged in what little water remains. It does fill its lungs with air, however, which contributes substantially to the gas exchange.

When the three kinds of lungfish are kept in water with access to air, the relative roles of gills and lungs are as shown in Figure 1.21. In the Austra-

lian lungfish the gills are the primary organs of gas exchange. In the African lungfish most of the oxygen is taken up by the lung and only a small fraction through the gills. In the South American lungfish virtually all the oxygen uptake is via the lung and only a minute amount via the gills. With carbon dioxide the situation is different; both the African and the South American lungfish exchange about half through the lung, the remainder through the gills. Hence, gills are important in carbon dioxide exchange, and the lung plays the dominant role in oxygen uptake.

The changes that take place as the lungfish are removed from water to air show a similar relationship among the three species (Figure 1.22). The oxygen consumption of the African and South American lungfish remains nearly unchanged in air. In contrast, the Australian lungfish shows a precipitous drop in oxygen consumption, because normally it is an aquatic breather and in air it cannot obtain sufficient oxygen.

The oxygen saturation of the arterial blood (lower part of Figure 1.22) shows that the two obligatory air breathers maintain or slightly increase the oxygen saturation of the blood when they are removed

FIGURE 1.21 The relative roles of gills (upper panel) and lungs (lower panel) in respiratory gas exchange in three kinds of lungfish when kept in water and with access to air. [Lenfant et al. 1970]

FIGURE 1.22 When African and South American lungfish are kept out of water, their rates of oxygen consumption remain nearly unchanged. The Australian lungfish shows a precipitous decrease in oxygen consumption when kept out of water. In this species the oxygen saturation in the arterial blood shows a corresponding drop. [Lenfant et al. 1970]

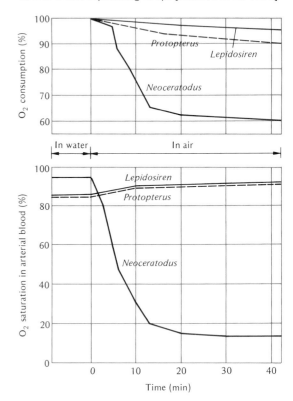

to air; the Australian lungfish's arterial oxygen content drops to nearly zero, a clear indication that the fish is becoming asphyxiated.

The lungfish are among the best studied of the air-breathing fish. The information on many others is inadequate or virtually nil. The study of these could be very rewarding, although often technically difficult, for many are quite small.

BIRD RESPIRATION

Structure of the respiratory system

The respiratory organs of birds are very different from their counterparts in mammals. The small, compact bird lungs communicate with voluminous, thin-walled *air sacs* and air spaces that extend between the internal organs and even ramify into the bones of the extremities and the skull. This extensive and intricate respiratory system has been

considered as an adaptation to flight. We can immediately say, however, that it is not *necessary* for flight, because bats (which have typical mammalian lungs) are good fliers and at times even migrate over long distances.

We may imagine that flying birds should have a very high oxygen consumption and that the avian respiratory system should be seen in this light. However, the rates of oxygen consumption of resting birds and mammals of equal body sizes are very similar, and although normal flight requires an

TABLE 1.8 Volumes of the respiratory systems of typical birds and mammals of 1 kg body size. [Estimated from data collected by Lasiewski and Calder 1971]

	Bird	Mammal
Lung volume (ml)	29.6	53.5
Tracheal volume (ml)	3.7	0.9
Air sac volume (ml)	127.5	—
Total respiratory system volume (ml)	160.8	54.4
Tidal volume (ml)	13.2	7.7
Respiratory frequency (min^{-1})	17.2	53.5

FIGURE 1.23 The smallest units of the mammalian lung are the saclike alveoli. In the bird lung the finest branches are tubes that are open at both ends and permit through flow of air. [Schmidt-Nielsen 1972]

8- or 10-fold increase in oxygen consumption, many mammals are capable of similar increases. Finally, during flight bats have an oxygen consumption very similar to that of birds of the same body mass. On the other hand, although the avian-type respiratory system is not a prerequisite for flight, it may still have considerable advantages.

The presence of air spaces in the body of a bird can be said to make the bird lighter, but only in a very limited sense. The bird needs a digestive system, a liver, kidneys, and so on, and merely adding large sacs of air to the abdominal cavity does not make the bird any lighter. If we removed this air, or doubled its volume, the bird would still have exactly the same weight to carry during flight. On the other hand, if the marrow of a bone is replaced by an equal volume of air, the bone weighs less. The air-filled bones, therefore, do contribute to making the bird lighter, but the other large air spaces do not.

If we compare the volumes of the respiratory systems of birds and mammals, we find some conspicuous differences (Table 1.8). The lung volume of a typical bird is only a little more than half that of a mammal of the same body size. In contrast, the tracheal volume of a bird is much larger than that of a mammal. This can easily be understood in view of the bird's long neck, but we shall later see that it has other implications as well. The air sacs of a bird are large, several times as large as

the lung, and mammals have no air sacs at all. Therefore, the total volume of the respiratory system of a bird is some three times as large as that of a mammal.

The difference between birds and mammals is not restricted to the air sacs; in structure the avian lungs differ radically from those of mammals. In mammals the finest branches of the bronchi terminate in saclike alveoli (Figure 1.23). In birds the finest branches of the bronchial system (known as *parabronchi*) permit through passage of air. Air can thus flow *through* the bird lung and continuously past the exchange surface; in mammals air must flow in and out. This is the most important difference between the respiratory systems of birds and mammals, and it has profound physiological consequences.

To understand how the avian respiratory system works, we must know a few additional facts about its complex anatomy; a diagram of the lungs and major air sacs will help (Figure 1.24). It is not necessary to learn the names of all the sacs, particularly as different investigators often use different names that merely confuse the nonspecialist. It is important, however, to know that the air sacs anatomically and functionally form two groups: a *pos-*

FIGURE 1.24 The body of a bird contains several large, thin-walled air sacs. The paired lungs are small and located along the vertebral column. The main bronchus, which runs through the lung, has connections to the air sacs as well as to the lung. Below the bird is a diagram of the system, simplified by combining all cranial sacs into one single space and all caudal sacs into another. [Schmidt-Nielsen 1972]

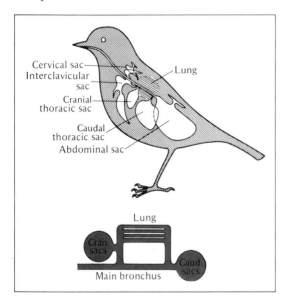

BIRD LUNG In cross section a bird lung shows cylindrical tubes (parabronchi) that allow the unidirectional through flow of air characteristic of bird lungs. The diameter of each tube in this picture, which is from a chick of domestic fowl, is slightly less than 0.5 mm. [Courtesy of Professor H.-R. Duncker, Giessen, Germany]

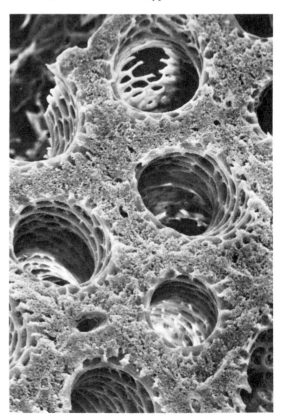

terior or *caudal group* that includes the large abdominal sacs, and an *anterior* or *cranial group* that consists of several somewhat smaller sacs.

The trachea divides into two bronchi; each bronchus runs to and then actually through one of the lungs, and terminates in the abdominal sac. The cranial sacs connect to this main bronchus in the anterior part of the lung; the caudal sacs connect to the posterior part of the main bronchus. The main bronchus also connects to the lung, and furthermore, some of the air sacs connect directly to the lung tissue.

Function of the respiratory system

What is the function of the air sacs? Do they function in gas exchange, or do they serve as bellows to move air in and out?

The morphology of the air sacs does not indicate that they have any major role in gas exchange between air and blood. Their walls are thin, flimsy, poorly vascularized, and there are no foldings or ridges to increase the surface area.

A simple experiment that excluded a direct role in gas exchange was made in the last century by a French investigator. He plugged the openings from the large abdominal air sacs to the rest of the respiratory system and then introduced carbon mon-

FIGURE 1.25 When an ostrich inhales a single breath of pure oxygen and then breathes ordinary air, the oxygen acts as a tracer gas. It is first found in the caudal air sacs. Only during the second or third respiratory cycle (indicated at top) does the gas show up in the cranial sacs. Respiratory rate about 6 cycles per minute. The usual gas composition in the air sacs is given at right. [Schmidt-Nielsen et al. 1969]

oxide into the sacs (Soum 1896). Birds are quite sensitive to carbon monoxide, but these birds showed no signs of carbon monoxide poisoning. Therefore, carbon monoxide had not been taken up by the blood from the air sacs. This conclusion can be generalized; it is valid for other gases that have similar diffusion properties – oxygen, for example (Magnussen et al. 1976).

A more plausible hypothesis for the function of the air sacs is that they serve as bellows to move air in and out. As an inspiration begins, there is a simultaneous pressure fall in both the cranial and caudal sacs as they expand (see diagram in Figure 1.24). This means that during inspiration air flows into all sacs but, as we shall see later, not all sacs fill with outside air. During exhalation the pressure in the sacs increases, and air flows out again.

To follow the flow of gas in greater detail, we could use a gas mixture of a composition different from the usual air and follow the movements of this gas through the respiratory system. A conve-

nient gas is pure oxygen, for it is harmless and can easily be measured with an oxygen electrode. For such experiments the ostrich has the advantage that it breathes quite slowly (a single respiration lasts about 10 seconds), and changes in gas composition can therefore be followed readily.

When an ostrich is permitted to inhale a single breath of pure oxygen instead of air, the oxygen shows up in the caudal air sacs toward the end of this breath (Figure 1.25). This must mean that the inhaled oxygen reaches these sacs directly through the main bronchus. In the cranial sacs, on the other hand, the oxygen concentration never increases during the inhalation of oxygen; however, these sacs do expand during inhalation, and this means that they receive air from somewhere else. Toward the end of the second inspiration, when the bird again breathes ordinary air, the oxygen in the cranial sacs begins to increase. This can only mean that the oxygen that now appears has, in the intervening time, been located elsewhere in the respiratory system, presumably in the caudal sacs, and has passed through the lung to reach the cranial sacs.

The gas concentrations found in the air sacs are interesting. The caudal sacs contain some 4% carbon dioxide (see Figure 1.25), and relative to atmospheric air the oxygen is depleted by a similar amount, from 21% to about 17%. In the cranial sacs, however, the carbon dioxide concentration is higher, between 6 and 7%, and the oxygen is correspondingly reduced to some 13 to 14%. These differences in gas composition might suggest that the caudal sacs are better ventilated and the cranial sacs contain more stagnant air, which therefore reaches a higher carbon dioxide concentration. This conclusion is not correct.

A marker gas can also serve to determine how rapidly air is renewed in a sac. After introduction directly into a sac, the concentration of the marker decreases stepwise in synchrony with each respiratory cycle, and the rate of washout indicates the extent of air renewal. The time required for the marker gas to be reduced to one-half the initial level can be designated as the *half time*. In the ostrich this half time was between two and five respiratory

FIGURE 1.26 The movement of a single inhaled volume of gas through the avian respiratory system. It takes two full respiratory cycles to move the gas through its complete path. [Bretz and Schmidt-Nielsen 1972]

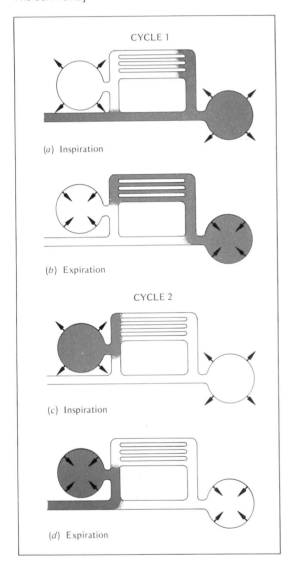

CYCLE 1

(a) Inspiration

(b) Expiration

CYCLE 2

(c) Inspiration

(d) Expiration

equal half times for cranial and caudal sacs (Bretz and Schmidt-Nielsen 1972). Because tracer gas experiments indicate that air flows into the cranial sacs from the lung (their carbon dioxide concentration is consistent with this conclusion), and because their air renewal is high, the evidence indicates that the air sacs serve as a holding chamber for air from the lung, to be exhaled on the next exhalation.

The flow of air can also be determined by placing small probes, sensitive to air flow, in the various passageways. Studies on ducks indicate the flow patterns illustrated in Figure 1.26, which shows how a single bolus of air would flow. During inhalation (a) most of the air flows directly to the caudal sacs. Although the cranial sacs expand on inhalation, they do not receive any of the inhaled outside air; instead, they receive air from the lung. On exhalation (b), air from the caudal sacs flows into the lung instead of out through the main bronchus. On the following inhalation (c), air from the lung flows to the cranial sacs. Finally, on the second exhalation (d), air from the cranial sacs flows directly to the outside. Two full respiratory cycles are required to move a single bolus of gas through the respiratory system. This does not mean that the two cycles differ in any way: They are completely alike, each bolus of gas being followed by another similar bolus, in tandem, on the next cycle.

The most notable characteristics of this pattern are that air always flows through the lung from the posterior in the anterior direction and that air moves through the lung during both inhalation and exhalation.

This flow pattern has an important consequence for gas exchange between air and blood that in principle is similar to the countercurrent flow in the fish gill (see Figure 1.7). It allows the oxygenated blood that leaves the lung to have a higher oxygen tension than the oxygen partial pressure in exhaled air. A highly simplified diagram explains this (Figure 1.27). Blood that is just about to leave the lung (right side of the diagram) is in exchange with air that has just entered the lung as it comes directly from the caudal sacs with a high oxygen partial

cycles for both cranial and caudal air sacs; this means that both sets of sacs are about equally well ventilated.

Similar determinations on ducks also gave about

FIGURE 1.27 Gas exchange in the bird lung. This highly simplified diagram shows the flow of blood and air through the lung, each represented by a single stream with opposite flow directions. This flow pattern permits the oxygenated blood to leave the lung with the highest possible oxygen tension. [Schmidt-Nielsen 1972]

FIGURE 1.28 The blood in the avian lung does not flow in parallel capillaries, but rather in an irregular, complex network. The simplified diagram below shows that blood leaving the lung is a mixture of blood flowing through different parts of the lung and having different degrees of oxygenation. This pattern can be described as cross-current flow. [Scheid and Piiper 1972]

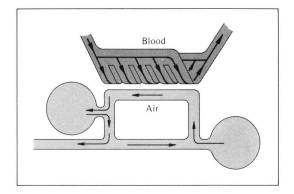

pressure. As the air flows through the lung (toward the left in the diagram), it loses oxygen and takes up carbon dioxide. All along, this air encounters blood with a low oxygen tension and therefore gives up more and more oxygen to the blood. Due to this type of flow, the blood can become well saturated with oxygen, and yet be able to extract more oxygen from the pulmonary air and deliver more carbon dioxide to it, than is the case in mammals.

The flow of air in the bird lung is in fact not an ideal countercurrent exchange system, but rather a *cross-current* type of flow (Figure 1.28). The result, with regard to arterial gas tensions, is similar to that described above, although a cross-current system is not as effective as a true countercurrent system in achieving maximum advantage in gas exchange.

The effectiveness of the unidirectional air flow in the avian lung is particularly important at high altitude. In experiments in which mice and sparrows were exposed to an atmospheric pressure of 350 mm Hg, corresponding to 6100 m or 20 000 ft altitude (slightly less than 0.5 atm), the mice were lying on their bellies and barely able to crawl, while the sparrows were still able to fly (Tucker 1968).

Mice and sparrows have the same body mass, their blood has the same affinity for oxygen, and their metabolic rates are similar, so the difference cannot be explained in terms of their rates of oxygen consumption or the chemistry of their blood. The flow pattern in the bird lung is the most plausible explanation, for it allows blood to take up oxygen from air that has a higher oxygen concentration than would be found in the mammalian system, and in addition, because of the unidirectional and continuous flow through the lung, to extract more oxygen from the air. It is consistent with these experiments that birds in nature have been seen in the high Himalayas, flying overhead at altitudes where mountain climbers can barely walk without breathing oxygen.

RESPIRATION IN EGGS

The problem of supplying oxygen to the growing embryo and chick within the hard shell of a bird's egg forms an interesting chapter in respiration physiology. Most of the work in this field has been concerned with hens' eggs, but the general principles apply to all birds' eggs, except for differences that result from the enormous differences

FIGURE 1.29 During the 21-day incubation period a 60-g hen's egg takes up oxygen while it loses both carbon dioxide and water. At the end of incubation, before the chick hatches, the egg weighs 51 g.

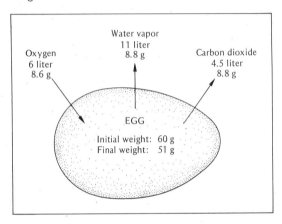

in size. The smallest eggs, those of hummingbirds, may weigh less than 0.3 g; those of the ostrich weigh over 1 kg. The largest bird eggs known are those of the extinct *Aepyornis,* which average about 10 kg. Thus, there is more than a 30 000-fold difference in weight between the smallest and largest bird egg.

The newly laid egg is a microcosm that contains everything that is needed to produce a living chick, with one exception – oxygen. Therefore the shell must be permeable to oxygen, and as a result it is also permeable to other gases. Water vapor and carbon dioxide therefore are lost during incubation.

The overall changes in a hen's egg during incubation are shown in Figure 1.29. The egg initially weighs 60 g, and during the 21 days of incubation it takes up 6 liters of oxygen. At the same time it loses 4.5 liters of carbon dioxide. As the density of carbon dioxide is higher, the weight of the two gases is almost identical. However, because the shell is permeable, some water evaporates. The amount is 11 liters of water vapor, weighing 8.8 g or 14.7% of the initial egg weight. Just before hatching the 60-g egg has been reduced to 51 g, almost the en-

tire loss being due to evaporation of water. The newly hatched chick that comes out of the egg weighs 39 g, the remaining 12 being shell and some membranes.

The shell of an egg consists of a hard outer layer of calcium carbonate, which on the inside has two soft membranes, called the outer and the inner shell membrane. The hard shell is much less permeable to gases than the shell membranes. When the egg is first laid, the permeability of the shell with its membranes is low, but it increases several-fold during the first few days after the egg is laid. This increase is correlated with a decline in the water content of the membranes, and most of the resistance to diffusion now resides in the hard shell (Kutchai and Steen 1971).

At the blunt end of the egg is an *air cell* that lies between the inner and outer shell membranes. This air space increases in size during incubation because water is lost from the egg by evaporation.

The hard shell of a typical hen's egg is perforated by about 10 000 pores. Because the surface area of the egg is about 70 cm^2, there is an average of 1.5 pores per square millimeter of shell. The pore diameter is about 0.017 mm, so that the total pore area of the egg is 2.3 mm^2 (Wangensteen et al. 1971). All gas exchange between the embryo and the surrounding atmosphere must take place through these pores.

The oxygen consumption and the carbon dioxide production increase throughout the incubation period, which lasts about 21 days. Because the diffusion characteristics of the eggshell during the later part of incubation remain constant, or nearly so, there must be a progressive increase in the gradients of gases between the ambient air and the embryo. The steeper gradients are produced by a progressive fall in the oxygen partial pressure within the egg and a concomitant increase in carbon dioxide during incubation (Figure 1.30).

During the last day of incubation, things change rapidly. About 28 hours before hatching the chick perforates the air cell membrane, lung respiration begins, and the chick rebreathes the gas of the air

FIGURE 1.30 As the incubation of a hen's egg progresses, the oxygen concentration in the air cell gradually decreases as the carbon dioxide shows a corresponding increase. [Romijn and Roos 1938]

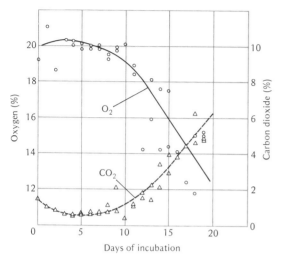

space (which because of evaporation has increased to about 10 cm^3 in volume). About 12 hours later the chick begins to break through the eggshell, the stage known as *pipping,* and the main barrier to gas exchange is now broken. During the following hours, while the chick struggles to break out completely, pulmonary respiration has taken over the whole task of gas exchange.

At high altitude, where the oxygen pressure is reduced, the oxygen supply to the embryo is severely challenged. This problem was studied in a colony of white leghorns maintained for many years at the White Mountain Research Station in California at 3800 m altitude (barometric pressure = 480 mm Hg, or 64 kPa). When the colony was started, only 16% of the fertile eggs hatched, compared with the usual 90% at sea level. As the colony became established the hatching rate gradually increased and after eight generations reached 60%.

One consequence of a reduced air pressure is an increase in the diffusion coefficient for gases. At 3800 m the increase is slightly more than 1.5-fold,

and this increase is beneficial to the delivery of oxygen to the embryo. However, the higher diffusion coefficient at low atmospheric pressure applies also to water vapor, and this heightens the danger of desiccation.

Eggs laid at high altitude are small, and the incubation period is longer than at sea level. The pore area per unit shell surface is decreased, though the thickness of the shell is unchanged. The reduction in pore area in combination with the increased gas diffusion coefficient results in an overall gas conductance that is similar at sea level and at high altitude. We can therefore see that the reduction in total pore area of an egg laid at high altitude is a necessary adaptation because the water loss otherwise would be excessive (Wangensteen et al. 1974).

Do the eggs of birds that are native to high altitude show similar diffusion characteristics? A study of six species that breed at altitudes above 2800 m (9300 feet) shows that the gas conductance of their eggs is significantly reduced as compared with similar species breeding at lower elevations. In this way the lower eggshell conductance compensates for the increased diffusion rates, and thus water loss (Carey 1980).

At very high altitude, however, there must be a compromise between reducing water loss on the one hand and the defense of the required oxygen supply to the embryo on the other (Monge C. et al. 1988).

One group of birds lay eggs that differ from the usual pattern. The mound-building birds of Australia, known as mallee fowl and brush turkeys, place their eggs in huge mounds of fermenting vegetation. The birds attend carefully to the mounds, open and rearrange them to maintain a nearly constant temperature of about 37°C. The eggs hatch after nearly two months and the chicks dig themselves out.

The atmosphere within the mounds is saturated with water vapor and the eggs therefore do not lose water during incubation. However, because of the fermentation the oxygen is as low as 14% and the CO_2 may reach 9%. To admit sufficient oxygen the eggs are thin shelled and have a large pore area, but this is irrelevant to water loss because the air in the mounds is saturated.

This information about the mound-building birds has permitted some conclusions about the nesting habits of dinosaurs. Fragments of dinosaur eggs collected in southern France were thinner and had a higher porosity than bird eggs of the same size, and their gas conductance ranged from 8 to 24 times the value for bird eggs. This similarity to the mound-building birds suggests that the dinosaur eggs were incubated under similar conditions of high humidity, probably underground. This conclusion is supported by what we know about the eggs of living reptiles. Those that deposit hard-shelled eggs in locations of high humidity have eggs with conductances several times higher than bird eggs (Williams et al. 1984).

Reptile eggs

The eggs of reptiles are much more variable than bird eggs. Their structure can range from thin, extensible, and membrane-like shells to hard calcareous shells more like bird eggs. These differences are related to the habitat and especially to the humidity conditions where the eggs are laid.

The eggs of egg-laying reptiles can be assigned to three broad categories. The shells of most lizard and all snake eggs are relatively thin and highly extensible. Such eggs are usually placed in a relatively dry environment, and if the environment is moist, they would tend to take up water and increase in size. Such eggs are quite sensitive to the proper humidity of the environment, and hatchability depends on proper conditions.

In contrast, turtles have somewhat flexible eggshells with a well-defined calcareous layer. They are relatively independent of humidity conditions, which do not unduly affect hatchability.

The eggs of all crocodilians, some turtles, and a few lizards are hard and rigid. These eggs will therefore not swell unduly in a moist environment, such as the nests used by alligators and crocodiles (Packard et al. 1982).

Marine turtles spend almost their entire lives at sea. After they enter the water as hatchlings, only the females will ever again leave the sea, and then only for a brief time to deposit eggs in nests on sandy tropical beaches. A female turtle places a clutch of about 100 eggs in a carefully excavated chamber above the high-tide mark, about 0.5 m below the sand surface. The soft, leathery eggs are round and look like Ping-Pong balls. After a 60-day incubation period the eggs hatch simultaneously, and the hatchlings collectively dig their way to the beach surface and head for the water.

The soft shell of the turtle egg is much more permeable to gases than is the bird egg shell. This does not matter to the loss of water, for the atmosphere within the turtle's nest is water-saturated, and a high gas permeability is an advantage to the supply of oxygen and loss of carbon dioxide.

What is the supply of oxygen in the turtle's nest when 100 eggs simultaneously consume oxygen and release carbon dioxide? Based on determinations of the metabolic rates of eggs near hatching, the metabolic demand for oxygen of an average clutch of 100 eggs is about 300 to 600 cm^3 oxygen per hour. As a result the oxygen concentration within the nest decreases and establishes a gradient between the gases in the nest and the atmosphere above the beach surface. During the 60-day incubation period the P_{O_2} in the nest gradually decreases from about 150 to between 80 and 100 mm Hg, and the P_{CO_2} at the same time rises to 40 to 60 mm Hg.

The diffusion characteristics of gases through moist sand (but not wet sand) are sufficient to supply the nest with oxygen at the rate it is used by the clutch of eggs. The spaces between the eggs are not filled with sand, and because diffusion in the gas phase is rapid, the eggs in the center are also adequately supplied with oxygen. In fact, the 100-egg clutch exchanges respiratory gases in a manner quite similar to that of a bird egg, where the diffusion resistance resides in the hard shell. For the turtle nest the diffusion resistance is in the sand, and the respiratory environment for the eggs is always saturated with water vapor. Hence, a highly permeable eggshell is no liability, but rather a necessity.

The characteristics of the nesting beach are very important for the gas exchange and probably for the successful hatching of sea turtle eggs. The similarity between the entire clutch of sea turtle eggs and the single bird egg is striking. The physical characteristics of the beach and the nest chamber dictate the gas exchange and play an important role in explaining why all species of sea turtles lay clutches of about 100 eggs in similar, very carefully constructed nests (Prange and Ackerman 1974; Ackerman 1980).

INSECT RESPIRATION

Terrestrial life poses a continual conflict between the need for oxygen and the need for water. Conditions that favor the entry of oxygen also favor the loss of water. Insects, the most successful terrestrial animals, have a hard cuticle, which is highly impermeable to gases and whose covering wax layer makes it virtually impermeable to water as well. Gas exchange takes place through a system of internal air-filled tubes, the *tracheae,* which connect to the outside by openings called *spiracles* (Figure 1.31). The spiracles usually have a closing mechanism that permits accurate control of the exchange between the air in the tracheal system and the outside atmosphere. The tubes branch and ramify and extend to all parts of the body. The finest branches, the *tracheoles,* are about 1μm or less in diameter and can even extend into individual cells, such as muscle fibers.

The tracheal system conveys oxygen directly to the tissues and carbon dioxide in the opposite direction. This makes insect respiration independent of a circulatory system. Insect blood has no direct role in oxygen transport, in contrast to the role of blood in vertebrates. The comparison with vertebrates is by no means irrelevant, for the rates of oxygen consumption of a large moth and a hummingbird in flight are similar; the oxygen supply mechanisms, however, differ radically.

FIGURE 1.31 The tracheal system of insects consists of fine air-filled tubes, *tracheae,* that convey respiratory gases to and from all parts of the body. This drawing shows only the main branches of the tracheal system of a flea. The openings to the outside air, the *spiracles,* are indicated by the row of small circles along the side of the body. [Wigglesworth 1972]

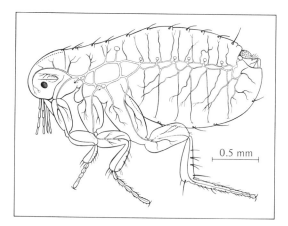

0.5 mm

We cannot conclude that the circulatory system in insects is unimportant; it has many other functions. One role is obvious: During flight the muscles have a high power output, and while oxygen is provided through the tracheal system, fuel must be supplied by the blood.

The tracheal system

Tracheal systems are suitable primarily for respiration in air and must have evolved in air. Insects that secondarily have become aquatic have in principle maintained air respiration, but their tracheal systems have many interesting modifications that make them suitable for gas exchange in water.

Most tracheal systems show variations from a basic pattern in which there characteristically are 12 pairs of spiracles, 3 pairs on the thorax and 9 on the abdomen. Often there are fewer spiracles, and there may be none. Some of the possible variations are shown in Figure 1.32.

In a typical pattern the larger tracheae connect

FIGURE 1.32 The basic pattern of the insect tracheal system (a) may be modified into a variety of other patterns (*b–f*). For details, see text. [Wigglesworth 1972]

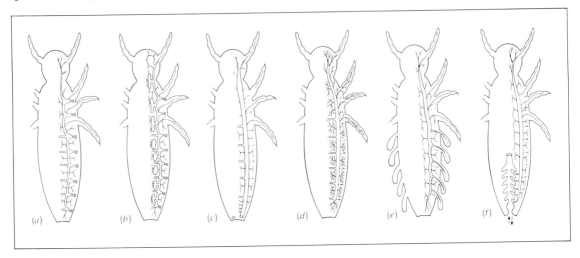

(*a*)   (*b*)   (*c*)   (*d*)   (*e*)   (*f*)

so that the system consists of tubes that all interconnect (Figure 1.32a). This pattern is commonly modified by the addition of enlarged parts of the tracheae, or air sacs, which are compressible (*b*). The volume of these sacs can be altered by movements of the body, thus pumping air in and out of the tracheal system. This is important, for in large and highly active insects diffusion alone does not provide sufficient gas exchange. The tracheae themselves have internal spiral ribs and are rather incompressible; the large air sacs are therefore necessary for ventilation. It is possible to achieve a unidirectional stream of air through the larger tracheal stems by having the spiracles open and close in synchrony with the respiratory movements, but out of phase with each other.

A modification characteristic of many aquatic insects is shown in Figure 1.32c. Most of the spiracles are nonfunctional, and only the two hindmost open to the outside. They are located so that the insect, by penetrating the water surface with the tip of the abdomen, can make contact with the atmosphere and obtain gas exchange, either by diffu-

sion, or more effectively, by respiratory movements.

The tracheal system may be completely closed, without any opening to the surface, although it is filled with air (Figure 1.32d). In this event gas exchange must be entirely by diffusion through the cuticle. Many small aquatic insects can obtain sufficient gas exchange in this way; their cuticle is relatively thin, and they can remain submerged without the need to make contact with the atmosphere. Yet the tracheal system is necessary for gas transport inside the animal because diffusion through the tissues would be too slow. As we saw earlier, diffusion of oxygen in air is some 300 000 times faster than in water; this clearly points out the advantage of a tracheal system in conveying gases when the circulation of blood is unsuited for this purpose.

A further development on the same theme again represents an aquatic insect (Figure 1.32e). The closed tracheal system extends into abdominal appendages or ''gills,'' which, with their relatively large surface and thin cuticle, permit effective gas

INSECT TRACHEOLES Tracheoles in the tymbal muscle of a cicada (*Tibicen* sp.). The tracheoles (Tr) are cut in cross section, except for one which is cut tangentially and shows the spirally ribbed wall. The diameter of these tracheoles is about 0.5 μm.

The distance between the two labeled tracheoles is about 6.5 μm (i.e., about the diameter of a human red blood cell). The light areas (M) are muscle fibers in cross section; the dark areas are mitochondria. [Courtesy of David S. Smith, University of Miami]

exchange between the water and air within the tracheal system. Such tracheal gills are found, for example, in mayfly larvae (Ephemeroptera).

Another variation is shown in Figure 1.32*f*. The tracheal gills are located within the lumen of the rectum, and ventilation takes place by moving water in and out of the rectum. This system is found in some dragonfly larvae. These larvae may also use the water in the rectum for locomotion by expelling it rapidly, thus moving by jet propulsion like squid and octopus.

Diffusion and ventilation

Diffusion alone suffices for gas exchange within the tracheal system of many small insects and in relatively inactive large insects. As an example we will use a large wood-boring larva of the genus *Cossus* (goat moth), which was studied by Krogh (1920). This larva weighs about 3.4 g and is 60 mm long. There are nine pairs of spiracles along the body, and careful measurements of the tracheae leading from these showed that the aggregate cross-sectional area of all the tracheae combined was 6.7 mm^2. The average length of the tracheal system was 6 mm. Curiously, as the tracheae branched and subdivided, the aggregate cross-sectional area of the system remained constant, not changing much with the distance from the spiracles.

Therefore, diffusion through the entire tracheal system can be represented by a single cylindrical tube of 6.7 mm^2 cross-sectional area and a length of 6 mm. The oxygen consumed by the animal, 0.3 μl O$_2$ s^{-1}, will diffuse through a tube of these dimensions if the partial pressure difference between the two ends is 11 mm Hg (1.5 kPa). This means that, with an oxygen pressure in the atmosphere of 155 mm Hg (20.7 kPa), the tissues could still have an oxygen tension of 144 mm Hg (19.2 kPa). Obviously, for the *Cossus* larva an adequate oxygen supply is secured by diffusion alone, even if the metabolic rate during activity should be increased several-fold.

Intuitively it may seem that the diffusion through 6 mm of air would be slow, but because the diffusion of oxygen in air is 300 000 times faster than in water, the diffusion through 6 mm of air is as fast as the diffusion through a water layer of 0.02 μm. The greatest barrier to oxygen reaching the tissues is therefore likely to be between the finest branches of the tracheoles and the cells. In very active tissues, such as insect flight muscles, electron micrographs reveal that the tracheoles are as close as 0.07 μm to a mitochondrion.

A quantitative consideration of the branched tracheal system gives information about another aspect of gas exchange. As the tracheae divide and subdivide, the aggregate wall area of the system increases. In the *Cossus* larva the largest tracheae are about 0.6 mm in diameter and the finest tracheoles 0.001 mm. The total cross-sectional area of the tubes remains constant, and the wall area of the tracheoles therefore increases to 600 times that of an equal length of the largest tracheae. Thus, the reduction in diameter alone increases the wall surface area 600-fold. This immediately tells us that virtually the entire area available for gas exchange

is in the finest branches. Consider now that the wall thickness in the tracheoles is less than one-tenth of the wall thickness of the large tracheae, and it becomes evident that diffusion through the walls of the larger branches must be insignificant.

The spiracles

The openings of the tracheal system to the outside, the spiracles, are highly complex structures that can be opened or closed to allow a variable amount of gas exchange. Their accurate control helps impede the loss of water.

The spiracles open more frequently and more widely at high temperature and when there is increased activity, in accord with the increased need for oxygen. The spiracles do not necessarily all open simultaneously; they are under control of the central nervous system, and out-of-phase opening and closing permit the control of air flow through the tracheal system.

The ventilation of the tracheal system, and especially the function of the spiracles, is influenced by both carbon dioxide and lack of oxygen. Carbon dioxide seems to be a primary stimulus for opening of the spiracles. If a tiny stream of carbon dioxide is directed toward one spiracle, only this spiracle will open; this shows that the single spiracle may respond independently. The necessary concentration of carbon dioxide is fairly small; in the cockroach, for example, 1% carbon dioxide in the air has a perceptible effect; 2% keeps the spiracles open; and 3% makes them remain widely open.

The effect of open spiracles on water loss is considerable. In the mealworm, which spends its entire life in the dry environment of flour, the control of the spiracles is very important. If the spiracles are made to remain open by adding carbon dioxide to the air, the water loss immediately increases several-fold (Figure 1.33).

Carbon dioxide is not the only agent that controls the spiracles; they can be made to open also by pure nitrogen (i.e., oxygen deficiency is a stimulus). This could be interpreted as an effect of carbon dioxide, for when there is a lack of oxygen

FIGURE 1.33 The water loss from a mealworm kept in dry air. During periods when the spiracles were made to remain open by adding carbon dioxide to the air, the water loss increased several-fold. [Mellanby 1934]

there is often an accumulation of acid metabolites (e.g., lactic acid), and acids increase the carbon dioxide tension. The anoxic effect could therefore be an indirect carbon dioxide effect. However, even a moderate oxygen deficiency stimulates respiration, and it is most likely that oxygen has a direct effect.

Ventilation

The tracheae are relatively rigid and do not easily collapse, but the spiral folding of the wall permits some accordion-like shortening. This can serve to reduce the volume sufficiently to contribute to the active ventilation of the system. If the tracheae are oval or flat in cross section, they are more compressible. However, effective ventilation depends on thin-walled dilations or air sacs that are connected with the larger tracheae, for when these are compressed, a large volume of air can be expelled. Expiration is usually the active phase, inspiration being passive.

In many insects the major respiratory movements are made by the abdomen. During flight, synchronous pressure changes and movements may increase the ventilation, and even the head may have some active ventilation owing to transmission of blood pressure. The ventilation volume may be quite large. As much as one-third or one-half of the total capacity of the air system can be emptied in one

expiration, giving a renewal of about half the volume of the respiratory system. This is much greater than in a mammal at rest, where the renewal for one breath is about one-tenth of the air contained in the system. In exercise the maximum renewal in a mammal is closer to two-thirds of the total volume of the respiratory system.

The breathing movements are synchronized with the opening and closing of spiracles and are similarly controlled by lack of oxygen or excess of carbon dioxide. Excess oxygen may cause complete arrest of respiratory movements. As high oxygen should have no influence on the carbon dioxide production, it probably affects the nervous system directly.

The breathing movements are directed by centers in the segmented ventral nerve chain, and isolated segments may perform respiratory movements. Coordination between the segments is brought about by higher centers located in the prothorax. The head, apparently, is not involved. Insects in general have fewer higher centers in the head, and decapitation has little effect on respiration and many other functions.

Respiration in aquatic insects

Insects that live in water have evolved from terrestrial forms and have retained many characteristics of their terrestrial ancestors. Almost all are freshwater forms; very few live in brackish water, and virtually no insect is truly marine.

Some aquatic insects have retained air breathing through spiracles. The spiracles are then usually reduced in number and located at the hind end of the body, which is brought into contact with the atmosphere. The spiracular openings are hydrophobic and often surrounded by hydrophobic hairs. The water surface is therefore easily broken and contact established with the air.

Small insects (e.g., mosquito larvae) depend on diffusion exchange between the atmosphere and the tracheal system, but larger larvae use respiratory movements to aid in the replacement of the tracheal air. The tracheae are often large and voluminous so they can contain enough air to serve as an oxygen store, permitting submersion for longer periods. There are strict limits to this solution, however, because an unlimited increase of air stores gives too much buoyancy and the insect will be unable to submerge.

External oxygen stores

Many insects carry air attached to the outside of the body, and the spiracles then open into this air mass. The air is held in place by nonwettable surfaces, often helped by hydrophobic hairs. The adult *Dytiscus* beetle has a large air space under the wings; when the beetle makes contact with the water surface, this air space is ventilated, and on submersion it serves as an air store into which the spiracles open.

A similar means is used by the back swimmer, *Notonecta*, which carries air attached to the ventral surface of the abdomen. The rather large air bubble gives it positive buoyancy, so that it must swim vigorously to descend from the surface and must attach itself to vegetation or other solid objects in order to remain submerged.

A simple experiment can demonstrate that the air carried by *Notonecta* has a respiratory role and is not related to buoyancy problems. If *Notonecta* is kept submerged in water that has been equilibrated with pure nitrogen instead of pure air, it will live for only 5 minutes. If it is kept submerged in water saturated with air, it will live for 6 hours. If it is submerged in water equilibrated with 100% O_2 instead of air, and is allowed to fill its air space with oxygen before submerging, it will live for only 35 minutes (Ege 1915). This seemingly paradoxical situation comes about because the bubble serves two purposes: It contains a store of oxygen that is gradually used up, but in addition it serves as a *diffusion gill* into which oxygen can diffuse from the surrounding water.

Let us examine what happens in each case. In the first case, the insect is allowed to carry with it a bubble of atmospheric air into oxygen-free water. The oxygen in the bubble disappears because the animal uses some; in addition, more oxygen diffuses rapidly into the surrounding oxygen-free water, and after 5 minutes all the oxygen is gone.

In the second case, the insect carries with it a bubble of air. After a while the animal's oxygen consumption has reduced the oxygen in the bubble to, say 5%, while the carbon dioxide has increased to 1%. The remainder of the gas in the bubble (94%) is nitrogen. Because the oxygen partial pressure in the bubble is lower than in the water, oxygen diffuses from the water into the bubble. The nitrogen in the bubble, on the other hand, is at 94%, and the water is in equilibrium with the normal 79% nitrogen in the atmosphere. The partial pressure of nitrogen in the bubble is, therefore, higher than in the water; nitrogen diffuses out; and the volume of the bubble gradually diminishes. In the end the bubble is lost, but in the case described it lasted for 6 hours.

In the third case, both bubble and water contained only oxygen. The bubble disappears at the rate the animal uses oxygen. Although the water is saturated with oxygen, this is of no help, for the partial pressures of oxygen in bubble and water are the same and no additional oxygen diffuses in from the water. Consequently, the bubble lasts a much shorter time than if nitrogen is also present, as in the second case.

An air bubble has one liability: How long it lasts depends on the depth to which the insect dives. With increasing depth in the water, the total pressure within the bubble increases, while the partial pressures of dissolved gases in the water remain unchanged. The bubble, therefore, is lost faster. For a 1-m increase in depth, the partial pressure of nitrogen in the bubble increases by about one-tenth, accelerating the rate of loss.

At a depth of 10 m, the total pressure within a submerged air mass is 2 atm. This high pressure causes a rapid loss of gas to the water, and an air mass at this depth could function only for a very short time. Therefore, there are limitations on the depth to which an insect with an air mass can dive. It has to return to the surface very often for renewal

AIR-BREATHING AQUATIC INSECTS Many aquatic insects breathe air. The mosquito larva (left) penetrates the surface film with a spiracular tube that opens at the tip of the abdomen. The leaflike structures to the right are anal papillae, whose primary function is in osmoregulation. The larva remains hanging still at the surface, while gas exchange takes place by diffusion. The water beetle, *Dytiscus* (right), makes contact briefly with the surface, only long enough to pull air into the space beneath its wings. The spiracles open into this air space, and gas exchange in the tracheae continues after the beetle has submerged again. [Courtesy of Thomas Eisner, Cornell University]

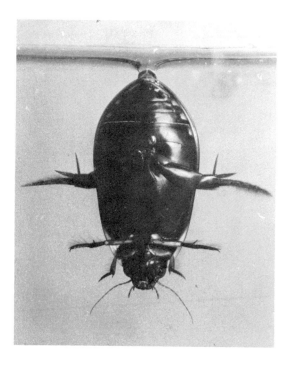

of the air; the distance to be traversed requires additional work and use of oxygen; exposure to predators is increased; and the total time spent at the bottom would be short.

For an insect that carries an air mass as a diffusion gill, the total obtainable oxygen far exceeds the initial supply carried in the mass when the insect submerged. Nitrogen is lost more slowly than oxygen diffuses in, primarily because nitrogen, because of its lower solubility, diffuses only half as fast as oxygen between air and water. Because of the initial nitrogen supply, an air mass will last long enough to obtain eight times as much oxygen by diffusion from the surrounding water as was originally in the bubble. This eightfold ratio remains constant and independent on the metabolic rate of the insect, of the exposed surface area, of the initial volume of the air mass, and of the thickness of the boundary layer of the water (meaning that turbulence in the water in the form of active ventilation will not influence the amount of obtainable oxygen) (Rahn and Paganelli 1968).

Plastron

Another type of respiratory device is known as a *plastron.* A plastron-breathing insect has part of the body surface densely covered with hydrophobic hairs that provide a nonwettable surface, the plastron, where air remains permanently. This device permits the insect to remain submerged without time limits. In the bug *Aphelocheirus,* there are 2 million single hairs per square millimeter of plastron surface, each single hair being 5 μm long (Thorpe and Crisp 1947).

The plastron breathers have a tremendous advantage above those that carry a single bubble as a diffusion gill, for water will not penetrate between the hydrophobic hairs. As a result the plastron serves as a noncompressible gill into which oxygen diffuses from the water. Because the volume remains constant, the net diffusion of nitrogen in this gill must be zero, and the total gas pressure within the air mass is negative relative to the gas tensions in the surrounding water. This is possible because the surface film of the water is supported by the hydrophobic hairs, and to force water into the air spaces between the hairs takes a pressure of 3.5 to 5 atm. As a result, plastron breathers can readily remain indefinitely at depths exceeding many meters.

Internal oxygen store

The hemipteran insects, *Buenoa* and *Anisops,* which are relatives of the back swimmer *Notonecta,* are remarkable for their ability to remain poised in midwater for several minutes, almost effortlessly. This is most unusual among insects, but it permits these predaceous animals to exploit the midwater zone in oxygen-poor waters. In small temporary bodies of water where fish are absent, the midwater zone is relatively free from predators. The bottom is dangerous because of voracious dragonfly and beetle larvae, and other predators hunt at the surface, but in midwater there is safety as well as little competition for food.

Both *Buenoa* and *Anisops* have large hemoglobin-filled cells in the abdomen. These serve to store oxygen, which is consumed during the dive. The insects carry with them only a small air mass, but this external supply is depleted only slowly while oxygen from the hemoglobin is being used. When the oxygen of the hemoglobin has been exhausted, the external oxygen is rapidly consumed, and the insects lose buoyancy and become heavier than water. Then they rapidly swim to the surface to replenish the oxygen stores.

The long-lasting internal oxygen store confers an advantage on these animals. After they have replenished their oxygen store at the surface, they swim actively to descend and then remain poised in midwater where they are in near-neutral buoyancy for an extended period while the oxygen stores are being used.

If *Anisops* is placed in air-free water, the duration of the dive is not much shorter. This shows that the relatively small air mass does not act as a gill, as it does in *Notonecta.* However, if the hemoglobin is poisoned with carbon dioxide, the dive is much shortened, for the hemoglobin now carries no oxygen. Even if the insect is permitted to take normal air with it, the situation is unchanged; the oxygen store is absent, and the external air mass does not help appreciably (Miller 1964).

Discontinuous or cyclic respiration

Many insects show a peculiar phenomenon characterized by a periodic or cyclic release of large amounts of carbon dioxide. Oxygen is taken up at a more or less constant rate, but carbon dioxide is released during brief periods or bursts. The carbon dioxide bursts may last for a few minutes and alternate regularly with long periods of minimal carbon dioxide release of several hours or even days.

A record of a single carbon dioxide burst from the pupa of a silkworm is shown in Figure 1.34. Such bursts may occur from once every week to

FIGURE 1.34 A record of the gas exchange of the pupa of a *Cecropia* silkworm (at 20 °C) shows steady oxygen uptake. The concurrent release of carbon dioxide is minimal until a major burst occurs after several hours. [Punt et al. 1957]

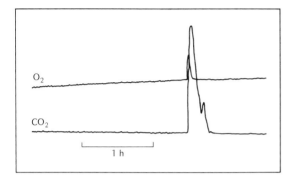

many times per hour, depending on temperature and metabolic rate. Frequencies as high as two bursts per minute have been recorded from ants (Lighton 1988).

Cyclic respiration has been observed in a variety of insect groups, including roaches, grasshoppers, beetles, and larvae and pupae as well as diapausing adult butterflies and moths. The phenomenon has been most carefully studied in the *Cecropia* moth because its large size makes it a convenient experimental object.

What happens between the bursts, during the interburst period? A minute amount of carbon dioxide is being released, but the concurrent oxygen uptake is many times greater, in some cases 20 or even 100 times as high. It is clear that gas exchange takes place through the spiracles and not through the cuticle, for if all the spiracles are sealed with wax, there is no detectable oxygen uptake or carbon dioxide release (i.e., the cuticle is highly impermeable to gases). This means that during the interburst period the spiracles permit inward passage of oxygen while virtually no carbon dioxide escapes. How is this possible? This and many other questions need answers.

Amount of carbon dioxide released

In a *Cecropia* pupa that weighs 6 g, more than 0.5 cm^2 CO_2 is released during a single burst. This volume exceeds the total volume of the tracheal system, so that even if the tracheal system were filled with carbon dioxide only, the carbon dioxide would still have to come from the blood or tissues during the short duration of the burst.

Accurate measurements of the carbon dioxide released from the pupa of another moth, *Agapema*, showed the tracheal volume to be 60 mm^3. Just before a burst, the tracheal air contained about 5.9% CO_2. This means that 3.5 mm^3 CO_2 was present within the tracheal system when the burst started. However, during the burst 30 mm^3 CO_2 was released; in other words, only 10% of the carbon dioxide could have come from the tracheal air and the remainder must have come from elsewhere.

Is metabolism a cyclic event?

The record in Figure 1.34 shows that the oxygen uptake, although continuing between the bursts, was intensified during the burst. This periodic nature of both carbon dioxide release and oxygen uptake raises the question of whether the metabolism of the pupa is cyclic, with short periods of sudden or cataclysmic carbon dioxide production interspersed with long quiet periods of virtual metabolic standstill.

This question can be answered by a simple experiment. If a single spiracle is kept permanently open by inserting a fine glass tube through the opening, the cyclic phenomenon disappears. We can, therefore, discount the hypothesis of cyclic biochemical events in the metabolism. It then becomes necessary to examine the function of the spiracles in relation to the cycles.

Movements of the spiracles

Direct observation of the spiracles with a microscope shows that they are widely open during the carbon dioxide burst. Shortly after the burst, the valve of each spiracle flutters for a short while, and then closes completely and remains closed for as

FIGURE 1.35 Record of the movements of the spiracular opening of a *Cecropia* pupa. With the pupa in pure oxygen (*a*), there is no flutter before the burst. WIth the pupa in atmospheric air (*b*), the burst is preceded by flutter. With the pupa in 10% oxygen (*c*), the spiracle flutters continuously. [Schneiderman 1960]

FIGURE 1.36 Composition of air obtained from within the tracheal system at various stages of cyclic respiration. A sudden rise in oxygen concentration occurs when the spiracles are fully open during the burst. Pupal weight, 8.4 g; temperature, 25 °C. [Levy and Schneiderman 1958]

long as an hour. Following this period the valve begins to pulse or flutter faintly, slightly enlarging the spiracular opening. The flutter continues, often for several hours, until the next period of wide opening and burst of carbon dioxide.

The movements of the spiracles of a *Cecropia* silkworm pupa are shown in Figure 1.35. The top record (*a*) was obtained with the insect in pure oxygen. After a long closed period, the spiracles opened, remained open for about 4 minutes, and then closed again. In atmospheric air (*b*), after a long closed period, the spiracles fluttered for about 10 minutes before they opened completely; then they fluttered for some time and gradually closed and remained closed until the next fluttering period started. The third graph (*c*) shows that 10% oxygen caused the spiracles to flutter continuously.

During the long period of flutter, when the spiracles are slightly open, the release of carbon diox-

ide is minimal, but oxygen uptake is almost continuous. We therefore are faced with the perplexing fact that, when the spiracle flutters, oxygen enters at a rate up to 100 times as high as that with which carbon dioxide leaves.

How can the spiracles, which are the only site of gas exchange, allow oxygen to enter but effectively prevent carbon dioxide from leaving? To answer this question we need information about the gas concentrations inside the tracheal system.

The insects we have discussed usually have no visible respiratory movements, and we might therefore assume that gas exchange between the tracheal system and outside air takes place by diffusion. However, the gas concentrations actually found in the tracheal system cannot be explained by diffusion processes only.

Schneiderman and his collaborators have succeeded in removing minute gas samples for analysis (Figure 1.36). As expected, they found the lowest carbon dioxide concentration, about 3%, immediately after a burst, and during the next 6 hours the carbon dioxide concentration rose very slowly to 6.5%. During the burst the oxygen concentration increased to nearly atmospheric concentrations, 18 to 20%, but after the burst it declined rapidly again and remained constant at 3.5% throughout the interburst period.

These amazing figures cannot be explained if diffusion is the only driving force for the movement of gases through the spiracles. However, gases can also be moved by pressure, and it was suggested by Buck (1958) that a slightly negative pressure within the tracheal system would cause a mass flow of air in through the spiracle.

It is possible to measure the pressure within the tracheal system by sealing a small glass capillary into one spiracle while leaving the others intact, and then connecting the tube to a manometer. During periods of flutter the pressure within the tracheal system is very close to atmospheric, although slightly negative. This slightly negative pressure causes an inward mass flow of air. When the spiracles close completely, the pressure in the tracheal system drops precipitously. As the spiracles remain closed, a considerable negative pressure develops, corresponding to the rapid decrease in oxygen concentration within the tracheal system immediately after the closing of the spiracles (Figure 1.36). When flutter begins again, the pressure rises stepwise, and when flutter continues regularly, the pressure has again returned to near-atmospheric (Levy and Schneiderman 1966b).

The rapid drop in pressure when the spiracles close is caused by the rapid consumption of oxygen from the closed tracheal system. Most of the carbon dioxide produced remains dissolved or buffered in blood and tissues, and therefore a negative pressure must develop. If the pupa is placed in pure oxygen before the spiracles close, so that the tracheal system is filled with oxygen only, the negative pressure becomes even greater. This is precisely what would be expected when the oxygen is consumed and no nitrogen is present.

If we combine all the information at hand, the facts are consistent with the following interpretation (Figure 1.37). The top graph shows flutter, followed by a period of open spiracles, and then a

FIGURE 1.37 Summary graph, showing the synchronous events in spiracular movements during cyclic respiration. [Levy and Schneiderman 1966a]

period of completely closed spiracles, after which the flutter is resumed. The carbon dioxide output (middle graph) is low during the closed and flutter periods; most of the carbon dioxide output takes place in a short burst during the open period. The oxygen uptake is recorded as being continuous; this corresponds to the actual consumption of oxygen by the tissues and can be measured by suitable manometric methods (Schneiderman and Williams 1955).

The composition of the tracheal air (Figure 1.37, bottom graph) shows that the carbon dioxide concentration falls during the burst from 6.5% to 3%, and again rises slowly during the interburst period. The changes in the oxygen concentration are more spectacular. When the spiracles open the oxygen rises rapidly to 18%, not far from atmospheric concentration. During the ensuing constricted period, the tracheal oxygen concentration falls steadily as the tissues use oxygen at a constant rate. When flutter begins, a negative pressure has already developed within the tracheal system, and inward bulk flow of air begins. This partly replenishes the oxygen, but nitrogen also enters and gradually increases. If there is, say 4% carbon dioxide and 4% oxygen in the tracheal system, the remainder must be nitrogen in a concentration of 92%.

Significance of cyclic respiration

Terrestrial animals are vulnerable to desiccation and insects often live in very dry environments. The diapausing pupa, which may remain encased for nearly a year from one summer to the next, has no possibility for replenishing lost water. It must be extremely economical in the use of water while permitting sufficient gas exchange for its metabolic processes. The spiracles must therefore be governed by a compromise between permitting oxygen uptake and preventing water loss.

We can now see a rationale for the periodic opening of the spiracles. The water loss depends on the vapor pressure within the tracheae (a function of temperature), the humidity in the outside atmosphere, and the spiracular opening and closing. It the spiracles are closed, water loss from the tracheal system is nil; when they are fully open, water loss is maximal.

As we have seen, partial closing (flutter) permits a sufficient oxygen uptake, for in addition to the inward bulk flow of air, the high oxygen concentration in the atmosphere gives a steep inward diffusion gradient. However, as the carbon dioxide concentration in the tracheal system builds up and reaches a limit of 6% or so, the spiracles open widely and permit this gas to escape (the carbon dioxide burst). If, instead, the carbon dioxide were permitted to leave continuously through open spiracles, as fast as it is formed, water would also be lost continuously. The cyclic nature of the gas exchange, which at first glance appears as a puzzling deviation, can be considered a clever device for water conservation.

We have now seen how gas exchange takes place in the respiratory organs and how special adaptations serve to provide for an adequate uptake of oxygen and the concomitant discharge of carbon dioxide. In the next chapter we shall discuss how the blood serves in the transport of these two important gases between the respiratory organs and the metabolizing tissues.

REFERENCES

Ackerman, R. A. (1980) Physiological and ecological aspects of gas exchange by sea turtle eggs. *Am. Zool. 20:*575–583.

Alkalay, I., Suetsugu, S., Constantine, H., and Stein, M. (1971) Carbon dioxide elimination across human skin. *Am. J. Physiol. 220:*1434–1436.

Bentley, P. J., Herreid, C. F., and Schmidt-Nielsen, K. (1967) Respiration of a monotreme, the echidna, *Tachyglossus aculeatus. Am. J. Physiol. 212:*957–961.

Bentley, P. J., and Shield, J. W. (1973) Ventilation of toad lungs in the absence of the buccopharyngeal pump. *Nature, Lond. 243:*538–539.

Berg, T., and Steen, J. B. (1965) Physiological mechanisms for aerial respiration in the eel. *Comp. Biochem. Physiol. 15:*469–484.

Bolin, B., Döös, B. R., Jäger, J., and Warrick, R. A. (eds.) (1986) *The Greenhouse Effect, Climate Change, and Ecosystems.* New York: John Wiley & Sons. 541 pp.

Bourbon, J. R., and Rieutort, M. (1987) Pulmonary surfactant: Biochemistry, physiology, and pathology. *News Physiol. Sci. 2:*129–132.

Bretz, W. L., and Schmidt-Nielsen, K. (1972) Movement of gas in the respiratory system of the duck. *J. Exp. Biol. 56:*57–65.

Brown, C. E., and Muir, B. S. (1970) Analysis of ram ventilation of fish gills with application to skipjack tuna (*Katsuwonus pelamis*). *J. Fish. Res. Bd. Can. 27:*1637–1652.

Buck, J. (1958) Cyclic CO_2 release in insects. 4. A theory of mechanism. *Biol. Bull. 114:*118–140.

Burggren, W. W. (1982) ''Air gulping'' improves blood oxygen transport during aquatic hypoxia in the gold fish *Carassius auratus. Physiol. Zool. 55:*327–334.

Burggren, W., Dunn, J., and Barnard, K. (1979) Branchial circulation and gill morphometrics in the sturgeon *Acipenser transmontanus*, an ancient Chondrosteian fish. *Can. J. Zool. 57*:2160–2170.

Cameron, J. N., and Mecklenburg, T. A. (1973) Aerial gas exchange in the coconut crab, *Birgus latro*, with some notes on *Gecarcoidea lalandii. Respir. Physiol. 19*:245–261.

Carey, C. (1980) Adaptation of the avian egg to high altitude. *Am. Zool. 20*:449–459.

Clements, J. A., Nellenbogen, J., and Trahan, H. J. (1970) Pulmonary surfactant and evolution of the lungs. *Science 169*:603–604.

Craig, A. B., Jr. (1961a) Causes of loss of consciousness during underwater swimming. *J. Appl. Physiol. 16*:583–586.

Craig, A. B., Jr. (1961b). Underwater swimming and loss of consciousness. *J.A.M.A. 176*:255–258.

Darden, T. R. (1972) Respiratory adaptations of a fossorial mammal, the pocket gopher (*Thomomys bottae*). *J. Comp. Physiol. 78*:121–137.

Dejours, P. (1966) *Respiration*. New York: Oxford University Press. 244 pp.

Dejours, P. (1981) *Principles of Comparative Respiratory Physiology,* 2d ed. Amsterdam: North Holland. 253 pp.

Dickinson, R. E., and Cicerone, R. J. (1986) Future global warming from atmospheric trace gases. *Nature 319*:109–115.

Dolk, H. E., and Postma, N. (1927) Ueber die Haut- und die Lungenatmung von *Rana temporaria. Z. Vergl. Physiol. 5*:417–444.

Ege, R. (1915) On the respiratory function of the air stores carried by some aquatic insects (Corixidae, Dytiscidae, and *Notonecta*). *Z. Allg. Physiol. 17*:81–124.

Feder, M. E., and Burggren, W. W. (1985) Cutaneous gas exchange in vertebrates: Design, patterns, control, and implications. *Biol. Rev. 60*:1–45.

Freadman, M. A. (1981) Swimming energetics of striped bass (*Morone saxatilis*) and bluefish (*Pomatomus saltatrix*): Hydrodynamic correlates of locomotion and gill ventilation. *J. Exp. Biol. 90*:253–265.

Gatz, R. N., Crawford, E. C., Jr., and Piiper, J. (1974) Respiratory properties of the blood of a lungless and gill-less salamander, *Desmognathus fuscus. Respir. Physiol. 20*:33–41.

Gehr, P., Mwangi, D. K., Ammann, A., Maloiy, G. M. O., **Taylor, C. R., and Weibel, E.** (1981) Design of the mammalian respiratory system. 5. Scaling morphometric pulmonary diffusing capacity to body mass: Wild and domestic mammals. *Respir. Physiol. 44*:61–86.

Graham, J. B. (1974) Aquatic respiration in the sea snake, *Pelamis platurus. Respir. Physiol. 21*:1–7.

Goerke, J., and Clements, J. A. (1986) Surfactant. Macklem, P. T., and Meads, J. (eds). *Handb. Physiol. 3*:247–261.

Gray, I. E. (1954) Comparative study of the gill area of marine fishes. *Biol. Bull. 107*:219–225.

Grigg, G. C. (1970) Use of the first gill slits for water intake in a shark. *J. Exp. Biol. 52*:569–574.

Hansen, J., Johnson, D., Lacis, A., Lebedeff, S., Lee, P., Rind, D., and Russel, G. (1981) Climate impact of increasing atmospheric carbon dioxide. *Science 213*:957–966.

Harvey, E. N. (1928) The oxygen consumption of luminous bacteria. *J. Gen. Physiol. 11*:469–475.

Hazelhoff, E. H., and Evenhuis, H. H. (1952) Importance of the ''counter-current principle'' for the oxygen uptake in fishes. *Nature, Lond. 169*:77.

Herreid, C. F., Bretz, W. L., and Schmidt-Nielsen, K. (1968) Cutaneous gas exchange in bats. *Am. J. Physiol. 215*:506–508.

Hughes, G. M. (1960) A comparative study of gill ventilation in marine teleosts. *J. Exp. Biol. 37*:28–45.

Hughes, G. M. (1966) The dimensions of fish gills in relation to their function. *J. Exp. Biol. 45*:177–195.

Hughes, G. M., Knights, B., and Scammell, C. A. (1969) The distribution of P_{O_2} and hydrostatic pressure changes within the branchial chambers in relation to gill ventilation of the shore crab *Carcinus maenas* L. *J. Exp. Biol. 51*:203–220.

Hughes, G. M., and Shelton, G. (1958) The mechanism of gill ventilation in three freshwater teleosts. *J. Exp. Biol. 35*:807–823.

Hutchinson, V. H., Whitford, W. G., and Kohl, M. (1968) Relation of body size and surface area to gas exchange in anurans. *Physiol. Zool. 41*:65–85.

Johansen, K. (1966) Air breathing in the teleost, *Symbranchus marmoratus. Comp. Biochem. Physiol. 18*:383–395.

Johansen, K. (1968) Air-breathing fishes. *Sci. Am. 219*:102–111.

Johansen, K., Lenfant, C., Schmidt-Nielsen, K., and Petersen, J. A. (1968) Gas exchange and control of breathing in the electric eel, *Electrophorus electricus. Z. Vergl. Physiol. 61:*137–163.

Johnston, A. M., and Jukes, M. G. M. (1966) The respiratory response of the decerebrate domestic hen to inhaled carbon dioxide-air mixture. *J. Physiol. 184:*38–39P.

Joshi, M. C., Boyer, J. S., and Kramer, P. J. (1965) Growth, carbon dioxide exchange, transpiration, and transpiration ratio of pineapple. *Botan. Gaz. 126:*174–179.

Kluge, M., and Ting, I. P. (1978) *Crassulacean Acid Metabolism.* Ecological Studies, vol. 30. New York: Springer Verlag. 209 pp.

Krogh, A. (1920) Studien über Tracheenrespiration. 2 Ueber Gasdiffusion in den Tracheen. *Pflügers Arch. 179:*95–112.

Krogh, A. (1941) *The Comparative Physiology of Respiratory Mechanisms.* Philadelphia: University of Pennsylvania Press. 172 pp. Reprinted by Dover Publications, New York, 1968.

Kuhnen, G. (1986) O_2 and CO_2 concentrations in burrows of euthermic and hibernating golden hamsters. *Comp. Biochem. Physiol.. 84A:*517–522.

Kutchai, H. and Steen, J. B. (1971) Permeability of the shell and shell membranes of hens' eggs during development. *Resp. Physiol. 11:*265–278.

Lasiewski, R. C., and Calder, W. A., Jr. (1971) A preliminary allometric analysis of respiratory variables in resting birds. *Respir. Physiol. 11:*152–166.

Lenfant, C., Johansen, K., and Hanson, D. (1970) Bimodal gas exchange and ventilation-perfusion relationship in lower vertebrates. *Fed. Proc. 29:*1124–1129.

Levy, R. I., and Schneiderman, H. A. (1958) An experimental solution to the paradox of discontinuous respiration in insects. *Nature, Lond. 182:*491–493.

Levy, R. I., and Schneiderman, H. A. (1966a) Discontinuous respiration in insects. 2. The direct measurement and significance of changes in tracheal gas composition during the respiratory cycle of silkworm pupae. *J. Insect Physiol. 12:*83–104.

Levy, R. I., and Schneiderman, H. A. (1966b) Discontinuous respiration in insects. 4. Changes in intratracheal pressure during the respiratory cycle of silkworm pupae. *J. Insect Physiol. 12:*465–492.

Lighton, J. R. B. (1988) Discontinuous CO_2 emission in a small insect, the formicine ant *Camponotus vicinus. J. Exp. Biol. 134:*363–376.

Machia, L., and Hughes E. (1970) Atmospheric oxygen in 1967 to 1970. *Science 168:*1582–1584.

Magnussen, H., Willmer, H., and Scheid, P. (1976) Gas exchange in air sacs: Contribution to respiratory gas exchange in ducks. *Respir. Physiol. 26:*129–146.

Margaria, R., Milic-Emili, G., Petit, J. M., and Cavagna, G. (1960) Mechanical work of breathing during muscular exercise. *J. Appl. Physiol. 15:*354–358.

McDonald, D. G., Wood, C. M., and McMahon, B. R. (1980) Ventilation and oxygen consumption in the Dungeness crab, *Cancer magister. J. Exp. Zool. 213:*123–136.

Mellanby, K. (1934) The site of loss of water from insects. *Proc. R. Soc. Lond. B, 166:*139–149.

Miller, P. L. (1964) The possible role of haemoglobin in *Anisops* and *Buenoa* (Hemiptera:Notonectidae). *Proc. R. Entomol. Soc. Lond. A39:*166–175.

Monge C. C., León-Velarde, F., and Gómez de la Torre, G. (1988) Laying eggs at high altitude. *News Physiol. Sci. 3:*69–71.

Nicol, J. A. C. (1960) *The Biology of Marine Animals.* New York: Interscience. 707 pp.

Otis, A. B. (1954) The work of breathing. *Physiol. Rev. 34:*449–458.

Otis, A. B. (1964) Quantitative relationships in steady-state gas exchange. In *Handbook of Physiology,* sect. 3, *Respiration,* vol. I (W. O. Fenn and H. Rahn, eds.), pp. 681–698. Washington, D.C.: American Physiological Society.

Packard, M. J., Packard, G. C., and Boardman, T. J. (1982) Structure of eggshells and water relations of reptilian eggs. *Herpetologia, 38:*136–155.

Prange, H. D., and Ackerman, R. A. (1974) Oxygen consumption and mechanisms of gas exchange of green turtle (*Chelonia mydas*) eggs and hatchlings. *Copeia 1974:*758–763.

Punt, A., Parser, W. J., and Kuchlein, J. (1957) Oxygen uptake in insects with cyclic CO_2 release. *Biol. Bull. 112:*108–119.

Rahn, H., and Paganelli, C. V. (1968) Gas exchange in gas gills of diving insects. *Respir. Physiol. 5:*145–164.

Rahn, H., Rahn, K. B., Howell, B. J., Gans, C., and Tenney, S. M. (1971) Air breathing of the garfish (*Lepisosteus osseus*). *Respir. Physiol. 11:*285–307.

Ramanathan, V. (1988) The greenhouse theory of climate change: A test by an inadvertent global experiment. *Science. 240:*293–299.

Randall, D. J. (1968) Fish physiology. *Am. Zool. 8:*179–189.

Roberts, J. L. (1975) Active branchial and ram gill ventilation in fishes. *Biol. Bull. 148:*85–105.

Romijn, C., and Roos, J. (1938) The air space of the hen's egg and its changes during the period of incubation. *J. Physiol. 94:*365–379.

Scheid, P., and Piiper, J. (1972) Cross-current gas exchange in avian lungs: Effects of reversed parabronchial air flow in ducks. *Respir. Physiol. 16:*304–312.

Schmidt-Nielsen, K. (1972) *How Animals Work.* London: Cambridge University Press. 114 pp.

Schmidt-Nielsen, K., Kanwisher, J., Lasiewski, R. C., Cohn, J. E., and Bretz, W. L. (1969) Temperature regulation and respiration in the ostrich. *Condor 71:*341–352.

Schneiderman, H. A. (1960) Discontinuous respiration in insects: Role of the spiracles. *Biol. Bull. 119:*494–528.

Schneiderman, H. A., and Williams, C. M. (1955) An experimental analysis of the discontinuous respiration of the cecropia silkworm. *Biol. Bull. 109:*123–143.

Schumann, D., and Piiper, J. (1966) Der Sauerstoffbedarf der Atmung bei Fischen nach Messungen an der narkotisierten Schleie (*Tinca tinca*). *Pflügers Arch. 288:*15–26.

Soum, J. M. (1896) Recherches physiologiques sur l'appareil respiratoire des oiseaux. *Ann. Univ. Lyon 28:*1–126.

Steffensen, J. F. (1985) The transition between branchial pumping and ram ventilation in fishes: Energetic consequences and dependence on water oxygen tension. *J. Exp. Biol. 114:*141–150.

Szarek, S. R., Johnson, J. B., and Ting, I. P. (1973) Drought adaptation in *Opuntia basilaris:* Significance of recycling carbon through crassulacean acid metabolism. *Plant Physiol. 52:*539–541.

Thomas, H. J. (1954) The oxygen uptake of the lobster (*Homarus vulgaris* Edw.). *J. Exp. Biol. 31:*228–251.

Thorpe, W. H., and Crisp, D. J. (1947) Studies on plastron respiration. The biology of *Aphelocheirus* [Hemiptera, Aphelocheiridae (Naucoridae)] and the mechanism of plastron retention. *J. Exp. Biol. 24:*227–269.

Tucker, W. A. (1968) Respiratory physiology of house sparrows in relation to high-altitude flight. *J. Exp. Biol. 48:*55–66.

Wangensteen, O. D., Wilson, D., and Rahn, H. (1971) Diffusion of gases across the shell of the hen's egg. *Respir. Physiol. 11:*16–30.

Wangensteen, O. D., Wilson, D., and Rahn, H. (1974) Respiratory gas exchange of high altitude adapted chick embryos. *Respir. Physiol. 21:*61–70.

Weibel, E. R. (1979) Oxygen demand and the size of respiratory structures in mammals. In *Evolution of Respiratory Process,* vol. 13 (S. C. Wood and C. Lenfant, 3ds.), pp. 289–346. New York: Dekker.

Weibel, E. R. (1984). *The Pathway for Oxygen.* Cambridge, Mass: Harvard University Press. 425 pp.

Wigglesworth, V. B. (1972) *The Principles of Insect Physiology,* 7th ed. London: Chapman & Hall. 827 pp.

Williams, D. L. G., Seymour, R. S., and Kerourio, P. (1984) Structure of fossil dinosaur eggshell from the Aix Basin, France. *Palaeogeogr. Palaeoclimatol. Palaeoecol. 45:*23–27.

ADDITIONAL READING

Altman, P. L., and Dittmer, D. S. (eds.) (1971) *Biological Handbooks: Respiration and Circulation.* Bethesda: Federation of American Societies for Experimental Biology. 930 pp.

American Society of Zoologists (1980) Symposium on physiology of the avian egg (C. Carey, organizer). *Am. Zool. 20(2):*325–484.

Bartels, H., Dejours, P., Kellogg, R. H., and Mead, J. (1973) Glossary on respiration and gas exchange. *J. Appl. Physiol. 34:*549–558.

Barth, M. C., and Titus, J. G. (eds.) (1984) *Greenhouse Effect and Sea Level Rise: A Challenge for This Generation.* New York: Van Nostrand Reinhold. 325 pp.

Butler, P. J. (ed.) (1982) Control and co-ordination of respiration and circulation. *J. Exp. Biol. 100:*1–319.

Dejours, P. (1981) *Principles of Comparative Respiratory Physiology,* 2d ed. Amsterdam: North Holland, 265 pp.

Dejours, P. (1988) *Respiration in Water and Air.* Amsterdam: Elsevier. 179 pp.

Duncker, H.-R. (1972) Structure of avian lungs. *Respir. Physiol. 14:*44–63.

Fenn, W. O., and Rahn, H. (eds.) (1964, 1965)

Handbook of Physiology, sect. 3, *Respiration,* vol. 1, pp. 1–926 (1964); vol. 2, pp. 927–1696 (1965). Washington, D.C.: American Physiological Society.

Heath, A. G., and Mangum, C. (eds.) (1973) Influence of the environment on respiratory function. *Am. Zool. 13:*446–563.

Hughes, G. M. (ed.) (1976) *Respiration of Amphibious Vertebrates.* New York: Academic Press. 402 pp.

Hughes, G. M., and Morgan, M. (1973) The structure of fish gills in relation to their respiratory function. *Biol. Rev. 48:*419–475.

Johansen, K. (1971) Comparative physiology: Gas exchange and circulation in fishes. *Annu. Rev. Physiol. 33:*569–612.

Johansen, K., et al. (1970) Symposium on cardiorespiratory adaptations in the transition from water breathing to air breathing. *Fed. Proc. 29*(3):1118–1153.

Krogh, A. (1939). *The Comparative Physiology of Respiratory Mechanisms.* Philadelphia: University of Pennsylvania Press, 172 pp. Reprinted by Dover Publications, New York, 1968.

Miller, P. L. (1964) Respiration: Aerial gas transport. In *The Physiology of Insecta,* vol. III (M. Rockstein, ed.), pp. 557–615. New York: Academic Press.

Mines, A. H. (1986) *Respiratory Physiology,* 2d ed. New York: Raven Press. 168 pp.

Piiper, J., and Scheid, P. (1977) Comparative physiology of respiration: Functional analysis of gas exchange organs in vertebrates. *Int. Rev. Physiol. 14:*219–253.

Ramanathan, V. (1988) The greenhouse theory of climate change: A test by an inadvertent global experiment. *Science. 240:*293–300.

Randall, D. J., Burggren, W. W., Farrell, A. P., and Haswell, M. S. (1981) *The Evolution of Air Breathing in Vertebrates.* New York: Cambridge University Press. 134 pp.

Taylor, C. R., Karas, R. H., Weibel, E. R., and Hoppeler, H. (1987) Adaptive variation in the mammalian respiratory system in relation to energetic demand. *Respir. Physiol. 69:*1–127.

Weibel, E. R. (1984) *The Pathway for Oxygen.* Cambridge, Mass: Harvard University Press. 425 pp.

Weibel, E. R. (1987) Scaling of structural and functional variables in the respiratory system. *Annu. Rev. Physiol. 49:*147–157.

Weibel, E. R. and Taylor, C. R. (eds.) (1981) Design of the mammalian respiratory system. *Respir. Physiol. 44*(1):i–viii; 1–164.

West, J. B. (1985) *Everest: The Testing Place.* New York: McGraw-Hill, 187 pp.

Wigglesworth V. B. (1984) *Insect Physiology.* 8th ed. London: Chapman and Hall. 191 pp.

Wood, S. C., and Lenfant, C. (1979) *Evolution of Respiratory Processes: A Comparative Approach.* Lung Biology in Health and Disease, vol. 13. New York: Dekker. 370 pp.

An excellent source of references in all fields of animal physiology is:

Prosser, C. L. (ed.) (1973) *Comparative Animal Physiology,* Neural and Integrative Animal Physiology. pp. 1–776 (1991); Environmental and Metabolic Animal Physiology, pp. 1–578 (1991). New York, N.Y.: Wiley-Liss.

The following volume provides the opportunity to further pursue several of the topics introduced in the present animal physiology text:

Taylor, C. R., Johansen, K., and Bolis, L. (1982) *A Companion to Animal Physiology.* Cambridge: Cambridge University Press. 365 pp.

CHAPTER TWO

Blood

We have seen that the transport of oxygen and carbon dioxide by diffusion is insufficient, except for very small animals. Nearly all large animals, unless they have very low oxygen demands, have a distribution system that is designed around the movement of a fluid, the blood.

We tend to think of *gas transport* as the primary function of blood, but blood has many other functions that may not be immediately apparent (see Table 2.1). Insects, for example, do not use blood for gas transport; they use air-filled tubes. Yet insects do have blood that is pumped around in the body, for many other substances need to be transported faster than diffusion alone can provide.

Thus, blood serves to *transport nutrients* that are absorbed from the digestive tract and to carry excretory products to the organ of excretion. A variety of intermediary metabolic products, including hormones and other important compounds, need a transport system between production site and the site of use.

An important but often overlooked function of blood

is the *transmission of hydraulic force*, which is used in many processes, such as ultrafiltration in the kidney, locomotion of the earthworm, breaking of the shell in molting crustaceans, erection of the penis, etc.

In large animals with a high metabolic rate the blood is essential for the *transport of heat;* otherwise their internal organs would rapidly become overheated.

The ability to *coagulate* is an inherent characteristic of blood, necessary to reduce the loss of this valuable fluid when the vascular system has been damaged.

Most of the transport functions, as well as the transmission of force and movement of heat, can be carried out by nearly any aqueous medium. Exceptions are gas transport and coagulation, which are associated with highly complex biochemical properties of the blood.

In this chapter we shall be concerned primarily with the role of blood in the transport of gases and the properties that serve this purpose.

TABLE 2.1 The most important functions of blood.

1. *Transport of nutrients* from digestive tract to tissues; to and from storage organs (e.g., adipose tissue, liver)
2. *Transport of metabolites* (e.g., lactic acid from muscle to liver), enabling metabolic specialization
3. *Transport of excretory products* from tissues to excretory organs; from organ of synthesis (e.g., urea in liver) to kidney
4. *Transport of gases* (oxygen and carbon dioxide) between respiratory organs and tissues; storage of oxygen
5. *Transport of hormones* (e.g., adrenaline [fast response]; growth hormone [slow response])
6. *Transport of cells* of nonrespiratory function (e.g., vertebrate leukocytes); insect blood lacks respiratory function, yet carries numerous types of blood cells
7. *Transport of heat* from deeper organs to surface for dissipation (essential for large animals with high metabolic rates)
8. *Transmission of force* (e.g., for locomotion in earthworms; for breaking shell during molting in crustaceans; for movement of organs such as penis, siphon of bivalve, extension of legs in spider; for ultrafiltration in capillaries of the kidneys)
9. *Coagulation,* inherent characteristic of many blood and hemolymph fluids; serves to protect against blood loss
10. *Maintenance of "milieu interieur"* suitable for cells in regard to pH, ions, nutrients, etc.

OXYGEN TRANSPORT IN BLOOD

Respiratory pigments

In many invertebrates oxygen is carried in the blood or hemolymph in simple physical solution. This aids in bringing oxygen from the surface to the various parts of the organism, for diffusion alone is too slow for any but the smallest organisms.

The amount of oxygen that can be carried in simple solution is small, however, and many highly organized animals (vertebrates almost without exception) have blood that can bind larger quantities of oxygen reversibly, thus greatly increasing the amount of oxygen carried. In mammalian blood, the amount of physically dissolved oxygen is about

0.2 ml O_2 per 100 ml blood, and the amount bound reversibly to hemoglobin is up to some 100 times as great, about 20 ml O_2 per 100 ml blood. Dissolved oxygen is therefore of minuscule importance compared with the oxygen in hemoglobin.

The substances we know as oxygen carriers in blood are proteins that contain a metal (commonly iron or copper). Usually they are colored, and therefore they are often called *respiratory pigments*. The commonest ones are listed in Table 2.2.

Blood corpuscles

In some animals the respiratory pigment occurs dissolved in the blood fluid; in others (such as vertebrates) it is enclosed in cells and the blood fluid contains no dissolved respiratory pigment. The distribution between cells and blood fluid for the most common respiratory pigments is listed in Table 2.3, which in addition gives the molecular weights of the pigments in representative animals.

This table immediately reveals that when the pigments occur enclosed in cells, their molecular weights are relatively low, ranging from 20 000 to 120 000; if the pigments occur dissolved in plasma, their molecular weights are much higher, from 400 000 to several million. The sole exception is the hemoglobin in the insect *Chironomus*.

The large molecules of respiratory pigments dissolved in the plasma are, in fact, aggregates of smaller molecules. They increase the total amount of the pigment without increasing the number of protein molecules dissolved in the blood plasma. A large increase in the number of dissolved particles would raise the colloidal osmotic pressure of the plasma, which in turn would influence many other physiological processes such as the passage of fluid through capillary walls and ultrafiltration (the initial process in the formation of urine in the kidney).

It has been said that enclosing a high concentration of hemoglobin within cells has the advantage of decreasing the viscosity of the blood as compared with a hemoglobin solution with the same oxygen-carrying capacity. Mammalian plasma

TABLE 2.2 Common respiratory pigments and examples of their occurrence in animals.

Pigment	Description	Molecular weight	Occurrence in animals
Hemocyanin	Copper-containing protein, carried in solution	300 000– 9 000 000	Molluscs: chitons, cephalopods, prosobranch and pulmonate gastropods, not lamellibranchs Arthropods, crabs, lobsters Arachnomorphs: *Limulus, Euscorpius*
Hemerythrin	Iron-containing protein, always in cells, nonporphyrin structure	108 000	Sipunculids: all species examined Polychaetes: *Magelona* Priapulids: *Halicryptus, Priapulus* Brachiopods: *Lingula*
Chlorocruorin	Iron-porphyrin protein, carried in solution	2 750 000	Restricted to four families of Polychaetes: Sabellidae, Serpulidae, Chlorhaemidae, Ampharetidae; prosthetic group alone found in starfish (*Luidia, Astropecten*)
Hemoglobin	Iron-porphyrin protein, carried in solution or in cells; most extensively distributed pigment	17 000– 3 000 000	Vertebrates: almost all, except leptocephalus larvae and some Antarctic fish (*Chaenichthys*) Echinoderms: sea cucumbers Molluscs: *Planorbis*, Pismo clam (*Tivela*) Arthropods: insects (*Chironomus, Gastrophilus*); crustacea (*Daphnia, Artemia*) Annelids: *Lumbricus, Tubifex, Arenicola, Spirorbis* (some species have hemoglobin, some chlorocruorin, others no blood pigment), *Serpula* (both hemoglobin and chlorocruorin) Nematodes: *Ascaris* Flatworms: parasitic trematodes Protozoa: *Paramecium, Tetrahymena* Plants: yeasts, *Neurospora*, root nodules of leguminous plants (clover, alfalfa)

contains about 7% protein, and the red blood cells contain about 35% hemoglobin. If all the hemoglobin from the red cells were carried dissolved in the blood instead of enclosed in the red blood cells, the total protein concentration would be about 20%, or three times as high as the typical mammalian plasma protein concentration.

The argument goes that a 20% protein solution is highly viscous or syrupy – far more viscous than blood. However, if we first measure the viscosity of a blood sample, and then disrupt the red cells with ultrasound so that the hemoglobin is set free, the resulting hemoglobin solution is less than half as viscous as blood with the hemoglobin carried in the cells (Figure 2.1). It is therefore clear that the reason for enclosing the hemoglobin within cells is not directly related to the question of reduced viscosity. However, the viscous properties of blood are very complex and will be discussed further in Chapter 3.

Aside from viscosity, other advantages are gained by enclosing hemoglobin in cells. The most important is probably that the chemical environment within the red cell can differ from that of the blood plasma. The reaction between oxygen and hemoglobin is greatly influenced by inorganic ions as well as by certain organic compounds, notably organic phosphates, and hemoglobin located within cells can be

TABLE 2.3 Molecular weights and locations (in cells or in plasma) of respiratory pigments in various animals.

Pigment	In cells		In plasma	
	Animal	Molecular weight	Animal	Molecular weight
Hemoglobin	Mammals	ca. 68 000[a]	Oligochaetes	
	Birds	ca. 68 000[a]	*Lumbricus*	2 946 000
	Fish	ca. 68 000[a]	Polychaetes	
	Cyclostomes		*Arenicola*	3 000 000
	Lampetra	19 100	*Serpula*	3 000 000
	Myxine	23 100	Molluscs	
	Polychaetes		*Planorbis*	1 539 000
	Notomastus	36 000	Insects	
	Echinoderms		*Chironomus*	31 400
	Thyone	23 600		
	Molluscs			
	Arca	33 600		
	Insects			
	Gastrophilus	34 000		
Chlorocruorin			Polychaetes	
			Spirographis	3 400 000
Hemerythrin	*Sipunculus*	66 000		
	Phascolosoma	120 000		
Hemocyanin			Gastropods	
			Helix	6 680 000
			Cephalopods	
			Rossia (squid)	3 316 000
			Octopus	2 785 000
			Eledone	2 791 000
			Arachnids	
			Limulus	1 300 000
			Crustacea	
			Pandalus	397 000
			Palinurus	447 000
			Nephrops	812 000
			Homarus	803 000

[a] The molecular weight of muscle hemoglobin (myoglobin) is 17 000.

provided with a separate well-adjusted environment different from the plasma.

Size and shape of the red cell

The red blood cells of mammals are round, slightly biconcave discs. There is one exception: All members of the camel family have oval red cells. The diameter of the red cell is characteristic for each species, the range for mammals being from about 5 to 10 μm. There is no particular relationship between the size of the red cell and the size of the animal: The smallest mammalian red cell is found in a small Asiatic deer, but a mouse and an elephant have red cells of approximately the same size.

Mammalian red cells lack nuclei, although they

FIGURE 2.1 Blood viscosity increases with increasing hemoglobin content. Blood viscosity is here expressed relative to the viscosity of pure water (water = 1.00). The viscosity of goat blood containing intact red cells is shown in the upper curve. The lower curve shows the relative viscosity of hemolyzed blood. The observed plasma viscosity of the same blood samples is shown by the dashed line. A comparison of points A, B, and C shows that a hemoglobin solution has a relative viscosity about half that of intact blood with the same total hemoglobin content. Normal human blood has a hemoglobin concentration of 15 g per 100 ml. [Schmidt-Nielsen and Taylor 1968]

have a life span of about 100 days and carry out complex metabolic functions. The red cells of all other vertebrates – birds, reptiles, amphibians, and fish – have nuclei and are almost universally oval. Frequently they are much larger than mammalian cells; the red cells of salamanders, for example, have 100 or 200 times as great a volume as a mammalian red cell. The importance, if any, of the nucleated versus a nonnucleated red cell is not understood. Neither is it clear whether the size of the red cell is of functional importance.

The only vertebrates that lack hemoglobin and red cells are the larvae of eels (leptocephalus larvae) and a few Antarctic fish of the family Chaenichthyidae.

OXYGEN DISSOCIATION CURVES

The property of blood that is of greatest importance to oxygen transport is the reversible binding of oxygen to the hemoglobin molecule. This binding can be written as an ordinary reversible chemical reaction:

$$Hb + O_2 \rightleftharpoons HbO_2$$

At high oxygen concentration the hemoglobin (Hb) combines with oxygen to form oxyhemoglobin (HbO_2) and the reaction goes to the right. At low concentration, oxygen is given up again and the reaction proceeds to the left. If the oxygen concentration is reduced to zero, the hemoglobin gives up all the oxygen it carries.

Each iron atom in the hemoglobin molecule binds one oxygen molecule, and when at high oxygen concentration (or pressure) all available binding sites are occupied, the hemoglobin is fully saturated and cannot take up any more oxygen. At any given oxygen concentration there is a definite proportion between the amounts of hemoglobin and oxyhemoglobin.

If we plot the amount of oxyhemoglobin present at each oxygen concentration, we obtain an *oxygen–hemoglobin dissociation curve*, which describes how the reaction depends on the oxygen concentration or partial pressure. Such a dissociation curve is depicted in Figure 2.2. This curve shows that pigeon blood is fully saturated with oxygen (100% HbO_2) at or above about 80 to 100 mm Hg (10 to 13 kPa). At still higher oxygen pressure, the hemoglobin takes on no more oxygen, and we say it is fully saturated. At lower oxygen pressure, however, it gives off oxygen, and at $P_{O_2} = 30$ mm Hg (4.0 kPa), half the hemoglobin is present as oxyhemoglobin, the other half as hemoglobin. This particular oxygen pressure is re-

FIGURE 2.2 Oxygen dissociation curve for pigeon blood: (a) curve determined for normal temperature (41 °C), blood P_{CO_2} (35 mm Hg), and pH (7.5) for this bird; (b) curve determined after shifting the pH from 7.5 to 7.2 with unchanged P_{CO_2}. [Lutz et al. 1973]

3. Carbon dioxide
4. Organic phosphate compounds
5. Inorganic ions, ionic strength

Because of the great functional importance of these factors, we need more information about their effects.

Effect of temperature

If the dissociation curve is determined for a sample of mammalian blood at room temperature and then at the normal body temperature of the animal, the results are entirely different. This is because increased temperature weakens the bond between hemoglobin and oxygen and causes an increased dissociation of the bond. As a consequence, at higher temperature the hemoglobin gives up oxygen more readily and the dissociation curve is shifted to the right.

This is of physiological importance, especially in cold-blooded animals, because increased temperature is usually accompanied by an increased metabolic rate (or need for oxygen); therefore, it is an advantage that hemoglobin delivers oxygen more readily at higher temperature. Even in warm-blooded animals this may be of some importance – in fever or in exercise, for example, when there is a rise in the rate of oxygen consumption.

Effect of carbon dioxide and pH

Another important influence on the dissociation curve is the pH of the blood plasma. Increase in carbon dioxide or other acids lowers the pH of the plasma, and this shifts the dissociation curve to the right (Figure 2.2). The significance of this shift is that a high carbon dioxide concentration causes more oxygen to be given up at any given oxygen pressure. This effect is known as the *Bohr effect*. In the capillaries of the tissues, as carbon dioxide enters the blood, the hemoglobin gives up a larger amount of oxygen than would be the case if there were no effect of carbon dioxide on the binding. The Bohr effect can therefore be said to facilitate the delivery of additional oxygen to the tissues.

The effect of carbon dioxide is not entirely due

ferred to as the *half-saturation pressure* (P_{50}). As the oxygen pressure is decreased further, more oxygen is given off, until the hemoglobin is fully deoxygenated at zero oxygen pressure.

The oxygen dissociation curves of other animals have a similar shape, but are not identical to the one shown in Figure 2.2. The blood of some animals has a higher affinity for oxygen (it gives up oxygen less readily), and the dissociation curve is located further to the left. The blood of others gives up oxygen more readily (the affinity for oxygen is lower), and the dissociation curve is located further to the right.

The oxygen dissociation curve that is observed for a given sample of blood is greatly affected by the conditions under which the determinations are made. This is because the binding of oxygen to hemoglobin is affected by several factors. The most important of these are:

1. Temperature
2. pH

FIGURE 2.3 The Bohr shift of hemoglobin in relation to body size. The hemoglobin of small mammals has a greater Bohr shift (i.e., is more acid-sensitive) than the hemoglobin of large mammals and, therefore, can release more oxygen at a given P_{O_2}. [Riggs 1960]

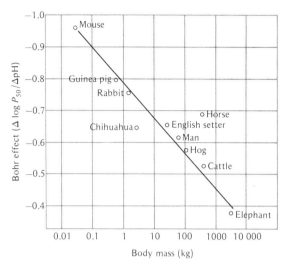

FIGURE 2.4 Oxygen dissociation curves of toadfish blood at 15 °C: (a) curve for whole blood in the absence of CO_2; (b) curve for whole blood in the presence of 25 mm Hg (3.3 kPa) CO_2. [Root et al. 1939]

to its effect on pH, however. If the pH of a blood sample is kept constant by using suitable buffers, an increase in carbon dioxide still shifts the curve to the right. In other words, the carbon dioxide lowers the oxygen affinity of hemoglobin even at a constant pH. This effect is caused by the binding of carbon dioxide directly to terminal amino groups of the hemoglobin molecule (i.e., this binding is not to the same site on the molecule where oxygen is bound).

The Bohr effect is not of equal magnitude in all mammals. We can express the Bohr effect as the shift in the P_{50} that occurs for a unit change in pH. We then find that the Bohr effect is dependent on body size and that the shift in P_{50} is greater in a mouse than in an elephant (Figure 2.3). In other words, mouse hemoglobin is more acid-sensitive than elephant hemoglobin. We shall later see that this is of importance to the delivery of oxygen to the tissues of small animals, which have a higher metabolic rate per gram than larger animals (Schmidt-Nielsen 1972).

The effect of acid on the dissociation curve is much more pronounced in some fish than in mammals; this stronger effect is known as the *Root effect* (Figure 2.4). The most significant consequence of the Root effect has to do with the function of the blood in the secretion of oxygen into the swimbladder (Chapter 10).

The peculiarity of the Root effect is that the blood is not fully saturated with oxygen at low pH, however high the oxygen pressure. This is explained by the fact that fish blood contains two different hemoglobin species, one highly sensitive to acid and the other acid-insensitive. In the presence of high carbon dioxide the former carries no oxygen, while the latter presents the usual dissociation curve for this hemoglobin species (Gillen and Riggs 1973).

Why do fish have two different hemoglobins? It is easy to see the value of the acid-sensitive species, but why have another kind that is acid-insensitive? Isn't this unnecessary?

The explanation is that when a fish in an emergency is making a maximal swimming effort, it produces a great deal of lactic acid. If the presence of this acid in the blood made all the hemoglobin incapable of binding oxygen, the fish would simply die of asphyxia. However, the presence of an acid-insensitive hemoglobin avoids this calamity.

Effect of organic phosphates

The function of the red cell in gas transport is intimately related to the metabolic activities of the cell itself, and we now understand many peculiarities of the oxygen dissociation curve that formerly were confusing.

The mammalian red cell has no nucleus, and for many years this cell was considered little more than an inert sac packed full of hemoglobin. The red cell, however, has an active carbohydrate metabolism that not only is essential to the viability of the cell, but also has profound effects on its function in oxygen transport. The red cell also has a high content of adenosine triphosphate (ATP) and an even higher level of 2,3-diphosphoglycerate (DPG).

In mammalian blood the important organic phosphate is diphosphoglycerate, but other vertebrates employ different compounds. In birds the dominant phosphate is inositol pentaphosphate (IPP), and in fish it is adenosine triphosphate (ATP) and/or guanosine triphosphate (GTP).

The presence of these organic phosphates helps explain why a purified solution of hemoglobin has a much higher oxygen affinity than whole blood (a hemoglobin solution has a dissociation curve far to the left of the curve for whole blood). If organic phosphates, notably DPB, are added to a hemoglobin solution, its oxygen affinity is greatly decreased and approaches that of intact cells. The decrease is attributable to a combination of the hemoglobin with the DPG, which alters the oxygen affinity (Chanutin and Curnish 1967; Benesch et al. 1968a,b).

The discovery of the DPG effect explains why, if we examine whole blood, we find many characteristics of the dissociation curve that remain unnoted if hemoglobin solutions are studied. For example, the DPG level of the red cell is higher in people living at high altitude than in people living at sea level. As a consequence, there is a shift to the right of the oxygen dissociation curve at high altitude. This shift to the right has been well known, but the mechanism remained unclear until it was discovered that the level of DPG increases in response to hypoxia (Aste-Salazar and Hurtado 1944; Lenfant et al. 1968).

The role of DPG and other phosphates explains the long-known fact that the oxygen affinity of blood stored in a blood bank increases with time; this is related to a gradual decrease in DPG during storage. It also helps explain why purified fetal and adult hemoglobins, when stripped of DPG, have similar oxygen affinities, although fetal blood has a higher oxygen affinity. Furthermore, a shift to the right of the maternal dissociation curve relative to the curve for a nonpregnant adult can now be understood, although the pregnant animal (as opposed to the fetus) has the same type of hemoglobin throughout.

Dissociation curve of fetal blood

For many mammals, including humans, the dissociation curve of fetal blood is located to the left of that of maternal blood (Figure 2.5). This is related to how the fetus obtains its oxygen by diffusion from the maternal blood. Because fetal blood has a higher affinity for oxygen than maternal blood, it can take up oxygen more readily. At a given oxygen pressure fetal blood contains more oxygen (has a higher percent saturation) than maternal blood at the same oxygen pressure; this facilitates the uptake of oxygen by the fetal blood in the placenta. The difference in dissociation curve between fetal and maternal blood is in part attributable to the fact that fetal hemoglobin is slightly different from maternal hemoglobin, and in part to a difference in the organic phosphate within the red cell. After birth fetal hemoglobin gradually disappears and is replaced by adult-type hemoglobin.

Displacement of the dissociation curve to the left, so that the blood more readily takes up oxygen, is found also in the chick embryo and frog tadpole.

FIGURE 2.5 Oxygen dissociation curves of fetal and maternal blood of goat. The higher oxygen affinity of the fetal blood helps in the transfer of oxygen to the fetal blood in the placenta. Broken lines indicate limits of nonpregnant adult goats. [Barcroft 1935]

FIGURE 2.6 The oxygen dissociation curve for tadpole blood is located to the left of that for adult frogs (i.e., tadpole blood has a higher oxygen affinity). Furthermore, tadpole blood is nearly acid-insensitive, whereas adult frog blood has a pronounced Bohr shift. [Riggs 1951]

Tadpole hemoglobin has a higher oxygen affinity than adult-type frog hemoglobin and, particularly interesting, it is insensitive to acid, whereas adult-type hemoglobin has a pronounced Bohr effect and is highly acid-sensitive (Figure 2.6).

If we consider the normal habitat of tadpoles, we can readily see the advantage; tadpoles often live in stagnant pools with low oxygen content and, at times, also high carbon dioxide content in the water. A high oxygen affinity therefore is an advantage. Because low oxygen and high carbon dioxide are likely to occur at the same time, a pronounced Bohr effect would be undesirable, for it would decrease the oxygen affinity of the blood when a high affinity is most needed.

Effect of altitude

The low atmospheric pressure at high altitude means that the partial pressure of oxygen is lower than at sea level. Animals that normally live in this environment of relative oxygen deficiency have corresponding adaptations in the oxygen dissociation curves of their blood. For example, the dissociation curve of the llama is located to the left of the usual range for mammals (Figure 2.7). The llama lives in the high Andes of South America, often above 5000 m, and the high oxygen affinity of its hemoglobin enables its blood to take up oxygen at the low pressures. This adaptation to altitude is an

FIGURE 2.7 Oxygen dissociation curves of llama and vicuña blood are located to the left of curves for other mammals (shaded area). The higher oxygen affinity of the blood of these animals aids the oxygen uptake at the low pressure of high altitude. [Hall et al. 1936]

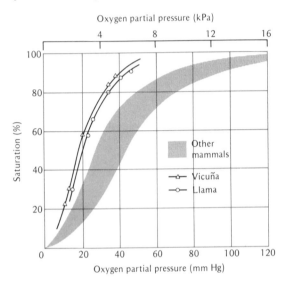

inborn characteristic, for the oxygen dissociation curve of a llama brought up in a zoo at sea level shows the same high oxygen affinity.

The same relationship has been reported for other animals: Species native to high altitude have blood with higher oxygen affinity. The bar-headed goose, which is known to fly over the high mountains in the Himalayas on its migration between the breeding grounds in Siberia and the wintering grounds in India, has a P_{50} about 10 mm Hg lower than close relatives from moderate altitudes; that is, its blood has a higher oxygen affinity (Black and Tenney 1980).

A study of deer mice (*Peromyscus maniculatus*) gave similar results. Ten different subspecies whose native habitats ranged from below sea level (-60 m in Death Valley) to the summits of the nearby Sierra Nevada (4350 m) showed a strong correlation between altitude and oxygen affinity of the

blood. To eliminate the possibility that the mice from high elevations were only temporarily altitude-acclimated, they had been kept for a prolonged period at low altitude. The difference from native low-altitude animals persisted. Furthermore, the progeny of wild-caught mice showed the same relationship: The subspecies from high altitude had greater oxygen affinity of the blood than animals from lower altitudes. The fact that the relationship persisted in second-generation animals indicates that the high-altitude adaptation in these animals is related to genetic factors (Snyder et al. 1982).

If a person moves to high altitude, he gradually becomes better adapted to the low oxygen pressure; he becomes acclimated to the altitude and performs better, both physically and mentally. However, the adaptation does not include an increased oxygen affinity of the blood. Paradoxically, the dissociation curve moves slightly to the right as altitude tolerance is improved. This is a result of other, more complex physiological adaptations (Lenfant et al. 1969).

The shift to the right of the dissociation curve of humans at high altitude (owing to increased DPG levels) is exactly the opposite of the change in the blood of the llama and other animals native to high altitude. This may be difficult to reconcile, but we should realize that the affinity of hemoglobin for oxygen is important at two different locations in the body: (1) when the hemoglobin takes up oxygen in the lung and (2) when the hemoglobin gives up oxygen to the tissues.

For uptake in the lung, a high oxygen affinity (shift to the left) increases the effectiveness of the uptake; in the tissues, on the other hand, a low oxygen affinity (shift to the right) facilitates the delivery of oxygen. It appears that the left-shift of the llama's dissociation curve is primarily an adaptation to improved oxygen uptake at the low pressures of high altitude, and that the right-shift in humans under similar conditions serves to augment the oxygen delivery to the tissues. Thus, a given dissociation curve can be regarded as a trade-off between a desirable high oxygen affinity in the lung and an equally desirable low affinity in the tissues.

The red cells of various mammals do not have equally high DPG levels, nor are their hemoglobins equally sensitive to the DPG level. The red cells of human, horse, dog, rabbit, guinea pig, and rat all have high levels of DPG, and their hemoglobins, when stripped of DPG, have high oxygen affinities. In contrast, sheep, goat, cow, and cat have low concentrations of DPG, and their hemoglobins interact only weakly with DPG (Bunn 1971). At present, these differences are difficult to understand. They seem to follow neither phylogenetic relationships nor any characteristic differences in the biology of the animals.

The situation appears even more confusing when we consider the other vertebrate classes. DPG seems to be absent from other vertebrates, except amphibians. Birds have no DPG, but instead have a high concentration of inositol pentaphosphate (IPP) in their red cells. IPP seems to be absent in all other vertebrates except turtles. Fish have high concentrations of adenosine triphosphate (ATP), as do snakes and lizards, and fish also have rather high concentrations of guanosine triphosphate (GTP), which is found in low concentrations in the red blood cells of all vertebrates (Bartlett 1980). Until now it had been difficult to establish any rational reason for this odd assortment, but it remains obvious that the organic phosphates are of great physiological importance.

The effect of a relative oxygen deficiency on fish shows an example where the response of the organic phosphates clearly is an advantage. When trout are exposed to lowered oxygen in the water, the ATP content of the red cell decreases (Table 2.4), and the oxygen affinity of the blood increases accordingly (i.e., P_{50} is decreased). As there is no change in the hemoglobin, the effect can be ascribed to the altered ATP concentration, which in this case appears as a useful adaptation.

TABLE 2.4 The half-saturation pressure (P_{50}) in the blood of rainbow trout acclimated to different oxygen levels in the water. The decrease in P_{50} (i.e., increased oxygen affinity) at lower oxygen in the water is correlated with the decrease in ATP in the red cells. [Tetens and Lykkeboe 1981]

	Oxygen in water (mm Hg)		
	50	80	150
P_{50} of blood (mm Hg)	16.8	21.7	24.1
ATP (mmol per liter red cells)	0.5	1.0	1.3

Diving animals

It might be expected that diving mammals, such as seals which are normally exposed to a relative oxygen deficiency when they dive, could gain an advantage by having blood with a high oxygen affinity. This is not so, however, and a little thought will clarify why they would gain no special advantage in this way.

When a seal breathes at the surface, it inhales ordinary atmospheric air. The uptake of oxygen into the blood in the lung therefore takes place under circumstances similar to those in other mammals, and the loading of the blood in the lung does not require any particularly high oxygen affinity. During diving, the unloading of oxygen in the tissues (primarily the central nervous system and the heart muscle) would not be aided by a high oxygen affinity of the blood; on the contrary, perhaps, a high oxygen affinity of the hemoglobin would tend to impede delivery of oxygen to the most vital organs. Accordingly, oxygen dissociation curves of diving animals generally differ little from those of similar nondiving animals.

Dissociation curve and body size

If the dissociation curves for various mammals are plotted together, we see that mammals of large body size have curves located to the left, and small animals have curves to the right (Figure 2.8).

FIGURE 2.8 Oxygen dissociation curves of blood of mammals of various sizes. At similar pH and CO_2 small mammals have a lower oxygen affinity. This helps in the delivery of oxygen in the tissues to sustain the high metabolic rate of a small animal. [Schmidt-Nielsen 1972]

globin (50% oxyhemoglobin and 50% hemoglobin), which is a convenient way of describing the location of the dissociation curve. The corresponding oxygen pressure, which we can consider an average unloading pressure, is about 22.5 mm Hg (3.0 kPa) for elephant blood, 28 mm Hg (3.7 kPa) for human blood, and about 45 mm Hg (6.0 kPa) for mouse blood.

These different unloading pressures of the blood can be related to the metabolic need for oxygen. Small animals have a higher rate of oxygen consumption per gram body weight than large animals. For example, the oxygen consumption of a horse weighing 700 kg is 1.7 μl O_2 g^{-1} min^{-1}, and that of a 20-g mouse is 28 μl O_2 g^{-1} min^{-1}. That is, the mean oxygen consumption of 1 g mouse tissue is about 16 times as high as that of 1 g horse tissue.

It follows that mouse tissue must be supplied with oxygen at a rate 16 times higher than horse tissue, and this can be achieved only by a diffusion gradient from capillary to cells 16 times as steep. This is shown diagrammatically in Figure 2.9. Each capillary supplies the surrounding tissue with oxygen, and the distance over which oxygen must diffuse is determined by the distance to the next capillary, the halfway point being the most distant point to which oxygen must diffuse. Horses and mice have approximately the same size red cells; therefore, their capillaries are also of approximately the same size, and this simplifies the analysis. To achieve a diffusion gradient in the mouse that is 16 times as steep as that in the horse, two variables can be adjusted. First, the diffusion distance can be reduced by reducing the intercapillary distance (i.e., an increase in capillary density in the tissue), and second, the unloading pressure for oxygen can be increased (i.e., a dissociation curve shifted to the right).

The capillary density is indeed larger in the small animal. In a mouse, for example, a 1-mm^2 cross section of a muscle may contain about 2000 capillaries; 1-mm^2 cross section of a horse muscle contains less than 1000 capillaries (Schmidt-Nielsen

As stated before, the loading–unloading reactions of hemoglobin are of importance at two locations: in the lung when oxygen is taken on and then in the tissues when oxygen is unloaded. Figure 2.8 shows that in all mammals the blood is practically 100% saturated at the normal oxygen pressure in the mammalian lung, which is about 100 mm Hg (13 kPa). The blood becomes fully loaded with oxygen in the lung, and there is no great difference among the species in this regard. Let us therefore examine the unloading of oxygen in the tissues.

The blood of a small mammal releases oxygen more readily because of its lower affinity for oxygen. Usually blood does not give up all its oxygen in the tissues; as an approximation we can say that on the average, about half the oxygen is given up. This corresponds to half-saturation of the hemo-

FIGURE 2.9 Each capillary is surrounded by an approximately cylindrical mass of tissue, which is supplied with oxygen by diffusion from the capillary. The steepness of the diffusion gradient depends on the radius of the tissue cylinder (R), the radius of the capillary (r), and the unloading pressure (tension) for oxygen.

and Pennycuik 1961). Figure 2.8 shows that the unloading pressure (P_{50}) for mouse blood is about 2.5 times higher than for horse blood. We can therefore conclude that the higher metabolic rate in the mouse and the necessary steep diffusion gradient for oxygen from capillary to tissue cells are sustained by a combination of a reduced diffusion path and a higher unloading pressure for oxygen.

Although the oxygen affinity of the blood of smaller mammals tends to be lower than in mammals of large size, we should remember that body size is only one of the many factors that affect the oxygen transport system. There are many deviations from the general trend (Bartels 1980). As we saw before, high-altitude animals as a group tend to have higher oxygen affinities (but within the group a decrease with size); burrowing animals, which often encounter low oxygen, likewise have high affinities; and there are deviations from the usual patterns that we just cannot account for.

Respiratory function of fish blood

When we compare the oxygen affinities of the blood of various fishes, we immediately meet some problems that are difficult to resolve. For example, at what temperature should the comparison be made? If we compare blood from two different fish, should they be examined at the same temperature, or should they be examined at the temperature at which each fish normally lives? The question is not easy to answer if we compare a tropical and an Arctic fish, but with fish from similar habitats we can use the same temperature.

The dissociation curves for fish hemoglobin show a wide range in oxygen affinity with characteristic differences related to the activity of the fish (Hall and McCutcheon 1938). The mackerel, an active, fast-swimming fish that normally moves in well-oxygenated water, has a dissociation curve to the right of other fish. This is in accord with what we have just learned: A low oxygen affinity facilitates the delivery of oxygen to the tissues, and in well-aerated water the arterial blood will still be 100% saturated.

The toadfish, whose curve lies to the extreme left, is a slow-moving and relatively sluggish bottom-living fish (Randall 1970). It is often found in less well oxygenated water, and it is highly tolerant of oxygen deprivation. The hemoglobin of the toadfish has a particularly high oxygen affinity, and this is in accord both with the environment in which the fish normally lives and with its relatively low metabolic rate, the demands on the loading mechanism apparently being more important than the unloading in the tissues.

Dissociation curves of invertebrate respiratory pigments

We saw earlier (Table 2.2) that invertebrates have a variety of respiratory pigments. In many, the oxygen affinity is very high, and the dissociation curve is located far to the left in the customary coordinate system. The hemoglobin from a bivalve mollusc, *Phacoides,* for example, has a P_{50} for oxygen of about 0.2 mm Hg (27 Pa) (Read 1962).

The hemoglobin of the *Chironomus* larva also has an extremely high oxygen affinity. The dissociation curve is influenced by both temperature and pH, and within a temperature range of 5 to 24°C, the P_{50} varies from 0.1 to 0.6 mm Hg (13 to 80 Pa). By mammalian standards this is an extremely high oxygen affinity. Characteristically, these animals live in environments that may periodically have very low oxygen concentrations, and it seems that their hemoglobins are of importance primarily when there is a lack of oxygen.

This suggestion is further supported by the fact that those invertebrates that do have hemoglobin often develop much higher hemoglobin concentrations when they are kept in water low in oxygen. An example is illustrated in Figure 2.10, which shows that *Daphnia*, a small planktonic fresh-water crustacean, develops relatively high hemoglobin concentrations in water low in oxygen. This has a profound effect on the survival of *Daphnia* in oxygen-deficient water. Survival is directly related to the hemoglobin concentration: The animals with

high hemoglobin survive in water where all low-hemoglobin *Daphnia* die from lack of oxygen (Kobayashi and Hoshi 1982). The same holds for the brine shrimp (*Artemia*), *Chironomus* larvae, and many others.

Hemerythrin differs chemically from hemoglobin in that it contains no porphyrin group. It is found only in a few marine forms (see Table 2.2), which show an interesting relationship between oxygen affinity and the ecological conditions under which the animals normally live.

The two sipunculid worms, *Dendrostomum* and *Siphonosoma*, have hemerythrin both in their blood and in their coelomic fluid. In *Dendrostomum* the tentacles around the mouth are richly supplied with blood; they are kept above the surface of the sand in which the animal lives and serve respiratory purposes.

Characteristically, in *Dendrostomum* the oxygen affinity of the coelomic fluid is higher than that of the blood (Figure 2.11). This means that oxygen taken up by the blood in the respiratory organ is readily transferred to the pigment of the coelomic fluid. In *Siphonosoma*, the blood and coelomic fluid do not differ much; in fact, the blood has a slightly

FIGURE 2.12 Oxygen dissociation curves of hemocyanin (from blood) and myoglobin (from radula muscle) of the amphineuran mollusc *Cryptochiton*. [Manwell 1958]

higher oxygen affinity. This is in accord with the habits of the worm: The oral tentacles are used for feeding but not for respiration; the worm remains in its burrow deep below the surface of the mud, and oxygen is taken up through the general body surface. Hence, both blood and coelomic fluids have a similar high oxygen affinity.

A similar relationship in the oxygen affinities of two different pigments is found in the amphineuran mollusc, *Cryptochiton*. This animal's blood contains hemocyanin, and the muscle of its tongue contains myoglobin. The blood has a relatively low oxygen affinity and the myoglobin a high oxygen affinity (Figure 2.12). This situation is completely analogous to what we find in vertebrates. In these, oxygen is transferred from the hemoglobin of the blood to the muscle hemoglobin (myoglobin), which has a far greater oxygen affinity and a dissociation curve displaced to the far left, with a P_{50} of about 0.5 mm Hg (0.07 kPa).

FACILITATED DIFFUSION

It is easy to see that hemoglobin carried in the blood vastly improves the transport of oxygen to the tissues. But hemoglobin occurs not only in blood but also in various tissues, in higher animals especially in muscle. Can this hemoglobin, lodged inside the cells and removed from circulation, serve any function?

An increased oxygen capacity of tissues can help smooth fluctuations when there are rapid increases in the demand for oxygen. For example, during muscle contraction the blood flow to a muscle may be reduced or momentarily stopped due to mechanical compression of the blood vessels by the contracting muscle. In this situation the myoglobin gives up its oxygen and thus acts as a short-time oxygen store (Millikan 1937). It is unlikely, however, that muscle contractions regularly cause complete cessation of blood flow. On the contrary, the circulation in heavily working muscles is usually greatly increased.

The discovery that hemoglobin greatly accelerates the diffusion rate of oxygen throws a different light on the presence of hemoglobin in muscle, as well as its wide occurrence in animals that live in environments that at least periodically subject them to low oxygen tensions. Some such animals are mentioned in Table 2.2, and it is probable that these also enjoy the advantages of accelerated or *facilitated diffusion* of oxygen.

Scholander (1960) discovered that although nitrogen diffuses through a hemoglobin solution a little more slowly than through water, the rate of oxygen transfer through the hemoglobin solution is much faster than through water.

Scholander arranged the following experiment. The solution to be tested (hemoglobin solution, blood plasma, or water) was absorbed in a highly porous membrane. One side of this membrane was then exposed to ordinary air and the other to vacuum. Both oxygen and nitrogen would now diffuse from the air through the membrane to the vacuum. If the membrane contained water or blood plasma, the ratio between the amount of oxygen and nitrogen diffusing through the liquid layer would be about 0.5. (This particular ratio could be predicted and comes about because the air contains about four

FIGURE 2.13 The diffusion of oxygen and nitrogen from air to a vacuum through a layer of blood plasma takes place at the constant ratio of 0.5. Through a hemoglobin solution the relative rate of oxygen diffusion is higher and increases with decreasing pressure. Shaded area at right indicates the water vapor pressure, which was identical on the two sides of the membrane. For further explanation, see text. [Scholander 1960]

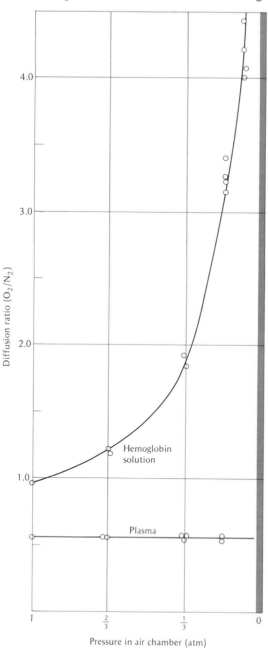

Pressure in air chamber (atm)

times as much nitrogen as oxygen, but nitrogen is only one-half as soluble as oxygen.)

If the air pressure above the membrane was lowered to 0.5 atm, only half as much air would diffuse across to the vacuum per unit time, but the ratio between the two gases remained unchanged at 0.5 volume oxygen per volume of nitrogen (Figure 2.13). If the plasma was replaced by a hemoglobin solution containing as much hemoglobin as normal blood (obtained by hemolyzing blood), the ratio of oxygen to nitrogen diffusion changed to nearly 1. This means that, relative to nitrogen, the diffusion of oxygen was speeded by the presence of hemoglobin. With further lowering of the pressure in the air chamber, the ratio increasingly favored the oxygen above the nitrogen, until the ratio indicated that the diffusion of oxygen was speeded up about eightfold relative to its diffusion in the absence of hemoglobin.

Facilitated diffusion is associated with the ability of hemoglobin to bind oxygen reversibly. Hemoglobin that has been changed to methemoglobin (achieved by oxidizing the divalent iron to trivalent iron) can no longer bind oxygen, and a solution of methemoglobin causes no facilitation of oxygen diffusion. However, myoglobin and various hemoglobins from vertebrates as well as invertebrates all show facilitated diffusion.

One important aspect of facilitated diffusion is that the enhanced diffusion occurs only when the oxygen pressure at the downhill end of the diffusion path is close to zero. In fact, maximal facilitation is found when the degree of saturation on the high and low sides of the hemoglobin layer are on the steepest part of the dissociation curve.

Do other respiratory pigments, such as hemocyanin, also facilitate diffusion? They do not. This seems related to the large molecular size of hemocyanin (several millions), which greatly reduces the motility of the hemocyanin molecule in solution.

Although the exact mechanism of oxygen transfer is inadequately understood, it seems that the normal thermal motility of the hemoglobin molecules in solution plays a role. If this motility is slowed (e.g., by adding gelatin to the hemoglobin solution), the facilitation is reduced. Myoglobin, which has a lower molecular weight (17 000) than hemoglobin (68 000), shows greater facilitation, roughly in inverse proportion to the square root of the molecular weights of the two pigments. This indicates that the difference results from the thermal motion of the pigment in solution (Hemmingsen 1965).

The main points of facilitated diffusion of oxygen in hemoglobin solutions can be summarized as follows: (1) the net diffusion of oxygen is accelerated when the exit pressure is low; (2) the transfer equals the sum of simple diffusion and the enhancement attributable to the pigment; (3) the facilitation increases with increasing hemoglobin concentration; and (4) the transfer rate is inversely related to the square root of the molecular weight of the pigment, so that a small hemoglobin molecule gives greater facilitation.

Physiological implications

In vertebrates hemoglobin is present in high concentration primarily in the red blood cells and secondarily in red muscles. It is doubtful that facilitated diffusion has much importance in the erythrocytes, for facilitation requires a near-zero oxygen pressure at the receiving end of the diffusion path. The red blood cells, which already have a very short diffusion distance, are rarely if ever exposed to near-zero oxygen pressures, except in the capillaries of heavily working muscles.

For myoglobin the situation is different. At some distance from a capillary the oxygen tension in a working muscle may readily approach zero. This establishes an ideal situation for facilitated diffusion, the oxygen tension in the capillary being somewhere between arterial and venous tension, and in the tissue (the receiving end of the diffusion path) nearly zero. A several-fold diffusion facilitation, such as was observed in laboratory experiments, would indeed be important in improving the

oxygen supply to the muscle. Experiments on pigeon breast muscle confirm that myoglobin may transport a significant fraction of the oxygen consumed by the muscle (Wittenberg et al. 1975).

It now appears particularly noteworthy that those invertebrate organisms that have hemoglobin frequently live in environments that, at least at times, have low oxygen tensions. When the oxygen supply is poor, the oxygen tension in the tissues is particularly low, and facilitation is most important. Hemoglobin is especially important for these animals, and many of them adapt to prolonged exposure to oxygen deficiency by increasing the amount of hemoglobin in their bodies. It seems that the major role of hemoglobin in these animals is to facilitate oxygen diffusion.

Indeed, some parasitic nematodes that live in the intestines of sheep respire actively at considerably higher rates than the calculated maximum rate of inward diffusion of oxygen, in the absence of facilitation (Rogers 1949). However, these animals have highly concentrated intracellular hemoglobin that would serve to augment the rate of oxygen diffusion at the very low oxygen partial pressure in the intestine of their host.

The role of hemoglobin in plants has been the subject of much speculation, but is now reasonably clear. Nitrogen-fixing plants, such as soy-beans and clover, have small nodules on the roots that contain bacteria that utilize atmospheric nitrogen to form nitrogen compounds (nitrogen fixation). These nodules contain a hemoglobin, called leghemoglobin, which is required for nitrogen fixation. It has a very high oxygen affinity ($P_{50} = 0.04$ mm Hg, 0.005 kPa), and the flux of leghemoglobin-bound oxygen exceeds the flux of dissolved oxygen by 10 000- to 100 000-fold. The oxygen delivered supports the formation of ATP, which is consumed in the complex reduction process involved in nitrogen fixation (Wittenberg 1977).

Evolutionary implications

It is curious that a molecule as complex as hemoglobin occurs so widely, yet in such a scattered or sporadic way. In addition to being nearly universally present in vertebrates, hemoglobin occurs here and there in a number of animal phyla, including protozoans, and in plants and bacteria. Often, it is found in only one or a few species within a large phylum, at times in a parasite and not in its free-living relative, and so on. Once the advantage of hemoglobin in facilitated diffusion has been established, however, the peculiar distribution of hemoglobin is easier to understand.

The basic structure of the heme part of the hemoglobin molecule is a porphyrin nucleus, which is universally present in plants and animals as part of cellular enzymes that belong to the cytochrome system. Relatively minor biochemical alterations can change this raw material into a hemoglobin-type pigment that binds oxygen reversibly. For such mutations to be permanently established requires some suitable evolutionary advantage that favors their maintenance.

The most obvious and immediate advantage of hemoglobin is the improvement in oxygen delivery due to facilitated diffusion. Once hemoglobin synthesis is established, further improvement in oxygen transport can be achieved by adding convective transport (i.e., blood and a circulatory system).

The establishment of a complete circulation system is a complex change that would require not only relatively minor biochemical changes in porphyrin, but also complex morphological and physiological alterations of the entire organism. Such drastic changes are, of course, possible, but can hardly be achieved as a single or few-step event. In contrast, an initial occurrence of hemoglobin in any tissue immediately endows the organism with the advantage of facilitated diffusion, giving a selective advantage to its maintenance. A later step would then be to permit this hemoglobin to enter an already existing blood system, thus further augmenting oxygen delivery by convection.

It follows from this hypothetical sequence of events that the scattered occurrence of hemoglobin is attributable to the universal presence of iron-porphyrin compounds, that facilitated diffusion gives the primary impetus to maintaining the biochemical mechanism for hemoglobin synthesis, and that adding hemoglobin to a preexisting convective system then becomes a logical later step in the evolutionary history.

CARBON DIOXIDE TRANSPORT IN BLOOD

As blood passes through the tissues and gives off oxygen, it concurrently takes up carbon dioxide; as it passes through the lungs, the reverse takes place. Let us examine the events in the lung as carbon dioxide is given off.

The venous blood of a mammal, as it reaches the lungs, contains about 550 cm^3 CO_2 per liter blood and has a carbon dioxide tension of about 46 mm Hg (6.1 kPa). As the blood leaves the lungs, it contains about 500 cm^3 CO_2 per liter blood at a tension of about 40 mm Hg. Thus, the blood has given up only a small amount of the total carbon dioxide it contains, the arteriovenous difference being about 50 ml CO_2 per liter blood. To understand carbon dioxide transport, we need to know more about carbon dioxide and its behavior in solution.

The total amount of carbon dioxide carried in the blood far exceeds the amount that would be dissolved in water at the carbon dioxide tension of the blood. This is because the largest part of the carbon dioxide is carried in chemical combination rather than as free dissolved carbon dioxide gas. The two most important such compounds are (1) bicarbonate ions and (2) a combination of carbon dioxide with hemoglobin. (Note that the carbon dioxide binds to terminal amino groups of the protein molecule, not to the binding site for oxygen.)

When carbon dioxide is dissolved in water, it combines with water to form carbonic acid (H_2CO_3). This can be described as follows:

$$CO_2 + H_2O \rightleftarrows H_2CO_3 \rightleftarrows H^+ + HCO_3^-$$
$$\rightleftarrows (H^+ + CO_3^{2-})$$

The H_2CO_3 dissociates as an acid into hydrogen ion and bicarbonate ion (HCO_3^-). Carbonic acid

(H_2CO_3) acts as a very weak acid with an apparent dissociation constant of 8.0×10^{-7} at 37°C ($pK = 6.1$).* The further dissociation of bicarbonate ion into one additional hydrogen ion and the carbonate ion (CO_3^{2-}) is included for completeness, but at the pH of living organisms the amount of carbonate ion present is minuscule and can usually be disregarded.

At the normal pH of mammalian blood, most of the total carbon dioxide present is in the form of bicarbonate ion. This can readily be calculated from the Henderson-Hasselbalch equation in the following way:

$$\frac{[H^+][A^-]}{[HA]} = K$$

$$[H^+] = K \cdot \frac{[HA]}{[A^-]}$$

$$pH = pK - \log \frac{[HA]}{[A^-]}$$

for $pK = 6.1$ and at $pH = 7.4$

$$-\log \frac{[HA]}{[A^-]} = pH - pK = 7.4 - 6.1 = 1.3$$

$$\log \frac{[HA]}{[A^-]} = -1.3 = 0.7 - 2$$

$$\frac{[HA]}{[A^-]} = \frac{[H_2CO_3]}{[HCO_3^-]} = 0.05 = \frac{1}{20}$$

*Carbonic acid is actually a stronger acid with a true pK value of 3.8. The reason for the discrepancy in pK values is that only about 0.5% of the dissolved carbon dioxide combines with water to form the acid. The bulk of the carbon dioxide remains as dissolved gas, and only the fraction 0.005 is hydrated to form H_2CO_3. In physiology it has been customary to consider that all carbon dioxide is in the form of carbonic acid; hence, the carbonic acid appears to be weaker than the molecular species, H_2CO_3, in fact is. If to the true pK for carbonic acid (3.8) we add the negative log of the hydration constant (log $0.005 = -2.3$), the sum is 6.1, which is the apparent pK for carbon dioxide as used in physiology. In this book we will follow physiological convention and use the apparent dissociation constant.

This means that at $pH = 7.4$ the ratio of carbonic acid to bicarbonate ion is 1 : 20; that is, for each one part of carbon dioxide present as acid, 20 times as much carbon dioxide is present as bicarbonate ion.

The hydrogen ion formed as a result of the ionic dissociation of the carbonic acid is buffered by various substances in the blood, and the effect of carbon dioxide on the pH of the blood is therefore only moderate. For example, arterial blood may have a normal pH of 7.45 and venous blood a pH of 7.42; the difference in pH caused by the uptake of carbon dioxide by the blood in the tissues thus is no more than 0.03.

The most important buffering substances in the blood are the carbonic acid–bicarbonate system, the phosphates, and the proteins in the blood. Proteins act as excellent buffers because they contain groups that can dissociate both as acids and as bases, the result being that proteins can either take up or give off hydrogen ions. Hemoglobin is the protein present in the largest amount in blood, and it has the greatest role in buffering; the plasma proteins are second in importance in this regard.

Carbon dioxide dissociation curve

The amount of carbon dioxide taken up by blood varies with its partial pressure. It is therefore possible to construct a *carbon dioxide dissociation curve* for blood, analogous to the oxygen dissociation curve. This is done by plotting on the ordinate the total amount of carbon dioxide in the blood and on the abscissa the various carbon dioxide partial pressures. The result is a curve as in Figure 2.14. This graph shows a slightly different dissociation curve for oxygenated and deoxygenated blood. This is because oxyhemoglobin is a slightly stronger acid than hemoglobin; hence, oxygenated blood binds slightly less carbon dioxide. This phenomenon is closely related to the Bohr effect, which merely shows the other side of the coin. When carbon dioxide (acid) is added to the blood, it pushes the equilibrium between oxyhemoglobin and hemoglobin in the direction of the weaker acid (hemo-

FIGURE 2.14 Carbon dioxide dissociation curves of mammalian blood. The curves for oxygenated and deoxygenated blood differ somewhat. The line A-V is the physiological dissociation curve in the body, describing the difference between arterial and venous blood. The dashed line indicates the amount of carbon dioxide held by physical solution in water. [Winton and Bayliss 1955]

globin). This tends to release more oxygen, which, in fact, is the Bohr effect.

In Figure 2.14 the point marked A stands for normal arterial blood, and the point marked V for mixed venous blood (venous blood is never completely deoxygenated). In the body the actual dissociation curve for carbon dioxide is the fully drawn line A–V, and this curve is said to be the *functional dissociation curve*.

Figure 2.14 also shows the dissociation curve for a sodium bicarbonate solution with the same total available base as in blood. The slope of the upper straight part of this curve represents the solubility of carbon dioxide in an aqueous medium. As carbon dioxide pressure is reduced, this dis-

solved carbon dioxide is given up. When the carbon dioxide pressure falls to about 10 mm Hg, the bicarbonate solution begins to give up carbon dioxide from the bicarbonate ion, leaving carbonate behind.

By exposing a solution of sodium bicarbonate to a vacuum, half the total carbon dioxide of the bicarbonate can be removed, leaving a solution of sodium carbonate (Na_2CO_3) behind. Sodium carbonate, however, does not give off carbon dioxide to a vacuum, and the final solution retains half the total carbon dioxide that was originally present as bicarbonate.

In contrast, whole blood exposed to a vacuum gives up all the carbon dioxide present. This is because other ions, primarily proteins, furnish the necessary anions. If separated plasma is subjected to a vacuum, it behaves more like a solution of sodium bicarbonate; the presence of red cells is necessary to obtain the normal dissociation curve shown in Figure 2.14.

Carbon dioxide and aquatic respiration

The amount of carbon dioxide produced in metabolism is closely related to the amount of oxygen consumed. The ratio between the respiratory exchange of carbon dioxide and oxygen, the *respiratory exchange ratio* (also called the *respiratory quotient*), normally remains between 0.7 and 1.0 (see Chapter 5). For the following discussion we shall consider an animal in which the exchange ratio is 1.0; that is, for each molecule of oxygen taken up, exactly one molecule of carbon dioxide is given off.

An air-breathing animal – a mammal, for example – with a gas-exchange ratio of 1.0 has an easily predicted carbon dioxide concentration in exhaled air. If the 21% O_2 in inhaled air is reduced to 16% in the exhaled air, the exhaled air of necessity contains 5% CO_2. If, instead, we consider the change in partial pressure of the two gases, we can say that if oxygen in the respired air is reduced by 50 mm Hg, the carbon dioxide is increased by the same amount, 50 mm Hg (6.7 kPa). Each milli-

FIGURE 2.15 Relation between P_{O_2} and P_{CO_2} in air and in water, when the respiratory exchange ratio (RQ) between oxygen and carbon dioxide is unity (RQ = 1). For further explanation, see text. [Rahn 1966]

meter change in the partial pressure of oxygen in the respiratory air results in an equal and opposite change in the carbon dioxide.

If the exchange ratio of unity is applied to gas exchange in water, the amounts of oxygen and carbon dioxide are, of course, equal, but the changes in the partial pressures of the two gases are very different because their solubilities are so different.

For example, if a fish removes two-thirds of the oxygen present in the water, thus reducing the P_{O_2} from 150 to 50 mm Hg (20 to 6.7 kPa), the increase in P_{CO_2} is about 3 mm Hg (provided the water is unbuffered; a buffered medium may give an even smaller increase). Depending on the temperature, the 100 mm Hg reduction in P_{O_2} may mean that, say, 5 cm^3 O_2 was removed from 1 liter of water. Returning 5 ml CO_2 increases the P_{CO_2} by only about 3 mm Hg because the solubility of carbon dioxide is very high, roughly 30 times as high as the solubility of oxygen.

This relationship in the carbon dioxide tensions resulting from respiration in air and in water is expressed in Figure 2.15. This graph shows the P_{CO_2} that results from any given decrease in oxygen in the respired medium. Even if a fish were to remove

all oxygen in the water, the carbon dioxide in the water would increase to only 5 mm Hg (0.67 kPa) partial pressure. Because carbon dioxide diffuses across the gill epithelium nearly as readily as oxygen, the blood P_{CO_2} of the fish can never exceed 5 mm Hg (in air-breathing vertebrates it is about 10 times as high).

As a consequence, in regard to the effect on acid–base regulation, air respiration and aquatic respiration are very different. This is particularly important for animals that shift between air and water respiration, such as amphibians and air-breathing fish. For these animals a successful shift between the two media requires substantial physiological adjustments in their acid–base regulation.

Blood carbon dioxide and pH

We have seen that the tension or partial pressure of carbon dioxide in the arterial blood is very different in aquatic and terrestial animals. The total amount of carbon dioxide present also varies greatly: It is usually low in fish (often less than 100 cm^3 CO_2 per liter blood) and may be very high in turtles (in excess of 1000 cm^3 CO_2 per liter blood). Clearly, there is uniformity neither in the carbon dioxide tension nor in the total carbon dioxide content of vertebrate blood. In contrast, we find a much greater uniformity in the pH of the blood of various vertebrates than would be expected from the large variation in carbon dioxide. This results from corresponding adjustments in the acid–base balance, achieved primarily through the amount of sodium (often called *free bicarbonate*) present in the blood.

The pH of the blood of a variety of vertebrates, both warm- and cold-blooded, usually varies around values between about 7.4 and 8.2. If the data are plotted against the temperature of animals, however, we obtain a more regular distribution than indicated by this rather wide pH range (Figure 2.16). At temperatures near 0°C, the pH often reaches values well over 8.0, and at 40°C it is closer to 7.5. At any given temperature, however, the blood pH of a variety of animals falls within a narrow range of about 0.2 pH units. It is important to remember

FIGURE 2.16 Blood pH values for vertebrates of all classes, determined at the normal body temperature for each animal. Shaded band indicates range of observed values; central fully drawn line is the pH of water at neutrality as a function of temperature. [Rahn 1966]

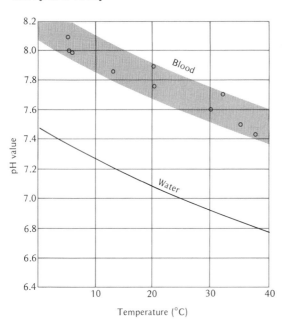

that the dissociation of water, and therefore the pH of neutrality, changes with temperature. The neutrality point is at 7.0 only at room temperature (25°C) and changes with temperature according to the line in the center of Figure 2.16, marked ''water.''

The blood pH of all vertebrates differs from neutrality by about 0.6 pH units, always in the alkaline direction (see Figure 2.16). Consequently, although vertebrates differ considerably in regard to blood pH value, they are amazingly uniform in the extent of deviation from neutrality when measured at the normal temperature for the animal. There is a similar uniformity in the maintenance of the intracellular pH, which is of great importance in providing suitable conditions for the activity of metabolic enzymes and so forth (Malan et al. 1976).

The very different conditions for achieving an appropriate acid–base balance in water and in air cause great physiological difficulties for animals that live in a transition between the two media. A fish that normally lives in water has its acid–base balance adjusted to a carbon dioxide tension of 2 or 3 mm Hg. If the fish changes to air respiration, the blood carbon dioxide tension rises sharply; this requires an adjustment in the blood buffering system, primarily of the sodium ion concentration, or some auxiliary mechanism for the elimination of carbon dioxide.

Amphibious animals lose substantial amounts of carbon dioxide through the skin, and this is of help to both lungfish and aquatic amphibians. If an animal emerges upon land and becomes completely terrestrial, it cannot retain a highly permeable skin because of the danger of desiccation. If it were to keep the P_{CO_2} low by lung ventilation, the ventilation would be vastly increased (relative to that needed for the oxygen supply), and evaporation would become a major problem. Thus, a radical readjustment in the acid–base regulatory system is a primary prerequisite for moving permanently onto land.

Carbonic anhydrase and velocity of carbon dioxide–water interaction

The events that take place when carbon dioxide is dissolved in water (see equations earlier in this chapter) are not instantaneous. When carbon dioxide enters water and becomes dissolved, the initial hydration of the carbon dioxide molecule to form H_2CO_3 is a relatively slow process. The reverse process – the release of carbon dioxide from H_2CO_3 – is also slow, requiring several seconds to a fraction of a minute. (The dissociation of H_2CO_3 into H^+ and HCO_3^- and the reverse process are, for physiological considerations, instantaneous.) If we consider that the time blood remains in a capillary is usually only a fraction of a second, how is it possible for the blood to take up carbon dioxide in the tissues as rapidly as it does, and again give off carbon dioxide during the short time the blood is in the lung capillary?

The problem appeared to be solved with the discovery of an enzyme that accelerates the formation of carbon dioxide from H_2CO_3 so that the process becomes very fast. The enzyme was given the name *carbonic anhydrase,* which is something of a misnomer because both the combination of carbon dioxide with water and the release of carbon dioxide from carbonic acid are speeded up (i.e., the process is catalyzed in both directions).

Carbonic anhydrase is not present in blood plasma, but is found in high concentration inside the red blood cells. It also occurs in other organs, notably in glands such as kidneys, the secretory epithelium of the stomach, pancreas, and salivary glands, but this is of no concern to us here. It should be noted, however, that carbonic anhydrase is highly specific for the hydration reaction of carbon dioxide and has no other known function in the organism.

Some years after the discovery of carbonic anhydrase, several highly potent carbonic anhydrase inhibitors were synthesized and became important as drugs, especially in the treatment of certain types of kidney malfunction. Much to the surprise of physiologists, it turned out that these enzyme inhibitors have a very low toxicity and that the complete inhibition of blood carbonic anhydrase has only minor effects on carbon dioxide transport. In other words, the enzyme is not a critical factor in carbon dioxide transport as such, although its inhibition causes some changes in the acid–base balance.

An examination of the amount of carbonic anhydrase found in various animals might be helpful in examining its possible role. We know that small animals have higher metabolic rates, per unit body weight, than large animals. They use oxygen and produce carbon dioxide more rapidly, and carbon dioxide must therefore be released in a shorter time in the lungs of a small animal.

An examination of the carbonic anhydrase of the blood in relation to body size shows that small animals indeed have significantly higher concentrations of the enzyme in their blood (Figure 2.17).

FIGURE 2.17 Carbonic anhydrase activity in mammalian blood is related to body size and is higher in small than in large mammals. [Magid 1967]

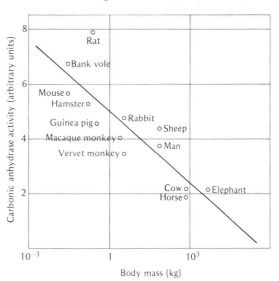

This seems to be in accord with the need for carbon dioxide transport, but the enzyme is present in amounts far in excess of what is needed. There is, however, another process where rapid hydration of carbon dioxide is important. As carbon dioxide in the tissue capillary enters the blood, the presence of carbonic anhydrase permits an immediate formation of carbonic acid, which in turn immediately affects the dissociation of the acid-sensitive oxyhemoglobin. This acid sensitivity, the Bohr effect, obviously would not have time to take effect while the blood still remains in the capillary unless the hydration of the carbon dioxide molecule were speeded up by the enzyme. In the absence of carbonic anhydrase, therefore, the Bohr effect could not be of much importance.

It seems, then, that carbonic anhydrase, an enzyme that specifically catalyzes the hydration of the carbon dioxide molecule, perhaps is of minor importance in carbon dioxide transport but contributes to the efficient delivery of oxygen to the tissues. As inhibition of the enzyme in the living animal does not cause great difficulties to the animal, the enzyme is probably important primarily under conditions of extreme demands on the gas-transport mechanism.

In other organs where carbonic anhydrase occurs, it seems always to be related to functions where ion transport processes are important, such as in glands (kidneys, pancreas, salivary gland) and in the ciliary body of the eye. The effect of carbonic anhydrase inhibition on these processes is profound and is the basis for the widespread and important use of inhibitor drugs in clinical medicine as well as in physiological research.

After having discussed the physiological properties of blood and the way in which it carries gases, we can move on to the next chapter, in which we will be concerned with the movement of blood within the organism.

REFERENCES

Aste-Salazar, H., and Hurtado, A. (1944) The affinity of hemoglobin for oxygen at sea level and at high altitudes. *Am. J. Physiol. 142:*733–743.

Barcroft, J. (1935) Foetal respiration. *Proc. R. Soc. Lond. B. 118:*242–263.

Bartels, H. (1980) Aspekte des Gastransports bei Säugetieren mit hoher Stoffwechselrate. *Verh. Dtsch. Zool. Ges. 1980:*188–201.

Bartlett, G. R. (1980) Phosphate compounds in vertebrate red blood cells. *Am. Zool. 20:*103–114.

Benesch, R., Benesch, R. E., and Enoki, Y. (1968a) The interaction of hemoglobin and its subunits with 2, 3-diphosphoglycerate. *Proc. Natl. Acad Sci. U.S.A. 61:*1102–1106.

Benesch, R., Benesch, R. E., and Yu, C. I. (1968b) Reciprocal binding of oxygen and diphosphoglycerate by human hemoglobin. *Proc. Natl. Acad. Sci. U.S.A. 59:*526–532.

Black, C. P., and Tenney, S. M. (1980) Oxygen transport during progressive hypoxia in high-altitude and sea-level waterfowl. *Respir. Physiol. 39:*217–239.

Bunn, H. F. (1971) Differences in the interaction of 2, 3-diphosphoglycerate with certain mammalian hemoglobins. *Science 172:*1049–1050.

Chanutin, A., and Curnish, R. R. (1967) Effect of organic and inorganic phosphates on the oxygen equilibrium of human erythrocytes. *Arch. Biochem. Biophys. 121:*96–102.

Gillen, R. G., and Riggs. A. (1973) Structure and function of the isolated hemoglobins of American eel, *Anguilla rostrata. J. Biol. Chem. 248:*1961–1969.

Hall, F. G., Dill, D. B., and Barron, E. S. G. (1936) Comparative physiology in high altitudes. *J. Cell. Comp. Physiol. 8:*301–313.

Hall, F. G., and McCutcheon, F. H. (1938) The affinity of hemoglobin for oxygen in marine fishes. *J. Cell. Comp. Physiol. 11:*205–212.

Hemmingsen, E. A. (1965) Accelerated transfer of oxygen through solutions of heme pigments. *Acta Physiol. Scand. Suppl. 246:*1–53.

Kobayashi, M., and Hoshi, T. (1982) Relationship between the haemoglobin concentration of *Daphnia magna* and the ambient oxygen concentration. *Comp. Biochem. Physiol. 72A:*247–249.

Lenfant, C., Torrance, J., English, E., Finch, C. A., Reynafarje, C., Ramos, J., and Faura, J. (1968) Effect of altitude on oxygen binding by hemoglobin and on organic phosphate levels. *J. Clin. Invest. 47:*2652–2656.

Lenfant, C., Ways, P., Aucutt, C., and Curz, J. (1969) Effect of chronic hypoxic hypoxia on the O_2–Hb dissociation curve and respiratory gas transport in man. *Respir. Physiol. 7:*7–29.

Lutz, P. L., Longmuir, I. S., Tuttle, J. V., and Schmidt-Nielsen, K. (1973) Dissociation curve of bird blood and effect of red cell oxygen consumption. *Respir. Physiol. 17:*269–275.

Magid, E. (1967) Activity of carbonic anhydrase in mammalian blood in relation to body size. *Comp. Biochem. Physiol. 21:*357–360.

Malan, A., Wilson, T. L., and Reeves, R. B. (1976) Intracellular pH in cold-blooded vertebrates as a function of body temperature. *Respir. Physiol. 28:*29–47.

Manwell, C. (1958) The oxygen-respiratory pigment equilibrium of the hemocyanin and myoglobin of the amphineuran mollusc *Cryptochiton stelleri. J. Cell Comp. Physiol. 52:*341–352.

Manwell, C. (1960) Histological specificity of respiratory pigments. 2. Oxygen transfer systems involving hemerythrins in sipunculid worms of different ecologies. *Comp. Biochem. Physiol. 1:*277–285.

Millikan, G. A. (1937) Experiments on muscle hae-moglobin *in vivo:* The instantaneous measurement of muscle metabolism. *Proc. R. Soc. Lond. B. 123:*218–241.

Rahn, H. (1966) Gas transport from the external environment to the cell. In *Development of the Lung* (A.V.S. de Reuck and R. Porter, eds.), pp. 3–23. London: CIBA.

Randall, D. J. (1970) Gas exchange in fish. In *Fish Physiology,* vol. 4 (W. S. Hoar and D. J. Randall, eds.), pp. 253–292. New York: Academic Press.

Read, K. R. H. (1962) The hemoglobin of the bivalved mollusc, *Phacoides pectinatus* Gmelin. *Biol. Bull. 123:*605–617.

Riggs, A. (1951) The metamorphosis of hemoglobin in the bullfrog. *J. Gen. Physiol. 35:*23–44.

Riggs, A. (1960) The nature and significance of the Bohr effect in mammalian hemoglobins. *J. Gen. Physiol. 43:*737–752.

Rogers, W. P. (1949) On the relative importance of aerobic metabolism in small nematode parasites of the alimentary tract. 2. The utilization of oxygen at low partial pressures by small nematode parasites of the alimentary tract. *Aust. J. Sci. Res. 2:*166–174.

Root, R. W., Irving, L., and Black, E. C. (1939) The effect of hemolysis upon the combination of oxygen with the blood of some marine fishes. *J. Cell. Comp. Physiol. 13:*303–313.

Schmidt-Nielsen, K. (1972) *How Animals Work.* London: Cambridge University Press, 114 pp.

Schmidt-Nielsen, K., and Pennycuik, P. (1961) Capillary density in mammals in relation to body size and oxygen consumption. *Am J. Physiol. 200:*746–750.

Schmidt-Nielsen, K., and Taylor, C. R. (1968) Red blood cells: Why or why not: *Science 162:*274–275.

Scholander, P. F. (1960) Oxygen transport through hemoglobin solutions. *Science 131:*585–590.

Snyder, L. R. G., Born, S., and Lechner, A. J. (1982) Blood oxygen affinity in high- and low-altitude populations of the deer mouse. *Respir. Physiol. 48:*89–105.

Tetens, V., and Lykkeboe, G. (1981) Blood respiratory properties of rainbow trout, *Salmo gairdneri:* Responses to hypoxia acclimation and anoxic incubation of blood in vitro. *J. Comp. Physiol. 145:*117–125.

Winton, F. R., and Bayliss, L. E. (1955) Blood and the transport of oxygen and carbon dioxide. In *Human Physiology* (F. R. Winton and L. E. Bayliss, eds.), pp. 78–118. Boston: Little, Brown.

Wittenberg, B. A., Wittenberg, J. B., and Caldwell, P. R. B. (1975) Role of myoglobin in the oxygen supply to red skeletal muscle. *J. Biol. Chem. 250:*9038–9043.

Wittenberg, J. B. (1977) Facilitation of oxygen diffusion by intracellular leghemoglobin and myoglobin. In *Oxygen and Physiological Function* (F. F. Jöbsis, ed.), pp. 228–246. Dallas: Professional Information Library.

ADDITIONAL READING

Altman, P. L., and Dittmer, D. S. (eds.) (1971) *Biological Handbooks: Blood and Other Body Fluids* (3d printing, with minor corrections). Bethesda: Federation of American Societies for Experimental Biology. 540 pp.

American Society of Zoologists (1980) Respiratory pigments. *Am. Zool. 20:*1–211.

Bonaventura, J., Bonaventura, C., and Sullivan, B. (1977) Non-heme oxygen transport proteins. In *Oxygen and Physiological Function* (F. F. Jöbsis, ed.), pp. 177–220. Dallas: Professional Information Library.

Brittain, T. (1987) The Root effect. *Comp. Biochem. Physiol. 86B:* 473–481.

Durliat, M. (1985) Clotting processes in Crustacea Decapoda. *Biol. Rev. 60:*473–498.

Esmon, C. T. (1987) The regulation of natural anticoagulant pathways. *Science 235:*1348–1352.

Howell B. J., Rahn, H., Goodfellow, D., and Herreid, C. (1973) Acid–base regulation and temperature in selected invertebrates as a function of temperature. *Am. Zool. 13:*557–563.

Jackson, C. M., and Nemerson, Y. (1980) Blood coagulation. *Annu. Rev. Biochem. 49:*765–811.

Linzen, B., et al. (1985) The structure of Arthropod hemocyanins. *Science 229:*519–524.

MacFarlane, R. G. (1970) *The Haemostatic Mechanism in Man and Other Animals.* Symposium of the Zoological Society of London, No. 27. London: Academic Press. 248 pp.

Perutz, M. F. (1983) Species adaptation in a protein molecule. *Mol. Biol. Evol. 1:*1–28.

Ratcliffe, N. A., and Rowley, A. F. (eds.) (1981) *Invertebrate Blood Cells,* vols. 1 and 2. New York: Academic Press, Vol. 1, 340 pp; vol. 2, 328 pp.

Reeves, R. B. (1977) The interaction of body temperature and acid–base balance in ectothermic vertebrates. *Annu. Rev. Physiol. 39:*559–586.

Riggs, A. F. (1988) The Bohr effect. *Annu. Rev. Physiol. 50:*181–204.

Toulmond, A. (1985) Circulating respiratory pigments in marine animals. *Soc. Exp. Biol. Symp. 39:*163–206.

Truchot, J. P. (1981) L'équilibre acido-basique extra-cellulaire et sa régulation dans les divers groupes animaux. *J. Physiol., Paris 77:*529–580.

Weber, R. E., and Jensen, F. B. (1988) Functional adaptations in hemoglobins from ectothermic vertebrates. *Annu. Rev. Physiol. 50:*161–179.

Wittenberg, B. A., and Wittenberg, J. B. (1989) Transport of oxygen in muscle. *Annu. Rev. Physiol. 51:*857–878.

CHAPTER THREE

Circulation

The major purpose of moving a fluid in the body is to provide rapid mass transport over distances where diffusion is inadequate or too slow. Circulation is therefore important in virtually all animals more than a few millimeters in size and a necessity for large animals with high metabolic rates.

In addition to transport of gases, the circulating blood serves (1) to transport other solutes, (2) to transport heat, and (3) to transmit force.

Functions that depend on the transmission of force are related mostly to the movement of the whole animals, movement of organs, and providing pressure for ultrafiltration in the capillaries of the kidney; these functions will be discussed elsewhere.

Those functions that pertain to the movement of solutes (including gases) and to heat will be discussed in this chapter.

To understand how blood moves in the circulatory system it is necessary to know some basic physical principles of the movement of fluid in tubes. Also, blood as a fluid has some unusual properties that affect its flow characteristics.

A system that is adequate for the convective transport of oxygen invariably suffices for transport of carbon dioxide and other solutes as well. Therefore, our main emphasis will be on the role of circulation in transport of gases, notably oxygen.

Circulatory systems of vertebrates are well known and will be discussed first. The function of invertebrate circulatory systems is less well known; in many cases the morphology has been adequately worked out, but information about function is often less satisfactory. As a result, there is little possibility for generalizations, and the main emphasis of this chapter will be on the highly developed and better understood circulatory systems of vertebrates.

GENERAL PRINCIPLES

Pumps

An adequate circulatory system depends on one or more pumps and on channels or conduits in which the blood can flow. The pump, or *heart,* is based on the ability of muscle to contract or shorten. By wrapping muscle around a tube or chamber, it is possible to achieve a reduction of volume. Two different types of pumps can be designed this way: *peristaltic pumps* or *chamber pumps* with valves (Figure 3.1). A chamber pump may have contractile walls (Figure 3.1*b*) or the reduction in volume may be achieved by external pressure from other body parts (Figure 3.1*c*).

Peristaltic hearts are found mostly in invertebrates; the vertebrate heart is almost without exception a chamber pump with contractile walls. A typical pump for which the force is provided by outside pressure is found in the larger veins in the human legs. The walls of these veins are relatively thin; the veins are provided with valves that prevent the backflow of blood, and when the leg muscles contract the veins are compressed and the valves ensure that blood is forced in the direction of the heart. This pumping action aids greatly in moving the blood out of the leg veins against the force of gravity, which tends to make blood accumulate in the legs.

Because muscle can provide force only by shortening, the heart cannot pull blood unless special arrangements are made. The elasmobranch heart is an example. It is enclosed in a stiff capsule, and when one chamber contracts, the other chamber is filled with blood by ''suction'' while valves prevent backflow from the arterial side. This is described in further detail later in this chapter.

Channels

The circulating fluid or blood is carried from the heart in channels or conduits and eventually returns to the heart. Vertebrate blood is carried in a system of elastic tubes or pipes (arteries, capillaries, veins). Blood returns to the heart without leaving this system of tubes, and because the blood

FIGURE 3.1 The three types of pumps that move blood in a circulatory system. In the peristaltic pump (*a*) a constriction in a tube moves along the tube and pushes blood ahead of it. In the common chamber pump (*b*) rhythmical contractions of the walls force blood out. Valves prevent backflow, and the blood is therefore expelled in one direction only. In another form of chamber pump (*c*) blood is pushed out of a collapsible tube by pressure from surrounding tissues. In this diagram contracting muscles provide the pressure and valves prevent backflow. The venous pump in the legs of humans is of this type.

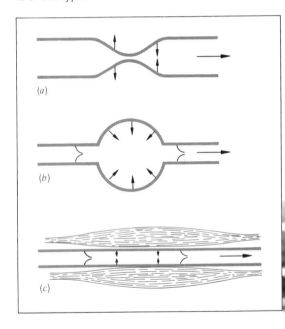

remains within this closed system, we refer to vertebrate circulation as a *closed circulation.* In many invertebrates (e.g., insects, most crustaceans, many molluscs) blood is pumped from the heart into the blood vessels, but these terminate and blood flows more or less freely between the tissues before it eventually returns to the heart. Such a system is called an *open circulation.* Some of the characteristics of open and closed circulatory systems are listed in Table 3.1.

Closed circulatory systems are found in verte-

TABLE 3.1 Major differences between closed and open circulatory systems.

Closed systems	Open systems
Usually high-pressure systems	Usually low-pressure systems
High pressure requires high peripheral resistance	Pressures moderate or low
High pressure sustained between heartbeats, requires elastic walls	Sustained pressure possible
Blood conveyed directly to organs	Similar to closed systems
Distribution to different organs well regulated	Distribution of blood less readily regulated
Blood return to heart rapid	Blood return to heart often slow

brates, cephalopods (octopus), and echinoderms; open systems are found in most arthropods, non-cephalopod molluscs, and tunicates. The circulatory system of tunicates is interesting in several ways. It is an open system, and the tube-shaped heart pumps by means of a peristaltic wave passing along from one end to the other. The heart has no valves. After a series of contractions, say several dozen or a hundred, the heart gradually slows and stops. After a pause the beat reverses direction and blood is now forced in the opposite direction (Krijgsman 1956). This is an unusual arrangement: In most circulatory systems the heart pumps, and the blood always flows, in one direction only.

Capacity

Let us examine the capacity of our pumping system from the viewpoint of an adequate oxygen supply. We can immediately see that the *volume* of blood that must be pumped for a given amount of oxygen will be reduced if the *capacity* of the blood to carry oxygen is high. The *amount* of oxygen carried is, of course, the product of the volume of blood and its oxygen content.

This simple principle is illustrated in Figure 3.2. If the oxygen capacity of the blood is 20 cm³ O_2 per 100 cm³ blood, as it is in many mammals, 1 cm³ of oxygen is contained in 5 cm³ of blood. If there is no hemoglobin or other respiratory pigment in the blood fluid, as in many invertebrates, the blood will contain only dissolved oxygen. The

FIGURE 3.2 The volume of blood that carries 1 cm³ of oxygen is plotted against the oxygen capacity of the blood. Note the rapidly increasing volume of blood needed to supply oxygen at low oxygen capacities (i.e., low concentration of respiratory pigment).

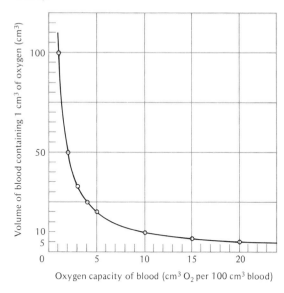

Oxygen capacity of blood (cm³ O_2 per 100 cm³ blood)

amount may be 0.5 to 1 cm³ O_2 per 100 cm³ blood, depending on the temperature. Consequently, some 100 or 200 cm³ of blood must be pumped to deliver 1 cm³ of oxygen. Thus, in the absence of hemoglobin, the pumping requirement on the heart is 20 or 40 times higher than it is for hemoglobin-containing mammalian blood.

The volume of fluid pumped by the heart can be adjusted (1) by increasing the frequency of the pump, (2) by increasing its stroke volume, or (3) by a combination of these two. We will return to the output of the heart in relation to changing needs for oxygen, especially during activity and exercise, but we will first discuss some other aspects of the body fluids.

VERTEBRATE CIRCULATION

Water compartments and blood volume

Roughly two-thirds of the vertebrate body consists of water. The exact figure varies a great deal, particularly as a result of variations in fat content. Adipose tissue contains only some 10% water, and a very fat individual, who may have half or more of the body weight as fat, therefore has a low overall water content. If the water content is expressed as a percentage of the fat-free body weight, the figure shows much less variation. Most of the water is located inside the cells (*intracellular water*); a smaller fraction is outside the cell membranes (*extracellular water*). The extracellular water is located partly in tissue spaces (*interstitial fluid*) and partly in the blood (*plasma water*). We can therefore consider the body water as being distributed among three compartments: intracellular, interstitial, and plasma (Figure 3.3).

The volumes of the body fluid compartments can be measured in various ways, often without harm to the organism. One of the most convenient ways is to introduce some suitable substance that is distributed uniformly within the compartment to be measured. The dilution of this substance can then be used for a calculation of the volume into which it is distributed. The quantities most readily measured in this way are plasma volume, extracellular volume, and total body water.

Total blood volume cannot be measured accurately by draining the blood from an animal, for a substantial amount of blood will remain in the vas-

FIGURE 3.3 The approximate distribution of water and solids in the vertebrate body, expressed as percentage of the fat-free body weight. More detailed information is given in Table 3.2.

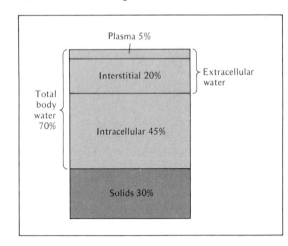

cular system. The dilution technique, however, gives reliable and reproducible results. One of the commonly used substances for this purpose is a blue dye, Evans blue (also known as T-1824). A measured small amount of this dye is injected as an aqueous solution into the bloodstream; the dye binds to the plasma proteins and remains within the vascular system. After complete mixing has taken place (after several minutes), a blood sample is withdrawn and the dye concentration in the plasma is measured. The *plasma volume* is then readily calculated from the dilution of the dye. The *total blood volume* can be calculated from the plasma volume if we also determine the percent red cells in the blood (the *hematocrit*), which is obtained by centrifugation of a blood sample.

The extracellular fluid volume is determined by injecting a substance that penetrates the capillary walls and becomes distributed in blood and interstitial spaces, but does not penetrate the cells. One such substance is the polysaccharide inulin (also used in renal function studies; see Chapter 9). Inulin, which has a molecular weight of about 5000, distributes itself uniformly in extracellular space, but is not metabolized and does not penetrate the

TABLE 3.2 Body fluid compartments in representative vertebrates. All values in percent of body weight. The minor differences shown are of uncertain significance and may be due to different techniques, to differences in the fat content of the animal body, etc. The blood volume of teleost fish, however, is consistently much lower than in other vertebrates.

Animal	Total body water (%)	Intracellular water (%)	Extracellular water (%)	Blood volume (%)
Lamprey[a]	76	52	24	8.5
Dogfish[b]	71	58	13	6.8
Carp (fresh water)[c]	71	56	15	3.0
Red snapper (marine)[c]	71	57	14	2.2
Bullfrog[d]	79	57	22	5.3
Alligator[e]	73	58	15	5.1
Gopher snake[e]	70	52	17	6.0
Pigeon[f]	—	—	—	9.2
Great horned owl[f]	—	—	—	6.4
Goat[g]	76	40	27	9.9

[a] Thorson (1959). [b] Thorson (1958). [c] Thorson (1961). [d] Thorson (1964). [e] Thorson (1968).
[f] Bond and Gilbert (1958). [g] Shkolnik et al. (1972).

cells. Its dilution, therefore, indicates the total extracellular volume. The volume of interstitial fluid is then calculated as the difference between total extracellular volume and plasma volume.

Total body water can, of course, be determined by killing the organism and determining the water content by drying. A dilution technique is, however, less dramatic and more convenient. Presently the commonest substance used is water labeled with the radioactive hydrogen isotope tritium. A small amount of tritiated water is injected; it diffuses through capillary walls and cell membranes without hindrance and rapidly becomes evenly distributed throughout all body water. A determination of the radioactivity of any sample of a body fluid can be used for a calculation of the volume into which the injected tritiated water was diluted. In fact, it is not even necessary to take a blood sample, for if the bladder has been emptied before the experiment, a freshly voided urine sample will be as representative of the body water as a water sample obtained from any other source.

Table 3.2 gives the total body water and its distribution in intracellular and extracellular volumes for some selected vertebrates. The total body water in aquatic and terrestrial vertebrates is of the same magnitude, about 70%, but there are variations in both directions. The distribution between intracellular and extracellular volume again is rather uniform, but with some inconsistencies.

The blood volume of vertebrates is usually between 5 and 10% of the body weight, except that teleost fish in general have blood volumes only 2 to 3% of body weight, and elasmobranchs and cyclostomes have volumes 6 to 10% of body weight. Air-breathing vertebrates (amphibians, reptiles, birds, and mammals) have blood volumes between 5 and 10%. We can therefore say that vertebrates on the whole have blood volumes between 5 and 10%, with the exception of teleost fish, which have substantially smaller blood volumes. At the present time the functional significance of this difference is uncertain.

Circulation patterns

Each class of vertebrates has a quite uniform type of circulation, but differences among the classes are substantial. As vertebrate life changes from

FIGURE 3.4 In the circulatory system of a fish, blood is pumped by the heart through the gills, from where it flows on to the tissues. In the mammalian system, blood is pumped through the lungs and then returns to the heart before being passed on to the tissues by a second pump. The mammalian arrangement allows a higher blood pressure in the tissues than in the lungs.

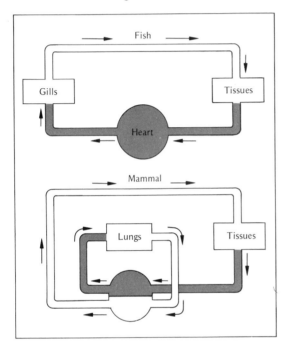

aquatic to terrestrial, the circulation becomes more complex.

There are two important consequences of the arrangement of mammalian circulation, both immediately apparent from Figure 3.4. First, the blood flow through the lungs must equal the blood flow through the entire remaining part of the body (except for minute transient variations that can be caused by slight changes in heart volume during one or a few strokes). Second, because both halves of the heart contract simultaneously, as blood is ejected from the heart, the entire ejected volume must be taken up by changes in the volume of the elastic blood vessels.

Fish and mammals represent two extremes in

vertebrate circulation. There is a gradual separation of the heart into two separate pumps as the vertebrates progress from aquatic life to fully terrestrial respiration. Each vertebrate class has certain characteristics, and a few comments will help explain these differences.

Cyclostomes

The circulatory system of hagfish differs from that of all other vertebrates. It is partly an open system with large blood sinuses, rather than a closed system as in other vertebrates. Its notable characteristic is that in addition to the regular heart (the *branchial heart*), the hagfish has several accessory hearts, especially in the venous system. There are three sets of such accessory hearts: the *portal heart,* which receives venous blood from the cardinal vein and from the intestine and pumps this blood to the liver; the *cardinal hearts,* which are located in the cardinal veins and help to propel the blood; and the *caudal hearts,* which are paired expansions of the caudal veins. In addition to these accessory hearts, all located in the venous system, the gills take active part in the forward propulsion of the arterial blood. This is accomplished by contraction of striated muscular elements in the gills and gill ducts that helps propel the blood in the arterial system (Johansen 1960).

The caudal hearts of the hagfish are particularly interesting because they differ in design from all other hearts (Figure 3.5). A longitudinal rod of cartilage separates two chambers, and alternate contractions of muscles on the two sides cause the rod to be flexed. As the muscles on one side contract, those on the opposite side provide pressure for expulsion of the blood on that side. Simultaneously, the volume on the contracting side increases, so that this chamber becomes filled with blood. By alternate contractions, the two chambers fill and empty in opposing phase, while appropriate valving assures a unidirectional flow.

Fish

Fish, both teleosts and elasmobranchs, have a heart that consists of two chambers in series, an

FIGURE 3.5 The caudal heart of hagfish differs from other vertebrate hearts. A central cartilage is flexed by alternate contractions of muscles on the two sides, which also serve as walls for the two chambers. As the muscles on one side contract, the chamber on this side is filled with blood, while blood is expelled from the other side. Valves help to ensure a unidirectional flow of blood. [Gordon 1972]

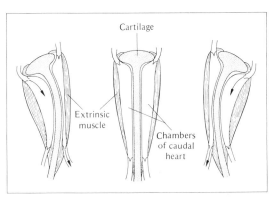

FIGURE 3.6 The two-chambered heart of fish is preceded by an enlarged venous sinus that supplies blood for filling the atrium. The heart ejects blood into a thickened part of the artery, the bulbus arteriosus (in teleosts) or conus arteriosus (in elasmobranchs). [Randall 1970]

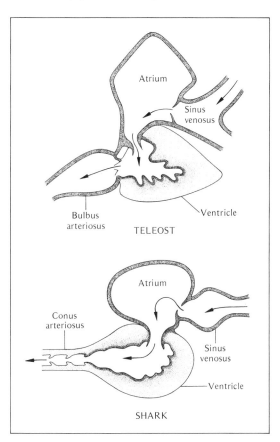

atrium and a ventricle. On the venous side the heart is preceded by an enlarged chamber or sinus on the vein, the *sinus venosus,* which helps assure a continuous flow of blood to the heart.

On the arterial side the teleost heart is immediately followed by a thickened muscular part of the ventral aorta, the *bulbus arteriosus* (Figure 3.6). The elasmobranch heart has a similarly located thickened part, the *conus arteriosus,* developed from the heart muscle. It is fibrous and is equipped with valves that prevent backflow of blood into the ventricle. This is particularly important because the elasmobranch heart, located in a rigid chamber, can produce negative pressures. A negative pressure in the heart facilitates the filling by "suction" of the atrium from the large sinus venosus.

The sequence of events in a fish heart during one contraction cycle can be followed on the pressure tracing shown in Figure 3.7. During contraction of the ventricle, the blood pressure rises and this pressure is transmitted to the bulbus arteriosus. As the ventricle relaxes, backflow from the bulbus is prevented by valves, and the high pressure therefore

persists in the bulbus after the ventricle begins to relax. The elastic properties of the filled bulbus serve to maintain a flow of blood in the ventral aorta during the diastolic period of the heartbeat.

During contraction of the ventricle, the systole, the heart decreases in volume, and because the elasmobranch heart is located in a rigid chamber, the decrease in volume causes a negative pressure to develop in this chamber. In the Port Jackson shark the pressure in the ventricle during its contraction rises to above 30 mm Hg (4.0 kPa), and the con-

FIGURE 3.7 Pressure changes in the heart of a lingcod (*Ophiodon elongatus*) during a single contraction cycle in relation to ventral aortic blood flow. [Randall 1970]

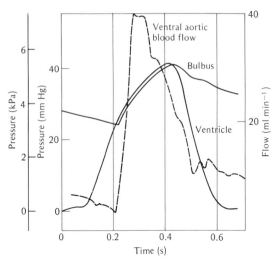

current negative pressure in the atrium and the sinus venosus may be as low as −4 mm Hg (−0.6 kPa) (Satchell 1970). Unless the large sinus and the adjoining veins could provide an equal volume of blood to flow into the atrium as the ventricle contracts, the negative pressure would only impede the contraction. With the inflow of blood, however, the negative pressure does not become excessive, but merely serves to fill the atrium. The filled atrium then contracts and forces blood into the ventricle that now is empty and relaxed, backflow into the sinus being prevented by valves. Thus, in this part of the cycle blood is merely shifted from atrium to ventricle, while the volume of the pericardial contents remains unchanged (Randall 1970).

In an air-breathing vertebrate, the sinus venosus and bulbus arteriosus decrease in importance, and in the mammalian heart they are not present as separate structures.

Lungfish

The major evolutionary change in lungfish is that, in addition to gills, they have lungs as respiratory organs. The gills in part receive blood that has already passed through the lungs. If the gills were similar to the gills of ordinary fishes, this might be a disadvantage, for a lungfish swimming in oxygen-depleted water would then lose oxygen from the blood to the water that flows over the gills. The lungfish gills, however, have degenerated, and some of the gill arches permit a direct through flow of blood (Johansen 1970).

The atrium of the heart is divided into two chambers by a septum, and the ventricle is partially divided. In this way the lungfish heart somewhat resembles the completely divided heart of mammals, birds, and crocodiles. The lungfish heart, in fact, shows an amount of structural division greater than that of any amphibian (Foxon 1955).

Blood from the lungs returns to the left atrium, and the right atrium receives blood from the general circulation. The partial division of the ventricle tends to keep the two bloodstreams separated, so that oxygenated blood tends to flow into the first two gill arches and supply the head with relatively oxygen-rich blood. The less well oxygenated blood from the right side of the heart flows through the posterior gill arches and passes on to the dorsal aorta and in part to the lungs. The lungfish represents the beginning of a complete separation between circulation to the lungs and to the remaining parts of the body (Johansen et al. 1968).

The anatomy of the lungfish does not explain how blood actually flows. Studies of live African lungfish (*Protopterus*) show that circulation to the lungs and circulation to the systemic vascular circuit have a high degree of functional separation, with preferential passage of oxygen-poor blood to the lungs and oxygen-rich blood to the systemic circulation. It is particularly significant that the functional separation is highest immediately after a breath of air when the oxygen in the lungs is highest, whereas later in the interval between breaths, the degree of separation diminishes. This is of obvious importance for the efficiency of gas exchange in a lung that only at intervals is filled with fresh air (Johansen et al. 1968).

FIGURE 3.8 The circulation in the bullfrog and the degree of oxygen saturation (*S*) of the blood. In spite of the one-chambered ventricle the blood coming from the lungs and from the rest of the body remains nearly unmixed. [Tazawa et al. 1979]

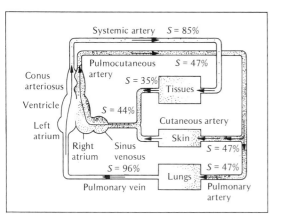

Amphibians

The heart of the modern amphibians (frogs, toads, salamanders) has two completely separate atria, but only one undivided ventricle. The left atrium receives oxygenated blood from the lungs, and the right atrium receives venous blood from the general systemic circulation. Although the ventricle is undivided, the two kinds of blood tend to remain unmixed, so that oxygenated blood enters the general circulation and oxygen-poor blood flows separately into the pulmonary circulation.

The pulmonary artery also sends branches to the skin; this is important because the moist amphibian skin is a major site of oxygen uptake. The anatomical arrangement of the heart includes a longitudinal ridgelike baffle in the conus arteriosus (known as the *spiral valve*), which seems important in keeping the bloodstreams separate (DeLong 1962).

Let us follow the circulating blood and its oxygen content in a bullfrog (see Figure 3.8). The single ventricle of the heart pumps blood into the conus arteriosus, which separates the flow into two nearly equal streams, one running in the systemic artery that supplies the general body with blood of high

saturation (*S* = 85%), and the other stream with low oxygen (*S* = 47%) running in the pulmocutaneous artery. This artery in turn supplies both skin and lungs.

When the frog breathes air, the blood leaving the lungs is well oxygenated (*S* = 96%), and the blood coming from the skin also has an increased oxygen content but is mixed with venous blood from the tissues. The two streams that return to the heart remain almost completely separate within the one-chambered ventricle. The result is that mixing is of little significance and the systemic artery can carry blood with a high oxygen saturation (Tazawa et al. 1979).

Is there any advantage to this arrangement? When the frog is submerged, there is no pulmonary gas exchange and the skin takes on the role as the only organ of gas exchange. There is now a further change; there is an increase in the systemic blood flow to the skin (Moalli et al. 1980), and as a result the mixed venous blood returning to the heart carries more oxygen.

Reptiles

In a noncrocodilian reptile the atria are completely separated, but the ventricle is only partially divided. Even so, the streams of oxygenated and nonoxygenated blood are kept well separated so that there is very little mixing of the blood and there is, in effect, a well-developed double circulation. Thus, the incomplete division of the ventricle in amphibians and noncrocodilian reptiles cannot be interpreted simply on the basis of anatomical appearance; the bloodstreams remain much more separated than anatomical considerations would indicate.

In crocodiles both chambers of the heart are completely divided. This complete separation is confounded by the peculiar fact that the left aortic arch originates from the right ventricle and thus should receive venous blood. There is, however, a hole or foramen connecting the two aortic arches.

On an anatomical basis this has been used to argue that crocodilian circulation allows mixing of oxygenated and nonoxygenated blood, but more careful study shows that the two types of blood remain separate and that both aortic arches carry unmixed oxygenated blood (White 1956). During diving, however, the circulation changes: Blood flow to the lungs is decreased, and a major part of the output of the right ventricle is ejected into the left aortic arch. This shunt permits a rerouting of the blood during diving and a partial or complete bypass of the lungs.

Birds and mammals

The division of the heart and separation of pulmonary and systemic circulation are complete in birds and mammals. This has one important consequence: The pressure can be different in the pulmonary and the systemic circulation.

The resistance to flow in the pulmonary system is much lower than in the systemic circulation, and the blood pressure in the pulmonary circulation is only a small fraction of the pressure in the systemic part. Such a difference is, of course, not possible if separation of the heart is incomplete. Because incomplete separation has been retained in both amphibians and reptiles, this arrangement may have other advantages that are not well understood.

There are some differences between the circulation in birds and mammals that are of great significance in comparative anatomy. For example, mammals retain the left aortic arch, whereas birds retain the right. A difference of physiological importance is that the kidneys of all nonmammalian vertebrates receive venous blood from the posterior part of the body (the *renal portal circulation*). Birds have retained this renal portal circulation, but it is absent in mammals. This difference is important to the understanding of renal function (Chapter 9).

Heart and cardiac output

As a rule, a small animal has a higher rate of oxygen consumption per unit body mass than a large animal, and thus the heart of the small animal must supply oxygen at a higher rate. We have already seen that the oxygen capacity of the blood of small

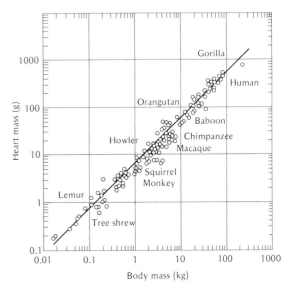

FIGURE 3.9 Heart size of mammals in relation to body size. The size of the heart is nearly proportional to body size and makes up approximately 0.6% of the body mass in small and large mammals alike. The heart size of a wide variety of primates falls within the range of other mammals. [Stahl 1965]

and large mammals is similar; as a consequence the heart of the small mammals must pump blood at a higher rate. Is the necessary increase achieved by a larger pump, by a larger stroke volume, or by a higher frequency?

Size of vertebrate hearts

The heart sizes of a number of mammals are plotted in Figure 3.9. As expected, heart size increases with body size, but surprisingly, relative to body size, small and large mammals have about the same heart size. If the heart mass were exactly proportional to body mass, the slope of the regression line in Figure 3.9 would be 1.0; the actual slope is 0.98, which is statistically indistinguishable from exact proportionality. The equation for the line is $M_h = 0.0059 M_b^{0.98}$, which says that the average mass (M_h) of the mammalian heart is 0.59%

of the body mass (M_b in kilograms), irrespective of the size of the mammal.

The heart size of birds can be described by a similar equation: $M_h = 0.0082\, M_b^{0.91}$ (Lasiewski and Calder 1971). This equation says that heart size is not strictly proportional to body size; the body mass exponent is significantly less than 1.0, which means that, relative to body size, larger birds tend to have slightly smaller hearts than birds of small body size. A 1-kg bird can be expected to have a heart of 8.2 g; for a mammal of the same size the expectation is a heart of 5.9 g.

The heart sizes of reptiles and amphibians are less well studied, but available data indicate that a reptile's heart size is about 0.51% of body weight; an amphibian's, 0.46%. These proportions are only slightly lower than in mammals, although the metabolic rates of reptiles and amphibians are about one-tenth the mammalian rate. Fish have smaller hearts again, about 0.2% of body weight.

To summarize, whether we compare the different classes, or animals of different body size within one class, the large differences in metabolic rates of vertebrates are not conspicuously reflected in the size of the heart. Differences in the need for oxygen must therefore be reflected primarily in pumping frequency, for the stroke volume depends on heart size, and the amount of oxygen contained in each volume of blood is not body-size-dependent.

Heartbeat frequency

The heartbeat frequency, or pulse rate, is usually given as the number of heartbeats per minute. The pulse rate for an adult human at rest is about 70 per minute; in exercise the rate increases several-fold.

The heart frequency is clearly inversely related to body size. An elephant that weights 3000 kg has a resting pulse rate of 25 per minute, and a 3-g shrew, the smallest living mammal, has a resting pulse rate of over 600 (Morrison et al. 1959). This means that in the shrew the heart goes through 10 complete contraction cycles per second, an almost unbelievable rate. During activity, the heart fre-

FIGURE 3.10 Relationship between heart rate and body size of mammals. The heart rate is higher in small than in large animals. If the observed data are plotted on logarithmic coordinates, they are grouped around a straight line, which can be represented by a logarithmic equation (see text). [Stahl 1967]

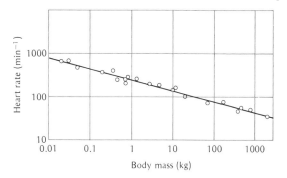

quency increases further; as much as 1200 beats per minute have been measured in hummingbirds (Lasiewski et al. 1967) and in small bats in flight (Studier and Howell 1969).

If heart frequency is plotted relative to body mass on logarithmic coordinates, the points are located close to a straight line (Figure 3.10). The equation for the line is $f_h = 241\, M_b^{-0.25}$. The slope of the regression line is negative; that is, the larger the body mass (M_b, in kilograms), the lower is the heart frequency (f_h).

In addition to the negative slope, the most significant information contained in this equation is the numerical value of the slope, which is 0.25. This is exactly the same as the slope of a regression line between body mass and resting oxygen consumption per unit body mass, the *specific rate of oxygen consumption* (see Chapter 5).*

We now have two important pieces of information: (1) the size of the pump, the heart, remains a constant percentage of the body mass, and (2) the increase in the pumping rate (the heart rate) in the smaller mammal increases in exact proportion to

---

*The word *specific*, as used here in front of a physical quantity, has the recommended meaning "divided by mass." For example, specific volume is the volume divided by the mass.

the need for oxygen. Let us turn to the volume of blood pumped by the heart.

Cardiac output

The volume of blood pumped by the heart per unit time is usually called the *cardiac output* or the cardiac minute volume. In hearts with a complete separation of the ventricles, the expression refers to the output of one side of the heart only.

The cardiac output can be determined in a number of different ways. One commonly used method is known as *Fick's principle*. It depends on the simple fact that all the oxygen consumed ($\dot{V}_{O_2}$) by the animal is carried by the blood ejected from the heart. Therefore, if we know the difference in the oxygen content of arterial and venous blood, we can calculate cardiac output ($\dot{Q}_h$) from the following equation.*

$$\dot{V}_{O_2} = \dot{Q}_h(C_{aO_2} - C_{vO_2})$$

To determine the cardiac output, we can determine (1) the rate of oxygen consumption, (2) the oxygen content in arterial blood, and (3) oxygen content in venous blood. An arterial blood sample can be obtained from any artery, but the venous blood must be obtained as mixed venous blood (i.e., from the right side of the heart or from the pulmonary artery). Venous blood from some other place will not suffice, for different organs remove different amounts of oxygen from the blood passing through them. A sample of mixed venous blood is most easily obtained from the right side of the heart by inserting a catheter in a vein and threading it in until it reaches the heart. This is a widely used and relatively harmless procedure.

During activity, when the oxygen consumption increases, the cardiac output also rises, but not in

proportion to the increased oxygen consumption. Part of the increase is covered by an increase in the arterio-venous oxygen difference (i.e., during activity the amount of oxygen extracted from the blood increases).

Determinations of cardiac output according to Fick's principle can be carried out only in animals that have a complete separation between arterial and venous blood in the heart, as do mammals and birds. In amphibians and reptiles, where there is a possibility for mixing arterial and venous blood in the heart, this principle cannot be used. However, in fish the principle applies: The venous blood sample can be taken from or near the heart (but before the gills) and arterial blood from any artery after the gills. This method may be misleading if there is any appreciable oxygen uptake through the skin.

Other methods for determining cardiac output are based on a variety of principles. One method consists of injecting an easily measured dye at a known moment in a vein leading to the heart. The dye will shortly afterward appear on the arterial side, and by integrating the curve representing the concentration of the dye as it rapidly increases and again disappears in the artery, it is possible to calculate the cardiac output.

In principle, the thermodilution method is similar. A known volume of saline is injected in the vein; if the saline is colder than the blood, there will be a transient decrease in arterial blood temperature. Again, the integrated curve gives information about the blood flow (i.e., the cardiac output).

The cardiac output ($\dot{Q}_h$, in milliliters per minute) is the product of the heart frequency (f_h, per minute) and the volume of blood ejected in one contraction, the stroke volume (V_h, in milliliters):

$$\dot{Q}_h = f_h \times V_h$$

An increase in cardiac output can therefore be attained by increasing the frequency, or by increasing the stroke volume, or both.

Let us compare two highly active animals, a flying

---

*In equations related to gas metabolism V usually stands for volume, and $\dot{V}$ (V-dot) for the time derivative of V (i.e., the volume per unit time). The subscript (in this case O_2) is chosen to aid easy recognition. Q stands for quantity of blood, $\dot{Q}$ for blood flow per unit time, and $\dot{Q}_h$ for blood flow from the heart (i.e., cardiac output). C stands for gas concentration in the blood, C_a for arterial blood, and C_v for venous blood.

TABLE 3.3 The increased demand for oxygen in a flying pigeon is met primarily by an increased heart rate while stroke volume actually drops a little. In a swimming trout, in contrast, the cardiac response is primarily an increased stroke volume with only a small increase in heart rate. Body size of pigeon, 0.44 kg; trout, 0.9–1.5 kg. [Butler et al. 1977; Kiceniuk and Jones 1977]

	Rest	Active	Increase
Flying pigeon:			
oxygen consumption increased 10-fold			
f_h (min^{-1})	115	670	5.8×
V_h (ml)	1.70	1.59	0.9×
$\dot{Q}_h$ (ml min^{-1})	195.5	1065	5.4×
Swimming trout:			
oxygen consumption increased 8-fold			
f_h (min^{-1})	37.8	51.4	1.36×
V_h (ml)	0.46	1.03	2.24×
$\dot{Q}_h$ (ml min^{-1})	17.4	52.9	3.04×

TABLE 3.4 Blood flow to the major organs of a 70-kg man at rest. [Data from Folkow and Neil 1971]

Organ	Organ size (kg)	Blood flow (liter min^{-1})	Blood flow (liter kg^{-1} min^{-1})
Kidneys	0.3 ⎫	1.2 ⎫	4.0
Liver	1.5 ⎬ 3.5	1.4 ⎬ 3.6	0.9
Heart	0.3	0.25	0.8
Brain	1.4 ⎭	0.75 ⎭	0.5
Skin	2.5 ⎫	0.2 ⎫	0.08
Muscle	29 ⎬ 66.5	0.9 ⎬ 2.0	0.03
Remainder	35 ⎭	0.9 ⎭	0.03
Total	70	5.6	

pigeon and a fast-swimming trout, to see how they adjust their cardiac outputs (Table 3.3). Both increase their oxygen consumption in about the same proportion: 10- and 8-fold, respectively. The pigeon increases its heart rate nearly 6-fold, but the trout swimming at maximum speed increases its heart rate only by one-third. The changes in stroke volume are different: The pigeon in fact shows a slight decrease in stroke volume, but the trout more than doubles the stroke volume. Neither animal increases the cardiac output in proportion to the oxygen consumption. This is because the venous oxygen content decreases during activity; in other words, more oxygen is taken out of each volume of blood before it returns to the heart. We shall return to this phenomenon later in this chapter.

Distribution of blood flow

The blood is not evenly distributed to all parts of the body; some organs receive much more than others, both in relative and in absolute terms. The distribution of blood to the major organs of a human at rest is shown in Table 3.4.

The four most heavily perfused organs – kidneys, liver, heart, and brain – make up only 5% of the body mass, but these organs receive more than half the total cardiac output. The relative blood flow to the kidneys is higher than to any other major organ (although blood flow to some endocrine organs, especially the thyroid and adrenals, may be even higher). Per unit weight the kidneys receive more than 100 times as much blood as muscles at rest.

During exercise the cardiac output is increased and the distribution of blood flow is changed. This is primarily because of the vastly increased oxygen demands of the muscles and because the work load on the heart is augmented due to the increase in pumping.

The blood flow to the skin is also very variable, changing with the need for temperature regulation. In the cold there is vasoconstriction in the skin, and the blood flow to the skin is at a minimum; when there is a need to dissipate heat, as a result either of high ambient temperatures or of increased heat production during exercise, blood flow to the skin is vastly increased. This subject will be further discussed in Chapter 7.

Regulation of heartbeat

The heart has the inherent capacity to contract rhythmically without any external stimulus. Even if the heart is completely removed from the body, it may continue to beat for a considerable time: An isolated frog or turtle heart beats rhythmically for hours. Mammalian hearts do not continue to beat for as long, for they are more sensitive to temperature and an adequate oxygen supply. Since the isolated heart continues beating by itself, its ability to contract rhythmically must be inherent. Excellent proof of the independent contractility of the heart is that the heart in a developing chick embryo begins to beat before any nerves have reached it, and that heart muscle cells grown in tissue culture also contract rhythmically without any external stimulus.

The contraction in the mammalian heart begins in a small piece of embryonic-type muscle located where the vena cava enters the right atrium. It is the vestige of the sinus venosus and is known as the *sinus node*. The contraction spreads rapidly through the muscle of the two atria, and, after a slight delay, to the muscle of the ventricles. When the wave of contraction reaches the partition between the atria and the ventricles, a patch of tissue known as the *atrioventricular bundle* conducts the impulse to the ventricles, which then, after the brief delay resulting from conduction, contract simultaneously.

The sinus node, where the contraction originates, is also known as the *pacemaker* of the heart. It starts the beat and sets the rate by its inherent rhythmicity. As we know, the heart rate can vary by several-fold. Two different mechanisms are involved in the control of the rate: nerve impulses to the pacemaker and hormonal influences. A branch of the vagus nerve, when stimulated, slows the rate of heartbeat. The nerve is of parasympathetic origin and acts by release of *acetylcholine*. Another nerve, the accelerans nerve, is of sympathetic origin. When it is stimulated, the heart accelerates because *noradrenaline* is released from the nerve endings at the pacemaker.

Noradrenaline is also known as a hormone. When it is released from the adrenal medulla into the blood, it accelerates the heart. This effect on the heart is one of the major effects of adrenaline, well known as part of the fight-or-flight syndrome. There is no corresponding release of the decelerating substance, acetylcholine, except from the endings of the vagus nerve. If acetylcholine were released elsewhere in the body, it would not affect the heart for the simple reason that the blood contains an enzyme that rapidly splits acetylcholine by hydrolysis.

The amount of blood pumped by the heart can be changed, not only by changing the frequency, but also by changing the volume of blood ejected in a single beat, the *stroke volume*. When the heart contracts, it does not eject all the blood contained in the ventricles. A substantial but variable volume of blood remains in the ventricles at the end of contraction.

Two major factors influence the stroke volume. One is the hormone adrenaline, which augments the force of the cardiac muscle contraction, thus forcing a larger amount of blood out of the heart in a single stroke. The other is the amount of blood in the ventricles when contraction begins. The latter effect results from an inherent characteristic of heart muscle, which will contract with greater force if it is more distended at the moment contraction begins.

If the return of venous blood to the heart is increased for some reason, the ventricles are filled with more blood, the muscle is stretched more than before, and the following contraction ejects a larger volume of blood. This relationship between increased venous return and increased cardiac output was discovered by the famous English physiologist Starling, and its known as *Starling's law of the heart*. In exercise the venous return to the heart increases, due partly to the increased pumping effect of the contracting muscles on the veins and partly to contraction in the venous system of the abdominal region, causing an increase in cardiac output.

In summary, the heartbeat originates in the sinus node or pacemaker. Cardiac output is under the influence of three major controlling systems: the ner-

TABLE 3.5 Geometry of the blood vessels in the mesentery of the dog. [Burton 1972]

Kind of vessel	Diameter (mm)	Number	Total cross-sectional area (cm²)	Length, approx. (cm)	Total volume (cm³)[a]
Aorta	10	1	0.8	40	
Large arteries	3	40	3	20	
Arterial branches	1	2 400	5	5	190
Arterioles	0.02	40 000 000	125	0.2	
Capillaries	0.008	1 200 000 000	600	0.1	60
Venules	0.03	80 000 000	570	0.2	
Veins	2	2 400	30	5	
Large veins	6	40	11	20	680
Vena cava	12.5	1	1.2	40	

[a] The estimates of total volume are based on a more detailed analysis than indicated by the figures for length given in the preceding column.

vous system, hormonal control, and the autoregulation attributable to the effect of the venous return on the heart muscle.

The conduits: blood vessels

The blood vessels are not merely tubes of different sizes. They have elastic walls, and a layer of smooth muscle within their walls enables them to change diameter.

There are characteristic differences among arteries, capillaries, and veins. Arteries have relatively thick walls that consist of heavy, strong layers of elastic fibers and smooth muscle. As the arteries branch and become smaller, and their walls thinner, the relative amount of muscle gradually increases in proportion to the elastic tissue in the wall. Capillaries, the smallest units of the system, consist of a single cell layer. Virtually all exchange of substances between blood and tissues takes place through the walls of the capillaries. Veins have thinner walls than arteries, but both elastic fibers and smooth muscle are found throughout the venous system.

Some measurements made on the vascular system of dogs are given in Table 3.5. Every time a larger vessel branches, the number of branches increases and the diameter decreases. As the number of branches increases, their total combined cross-sectional area increases, and at the level of the capillaries the aggregate cross section of all the capillaries in the mesentery alone is some 800 times that of the aorta. From that point on, the vessels converge into larger and larger veins, their number decreases, and so does the total cross-sectional area, until in the vena cava the cross section slightly exceeds that of the aorta. Another interesting fact is evident from the table: The amount of blood located in the venous system exceeds by several-fold the amount of blood in the arterial system.

In addition to the volume of the vascular system, we are interested in the pressure and the flow velocity in the various parts. Such information is given in Figure 3.11, which refers to measurements on humans. The first column shows that most of the blood at any given moment is located on the venous side.

The last column shows that the velocity of the blood is highest in the aorta. As the total cross-sectional area increases, the velocity decreases drastically, until in the capillaries it is between one-hundredth and one-thousandth of the velocity in the aorta. On the venous side, the velocity increases

FIGURE 3.11 Distribution of blood volume, blood pressure, and velocity in the various parts of the vascular system of humans (1 mm Hg = 0.13 kPa).

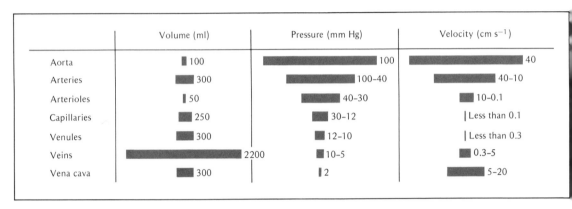

again, but does not reach the high velocity of the arterial system.

The center column in Figure 3.11 shows the pressure throughout the system. The pressure must, of course, gradually decrease throughout, from the aorta to the larger veins. The greatest pressure drop takes place in the smallest arteries, called *arterioles,* which have a high proportion of smooth muscle in their walls and, by changing their diameter, are the most important factor in changing the resistance to flow and thus in regulating the distribution of blood flow to various organs.

The pulmonary circulation differs from the systemic in that the total resistance is less; the blood pressure can therefore be lower and the arteries have thinner walls. In cross section the pulmonary artery is not circular, but oval. During the contraction of the heart, as blood is injected into the pulmonary artery, the flattened vessel absorbs a considerable increase in volume as the cross section becomes circular; only then does the wall become stretched. In this way, much of the ejected blood is taken up in the increased volume of the artery, which thus serves as a damping device so that the blood does not flow through the lung capillaries in spurts with each heartbeat (Melbin and Noordergraaf 1971).

THE PHYSICS OF FLOW IN TUBES

To understand how blood moves in the vascular system it is necessary to understand the basic physics of flow in tubes and also some of the peculiarities of blood as a fluid.

The flow of a fluid (whether gas or liquid) may be smooth and regular so that in a straight tube each particle of fluid moves in a straight line. This is called *laminar flow.* If in a curved tube the direction of flow at any given point remains constant, the flow is also said to be laminar. The path followed by a fluid particle in laminar flow is called a *streamline.* If the fluid particles flow irregularly and swirl about, however, the flow is *turbulent.* In a given tube the flow changes from laminar to turbulent if the velocity increases above a critical point. The following discussion will refer specifically to laminar flow; the fluid dynamics of turbulent flow is more complex and has little application to the flow in blood vessels.

Bernoulli's theorem

The steady flow in an ideal fluid along a streamline is described by *Bernoulli's theorem,* named after a Swiss physician (1700–1782) who at the age of 25 became professor of mathematics. First we must note that flow in a tube is called *steady* if the velocity at every point of the tube remains constant, even if the pipe gets narrower or wider and the fluid therefore moves faster or slower as it flows along the tube.

Bernoulli's theorem states that the total fluid energy (E) is the sum of (1) the potential energy attributable to internal pressure, (2) the potential energy attributable to gravity, and (3) the kinetic energy of the moving fluid.

$$E = (p\,v) + (m\,g\,h) + (\tfrac{1}{2}\,m\,u^2) \qquad (1)$$

On the right side of the equation the first term is the pressure energy of the fluid or the product of pressure (p) and volume (v). The second term is the gravitational potential energy or the product of the mass (m), gravitational constant (g), and the height (h). The third term is the kinetic energy attributable to the velocity (u) of the fluid. The sum of these three is the total fluid energy. The energy content per unit volume (E') can be obtained by dividing both sides of the equation by the volume:

$$E' = p + \rho\,g\,h + \tfrac{1}{2}\,\rho\,u^2 \qquad (2)$$

in which ρ (rho) is the density of the fluid.

In a horizontal tube there is no change in gravitational potential energy ($\rho\,g\,h$) from one end of the tube to the other, and if the flow is in a frictionless tube, the energy content of the fluid (E') also remains constant. This gives:

$$p + \tfrac{1}{2}\rho\,u^2 = E' - \rho\,g\,h = k \qquad (3)$$

Because the sum of the pressure term and the velocity term is constant, if the velocity (u) changes due to a change in diameter, the pressure (p) must change in the opposite direction.

Let us now look at a tube of uniform diameter in which there is friction. With constant diameter, the velocity (u) remains constant. Due to resistance to flow, energy is used to drive the fluid through the tube, and the energy loss is therefore expressed as a decrease in pressure (p), as indicated in Figure 3.12. The energy loss from frictional forces is degraded as heat; this shows up as a temperature increase in the fluid, but in the bloodstream this is of no significance.

FIGURE 3.12 (Top) When a fluid flows in a horizontal tube, the frictional resistance leads to a drop in pressure along the tube.

FIGURE 3.13 (Bottom) When a fluid flows through a narrow part of a tube, the velocity of the fluid is increased at this point. The increased velocity is accompanied by a decrease in pressure. For further explanation, see text.

If the diameter of the tube changes, there must be a change in the velocity (u) as well. Again, in a horizontal tube the gravity term of equation 2 can be disregarded. Let us, for the moment, assume that no energy is dissipated by frictional forces and the energy content remains constant. We now see that velocity and pressure must change inversely according to expression 3, which states that if velocity increases, pressure in the fluid must decrease. This is illustrated in Figure 3.13 for a tube with frictional resistance.

It is often stated that fluid always moves from higher pressure to lower pressure. Figure 3.13 shows that this is not necessarily so, for the pressure is lower in the narrow tube than further downstream.

FIGURE 3.14 A fluid in a U-shaped tube remains at rest. At the bottom of the U the pressure is higher than in the arms, due to the effect of gravity on the fluid.

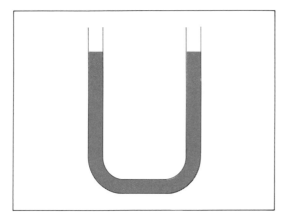

The correct statement is that fluid always moves from a point where the total fluid energy is higher to a point where it is lower.

In the preceding discussion of flow in a horizontal tube the gravitational term in equation 2 was disregarded. If we have a U-tube, as in Figure 3.14, a liquid at rest has the same energy content per unit volume throughout the tube. Although the pressure at the bottom of the U is higher than in the arms, the fluid remains still. Reference to equation 2 shows that, as the height (h) decreases, the gravitational potential energy per unit volume ($\rho\,g\,h$) decreases, and the pressure energy (p) must increase accordingly. We refer to this increase in pressure energy as the *hydrostatic pressure* of a fluid, an expression used to describe the increase in pressure with depth relative to the pressure at the surface.

Blood pressure

Let us see how these principles can be applied. In the foot of a person the venous pressure is increased due to the height of the column of blood (Figure 3.15). The arterial pressure at the foot actually exceeds the arterial pressure at the heart, again due to the weight of the column of blood. The high pressure in the venous system of the legs distends the veins, and contractions of the leg muscles and valves in the veins are main aids in returning the venous blood to the heart.

Another consequence of the gravitational term is that a certain arterial pressure is necessary to drive the blood to the head. Because of the gravity effect, the arterial blood pressure in the head of a man is reduced to about 50 mm Hg (7 kPa). For a giraffe, which carries its head about 2 m above the level of the heart, sufficient pressure to supply the brain with blood requires a much higher blood pressure than in humans (Figure 3.16).

Measurements of blood pressures in the giraffe have shown a systolic blood pressure as high as 260 mm Hg (35 kPa) (Van Citters et al. 1968). This is more than twice as high as the normal systolic blood pressure in humans, which is about 100 mm Hg (13 kPa). To withstand the high pressures, the giraffe's arterial system has exceptionally thick walls, and the venous system is equipped throughout with valves that facilitate the return of blood from the limbs. When a giraffe that stands 4 m tall lowers its head to drink, valves in the neck veins help impede backflow of blood to the head and presumably reduce the increase in hydrostatic pressure in the brain that otherwise would be caused by swinging the head down. (Hargens et al. 1987).

Wall thickness and tube diameter

In a hollow cylinder the tension in the wall (T) equals the product of the pressure across the wall (p) and the radius (r):

$$T = p \times r$$

This relationship was derived by Bernoulli, but is usually known as *Laplace's law*. It tells us that, for a given pressure, the tension in the wall increases in direct proportion with increasing radius. For the wall to withstand the tension, the thickness must therefore be increased accordingly. This is the reason a large artery must have a thicker wall than a small artery.

A vein, which carries much lower blood pressure, has a thinner wall. Again, a smaller vein can withstand the venous pressure with a thinner wall

FIGURE 3.15 Arterial and venous pressures in a man as he assumes different postures. The figures indicate the pressures at various points in relation to the pressure in the right atrium of the heart. (1 mm Hg = 0.13 kPa). [Modified from Burton 1972]

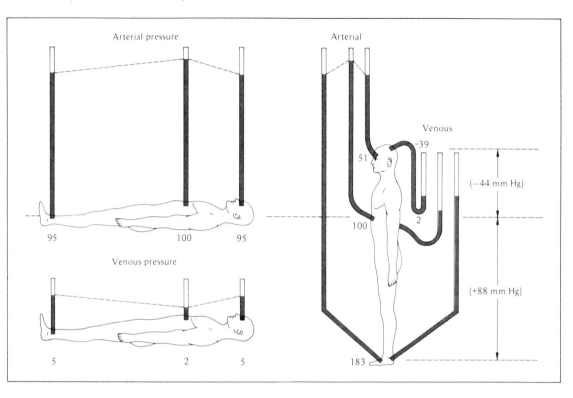

FIGURE 3.16 Because the head of the giraffe is as much as 2 m above the level of the heart, the brain can be supplied with blood only if the blood pressure at the heart is at a level so high that it would be extremely dangerous in humans.

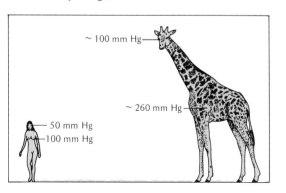

than a larger vein. A capillary, of course, has a higher blood pressure than the veins, but nevertheless, because of its very small radius, a wall consisting of a single layer of cells has sufficient strength. Thus, the smallness of the capillary is a prerequisite for a capillary wall thin enough to permit rapid exchange of material between the blood and the tissues.

Viscosity

The resistance to flow in a tube results from inner friction in the fluid, the *viscosity* of the fluid. Everybody has some experience with viscosity – water and syrup do not flow equally fast out of a bottle. We say that water has a low viscosity and syrup a high viscosity.

The flow through a tube $(\dot{Q})$ is proportional to the pressure drop through the tube and inversely proportional to the resistance:

$$\dot{Q} = \frac{p}{R}$$

The resistance term (R) is a function of the dimensions of the tube and the nature of the fluid (viscosity). This relationship was clarified by the French physician Poiseuille (1799–1869), who became interested in the flow of blood in capillaries. Poiseuille worked with tubes of different diameters and found that the flow through a tube is proportional to the pressure and proportional to the fourth power of the radius of the tube, but inversely proportional to the length of the tube and to the viscosity of the fluid. This is formally expressed in *Poiseuille's equation,* which can be written as follows:

$$\dot{Q} = \Delta p \, \frac{r^4}{l\eta} \times \frac{\pi}{8}$$

in which $(\dot{Q})$ = rate of blood flow, Δp = pressure drop, r = radius, l = length of tube, and η (eta) = viscosity.*

The most important aspect of this equation is the striking effect the radius of the tube has on the flow. For our purposes here, however, we shall be more concerned with the viscosity. The equation shows that as viscosity increases, the fluid flow decreases proportionately.

For many purposes it is convenient to express the viscosity of a fluid relative to that of water, assigning unity to the viscosity of water. Blood plasma is more viscous than water and has a *relative viscosity* of about 1.8, mostly as a result of the 7% dissolved protein. Assume that we measure the flow of water and of plasma through a given tube: If we assign the value of 1.0 to the flow of water, the flow of plasma will be 1/1.8, or 0.55.

Let us now reduce the radius of the tube by one-half, keeping pressure and length of the tube constant. According to Poiseuille's equation the flow of water will be reduced to $(\frac{1}{2})^4$ or 1/16. So will the flow of plasma, and the ratio of the flows will still be 1/1.8; that is, the viscosity of plasma relative to water (the relative viscosity) is independent of the size of the tube.

If instead of plasma we use whole blood, we find that its viscosity is higher than the viscosity of plasma and increases with the concentration of red cells (Figure 2.1). Blood, however, does not behave quite as expected. Its relative viscosity changes with the radius of the tube through which it flows. With decreasing radius, the relative viscosity of blood decreases (i.e., it becomes more similar to water and flows more easily).

Such an anomalous fluid is called a *non-Newtonian fluid;* its behavior makes the flow of blood in the body difficult to describe, particularly in view of the fact that blood vessels are elastic and by no means maintain a constant radius. The end result, however, is important: Flow of blood through the capillary bed is easier, and the apparent viscosity of blood is lower than expected merely from the dimensions of the blood vessels.

We can now return to the question of blood viscosity that was raised in Chapter 2: What is the effect on blood viscosity of enclosing the hemoglobin in red cells? Because of its non-Newtonian behavior, whole blood flows more easily than corresponding hemoglobin solutions in the arterioles and capillaries (which are the major resistance elements of the circulatory system) (Snyder 1973).

Another peculiar aspect of blood flow through capillaries is that the capillary diameter is often considerably less than the size of a red blood cell. Unexpectedly, this seems to be no impediment to flow in the capillary: The red cell is easily deformed and passes readily through the capillary (Skalak and Branemark 1969). This gives rise to a very different type of flow (*bolus flow*) in which the red cell acts as a plug that causes a rapid renewal of liquid in the boundary layer along the

---

*The viscosity of a fluid changes with temperature. A reduction in temperature from 37 to 0 °C causes the viscosity of water to increase about 2.6-fold.

capillary wall, thus facilitating the renewal of diffusible substances in this layer.

The capillary system

The total number of capillaries in the body is enormous. Merely in the mesentery of a dog the number of capillaries exceeds one thousand million (Table 3.5). Muscle is particularly well suited for an exact count of the number of capillaries, for here the capillaries run between and parallel to the muscle fibers. In a cross section of a muscle it is therefore relatively easy to count the number of capillaries per unit cross-sectional area. The number is very great, but the capillaries usually are not all open and filled with blood.

In the resting muscle of a guinea pig, a cross section of 1 mm^2 contains about 100 open capillaries through which blood flows. If the muscle is performing work, however, the constricted arterioles open, more capillaries carry blood, and in maximal exercise a 1-mm^2 cross section may have more than 3000 open capillaries. In an ordinary pencil the cross section of the lead is about 3 mm^2. Imagine nearly 10 000 tiny tubes running in parallel down the lead of the pencil!

A mammal of small body size has a higher capillary density than a mammal of large body size. This is consistent with the higher metabolic rate and need for oxygen in the small animal, as discussed in Chapter 2.

Two major processes account for the exchange of material between the fluid within the capillary and the interstitial spaces of the tissues. The capillary wall is quite thin, about 1 μm thick, and consists of a single layer of cells. Water and dissolved substances of small molecular weight (gases, salts, sugars, amino acids, etc.) can *diffuse* relatively unhindered across the capillary wall, but in addition, pressure within the capillary forces some fluid out through the wall by *bulk filtration*. Molecules as large as most proteins (molecular weight about 70 000 or larger), however, do not pass through the capillary wall. Proteins are withheld within the capillary, and a relatively protein-free filtrate is forced out through the wall. The concentrations of the small-size solutes in this filtrate are similar to those in the blood plasma, but not quite identical because the proteins have some influence on the distribution of ions (the Gibbs–Donnan effect).

If we regard the capillary wall as a semipermeable membrane, we can see that because the nonprotein solutes penetrate freely, they have no osmotic effect. The proteins, however, are retained inside the capillary and exert a certain osmotic effect, called the *colloidal osmotic pressure*. In mammalian plasma this pressure is about 25 mm Hg and tends to draw water back into the capillary from the surrounding tissue fluid.

When the hydrostatic pressure (blood pressure) within the capillary exceeds the colloidal osmotic pressure, fluid is forced out through the capillary wall; when the hydrostatic pressure within falls below the colloidal osmotic pressure, fluid is drawn in. The blood pressure within the capillary is variable, but in the arterial end it is often higher and in the venous end lower than the colloidal osmotic pressure. As a result, fluid is filtered out at the arterial end and reenters at the venous end of the capillary (Figure 3.17). This theory for bulk filtration and reentry of fluid through the capillary wall was originally proposed by Starling (1896).

The amount of fluid filtered through the capillary wall, and the amount withdrawn again because of the colloidal osmotic pressure, fluctuate greatly. Usually outflow exceeds inflow, and the excess fluid remains in the interstitial spaces between the tissue cells. This fluid, the *lymph*, slowly drains into tiny lymph vessels or lymph capillaries. It is now accepted that this flow is not an entirely passive drainage. The small lymph vessels undergo continual rhythmic contractions whose pumping action causes a slight negative pressure to develop, and as long as the amount of fluid filtering through the capillary wall remains slight, the lymphatic system is capable of removing all the excess fluid (Guyton 1976).

The smaller lymph vessels gradually merge with other lymph vessels and eventually join into larger

FIGURE 3.17 The capillary wall is semipermeable, and the blood pressure forces fluid out by ultrafiltration (top). The plasma proteins remain within the capillary and oppose the ultrafiltration process. The blood pressure decreases along the length of the capillary, and when the blood pressure falls below the colloidal osmotic pressure exerted by the proteins, these cause an osmotic flow of fluid back into the capillary (bottom). The wall is freely permeable to salts and other small-molecular-size solutes, and the fluid movements therefore take place as if these were not present.

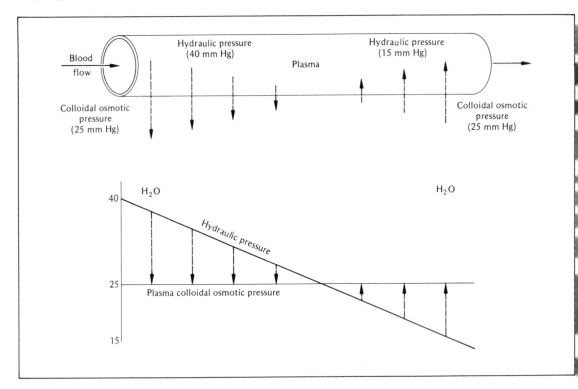

lymph ducts. These, in turn, empty the lymph into the larger veins, returning the fluid to the blood system. The lymphatic system can be considered part of the vascular system, but there is no direct circulation of the lymph as such; it is, rather, a drainage system that returns to the blood excess fluid lost from the capillary system.

If plasma proteins were absent, any hydrostatic pressure in the capillary in excess of that in the surrounding tissue spaces would cause fluid to filter out, and no force would be available to cause fluid to reenter. The plasma proteins are therefore essential to retain the blood fluid within the vascular system.

Plasma colloidal osmotic pressure is higher in mammals than in other vertebrates (Table 3.6). Surprisingly, birds have a rather low colloidal pressure, although their arterial blood pressure usually far exceeds the common values for mammals. Some birds have blood pressures above 200 or even 250 mm Hg (27 to 33 kPa), against about 100 mm Hg (13 kPa) commonly found in mammals. These differences are not well understood. It seems either that birds must have a greater resis-

TABLE 3.6 Colloidal osmotic pressure of vertebrate blood plasma (1 mm Hg = 0.13 kPa). [Data from Altman and Dittmer 1971]

Animal	Pressure (mm Hg)
Cow	21
Sheep	22
Dog	20
Cat	24
Chicken	11
Dove	8.1
Alligator	9.9
Turtle	6.4
Frog	5.1
Toad	9.8
Codfish	8.3
Flounder	8.5

tance to blood flow at the level of the arterioles so that the capillary pressure is not very high, or that the avian capillary wall has properties that differ from those of mammalian capillaries. A higher resistance in the arterioles seems meaningless, for why should birds then have the high arterial pressure to begin with?

Other vertebrates – reptiles, amphibians, and fish – also have relatively low colloidal osmotic pressure. This is consistent with the fact that these groups in general have lower arterial blood pressures than mammals.

Circulation during exercise

Muscular activity increases the demand for oxygen and thus the amount of oxygen the heart must deliver to the tissues. Two avenues are open to meet this increased demand: increasing the volume of blood pumped by the heart (the cardiac output) or increasing the amount of oxygen delivered by each volume of blood. Arterial blood is already fully saturated and cannot take up more oxygen, but venous blood normally still contains more than half the oxygen content of arterial blood. Increasing the

oxygen extraction from the blood is an obvious way of obtaining more oxygen from each volume of blood.

Let us first examine oxygen extraction. The total muscle mass of a lean man, which makes up nearly half his weight, uses about 50 ml O_2 per minute. This oxygen is supplied by a blood flow of about 1 liter (i.e., the oxygen content of arterial blood, 200 ml O_2 per liter blood, is reduced to 150 ml O_2 per liter in venous blood). Because a quarter of the oxygen in arterial blood is removed, we say that the oxygen extraction is 25%. In a normal man during heavy exercise the blood flow to the muscles may be 20 liters per minute (in well-trained athletes, even higher), and the oxygen extraction in the muscles increases to 80 or 90%; in other words, the venous blood coming from heavily working muscles has very little oxygen left in it (Folkow and Neil 1971).

The second avenue available for increasing the delivery of oxygen is increased cardiac output. This can be achieved by increasing both heart rate and stroke volume. Because of the interest in humans, both for medical reasons and in relation to athletic performance, much more information is available for humans than for other animals.

At rest the human heart beats at a rate of about 70 per minute, with a stroke volume of about 70 ml (from each side), giving a total cardiac output of about 5 liters per minute. In heavy exercise, the cardiac output can readily increase fivefold or more (if the oxygen extraction is tripled, this corresponds to a 15-fold increase in oxygen delivery). Most of the increase in cardiac output is attributable to increased pulse rate, which may rise to 200 strokes per minute, but there is also an increase in stroke volume, which may exceed 100 ml.

The distribution of the increased cardiac output to the muscles and to the remainder of the body in rest and exercise is shown in Figure 3.18. The blood flow to the muscles of an athlete in top condition may increase as much as 25- or 30-fold, while there is a slight decrease in circulation to the remainder

FIGURE 3.18 Distribution of total blood flow (cardiac minute volume) (*a*) and of oxygen consumption (*b*) between the muscles (shaded bars) and all other parts of the body (unshaded bars). Data for resting man, heavy exercise in a normal man, and heavy exercise in a top athlete. All values are in liters per minute. [Folkow and Neil 1971]

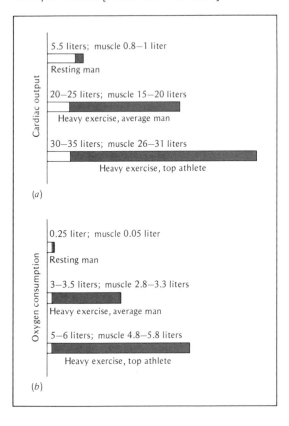

(*a*)

(*b*)

FIGURE 3.19 In a trout the oxygen extraction from the blood increases with increasing swimming speed. At the highest sustainable speed each volume of blood delivers about three times as much oxygen as at rest. (1 mmol O_2 per liter = 2.24 cm^3 O_2 per 100 cm^3 blood.) [Data from Kiceniuk and Jones 1977]

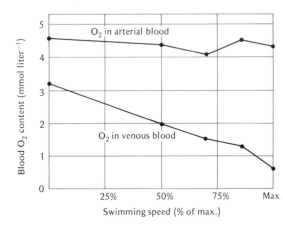

of the body. In the athlete the oxygen consumption of the muscles may be increased as much as 100-fold; this is possible only because of an approximately threefold increase in the oxygen extraction.

It would be interesting to compare the human athlete and a fast-swimming fish. Let us therefore return to the trout that we compared with a pigeon in Table 3.3 (p. 105).

Trout of about 1-kg size were trained to swim against a water current and were studied at various speeds up to the maximal speed they could sustain. As expected, the rate of oxygen consumption increased with increasing swimming speed and reached a sustainable level of about eight times the resting rate of oxygen consumption – a rather good performance for a fish.

How does the oxygen transport system of the trout adjust to the need for supplying oxygen at eight times the resting rate? The heart rate was increased only by about 35%, but the stroke volume was more than doubled, giving a threefold increase in cardiac output. This by itself is far from sufficient to sustain an eightfold increase in rate of oxygen consumption. Accordingly, each volume of blood must deliver more oxygen (Figure 3.19). We see that the amount of oxygen carried in the arterial blood is nearly unchanged in exercise, but the venous blood gives up increasing amounts of oxygen. At rest, the oxygen extraction is 32%, but in maximum exercise it was increased to 85%. Thus, the increased need for oxygen in exercise is sustained by a combination of three factors: a moder-

FIGURE 3.20 The circulation in the earthworm. The dorsal vessel is contractile and pumps blood into the larger vessels. The integument serves in gas exchange and is heavily vascularized. Special vessels from the integument supply the ventral nerve cord with oxygen-rich blood directly from the skin. [Meglitsch 1972]

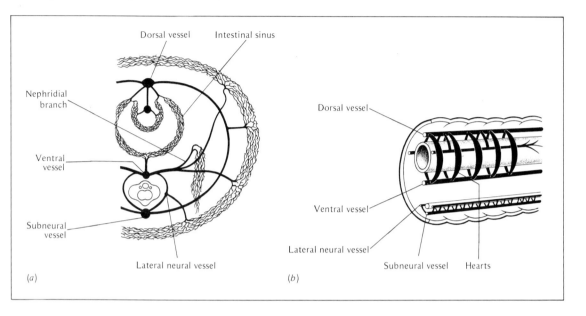

ate increase in heart rate (36%), a more than twofold increase in stroke volume, and a nearly threefold increase in the oxygen extraction from the blood. In other words, the adjustments to athletic performance in a fish and a human are similar, the difference being that the fish adjusts cardiac output primarily through increasing stroke volume and the human relies more on heart rate, while both greatly increase the oxygen extraction from the blood. The increased oxygen extraction, of course, reduces the work load on the heart for both of them.

INVERTEBRATE CIRCULATION

Many invertebrates have well-developed circulatory systems; examples are annelids, echinoderms, arthropods, and molluscs. Most of these have open systems. Cephalopods (octopus and squid) have closed circulations, although the other classes of molluscs have open systems. It is worth noting that insects, which are highly organized and complex animals and can sustain exceptionally high metabolic activity, have open circulatory systems.

Annelids

Annelids have closed, well-organized circulatory systems and their blood often contains respiratory pigments dissolved in the plasma. Hemoglobin is the most common pigment, but some annelids have chlorocruorin or hemerythrin.

For the earthworm, *Lumbricus,* the general body surface serves as a respiratory organ and is supplied with a dense capillary network (Figure 3.20). The circulatory system has two longitudinal vessels: a dorsal vessel in which the blood is pumped forward, and a ventral vessel in which it flows in the opposite direction. The nerve cord of many annelids, including the earthworm, is supplied through additional longitudinal vessels, and some of the oxygenated blood from the integument is routed directly to this important organ. The closed system of blood vessels thus has a distinct advantage in controlling precisely where the blood will flow.

An annelid does not have one single distinct heart, although several of the blood vessels have dilations that are contractile. The dorsal blood vessel is the most important vessel, in which peristaltic waves propel blood forward. In addition, blood vessels that on each side connect the main dorsal with the main ventral longitudinal vessel are contractile and serve as accessory hearts.

In the giant earthworm (*Glossoscolex giganteus*) the dorsal blood vessel contracts about six to eight times per minute, raising the systolic pressure to nearly 20 mm Hg (2.7 kPa). This contraction is responsible for filling five pairs of segmental lateral hearts in the front part of the animal. These in turn contract and cause the pressure in the ventral vessel to rise as high as 75 mm Hg (10 kPa) (Johansen and Martin 1965).

The giant earthworm is found in the tropical and subtropical regions of Brazil. It is the largest terrestrial animal without specialized respiratory organs known to us. It may be as long as 120 cm with a diameter of 2 to 3 cm and a weight of more than 500 or 600 g. It depends entirely on the skin for gas exchange, as other earthworms do. Its blood contains hemoglobin and has a rather high oxygen affinity with a half-saturation pressure (P_{50}) of about 7 mm Hg (Johansen and Martin 1966). In combination with the highly developed circulatory system, the giant earthworm obtains an adequate supply of oxygen, using only the skin as respiratory organ.

Echinoderms

Echinoderms (e.g., starfish, sea urchins, and sea cucumbers) have three fluid systems: the *coelomic system,* the blood or *hemal system,* and the *water vascular system.* The last functions primarily in locomotion; it is a hydraulic system used in the movement of the tube feet and is filled with a fluid similar to sea water.

An echinoderm has a large coelom between the body wall and the digestive tract, filled with a coelomic fluid that seems to be important in transporting nutrients between the digestive tract and other parts of the body (Farmanfarmaian and Phillips 1962).

The hemal or blood vascular system of the echinoderm contains a separate fluid that, in some sea cucumbers (holothurians), contains hemoglobin. The hemoglobin from the sea cucumber *Cucumaria* has a rather high oxygen affinity, with half-saturation pressures between 5 and 10 mm Hg (depending on temperature) and a striking absence of a Bohr effect between pH 7.5 and 6.5 (Manwell 1959). These characteristics are apparently well suited to the often oxygen-deficient mud environment in which sea cucumbers live.

The relationship between the hemal system and the coelomic system is not well understood, and the role of the hemal system in respiration and oxygen supply to the animal is not clear (Martin and Johansen 1965). The role of the water vascular system is primarily in locomotion and is better understood.

Molluscs

With the exception of octopus and squid, molluscs have open circulatory systems. The blood of many molluscs contains hemocyanin, and a few have hemoglobin. The oxygen-carrying capacity of the blood tends to be correlated with the size of the animal and, in particular, with its activity.

Molluscs, in general, have a well-developed heart, and the heartbeat is adjusted to meet the physiological demands for oxygen. An increased venous return of blood to the heart causes an increase in both amplitude and frequency of the heartbeat. The mollusc heart continues to beat if it is separated from the nervous system; it has an inherent rhythmicity with a pacemaker. However, the heart is also under the influence of neurosecretions that modify its beat. Acetylcholine inhibits the heart, and serotonin acts as an excitatory substance.

In clams, active movement of the foot is based on the use of blood as a hydraulic fluid. Large blood sinuses in the foot are controlled by valves that help in adjusting both the size and the movements of the foot. This is particularly important in burrowing species, which use the foot in moving through the bottom material.

Cephalopods (octopus and squid) have highly developed closed circulatory systems with distinct arteries, veins, and capillary networks. This is related to the high organization and activity of these animals and to the importance of the blood in respiratory exchange in the gills as well as in the function of the kidney.

Both octopus and squid have relatively high blood pressures, which during activity may exceed 75 mm Hg (10 kPa) (Wells 1979; Bourne 1982). As in vertebrates the arteries have elastic walls which distend during systole. They therefore serve as an elastic reservoir that maintains the blood pressure during diastole, serving to smooth out the pulsatile flow of blood from the heart. Isolated elastic fibers from the arteries of octopus have rubberlike properties similar to the elastic fibers in arterial walls of vertebrates (Shadwick and Gosline 1981). The amino acid composition of these rubberlike proteins differs, suggesting that each protein arose independently during evolution.

The fact that cephalopods have a closed circulatory system means that the blood remains confined within the blood vessels. Therefore, the blood remains distinctly separate from the interstitial fluid. The blood makes up about 6% of the body weight, and the interstitial fluid, about 15%. These magnitudes are strikingly similar to those of the corresponding fluid volumes in vertebrate animals.

The open system of a noncephalopod mollusc does not separate blood from interstitial fluid, and blood flows freely throughout the extracellular space; the blood volume is therefore quite large and may constitute as much as 50% of the wet body weight (without the shell) (Martin et al. 1958).

The open circulatory systems lack discrete peripheral blood vessels, and this has created the impression that the flow of blood in the body spaces is undirected and not carefully partitioned, as it is in vertebrates. Studies of the abalone (the gastro-

pod *Haliotis cracherodii*) suggests that the blood is distributed to the various organs, not in direct proportion to their mass, but in proportion to their oxygen consumption. For example, the foot muscle makes up 66% of the total body weight (exclusive of the shell) but receives only 27% of the cardiac output. In contrast, the digestive gland represents only 6% of the body weight but receives 13% of the cardiac output (Jorgensen et al. 1984). Figure 3.21 shows that per unit weight, the blood flowing to the right kidney is about eight times as great as the flow to the foot muscle. Clearly, open systems may be far more effective in distributing the blood than previously thought.

Insects

An insect usually has a major blood vessel running along the dorsal side. The posterior part of this vessel functions as a "heart" and is provided with a series of valved openings through which blood can enter. The anterior part of the blood vessel, which could be called the "aorta," is contractile and may have peristaltic waves propelling the blood forward. This main blood vessel branches and continues to the head, where it ends. The branches from the aorta supply most of the body, where blood flows freely among the tissues and slowly percolates back toward the heart.

In many insects the blood that flows among the tissues is given some direction by longitudinal membranes. In the antennae and the limbs, the blood may enter on one side and leave on the other. In this way the blood is led into certain pathways, although it does not run in discrete blood vessels (Figure 3.22).

Because the blood system is open, the blood pressure of an insect can barely exceed the tissue pressure. Insect blood plays no particular role in respiration; its main functions are to carry nutrients and metabolites around in the body and to provide a transport system for hormones, which in insects are of great importance in physiological coordination (growth, molting, etc.). In some insects the circulation of blood is also essential in the distribution of heat and, in particular, in the regulation

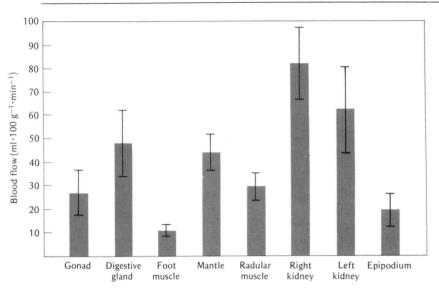

FIGURE 3.21 Blood flow rates in selected tissues of the black abalone. Vertical bars are ± 1 SE.

FIGURE 3.22 The well-developed circulatory system in an insect has a main dorsal vessel that carries blood from the heart. Although the circulation is open, the blood follows certain channels because of the presence of longitudinal membranes, especially in the legs. Arrows indicate course of circulation. D, dorsal diaphragm with muscles; N, nerve cord; PO, pulsatile organs; S, septa dividing appendages. [Wigglesworth 1972]

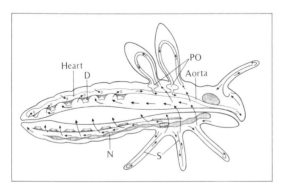

of heat loss in highly active flying insects (Heinrich 1971).

Although the heart and the dorsal vessel are the principal components of the circulatory system of insects, many also have accessory pumping organs or hearts. These are particularly important for maintaining circulation in the appendages (wings, legs, and antennae). The flow of blood in the wings is aided by pulsating organs within the wings themselves, located in the channels that return the blood to the body (Thomsen 1938).

Blood flow into the antennae is aided by contractile organs located at their bases. These function by aspiration of blood from blood sinuses in the head. Many insects also have leg hearts that aid in circulating blood to these important appendages. The accessory hearts of an insect and their contraction are independent of the dorsal heart: They may beat at different rates; one may stop while others continue to beat, and none is dependent on the dorsal heart.

The filling of the dorsal heart of an insect takes place by suction. A set of muscles (the *aliform muscles*) radiate from the heart; when they contract, the expansion of the heart pulls blood through valved openings (*ostia*) into the lumen of the heart. The subsequent contraction of the heart propels blood forward, in the direction of the head, and the valves prevent the blood from flowing out through the ostia (Figure 3.23).

FIGURE 3.23 The insect heart consists of a contractile portion of the main dorsal blood vessel. Filling of the heart with blood is achieved by suction, provided by muscles that pull externally on the walls of the heart. This figure shows the dissection of a cockroach (*Blaberus*). [Modified from Nutting 1951]

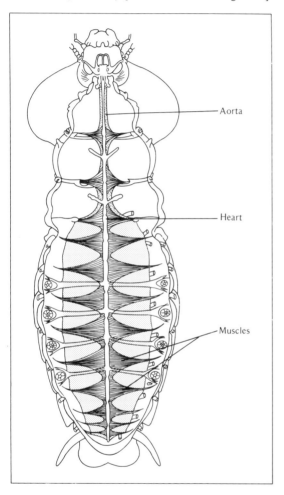

— Aorta

— Heart

— Muscles

Arachnids

The circulatory system of spiders and scorpions is similar to that of insects, but it may have a greater role in respiration than is usual in insects. Hemocyanin is found in the blood of some scorpions, and both scorpions and spiders have definite respiratory organs that are perfused by blood. This should not be surprising, for arachnids lack the tracheal system that is characteristic of insects.

The arachnid heart is located dorsally in the abdomen. It is filled through valved openings (ostia) and empties into arteries. In spiders, distinct arteries lead to the legs, where a relatively high blood pressure is of importance in locomotion. Spider legs lack extensor muscles, and blood is used as a hydraulic fluid for leg extension. In jumping spiders the blood pressure may reach an amazing 400 mm Hg (50 kPa), or about 0.5 atm (see Chapter 10).

Crustaceans

The circulatory system of crustaceans is extremely variable. Small crustaceans have poorly developed circulatory systems, often without any heart; large crustaceans, in particular decapods (lobster, crabs, and crayfish), have well-developed circulatory systems and blood with respiratory pigments (hemocyanin).

The crustacean circulatory system is based on the same morphological structures as in other arthropods, but it differs from insect circulation in one important physiological aspect: The crustaceans have gills and, therefore, also well-defined circulation to the gills.

The dorsal heart of a large crustacean lies in a pericardial sinus from which blood enters the heart through valved ostia. From the heart there is usually a main artery running in the anterior and another in the posterior direction. As the arteries branch, the blood leaves the vessels and flows between the tissues to a system of ventral sinuses. From these the blood flows into the gills and then in discrete vessels back to the heart.

As a result of this arrangement the decapod heart is directly supplied with oxygenated blood, which is then pumped to the tissues (Figure 3.24). This is the opposite of the arrangement in fish; in a fish the heart receives deoxygenated venous blood, which is then pumped to the gills and on to the tissues. The fish heart itself, however, is supplied with oxygenated blood through a branch of the gill circulation from which oxygenated blood reaches the heart muscle directly.

FIGURE 3.24 Circulation in the lobster (a). In large crustaceans the oxygenated blood from the gills is conveyed directly to the heart, which pumps it on to the tissues (b). In fish the heart is located differently, before the gills (see Figure 3.4). [Meglitsch 1972]

FIGURE 3.25 The volume of blood required to supply 1 ml of oxygen to an animal, plotted against the hemocyanin concentration in its blood. For each of 10 species of crabs the measured blood flow is related to the consumption, in the same animal, of 1 ml of oxygen. [Burnett 1979]

Earlier in this chapter we discussed how an increased oxygen capacity of the blood, caused by the presence of a respiratory pigment, reduces the volume of blood that must be pumped to supply oxygen to the tissues (see Figure 3.2). Crabs provide a good demonstration of this principle because of the wide species-to-species variations in the hemocyanin content of their blood. Figure 3.25 shows that the higher the hemocyanin content of the blood is, the less blood is pumped to supply the tissues with a given amount of oxygen.

The same principle applies, of course, to the demands on the circulation in all animals that have a respiratory pigment: The higher the oxygen capacity of the blood, the less volume needs to be pumped. There is a trade-off here between the cost of providing the respiratory pigment and the cost of pumping, and the question is, Which strategy pays best? It seems that for highly active animals

a high oxygen capacity is most important; for slow and sluggish animals it may be more economical to avoid a heavy investment in the synthesis of high concentrations of a respiratory pigment.

BLOOD COAGULATION AND HEMOSTASIS

Several mechanisms help prevent loss of blood from ruptured blood vessels. Severe blood loss leads to a decrease in blood pressure and thus reduces the flow of blood from the damaged area. Damaged blood vessels constrict and thereby decrease blood flow. The most important mechanism, however, is the closing of blood vessels at the site of injury by a plug consisting of coagulated protein and blood cells. Such a plug or *clot* is important in the complete arrest of bleeding from minor injuries, but if major blood vessels have been ruptured, it does not suffice.

The clotting or *coagulation* mechanism has been well studied in mammals, particularly humans, because blood clotting is of great medical importance. To be effective, a clotting mechanism must act rapidly; yet blood must not clot within the vascular system. Blood must therefore have the inherent ability to clot, and the clotting mechanism should be ready to be turned on when needed. On the other hand, this mechanism must not go off inadvertently.

In vertebrates the blood clot consists of the protein *fibrin*, an insoluble fibrous protein formed from *fibrinogen*, a soluble protein present in normal plasma in an amount of about 0.3%. The transformation of fibrinogen to fibrin is catalyzed by the enzyme *thrombin*, and the reason blood does not clot in the vascular system is that thrombin is absent from the circulating blood. Thrombin, however, can be formed rapidly because its precursor, *prothrombin*, is already present in the plasma.

What is necessary to initiate coagulation is the formation of thrombin from prothrombin. This is only the final step in a complex sequence of biochemical events that has been slowly unraveled in studies of human patients with various deficiencies in the clotting mechanism (e.g., hemophilia). A total of 12 clotting factors have been identified, numbered I through XIII (factor VI is a term no longer used). A few of the final steps of the clotting mechanism are shown in the following diagram.

The many steps in the clotting mechanism may seem an unnecessary complication, and indeed they have been a severe hindrance in the clarification of

RED BLOOD CELL A human red blood cell caught in the meshwork of fibrin in a blood clot. The cell, whose diameter is about 6 μm, has the biconcave shape characteristic of mammalian red blood cells. [Courtesy of Emil Bernstein, the Gillette Company Research Institute; from *Science,* 1971, vol. 173, No. 3993, cover picture]

what takes place. Biologically, the importance of the many steps seems to be explained by the fact that the clotting mechanism works as a biochemical amplifier (MacFarlane 1964). Clotting is normally initiated when blood contacts "foreign" surfaces or damaged tissues. This initiates a series of enzymatic steps in which the enzyme formed in the first step serves as a catalyst or activator for the next step, and so on. The series of steps thus forms an *enzyme cascade*, which ends with the final formation of the clot in which the soluble fibrinogen is changed to the insoluble fibrin.

The enzymatic amplification allows clotting to take place rapidly, yet provides a considerable safety margin to prevent spontaneous coagulation within the vascular system. The analogy to an electronic amplification system is obvious. If we want an amplification system with a low noise level, we use several steps with low amplification in each step, rather than a single step with a high amplification. This minimizes the chance that random noise in the system may set off the final step and thus ensures an adequate margin of safety.

A hemostatic mechanism is as essential for most invertebrates as for vertebrates. The fact that many have open circulatory systems makes the situation more difficult, for in such a system the contraction of blood vessels is of no help. However, open systems have lower blood pressures, and this decreases the tendency for loss of large volumes of fluid.

The two distinct hemostatic mechanisms of vertebrates – clotting of blood and local vasoconstriction – have their counterparts in invertebrates. The simplest invertebrate mechanism is the agglutination of blood corpuscles without participation of plasma proteins (Grégoire and Tagnon 1962). Agglutination is followed by the formation of a cellular meshwork that shrinks and helps close a wound. This is often helped by the contraction of the muscles of the body wall, thus aiding in wound closure.

A true clotting, induced by enzymatic changes of unstable blood proteins, has been described for many arthropods, especially crustaceans. The clotting mechanism of invertebrates, where it occurs, is biochemically distinctly different from the vertebrate mechanism. For example, vertebrate clotting is inhibited by heparin, a mucopolysaccharide that can be isolated from mammalian liver. Heparin has no effect on the clotting system of the horseshoe crab *Limulus* (an arachnid, not a crab) and virtually no effect on crustacean blood (Needham 1970).

Knowledge of the clotting mechanisms of invertebrates is very incomplete, but present information indicates that such mechanisms must have evolved independently many times in the course of evolution.

In these first three chapters we have discussed respiratory gases and their transport. We shall now turn to questions concerning the supply of energy and related matters.

REFERENCES

Altman, P. L., and Dittmer, D. S. (eds.) (1971) *Biological Handbooks: Respiration and Circulation.* Bethesda: Federation of American Societies for Experimental Biology. 930 pp.

Bond, C. F., and Gilbert, P. W. (1958) Comparative study of blood volume in representative aquatic and nonaquatic birds. *Am. J. Physiol. 194:*519–521.

Bourne, G. B. (1982) Blood pressure in the squid, *Loligo pealei. Comp. Biochem. Physiol. 72A:*23–27.

Burnett, L. E. (1979) The effects of environmental oxygen levels on the respiratory function of hemocyanin in the crabs, *Libinia emarginata* and *Ocypode quadrata. J. Exp. Zool. 210:*289–300.

Burton, A. C. (1972) *Physiology and Biophysics of the Circulation: An Introductory Text.* 2d ed. Chicago: Year Book. 217 pp.

Butler, P. J., West, N. H., and Jones, D. R. (1977) Respiratory and cardiovascular responses of the pigeon to sustained level flight in a wind-tunnel. *J. Exp. Biol. 71:*7–26.

DeLong, K. T. (1962) Quantitative analysis of blood circulation through the frog heart. *Science 138:*693–694.

Farmanfarmaian, A., and Phillips, J. H. (1962) Digestion, storage, and translocation of nutrients in the purple sea urchin (*Strongylocentrotus purpuratus*). *Biol. Bull. 123:*105–120.

Folkow, B., and Neil, E. (1971) *Circulation.* New York: Oxford University Press. 593 pp.

Foxon, G. E. H. (1955) Problems of the double circulation in vertebrates. *Biol. Rev. 30:*196–228.

Gordon, M. S. (1972) *Animal Physiology: Principles and Adaptations,* 2d ed. New York: Macmillan. 592 pp.

Grégoire, C., and Tagnon, H. J. (1962) Blood coagulation. *Comp. Biochem. 4:*435–482.

Guyton, A. C. (1976) Symposium: Interstitial fluid pressure and dynamics of lymph formation. *Fed. Proc. 35:*1861–1885.

Hargens, A. R., Millard, R. W., Petterson, K., and Johansen, K. (1987) Gravitational haemodynamics and oedema prevention in the giraffe. *Nature, Lond. 329:*59–60.

Heinrich, B. (1971) Temperature regulation of the

sphinx moth, *Manduca sexta.* 2. Regulation of heat loss by control of blood circulation. *J. Exp. Biol. 84:*153–166.

Johansen, K. (1960) Circulation in the hagfish, *Myxine glutinosa* L. *Biol. Bull. 118:*289–295.

Johansen, K. (1970) Air breathing in fishes. In *Fish Physiology,* vol. 4 (W. S. Hoar and D. J. Randall, eds.), pp. 361–411. New York: Academic Press.

Johansen, K., Lenfant, C., and Hanson, D. (1968) Cardiovascular dynamics in the lungfishes. *Z. Vergl. Physiol. 59:*157–186.

Johansen, K., and Martin, A. W. (1965) Circulation in a giant earthworm, *Glossoscolex giganteus.* 1. Contractile processes and pressure gradients in the large blood vessels. *J. Exp. Biol. 43:*333–347.

Johansen, K., and Martin, A. W. (1966) Circulation in a giant earthworm, *Glossoscolex giganteus.* 2. Respiratory properties of the blood and some patterns of gas exchange. *J. Exp. Biol. 45:*165–172.

Jorgensen, D. D., Ware, S. K., and Redmond, J. R. (1984) Cardiac output and tissue blood flow in the abalone, *Haliotis cracherodii* (Mollusca, Gastropoda). *J. Exp. Zool. 231:*309–324.

Kiceniuk, J. W., and Jones, D. R. (1977) The oxygen transport system in trout (*Salmo gairdneri*) during sustained exercise. *J. Exp. Biol. 69:*247–260.

Krijgsman, B. J. (1956) Contractile and pacemaker mechanisms of the heart of tunicates. *Biol. Rev. 31:*288–312.

Lasiewski, R. C., and Calder, W. A., Jr. (1971) A preliminary allometric analysis of respiration variables in resting birds. *Respir. Physiol. 11:*152–166.

Lasiewski, R. C., Weathers, W. W., and Bernstein, M. H. (1967) Physiological responses of the giant hummingbird, *Patagona gigas. Comp. Biochem. Physiol. 23:*797–813.

MacFarlane, R. G. (1964) An enzyme cascade in the blood clotting mechanism, and its function as a biochemical amplifier. *Nature, Lond. 202:*498–499.

Manwell, C. (1959) Oxygen equilibrium of *Cucumaria miniata* hemoglobin and the absence of the Bohr effect. *J. Cell. Comp. Physiol. 53:*75–84.

Martin, A. W., Harrison, F. M., Huston, M. J., and Stewart, D. M. (1958) The blood volumes of some representative molluscs. *J. Exp. Biol. 35:*260–279.

Martin, A. W., and Johansen, K. (1965) Adaptations of the circulation in invertebrate animals. In *Handbook of Physiology,* sect. 2, *Circulation,* vol. 3 (W. F. Ham-ilton and P. Dow, eds.), pp. 2545–2581. Washington, D.C.: American Physiological Society.

Meglitsch, P. A. (1972) *Invertebrate Zoology,* 2d ed. New York: Oxford University Press. 834 pp.

Melbin, J., and Noordergraaf, A. (1971) Elastic deformation in orthotropic vessels: Theoretical and experimental results. *Circulation Res. 29:*680–692.

Moalli, R., Meyers, R. S., Jackson, D. C., and Millard, R. W. (1980) Skin circulation of the frog, *Rana catesbeiana:* Distribution and dynamics. *Respir. Physiol. 40:*137–148.

Morrison, P., Ryser, F. A., and Dawe, A. A. (1959) Studies on the physiology of the masked shrew *Sorex cinereus. Physiol. Zool. 32:*256–271.

Needham, A. E. (1970) Haemostatic mechanisms in the invertebrata. In *The Haemostatic Mechanism in Man and Other Animals.* (R. G. MacFarlane, ed.), pp. 19–44. Symposium of the Zoological Society of London, No. 27. London: Academic Press.

Nutting, W. L. (1951) A comparative anatomical study of the heart and accessory structures of the orthopteroid insects. *J. Morphol. 89:*501–597.

Randall, D. J. (1970) The circulatory system. In *Fish Physiology,* vol. 4 (W. S. Hoar and D. J. Randall, eds.), pp. 133–172. New York: Academic Press.

Satchell, G. H. (1970) A functional appraisal of the fish heart. *Fed. Proc. 29:*1120–1123.

Shadwick, R. E., and Gosline, J. M. (1981) Elastic arteries in invertebrates: Mechanics of the octopus aorta. *Science 213:*759–761.

Shkolnik, A., Borut, A., and Choshniak, J. (1972) Water economy of the Beduin goat. *Symp. Zool. Soc. London. 31:*229–242.

Skalak, R., and Branemark, P. I. (1969) Deformation of red blood cells in capillaries. *Science 164:*717–719.

Snyder, G. K. (1973) Erythrocyte evolution: The significance of the Fåhraeus-Lindqvist phenomenon. *Respir. Physiol. 19:*271–278.

Stahl, W. R. (1965) Organ weights in primates and other mammals. *Science 150:*1039–1042.

Stahl, W. R. (1967) Scaling of respiratory variables in mammals. *J. Appl. Physiol. 22:*453–460.

Starling, E. H. (1896) On the absorption of fluids from the connective tissue spaces. *J. Physiol., Lond. 19:*312–326.

Studier, E. H., and Howell, D. J. (1969) Heart rate of female big brown bats in flight. *J. Mammal. 50:*842–845.

Tazawa, H., Mochizuki, M., and Piiper, J. (1979) Respiratory gas transport by the incompletely separated double circulation in the bullfrog, *Rana catesbeiana*. *Respir. Physiol. 36:*77–95.

Thomsen, E. (1938) Ueber den Kreislauf im Flügel der Musciden, mit besonderer Berücksichtigung der akzessorischen pulsierenden Organe. *Z. Morphol. Oekol. Tiere 34:*416–438.

Thorson, T. B. (1958) Measurement of the fluid compartments of four species of marine Chondrichthyes. *Physiol. Zool. 31:*16–23.

Thorson, T. B. (1959) Partitioning of body water in sea lamprey. *Science 130:*99–100.

Thorson, T. B. (1961) The partitioning of body water in Osteichthyes: Phylogenetic and ecological implications in aquatic vertebrates. *Biol. Bull. 120:*238–254.

Thorson, T. B. (1964) The partitioning of body water in Amphibia. *Physiol. Zool. 37:*395–399.

Thorson, T. B. (1968) Body fluid partitioning in Reptilia. *Copeia 1968 (3):*592–601.

Van Citters, R. L., Kemper, W. S., and Franklin, D. L. (1968) Blood flow and pressure in the giraffe carotid artery. *Comp. Biochem. Physiol. 24:*1035–1042.

Wells, M. J. (1979) The heartbeat of *Octopus vulgaris. J. Exp. Biol. 78:*87–104.

White, F. N. (1956) Circulation in the reptilian heart (*Caiman sclerops*). *Anat. Rec. 125:*417–431.

Wigglesworth, V. B. (1972) *The Principles of Insect Physiology,* 7th ed. London: Chapman & Hall. 827 pp.

ADDITIONAL READING

Altman, P. L., and Dittmer, D. S. (eds.) (1971) *Biological Handbooks: Respiration and Circulation.* Bethesda: Federation of American Societies for Experimental Biology. 930 pp.

American Society of Zoologists, Symposium. (1987) Cardiovascular adaptation in reptiles. *Amer. Zool. 17:*3–131.

Berne, R. M., and Levy, M. N. (1981) *Cardiovascular Physiology,* 4th ed. St. Louis: Mosby. 286 pp.

Butler, P. J. (ed.) (1982) Control and co-ordination of respiration and circulation. *J. Exp. Biol. 100:*1–319.

Caro, C. G., Pedley, T. J., Schroter, R. C., and Seed, W. A. (1978) *The Mechanics of Circulation.* Oxford: Oxford University Press. 540 pp.

Folkow, B., and Neil, E. (1971) *Circulation.* New York: Oxford University Press. 593 pp.

Goerke, J., and Mines, A. H. (1988) *Cardiovascular Physiology.* New York: Raven Press. 232 pp.

Hamilton, W. F., and Dow, P. (eds.) (1962, 1963, 1965) *Handbook of Physiology,* sect. 2, *Circulation,* vol. 1, pp. 1–758 (1962); vol. 2, pp. 759–1786 (1963); vol. 3, pp. 1787–2765 (1965).

MacFarlane, R. G. (ed.) (1970) *The Haemostatic Mechanism in Man and Other Animals.* Symposium of the Zoological Society of London, No. 27. London: Academic Press. 248 pp.

Martin, A. W. (1974) Circulation in invertebrates. *Annu. Rev. Physiol. 36:*171–186.

Nadel, E. R. (1980) Circulatory and thermal regulations during exercise. *Fed. Proc. 39:*1491–1497.

Satchell, G. H. (1971) *Circulation in Fishes.* New York: Cambridge University Press. 132 pp.

Taylor, M. G. (1973) Hemodynamics. *Annu. Rev. Physiol. 35:*87–116.

White, F. N. (1976) Circulation. In *Biology of the Reptilia,* vol. 5 (C. Gans, ed.), pp. 275–334. New York: Academic Press.

PART TWO

FOOD AND ENERGY

CHAPTER FOUR

Food and fuel

Animals need food (1) to provide the energy needed to keep alive and to maintain body processes, for muscle contraction, and for many other processes, (2) as raw materials for building and maintaining cellular and metabolic machinery, and (3) for growth and reproduction.

Plants use the energy from sunlight and carbon dioxide from the atmosphere to synthesize sugars and, indirectly, all the complicated compounds that constitute a plant.

All animals use chemical compounds to supply energy and building materials. They must obtain these either directly by eating plants or by eating other organic material. Therefore, the organic compounds animals need are ultimately derived from plants and thus indirectly from sunlight.

There are exceptions to this universal dependence on sunlight. In a few locations in the deep ocean where no light penetrates, there are rich animal communities that utilize the peculiar chemical characteristics of geothermally heated water.

The subject of food has three major aspects, *feeding, digestion,* and *nutrition.*

We shall first deal with *feeding*, which refers to the acquisition and ingestion of food. Virtually all food, whether of plant or animal origin, consists of highly complex compounds that cannot be used without first being broken down to simpler compounds. We refer to these processes as *digestion*.

A variety of organic compounds can provide energy, but in addition, animals have specific needs for compounds they cannot synthesize, such as amino acids and vitamins. Both the need for food to provide energy and the need for specific food components belong to the subject of *nutrition*.

We shall first discuss food and food intake, proceed to the processes of digestion that make the food compounds available to the organism, and then briefly mention some specific nutritional requirements. Finally, we shall discuss how organisms may defend themselves by producing toxic substances that make them inedible to many animals.

TABLE 4.1 Methods used by various animals to obtain food, classified by character of the food.

Type of food	Feeding method	Animals using method
Small particles	Formation of digestive vacuoles	Amebas, radiolarians
	Use of cilia	Ciliates, sponges, bivalves, tadpoles
	Formation of mucus traps	Gastropods, tunicates
	Use of tentacles	Sea cucumbers
	Use of setae, filtering	Small crustaceans (e.g., *Daphnia*), herring, baleen whales, flamingos, petrels
Large particles or masses	Ingestion of inactive masses	Detritus feeders, earthworms
	Scraping, chewing, boring	Sea urchins, snails, insects, vertebrates
	Capture and swallowing of prey	Coelenterates, fish, snakes, birds, bats
Fluids or soft tissues	Sucking plant sap, nectar	Aphids, bees, hummingbirds
	Ingestion of blood	Leeches, ticks, insects, vampire bats
	Sucking of milk or milklike secretion	Young mammals, young birds
	External digestion	Spiders
	Uptake through body surface	Parasites, tapeworms
Dissolved organic material	Uptake from dilute solutions	Aquatic invertebrates
Symbiotic supply of nutrients	Action of intracellular symbiotic algae	Paramecia, sponges, corals, hydras, flatworms, clams

FEEDING

Food is obtained by a diversity of mechanical means, and these determine the nature of the food a given animal can obtain and utilize. Major feeding mechanisms and examples of animals that use the different means to obtain food are listed in Table 4.1.

Small particles

Microscopic algae and bacteria can be taken directly into a cell (e.g., into the digestive vacuole of an ameba). A variety of organisms, even some quite large ones, catch microscopic organisms with the aid of cilia. A few animals, notably tunicates and some gastropods, use a sheet of mucus that traps fine suspended particles; the mucus sheet is then ingested and the food organisms digested.

Tentacular feeding is used by sea cucumbers, which live burrowed in the mud with their tentacles extended above the mud surface. The tentacles entangle fine particles and at intervals are retracted into the animal's mouth, where digestible material is removed.

Various filtering devices are common in a wide variety of small planktonic crustaceans such as copepods and in amphipods, sponges, and bivalves. Quite a few vertebrates, ranging from fish to mammals, also use filtering to obtain food, but their filters usually strain off organisms larger than the microscopic food commonly utilized by invertebrates. A little reflection will tell us that even spiders are filter feeders; their filters are familiar to all of us as spider webs.

Many pelagic fish are plankton eaters. Herring and mackerel have gill rakers structured so that they function as a sieve that catches plankton, predominantly small crustaceans. Some of the largest sharks, the basking sharks and whale sharks, feed exclusively on plankton, which is strained from the water that enters the mouth and flows over the gills. It has been estimated that a large basking shark, in 1 hour, strains the plankton from 2000 tons of water (Matthews and Parker 1950).

A few marine birds also feed on plankton. One of the petrels, the whale bird or prion, has a series of lamellae extending along the edge of the upper beak. It feeds by straining crustaceans from the surface water in a way similar to what baleen whales do on a much larger scale. The flamingo, also, is a plankton eater, using its beak to strain small organisms from the water.

The baleen whales are entirely specialized for feeding on plankton. Their filtering apparatus consists of a series of horny plates attached to the upper jaw and hanging down on both sides. As the whale swims, water flows over and between the plates, and plankton is caught in the hairlike edges of the plates. The plankton-eating whales include the largest living animal, the blue whale, which may reach a weight of over 100 tons. It is interesting that both the largest whales and the largest sharks are plankton feeders rather than carnivores and thus avoid any extra links in the food chain.

Large food masses

Methods for dealing with large particles or masses of food include a tremendous variety of mechanisms and structures. Some animals ingest the medium in which they live and digest the organic material it contains, whether small living organisms or dead organic matter. Such animals frequently pass large amounts of relatively inert material through their digestive tracts and utilize whatever organic material it contains.

Many animals use mechanical methods such as chewing and scraping to obtain their food, which is frequently of plant origin. The range of these animals includes such a variety that to list them would be pointless. We need only think of the large numbers of insects and other invertebrates, as well as the many vertebrates, that are herbivores. Some carnivores are also included in this group. They capture prey and tear, shred, or chew it before swallowing.

Some carnivores capture and swallow their prey whole. Relatively few invertebrates feed this way, but many vertebrates do, including representatives of all classes: fish, amphibians, reptiles, birds, and mammals.

The mouth and oral cavity may have various kinds of jaws or teeth that serve in the mechanical breakdown of food into smaller particles (*mastication*), making it more accessible to the digestive enzymes.

In some animals the mechanical breakdown of food occurs in a special compartment with thick muscular walls, the *gizzard,* where a grinding action takes place. An earthworm, for example, has a gizzard for such grinding, called *trituration.* A bird, which has no teeth, also has a gizzard and often swallows small stones that help in the trituration of food. This is particularly important in a seed-eating bird, for its food is hard and would, in the absence of grinding, take a long time to soften in the digestive tract. The gizzard may be preceded by a simple storage chamber, a *crop,* in which the moisturizing or softening begins, but without any important grinding action.

At the beginning of the digestive tract there is usually a special larger compartment, a *stomach,* which serves as a storage organ and in which digestive action also takes place. In all vertebrates, with a few exceptions, the stomach has a major role in the digestion of proteins, and the most important enzyme in the stomach is the protein-digesting pepsin, which acts best in highly acidic solution. The stomach of a vertebrate also has special cells that secrete acid (hydrochloric acid).

The stomach is followed by a hollow tube, the *intestine,* in which further digestion takes place and which is particularly important for the absorption of the products of digestion. These products are mainly simple sugars, glycerol and fatty acids, and amino acids. The length of the intestine varies a great deal, and it can usually be subdivided into several parts. The posterior portion is of little importance in digestion and serves mainly in absorption. Periodically the remaining material is expelled as feces through the anus.

Liquids

Many liquids are of value as foods. Animals feeding on liquids are often highly specialized and adapted to their food sources, and in some cases there is a mutual benefit to the feeding animal and the provider. For example, insects obtain nectar from

TABLE 4.2 Composition of "milk" from pigeons and emperor penguins, as percentage of total dry matter. The composition of rabbit milk is given for comparison. [Prévost and Vilter 1962]

Constituent	Pigeon	Penguin	Rabbit
Protein	57.4	59.3	50.6
Lipid	34.2	28.3	34.3
Carbohydrate	0	7.8	6.4
Ash	6.5	4.6	8.4
Total	98.1	100.0	99.7

flowers and in turn provide pollination by moving from flower to flower. In many cases, fluid feeders are parasitic; this is true of aphids, which suck plant sap, and of animals such as leeches, mosquitoes, and ticks that suck the blood of other animals.

All mammals begin their lives as fluid feeders when for a period they live exclusively on the milk produced by their mothers. It is less well known that some birds feed their young on a milklike secretion. In the pigeon this secretion is formed in the crop; it is known as *crop milk* and is regurgitated to feed the nestlings. Curiously, the formation of crop milk is stimulated by the same hormone, prolactin, that in mammals stimulates the mammary glands to produce milk. The biological advantage of feeding the young this way is that it allows the parents to be opportunistic in their own feeding and frees them from the need to find special kinds of food (e.g., insects) for the young. It also buffers the young against fluctuations and shortages in the food supply.

This advantage is particularly striking in the emperor penguin, which can feed its young on "milk" secreted by the esophagus (Prévost and Vilter 1962). The emperor penguin reproduces in midwinter. When the Antarctic winter approaches in March, the birds leave the sea and walk over the ice to the rookeries, where the female leaves a single egg with her mate to be incubated for over 2 months while he stands on the ice in the cold and dark of winter. In the meantime, the female walks back to the sea

to feed and returns later to the fasting male to feed the young. If she is delayed, however, the still-fasting male begins to feed "milk" to the young, which is kept alive and even may gain in weight. In protein and fat content pigeon and penguin milk (Table 4.2) is similar to mammalian milk (although the milk of many mammals has a much higher carbohydrate content). Note that the numbers in this table refer to percentages in the dry matter.

The composition of mammalian milk varies a great deal. The protein content is related to whether the young grow slowly or rapidly. It may vary from about 1% in human milk to as much as 15% in rabbit milk. The fat content varies even more, from a fraction of 1% in the donkey to more than 50% in many seals and whales (see Table 8.19, p. 348).

The carbohydrate content may be zero or as high as 7%, as in horse milk. Milk never seems to exceed about 7% milk sugar, or lactose. Why is this so? It would seem simple to increase the amount of this easily digested and metabolized compound. The limit seems to have an osmotic explanation. A lactose concentration much higher than 7% would give the milk an osmotic concentration higher than the blood plasma, and this seems incompatible with the secretory processes in the mammary glands (Schmidt-Nielsen 1986).

Spiders provide a special case of fluid feeding. A spider's prey is often as large as, or larger than, its own body, usually covered with hard chitin, and not easily torn apart. Spiders have solved the problem of obtaining food from their insect prey by piercing it with their hollow jaws and pumping digestive juices into its body. These liquefy the tissues and the spider then sucks the prey empty.

Some parasites (e.g., tapeworms) live in highly nutritive media and have completely lost their digestive tracts. Because their hosts provide the digestive system, the parasites lack both digestive tract and the digestive enzymes that are otherwise necessary before food can be absorbed, and all their nutrients are taken up directly through the body surface.

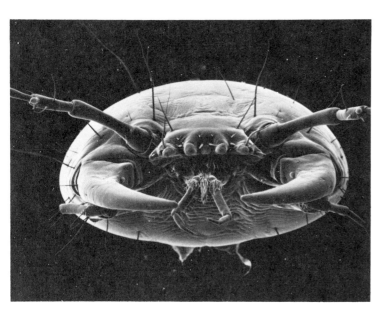

FLUID-FEEDING INSECT Rove beetles (Staphylinidae) are predators that feed on various invertebrates. They pierce their prey with sharp fangs (mandibles) and pump it full of enzymes that digest the prey from inside. The beetles then suck the liquefied food through a strainer composed of hair at the entrance to the mouth. [Courtesy of John F. Lawrence, Museum of Comparative Zoology, Harvard University]

Dissolved organic material

Whether organic matter dissolved in the water can be utilized by aquatic animals is a question that remains controversial. It was suggested early in this century by the German biologist Pütter that dissolved organic matter in the sea, although present in a highly dilute state, could be used directly by aquatic animals. Pütter even suggested that such dissolved organic matter was more important than plankton, for plankton organisms are scattered so far apart in the water that often their capture is not feasible (Jørgensen 1976).

Whether the uptake of dissolved organic material is important has been difficult to study experimentally because of the extremely dilute solutions to be examined. This changed with the introduction of radioisotopes to label organic compounds, and it became possible to demonstrate clearly that many aquatic invertebrates indeed can absorb both glucose and amino acids from exceedingly dilute solutions.

In such experiments it is not difficult to show that some uptake has taken place. This is a fact

beyond doubt; the difficulty is to show that the amount of a compound taken up exceeds the amount the organism loses in the same period of time.

Another difficulty is that the animals in question might not have a physiological mechanism for the uptake of organic substances from dilute solutions, and that bacteria could play a role by absorbing organic solutes and later be ingested by the animals. Bacteria can effectively remove glucose or acetates from solutions containing only 1 to 10 μg per liter, which is impossible for common freshwater algae (Wright and Hobbie 1966). Recent investigators have taken this possibility into consideration and have apparently been successful in eliminating the role of bacteria.

Some studies have reported that losses exceed uptake, but more recent investigations that have used increasingly refined techniques indicate that uptake often by far exceeds loss (Wright and Stephens 1978). It is increasingly certain that net uptake can take place from natural sea water, which indicates that the process is of importance under natural conditions (Manahan et al. 1982).

TABLE 4.3 Release of carbohydrate from symbiotic algae to host animal. [Smith et al. 1969]

Type of alga	Host phylum	Host species	Carbohydrate released to host
Zoochlorellae	Protozoa	*Paramecium bursaria*	Maltose
	Porifera	*Spongilla lacustris*	Glucose
	Coelenterata	*Chlorohydra viridissima* (wild type)	Maltose
		Chlorohydra viridissima (mutant)	Glucose
	Platyhelminthes	*Convoluta roscoffensis*	Unknown
	Mollusca	*Placobranchus ianthobapsus*	Unknown
		Tridacna crispata	
Zooxanthellae	Coelenterata	*Pocillopora damicornis*	Glycerol
		Anthopleura elegantissima	Glycerol
		Zoanthus confertus	Glycerol
		Fungia scutaria	Glycerol
	Mollusca	*Tridacna crocea*	Glycerol

In one study that involved 21 species from seven different phyla of marine invertebrates, all animals except arthropods showed influx rates of free amino acids that were dependent on the concentration in the water, while the loss was at low constant rates. There must, therefore, always be a certain concentration at which uptake and loss are equal, and this concentration was in all cases lower than the concentration in the natural bay water tested. This means that a net gain should always be possible, and the estimate was that all the animals tested, except the arthropods, commonly support 3 to 10% or more of their metabolism this way (Ferguson 1982). Whether any aquatic invertebrate depends primarily on the uptake of dissolved organic material still remains uncertain, but it seems established beyond doubt that many do derive a net benefit from such uptake.

Symbiotic supply of nutrients

Some invertebrate animals contain within their cells symbiotic algae. These algae are of importance in supplying nutrients to the host, providing a shorter pathway between photosynthesis and animal than is possible in any other way. The algae, in turn, utilize ammonia given off by the animal host for synthesis of protein. Such symbiotic relationships between animal and plant occur among protozoans, sponges, corals, hydras, flatworms, and clams.

In this relationship the animal obtains organic compounds (primarily carbohydrates) from the living cells of the plant, and not from the destruction and digestion of the plant cells. In this respect the symbiotic supply of nutrients differs in principle from all the feeding methods discussed above, which always involve destruction of part or the whole of the organism that provides the food.

The importance of the symbiosis has been demonstrated in hydras by comparing animals that carry algae (green hydras) with animals that do not. The presence of symbiotic algae greatly prolongs the survival of hydras under conditions of starvation. If a limited amount of food is given, the green hydras grow faster than pale ones, again showing that the green algae contribute organic material. The oxygen consumption of the green hydras is lower in light than in the dark and lower than in animals without symbiotic algae (Muscatine and Lenhoff 1965; Pardy and Dieckmann 1975).

Two major types of algae occur in the host animals: green algae (Zoochlorellae) or brown algae (Zooxanthellae) (Table 4.3). If we include in the term carbohydrate not only sugars, but also derivatives such as sugar alcohols and glycerol, we can say that the bulk of material transferred is of carbohydrate nature. The soluble carbohydrate given off to the host animal is invariably different from the major intracellular carbohydrate of the algae. Zoochlorellae, for example, synthesize and retain in their cells sucrose, but they excrete maltose or glucose to the host animal.

The amount of organic material obtained in a symbiotic relationship is substantial. The algae in a green hydra release as much as 45 to 50% of the carbon fixed in photosynthesis to their host. Isolated algae release carbohydrate, mainly maltose, to the medium, and the amount depends on the pH of the solution. At pH 4.5 these algae release as much as 85% of their photosynthesized carbon, but in alkaline medium the release declines to a few percent. This opens the possibility that the host animal may be able to control the amount of carbohydrate released by the algae by adjusting its intracellular pH.

When the Zooxanthellae are isolated from their host and incubated in sea water, they release some carbohydrate, but the amount increases if homogenized host tissue is added. In the case of Zooxanthellae from *Tridacna* (the giant clam) a 16-fold increase in carbohydrate release was obtained in the presence of homogenate of the host. This is not a universal phenomenon, however; Zoochlorellae from a hydra are unaffected by homogenates of their own host animal (Smith et al. 1969).

Food from the earth's interior

Until a few years ago it was thought that all life on earth directly or indirectly depends on energy from the sun. Recent discoveries have shaken this universal principle, and an exciting new chapter has been added to biology.

Deep-sea biologists have found rich animal communities in several places of the deep ocean, where their existence is completely independent of the usual food chains that are based on photosynthesis. They occur in places of undersea volcanic activity where the rigid plates of the earth's crust

pull apart. The sea floor in these areas is marked by rifts where the thin rock layer is heated to high temperatures by the underlying hot magma. As water seeps into the rock, it is heated and emerges again as hot springs, or *hydrothermal vents*. In some of these the water temperature is only a few degrees higher than the surrounding water, which is at about $+2\,°C$. In others the water is heated to as much as $350\,°C$, but is kept from boiling by the high pressure, which at a depth of 2500 m is about 250 atm.

These vents are surrounded by rich communities of mussels, clams, limpets, crabs, tubeworms, and coelenterates. Some of the animals grow to enormous size; the clams may be 30 cm and the tubeworms more than 2 m long. Most of these animals are unique, previously unknown to science (Corliss et al. 1979).

Additional vent communities have been discovered at great depth near the Japanese coast, at the Mid Atlantic Ridge, and in the Gulf of Mexico. At least 16 previously unknown families of invertebrates have been discovered in these communities (Grassle 1985).

How is it possible to support a rich animal life in total darkness at a depth of several thousand meters? No light penetrates to this depth, and no green plants can grow there. Of the organic material that sinks down from the surface layers, only minimal amounts reach such depths. Everywhere else at the bottom of the deep ocean animal life is very dilute, and the higher temperature of the water at the hydrothermal vents cannot by itself make animals grow and thrive. The animals need food; where does it come from? The answer is not only interesting, it is of the greatest fundamental importance to animal physiology.

The water from the hydrothermal vents contains a variety of dissolved minerals and high concentrations of hydrogen sulfide (H_2S). Where this water mixes with oxygen-rich ocean water, there is a tremendous production of bacteria that obtain chemical energy from the oxidation of hydrogen sulfide (Karl et al. 1980). Filter-feeding animals such as clams and mussels thrive on filtering out these bacteria, and the crabs seem to be predatory and live on other animals. Thus the food chain begins with the bacteria and is based on chemical energy (oxidation of H_2S) rather than on light from the sun.

It has been possible to estimate the age of some of these animals and thus their growth rates by examining the ratios of radioactive isotopes found in their shells. One of the large clams was estimated to grow at a yearly rate of 4 cm and to deposit 20 g of calcium carbonate in the shell in one year. The estimated age of a 22-cm-long shell was 6.5 years (Turekian et al. 1979). Compared with other deep-sea animals this is a striking growth rate. For example, a deep-sea clam from the Atlantic Ocean, *Tindaria callistiformis,* reaches a total length of 0.84 cm when mature and takes 100 years to reach this size (Turekian et al. 1975). Evidently, the rift animals not only occur in dense populations, but grow at extraordinarily fast rates.

It is easy to understand that filter feeders ingest bacteria and that crabs make a living as scavengers. But how do the large tubeworms make a living, for these have neither mouth nor intestinal tract to take advantage of a rich food supply?

The tubeworms at the Galápagos rift grow to giant size: 2 or 3 cm in diameter and more than 2 m long. They belong to the Pogonophora and are named *Riftia pachyptila.* It is reasonable to ask how these worms are capable of living there at all, and the answer is interesting. Large organs, known as *trophosomes,* fill the greater part of the coelom and contain virtually solid masses of symbiotic bacteria. Evidently these bacteria provide their host with energy obtained from oxidation of H_2S.

The strongest evidence for this conclusion comes from tests of the enzymes present in the trophosome. Two sets of enzymes are found in high concentration. One set is involved in sulfur metabolism and the formation of adenosine triphosphate (ATP) from the chemical energy of oxidation of H_2S. This can provide chemical energy for the tubeworms, but not the carbon compounds needed for growth.

The latter are provided by the other set of en-zymes, which have functions identical to those of two key enzymes in the so-called Calvin-Benson cycle (i.e., the group of enzymes that makes it possible for green plants to carry out the photosynthetic CO_2 fixation that forms the basis for all green plant life). With the aid of these enzymes the trophosome can use inorganic carbon dioxide to synthesize organic compounds just as green plants use the carbon dioxide from the atmosphere. The difference is that the energy for photosynthesis comes from light quanta and the energy for CO_2 fixation in the trophosomes comes from bacterial ATP (Cavanaugh et al. 1981; Felbeck 1981).

In summary, the rich animal life in the deep sea rift communities depends, not on energy from the sun, but from the interior chemistry of the earth. Energy is derived from the symbiotic oxidation of hydrogen sulfide. Organic compounds are obtained through CO_2 fixation carried out, not with the aid of sunlight, but with the chemical energy from sulfide oxidation.

There are many other interesting problems connected with the unusual physiology of the rift animals. The large tubeworms, *Riftia,* have blood as well as coelomic fluid that contains hemoglobin. Although these animals depend on hydrogen sulfide, sulfide is a threat to aerobic respiration due to its capacity to inhibit the reversible binding of oxygen to hemoglobin and to inhibit the activity of cytochrome-*c*-oxidase. However, both oxygen and hydrogen sulfide must reach the trophosome, and both are carried by the blood, which does contain hemoglobin. The hydrogen sulfide is indeed bound to the hemoglobin, but in the tubeworms the binding site is not the oxygen-binding site of the hemoglobin molecule (Arp et al. 1987).

The oxygen capacity of the blood is high: 8.4 ml O_2 per 100 ml blood, about one-half of what is common in mammals. Both blood and coelomic fluid have high oxygen affinities; at 3 °C the P_{50} in the blood is 1.2 mm Hg and in the coelomic fluid 0.7 mm Hg. The higher affinity in the coelomic fluid makes the system well suited for oxygen transport to the tissues (Arp and Childress 1981a).

RED-TIPPED TUBEWORMS These tubeworms (*Riftia pachyptila*) live in the unique, dense animal communities that surround deep-ocean springs where hot, sulfur-loaded water emerges from volcanic rifts. The worms, which are up to 2 m long, bear a bright red crown where oxygen is taken up by the hemoglobin-containing blood. [Courtesy of Jack Donnelly, Woods Hole Oceanographic Institution]

Some of the other animals in the rift communities have very interesting physiological characteristics. The crabs (*Bythograea*) that live there have hemocyanin in their blood. The oxygen affinity is rather high; the P_{50} is 6.6 mm Hg (at 2.6 °C and pH 7.5) and remains the same throughout the temperature range of 2 to 30 °C. The oxygen capacity is high for a hemocyanin-containing blood: 4.5 ml O_2 per 100 ml blood. The independence of temperature and the high oxygen capacity make the blood well suited as an oxygen reservoir to permit the crabs to range as predators on long excursions into warmer and oxygen-deficient water (Arp and Childress 1981b).

After oceanographers became aware of the unusual vent communities, other extraordinary discoveries were made. Certain clams (*Solemya reidi*) that lack a digestive system are found exclusively in habitats rich in hydrogen sulfide near sewage outfall areas on the California coast near Los Angeles. These clams are able to absorb dissolved organic molecules from sea water, but it is likely that they are nourished also by symbiotic bacteria in

The crown of the tubeworm *Riftia* serves to take up hydrogen sulfide from the hot water emerging from deep sea vents. With the help of symbiotic bacteria in the trophosome they cover their energy needs from the oxidation of hydrogen sulfide. [Courtesy John A. Baross, Oregon State University]

their gills, similar to the deep sea vent organisms (Felbeck, 1983).

In the Gulf of Mexico, in areas where there is no volcanic activity, mussels, clams, and tube-worms are found in large numbers near seeps of oil and gas. Their dietary carbon seems to be derived from the seeping hydrocarbons (Brooks et al. 1987).

A mussel of the genus *Mytilus* from this area contains methanotrophic symbiotic bacteria in its gills. It will grow with methane as its only carbon and energy source, while its growth rate is zero in the absence of methane (Cary et al. 1988).

Because these unusual animals occur at great depth in the ocean, it is difficult to study their physiology under conditions normal for these animals.

DIGESTION

Animal food consists of organic material, most of which belongs to three major groups: proteins, fats, and carbohydrates. These three kinds of organic compounds completely dominate the makeup of virtually all plants and animals. Most have fairly large molecules. Even the smallest (sugars) have molecular weights of a few hundred; fats are larger; many proteins have molecular weights of 100 000 to several million; and starch and cellulose are of indeterminate size, being polymers of small carbohydrate units.

Whether the food is used for fuel or for building and maintenance, the large molecules of the food are first broken down into simpler units; these are then absorbed and either incorporated into the body or metabolized to provide energy.

The main function of digestion is to break up the large and complex molecules in the food so that they become absorbable and available for use in the body. This breakdown is achieved in the digestive tract with the aid of enzymes.

Intracellular and extracellular digestion

Digestion can take place within cells or outside. In a unicellular animal digestion is usually, of ne-

cessity, inside the cell. A protozoan takes food into the digestive vacuole, and enzymes that aid in the digestion of carbohydrates, fats, and proteins are secreted into the vacuole. Similar intracellular digestion occurs in sponges and to some extent in coelenterates, ctenophores, and turbellarians. It is found also in a number of more complex animals in conjunction with extracellular digestion. For example, a clam often takes small food particles into the cells of the digestive gland, where they are digested intracellularly.

Some animals, such as coelenterates, that feed on larger pieces of food have partly extracellular and partly intracellular digestion. Digestion begins in the digestive cavity, the coelenteron; and fragments of partly digested food are then taken into the cells of the cavity wall where they are digested completely.

Extracellular digestion has one obvious advantage: It permits the ingestion of large pieces of food, whereas intracellular digestion is limited to particles small enough to be taken into the individual cells of the organism.

Extracellular digestion is usually associated with a well-developed digestive tract that allows secreted enzymes to act on the food material. The digestive tract may have one opening, as in coelenterates, brittlestars, and flatworms. In these animals, any undigested material is expelled through the same opening that served as a mouth.

In more complex animals the digestive tract has two openings, a mouth and an anus. This permits an assembly-line type of digestion. Food is ingested through the mouth and is passed on and acted upon by a series of digestive enzymes; the soluble products of digestion are absorbed, and in the end undigested material is expelled through the anus without interference with food intake. In this way food intake can continue while digestion takes place, and the passage of food through the digestive tract can continue uninterrupted.

All coelenterates are carnivorous. They have capture devices, tentacles, that help them catch and paralyze prey. The tentacles have specialized cells, *nematocysts,* which, when suitable prey comes in contact with them, fire tiny hollow spines that penetrate the prey. A poison is ejected from the cell through the spine, and this paralyzes the prey. The tentacles then draw the prey into the coelenteron for digestion.

The mouth of a flatworm (planarian) leads into a gastrovascular cavity that is widely branched and ramifies throughout the body. This cavity, because of its distribution, not only serves in digestion, but at the same time can transport food to all parts of the body. The extensive branching also increases the total surface of the gastrovascular cavity, aiding the absorption of digested food. In planarians extracellular digestion aids in breaking down the food, but most of the food particles are taken up by the cells that line the cavity and are digested intracellularly.

Enzymatic digestion

Most food compounds are very large molecules (proteins), highly insoluble in water (fats), or large as well as insoluble (starch, cellulose). Before the elements of the food can be absorbed and utilized, they must be brought into soluble form and broken down into smaller component units.

Proteins, starches, and cellulose are polymers formed from simpler building blocks by removal of water; fats are esters of fatty acids and glycerol, again formed by removal of water. The breakdown of these food compounds into simpler components involves the uptake of water and therefore called *hydrolysis.*

Hydrolysis is a spontaneous reaction, and energy in the form of heat is released in the process. Spontaneous hydrolysis of food compounds proceeds at an immeasurably slow rate, but can be speeded up by catalysts. Catalysts produced by living organisms are called *enzymes,* and enzymes are essential to all digestive processes. Virtually all metabolic processes within the cells are also catalyzed by enzymes.

All enzymes are proteins; many have been isolated in pure form, and their action has been studied in great detail. Many enzymes have been named according to the substrates on which they act, with the attachment of the ending *-ase.* Digestive enzymes frequently act on a broad group of substrates, and many are best known by common or trivial names, although they possess more precise names according to internationally accepted nomenclature. Because the digestive enzymes from only a few organisms are known in sufficient detail, we shall discuss most of them under their familiar common names, rather than following the technical nomenclature.

After being described by the substrate it acts on or the reaction it catalyzes, the most important measure of an enzyme is its activity, that is, the rate at which the catalyzed reaction proceeds. Many factors influence the activity, and if the protein nature of the enzyme is changed, it usually loses its activity. High temperature, strong chemicals, other enzymes that split proteins, and heavy metals that bind to the protein may all inactivate the enzymatic activity.

The activity of all enzymes is greatly influenced by the *pH of the solution.* Most enzymes show their highest activity in a narrow range of pH, called the *optimum pH* for the enzyme in question. Two well-known protein-digesting enzymes are *pepsin,* found in the vertebrate stomach, and *trypsin,* produced in the vertebrate pancreas. Pepsin has a pH optimum in highly acidic solutions, about pH 2, and trypsin has its optimum in slightly alkaline solutions, about pH 8.

The optimum pH for a given enzyme is not necessarily identical for all substrates on which it acts. Pepsin, for example, shows an optimum pH of 1.5 for digestion of egg albumin, 1.8 for casein, and 2.2 for hemoglobin (White et al. 1964). On both sides of the optimum pH the activity is decreased, and at a very different pH the activity is nil. Pepsin still has some activity at pH 4, and resumes full activity if the pH is returned to the optimum. At pH 8, however, pepsin is destroyed and permanently inactivated.

The rate of enzymatic reactions is greatly influ-

FIGURE 4.1 The effect of temperature on a protein-splitting enzyme (protease) from the ascidian *Halocynthia*. The enzyme effect seems to have a temperature optimum, but this optimum is lower the longer the duration of the experiment. This is explained in the text. [Berrill 1929]

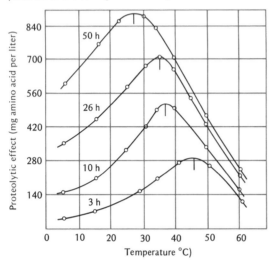

enced by the *temperature*. An increased temperature speeds the rate. Heat usually also inactivates enzymes; at temperatures above 50°C or so, most enzymatic action is completely destroyed. Therefore, a moderate increase in temperature gives an increased rate of reaction, but when the temperature increases further, thermal destruction of the enzyme catches up with the accelerated reaction rate. At some temperature we will observe a maximum, but because thermal destruction increases with time, we will find that the optimum temperature depends on the duration of the experiment. In a long-lasting experiment the temperature effect on the enzyme will have proceeded further, and the observed temperature optimum will therefore be lower; in a short experiment with the same enzyme the temperature optimum will be higher. Thus the temperature optimum is not a specific characteristic of an enzyme; it depends on the duration of an experiment (Figure 4.1).

Most enzymes are rapidly inactivated by temperatures above 45 or 50°C. This is the temperature range that for most complex organisms forms the upper limit for life, except for some resting stages and certain thermophilic bacteria. A few enzymes, however, are heat-resistant; examples are the protein-splitting enzyme papain from the papaya fruit and a similar enzyme from pineapple. These enzymes are used as meat tenderizers because their proteolytic action continues at cooking temperatures as high as 70 or 80°C, although at boiling temperature they are rapidly inactivated.

Protein digestion

Proteins are polymers consisting of amino acids bound together by peptide bonds. An amino acid is a relatively simple organic acid that on the carbon atom next to the acid group (carboxyl group, $-COOH$) carries an amino group ($-NH_2$). A peptide bond is formed by removal of water between the carboxyl of one amino acid and the amino group of another.

There are about 20 common amino acids that in various proportions occur in virtually all living organisms. This means that when proteins are hydrolyzed into amino acids, these amino acids are the building blocks required for synthesis of nearly any specific protein.

The enzymes that digest proteins are divided into two groups, according to where on the protein molecule they act. *Exopeptidases* hydrolyze a terminal peptide bond in a long peptide chain; *endopeptidases* act in the interior of the peptide chain. The two best known endopeptidases in vertebrate digestion are pepsin and trypsin.

Pepsin is secreted by the stomach of vertebrates as an inactive precursor, *pepsinogen*. The stomach also secretes hydrochloric acid, which gives its contents a low pH. In an acid medium, below pH 6, pepsinogen is activated autocatalytically to form the active enzyme, pepsin.

Pepsin specifically hydrolyzes peptide bonds between an amino acid that carries a phenyl group (tyrosine or phenylalanine) and a dicarboxylic acid (glutamic or aspartic acid). Pepsin therefore at

FIGURE 4.2 The protein-digesting enzyme pepsin splits peptide bonds between a dicarboxylic and an aromatic amino acid. Trypsin acts on peptide bonds adjacent to an amino acid with two amino groups. Arrows show points of attack. See text for further details.

tivated more and more rapidly as more trypsin is formed). This is known as *autocatalytic activation*.

Trypsin acts best at a slightly alkaline pH, between 7 and 9. It hydrolyzes peptide bonds that are adjacent to a basic amino acid (i.e., one that carries two amino groups, lysine or arginine) (Figure 4.2).

The protein fragments and peptide chains formed by the action of pepsin and trypsin are further digested with the aid of exopeptidases, which act only on terminal peptide bonds. Carboxypeptidase, secreted by the pancreas, hydrolyzes terminal peptide bonds next to a free carboxyl group, and aminopeptidase, secreted by the intestine, hydrolyzes peptide bonds next to a free amino group. Finally, dipeptidases hydrolyze the peptide bond of fragments that consist of only two amino acids.

Protein digestion in invertebrates, whether extracellular or intracellular, in principle proceeds as in vertebrates, except that invertebrates generally have no pepsinlike enzymes that act in acid solution. Otherwise, although the enzymes are not identical in their structure to the vertebrate enzymes, their action is similar.

Pepsinlike enzymes are not an exclusive characteristic of vertebrates, however. The maggot of the ordinary housefly *(Musca domestica)* secretes a protease that is active in the pH range of 1.5 to 3.5 with an optimum at 2.4. Whether this enzyme is of importance in the maggot's digestion depends on a suitably acid medium (Greenberg and Paretsky 1955; Lambremont et al. 1959).

Fat digestion

Digestion of fats is similar in vertebrates and invertebrates. Ordinary fats, whether of plant or animal origin, consist of esters between one molecule of glycerol (a trivalent alcohol) and three molecules of long-chain fatty acids. They are highly insoluble in water and as a result are not easily hydrolyzed by enzymes.

The vertebrate pancreas secretes a fat-hydrolyzing enzyme, *lipase,* but to bring it into contact with

tacks only a few of the peptide bonds in a large protein and not others, and the result of pepsin digestion is a series of shorter chains or fragments that are not further broken down by pepsin.

Trypsin is secreted by the pancreas in an inactive form, *trypsinogen*. It is activated in the intestine by the enzyme enterokinase, which is secreted by glands in the intestinal wall. Trypsinogen is also activated by active trypsin (i.e., trypsinogen is ac-the fat, the aid of a detergent is needed. The bile acids, secreted by the liver, serve this function. With the aid of the bile acids and the mechanical movements of the intestine, the fats are emulsified, and as hydrolysis proceeds, the resulting fatty acids (also relatively insoluble) are kept in solution with the aid of the bile salts and thus can be absorbed. The glycerol (which is a normal intermediary in virtually all cell metabolism) is water-soluble and is easily absorbed and metabolized.

Fats can, however, to some extent be absorbed by the intestinal epithelium without preceding hydrolysis. The tiny droplets formed when fats are emulsified with the aid of the bile are taken up directly by the epithelial cells.

Some fatty substances, among them *waxes,* are not hydrolyzed by ordinary lipases. Waxes are esters of one molecule of a long-chain fatty alcohol with one molecule of a fatty acid; if they could be hydrolyzed, the components would be metabolized with a high yield of energy. What we usually think of as wax is beeswax. With one known exception, beeswax remains undigested by vertebrates and therefore has no nutritive value.*

Is wax good food?

To humans wax is indigestible. Vertebrate pancreatic lipase has virtually no effect on waxes, which are esters of one molecule of a long-chain fatty alcohol and one molecule of fatty acid. If waxes could be hydrolyzed, ordinary beeswax would be an excellent energy source with a caloric value similar to ordinary fats.

There are a few known examples of wax digestion that have been considered as physiological curiosities. One is the larva of the wax moth, a common parasite in beehives and a nuisance to beekeepers. Another is a bird, the curious wax-eating South African honey guide. This bird is known for its habit of guiding humans (or the wild honey-

badger) to the nest of wild bees. It attracts attention by chattering and noisy behavior, and, if it is followed, it leads the way to a bees nest. It sits patiently while the nest is plundered, waiting to enjoy the leftover spoils. It is clear that the honey guide prefers to eat wax rather than honey; it has even happened that the bird has entered a chapel to attack the wax candles.

The honey guide does indeed digest wax, aided by symbiotic bacteria in the digestive tract. These bacteria have been isolated and grown in culture, and their essential role in wax digestion has been demonstrated by feeding the bacterial culture along with wax to domestic chicks. Normal chicks do not digest wax, but if wax is fed with the microorganisms from honey guides, the wax is digested and metabolized (Friedmann et al. 1957).

Although waxes are insignificant as food for terrestrial animals, they are extremely important in the food chain of marine animals, side by side with ordinary fats and oils.

Waxes occur in a wide variety of marine organisms: molluscs, cephalopods, shrimps, sea anemones, corals, and many fish. The primary producers of waxes appear to be small planktonic crustaceans, especially copepods. In some of these, waxes may constitute as much as 70% of the animal's dry weight. The copepods feed on phytoplankton, which contains no wax. Diatoms and dinoflagellates accumulate oil globules, which are largely triglycerides. However, the fatty acids of the waxes in the copepods closely resemble characteristic fatty acids found in the phytoplankton, and it is reasonable to conclude that these are used directly for wax production by the copepods.

The role of waxes in the marine food chain is overwhelmingly more important than was imagined only a few years ago, for the crustacean plankton is the major step between the photosynthetic microscopic algae and the consumers in the sea. It has been estimated that because of this step, at least half of the earth's photosynthetic production is for a time converted to wax (Benson et al. 1972).

Fish that prey on copepods (e.g., herring, an-

---

*The "wax" often used in the household and for candle making is a hydrocarbon fraction of crude mineral oil. It is chemically inert, is attacked only by certain strains of bacteria, and is not known to be utilized by any animal.

chovies, sardines, and pilchards) have wax lipases in their digestive tracts (Sargent and Gatten 1976). The fatty alcohols are oxidized to fatty acids, which then enter into ordinary neutral fats as triglycerides. In some other fish the amount of wax lipases is much smaller, and this leaves open the question of how well they can digest waxes.

Because waxes are found in the fats and oils of such a variety of marine animals, even in whales, it is difficult to say whether these waxes can be metabolized and serve as an energy reserve or are merely stored because they are ingested and not easily metabolized. This important question is the subject of intensive studies by marine biologists.

Marine birds also prey on animals that contain large amounts of wax. Petrels and auks are major predators of and feed their young on crustacean plankton that has a high wax content. Up to two-thirds of the digestible energy in the food that these birds give to their chicks consists of wax esters. The waxes are not excreted with the feces and are not deposited in the stored fat of the young. Therefore the wax in the food must either be metabolized directly or be converted to triglycerides for storage.

Feeding experiments with ^{14}C-labeled cetyl palmitate have shown that chicks of petrels and auks can digest wax esters with less than 1% of the ingested label excreted. The labeled fatty acids were assimilated and deposited, primarily as triglycerides in storage fat depots (Roby et al. 1986). It remains to be seen whether symbiotic bacteria are involved in wax digestion in these birds.

The conclusion is obvious, however. Marine birds are able to utilize effectively the large pool of high-energy waxes in the marine food chain. It is highly plausible that many other animals can also utilize the waxes in the marine food chain. Wax, after all, is not universally indigestible.

Carbohydrate digestion

There are no great differences between vertebrates and invertebrates in how carbohydrates are digested. Simple sugars, such as glucose and fructose, are absorbed without any change and utilized

directly in ordinary metabolic pathways. Disaccharides, such as sucrose (of plant orgin) or lactose (in milk), are broken down to monosaccharides before they are absorbed and can be utilized. The enzyme *sucrase* is secreted in the intestine, but it is absent from the cellular machinery of animals. Therefore, if sucrose is injected into the body of vertebrates, it is completely excreted again in the urine without any change.

Large numbers of plants store starch as a major energy reserve. Starch is a polymer of single glucose units. It is relatively insoluble, but it is hydrolyzed by the enzyme *amylase* (from Latin *amylum* = starch), secreted in the saliva of humans (and some, but not all, other mammals) and in larger amounts by the pancreas.

Although starch is relatively insoluble, it is well digested by many animals. It is more easily attacked by amylase, however, if it first has been heated, and this is the major reason that humans prepare starchy foods by cooking (potatoes, bread, etc.). It is quite probable that in many animals that readily digest uncooked starch, bacteria aid in the initial breakdown of the starch, which is then more easily attacked by the amylase.

Wood as food: Symbiotic digestion

The most important structural material of plants is cellulose, a glucose polymer that is extremely insoluble and refractory to chemical attack. Cellulose-digesting enzymes, *cellulases,* are absent from the digestive secretions of vertebrates, and yet many vertebrates depend on cellulose as their main energy source. True cellulases have been reported from the intestinal tracts of several invertebrates that feed on wood and similar plant products, but in many cases cellulose digestion is carried out by *symbiotic microorganisms* that live in the digestive tract of their host.

Invertebrate cellulose digestion

The question of whether a cellulase is produced by an animal or by symbiotic microorganisms is often difficult to establish with certainty. A few examples will illustrate the difficulties.

Snails. It has long bee accepted that the garden snail, *Helix pomatia,* digests cellulose and actually secretes a cellulase. However, careful examination of extracts of the snail's digestive gland and intestinal wall has failed to reveal any cellulase activity, although cellulase is readily demonstrated in the intestinal contents (Florkin and Lozet 1949). This seems to indicate a major role of intestinal microorganisms, but the fact that an extract of digestive organs does not reveal the enzyme does not necessarily exclude the possibility that the animal may produce the enzyme.

Shipworms. Another organism that has been accepted as able to digest cellulose is the shipworm *(Teredo),* actually a wood-boring bivalve mollusc. True cellulase has been found in the gut of the shipworm, and gut extracts liberate sugar from cellulose, although no cellulose-digesting bacteria or protozoans could be isolated from the digestive system (Greenfield and Lane 1953). However, the apparent absence of cellulose-digesting microorganisms is no proof that these do not exist in the living animal. In fact, bacteria known to digest cellulose and also fix nitrogen have been cultured from shipworms (Waterbury et al. 1983).

Silverfish. There are, however, reports of true cellulases obtained from some animals under conditions that seem to exclude any possibility of symbiotic microorganisms. The silverfish *(Ctenolepisma lineata)* digests cellulose and can survive on a diet of cellulose alone, although this is not satisfactory for prolonged feeding.

The gut of the silverfish contains many microorganisms, but no cellulose-digesting species has been isolated. More importantly, bacteria-free silverfish have been obtained by washing the eggs in a solution of mercuric chloride and ethanol and raising the nymphs on sterilized rolled oats and vitamins under aseptic conditions. Such bacteria-free silverfish digest cellulose marked with ^{14}C and exhale ^{14}C-labeled carbon dioxide. Finally, a cellulase has been demonstrated in extracts of the midgut, and it therefore seems certain that silverfish do produce the cellulase (Lasker and Giese 1956).

Termites. In the case of termites, which live virtually exclusively on wood, the role of symbiotic organisms is beyond doubt. The intestinal tract of a wood-digesting termite is virtually crammed full of several kinds of flagellates and bacteria. Several of the flagellates have been isolated and kept in pure culture, and this has permitted a careful study of their role in cellulose digestion.

The symbiotic flagellates from termites are obligate anaerobic organisms (i.e., they can live only in the absence of free oxygen). This sensitivity to oxygen can be used to remove the flagellates from the host animals and thus to obtain protozoa-free termites. If termites are exposed to oxygen at 3.5 atm pressure, the flagellates are selectively killed in about half an hour; the termites survive unharmed (Cleveland 1925). Such defaunated termites cannot survive if they are fed as usual on wood. They still have bacteria in the intestine, and this indicates that the protozoan symbionts, rather than the bacteria, are responsible for cellulose digestion. When the termites are reinfected with the proper flagellates, they can again digest cellulose (Hungate 1955).

The hypothesis that the flagellates carry out the cellulose digestion is further supported by studies of these organisms in isolated culture. One such flagellate *(Trichomonas termopsidis,* from the termite *Termopsis),* which has been kept in culture for as long as 3 years, yields extracts that contain cellulase (Trager 1932). In fact, this flagellate cannot utilize any carbon source other than cellulose, so there is no doubt about its role in the cellulose digestion of termites (Trager 1934).

Although the cellulose from wood can give a substantial amount of energy, wood has a very low nitrogen content. Because termites thrive on wood, it has been suggested that the intestinal microbes could be nitrogen-fixing bacteria, and thus be of aid in the protein supply. This hypothesis seems plausible.

Many earlier attempts were unsuccessful at demonstrating nitrogen fixation in the overall nitrogen balance of termites, but it is now clear that nitrogen fixation does take place with the aid of microorganisms (French et al. 1976).

Nitrogen fixation may, in fact, be more widespread among wood-boring animals than was formerly expected. In several species of shipworms (*Teredo* and others) nitrogen fixation is associated with bacteria in the gut, and these bacteria have been isolated and shown to be capable of nitrogen fixation (Carpenter and Culliney 1975).

In the shipworm *Bankia* (which is closely related to *Teredo*) the gills harbor intracellular bacteria. These appear to be responsible for the synthesis of essential amino acids that occurs when isolated gills are incubated with glucose labeled with radioactive carbon (Trytek and Allen 1980).

Vertebrate cellulose digestion

Many mammals are herbivores, and most of them live on food that make cellulose digestion essential. The ruminants, which include some of our most important domestic meat- and milk-producing animals (cattle, sheep, and goats) have specialized digestive tracts that are highly adapted to symbiotic cellulose digestion. However, many nonruminant mammals also depend on symbiotic microorganisms for cellulose digestion, although their anatomical adaptations differ from those of the true ruminants.

Ruminants. The stomach of a ruminant consists of several compartments (Figure 4.3), or, to be more precise, the true digestive stomach, the *abomasum*, is preceded by several large compartments. The first and largest is called the *rumen*. The rumen serves as a large fermentation vat in which the food, mixed with saliva, undergoes heavy fermentation. Both bacteria and protozoans are found in the rumen in large numbers. These microorganisms break down cellulose and make it available for further digestion. The fermentation products (mostly acetic, propionic, and butyric acids) are absorbed and utilized; carbon dioxide and methane (CH_4) formed in the fermentation process are permitted to escape by belching (eructation).

FIGURE 4.3 In the ruminant animal the true stomach (abomasum) is preceded by several other compartments. The first and largest of these, the rumen, serves as a giant fermentation vat that aids in cellulose digestion.

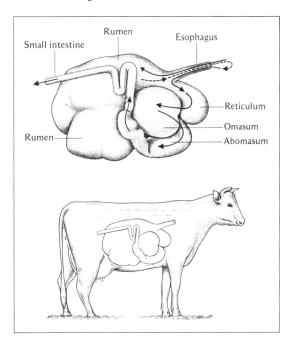

Rumination, or chewing the cud, is regurgitation and remastication of undigested fibrous material, which is then swallowed again. This process has given the ruminant group its name. As the food reenters the rumen, it undergoes further fermentation. Broken-down food particles gradually pass to the other parts of the stomach, where they are subjected to the usual digestive juices in the abomasum (which corresponds to the digestive stomach of other mammals).

The fermentation products from the rumen are a major source of energy. It has been estimated that the energy available from the organic acids produced in the rumen makes up 70% of the total energy requirement of cattle. The methane, on the other hand, is completely lost to the ruminant, and the loss is appreciable. A cow fed 5 kg of hay each day gave off 191 liters of methane each day; the energy value of this methane was more than 10% of the daily digested food energy (Kleiber 1961).

The fermentation products are mostly short-chain organic acids, and the copious amounts of saliva secreted by ruminants serve to buffer these fermentation products in the rumen. The saliva of a ruminant is little more than a dilute solution of sodium bicarbonate, which serves both as a buffer and as a proper fermentation medium for the microorganisms.

The amount of saliva secreted by the ruminant is impressive. The total secretion of saliva per day has been estimated at 6 to 16 liters in sheep and goats, and at 100 to 190 liters in cattle (Bailey 1961; Blair-West et al. 1967). Because sheep and goats weigh perhaps 40 kg, and cattle some 500 kg, the daily production of saliva may reach roughly one-third of the body weight. As two-thirds of the body weight is water, about half of the total body water passes through the salivary glands (and the rumen) each day.

The protozoans of the rumen include ciliates that superficially resemble free-living forms such as paramecia. They occur in numbers of several hundred thousand per milliliter of rumen contents. Many of the rumen organisms have been cultured in the laboratory, and extracts from pure cultures show cellulase activity (Hungate 1942).

The ciliates are obligate anaerobic organisms and must meet their energy requirements through fermentation processes. This gives a relatively low energy yield for the microorganisms, but it means that the fermentation products become available to the host animal, which in turn uses them in oxidative metabolism. Because symbiotic cellulose digestion is the only way cellulose becomes available to mammals, the relatively small loss to the symbiotic microorganisms is quantitatively unimportant. What is more important is that the rumen microorganisms contribute in several other ways to the nutrition of the host.

The rumen microorganisms can synthesize protein from inorganic nitrogen compounds, such as ammonium salts (McDonald 1952). It is particularly useful that urea, which is normally an excretory product eliminated in the urine, can be added to the feed of ruminants and increase the protein synthesis. This has been used in the dairy industry, for urea can be synthesized cheaply, and it is therefore less expensive to supplement the diet of milk cows with urea than to use more expensive high-protein feed.

The rumen contents of a cow may weight as much as 100 kg, with a total weight of protozoans of about 2 kg, and these again contain 150 g protozoan protein. The propagation time for these protozoans makes it possible to estimate that 69% of the protozoan population passes into the omasum each day, and that this contributes a protein supply of over 100 g per day (Hungate 1942).

The microbial protein synthesis in the rumen is of special importance when the animal is fed on low-grade feed. It has been found that a camel fed a nearly protein-free diet (inferior hay and dates) excretes virtually no urea in the urine. Urea continues to be formed in metabolism, but instead of being excreted, this ''waste'' product reenters the rumen, in part through the rumen wall and in part with the saliva.

In the rumen, urea is hydrolyzed to carbon dioxide and ammonia, the latter being used by the microorganisms for resynthesis of protein. In this way a camel on low-grade feed can recycle much of the small quantity of protein nitrogen it has available (Schmidt-Nielsen et al. 1957) (see Figure 4.4).

Similar reutilization of urea nitrogen in animals fed low-protein diets has been observed in sheep (Houpt 1959). Under certain conditions the rabbit (a nonruminant) can utilize urea to a significant extent in its nitrogen metabolism (Houpt 1963).

If inorganic sulfate is added to the diet of a ruminant, the microbial synthesis of protein is improved and, particularly important, the sulfate is incorporated in the essential amino acids cysteine and methionine (Block et al. 1951). Thus, the rumen microorganisms contribute both to protein synthesis and to the quality of the protein. Because

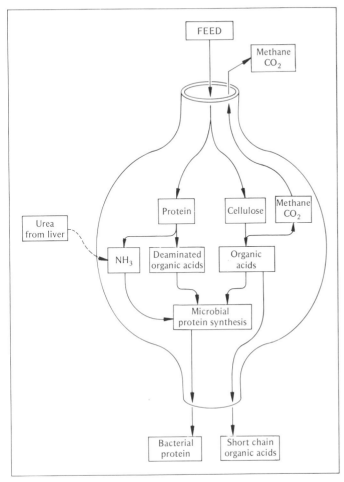

FIGURE 4.4 Major events in ruminant digestion. Cellulose from the feed undergoes microbial fermentation, forming short-chain organic acids that are used by the animal, while methane and carbon dioxide are lost to the atmosphere. The protein from the feed may be partly deaminated, forming organic acids and ammonia. Bacteria use ammonia, and ammonia from ammonium salts or from urea (formed in the liver or supplied in the food), together with organic acids for protein synthesis. The bacterial protein passes further down the digestive tract, is digested, and is used by the animal.

the microbes in the rumen can synthesize all the essential amino acids, ruminants are nutritionally independent of these, and the quality of the protein they receive in their feed is of minor importance.

Another nutritional advantage of ruminant digestion is that some important vitamins are synthesized by the rumen microorganisms. This applies to several of the vitamin B group; in particular, the natural supply of vitamin B_{12} for ruminants is obtained entirely from microorganisms.

Nonruminant animals. Cellulose digestion in many nonruminant herbivorous mammals is also aided by microorganisms. Cellulose-containing feed

is usually bulky, and fermentation is relatively slow and takes time. Much space is required, and the part of the digestive tract used for fermentation is therefore of considerable size. In some animals the stomach is large and has several compartments, and the digestion has obvious similarities to ruminant digestion. In others the major fermentation of cellulose takes place in a large diverticulum from the lower intestine, the *cecum*.

Multiple-compartment stomachs are found not only in some nonruminant ungulates, but in very far removed animals such as the sloth (Denis et al. 1967) and the langur monkey (Bauchop and Mar-

FIGURE 4.5 The stomach of the sloth (*Bradypus tridactylus*) is complex and highly reminiscent of the stomach of a ruminant. [Grassé 1955]

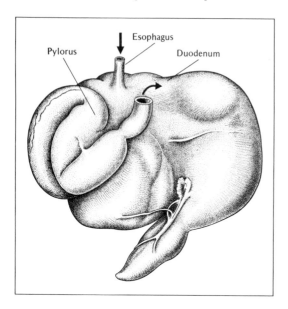

tucci 1968) (Figure 4.5). Even among marsupials there are animals that have a rumen-like stomach. One is the rabbit-sized quokka, which weighs 2 to 5 kg. Its large stomach, which harbors microorganisms that participate in cellulose digestion (Moir et al. 1956), may contain nearly 0.5 kg of moist material. For an animal weighing about 3 kg, this equals 15% of the body weight, an amount similar to that which is often found in ruminants.

Microbial fermentation in the cecum has considerable similarity to fermentation in the rumen, but the rumen has three definite advantages over the cecum.

One advantage is that rumen fermentation takes place in the anterior portion of the gastrointestinal tract, so that the products of fermentation can pass through the long intestine for further digestion and absorption.

The second advantage of the ruminant system is that the mechanical breakdown of the food can be carried much further, for coarse and undigested

particles can be regurgitated and masticated over and over again. This difference is clearly visible if we compare the fecal material of cattle (ruminants) and horses (nonruminants). Horse feces contain coarse fragments of the food intact; cow feces are a well-ground-up, smooth mass with few large visible fragments.

The third, and perhaps greatest, advantage of the ruminant system with microbial fermentation in the anterior portion of the gastrointestinal tract is the opportunity to recycle the urea nitrogen that otherwise would be excreted and lost to the organism. As we saw earlier, this is important for ruminants eating low-grade feed.

The ruminant-like large marsupials – kangaroos and wallabies – in which microbial fermentation also takes place anterior to the digestive stomach, utilize the same mechanism. When the nitrogen content in the vegetation declines at the beginning of the dry season, wallabies begin to recycle urea and continue to do so throughout the prolonged dry season, thus gaining a relative independence of the low quality of the available feed (Kinnear and Main 1975).

An improvement in the digestion of plant materials through fermentation is not restricted to mammals. For example, the willow ptarmigan in Alaska subsists during several winter months on a diet of only willow buds and twigs. Most gallinaceous birds have two large ceca suitable for cellulose fermentation. In the willow ptarmigan the fermentation products are ethanol and acetic, propionic, butyric, and lactic acids in various proportions. The contribution of cecal fermentation to the basal energy requirement is up to 30%, a substantial contribution indeed (McBee and West 1969).

Several herbivorous reptiles likewise benefit from symbiotic microorganisms. The marine green turtle *(Chelonia mydas)* feeds on sea grasses and is able to digest 90% of the cellulose it eats, aided by microbes that are present in numbers similar to those in ruminant mammals (Bjorndal 1979).

The common iguana *(Iguana iguana),* which is herbivorous throughout life, also benefits from microbial fermentation in the hindgut. If iguana hatchlings are reared in captivity, they fail to develop the same complex microbial population as wild hatchlings, and they do not grow as well as hatchlings in the wild. However, if the captive hatchlings are fed fresh fecal material from adult iguanas, they develop the complex microbial flora and grow as rapidly as wild hatchlings. The acquisition of the fermentation flora requires contact with animals that already possess the microbes; otherwise the young are deprived of the benefits of microbial digestion (Troyer 1982).

Coprophagy

The disadvantage of locating cellulose fermentation in the posterior part of the intestinal tract can be circumvented in a peculiar way. Many rodents as well as rabbits and hares form a special kind of feces from the contents of the cecum, and these are reingested so that the food passes through the entire digestive tract a second time. There are, in fact, two kinds of feces: the well-known ordinary, firm, dark rodent fecal pellets, and softer, larger, and lighter feces that are not dropped by the animal but are eaten directly from the anus. This latter type is kept separate from the ordinary droppings in the rectum, and reingestion permits a more complete digestion and utilization of the food.

Coprophagy (from the Greek *copros* = excrement and *phagein* = to eat) is common in rodents and is of great nutritional importance. For example, if coprophagy is prevented, rats require supplementary dietary sources of vitamin K and biotin. In addition, deficiencies of other vitamins develop more rapidly. What is more interesting is that, if the rats are provided with a variety of dietary supplements, their growth rate is nevertheless depressed by about 15 to 25%. Even if the rats are given access to eating normal rat feces, normal growth is not attained by rats prevented from coprophagy (Barnes et al. 1963).

In rabbits prevention of coprophagy leads to a decrease in the ability to digest food, as well as a decrease in protein utilization and nitrogen reten-

FIGURE 4.6 In the stomach of the rabbit ingested food is located in the pyloric part (left), which contains digestive glands. Reingested fecal pellets are located in the large fundus (right), where they remain separate from the food material while fermentation continues. [Harder 1949; Grassé 1955]

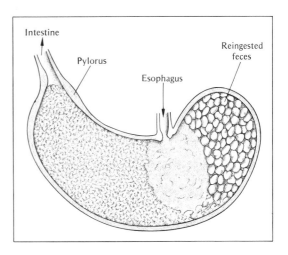

tion. When coprophagy is permitted again, there is a corresponding increase in the ability to digest cellulose (Thacker and Brandt 1955).

The special soft feces that a rabbit reingests originate in the cecum. When ingested, these feces are not masticated and mixed with other food in the stomach; they tend to lodge separately in the fundus of the stomach (Figure 4.6). The soft feces are covered by a membrane, and they continue to ferment in the stomach for many hours, one of the fermentation products being lactic acid (Griffiths and Davies 1963). In this way the fundus of the stomach serves as a fermentation chamber, analogous to the rumen of sheep and cattle, and thus provides essential nutritional advantages to the animal.

In the lagomorphs (rabbits and hares) the reingested feces are strikingly different, but in rodents reingested feces differ less sharply from the regular droppings. Furthermore, rodents chew the reingested feces carefully before swallowing, which again is different from rabbits. Whether these differences are reflected in the nutritional benefits of coprophagy has not been well studied. As coprophagy affects the yields of energy, vitamins, essential amino acids, and other nutrients, it would be interesting to know whether the difference between rodents and lagomorphs has any functional significance (Kenagy and Hoyt 1980).

Reingestion of feces has been reported for several other orders of mammals: shrews, marsupials, and prosimian primates. How widespread this nutritionally important process is deserves further study.

NUTRITION

The mechanical properties of the food must correspond to the animal's feeding apparatus, and the chemical nature of the food must be suited to the available digestive mechanisms. Once these requirements have been met, the next question is: Is the food adequate in regard to the needs of the organism? Nutrition deals with this question.

The problem can conveniently be divided into two parts: the quantity of food and the quality of food. The organism needs (1) energy for external activity and internal maintenance and (2) a supply of specific substances for upkeep and growth. These specific substances include amino acids, vitamins and certain other essential nutrients, and various minerals and other elements, some of them in such minute quantities that it has been difficult to establish whether they are really needed.

The following discussion will first deal with the need for energy and energy-yielding components of the diet. We will then proceed to specific substances needed by the organisms. Finally, we will mention some less desirable compounds that may occur in food and their biological importance.

Energy supply, fuel

The bulk of organic material used for food consists of proteins, carbohydrates, and fats. The main digestion products of these compounds are amino acids (from proteins), various simple sugars (present in the food or derived from starch digestion), short-chain fatty acids (primarily from cellulose fermentation), and long-chain fatty acids (from fat digestion).

The oxidation of these digestion products yields virtually all the chemical energy needed by animal organisms. All are quite simple organic compounds, consisting mostly of carbon, hydrogen, and oxygen in various proportions, except that the amino acids also contain nitrogen (in the amino group, $-NH_2$). If amino acids are used to supply chemical energy, the amino group is first removed by deamination, and the remaining relatively small fragments (various short-chain organic acids) enter the usual pathways of intermediary metabolism.

These various organic compounds are, within wide limits, interchangeable in energy metabolism, although there are certain limitations. For example, the human brain needs carbohydrate (glucose), but most other organs can use fatty acids, which constitute the main fuel for mammalian muscle metabolism. In contrast, the flight muscles of the fruit fly, *Drosphila,* require carbohydrate, and when the supply is exhausted, the fruit fly cannot fly, although stored fat is used for other energy-requiring processes (Wigglesworth 1949). Not all insects need glucose for flight, however; the migratory locust uses primarily fat for its long migratory flights (Weis-Fogh 1952).

For an animal to remain in a steady state, the total energy expenditure must be covered by an equal intake of food. If the amount of food is insufficient, the remaining energy requirement is covered by the consumption of body substance, primarily stored fat. If the food intake exceeds the energy used, most animals store the surplus as fat, irrespective of the nature of the food. This is the reason domestic livestock, such as beef and pigs, can be fattened by feeding on grain or corn, which contain mostly starch. Some animals, however, store primarily carbohydrate in the form of glycogen. This is important, for glycogen can provide energy in the absence of oxygen (Chapter 5).

Regulation of food intake

In many animals intake of food is remarkably well adjusted to energy expenditure (Mayer and Thomas 1967). If the energy needs are increased because of physical activity, the food intake is adjusted accordingly. The regulatory mechanisms for the food intake of mammals are located in the hypothalamic area of the brain, but specific peripheral influences play a major role in the momentary level of hunger and food intake. One of the important influences is the blood sugar level. It has been demostrated that hunger pangs coincide with contractions of the stomach and that these occur especially when the level of blood glucose decreases. Also, the degree of filling of the stomach is of obvious importance in inducing a feeling of satiation and termination of food intake.

These factors can be of only minor importance in the regulation of long-term food intake and maintenance of body weight. Increasing the bulk or volume of the food by adding nonnutritive material to the food has a considerable effect on both gastric filling and motility, but the animal nevertheless rapidly adjusts by increasing food intake. Rats fed a calorically dilute diet maintain their caloric intake by ingesting larger amounts. In rats fed a calorically concentrated diet, the caloric intake remains too high for several days, but then decreases to the basic level (Figure 4.7).

The exact mechanism for this control is not fully understood, but it is known that the hypothalamic region of the brain is important. If specific areas in the hypothalamus are destroyed, the food intake is greatly altered and may be either reduced or increased. Carefully placed hypothalamic lesions may lead to such tremendous increases in food intake that an animal eats continuously and grows so fat it is unable to move.

Regulation of food intake is not limited to higher vertebrates. Goldfish, which continue to grow throughout their lifetimes (although at a decreasing rate), have been used in similar experiments with caloric dilution of their diet. If permitted to control their food intake, goldfish respond to caloric dilu-

FIGURE 4.7 Changes in amount of food eaten per day by adult rats when the energy content per unit weight of food was (a) reduced and (b) increased. Solid line, weight of food consumed daily; broken line, energy content of food eaten, expressed as equivalent weight of initial diet (1 kilocalorie = 4.184 kilojoule). [Hervey 1969]

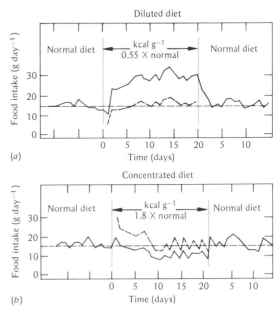

tion by increased food intake. Also, a decrease in the water temperature from 25°C to 15°C causes a decrease in food intake by one-half to one-third. Goldfish therefore seem to eat for calories, for a 10°C drop in temperature causes roughly a two- to three-fold decrease in metabolic rate and energy requirement (Rozin and Mayer 1961).

SPECIFIC NUTRITIONAL NEEDS

A diet that furnishes adequate fuel, measured in terms of energy units, can be utterly inadequate in regard to raw materials for growth and building and maintaining the cellular and metabolic machinery.

For most animals the energy-requiring processes can be fully satisfied by glucose alone, but glucose contributes nothing toward other needs. There are three major categories of such needs: proteins and amino acids, vitamins and related compounds, and minerals and trace elements.

Proteins and amino acids

When an animal grows, protein is continually synthesized and added to the organism. In the adult organism, however, the amount of protein remains much the same throughout life. It might therefore seem that once an organism has reached its adult size, dietary protein would be less important. This is not true: An inadequate supply of protein leads to serious malnutrition.

Why is this so? It would appear that, once proteins have been incorporated into an organ (a muscle, for example), they would remain as part of the permanent structure. It was, therefore, a great surprise when the biochemist Rudolf Schoenheimer found that the body proteins are constantly broken down and resynthesized. Schoenheimer introduced in the diet of adult rats amino acids labeled with the heavy nitrogen isotope ^{15}N, and these amino acids were invariably found incorporated in the tissue proteins. Thus, it was shown that cell components once regarded as stable structures have a high turnover rate of proteins (Schoenheimer 1964).

Not all proteins have the same turnover rate. In humans, for example, the half-life of serum proteins is approximately 10 days, but connective tissue has a low rate of protein turnover. Once these latter structural proteins are formed, they remaiin relatively stable compared with proteins of metabolically active tissues such as blood, liver, muscle, and so on.

New protein is synthesized from single amino acids, not from partially degraded fragments of other protein molecules. An adult man synthesizes and degrades about 400 g protein per day, and this far exceeds his total protein intake, which on an average diet may be between 100 and 150 g per day. Thus, the synthesis utilizes not only amino acids from the food, but also, in part, those available from protein breakdown in the body.

Each organism contains a large number of different proteins as cell components, enzymes, and so on. Furthermore, every animal or plant species has proteins that are specific for that species, and even individuals within a species may have characteristically different proteins. For example, the blood groups of humans are related to the different structures of some of the blood proteins.

Only about 20 different amino acids occur commonly in proteins, but because a protein molecule contains from less than 100 to many thousand amino acids, there is virtually no limit to the number of different proteins that can be built from the 20 basic building blocks.

The amino acids needed by the body are usually supplied by the protein of the food, but they can be replaced by chemically purified amino acids in suitable proportions. However, it is not necessary to supply every one of the 20 different amino acids. Some can be formed in the body from other amino acids, but others cannot be synthesized by the organism and must be supplied in the diet. These latter are called *essential amino acids*.

The terms *essential* and *nonessential* describe the dietary requirements for amino acids, but say nothing about how important these amino acids are. The nonessential amino acids are just as important as the others as structural components of body proteins. In fact, they are probably so important that the dietary supply cannot be trusted, and the organism cannot afford to be without a mechanism for their synthesis.

Of the 20 or so common amino acids, 10 are essential to rats (Table 4.4) (Rose 1938). The exclusion of any one of them (except arginine) produces profound nutritional failure. The absence of arginine gives a decreased rate of growth, but arginine can be synthesized, although not at a rate sufficient for optimal growth. Humans and other mammals have the same requirements, except that histidine is not needed by adult humans although it seems to be required by all other higher animals tested (Rose 1949). It is particularly striking that fish, many insects, and even protozoans have prac-

TABLE 4.4 Classification of the amino acids with respect to dietary needs of rats.

Essential	Nonessential
Lysine	Glycine
Tryptophan	Alanine
Histidine	Serine
Phenylalanine	Norleucine
Leucine	Aspartic acid
Isoleucine	Glutamic acid
Threonine	Proline
Methionine	Citrulline
Valine	Tyrosine
Arginine[a]	Cystine

[a] Arginine can be synthesized, but not at a sufficiently rapid rate to meet the demands of normal growth.

tically overlapping needs for essential amino acids (Tarr 1958; Taylor and Medici 1966).

Plants, in contrast to animals, do synthesize protein from inorganic nitrogen (ammonia salts or nitrate), and some microbes can even use molecular nitrogen. Such *nitrogen-fixing bacteria* are found in the root nodules of clover and other leguminous plants, which therefore are particularly important in increasing the protein yield from agriculture.

We have already mentioned that termites and other wood-boring insects use nitrogen-fixing bacteria in the gut to supplement the extremely low protein content of the wood they eat. Probably many more animals, themselves unable to utilize molecular nitogren, take advantage of nitrogen-fixing microbes.

A case in point is the high rate of biological productivity of coral reefs that has puzzled marine biologists for years. Tropical water masses characteristically have very low levels of disolved nutrients, and especially the low level of fixed nitrogen is often the major limiting factor in phytoplankton production. It appears that, as the ocean water flows over shallow coral reefs, there is a marked increase in its nitrogen content, caused mainly by the nitrogen-fixing blue-green alga *Calothrix*. Without this aid to the supply of nitrogen,

the high productivity of the reef community would be difficult to explain (Webb et al. 1975).

Ultimately, all animal proteins are of plant origin. This holds even for the protein synthesized from ammonia or urea in the rumen of a ruminant, for without microbes this synthesis does not take place. Microbial protein synthesis in the rumen has an interesting consequence; it makes the ruminant animal nutritionally independent of essential amino acids, for these are synthesized by the microorganisms. Ruminants therefore have no dietary need for the essential amino acids, although metabolically they are indispensable (Loosli et al. 1949; White et al. 1964).

Accessory foodstuffs: vitamins

Early in the twentieth century it was discovered that a diet of carbohydrates, fats, and proteins is insufficient in regard to a number of minor components that have become known as *vitamins*. The quantities needed are small, in the range of milligrams or micrograms. For this reason these compounds are also known as *accessory foodstuffs*.

The initial discovery of a vitamin was associated with the human disease beri-beri, which is caused by the absence of thiamine (vitamin B_1) in the diet. Originally, the vitamins were given letters of the alphabet, but as their chemical structure and exact biochemical roles have been clarified, it has become increasingly common to use their chemical names instead. Detailed lists of the vitamins and their metabolic roles can be found in any adequate textbook on nutrition and will therefore be omitted here.

The vitamin requirements are not the same for all animals. For example, ascorbic acid (vitamin C) can be synthesized by most mammals, but humans and a few other mammals (some primates, bats, and the guinea pig) must have it supplied in the diet. Many invertebrates, teleost fish, and some birds also need dietary ascorbic acid, but lungfish, amphibians, reptiles, and most birds are able to synthesize their own. These differences as well as the different organ of synthesis (liver or kidney) have interesting evolutionary implications (Birney et al. 1980).

A striking difference in synthesizing ability exists in regard to cholesterol, which vertebrates readily synthesize. For humans, cholesterol is, in fact, considered undesirable because it is a major factor in the development of arteriosclerosis. Insects, however, cannot synthesize cholesterol, and it must therefore be supplied in the food; for insects, cholesterol is a vitamin (Horning 1958).

All insects so far studied are incapable of synthesizing cholesterol from nonsteroid precursors and need a dietary source of sterols. The same applies to crustaceans. Several other invertebrates need a dietary source of sterols; among protozoans some require sterols and others do not.

Symbiotic organisms are often important sources of vitamins. For example, vitamin K, which is necessary for normal blood coagulation in vertebrates, is synthesized by bacteria in the digestive tracts of mammals. Its physiological role was discovered in birds fed a cholesterol-free diet for the purpose of studying cholesterol synthesis. The birds developed severe bleedings, which were traced to a vitamin deficiency. Only then was it established that this vitamin is needed also by mammals. Its discovery, however, depended on the accidental observation of its importance in birds, which require it in their food.

Symbiotic synthesis of vitamins is particularly important for ruminants, which appear not to need several of the vitamins in the B group. Again, it is the microbial synthesis that makes these animals independent of a dietary source.

Information about the vitamin requirements of lower vertebrates and invertebrates is inadequate. Studies are difficult, for they require the use of diets composed of purified components, and many animals will not feed on such diets in the laboratory. Broadly based nutritional studies have therefore mostly been confined to protozoans, insects, and higher vertebrates. Also, to establish nutritional requirements with certainty, it is necessary to know whether symbiotic microoganisms contribute to meet

TABLE 4.5 Approximate elementary composition of the body of a 70-kg man.

Element	Weight (g)	Percent body weight
Oxygen	45 500	65.0
Carbon	12 600	18.0
Hydrogen	7 000	10.0
Nitrogen	2 100	3.0
Calcium	1 350	1.93
Phosphorus	785	1.12
Potassium	245	0.35
Sulfur	175	0.25
Sodium	105	0.15
Chlorine	105	0.15
Magnesium	35	0.05
Total	70 000	100.00

the vitamin needs. This can be done with the aid of germ-free animals, but the procedures then become very complicated and costly.

Minerals and trace elements

How many elements are essential to life? The answer to this question is known for no more than a few laboratory animals. Of the over 90 naturally occurring elements, only 26 are known to be necessary for rats and/or chicks.

The four commonest elements – *oxygen, carbon, hydrogen,* and *nitrogen* – make up 96% of the total body weight of a mammal (Table 4.5). The seven next most abundant elements make up very close to the remaining 4%; they are calcium, phosphorus, potassium, sulfur, sodium, chlorine, and magnesium, in that order. Fifteen additional elements are required, but their total combined amount in the mammalian body is less than 0.01% of the body mass.

Water and the organic building blocks of the body – proteins, nucleic acids, fats, and carbohydrates – contain most of the oxygen, hydrogen, carbon, and nitrogen, plus part of the sulfur and phosphorus. The mineral compounds of the skeleton together with dissolved ions make up the remaining few percent. The quantitatively most important ions in animals are sodium, potassium, chloride, and bicarbonate. The maintenance of normal concentrations of sodium and chloride ions is necessary for the osmotic balance of animals, as will be discussed in Chapter 8.

Although sodium is usually the major cation in blood and extracellular fluids, potassium is the dominant cation inside the cells, where the sodium concentration is correspondingly lower. Chloride is usually higher in the extracellular and lower in the intracellular fluid. Bicarbonate ion is intimately connected with respiration and carbon dioxide transport and is of major importance in the regulation of the pH of the organism, as described in previous chapters.

Calcium, magnesium, phosphate, and sulfate are also important ions; all are necessary for many physiological processes. Any great change in their concentration is deleterious or fatal. For example, calcium ions are necessary for normal cell function, for nerve conduction, for muscle contraction, and for blood coagulation. If the calcium concentration in the blood of a mammal falls to about half its normal value (about 5 mEq per liter) the result is severe or fatal tetanic cramps.

Calcium is also an important constituent of skeletons and many other rigid mechanical structures. Vertebrate bones and teeth consist mainly of calcium phosphate (hydroxyapatite). Invertebrate mineralized structures, such as clam shells and coral, are mainly calcium carbonates.

Not all structural substances in animals are mineralized. The exoskeleton of an insect consists of the organic material chitin. In crustaceans the exoskeleton may be entirely organic, but in many larger forms (such as crabs and lobsters), it is encrusted with calcium carbonate.

Invertebrate skeletons frequently contain some magnesium carbonate in addition to the calcium carbonate, and often minor amounts of calcium sulfate as well. A peculiar exception is the skeletal material of the radiolarian *Acantharia* (a proto-zoan), which consists entirely of strontium sulfate (Odum 1957).

The supporting structures of many plants, in addition to cellulose, often contain amorphous silicon oxide (silica). Of animals, some sponges have skeletons composed of spicules of silica (SiO_2).

Trace elements

Let us now consider the 15 elements that make up less than 0.01% of the body of a mammal. These elements occur in such small amounts that, although early investigators could demonstrate their presence in the body, analytical methods were not sufficiently accurate for a precise analysis. These elements therefore were mentioned as occurring in ''trace'' amounts, and they became known as *trace elements*. With better analytical methods, and in particular with the aid of radioisotopes, it has become possible not only to determine these elements more accurately, but also to clarify their physiological and biochemical roles.

The trace elements can be divided into three groups: (1) those known to be essential, (2) those that have metabolic effects but have not been shown to be indispensable, and (3) those that occur widely in living organisms but seem to be only incidental contaminants. Some trace elements are needed in such small amounts that it is difficult to devise a diet completely free of them, and it is probable that as research progresses, some elements of the second and perhaps even of the third group will be moved to the first group (Mertz, 1981).

The mere occurrence of a trace element gives no information about its function, and the chemical demonstration of its presence is therefore rather uninformative, for virtually every element in the periodic table can be found in some minute amount.

An understanding of the role of trace elements is obtained mainly in two ways: (1) feeding of synthetic diets composed of highly purified components and designed to lack the element under study, and (2) the study of diseases, in both humans and animals, which are believed to result from a deficiency of a trace element. Such deficiency diseases

FIGURE 4.8 Periodic system of the naturally occurring elements. Elements known to be essential for animals are emphasized by shading.

1a	2a	3a	4a	5a	6a	7a	8a			1b	2b	3b	4b	5b	6b	7b	8b
1 H																	2 He
3 Li	4 Be											5 B	6 C	7 N	8 O	9 F	10 Ne
11 Na	12 Mg				Transition elements							13 Al	14 Si	15 P	16 S	17 Cl	18 Ar
19 K	20 Ca	21 Sc	22 Ti	23 V	24 Cr	25 Mn	26 Fe	27 Co	28 Ni	29 Cu	30 Zn	31 Ga	32 Ge	33 As	34 Se	35 Br	36 Kr
37 Rb	38 Sr	39 Y	40 Zr	41 Nb	42 Mo	43 Tc	44 Ru	45 Rh	46 Pd	47 Ag	48 Cd	49 In	50 Sn	51 Sb	52 Te	53 I	54 X
55 Cs	56 Ba	57-71 ΣLa	72 Hf	73 Ta	74 W	75 Re	76 Os	77 Ir	78 Pt	79 Au	80 Hg	81 Tl	82 Pb	83 Bi	84 Po	85 At	86 Rn
87 Fr	88 Ra	89 Ac	90 Th	91 Pa	92 U												

are often associated with exceptionally low concentrations of certain elements in the soil in specific geographical areas.

The first method has the disadvantage that it is difficult to eliminate every trace of a given element from the diet, and a highly purified diet also readily becomes deficient in other respects. For example, should a diet be inadequate with regard to some vitamin as well as the trace element under study, the animal will not show the expected improvement if the trace element alone is added. Therefore, progress in the study of trace elements was difficult before vitamin research was sufficiently advanced to design adequate test diets.

Iron [26].* The presence of iron in living animals has been known for so long that usually iron is not considered a trace element. Nevertheless, it

*The number in brackets after the name of an element is the atomic number of that element in the periodic table.

is essential. It is grouped in the periodic system among the ''transition elements'' (see Figure 4.8), which encompass a total of nine known trace elements. Another six trace elements are located in the periodic table in close proximity to those nonmetallic elements that are the most abundant in living organisms.

Iron is a constituent of hemoglobin and of several intracellular enzyme systems, notably the cytochromes. The total amount of iron in the organism is not very great; in an adult human it is about 4 g. Of this amount 70% is found in hemoglobin, 3.2% in myoglobin, 0.1% in cytochromes, 0.1% in catalase, and the remainder in storage compounds mostly in the liver (O'Dell and Campbell 1971). The dietary requirement for an adult mammal is very small, for iron from the breakdown of hemoglobin is stored in the liver and used again for hemoglobin synthesis. Growing animals need more iron, and adult females need to replace that which is lost in reproductive processes (growth of fetus, menstruation).

Cobalt [27]. Cobalt is unique among the trace elements because the cobalt ion as such is not known to be needed by any animal organism. All higher animals, however, need cobalt in the specific form of vitamin B_{12}, which is required for blood formation. Deficiency leads to severe anemia. In ruminants vitamin B_{12} is formed in the rumen by the microbial flora, provided a sufficient amount of cobalt is present in the diet.

The first conclusive evidence that cobalt is an essential trace element came from Australia, where a serious disease of cattle and sheep caused great economic losses. Addition of small amounts of cobalt prevents this disease. The use of cobalt in fertilizers is expensive because fairly large amounts are needed. Instead, it has been found practical to make each sheep swallow a small cobalt-containing ceramic ball that remains in the rumen and slowly releases cobalt over several years.

Nickel [28]. Nickel is adjacent to cobalt in the periodic table and is thought to be essential. Although nickel is known to influence the growth of feathers in chickens, it is not certain that the element is essential for animal life. The problem is that small amounts of nickel are nearly always associated with iron and therefore extremely difficult to eliminate completely from a diet. There is, however, evidence that nickel is essential for leguminous plants, and possibly for all higher plants (Eskew et al. 1983).

Copper [29]. Copper occurs widely in soil, but in certain areas the concentration is so low that both plants and animals suffer a deficiency. One of the most conspicuous symptoms is anemia. Copper is not part of the hemoglobin molecule, but copper deficiency prevents the formation of hemoglobin. The exact mechanism of this effect is not known.

Copper is also a constituent of a dozen or more enzymes, one of the most important being cytochrome oxidase, which contains both iron and copper.

Zinc [30]. Zinc is a constituent of many important enzymes, including carbonic anhydrase and several peptidases. Deficiency symptoms can be produced by feeding a purified diet.

Vanadium [23]. Vanadium exerts a pronounced growth-promoting effect on rats kept in an all-plastic isolated environment on a highly purified amino acid diet. The amounts needed, around 1 μg per day for a rat, are within the range of concentrations found naturally in food and in tissues of higher animals (Schwarz and Milne 1971).

It is peculiar that some species of tunicates accumulate very high concentrations of vanadium in their blood, although other tunicates have no unusual concentrations of this element. The function, if any, of this vanadium enrichment remains uncertain.

Chromium [24]. Chromium is a fairly recent addition to the list of essential trace elements. It is needed for insulin to exert its maximal effect in promoting glucose uptake by the tissues (Mertz 1969). So far there is no evidence that chromium is a component of any enzyme system or serves as an enzyme activator.

Manganese [25]. Manganese is widespread in soil and is found in all living plants and animals. It was established rather late that manganese is an essential dietary component. Deficiency symptoms include abnormal development of the bones, and the element is also needed for the function of ovaries and testes. Its biochemical effect has become known only recently. It functions as an activator of certain enzyme systems, but the connection with the deficiency symptoms is not entirely clear.

Molybdenum [42]. Molybdenum is required by all nitrogen-fixing microorganisms, by higher plants, and by animals. Molybdenum is a constituent of the enzyme xanthine oxidase and thus has a role in the oxidation of purine to uric acid.

Silicon [14]. It has long been known that silicon, which next to oxygen is the most abundant element in the earth's crust, is essential for plants. Certain knowledge that it is essential for growth and normal development in animals is, however, more recent. Silicon-deficient chicks develop bone and skull abnormalities, apparently due to an effect on the collagen matrix of the bone (Carlisle 1977).

Tin [50].   Located in the same group of the periodic table as carbon and silicon, tin is among the most recent additions to the list of essential trace elements. Inorganic tin salts are necessary for normal growth of rats kept on highly purified diets with other known trace elements added (Schwarz et al. 1970).

Arsenic [33].   This element probably is essential for rats and, by implication, for other animals. When fed on diets with extremely low arsenic content, rats develop rough hairy coats and grow more slowly than control animals. They also show other disturbances, with enlarged spleens and low hematocrit values. For some unknown reason males are consistently more affected than females (Nielsen et al. 1975).

Selenium [34].   Selenium has, because of its toxicity, been of greater practical importance in animal nutrition than as an essential element. In some geographical areas selenium is found in the soil in such high concentrations that plants become toxic to livestock, and grazing animals then may suffer acute selenium poisoning.

Selenium, however, is also an essential trace element; deficiency symptoms include liver necrosis and muscular dystrophy. The biochemical role of selenium seems to be related to the function of vitamin E, which is necessary for growth and normal reproduction in rats.

Fluorine [9].   Until recently fluorine was not classed as essential, although some of its physiological effects were well known. This changed, however, when fluorine was found to be essential for the normal growth of rats (Schwarz and Milne 1972).

It has long been known that fluorine is present in small amounts in teeth and bone, where the fluoride ion is incorporated into the crystal lattice of calcium phosphate (apatite), the main mineral component. Whether fluorine is essential in this role

is doubtful, but it makes the teeth more resistant to dental caries. The addition of fluorine to drinking water has therefore been recommended as a public health measure to reduce the incidence of tooth decay. The effectiveness is beyond doubt, and the amounts added are so small that there are no known toxic effects.

In some geographical areas, however, fluorine is present in the natural water supply in such high concentrations that it causes abnormalities of the tooth enamel, known as ''mottled enamel.'' Cosmetically, mottled enamel is undesirable, but mottled teeth are particularly resistant to decay. Even higher intake of fluorine leads to deformation of bones and serious poisoning; these have occurred near industrial plants and smelteries where fluorine is discharged to the atmosphere.

Iodine [53].   The physiological significance of iodine relates to its role as a constituent of the thyroid hormones. These accelerate the cellular metabolism and are necessary for the control of metabolic rate and for normal growth of vertebrates.

It has long been known that the incidence of human goiter (an enlargement of the thyroid gland) is concentrated in geographical areas where the iodine content of the soil is particularly low. This enlargement of the thyroid appears to be a response to iodine deficiency, and the addition of small amounts of iodine to the food (usually by adding a minute quantity of sodium iodide to ordinary table salt) has virtually eliminated this rather serious public health problem.

Iodine-deficient regions are mostly inland and mountainous areas far from the ocean. Sea water contains a small amount of iodine, and spray that enters the atmosphere from wave action is precipitated with rain and meets the needs of both animals and humans, even several hundred kilometers from the coast.

Other trace elements.   Many additional elements (e.g., lithium, rubidium, beryllium, strontium, aluminum, boron, germanium, bromine) are widely present in living material and may have functional significance, although definite evidence is lacking. Some are highly toxic, but even so may be needed in trace amounts. Furthermore, at present there is no reason to believe that all animals have exactly the same requirements for trace elements.

CHEMICAL DEFENSE

One way to avoid being eaten is to make oneself inedible. This simple principle is widely used by both plants and animals. Plants are the primary producers in the food chain, and because they cannot run away, they remain in place, ready to be eaten.

Making oneself inedible with the aid of toxic compounds is a much used defense, and an amazing number of plants are poisonous to animals. Their "predators," in turn, have developed mechanisms to circumvent or counteract these defenses, so that hardly any plant is completely without enemies (Whittaker and Feeny 1971). An example of such adaptation is the cabbage butterfly, whose larva feeds on a number of cruciferous plants (cabbage, mustard, etc.). These plants contain mustard oils that are toxic or lethal to other butterflies. The oils are not only tolerated by the cabbage butterfly larva, but even act as attractants that lure the adult butterfly to deposit its eggs on these plants.

Animals likewise use venoms and toxins for defense; for some (e.g., snakes, spiders, scorpions), the venom serves both for defense and as an attack mechanism for catching prey.

Some animals have not only circumvented the chemical defenses of plants, but have even exploited the toxic principles of the plants on which they feed to make themselves distasteful or inedible to their predators.

Plant poisons

Various plant compounds affect almost every physiological system. They may act on the skin, on the digestive system, on the blood, on various parts of the nervous system, on the metabolism, on the nutritional value of food, and on the hormonal and reproductive systems of animals. The variety of compounds and effects makes them difficult to classify. Nevertheless, they can be placed in some major groups, in part because of chemical similarities, and in part because of similarities in effects. A few examples will illustrate the immense variety of compounds and actions.

Alkaloids

Alkaloids belong to a group of organic compounds that have little in common except that they all contain nitrogen. Among these are toxic compounds that have profound effects on the central nervous system of vertebrates (e.g., opium, which is a crude product of the opium poppy containing the medically useful alkaloids morphine and codeine). A widely used alkaloid is nicotine; other well-known and highly toxic alkaloids are strychnine and the arrow poison curare.

Glycosides

Glycosides are compounds that on hydrolysis yield one or more sugar molecules and a variety of other compounds. One of the most widely known is digitalis, an important drug in the treatment of certain afflictions of the heart. It is found in the foxglove and is lethal in overdose.

Several plants contain glycosides that upon hydrolysis yield cyanide (HCN). One of the best known is the glycoside amygdalin, which occurs in bitter almonds and in the kernels of apricots, prunes, and cherries. Apple seeds also contain small amounts of this highly toxic compound. For most of us this is unimportant, but it has been reported that a man who was fond of apple seeds as a delicacy saved a cupful of them and, after eating them all at one time, died from cyanide poisoning (Kingsbury 1964).

A major staple food in many tropical countries, cassava (manioc), releases so much cyanide that even the preparation of the tubers for eating involves a risk of fatal poisoning. The tubers are harvested and shredded with quantities of water, and after the starch granules have settled, the poisonous liquid is poured off. Among some Central American Indians, the job of preparing cassava is given only to older people, presumably because they are considered expendable.

It is interesting that cyanide smells strongly of bitter almonds to some persons, and to others is completely odorless. This difference is not the result of an inability to smell "bitter almond," for benzaldehyde, which to cyanide smellers smells very similar to cyanide, has the characteristic "bitter almond" smell to cyanide nonsmellers. Cyanide nonsmellers, who are quite common, should be careful if they use this dangerous substance, for their sense of smell will not give them any warning.

Oils and resins

Oils and resins of various kinds are widely distributed in plants and have many different effects. One of the popularly known such resins is the active principle in marijuana, cannabis, which through its effect on the central nervous system distorts the conscious mind and causes hallucinations.

Other oils and resins are irritants to the skin or act on the gastrointestinal tract, causing vomiting, diarrhea, and a variety of other symptoms. Several compounds act as irritants to the nose and mouth and thus produce their deterrent effects more promptly.

Oxalic acid

Oxalic acid is found in a large number of plants from widely different families. It owes its acute toxicity to the fact that it precipitates calcium in the form of insoluble calcium oxalate. In acute poisoning the calcium concentration in the blood is reduced, causing nervous symptoms, muscle cramps, and acute damage to the kidneys.

One plant that is important because of its oxalate content is *Halogeton,* which is increasingly common on the ranges of the western United States. Over 30% of the dry weight of the plant may be soluble oxalate. It is highly toxic to grazing livestock and has killed virtually thousands of sheep.

Two plants commonly eaten by man, spinach and rhubarb, belong to the same family, the Chenopodiaceae. Neither is usually eaten in sufficient quantities to cause acute poisoning. However, the leaves of rhubarb, which have occasionally been used as a vegetable, contain more oxalic acid than the stems and have caused several deaths. If large amounts of spinach are eaten daily, the precipitation of calcium oxalate in the intestine will reduce the calcium available for absorption and may over a period of time lead to calcium deficiency.

Enzyme inhibitors

Some plants, including several of the pea family, contain substances that inhibit the action of trypsin, the important proteolytic enzyme of the pancreas. Such tryspin inhibitors are found in soybeans, lima beans, peanuts, and a number of plants from other families.

Most of the trypsin inhibitors are heat-sensitive, and because humans mostly eat cooked food, they are of minor importance in human nutrition. However, trypsin inhibitors reduce the protein utilization of animals that eat raw or untreated soybeans and related products, and are therefore of importance in animal husbandry. The deleterious effects can be reduced or eliminated by heat treatment (Rackis 1965).

Compounds with hormonal effects

Cabbage, mustard, radishes, and a number of other plants of the family Cruciferae owe their sharp taste to mustard oils, which are irritants and may even cause fatal poisoning. In addition, many contain substances that suppress the function of the thyroid gland and cause an enlargement of the gland, known as goiter. When laboratory animals are fed diets containing large amounts of cabbage leaves or turnips, they develop goiter.

The antithyroid substances are chemically related to drugs such as thiourea that are used in the treatment of thyroid hyperactivity. It is well known that some persons find the compound phenylthiourea extremely bitter; others find it tasteless. The nontasting condition is inherited as a recessive gene.

One of the antithyroid substances from turnips and cabbage is similarly tasteless to some and bitter to others, with tasters and nontasters matching the tasters and nontasters of phenylthiourea (Boyd 1950). This suggests a possible physiological role for the gene responsible for the tasting ability because it could help avoid eating plants that contain antithyroid agents.

Compounds that affect animal reproduction

One of the defense strategies used by plants is to interfere with the reproduction of their enemies through compounds that foul up the animals' endocrine systems. Both insects and vertebrates suffer the effects.

One way is to interfere with the normal development of insects which as larvae go through a number of molts before they change into adult form. A hormone, the juvenile hormone, is essential in directing insect larval development, but the change into the mature adult form depends on the *absence* of this hormone. Several American evergreen trees produce substances that act as juvenile hormone (see Chapter 13), and the effect is simple: The insects do not develop into the mature form and thus are prevented from reproducing.

Many synthetic compounds with juvenile hormone effects have been developed, some of them far more powerful than the naturally occurring hormones. Under certain conditions they can be used as potent insecticides with minimal effects on other animals, including domestic animals, which are insensitive to insect hormones. One disadvantage of these insecticides is that they are effective only during the brief period when the immature insect develops into the adult form.

A broader effect is produced by plant substances that have anti-juvenile-hormone activity (Bowers et al. 1976). They are effective over a much longer time and upset normal development as well as a number of other processes regulated by juvenile hormone. The action is to stop or depress the pro-

duction of juvenile hormone; this not only leads to precocious metamorphosis, but also interferes with induction of diapause, development of eggs, production of sex attractants, and so on. The substances promise to be of great interest as more information is obtained about their physiological effects.

Vertebrate reproduction is also affected by plant compounds that act on the endocrine system. Several plants that commonly are eaten by grazing animals, especially plants of the pea family, contain biologically active compounds with estrogenic effects that mimic the normal activity of animal estrogens. These estrogens cause reproductive disturbances in sheep and cattle and may be a factor in controlling wild rodent populations.

The harmful effects of plant estrogens have been extensively studied in Australia, where they are important in sheep production. Animals that eat subterranean clover show declined fertility, caused by inability to conceive, and the number of pregnant ewes therefore decreases drastically (Shutt 1976).

Plant estrogens are not only important to animal husbandry, they have a much broader role in nature. In the arid regions of California the annual plants produce phytoestrogens in dry years. When the leaves are eaten by quail, these compounds inhibit reproduction and prevent the production of young that during the drought would have inadequate food to eat. In wet years, the plants grow vigorously and estrogen concentration is low, and in such years the quail reproduce prolifically. The abundant crop of plants and seeds then assures the survival of the enlarged population. The close relationship between animal reproduction and plant productivity has long been known, and the mechanism is, at least in this instance, a well-adjusted natural biological control of population size (Leopold et al. 1976).

Animal use of plant poisons

The common milkweed, *Asclepias,* and other plants that belong to the same family produce car-

POISON MEAL An inexperienced blue jay eats a monarch butterfly (left). The poisonous cardiac glycosides that the monarch larva has obtained from its food plant make the blue jay violently ill (right). The bird will probably never again try to eat a monarch. [Courtesy of Lincoln Brower, University of Florida]

diac glycosides. In addition to affecting the heart of the animal that ingests these toxic plants, they also affect the nerve center in the brain that controls vomiting. When an animal eats this plant, vomiting usually occurs before a lethal amount of the poison has been absorbed, and the effect is limited to a very unpleasant gastric upset.

In spite of their toxicity, milkweeds are the exclusive food of the larva of a group of insects that includes the monarch butterfly. In nature various insect-eating vertebrates avoid these butterflies and their larvae for the simple reason that they have accumulated the poisonous substances from the milkweed. If a bird eats only one of the strikingly colored monarch butterflies, it will be violently ill and never try again. The butterflies have thus developed the ability not only to feed on the poisonous plants, but also to use the poisons as protection against their own predators (Brower et al. 1967).

Such use of toxic substances obtained from the feed is widespread among insects, but is not restricted to this class. Some marine animals feed on various coelenterates that otherwise are well protected by their stinging cells, or nematocysts.

One coelenterate that is painfully known to people is the Portuguese man-of-war, *Physalia,* which may even be lethal to swimmers who come in close contact with them. Certain sea slugs (nudibranch snails), which feed on the Portuguese man-of-war, ingest the tissues that contain nematocysts and di-

BOMBARDIER BEETLE These insects defend themselves with a chemical spray, ejected at a temperature of 100 °C. The spray is aimed directly at the attacker (e.g., an ant biting at one of the legs). This photograph shows a bombardier beetle kept in place for photography by a metal wire cemented to its back. The spray is directed toward the "attacker," a pair of forceps used to hold one of the legs. Bombardier beetles occur in all major faunal regions of the world, except Oceania. The species shown here, *Stenaptinus insignis,* is from Kenya and is about 20 mm long. [Courtesy of Thomas Eisner and D. Aneshansley, Cornell University]

gest them without triggering the release of the poison. The nematocysts are instead incorporated in the tissues of the snail and serve for defense against would-be predator fish (Thompson and Bennett 1969).

Animal use of their own poisons

Animals also use a wide variety of venoms and toxins that they produce themselves. Well-known examples are snakes, scorpions, spiders, bees, and wasps. More unusual types of defense include the hydrogen cyanide produced by millipedes when they are disturbed (Eisner et al. 1963) and, even more dramatic, the chemical spray ejected at a temperature of 100°C by bombardier beetles (Aneshansley et al. 1969).

The bombardier beetles produce their defensive spray from a pair of glands that open at the tip of the abdomen. The spray is visible as a fine mist, and the discharge occurs with an audible detonation. The secreting gland is a two-chambered apparatus. The larger, inner compartment contains a solution of hydroquinones and hydrogen peroxide; the outer compartment contains an enzyme mixture of catalases and peroxidases. When the beetle discharges, it squeezes fluid from the inner reservoir into the outer chamber. An instantaneous explosive reaction takes place as the hydrogen peroxide is decomposed by the enzymes and the hydroquinones are oxidized with a tremendous release of heat. Under the pressure of the evolved gas, the mixture is ejected as a fine spray that has a strongly repellent effect on predators. The beetle can spray repeatedly and in quick succession and can aim the abdominal tip in virtually any direction from which it is attacked.

In summarizing this section we can conclude that an overwhelming number of animals and plants use chemical defense mechanisms, either producing their own poisons, or taking advantage of compounds produced by others. In fact, probably only a minority of organisms is not protected in some such way against at least some potential enemies.

REFERENCES

Aneshansley, D. J., Eisner, T., Widom, J. M., and Widom, B. (1969) Biochemistry at 100°C: Explosive secretory discharge of bombardier beetles *(Brachinus)*. *Science 165:*61–63.

Arp, A. J., and Childress, J. J. (1981a) Blood function in the hydrothermal vent vestimentiferan tube worm. *Science 213:*342–344.

Arp, A. J., and Childress, J. J. (1981b) Functional characteristics of the blood of the deep-sea hydrothermal vent brachyuran crab. *Science 214:*559–560.

Arp, A. J., Childress, J. J., and Vetter, R. D. (1987) The sulphide-binding protein in the blood of the vestimentiferan tube-worm, *Riftia pachyptila*, is the extracellular haemoglobin. *J. Exp. Biol.* 128:139–158.

Bailey, C. B. (1961) Saliva secretion and its relation to feeding in cattle. 3. The rate of secretion of mixed saliva in the cow during eating, with an estimate of the magnitude of the total daily secretion of mixed saliva. *Br. J. Nutr.* 15:443–451.

Barnes, R. H., Fiala, G., and Kwong, E. (1963) Decreased growth rate resulting from prevention of coprophagy. *Fed. Proc.* 22:125–133.

Bauchop, T., and Martucci, R. W. (1968) Ruminant-like digestion of the langur monkey. *Science 161:*698–700.

Benson, A. A., Lee, R. F., and Nevenzel, J. C. (1972) Wax esters: Major marine metabolic energy sources. *Biochem. Soc. (Lond.) Symp.*, no. 35, pp. 175–187.

Berrill, N. J. (1929) Digestion in ascidians and the influence of temperature. *J. Exp. Biol.* 6:275–292.

Birney, E. C., Jenness, R., and Hume, I. D. (1980) Evolution of an enzyme system: Ascorbic acid biosynthesis in monotremes and marsupials. *Evolution 34:*230–239.

Bjorndal, K. A. (1979) Cellulose digestion and volatile fatty acid production in the green turtle, *Chelonia mydas*. *Comp. Biochem. Physiol.* 63A:127–133.

Blair-West, J. R., Coghlan, J. P., Denton, D. A., and Wright, R. D. (1967) Effect of endocrines on salivary glands. In *Handbook of Physiology*, sect 6, *Alimentary Canal*, vol II, *Secretion* (C. F. Code, ed.), pp. 633–664. Washington, D.C.: American Physiological Society.

Block, R. J., Stekol, J. A. and Loosli, J. K. (1951) Synthesis of sulfur amino acids from inorganic sulfate by ruminants. 2. Synthesis of cystine and methionine from sodium sulfate by the goat and by the microorganisms of the rumen of the ewe. *Arch. Biochem. Biophys.* 33:353–363.

Bowers, W. S., Ohta, T., Cleere, J. S., and Marsella, P. A. (1976) Discovery of insect anti-juvenile hormones in plants. *Science 193:*542–547.

Boyd, W. C. (1950) Taste reactions to antithyroid substances. *Science 112:*153.

Brooks, J. M., Kennicutt II, M. C., Fisher, C. R., Macko, S. A., Cole, K., Childress, J. J., Bidigare, R. R., and Vetter, R. D. (1987) Deep-sea hydrocarbon seep communities: Evidence for energy and nutritional carbon sources. *Science 238:*1138–1142.

Brower, L. P., Brower, J. V. Z., and Corvino, J. M. (1967) Plant poisons in a terrestrial food chain. *Proc. Natl. Acad. Sci. U.S.A.* 57:893–898.

Carlisle, E. M. (1977) A silicon requirement for normal skull formation. *Fed. Proc.* 36:1123.

Carpenter, E. J., and Culliney, J. L. (1975) Nitrogen fixation in marine shipworms. *Science 187:*551–552.

Cary, S. C., Fisher, C. R., and Felbeck, H. (1988) Mussell growth supported by methane as sole carbon and energy source. *Science 240:*78–80.

Cavanaugh, C. M., Gardiner, S. L., Jones, M. L., Jannasch, H. W., and Waterbury, J. B. (1981) Prokaryotic cells in the hydrothermal vent tube worm *Riftia pachyptila* Jones: Possible chemoautotrophic symbionts. *Science 213:*340–342.

Cleveland, L. R. (1925) Toxicity of oxygen for protozoa in vivo and in vitro: Animals defaunated without injury. *Biol. Bull.* 48:455–468.

Corliss, J. B., Dymond, J., Gordon, L. I., Edmond, J. M., von Herzen, R. P., Ballard, R. D., Green, K., Williams, D., Bainbridge, A., Crane, K., and van Andel, T. H. (1979) Submarine thermal springs on the Galápagos rift. *Science 203:*1073–1083.

Denis, G., Jeuniaux, Ch., Gerebtzoff, M. A., and Goffart, M. (1967) La digestion stomacale chez un paresseux: L'unau *Choloepus hoffmanni* Peters. *Ann. Soc. R. Zool. Belg.* 97:9–29.

Eisner, T., Eisner, H. E., Hurst, J. J., Kafatos, F. C., and Meinwald, J. (1963) Cyanogenic glandular apparatus of a millipede. *Science 139:*1218–1220.

Eskew, D. L., Welch, R. M., and Cary, E. E.

(1983) Nickel: An essential micronutrient for legumes and possibly for higher plants. *Science 222:*621–623.

Felbeck, H. (1981) Chemoautotrophic potential of the hydrothermal vent tube worm, *Riftia pachyptila* Jones (Vestimentifera). *Science 213:*336–338.

Felbeck, H. (1983) Sulfide oxidation and carbon fixation by the gutless clam *Solemya reidi:* An animal–bacteria symbiosis. *J. Comp. Physiol. 152:*3–11.

Ferguson, J. C. (1982) A comparative study of the net metabolic benefits derived from the uptake and release of free amino acids by marine invertebrates. *Biol. Bull. 162:*1–17.

Florkin, J., and Lozet, F. (1949) Origine bactérienne de la cellulase du contenu intestinal de l'escargot. *Arch. Int. Physiol. 57:*201–207.

French, J. R. J., Turner, G. L., and Bradbury, J. F. (1976) Nitrogen fixation by bacteria from the hindgut of termites. *J. Gen. Microbiol. 95:*202–206.

Friedmann, H., Kern, J., and Hurst, J. H. (1957) The domestic chick: A substitute for the honeyguide as a symbiont with cerolytic microorganisms. *Am. Nat. 91:*321–325.

Grassé, P.-P. (1955) *Traité de Zoologie: Anatomie, Systématique, Biologie*, vol. 17, fasc. 2; *Mammifères*, pp. 1173–2300. Paris: Masson.

Grassle, J. F. (1985) Hydrothermal vent animals: Distribution and biology. *Science 229:*713–717.

Greenberg, B., and Paretsky, D. (1955) Proteolytic enzymes in the house fly, *Musca domestica* (L.). *Ann. Entomol. Soc. Am. 48:*46–50.

Greenfield, L. J., and Lane, C. E. (1953) Cellulose digestion in *Teredo. J. Biol. Chem. 204:*669–672.

Griffiths, M., and Davies, D. (1963) The role of the soft pellets in the production of lactic acid in the rabbit stomach. *J. Nutr. 80:*171–180.

Harder, W. (1949) Zur Morphologie und Physiologie des Blinddarmes der Nagetiere. *Verh. Dtsch. Zool. 2:*95–109.

Hervey, G. R. (1969) Regulation of energy balance. *Nature, Lond. 222:*629–631.

Horning, M. G. (1958) The sterol requirements of insects and of protozoa. In *Cholesterol: Chemistry, Biochemistry, and Pathology* (R. P. Cook, ed.), pp. 445–455. New York: Academic Press.

Houpt, T. R. (1959) Utilization of blood urea in ruminants. *Am. J. Physiol. 197:*115–120.

Houpt, T. R. (1963) Urea utilization by rabbits fed a low-protein ration. *Am. J. Physiol. 205:* 1144–1150.

Hungate, R. E. (1942) The culture of *Eudiplodinium neglectum,* with experiments on the digestion of cellulose. *Biol. Bull. 83:*303–319.

Hungate, R. E. (1955) Mutualistic intestinal protozoa. In *Biochemistry and Physiology of Protozoa,* vol. II (S. H. Hutner and A. Lwoff, eds.), pp. 159–199. New York: Academic Press.

Jørgensen, C. B. (1976) August Pütter, August Krogh, and modern ideas on the use of dissolved organic matter in aquatic environments. *Biol. Rev. 51:*291–328.

Karl, D. M., Wirsen, C. O., and Jannasch, H. W. (1980) Deep-sea primary production at the Galápagos hydrothermal vents. *Science 207:*1345–1347.

Kenagy, G. J., and Hoyt, D. F. (1980) Reingestion of feces in rodents and its daily rhythmicity. *Oecologia (Berl.) 44:*403–409.

Kingsbury, J. M. (1964) *Poisonous Plants of the United States and Canada.* Englewood Cliffs, N.J.: Prentice-Hall. 626 pp.

Kinnear, J. E., and Main, A. R. (1975) The recycling of urea nitrogen by the wild tammar wallaby *(Macropus eugenii),* a "ruminant-like" marsupial. *Comp. Biochem. Physiol. 51A:*793–810.

Kleiber, M. (1961) *The Fire of Life. An Introduction to Animal Energetics.* New York: Wiley. 454 pp.

Lambremont, E. N., Fisk, R. W., and Ashrafi, S. (1959) Pepsin-like enzyme in larvae of stable flies. *Science 129:*1484–1485.

Lasker, R., and Giese, A. C. (1956) Cellulose digestion by the silverfish *Ctenolepisma lineata. J. Exp. Biol. 33:*542–553.

Leopold, A. S., Erwin, M., Oh, J., and Browning, B. (1976) Phytoestrogens: Adverse effects on reproduction in California quail. *Science 191:*98–100.

Loosli, J. K., Williams, H. H., Thomas, W. E., Ferris, F. H., and Maynard, L. A. (1949) Synthesis of amino acids in the rumen. *Science 110:*144–145.

Manahan, D. T., Wright, S. H., Stephens, G. C., and Rice, M. A. (1982) Transport of dissolved amino acids by the mussel *Mytilus edulis:* Demonstration of net uptake from natural seawater. *Science 215:*1253–1255.

Matthews, L. H., and Parker, H. W. (1950) Notes on the anatomy and biology of the basking shark (*Cetorhinus maximum* (Gunner)). *Proc. Zool. Soc. Lond. 120:*535–576.

Mayer, J., and Thomas, D. W. (1967) Regulation of food intake and obesity. *Science 156:*335–337.

McBee, R. H., and West, G. C. (1969) Cecal fermentation in the willow ptarmigan. *Condor 71:*54–58.

McDonald, I. W. (1952) The role of ammonia in ruminal digestion of protein. *Biochem. J. 51:*86–90.

Mertz, W. (1969) Chromium occurrence and function in biological systems. *Physiol. Rev. 49:*163–239.

Mertz, W. (1981) The essential trace elements. *Science 213:*1332–1338.

Moir, R. J., Somers, M., and Waring, H. (1956) Studies on marsupial nutrition. 1. Ruminant-like digestion in a herbivorous marsupial (*Setonix brachyurus* Quoy and Gaimard). *Aust. J. Biol. Sci. 9:*293–304.

Muscatine, L., and Lenhoff, H. M. (1965) Symbiosis of hydra and algae. 2. Effects of limited food and starvation on growth of symbiotic and aposymbiotic hydra. *Biol. Bull. 129:*316–328.

Nielsen, F. H., Givand, S. H., and Myron, D. R. (1975) Evidence of a possible requirement for arsenic by the rat. *Fed. Proc. 34:*923.

O'Dell, B. L., and Campbell, B. J. (1971) Trace elements: Metabolism and metabolic function. *Comp. Biochem. 21:*179–266.

Odum, H. T. (1957) Biogeochemical deposition of strontium. *Univ. Texas Publ. Inst. Mar. Sci. 4:*38–114.

Pardy, R. L., and Dieckmann, C. (1975) Oxygen consumption in the symbiotic hydra *Hydra viridis. J. Exp. Zool. 194:*373–378.

Prévost, J., and Vilter, V. (1962) Histologie de la sécrétion oesophagienne du Manchot Empereur. In *Proceedings, 13th International Ornithologic Congress,* vol. 2. pp. 1085–1094. Baton Rouge, La.: American Ornithologists' Union.

Rackis, J. J. (1965) Physiological properties of soybean trypsin inhibitors and their relationship to pancreatic hypertrophy and growth inhibition of rats. *Fed. Proc. 24:*1488–1500.

Roby, D. D., Place, A. R., and Ricklefs, R. E. (1986) Assimiliation and deposition of wax esters in planktivorous seabirds. *J. Exp. Zool. 238:*29–41.

Rose, W. C. (1938) The nutritive significance of the amino acids. *Physiol. Rev. 18:*109–136.

Rose, W. C. (1949) Amino acid requirements of man. *Fed. Proc. 8:*546–552.

Rozin, P., and Mayer, J. (1961) Regulation of food intake in the goldfish. *Am. J. Physiol. 201:*968–974.

Sargent, J. R., and Gatten, R. R. (1976) The distribution and metabolism of wax esters in marine invertebrates. *Biochem. Soc. (Lond.) Trans. 4:*431–433.

Schmidt-Nielsen, B., Schmidt-Nielsen, K., Houpt, T. R., and Jarnum, S. A. (1957) Urea excretion in the camel. *Am. J. Physiol. 188:*477–484.

Schmidt-Nielsen, K. (1986) Why milk? *News Physiol. Sci. 1:*140–142.

Schoenheimer, R. (1964) *The Dynamic State of Body Constituents.* New York: Hafner. 78pp.

Schwarz, K., and Milne, D. B. (1971) Growth effects of vanadium in the rat. *Science 174:*426–428.

Schwarz, K., and Milne, D. B. (1972) Fluorine requirement for growth in the rat. *Bioinorg. Chem. 1:*331–338.

Schwarz, K., Milne, D. B., and Vinyard, E. (1970) Growth effects of tin compounds in rats maintained in a trace-element-controlled environment. *Biochem. Biophys. Res. Commun. 40:*22–29.

Shutt, D. A. (1976) The effects of plant oestrogens on animal production. *Endeavour 35:*110–113.

Smith, D., Muscatine, L., and Lewis, D. (1969) Carbohydrate movement from autotrophs to heterotrophs in parasitic and mutualistic symbiosis. *Biol. Rev. 44:*17–90.

Tarr, H. L. A. (1958) Biochemistry of fishes. *Annu. Rev. Biochem. 27:*223–244.

Taylor, M. W., and Medici, J. C. (1966) Amino acid requirements of grain beetles. *J. Nutr. 88:*176–180.

Thacker, E. J., and Brandt, C. S. (1955) Coprophagy in the rabbit. *J. Nutr. 55:*375–385.

Thompson, T. E., and Bennet, I. (1969) Physalia nematocysts: Utilized by mollusks for defense. *Science 166:*1532–1533.

Trager, W. (1932) A cellulase from the symbiotic intestinal flagellates of termites and of the roach, *Cryptocerus punctulatus. Biochem. J. 26:*1762–1771.

Trager, W. (1934) The cultivation of a cellulose-digesting flagellate, *Trichomonas termopsidis,* and of certain other termite protozoa. *Biol. Bull. 66:*182–190.

Troyer, K. (1982) Transfer of fermentative microbes between generations in a herbivorous lizard. *Science 216:*540–542.

Trytek, R. E., and Allen, W. V. (1980) Synthesis of essential amino acids by bacterial symbionts in the gills of the shipworm *Bankia setacea* (Tryon). *Comp. Biochem. Physiol. 67A:*419–427.

Turekian, K. K., Cochran, J. K., Kharkaar, D. P., Cerrato, R. M., Vaišnys, J. R., Sanders, H. L.,

Grassle, J. F., and Allen, J. A. (1975) Slow growth of a deep-sea clam determined by ^{228}Ra chronology. *Proc. Natl. Acad. Sci. U.S.A. 72*:2829–2832.

Turekian, K. K., Cochran, J. K., and Nozaki, Y. (1979) Growth rate of a clam from the Galápagos Rise hot spring field using natural radionuclide ratios. *Nature, Lond. 280*:385–387.

Waterbury, J. B., Calloway, C. B., and Turner, R. D. (1983) A cellulolytic nitrogen-fixing bacterium cultured from the gland of Deshayes in shipworms (Bivalvia: Teredinidae). *Science 221*:1401–1403.

Webb, K. L., DuPaul, W. D., Wiebe, W., Sottile, W., and Johannes, R. E. (1975) Enewetak (Eniwetok) atoll: Aspects of the nitrogen cycle on a coral reef. *Limnol. Oceanogr. 20*:198–210.

Weis-Fogh, T. (1952) Fat combustion and metabolic rate of flying locusts (*Schistocerca gregaria* Forskål). *Philos. Trans. R. Soc. Lond. B. 237*:1–36.

White, A., Handler, P., and Smith, E. L. (1964) *Principles of Biochemistry*, 3d ed. New York: McGraw-Hill. 1106 pp.

Whittaker, R. H., and Feeny, P. P. (1971) Allelochemics: Chemical interactions between species. *Science 171*:757–770.

Wigglesworth, V. B. (1949) The utilization of reserve substances in *Drosophila* during flight. *J. Exp. Biol. 26*:150–163.

Wright, R. T., and Hobbie, J. E. (1966) Use of glucose and acetate by bacteria and algae in aquatic ecosystems. *Ecology 47*:447–464.

Wright, S. H., and Stephens, G. C. (1978) Removal of amino acid during a single passage of water across the gill of marine mussels. *J. Exp. Zool. 205*:337–352.

ADDITIONAL READING

American Society of Zoologists (1982) The role of uptake of organic solutes in nutrition of marine organisms: Symposium. *Am. Zool. 22*:609–733.

Bücherl, W., Buckley, E. E., and Deulofeu, V. (eds.) (1968, 1971) *Venomous Animals and Their Venoms*, vol. 1, 707 pp. (1968); vol. 2, 687 pp. (1971); vol. 3, 537 pp. (1971). New York: Academic Press.

Code, C. F., and Heidel, W. (eds.) (1967–1968) *Handbook of Physiology*, sect. 6, *Alimentary Canal*, vols. I–V, pp. 1–2896. Washington, D.C.: American Physiological Society.

Crawford, M. A. (ed.) (1968) *Comparative Nutrition of Wild Animals*. Symposium, Zoological Society of London, no. 21. London: Academic Press. 429 pp.

Eisner, T. (1970) Chemical defense against predation in arthropods. In *Chemical Ecology* (E. Sondheimer and J. Simeone, eds.), pp. 157–217. New York: Academic Press.

Fenchel, T. M., and B. B. Jørgensen (1976) Detritus food chains of aquatic ecosystems: The role of bacteria. *Adv. Microbl. Ecol. 1*:1–58.

Florkin, M., and Stotz, E. H. (eds.) (1971) *Comprehensive Biochemistry*, vol. 21, *Metabolism of Vitamins and Trace Elements*. Amsterdam: Elsevier. 297 pp.

Fowden, K., Garton, G. A., and Mills, C. F. (eds.) (1981) Metabolic and physiological consequences of trace element deficiency in animals and man. *Philos. Trans. R. Soc. Lond. B 294*:1–213.

Goreau, T. F., Goreau, N. I., and Yonge, C. M. (1973) On the utilization of photosynthetic products from zooxanthellae and of a dissolved amino acid in *Tridacna maxima* f. *elongata. (Mollusca: Bivalvia). J. Zool. Lond. 169*:417–454.

Halver, J. E. (ed.) (1972) *Fish Nutrition*. New York: Academic Press. 714 pp.

Harborne, J. B. (ed.) (1978) *Biochemical Aspects of Plant and Animal Coevolution*. New York: Academic Press. 436 pp.

Hume, I. D. (1982) *Digestive physiology and nutrition of marsupials*. Cambridge: Cambridge University Press. 256 pp.

Jannasch, H. W. (1985) The chemosynthetic support of life and the microbial diversity at deep-sea hydrothermal vents. *Proc. R. Soc. Lond. B 225*: 227–297.

Jørgensen, C. B. (1975) Comparative physiology of suspension feeding. *Annu. Rev. Physiol. 37*:57–79.

Labov, J. B. (1977) Phytoestrogens and mammalian reproduction. *Comp. Biochem. Physiol. 57A*:3–9.

Leschine, S. B., Holwell, K., and Canale-Parola, E. (1988) Nitrogen fixation by anaerobic cellulolytic bacteria. *Science 242*:1157–1159.

Liener, I. E. (1969) *Toxic Constitutents of Plant Foodstuffs*. New York: Academic Press. 500 pp.

McDonald, P., Edwards, R. A., and Greenhalgh, J. F. D. (1973) *Animal Nutrition*, 2d ed. New York: Hafner. 487 pp.

Powell, M. A., and Somero, G. N. (1986) Adaptations to sulfide by hydrothermal vent animals: Sites and mechanisms of detoxification and metabolism. *Biol. Bull. 171*:274–290.

Ruckebusch, Y., and Thivend, P. (eds.) (1980) *Digestive Physiology and Metabolism in Ruminants.* Westport, Conn.: AVI. 854 pp.

Ryan, P. R. (ed.) (1984) Deep-sea hot springs and cold seeps. *Oceanus 27:*1–84.

Southward, E. C. (1987) Contribution of symbiotic chemoautotrophs to the nutrition of benthic inverte-brates. In *Microbes in the Sea* (M. A. Sleigh, ed.) pp. 83–118. New York: Wiley. 241 pp.

Stephens, G. C. (1975) Uptake of naturally occurring primary amines by marine annelids. *Biol. Bull. 149:*397–407.

Stevens, C. E. (1988) *Comparative physiology of the vertebrate digestive system.* Cambridge: Cambridge University Press. 300 pp.

Underwood, E. J. (1977) *Trace Elements in Human and Animal Nutrition,* 4th ed. New York: Academic Press. 545pp.

Vonk, H. J., and Western, J. R. H. (1984) *Comparative biochemistry and physiology of enzymatic digestion.* London: Academic Press. 512 pp.

Wright, S. H., and Manahan, D. T. (1989) Integumental nutrient uptake by aquatic organisms. *Annu. Rev. Physiol. 51:*585–600.

CHAPTER FIVE

Energy metabolism

The preceding chapter dealt with food and feeding; this chapter will deal with the use of food to provide energy. Animals need chemical energy to carry out their various functions, and their overall use of chemical energy is often referred to as their *energy metabolism.*

Animals obtain energy mostly through the oxidation of foodstuffs. The amount of oxygen they consume can therefore be used as a measure of their energy metabolism. Much of this chapter will be concerned with the *rate of oxygen consumption,* which we often take to mean the *rate of energy metabolism.*

However, some animals can live in the absence of free oxygen. They still utilize chemical energy for their energy needs, although the metabolic pathways are different. Such metabolic processes are known as *anaerobic metabolism.* This situation is normal for quite a few animals that live in oxygen-poor environ-ments or tolerate prolonged or permanent lack of oxygen. They obtain energy through processes that do not utilize molecular oxygen.

The energy-requiring processes and reactions in the living organism use a common source of energy, *adenosine triphosphate* (ATP). This ubiquitous com-pound, through the hydrolysis of an ''energy-rich'' phosphate bond, is the immediate energy source for processes such as muscle contraction, ciliary move-ment, firefly luminescence, discharge of electric fish, cellular transport processes, all sorts of synthetic re-actions, and so on.

ATP is formed at the various energy-yielding steps in the oxidation of foodstuffs, and also in anaerobic energy-yielding processes. ATP is the universal in-termediate in the flow of the food's chemical energy to energy-requiring processes in both aerobic and an-aerobic organisms.

METABOLIC RATE

Metabolic rate refers to the energy metabolism per unit time.* It can, in principle, be determined in three different ways.

The first is by calculating the difference between *the energy value of all food* taken in and the energy value of all excreta (primarily feces and urine). This method assumes that there is no change in the composition of the organism. It therefore cannot be used for growing organisms or in organisms that have an increase or a decrease in storage of fat or other material. The method is technically cumbersome and is accurate only if carried out over a sufficiently long period of observation to assure that the organism has not undergone significant changes in size and composition.

The second method of determining metabolic rate is from *the total heat production* of the organism. This method should give information about all fuel used, and in principle it is the most accurate method. In practice, determinations are made with the organism inside a calorimeter. This can yield very accurate results, but technically it is a complex procedure. Such items as heating of ingested food and vaporization of water must be entered into the total heat account. Also, if the organism performs any external work that does not appear as heat, this work must be added to the account. Assume, for example, that a person located inside a large calorimeter lifts a ton of bricks from the floor onto the table. The increase in potential energy of the bricks is derived from the energy metabolism of the person and could be released as heat if the bricks were permitted to slide down again.

The third measure that can be used to determine the metabolic rate is *the amount of oxygen used* in oxidation processes, provided information is available about which substances have been oxidized (and there is no anaerobic metabolism).

The determination of oxygen consumption is technically easy and is so commonly used for estimation of metabolic rates that the two terms often are used interchangeably. This is obviously not correct; for example, a fully anaerobic organism has zero oxygen consumption but certainly does not have a zero metabolic rate!

The reason oxygen can be used as a practical measure of metabolic rate is that the amount of heat produced for each liter of oxygen used in metabolism remains nearly constant, irrespective of whether fat, carbohydrate, or protein is oxidized (Table 5.1, column c). The highest figure (5.0 kcal per liter O_2 for carbohydrate metabolism) and the lowest (4.5 kcal per liter O_2 for protein) differ by only 10%, and it has become customary to use an average value of 4.8 kcal per liter O_2 as a measure of the metabolic rate.† The largest error resulting from the use of this mean figure would be 6%, but if the metabolic fuel is a mixture of the common foodstuff (carbohydrate, fat, and protein), the error is usually insignificant. Most determinations of metabolic rates are not very precise anyway, not because the available apparatus is inherently inaccurate, but because the amounts of animal activity and other physiological functions tend to vary a great deal.

Although 1 liter of oxygen gives similar amounts of energy for the three major foodstuffs, each gram

---

*The word *metabolism* does not always refer to energy metabolism. For example, the term *iron metabolism* refers to the total physiological role of iron – its intake, absorption, storage, hemoglobin synthesis, excretion, etc. Another example, the term *intermediary metabolism* refers to all the biochemical reactions that take place in the living organism, e.g., the Krebs cycle. In this chapter the term will mean energy metabolism.

†The traditional unit for heat, the calorie (cal), is not a part of the International System of Units (the SI System). However, it is in universal use in physiological tables, books, and other publications, and it is impractical not to retain it in this book. The energy unit of the SI System is the joule (J), and the calorie is defined as 1 cal = 4.184 J (1 kcal = 4.184 kJ). Further, 1 cal is the amount of heat required to heat 1 g water from 14.5°C to 15.5°C, and 1 kcal = 1000 cal. Often, especially in elementary books and popular writings, the calorie and kilocalorie are confused. At times, a kilocalorie is called a ''large calorie'' and is written ''Cal.'' Because the calorie is a precisely defined unit it is meaningless to refer to ''large'' and ''small'' calories. For additional information about the use of symbols and units, see Appendix A.

TABLE 5.1 Heat produced and oxygen consumed in the metabolism of common foodstuffs. The values for protein depend on whether the metabolic end product is urea or uric acid. The ratio between carbon dioxide formed and oxygen used is known as the respiratory quotient (RQ). [Data based on Lusk 1931; King 1957]

Food	(a) kcal g^{-1}	(b) liter $O_2\, g^{-1}$	(c) kcal per liter O_2	(d) $RQ = \dfrac{CO_2 \text{ formed}}{O_2 \text{ used}}$
Carbohydrate	4.2	0.84	5.0	1.00
Fat	9.4	2.0	4.7	0.71
Protein (urea)	4.3	0.96	4.5	0.81
Protein (uric acid)	4.25	0.97	4.4	0.74

of foodstuff gives a very different amount of energy. The energy derived from the oxidation of 1 g fat is more than twice as high as from the oxidation of 1 g carbohydrate or protein (Table 5.1, column a).

The high energy value of fat is widely known. The biologically most important consequence is that energy stored as fat adds less to the weight and bulk of an animal than would otherwise be the case.

Many of you may be interested in knowing that the energy value of alcohol is also quite high, nearly 7 kcal g^{-1}. This is one reason that persons who wish to lose weight by dieting usually avoid alcohol; another reason is that alcohol readily weakens the resolution to restrict food intake and thus contributes further to the lack of success in controlling energy intake.

The energy value given for protein, 4.3 kcal g^{-1}, differs from the combustion value determined for proteins in the chemical laboratory, which is usually given as 5.3 kcal g^{-1}. The reason is that protein is not completely oxidized in the body. Mammals excrete the nitrogen from amino acids as urea, $CO(NH_2)_2$, an organic compound whose combustion value accounts for the difference. This difference must be considered if the energy balance of an animal is calculated from combustion values determined by employing a bomb calorimeter.

The amount of oxygen needed to oxidize 1 g fat is more than twice the amount needed for oxidation of carbohydrate or protein (Table 5.1, column b). In combination with the high energy yield from fat, the result is that the energy per liter oxygen is approximately the same for fat as for carbohydrate and protein (see above).

The last column in Table 5.1 shows the ratio between the carbon dioxide formed in metabolism and the oxygen used. This ratio, known as the *respiratory quotient (RQ)* or *respiratory exchange ratio (R)*, is an important concept in metabolic physiology.

The RQ gives information about the fuel used in metabolism. Usually the RQ is between 0.7 and 1.0. An RQ near 0.7 suggests primarily fat metabolism; an RQ near 1.0 suggests primarily carbohydrate metabolism. For an intermediate RQ, it is more difficult to say what foodstuffs have been metabolized; it could be protein, a mixture of fat and carbohydrate, or a mixture of all three. It is possible, however, to determine the amount of protein metabolized from the nitrogen excretion (urea in mammals, uric acid in birds and reptiles), and from this knowledge we can calculate that fraction of the oxygen consumption and carbon dioxide production that is attributable to protein metabolism. The remainder of the gas exchange is attributable to fat and carbohydrate metabolism, and the amount of each can then be calculated. From determinations of oxygen consumption, carbon dioxide elimination, and nitrogen excretion, it is thus possible to calculate the separate amounts of protein, fat, and carbohydrate metabolized in a given period of time.

Could carbon dioxide production be used for determinations of metabolic rate equally as well as oxygen consumption? In theory the answer is yes, but there are difficulties that make it less practical and in fact far less accurate. There are two main reasons: One is that the body contains a large pool of carbon dioxide; the other has to do with the caloric equivalent of 1 liter of carbon dioxide.

The large pool of carbon dioxide that is always present in the body changes easily. For example, hyperventilation of the lungs causes large amounts of carbon dioxide to leave in the exhaled air. This carbon dioxide, of course, is not a measure of the metabolic rate. Hyperventilation is followed by a period of reduced ventilation during which the carbon dioxide again builds up to its normal level, and during this period the reduced amount of carbon dioxide given off again is not a measure of metabolic rate.

Other circumstances also influence the pool of carbon dioxide. In heavy exercise, for example, lactic acid is formed in the muscles as a normal metabolic product. As the lactic acid enters the blood, it has the same effect as pouring acid on a bicarbonate solution: Carbon dioxide is driven out of the blood and released in the lung in excessive amounts. Again, the amount of carbon dioxide is not a measure of the metabolic rate.

The other reason carbon dioxide is less suitable than oxygen for measuring metabolic rates is that the different foodstuffs give rather different amounts of energy for each liter of carbon dioxide produced. Carbohydrate gives 5.0 kcal per liter CO_2 formed; fat gives 6.7 kcal per liter CO_2. The value for fat is therefore one-third higher than for carbohydrate, and unless we know whether fat or carbohydrate is metabolized, the error can be substantial.

For these reasons the carbon dioxide production is not a practical measure of metabolic rate.

ENERGY STORAGE: FAT AND GLYCOGEN

For most adult animals food intake and energy expenditure remain approximately equal. If energy expenditure exceeds food intake, the excess energy is covered by the consumption of body substance, primarily *fat*. If food intake exceeds energy used, the surplus is stored as fat, irrespective of the composition of the food. For example, pigs are fattened on feed consisting mostly of carbohydrate (grains).

During fattening, if large amounts of carbohydrate are changed into fat, the RQ will exceed 1.0. This is because fats contain relatively less oxygen than do carbohydrates (i.e., some of the ''excess'' oxygen from the carbohydrate can be used in metabolism, thus reducing the respiratory oxygen uptake and increasing the respiratory carbon dioxide:oxygen ratio.

Because fat yields more than twice as much energy as carbohydrate, it is better suited for energy storage. However, there are exceptions to this rule. For animals that do not move about, such as oysters and clams, weight is of minor consequence, and *glycogen* is used for storage. Also, many intestinal parasites, such as the roundworm *Ascaris,* store glycogen.

In these animals glycogen is probably a more suitable storage substance than fat, for they are frequently exposed to conditions of low or zero oxygen, and in the absence of oxygen, glycogen can yield energy by breakdown to lactic acid. Therefore, as bivalves frequently close their shells for long periods and intestinal parasites live in environments virtually devoid of oxygen, it is advantageous for these animals to store glycogen rather than fat. Sessile animals and parasites have little need for weight economy, and from this viewpoint the use of glycogen instead of fat is no disadvantage.

For animals that move about, however, weight economy is of great importance. Just think of how important this is for migrating birds, which may fly nonstop for more than 1000 km. To have enough fuel, migrating birds may carry as much as 40 to 50% of their body weight as fat; if the fuel were heavier, the weight would be excessive, and long-distance migration would be impossible because much of the available energy would merely go into carrying the heavier body aloft.

Storage of glycogen involves more weight, not only because of the lower energy content of carbohydrate compared with fat, but also because glycogen is deposited in the cells with a considerable amount of water. It has been estimated that the deposition of glycogen in the cells of liver and muscle is accompanied by about 3 g water for each gram of glycogen stored. Some investigations even indicate that the amount of water may be as high as 4 or 5 g per gram glycogen (Olsson and Saltin 1968).

The simplest way of expressing the weight of fuel relative to the energy value is to compute the weight per energy unit, or the *isocaloric weight*. For the fuels we have discussed, the isocaloric weights, in grams of fuel per kilocalorie, are as follows:

Fat	0.11
Protein	0.23
Starch	0.24
Glycogen + H$_2$O	~1.0

Expressed in this way, it is immediately apparent that, with the extra weight of water, glycogen is some 10 times as heavy to carry as the same energy in fat.

In spite of this weight difference, glycogen is an important storage form for energy. Its advantage is twofold: It can provide fuel for carbohydrate metabolism very quickly, whereas the mobilization of fat is slow; and, perhaps more importantly, glycogen can provide energy under anoxic conditions. This is common during heavy muscular exercise when the blood does not deliver sufficient oxygen to meet demands. For long-term storage of large amounts of energy, however, glycogen is unsuited, and fat is the preferred substance.

The fact that water is stored with the glycogen probably explains a puzzling observation made by many persons when they begin reducing diets. During the first few days they lose weight rapidly, often far in excess of what can be attributed to the loss of fat alone. In this initial period the glycogen stores become depleted, and the excess water is excreted. Much of the weight loss is therefore only loss of water. After a few days, even if food intake is less than what is metabolized, the glycogen stores are replenished.

When metabolism turns to the use of stored fat, there is no longer a rapid decrease in body weight; on the contrary, body weight may increase again, although the person remains in negative calorie balance and continues to lose fat. These unexpected changes in weight, which are rarely adequately explained, cause many dieters to be overoptimistic during the first days and then rapidly become discouraged when they lose weight less quickly or even gain weight despite their continued sacrifice of food intake.

EFFECT OF OXYGEN CONCENTRATION

It is often assumed that oxygen consumption (or metabolic rate) within wide limits is independent of the oxygen concentration. For example, if we replace the ordinary air with pure oxygen, a mammal continues to consume oxygen at the same rate, although oxygen is present in nearly five times the usual concentration. The reverse is also true; if we reduce the oxygen concentration to half, which can be done by reducing the total atmospheric pressure to half (equivalent to about 6000 m altitude), the rate of oxygen consumption likewise remains virtually unaffected.

This independence of oxygen concentration is not universal. The situation may be quite different. Many invertebrates respond to changing oxygen concentrations, as shown for the common lobster in Figure 5.1.

FIGURE 5.1 The oxygen uptake of lobsters is higher at high temperatures, but at any given temperature it varies linearly with the oxygen concentration in the water. [Thomas 1954]

FIGURE 5.2 At low oxygen concentrations the oxygen consumption of goldfish varies linearly with the oxygen content of the water, but at higher oxygen tension it is independent of the oxygen content. [Fry and Hart 1948]

At high temperature the oxygen consumption is higher than at low temperature, but at any one temperature the oxygen uptake is linearly related to the available oxygen.*

If the oxygen uptake is studied over a wider temperature range in an animal such as the goldfish, we obtain the relationship shown in Figure 5.2. At relatively high oxygen tension the oxygen uptake of goldfish is independent of the available oxygen, but at lower oxygen tensions there is a linear relationship. The inflection point, which indicates where

*In this book amounts of oxygen are given in volumes (i.e., volume of dry gas at 0°C and 760 mm Hg pressure). The SI unit for amount of a substance is the mole (symbol: mol), and an amount of oxygen should be given not in liters or milliliters, but in moles and millimoles (mmol). For example, 1 liter O_2 at 0°C and 760 mm Hg pressure = 44.64 mmol. We are presently in a transition period with increasing use of SI units, but many traditional units are more familiar and will persist for some time to come. Whenever practical, a change to SI units is preferred (see Appendix A).

the oxygen uptake changes from concentration dependence to independence, is lower at low temperature than at high temperature. Thus, in Figure 5.2 the inflection point changes from less than 20 mm Hg at 5°C to about 40 mm Hg at 35°C.

If we return to Figure 5.1, we can see that the straight-line relationships observed for lobsters may indicate that we are looking at only the lower part of curves of the same general shape as the goldfish curves. If this is so, we can expect that the curves for lobsters at some unknown higher oxygen concentration would level off and become independent of further increases in oxygen.

Even mammals, whose metabolic rates show little dependence on available oxygen, may be more dependent on oxygen concentration than we usually assume. The oxygen consumption of isolated muscles of rats and rabbits is reversibly depressed at low oxygen concentrations, without alteration of other cell functions (Whalen 1966).

This is of importance to our evaluation of the observation that diving animals, after a dive, do not increase their oxygen consumption enough to equal the oxygen debt they are expected to have acquired during the time of the dive (see later in this chapter). In the swimming animal the muscle mass is the major oxygen-consuming tissue, and an oxygen-dependent metabolic rate in this tissue would yield exactly the result observed in many diving animals. Perhaps an oxygen-independent metabolism as we usually think of it is a special case and oxygen dependence is the general rule.

Acclimation to low oxygen levels

Many fast-swimming, active fish are quite sensitive to low oxygen in the water. This is true of most salmonids – for example, the speckled trout (*Salvelinus fontinalis*). For this fish the lethal level of oxygen in the water depends on previous exposure to low but tolerable oxygen levels. If the fish has been kept in relatively oxygen-poor water, it becomes more tolerant to low levels of oxygen (Figure 5.3).

This improved tolerance is probably attributable to an enhancement of the ability to extract oxygen from the water. If there was a true increase in the ability of the tissues to tolerate anoxic conditions, a fish acclimated to low levels of oxygen could be expected to survive longer than a nonacclimated fish when placed in completely anaerobic conditions. There is no evidence of such a difference (Shepard 1955).

The improved oxygen extraction would result from increased volume of water pumped over the gills, increased ability to remove oxygen from the water flowing over the gills, improved oxygen transport capacity in the blood, improved function of the heart, or any combination of these.

Anaerobic metabolism

Animals that tolerate prolonged exposure to zero oxygen must of necessity obtain their metabolic energy from nonoxidative reactions. This holds true for many intestinal parasites, for animals that live

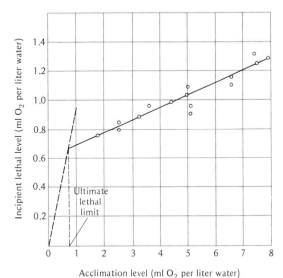

FIGURE 5.3 If speckled trout have already been exposed to a reduced oxygen concentration in the water (abscissa), they become more tolerant to low oxygen and have a lowered lethal oxygen limit (ordinate). [Shepard 1955]

in the oxygen-free mud of lakes and ponds, for bivalves that remain closed during long periods, and for many others.

Animals that can intermittently survive long periods of anoxic conditions are called *facultative anaerobic organisms*. An example is the sea anemone *Bunodosoma*, which is found along the Texas Gulf Coast. These sea anemones are often buried in the sand, and some are found underneath rocks where the sand is black from anaerobic sulfur bacteria. The survival of these sea anemones in the absence of oxygen is impressive. Animals that were placed individually in sealed chambers containing sea water that had been flushed with nitrogen were left undisturbed in sealed jars for 6 weeks. When the jars were opened, a pungent odor of sulfur was released, and there was no measurable oxygen left in the water. When the sea anemones were placed in aerated water, they all opened quickly and showed normal activity. Their neuromuscular responses were the same and showed the same threshold for stimulation whether or not oxygen was present (Mangum 1980).

Lactic acid, glycolysis

How does an animal obtain energy to remain alive in the total absence of oxygen? One common process for obtaining energy under anaerobic conditions is the breakdown of carbohydrate into lactic acid. For example, 1 mol glucose can anaerobically be broken down to 2 mol lactic acid ($C_6H_{12}O_6 \rightarrow 2\ C_3H_6O_3$).

This process, known as *glycolysis,* occurs commonly in vertebrate muscle when the energy demand in heavy exercise exceeds the available oxygen.

It will be convenient to discuss the needs for energy by using the common currency of energy metabolism, adenosine triphosphate (ATP). ATP is the immediate source of energy for nearly every energy-requiring process in the living organism. If we consider oxidative and anoxic metabolism from this viewpoint, the situation looks as follows.

When 1 mol glucose, through intermediate biochemical reactions, is degraded to 2 mol lactic acid, 2 mol ATP is synthesized and made available to supply energy. In contrast, the complete oxidation of 1 mol glucose to carbon dioxide and water yields 36 mol ATP (6 mol in the formation of 2 mol pyruvic acid and 30 mol in the complete oxidation of the pyruvic acid via the Krebs cycle). The yield from the anaerobic breakdown of glucose is therefore small, not quite 6% of what would be available from the complete oxidation of glucose.

Most anaerobic glycolysis depends on glycogen, rather than glucose, as the substrate. Glycogen is a polymer of glucose, and the single glucose unit in the polymer is called a glycosyl unit. If glycogen is the substrate in anaerobic metabolism, the yield is 3 mol ATP from the formation of 2 mol lactic acid from 1 mol glycosyl. The yield from anaerobic glycolysis, if glycogen is the substrate, therefore yields about 8% of what would be available from the complete oxidation of the substrate

to carbon dioxide and water. The reason a glycosyl unit yields more energy than glucose is that some energy was used in the synthesis of the glycogen polymer from glucose, and this eliminates one energy-requiring step in the breakdown to lactic acid.

It may seen uneconomical to use anaerobic glycolysis when the yield of ATP is so low, but an intestinal parasite probably doesn't care much about economy. However, animals that use glycolysis only intermittently can be expected to use lactic acid as a substrate for further oxidation (in the Krebs cycle) and thus in the end gain the full energy value of the original carbohydrate substrate.

The tolerance to anoxic conditions sometimes is amazing. It has been reported that the crucian carp (*Carassius carassius*) can live for 5.5 months under the ice of a frozen lake when the water under the ice is oxygen-free, as shown by the presence of hydrogen sulfide formed during fermentation of dead plant material (Blazka 1958). (Hydrogen sulfide is poisonous because it binds to and inactivates cytochrome oxidase, but because no oxygen is present to permit oxidative metabolism, this does not poison the fish.) These carp produce practically no lactic acid in the tissues, and other metabolic processes must therefore take place.

The anaerobic metabolism of fish has an interesting consequence that is related to the low carbon dioxide concentration in fish blood. The low carbon dioxide concentration means that the amount of sodium bicarbonate that serves as the main buffering substance in the blood is also very low. The addition of large amounts of lactic acid would therefore severely disturb the normal acid–base balance of the poorly buffered blood.

The observations illustrated in Figure 5.4 show one way in which a fish can counteract some of the effects of lactic acid. A trout that was made to swim vigorously for 15 minutes showed a considerable increase in the lactic acid concentration in the blood, but after it was allowed to rest the lactic acid surged to much higher levels (Black et al. 1966). Apparently, the extremely high demands on the swim-

FIGURE 5.4 The lactic acid concentration in a trout increased substantially during 15 minutes vigorous swimming. After it was allowed to rest, the blood lactic acid surged to much higher levels. Note the different time scales for swimming and recovery. [Black et al. 1966]

bursts of glycolytic function or prolonged, low-level anaerobic activity. Among the fish studied, warm-bodied species such as tuna had the greatest buffering capacities. Deep-sea fishes and shallow-living fishes with sluggish swimming habits had low buffering capacities. Marine mammals, on the other hand, had high muscle buffering capacities, and as we shall see later, diving mammals depend to a great extent on anaerobic lactic acid formation for the supply of energy to the muscles during underwater swimming (Castellini and Somero 1981).

Other pathways

We have so far discussed only glycolysis and formation of lactic acid in anaerobic metabolism; other reactions are available. Several different pathways have been established for a variety of organisms, and those invertebrates that can tolerate long periods of anoxia utilize a number of interesting processes (Hochachka 1980).

A study of goldfish, a close relative of carp, has given us further understanding of anaerobic metabolism in fish. Goldfish tolerate anoxia very well and can survive for several days in the absence of oxygen, especially if the temperature is low. The amount of glycolysis, as indicated by an increase in lactic acid in muscle and blood, is insufficient to support the metabolic rate. Furthermore, even if there is a complete lack of oxygen, some metabolic carbon dioxide is formed (Shoubridge and Hochachka 1980).

To study these problems, goldfish were first poisoned with carbon monoxide to block the oxygen transport in blood and to inhibit cytochrome oxidase so that any oxygen remaining in the body (such as in the swimbladder) would be unavailable for oxidative metabolism. Afterward the fish were moved to oxygen-free water and kept under nitrogen for 12 hours. After 12 hours the results were as shown in Table 5.2. Lactate had accumulated in the fish, but there was also a similar concentration of ethanol (ordinary ethyl alcohol) in the tissues. Furthermore, a substantial amount of ethanol was found in the water in which the fish were kept.

ming muscles were not met through oxidative metabolism, and lactic acid was formed. However, all lactic acid was not immediately released into the circulation but accumulated in the muscle, and only after the fish was allowed to rest and recover was this lactic acid released. It could then be used by other organs for oxidative metabolism and for resynthesis of glucose and glycogen, a process that normally takes place in the liver.

Measurements of the buffering capacity of the muscle itself have given some interesting results. Muscles from a variety of fish and marine mammals showed that the highest buffering capacities were found in muscles capable of either intense

Several further observations indicated that the ethanol had been formed from lactate. After the first 12 hours the lactate concentration in the fish increased only slowly, but no lactate was found in the water, and if additional lactate had been formed, it apparently had been metabolized to ethanol. To confirm this suggestion, lactate labeled with radioactive carbon (^{14}C) was injected, and 88 to 92% of the injected lactic acid was collected as ethanol.

Is there any advantage to the formation of ethanol? The ethanol is easily removed from the body because it readily diffuses out through the gills. This is wasteful because ethanol has a considerable fuel value, but on the other hand it is important for fish that the lactic acid concentration should not reach too high levels. As we saw before, fish blood has a poor bicarbonate buffer system, which in turn is a consequence of the high solubility of CO_2 in water. Transformation of lactate to ethanol, which is readily lost to the water, apparently serves as a means of avoiding acidosis, although the fish pays for this privilege in the form of excessive loss of metabolic fuel.

Many will wish to compare the ethanol concentration found in the goldfish with concentrations commonly found in humans after alcohol consumption. A human who has a blood alcohol level of 0.25% ethanol is intoxicated, and in many countries (but not all) the legal limit for automobile driving is one-fifth of this, or 0.05%. Any value lower than this should then not constitute being under the influence of alcohol; the concentration in the goldfish described in Table 5.2 was 0.024%, or one-half the legal limit.

The oxygen minimum layer

Most of the world's oceans contain, at intermediate depth, zones where the oxygen content is very low. There is a large such area in the northeastern tropical Pacific, which is known as the *oxygen minimum layer*. Off the coast of central and southern Mexico the layer extends from approximately 100 to 900 m in depth, and the dissolved oxygen content is generally less than 0.1 cm^3 per liter sea water, with relatively little variation throughout the

TABLE 5.2 Concentrations of lactate and ethanol in goldfish after 12 hours anoxia at 4 °C. [Shoubridge and Hochachka 1980]

| | Metabolite concentration (mmol per kg fish) | | |
	Tissue lactate	Tissue ethanol	Ethanol in water
Control	0.18	0	0
Anoxia	5.81	4.58	6.63

year. Nevertheless, a surprisingly abundant population of animals lives in the oxygen-poor water.

Some of the animals, particularly the fish, make daily migrations to the surface, ascending at dusk and returning to depth before sunrise, in vertical movements that amount to 300 m or more. Paradoxically, the fish present in the low-oxygen region often have gas-filled swimbladders that contain a high percentage of oxygen.

During the day, when the fish are at depth, it seems likely that they obtain energy by anaerobic pathways. Their blood has a low oxygen capacity and their hemoglobin such a strikingly low affinity for oxygen that no more than 1% of the hemoglobin would be oxygenated while the fish are at depth (Douglas et al. 1976).

Small crustaceans that live in the oxygen minimum layer are very effective at removing oxygen from the water. A study of 28 species from the oxygen minimum layer off southern California found that all the species are able to take up oxygen at least to the lowest P_{O_2} they encounter in nature, in some cases as low as 4 mm Hg. The exceptions are two parasites and one animal that makes daily vertical migrations to oxygen-rich water (Childress 1975).

At the present time, it appears that the existence of a fairly rich fauna in the oxygen minimum layer depends on the utilization of very low oxygen concentrations and, therefore, largely on aerobic metabolism. However, the exploration of the interesting biological problems of the oxygen minimum layer is just beginning, and it is not yet possible to

TABLE 5.3 Most of the major orders of mammals have a few aquatic representatives, and some are exclusively aquatic.

Mammalian order	Aquatic representatives
Monotremata	Platypus (*Ornithorhynchus*)
Marsupialia	Water opossum
Insectivora	Water shrew
Chiroptera	None
Primates	None
Rodentia	Beaver, muskrat
Lagomorpha	None
Cetacea	Whales (all)
Carnivora	
Fissipedia	Sea otter
Pinnipedia	Seals (all)
Sirenia	Sea cows (all)
Ungulata	
Perissodactyla	Tapir
Artiodactyla	Hippopotamus

TABLE 5.4 Many bird families are almost completely aquatic (except for their reproductive activities); others are primarily or exclusively nonaquatic.

Primarily aquatic	Primarily nonaquatic
Penguins	Ratites (ostrich, rhea, etc.)
Loons	Hawks
Grebes	Fowls and pheasants
Petrels and albatrosses	Pigeons
Pelicans	Parrots
Herons	Owls
Ducks	Nightjars
Waders	Hummingbirds
Gulls	Woodpeckers
Auks	Passeriformes (finches, starlings, crows, etc.)

say whether the fauna depends mostly on unique abilities to utilize low oxygen concentrations or on tolerance to long periods of anaerobic metabolism.

The latter option is not unthinkable, for several anaerobic biochemical mechanisms are potentially available. We have become conditioned to thinking of anaerobic glycolysis and lactic acid formation as the typical pathway for anaerobic metabolism because we know it is used in higher vertebrates.

However, glycolysis is a fundamentally inefficient metabolic strategy, which makes available only a few percent of the energy obtainable from complete oxidation of the substrate. Many other pathways are available for production of ATP during extended periods of anoxia. The large free amino acid pool of many invertebrates is an important potential source of energy that represents astonishing energetic advantages over lactic acid formation (Hochachka et al. 1973).

PROBLEMS OF DIVING

All aquatic mammals and birds have lungs and breathe air, but successful aquatic life requires a number of modificiations in the usual terrestrial pattern. The most obvious is that lengthy dives require effective utilization of a limited oxygen supply, but other modifications are also important.

Adaptation to an aquatic mode of life is found in all major orders of mammals, except the Chiroptera (bats), Lagomorpha (rabbits and hares), and primates (Table 5.3). Three orders are entirely aquatic: the seals, the whales, and the sea cows (which include the manatee and the dugong). A list of aquatic and nonaquatic birds shows that the successful ones are phylogenetically quite diverse (Table 5.4). However, the physiological adaptations these animals employ to cope with the special demands of their environment are similar. The facts that all divers use common mechanisms and that to a lesser degree such mechanisms are present in nondiving vertebrates suggest that the successful divers have taken advantage of and developed already existing mechanisms.

The problems that must be solved by diving animals can readily be explained by reference to the problems encountered by a person who dives, either with or without equipment. Some of the person's difficulties have no direct bearing on diving animals; others are directly applicable to them. All air-breathing animals that dive must cope with the problem of oxygen supply, but those that make fairly deep dives encounter additional problems. We can arrange these problems in five major groups, which we shall discuss in the following order:

1. The bends or diver's disease
2. Oxygen toxicity
3. Narcotic effects of gases
4. Direct effects of high pressure
5. Oxygen supply

The bends

This dangerous syndrome is known under several names: diver's disease, caisson disease, and aeroembolism. It occurs when a human diver returns to the surface after spending a prolonged time at considerable depth, below 20 m or so. It becomes increasingly severe with increasing depth and diving time. It is caused by gas bubbles in the tissues and the bloodstream, formed in the same way that bubbles form in a bottle of soda water when the cap is removed. In both cases bubbles appear when the pressure is reduced over a liquid that is saturated with gas at a higher pressure.

The bends may also occur during rapid ascent in aircraft (balloons or airplanes) with nonpressurized cabins; in these cases the danger of bends occurs when the pressure is reduced to about 0.5 atm (above 6000 m).

In soda water the dissolved gas is carbon dioxide. The gas causing the bends is always nitrogen (unless the diver has been breathing an artificial gas mixture that contains other inert gases such as helium). For the bends to develop when a diver returns to the surface requires a supersaturation with a gas at about 2 atm or more. This excludes carbon dioxide as a cause, for even when the diver is at great depth, the carbon dioxide tension does not increase much above the physiologically normal level of about 40 mm Hg. Oxygen does not cause the bends because it is rapidly used up by the tissues, and, as we shall see later, oxygen at a pressure of more than 1 atm is so toxic that higher concentrations must be avoided.

The pressure in water increases by about 1 atm

for each 10 m increase in depth. To keep a diving suit inflated with air at 10 m depth requires 1 atm pressure above normal atmospheric pressure, or a total air pressure of 2 atm. If a diver at 10 m depth is supplied with ordinary air, the air he breathes has 0.4 atm oxygen pressure and 1.6 atm nitrogen pressure. The blood dissolves nitrogen at the higher pressure, and gradually all the tissue water becomes equilibrated with nitrogen at this pressure. Furthermore, the body fat, which in a normal lean male constitutes about 15% of body weight, dissolves about five times as much nitrogen per unit weight as does water. Our standard 70-kg man contains nearly 50 liters water and 10 kg fat; the total amount of nitrogen dissolved in the fat, therefore, about equals the amount dissolved in the body water.

It takes considerable time before a person at a given depth reaches full saturation with nitrogen. During short dives this is an advantage, but after a long dive, when the tissues are nearly saturated with nitrogen at high pressure, the danger increases. When the diver returns to the surface, several hours are required to eliminate the nitrogen, for it is only slowly carried from the tissues, dissolved in the blood, and released in the lungs.

The danger of bubble formation is greater when gas is dissolved at a higher pressure. Increased movements, muscular effort, and increased circulation tend to increase bubble formation, analogous to the effect of shaking a beer can or a soda-water bottle before opening. Bubbles often form in the joints, which causes considerable pain. If bubbles occur in the bloodstream, they block the finer blood vessels, and when this happens in the central nervous system it is particularly dangerous and may cause rapid death.

The only possible treatment for the bends is to increase the pressure quickly so that the bubbles redissolve. This can be done either by returning the diver to the depth from which he ascended or by transferring him to a pressure chamber where the air can be recompressed to the desired level.

The way to prevent the bends from developing

is to return a diver slowly to the surface by stages. It is reasonably safe to let a diver ascend to a point where the pressure is half that at which he was working, and keep him there until a considerable amount of the nitrogen has been eliminated, say 20 to 30 minutes. It is then feasible to let him ascend part of the way again and continue the process of staging. Staging tables that give precise schedules for the fastest possible safe ascent from a given depth are widely used.

If a diver uses a tank of compressed air, he must breathe air at the pressure of the surrounding water or his chest would be collapsed by the water pressure. The scuba diver is therefore in the same danger of developing the bends as the diver using an inflated diving suit.

The danger of bends can to some extent be reduced by using gases other than nitrogen. One widely used gas is helium, which has the advantages of being less soluble than nitrogen in water and fat and of diffusing several times faster. The faster diffusion reduces the time necessary for staging. However, it also speeds up the saturation of the tissues with helium, and for short dives this counteracts some of the advantages.

Our discussion so far has dealt only with divers supplied with air under pressure. A skin diver without equipment who repeatedly descends to some considerable depth is also in danger of developing the bends. Each time he fills his lungs and descends, the water pressure on his chest compresses the rib cage so that the air pressure in the lungs increases. If he descends to 20 m, the air pressure in the lung will be 3 atm. Although one or a few dives to 20 m depth will not be of any significance, many repeated dives are risky.

A well-documented case of a skin diver developing the bends occurred when a Danish physician practiced underwater escape techniques in a 20-m-deep training tank. He used the so-called bottom drop technique, in which the diver after a deep breath pushes himself downward from the edge of the tank. After descending 2 or 3 m the chest is compressed so much that he continues to drop to the bottom

with increasing speed. After 5 hours of spending about 2 minutes at the bottom during each of 60 dives, he developed pain in the joints, breathing difficulties, blurred vision, and abdominal pains. He was placed in a recompression chamber at 6 atm pressure, whereupon the symptoms rapidly disappeared. He was then slowly staged to 1 atm pressure, according to the U.S. Navy staging tables. This required 19 hours and 57 minutes. The treatment was completely successful, but many other divers have not been so lucky (Paulev 1965).

If a person develops the bends from performing repeated breath-hold dives, how can seals and whales, which dive over and over again, avoid this danger? Often they come to the surface for seconds only, and they may dive to considerable depths.

The best diver among the seals is the Antarctic Weddell seal, which voluntarily dives to below 600 m and stays under water for more than 60 minutes (Kooyman et al. 1980). The deepest diving whale is the sperm whale (*Physeter catadon*), for which the known record is 1134 m. This curious record was documented because a trans-Atlantic cable was broken at this depth, and when the cable was taken up for repair, a dead sperm whale had its lower jaw entangled in the cable. The way the jaw was caught showed that the whale had been swimming along the bottom when it hit the cable and had tried to free itself.

Why can seals and whales tolerate deep dives and rapid ascent without developing the bends? There are three possibilities: (1) they are tolerant to bubbles, (2) they have mechanisms to avoid bubble formation in spite of supersaturation, or (3) there is no supersaturation. The information we have, although sparse, points to the third possibility as the likely answer.

The most important difference between seals and whales on the one hand and a human diving with equipment on the other is that the animals do not receive a continuous supply of air. However, as in humans, many repeated dives with air-filled lungs should eventually produce the bends. In this regard it is significant that a seal exhales at the beginning of each dive, rather than diving with filled lungs as a person usually does. Whether a whale also exhales before diving is uncertain, but it has an exceptionally large and wide trachea. This may be an adaptation to enable the whale to ventilate its lungs rapidly when it briefly surfaces to breathe.

Equally or more important, the lungs are small relative to the large volume of the upper airways (Scholander 1940). As a whale dives, the increasing water pressure compresses the lungs and forces air into the large trachea, which is rigidly reinforced with circular bone rings. If the volume of the noncompressible trachea is one-tenth of the lung volume, the lung will be completely collapsed and all the air forced into the trachea at 100 m depth. With no air in the lung, nitrogen cannot enter the blood, as it does in humans. Another reason nitrogen does not enter the blood is that the blood flow to the lung is minimal during the dive; thus, even if the lung contains some air before the animal reaches the depth where the lung is collapsed, little nitrogen enters the blood.

Oxygen toxicity

Pure oxygen at 1 atm pressure is harmful to most warm-blooded animals. A human can safely breathe 100% O_2 for 12 hours, but after 24 hours he feels substernal distress and increasing irritation of the lungs. Rats kept in pure oxygen die with symptoms of severe lung irritation after a few days. If the oxygen pressure is higher than 1 atm, nervous symptoms develop before the general irritation of the respiratory organs sets in. Continued breathing of oxygen at 2 atm pressure causes convulsions, and exercise reduces the tolerance. A pressure of 3 atm oxygen can be tolerated by humans at most for a few hours, and longer exposure must be avoided.

The importance of oxygen toxicity to diving is obvious. If a diver descends to 40 m and is supplied with compressed air, he breathes at a total pressure of 5 atm. Because one-fifth of the air is oxygen, the partial pressure of oxygen is 1 atm (i.e., near the toxic limit). Dives with compressed air at even greater depths are, of course, increasingly

harmful as the oxygen partial pressure exceeds 1 atm.

The only practical way for a diver working at 40 m or more to avoid oxygen toxicity is to decrease the oxygen content of the air he breathes. A diver working at 40 m (5 atm) could be supplied with a gas containing 10% O_2 and 90% N_2. He would then breathe oxygen at a partial pressure of 0.5 atm, which is completely safe. At greater depth the oxygen content of the gas mixture should be further reduced, and to reduce the danger of bends during the decompression, nitrogen may be partly or completely replaced by other gases such as helium.

Narcotic effects of inert gases

The depth to which a person can dive safely is also limited by the direct physiological effects of inert gases. Nitrogen at several atmospheres pressure has a narcotic effect, similar to that of nitrous oxide (laughing gas), but nitrogen requires higher pressures.

The narcotic effect takes some time to develop, but at a depth of about 100 m it is so severe that nitrogen–oxygen mixtures cannot be used. It is therefore necessary to use other gases, helium again being favored. Hydrogen is another possibility, although much more hazardous to handle because of the danger of explosions. Eventually, helium under pressure also has serious physiological side effects, thus limiting the depth to which it is possible to dive, even for the well-equipped person. One possibility is to use a completely rigid structure, such as a reinforced rigid diving suit or a bathyscaphe, in which a low pressure can be maintained.

Diving animals are not subject to the narcotic effects of inert gases (or to oxygen toxicity) for the simple reason that they do not breathe a continuous supply of air during the dive.

Oxygen supply during diving

For a person who swims under water without equipment, the most immediate need is to obtain oxygen. He usually fills his lungs to increase the

TABLE 5.5 Possible solutions for extending underwater time for diving animals. Solutions in brackets are not widely used or are improbable or physiologically impossible.

Increased oxygen storage
  [Increased lung size]
  Blood: larger blood volume and higher hemoglobin
    content
  Tissues: higher muscle hemoglobin
  [Unknown oxygen storage]

Use of anaerobic processes
  Lactic acid formation
  [Unknown hydrogen acceptors]

Decreased oxygen consumption
  Decreased metabolic rate

Aquatic respiration
  Cutaneous respiration (frog, sea snakes)
  Esophageal or rectal respiration (some turtles)
  [Breathing water (only tried experimentally)]

TABLE 5.6 Oxygen stores in seal and human. The figures are approximate (e.g., blood oxygen content will be lower than listed because a large part of the blood is located in the venous system and not fully saturated with oxygen). [Data for seal from Scholander 1940]

Location of oxygen	Amount of oxygen (ml)
Seal (30 kg)	
Air in lungs (350 ml, 16% O_2)	55
Blood (4.5 liters, 25 ml O_2 per 100 ml)	1125
Muscle (6 kg, 45 ml O_2 per kg)	270
Tissue water (20 liters, 5 ml O_2 per liter)	100
Total	1550
Milliliters O_2 per kilogram body weight	52
Human (70 kg)	
Air in lungs (4.5 liters, 16% O_2)	720
Blood, arterial (1½ liters, 20 ml O_2 per 100 ml)	300
Blood, venous (3½ liters, 15 ml O_2 per 100 ml)	525
Muscle (16 kg, 15 ml O_2 per kg)	240
Tissue water (40 liters, 5 ml O_2 per liter)	200
Total	1985
Milliliters O_2 per kilogram body weight	28

amount of air he carries with him, postponing the moment he must surface to breathe again. There are, however, many theoretical possibilities for extending diving time (Table 5.5). Some of these are physiologically not feasible or are not known to be used by animals. Most diving animals have adopted a combination of several possible methods, often resulting in diving performances that are striking when measured against human standards.

Oxygen storage

Increasing the amount of oxygen in the lung seems a simple solution to the problem of carrying more oxygen during the dive. However, diving animals have no greater lung volumes than other mammals, and many of the best divers have rather small lung volumes. A moment's thought helps to explain why this is so. First, a large amount of air in the lung may permit more nitrogen to invade the body fluids and thus increase the hazards of the bends. Furthermore, if the lungs are particularly large, an animal will have difficulty in initially becoming submerged. This last argument may be a minor matter, however, for once an animal with large lungs is successfully submerged, the pressure of the water compresses the lungs and decreases the buoyancy so that there would be no difficulty in remaining at depth.

Let us examine what other oxygen supplies are available to a seal at the beginning of a dive and how long this oxygen may last. Such an account has been given by the Norwegian physiologist P. F. Scholander, who compared the oxygen stores in a small seal and a human (Table 5.6).

One striking difference is the small amount of oxygen the seal carries in its lungs. Instead of filling the lungs at the beginning of a dive, seals seem to exhale and begin the dive with a minimal volume of air in the lungs. This is the reason for the very small amount of oxygen in the lungs of the seal compared with a person, who fills his lungs with air when he dives.

The amount of oxygen in the blood is a very different matter. Although the seal weighs less than half as much as the person, its total blood volume is almost as great. Furthermore, seal blood has a higher oxygen-carrying capacity than human blood, and as a result the total amount of oxygen in the seal's blood exceeds that in the person's.

A high oxygen-carrying capacity of the blood is characteristic of good divers. Although the oxygen capacity of human blood (20 ml O_2 per 100 ml blood) is in the higher range among terrestrial mammals, seals and whales have oxygen capacities ranging up to between 30 and 40 ml per 100 ml blood. There is a limit to how much the oxygen capacity can be increased, however. The red blood cells are already about as densely packed with hemoglobin as possible, and if the red cells make up much more than 60% of the total blood volume, the blood becomes so viscous that the heart must work excessively to pump the blood through the vascular system.

Increasing the blood volume is one way of increasing the total amount of hemoglobin in the body without increasing blood viscosity. This solution is common among successful divers, but again, there are limits to how much blood it is feasible to have in the body.

All body water contains some dissolved oxygen. The amount depends on the water content of the body, which is similar in human and seal, but the total amount is small. Some oxygen is dissolved in the body fat, but it is unimportant as an oxygen store because fat is poorly vascularized and probably excluded from circulation during diving.

The second largest amount of oxygen in the seal is stored in the muscle mass. The red color of mammalian muscle is caused by muscle hemoglobin (myoglobin). It is located within the muscle cells, and therefore remains even after all blood has been drained out. The hemoglobin content of muscle is much lower than that of blood, but because of the large muscle mass of the body, about 20%

of the body weight,* the muscle hemoglobin carries a substantial amount of oxygen.

Seal and whale muscles have much higher myoglobin contents and oxygen capacities than muscles of terrestrial mammals, and the oxygen held by the myoglobin constitutes a substantial reserve (Table 5.7). Because of the myoglobin content seal and whale meat is very dark, and when cooked it is almost black, nearly the same color as blood sausage.

Adding up all the oxygen we can account for in Table 5.6 shows that a seal may have, per kilogram body weight, about twice as much oxygen as a person.

The stored oxygen, if used up at the normal metabolic rate of the seal at rest, would last at most 5 minutes. However, the seal is able to dive for at least three times as long. The total oxygen store in a human would be sufficient to supply his metabolic rate at rest for perhaps 4 minutes, but this exceeds the maximum diving time for humans (about 2 minutes, except for unusual record performances). The reason is that a person cannot tolerate full depletion of all free oxygen in the body, and the maximum diving time is therefore shorter than that calculated from the total oxygen stores. Why, then, can a seal dive for much longer?

We cannot a priori exclude some unknown oxygen store in the seal, but on the basis of present knowledge, it seems most unlikely. What we know about diving mammals indicates that the performance of even the best divers can be fully accounted for without invoking new and so far unknown mechanisms.

Circulatory changes

One of the most characteristic physiological responses observed in a diving seal is that the heart rate at the beginning of submersion decreases precipitously (Figure 5.5). This sudden drop in heart

---

*This figure is low because of the large amount of fat in the seal's body. The muscles make up a much larger fraction of the fat-free body mass.

TABLE 5.7 Myoglobin concentration and oxygen capacity of skeletal muscle of terrestrial versus diving mammals. [Snyder 1983]

		Myoglobin (g/kg muscle)	Oxygen capacity (ml O_2/kg muscle)
Terrestrial	Human	6.0	8.0
	Rat	3.0	4.0
	Dog	6.7	9.0
Whales	Bottlenose dolphin	32.5	43.6
	Harbor porpoise	41.0	56.0
	Sperm whale	56.7	76.0
Seals	Harbor seal	52.1	69.8
	Weddell seal	44.6	59.8
	Bladdernose seal	41.0	55.0

FIGURE 5.5 When a trained harbor seal dived voluntarily in a pool, its heart rate dropped precipitously from the normal rate of about 150 to 40 beats per minute, and during the dive the heart rate decreased further to about 20 beats per minute. Beginning and end of the dive are indicated by the broken vertical lines. [Elsner 1965]

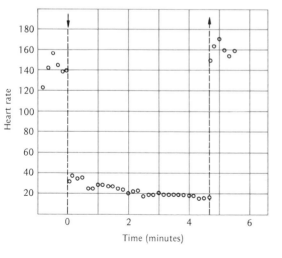

rate must be caused by a nerve reflex, for if it were a response to a gradual depletion of oxygen, it would develop more slowly.

The decrease in heart rate during submersion has become known as the *diving reflex*. It has been observed in a large number of animals, including the seal, whale, muskrat, and hippopotamus. It is very pronounced in laboratory experiments when the animals are forcibly submerged, but it also occurs in animals that are free to dive voluntarily. For example, muskrats in large holding tanks had heart rates that fell from 320 before a dive to 34 beats per minute after 20 seconds of voluntary submersion (Jones et al. 1982).

The slowing of the heart is not as pronounced in all animals, and in free-ranging birds the decrease in heart rate may be absent (Butler 1979; Kanwisher et al. 1981). There is an important difference between mammals and birds in regard to diving because birds have rather large volumes of air in their air sacs and it is probable that this air can be passed through the lung and used for oxygen supply during submersion. This in turn would require maintained blood flow to transport the oxygen to the organs where it is used.

A study of sea snakes diving under natural conditions in the sea has failed to show a sudden drop in heart rate. During dives lasting more than an hour, the heart rate decreased from the usual predive value of 30 to around 20 beats per minute, where it remained fairly constant during the dives

(Heatwole et al. 1979). However, in these observations the initial high value was probably connected with handling before the animals were released into the sea. It is likely that the maintained heart rate and circulation in the sea snake is related to the fact that a substantial part of the gas exchange during diving occurs through the skin, and that circulation therefore is maintained.

In spite of the decrease in heart rate during submersion of a diving mammal, the arterial pressure is maintained. This can happen only if the peripheral resistance in the vascular system is increased through vasoconstriction. With the reduction in cardiac output, only the most vital organs receive blood; these are the central nervous system and the heart itself. There is a complete redistribution of blood flow, and muscles as well as abdominal organs receive little or no blood during the dive. In a way, this makes the animal much smaller: The oxygen is reserved for a small part of the body – the organs for which oxygen is most essential. The oxygen present in the circulating blood at the beginning of the dive is therefore used up only slowly, and the duration of a dive approximately coincides with the time needed to deplete fully the oxygen in the blood.

Let us examine how the reduced blood flow is distributed to the major organs. Such studies cannot readily be made in free-ranging animals, and we must therefore use results obtained during submersion in the laboratory. Observations on Weddell seals (Figure 5.6) show virtually no change in the blood flow to the brain during submersion, while flow to the retina and to the lung is reduced by about one-third.

In other organs, however, the situation is very different. The blood flow to the heart muscle itself is greatly reduced. The flow to both atria and ventricles is decreased to about one-sixth of the presubmersion value. This is reasonable in view of the precipitous decrease in heart rate and thus in the work the heart must perform during submersion. The blood flow is decreased even more to the diaphragm, which of course does no work when breathing is suspended. The decrease is to onetwentieth of the normal value, similar to the drop for circulation to the pancreas, the liver, and skeletal muscle.

Most muscles, whether used vigorously or not, receive virtually no blood during a dive. After the oxygen that was present in the muscle to begin with has been used up, the work of the muscles is based on anaerobic processes, which lead to the formation of lactic acid in the muscles. Because there is no blood flow to the muscles, this lactic acid accumulates in the muscle during the dive, and the lactate concentration in the blood remains low. Immediately after the dive, when circulation to the muscles is resumed, the blood lactic acid concentration shoots up. (The ability of seal muscle to work anaerobically is not unique; lactic acid is normally formed in the muscles of people during strenuous exercise or athletic performances.)

The muscle hemoglobin in a diving seal provides some oxygen, but this oxygen is depleted early in the dive within a period much shorter than the normal duration of a dive. The amount of oxygen stored in the muscle hemoglobin is much too small to explain the long duration of a dive; the circulatory adjustments and the ability of muscle to work anaerobically are the major factors.

Experimental observations on alligators indicate that submersion causes a drop in their heart rate from more than 40 to 3 beats per minute. However, if the signals are transmitted from the animal to the recording apparatus by radio telemetry, we can obtain information from completely undisturbed animals, free of restraints and human presence.

In telemetric studies of the caiman (*Caiman crocodilus*) the animals showed typical diving bradycardia when handled, with drops from 28 to 5 beats per minute. However, in the absence of humans, the heart rates of resting animals gradually dropped over a period of several hours to 10 to 12 beats per minute. When such animals submerged voluntarily, their heart rates dropped by only 1 or 2 beats per minute, or not at all. If the investigator entered

FIGURE 5.6 Changes in blood flow to the major organs of a Weddell seal during submersion. Lightly shaded columns indicate normal blood flow; dark columns indicate blood flow during submersion. Numbers in parentheses indicate the blood flow during diving in percent of predive flow. [Data from Zapol et al. 1979]

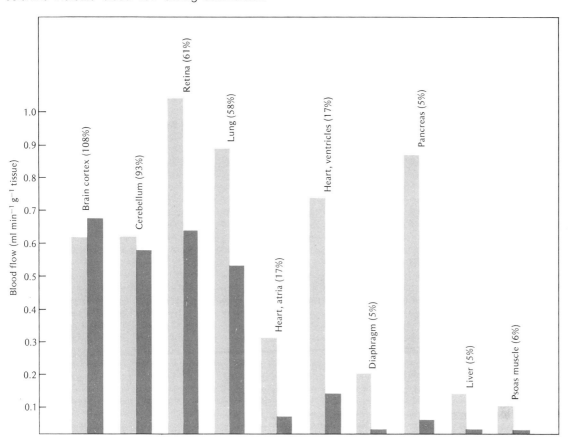

the room, on the other hand, the animals dived and their heart rates dropped to as little as 5 beats per minute (Gaunt and Gans 1969).

These results have been confirmed on alligators (*Alligator mississippiensis*) roaming free in a lake; voluntary short dives did not result in any reduction in heart rate (Smith et al. 1974). It is therefore important to make observations on animals under conditions that ensure a minimum of external disturbance.

Interestingly, a reverse "diving" bradycardia has been observed in fish. When a fish is taken out of water, it cannot obtain enough oxygen, and its heart rate immediately decreases. This has been observed in both teleosts and elasmobrachs. An air-breathing fish, such as the mud skipper (*Periophthalmus*) (see Chapter 1), differs from fully aquatic fish in that its heart rate decreases when it escapes from air into its water-filled burrow. In this respect the mud skipper reacts as a terrestrial air breather that performs a dive (Gordon et al. 1969).

A record of respiration and blood gases in a seal

subjected to an experimental submersion is shown in Figure 5.7. The top graph shows that the blood oxygen gradually is depleted during the dive, while the carbon dioxide correspondingly increases. The blood lactic acid changes only a little during submersion, but a wave of lactic acid appears in the blood as soon as air breathing is resumed. This lactic acid, which signifies an oxygen debt, is later gradually removed from the blood. Part of the lactic acid is resynthesized into glycogen by the liver and muscles, and a smaller fraction is completely oxidized to carbon dioxide and water.

The center graph (Figure 5.7*b*) shows the ventilation volume, which during submersion is zero, followed by increased ventilation of the lung immediately afterward.

The lower graph (Figure 5.7*c*) shows the corresponding respiratory oxygen uptake, which, immediately after breathing is resumed, shows a sharp increase. Before submersion the elimination of carbon dioxide is similar to but a little lower than the oxygen uptake. Immediately after the submersion period there is a surge in the output of carbon dioxide. This is because the lactic acid that enters the circulation from the muscles drives out carbon dioxide from the bicarbonate in the blood, as vinegar poured on baking soda drives off carbon dioxide.

Decreased metabolic rate

Figure 5.7*c* shows that the oxygen debt acquired during the dive is followed by an increased oxygen uptake. The expected oxygen debt is represented by the shaded area (which indicates the normal metabolic rate of the seal). The increased oxygen uptake after the dive indicates that the oxygen debt acquired during the dive is being paid off, and the after-dive increase should equal the debt. If the area under the after-dive oxygen uptake curve is compared with the oxygen debt, however, the amount repaid seems smaller. This could be caused by a long extension in time of the processes used to repay the oxygen debt that cannot easily be distinguished from normal small variations in resting metabolic rate. Or there may be a real decrease in metabolic rate during the dive, so that the true oxygen debt is smaller than suggested by the graph.

Further comparisons of the after-dive oxygen uptake and the expected oxygen debt show that a reduced metabolic rate is the probable explanation. In the duck the observed after-dive increase in oxygen consumption is about one-quarter or one-third of the expected oxygen debt; even if a duck is exposed to a series of repeated dives, it shows no signs of an accumulationg oxygen debt.

A decrease in metabolism to values below the normal resting rates is unexpected, for we are accustomed to thinking of metabolic rate as being very constant. This is not necessarily so. We should remember that during the dive many organs are poorly perfused with blood or not perfused at all. The kidneys, which normally have a high blood flow and a high oxygen consumption, have virtually no blood flow during a dive, and there is an immediate cessation of glomerular filtration and of urine production. Other organs are also virtually excluded from circulation during diving, and we can expect that their activity practically ceases, thus decreasing the amount of oxygen needed to repay an oxygen debt after the dive.

It is not easy to measure the *true* metabolic rate during the dive of a mammal or a bird. The three methods we have for determining metabolic rates are all inapplicable. The respiratory oxygen uptake is zero, determination of food uptake is useless, and the rate of heat production cannot be measured within the short time period of the dive. However, aquatic turtles are highly tolerant to prolonged submersion and decreased oxygen, and their heat production can more readily be measured.

In a diving turtle the metabolic rate gradually decreases, as shown by the dashed curve in Figure 5.8. After 2 hours' submersion, metabolic heat production has reached a minimum value that is similar to the level obtained if the turtle is submerged after breathing nitrogen. This heat production indicates the level of totally anaerobic processes sustained in the absence of oxygen.

FIGURE 5.7 Record of respiratory exchange in a 29-kg seal during a 12-minute experimental submersion: (a) blood concentrations of oxygen, carbon dioxide, and lactic acid; (b) ventilation volume; (c) respiratory oxygen uptake and carbon dioxide release. [Scholander 1940]

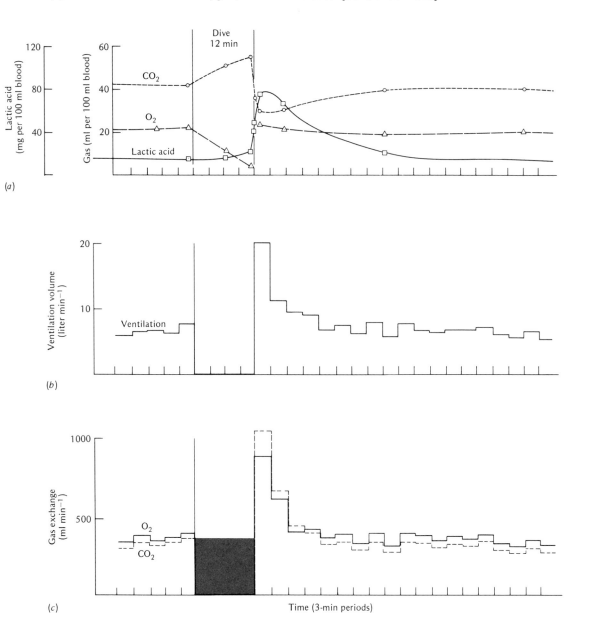

FIGURE 5.8 A diving turtle has a true metabolic rate that is indicated by its heat production. During the dive, the metabolic rate of a turtle that has been breathing air gradually decreases to a level of about 0.06 kcal per kilogram body weight per hour, the same level as when the turtle has been breathing pure nitrogen before the dive. [Jackson and Schmidt-Nielsen 1966]

If, however, the turtle is given pure oxygen to breathe before a dive, its heat production during submersion is initially very high and gradually decreases as the oxygen is depleted. In the turtle, then, metabolic rate is clearly dependent on the available oxygen and gradually decreases as the oxygen supply is depleted. These results are fully in accord with the fact that the oxygen consumption of resting muscle is not constant, but depends on blood flow (i.e., the oxygen supply) (Whalen et al. 1973).

Cutaneous and rectal respiration

We have already seen how important cutaneous respiration is in frogs and how lungless salamanders obtain all their oxygen through the skin (Chapter 1). There is no reason to believe that a diving bird or mammal uses a similar route to cover any substantial part of its need for oxygen. For one thing, these are large animals, and their relative surface area is therefore small. Further, they are warm-blooded animals and their need for oxygen is high. Finally, the nature of their skin is very different from the thin, moist, highly vascularized integument of amphibians.

Diving reptiles may be a different matter. Aquatic turtles have a thick and probably impermeable carapace, but they may use the oral mucosa as an aid in gas exchange when they remain submerged for long periods of time. Some turtles take water into their rectum, but how important this is for their gas exchange is uncertain; it is more probable that they use the water to adjust their buoyancy. To reduce buoyancy the animal would expel air from the lungs, but as the carapace is rigid, the reduction in volume can be compensated for by taking in water (or by drawing the limbs into the carapace).

Marine turtles are, of course, excellent swimmers and divers. Nevertheless, it was a surprise when it was reported that green turtles (*Chelonia mydas*) can overwinter, submerged and in a dormant state, at the bottom of the Gulf of California (Felger et al. 1976). The turtles are found at a depth of 10 to 15 m or more and may remain immobile for several months. Once fishermen have discovered the existence of these turtles and how easily they can be located with modern gear, the inevitable result will be overfishing and rapid depletion of these unique animals. Whether the green turtle, while submerged, depends on anaerobic processes or can obtain enough oxygen for its low needs during dormancy is a subject that should be studied.

Sea snakes (*Pelamis platurus*), which prey on small fish and dive to depths of perhaps 20 m, remain for long periods of time floating at the surface, but with the head submerged while looking for prey. In this animal up to one-third of the oxygen uptake can take place through the skin, and over 90% of the carbon dioxide leaves via this route. This not only prolongs the time the snake can remain immobile with the head submerged, but also is useful when the snake subdues and swallows large prey, a process that may require up to 20 minutes and prevents air respiration (Graham 1974).

Liquid breathing

What would happen if a terrestrial animal attempted to breathe water? We know that for a human this is impossible; he drowns. We also know that 1 liter of water contains only a small fraction of the amount of oxygen contained in 1 liter of air. It is therefore reasonable to ask what would happen if we were to increase the amount of oxygen dissolved in water, which can be achieved by equilibrating the water with oxygen under high pressure.

This has been tried. If water is equilibrated with pure oxygen at 8 atm pressure, it will dissolve about 200 ml O_2 per liter (at 37°C); in other words, 1 liter of this water contains about the same amount of oxygen as 1 liter of ordinary air.

Another measure must also be taken. Instead of using water to breathe, we must use a balanced salt solution similar to blood plasma. If just water were used, rapid osmotic uptake of water into the blood in the lung would occur and cause hemolysis of the red corpuscles and serious loss of salts. In fact, the actual cause of death in fresh-water drowning is usually this effect of inhaled water. For this reason, artificial respiration is often ineffective in saving victims of fresh-water drowning and more successful in cases of salt-water drowning.

Complete submersion in oxygen-supersaturated water has been tried with success in both mice and dogs, and the animals have survived for several hours. Mice introduced into a pressure chamber and submerged in water supersaturated with oxygen first try to reach the surface. But if they are kept under, they inhale the oxygenated liquid and instead of drowning continue to breathe the oxygenated saline solution.

A major drawback to liquid breathing is that water is approximately 50 times more viscous than air, and the work of breathing is correspondingly increased. Another problem is that the normal surfactants that line the lungs are washed away during liquid breathing; this causes no difficulty during the experiment, but after return to air breathing the lungs tend to collapse.

A more important limitation on liquid breathing is caused by the need to eliminate carbon dioxide at the same rate as oxygen is taken up. If the P_{CO_2} of blood is to be maintained at the normal 40 mm Hg, several times as much liquid must be moved in and out of the lungs. Animals used in experiments with liquid breathing show this quite clearly. Although the arterial blood is fully saturated with oxygen, the carbon dioxide concentration increases greatly. This problem will be difficult to overcome if an aqueous solution is used (Kylstra 1968).

Another approach to liquid breathing involves the use of certain synthetic liquids, such as fluorocarbons, instead of water. Oxygen is extremely soluble in some of these organic liquids and this alleviates the problem of saturation at high oxygen pressure. The solubility of carbon dioxide, however, is not as great, and this will cause difficulties in the practical application of fluorocarbons. Such liquids have been tried with animals, which have survived several hours' breathing without complications. For humans, however, the scheme does not seem promising for applications to diving.

Liquid breathing has one potential advantage that under certain very limited circumstances could be of value. Because only liquid and no gas is inhaled, a carrier gas such as nitrogen or helium is not needed. Theoretically, this could allow liquid breathing at very high pressures without supersaturating the body fluids with an inert gas, and the danger of bends could therefore in theory be avoided. The practical use of this possible advantage should be very limited.

Effects of high pressure

Human divers can operate fairly routinely down to about 200 m depth (20 atm), provided they use special gas mixtures such as helium–oxygen. Experimental dives are being carried out down to about 600 m, and this range may be extended by the use of other gas mixtures, including hydrogen.

We have discussed the long periods of decompression necessary to avoid the bends and the narcotic and other neurological effects of gases that are inert at moderate pressures. In addition, it is possible that high pressure per se may have deleterious effects, unrelated to the effects of the gases. For an organism breathing air, there is no simple way to distinguish direct pressure effects from the effects of the gases, for the two variables cannot be varied independently. The question will therefore not be simple to resolve.

How important is the problem of high pressure for diving seals and whales, which, as we have seen, may attain depths exceeding 600 m (Weddell seal) and 1100 m (sperm whale). Based on what we know, these animals do not suffer from bends because their tissues do not become heavily supersaturated with nitrogen. However, chemists and biochemists are well aware that chemical reactions and the structure of complex molecules are greatly influenced by pressure, and it was earlier believed that the immense pressures in the deeper parts of the oceans precluded all animal life. As exploration of the ocean bottom has proceeded, animal life has been found at even the greatest depths, in excess of 10 000 m (1000 atm).

If shallow-water animals are exposed to high pressures under otherwise normal conditions of temperature and oxygen supply, responses vary greatly from species to species. Many are stimulated to increased oxygen consumption, and often increased activity, by pressures of 50 or 100 atm. Higher pressures, of a few hundred atmospheres, often are inhibitory and may cause death (MacDonald 1975). It is quite conceivable that seals and whales are unaffected by hydrostatic pressures per se, for the deepest recorded dives are exceptional and probably rare events. When human divers operate close to the present limit, around 60 atm, it is possible that direct pressure effects may become important.

Many of the basic effects of pressure on biological systems can be referred to as the effects of pressure on protein structure and to the effect on the ionization of weak acids. Increases in pressure tend to favor the dissociation of weak acids and bases, and pressure also affects the velocity constants of chemical reactions.

The biologically most important effect on proteins is probably on their tertiary structure – the way in which a protein molecule folds upon itself, with weak bonds and other weak interactions providing stabilization. Pressure has a considerable effect on the tertiary structure, which in turn is responsible for many of the biologically important characteristics of protein molecules, enzymes, contractile proteins of muscle, structural proteins in membranes, and so on (Somero and Hochachka 1976).

In the intact organism these effects are likely to be expressed as derangements in nerve transmission and the nervous system, muscle contraction, oxygen transport in the blood, membrane function, and ion transport mechanisms. The complex biochemical processes of gene expression and transcription through the synthesis of messenger RNA are extremely pressure-sensitive, and embryonic development will therefore be seriously affected.

Organisms that live at abyssal depth in the oceans are physiologically adapted to function normally at the high pressures at which they normally live. The functional adaptations of their component proteins, enzymes, and other systems are currently the subject of intensive studies.

Experiments on live animals are technically very demanding because organisms from great depth are difficult to collect and must be kept cool and if possible at high pressure throughout the transportation to the surface and during experimentation. It is not easy to carry out experiments inside a closed container that must be able to resist pressures of several hundred atmospheres!

METABOLIC RATE AND BODY SIZE

Mammals

Let us return to the discussion of diving mammals. Large whales can remain submerged for a couple of hours or so, but a very small diver, such as the water shrew, makes dives that rarely exceed half a minute. On the whole, the larger a diving mammal, the longer lasting are the dives it can perform. Why this difference? The answer is that the rate of oxygen consumption relative to the body

TABLE 5.8 Observed rates of oxygen consumption in mammals of various body sizes.

Animal	Body mass, M_b (kg)	Total O_2 consumption, $\dot{V}_{O_2}$ (liter O_2 h^{-1})	O_2 consumption per kilogram, $\dot{V}_{O_2}/M_b$ (liter O_2 kg^{-1} h^{-1})
Shrew[a]	0.004 8	0.035 5	7.40
Harvest mouse[b]	0.009 0	0.022 5	2.50
Kangaroo mouse[c]	0.015 2	0.027 3	1.80
Mouse[d]	0.025	0.041	1.65
Ground squirrel[e]	0.096	0.09	1.03
Rat[d]	0.290	0.25	0.87
Cat[d]	2.5	1.70	0.68
Dog[d]	11.7	3.87	0.33
Sheep[d]	42.7	9.59	0.22
Man[d]	70	14.76	0.21
Horse[d]	650	71.10	0.11
Elephant[d]	3 833	268.00	0.07

[a] Hawkins et al. (1960). [b] Pearson (1960). [c] Bartholomew and MacMillen (1961). [d] Brody (1945). [e] Hudson (1962).

size is much higher in the small mammal than in the larger mammal. Let us compare small and large animals in this regard.

The rates of oxygen consumption of a variety of mammals are listed in Table 5.8. The largest land animal, the elephant, is a million times larger than the smallest, the shrew, and its total oxygen consumption obviously must be much higher. However, we do not get a fair assessment of the two animals by comparing their total oxygen consumption. If instead we calculate the rate of oxygen consumption per unit body mass, the *specific oxygen consumption** (last column), we get a striking picture of the relationship between body size and oxygen consumption.

We see that the rate of oxygen consumption per gram decreases consistently with increasing body size. This is even more apparent if the data are plotted graphically (Figure 5.9). In this graph the abscissa has a logarithmic scale; otherwise all val-

*The word *specific* used before the name of a physical quantity means "divided by mass." Thus, *specific oxygen consumption* means "oxygen consumption per unit mass."

ues for the medium-size to small animals would be crowded together at the left part of the graph. The ordinate, on the other hand, has a linear scale, and we can see that 1 g of shrew tissue consumes oxygen at a rate some 100-fold as great as 1 g elephant tissue. This tremendous increase in oxygen consumption of the small animal necessitates that the oxygen supply, and hence the blood flow, to 1 g tissue be about 100 times greater in the shrew than in the elephant. Other physiological variables – heart function, respiration, food intake, and so on – must be similarly affected.

If we plot the data from Table 5.8 on a graph with a logarithmic scale on both coordinates, the points fall more or less along a single straight line (Figure 5.10). This regression line represents the generalization that the oxygen consumption of mammals per unit body mass decreases regularly with increasing body size; it also gives a quantitative expression of the magnitude of the decrease.

We can use this graph to read off the "expected" oxygen consumption for a mammal of any given size. For example, a typical mammal of 1 kg size can be expected to have an oxygen consump-

FIGURE 5.9 Observed rates of specific oxygen consumption of various mammals. The oxygen consumption per unit body mass increases rapidly with decreasing body size. Note that the abscissa has a logarithmic scale and the ordinate an arithmetic scale. [Data from Table 5.8]

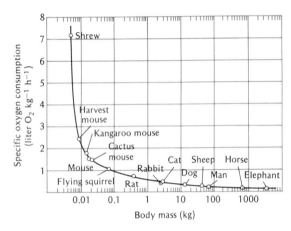

FIGURE 5.10 The rates of oxygen consumption of various mammals, when calculated per unit body mass and plotted against body mass on logarithmic coordinates, tend to fall along a straight regression line with a slope of −0.25.

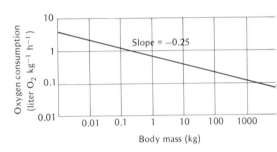

but these differences are too small to be statistically valid (Kleiber 1961).

It is easier to understand the meaning of the above equation if it is written in the logarithmic form:

$$\log \dot{V}_{O_2}/M_b = \log 0.676 - 0.25 \log M_b$$

This equation is of the general form $y = a + bx$, which represents a straight line with the slope b. In the metabolic equation above the slope is negative ($b = -0.25$) (see also Appendix C).

If we return to Table 5.8 and consider the oxygen consumption of the entire animal (column 2) we find, of course, that a whole elephant consumes more oxygen than a mouse. If the data are plotted on logarithmic coordinates, we obtain a straight line with a positive slope of 0.75. The equation for the regression line is:

$$\dot{V}_{O_2} = 0.676 \, M_b^{0.75}$$

This equation can be derived from the preceding equation by multiplying both sides by body mass ($M_b^{1.0}$). In its logarithmic form the equation is:

$$\log \dot{V}_{O_2} = \log 0.676 + 0.75 \log M_b,$$

Data that relate oxygen consumption to body size have been compiled for all sorts of animals, both vertebrate and invertebrate. Oxygen consumption is mostly lower for the cold-blooded than for the

tion of about 0.7 liter O_2 per hour. If we find that a mink which weighs 1 kg has a rate of oxygen consumption twice as high as indicated for a 1-kg mammal on the graph, we say that the mink has a higher oxygen consumption than expected for a typical mammal of its size.

Instead of using the graph, the regression line can be represented by the equation:

$$\dot{V}_{O_2}/M_b = 0.676 \times M_b^{-0.25}$$

in which $\dot{V}_{O_2}/M_b$ is the specific oxygen consumption in liters O_2 per kilogram body mass per hour (liter O_2 kg^{-1} h^{-1}) and M_b is body mass in kilograms. This general equation is based on numerous observations on a much greater variety of mammals than is listed in Table 5.8 and ranging in size from a few grams to several tons. Observations on a single species may deviate more or less from the line, which only represents the best fitting straight line calculated from all available data. Various authors have arrived at slightly different numerical constants (e.g., an exponent or slope of −0.27),

BODY SIZE This high-speed photograph of a shrew, attacking a cockroach thrown to it for food, gives a good impression of the relative sizes of one of the smallest mammals and a moderately large insect. [Courtesy of Peter Morrison, University of Alaska; from *Physiological Zoology* 32:263, 1959, © University of Chicago Press]

warm-blooded vertebrates, but the data again tend to fall on straight lines with slopes of about 0.75 (Figure 5.11).* Many invertebrate animals also have rates of oxygen consumption that fall on the same or similar regression lines, although there are some exceptions to this general rule. For example, some insects, pulmonate snails, and a few other groups have oxygen consumption rates that fall on regression lines with a slope closer to 1.0. A slope of 1.0 means that the rate of oxygen consumption is directly proportional to the body mass (i.e., an animal twice as big consumes twice as much oxygen, etc.).

The oxygen consumption rates of microorganisms also fall on lines with a similar slope, and even some trees show the same relationship between oxygen consumption and size (Hemmingsen 1960). The fact that oxygen consumption bears the same relationship to size in so many different organisms suggests that the phenomenon represents a general biological rule.

Several physiologists who have worked with these problems have attempted to reach a rational explanation for the regular relationship between oxygen consumption, or metabolic rate, and body size.

Nearly 100 years ago the German physiologist Max Rubner examined the metabolic rates of dogs of various sizes. He found that smaller dogs had a higher metabolic rate per unit body mass than larger dogs, a finding entirely in accord with the above discussion. Rubner had the clever idea that this relationship might be attributable to the fact that a smaller animal has a larger body surface relative to the body mass than a large animal.

Because small and large dogs have the same body temperature, in order to keep warm they must produce metabolic heat in relation to their heat loss. Small dogs, because of their larger relative surface, must therefore produce a greater amount of heat per unit body mass. Rubner then calculated the heat production per square meter body surface and found that it was approximately 1000 kcal per 1 m^2 body surface per day, in both large and small dogs. He therefore thought that his theory had been confirmed and that metabolic rate was determined

---

*Although the metabolic rate of an organism is most often measured as the rate of oxygen consumption, many authors recalculate their observations to heat production, most frequently by equating 1 liter oxygen (STP) to 4.8 kcal. The metabolic rates shown in Figures 5.11, 5.12, and 5.13 carry on the left ordinate the notation in kilocalories per hour and on the right ordinate the corresponding scale in liters of oxygen per hour.

The SI unit for energy (in this case heat) is the joule (J), and 1 cal = 4.184 J. A heat production of 1 kcal h^{-1} = 1.1622 J s^{-1} or (because 1 J s^{-1} = 1 W) 1.1622 W. A metabolic rate of 1 liter O_2 per hour, if we assume 4.8 kcal per liter O_2, equals 5.5787 W. For further use of conversion factors, see Appendix A.

FIGURE 5.11 The rates of oxygen consumption for a wide variety of organisms when plotted against body mass (log coordinates) tend to fall along regression lines with a slope of 0.75. Note that each division on the coordinates signifies a 1000-fold change. [Hemmingsen 1960]

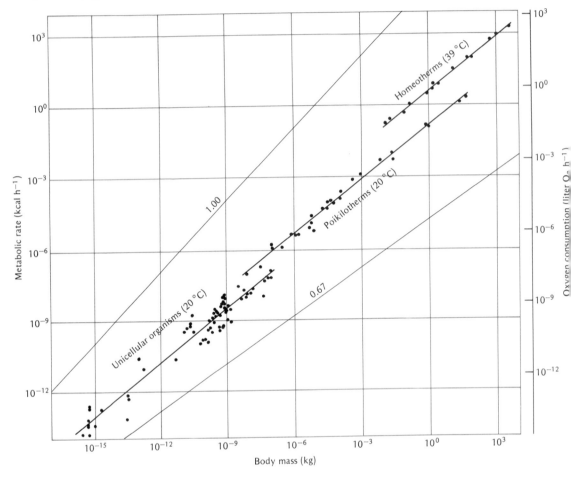

by surface area and the need to keep warm, a conclusion that became known as *Rubner's surface rule.*

Unfortunately, the slopes of the regression lines in Figure 5.11 cannot be explained by the need to compensate for heat loss, for temperature regulation is not a problem for fish or crabs or beech trees, which show the same relationship between metabolic rate and body size as mammals do. Furthermore, if the metabolic rate were really proportional to surface, the slopes of the regression lines

in Figure 5.11 should be 0.67.* Actually the slopes are closer to 0.75. The fact that cold-blooded ver-

---

*If two bodies of different size are geometrically similar their surface areas will be related as the square of a corresponding linear dimension, and their volumes as the cube of the linear dimension. Their areas will therefore be related as their volumes raised to the power 2/3, or 0.67. A few minutes spent in relating edges, surfaces, and volumes of cubes of different size will clarify this rule, which applies to any geometrically similar bodies.

tebrates as well as many invertebrates (and at least some plants) have regression lines with the same slope excludes the possibility that temperature regulation is a primary cause of the regularity of the regression lines.

The regular relationship between metabolic rate and body size, and the value of 0.75 for the slope of the regression lines, are not easy to explain. We can say, however, that it would be virtually impossible to design mammals of widely different body sizes that would follow a metabolic regression line with a slope of 1.0 (i.e., with metabolic rates directly proportional to body mass). Kleiber (1961) has calculated that if a steer were designed with the weight-specific metabolic rate of a mouse, to dissipate heat at the rate it is produced, the steer's surface temperature would have to be well above the boiling point. Conversely, if a mouse were designed with the weight-specific metabolic rate of a steer, to keep warm it would need as insulation a fur at least 20 cm thick.

It is important to realize that many physiological processes, not only heat loss, are functions of the surface area. In fact, to design a workable organism it would be necessary to include a careful consideration of surface areas. A large number of physiological processes are surface-related: The uptake of oxygen in the lungs or in the gills depends on the area of these organs; the diffusion of oxygen from the blood to the tissues takes place across the capillary walls, again a surface function; food uptake in the intestine depends on the surface area of the intestine; and so on. In fact, all cells have surfaces through which nutrients and oxygen must center the cell and metabolic products leave.

The cells maintain a different ionic composition from the extracellular fluid, and the difference must again be maintained by surface- or area-related processes. It is therefore easy to understand that overall metabolism cannot be independent of surface considerations; it is more difficult to explain why it deviates in such a regular fashion that the slope of the regression lines we have discussed often is 0.75 or very close to this value.

Birds

We have discussed the metabolic rates of mammals, and it would be interesting to know whether the other major group of warm-blooded vertebrates, the birds, have similar metabolic rates. Passerine birds (sparrows, finches, crows, etc.) mostly have somewhat higher metabolic rates that nonpasserine birds, and we will therefore divide the birds into two separate groups. Metabolic rates have been compiled for 58 species of nonpasserine birds, ranging in size from a 0.003-kg hummingbird to the 100-kg ostrich (Figure 5.12). The equation for the regression line corresponds to

$$\dot{V}_{O_2} = 0.679 \times M_b^{0.723}$$

This equation is similar to that for mammals, $\dot{V} = 0.676 \times M_b^{0.75}$ ($\dot{V}_{O_2}$ in liters O_2 per hour and M_b in kilograms). This similarity between mammals and birds means that a mammal and a nonpasserine bird of the same body size are likely to have nearly the same metabolic rates.

The available data for 36 species of passerine birds, ranging in size from 0.006 to 0.866 kg, give the equation

$$\dot{V}_{O_2} = 1.11 \times M_b^{0.724}$$

The slope of the regression lines for passerine and nonpasserine birds thus is the same, but the metabolic rate of a passerine bird can be expected to be higher than that of a nonpasserine bird of the same size by an amount of about 65% (the difference between 1.11 and 0.679). Of course, any individual species may deviate more or less from the regression lines, which represent the best fit for all available data (Lasiewski and Dawson 1967).

Marsupials and monotremes

The body temperatures of most marsupials (about 35°C) are somewhat lower than those of eutherian mammals (about 38°C). This fact has been widely quoted and is often taken to mean that marsupials are physiologically intermediate between the "lower" monotremes (which have even lower body

FIGURE 5.12 The resting oxygen consumption (metabolic rate) of birds increases with body size. The slope shown here indicates that for a non-passerine bird of any given size, oxygen consumption is similar to that of a mammal of the same size. Passerine birds as a group have higher metabolic rates. [Lasiewski and Dawson 1967]

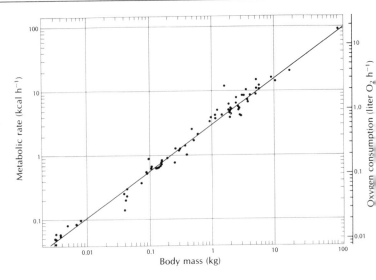

temperatures, about 30°C) and the ''higher'' eutherian mammals.

There is no a priori reason to regard a lower body temperature as physiologically inferior, for marsupials regulate their body temperature effectively over a wide range of external conditions. If a low body temperature indicated ''lower'' in the sense of ''inferior,'' birds would be ''superior'' to mammals because they have higher body temperatures (mostly about 40 to 42°C).

These differences in body temperature make it interesting to compare the energy metabolism of the various groups and see whether marsupials as a whole adhere to a pattern similar to that of other warm-blooded vertebrates. A study of Australian marsupials, ranging in size from 0.009 to 54 kg, showed that their metabolic rates varied with body size in a way similar to eutherian mammals, but at an overall somewhat lower level (Dawson and Hulbert 1970).

Expressed with the same units used for eutherian mammals ($\dot{V}_{O_2}$ in liters O_2 per hour and M_b in kilograms), the equation for marsupials is

$$\dot{V}_{O_2} = 0.409 \times M_b^{0.75}$$

FIGURE 5.13 The resting metabolic rate of marsupials is consistently somewhat lower than that of eutherian mammals. The increase with increasing body size, however, follows regression lines of identical slope. [Dawson and Hulbert 1970]

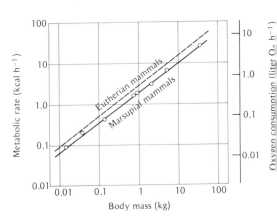

The data fall on a line with the same slope as the general equation for eutherian mammals ($\dot{V}_{O_2} = 0.676 \times M_b^{0.75}$), as shown in Figure 5.13. The consistently lower metabolism of the marsupials is evident. It is interesting that this lower metabolism of marsupials is associated with a 3°C lower body temperature. However, it is not possible to say whether the lower body temperature is a result of the lower metabolism or vice versa.

If we estimate what the metabolism of marsupials would be in the event their body temperature were raised to 38°C, we find that their "corrected" metabolic rate would be similar to that of eutherian mammals. This calculation,* carried out for the three major groups of mammals and for birds, shows that monotremes, marsupials, eutherian mammals, and nonpasserine birds all have similar temperature-corrected metabolic rates. Passerine birds, however, fall distinctly outside the common range; their metabolic rates are, on the whole, more than 50% higher, even if recalculated to a lower body temperature than they normally have (Dawson and Hulbert 1970).

SIZE AND PROBLEMS OF SCALING

In the preceding discussions we have related metabolic rates to body size, and we have derived general equations that represent this relationship. These equations express the consequences of a change in size, or in scale, for the metabolic rates of animals.

The study of the consequences of a change in size or scale is an important aspect of engineering science – so important that scaling is considered a separate discipline. The engineer meets problems of scaling every day when he builds taller buildings, longer bridges, and bigger ships, or changes from a small-scale working model to a full-size product.

In physiology, we can define *scaling* as the structural and functional consequences of a change in size or in scale among similarly organized animals (Schmidt-Nielsen, 1984). As we examine available information, we find a multiplicity of functional variables that are related to body size. We have already discussed some of the constraints

*For information on how to calculate the effect of temperature change on metabolic rate, see Chapter 6.

on the rate of metabolism that prevail for a mammal of a given size; many other constraints exist on how an animal of a given size must be designed in order to function well. For example, it is familiar to all of us that the bones of a large animal, such as an elephant, are proportionately heavier than those of a small mammal, say a mouse. The reason is simple, the bones must be able to support the weight of an animal, which increases with the third power of the linear dimensions. To support the increased weight, the diameter of the bones of the large animal must therefore be given a disproportionate increase.

Other structural and functional variables are also scaled in relation to body size. For example, although the heart of a horse weighs far more than that of a mouse, in both species the heart is about 0.6% of the total body mass (M_b). If we express this in a general equation, it would be:

$$\text{heart weight} = 0.006 \, M_b^{1.0}$$

This equation, which is based on the measurement of hundreds of heart weights of different mammals, shows that the weight of the mammalian heart is proportional to the body weight (exponent of 1.0). There is, of course, still room for a certain margin of variation, but the equation gives the general trend.

Similarly, the lung volume of mammals is scaled in simple proportion to body size. The regression line has a slope of 1.02, which means that in all mammals the lungs constitute very nearly the same proportion of the body. As a general rule, then, all mammalian lungs are scaled similarly and have a volume of 63 ml per kilogram body weight (again with a certain margin of variation).

Table 5.9 shows a number of physiological variables scaled to body size by similar equations. As we saw above, lung volume is directly related to body size. The tidal volume at rest is also directly related to body size (Table 5.9, line 5). If we divide the equation for tidal volume by the equation

TABLE 5.9 Relationship for mammals between physiological variables and body mass (M_b in kilograms).[a] [Data selected from Adolph 1949; Drorbaugh 1960; Stahl 1967]

O_2 consumption (liter h^{-1})	=	$0.676 \times M_b^{0.75}$
O_2 consumption per kilogram (liter h^{-1} kg^{-1})	=	$0.676 \times M_b^{-0.25}$
Lung ventilation rate (liter h^{-1})	=	$20.0 \times M_b^{0.75}$
Lung volume (liter)	=	$0.063 \times M_b^{1.02}$
Tidal volume (liter)	=	$0.0062 \times M_b^{1.01}$
Blood volume (liter)	=	$0.055 \times M_b^{0.99}$
Heart weight (kg)	=	$0.0058 \times M_b^{0.99}$
Respiration frequency (min^{-1})	=	$53.5 \times M_b^{-0.26}$
Heart rate (min^{-1})	=	$241 \times M_b^{-0.25}$

[a] If equations listed in this text are compared with similar equations given elsewhere, it is necessary to pay close attention to the units used. In this chapter the units are consistently liters O_2, hours, and kilograms. Recalculations between liters and milliliters and between hours, minutes, and seconds are a matter of simple arithmetic. However, if the body mass is expressed in grams instead of kilograms, the conversion is more complex. As an example, consider the equation for metabolic rate $\dot{V}_{O_2} = 3.8\,M_b^{0.75}$ with the units milliliters, hours, and grams. If we convert the equation for use with kilograms, what will the coefficient 3.8 become? To insert kilograms, we must *divide* the gram mass (M_b) by 1000, but because the number 1000 is afterward raised to the 0.75 power, we must *multiply* the coefficient 3.8 by the same number ($1000^{0.75}$ or 177.83). This gives the equation $\dot{V}_{O_2} = 3.8 \times 177.83 \times M_b^{0.75} = 676\,M_b^{0.75}$ (ml, h, kg), or $0.676\,M_b^{0.75}$ (liter, h, kg). These last units are those used in all equations in this chapter.

for lung volume, we obtain an expression for the tidal volume relative to lung volume. The body size exponents are so close that the residual exponent of 0.01 obtained in this division is insignificant compared with the inherent inaccuracy of the measurements used to establish the original equations. We thus obtain the value 0.0062:0.063, or very nearly 0.1.

This dimensionless number predicts that mammals in general will have a tidal volume at rest of one-tenth their lung volume, irrespective of their body size. Although there may be deviations from this general pattern, the statement has a predictive value that is useful. We can consider a rat or a dog

or any other mammal and predict expected physiological parameters for them. We can then examine our measurements to see how they fit the general pattern or deviate from it.

When we continue to examine the equations in Table 5.9 we can see other interesting relationships of predictive value. The blood volume of mammals, for example, is 5.5% of the body mass, irrespective of body size (the exponent 0.99 can be considered as insignificantly different from 1.0). As we saw before, heart weight (the size of the circulatory pump) is also proportional to body mass.

Because the blood supplies oxygen to the tissues at the rate required by metabolism, which is scaled to the 0.75 power of body size, the rate of the pump's beat (heart rate) must be adjusted accordingly, and we do indeed find this to be the case. The heart rate of the small animal is very high and decreases with body size as indicated by the negative slope of -0.25 of the regression line (Table 5.9, last line). This shows that heart rate, and thus cardiac output, is indeed scaled to pump blood at the rate required to supply the oxygen needed for the metabolic rate.

The equations listed in Table 5.9 relate to variables that are either directly proportional to body size (exponent = 1.0) or are scaled to the metabolic rate (exponent = 0.75). We cannot assume that all anatomical and physiological variables are scaled in one of these two ways; indeed, some differ substantially. For example, for the size of kidney and liver the equations are:

$$\text{kidney mass} = 0.021 \times M_b^{0.85}$$
$$\text{liver mass} = 0.082 \times M_b^{0.87}$$

Both these organs are metabolically very important but they are not scaled in proportion to body mass; they are relatively smaller in the larger animal. However, they are not scaled as metabolic rate either, for their size increases with an exponent that exceeds the 0.75 power for metabolic rate. We can therefore conclude that the scaling of animals, their organs, and their physiological functions is not always a simple function of body size and that more complex considerations enter into what is necessary to form an integrated and well-functioning organism.

ENERGY COST OF LOCOMOTION

Swimming, running, and flying require more energy than sitting still, but how do the different kinds of locomotion compare? Walking and running are more familiar to us than swimming and flying, and of more immediate interest, and we therefore know more about these ways of moving. Also, humans are our best experimental animal, for humans are highly cooperative and at times even highly motivated. This is of great help in studies of maximum performance, for when an animal refuses to run faster or longer, is it because he will not or cannot? Dogs have been extensively studied, for they are often cooperative and easy to work with – at times they even seem to be highly motivated. Flying is a different matter, for man has minimal experience in flying under his own power, and he is a clumsy and ineffective swimmer compared with seals and whales and fish.

When we compare the various kinds of locomotion, we should realize that the greatest differences between moving in water, on land, and in the air are attributable to the different physical qualities of these media. The two most important differences are (1) in the support of the animal body provided by air and by water, and (2) in the different resistance to movement that the two media present.

Most swimming animals are in near-neutral buoyancy; their weight is fully supported by the surrounding water, and no effort goes into supporting the body. Running and flying animals are in a very different situation; they must support the full weight of their bodies. The running animal has solid support under its feet, but the flying animal must continuously support its weight against a fluid of low density and low viscosity.

The resistance of the medium is of great importance to the swimming animal, for water has a high viscosity and density. In contrast, running and flying animals have the advantage of moving through a medium of low viscosity and low density. The physical characteristics of the medium have profound effects on the structural adaptations animals show in their modes of locomotion. Aquatic animals have streamlined bodies and propel themselves with fins and tail or modifications of these structures. Birds have streamlined bodies and wings, which basically function on the same principles of fluid dynamics as the fish tail. The smaller the size of the animal, however, the less effective is streamlining in reducing drag, and small insects are not obviously streamlined.

Running animals use their extremities as levers for moving over the solid substratum. Air resistance is of little importance, and running animals are not particularly streamlined. Streamlining is much more important for birds, for they move much faster than running mammals, and air resistance increases approximately with the square of the speed. Most mammals are quadrupedal, but an assortment of oddities such as humans and kangaroos are primarily bipedal.

Running

Let us first consider walking and running as we know it from our own experience. We know that it is more expensive, in terms of energy, to walk or run faster, and that running costs more than walking.

Studies of human athletes walking and running on a treadmill have given the results shown in Figure 5.14. As expected, with increasing speed both walking and running become more expensive. When the speed of walking exceeds about 8.5 km h^{-1}, the curve for walking crosses and exceeds the curve for running. The point of interesection of the two curves is where most humans naturally change to running when they decide to move faster. This corresponds to the fact that above this speed it is more expensive to walk than to run.

Figure 5.14 also shows the well-known fact that it is more expensive to move on an uphill grade

FIGURE 5.14 The energy expenditure of a person during walking and running, at three different grades. [Margaria et al. 1963]

and less expensive to move downhill. Within the range of running speeds recorded here, the cost of running increases linearly with running speed, but at very high speeds the curves tend to swing upwards.

Let us now look at a horse, an animal that can move in three different gaits (Figure 5.15). We can see that the cost of moving increases approximately linearly with increasing speed, but a careful examination shows that within each speed the increase is not completely linear; it is slightly curvilinear. The natural change from one gait to another for this horse occurred at the speeds where the curves intersected. This means that a horse for any given speed will choose to move in the most economical gait.

This particular horse had been trained to extend each gait beyond the natural limits. The extended gaits were found to be more expensive than the gait the horse would naturally select for a given speed, thus exaggerating the curvilinear relationship.

The next question that these studies raised gave an unexpected result. Remember that when a horse runs faster, it uses more energy, but it also covers a greater distance. Let us therefore examine how expensive it is to cover a given distance in the different gaits. The results in Figure 5.16 show the strongly exaggerated U-shape of the curves that were obtained when the gaits were extended artificially. However, the minimum cost to move a given distance was the same in each gait.

When a horse runs on a treadmill, it must run at the speed of the tread. However, when it runs free and can select its own speed, it chooses to run within a narrow range of speeds, and these speeds coincide with the most economical speeds for each gait (Figure 5.16).

We saw in Figure 5.15 that, when the horse was not moving at the artificially extended gaits, the relationship between running speeds and oxygen consumption was nearly linear. Such a linear relationship holds for a wide variety of animals. For example, consider a running rat. Over a wide range of speeds the increase in oxygen consumption is nearly linear (Figure 5.17). If we extend the straight line back to the ordinate, corresponding to zero running speed, the intersect is higher than the resting oxygen consumption. Therefore, if we draw a complete curve, covering all running speeds from zero to the maximum, we would not obtain a straight line. However, in the intermediate range of speeds, the curves for a wide variety of running mammals, including humans, follow straight lines.

Let us examine the running rat and the straight-line relationship between cost and speed. For any increment in running speed there is a proportional increase in cost. Thus, the increment in the metabolic cost for an increment in speed is given by the slope of the line. As the line is straight, the cost (slope) remains constant and does not change with the speed of running. This is an important observations that has broad implications, as we shall see when we compare animals of different body sizes.

The rate of oxygen consumption in relation to running speed for a number of animals is shown in

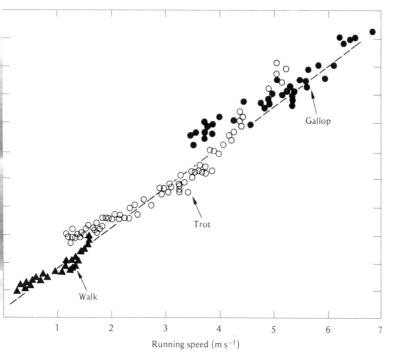

FIGURE 5.15 The oxygen consumption of a horse increases almost linearly with the running speed. Triangles – walk. Open circles – trot. Filled circles – gallop. This horse had been trained to extend the speed in each gait beyond the normal range. For example, it could trot at speeds where it would normally move in a gallop. [Hoyt and Taylor, 1981]

Gallop

Trot

Walk

Running speed (m s^{-1})

Figure 5.18. We can immediately see a body size relationship, the smaller animals having steeper regression lines than larger animals (i.e., their oxygen consumption increases more rapidly with increasing running speed).

A simple way of comparing the cost of running for different animals is to calculate how much energy it takes to move one unit of body mass over one unit of distance: for example, to move 1 kg body mass over 1 km (given as liters O_2 kg^{-1} km^{-1}). If we now return to Figures 5.17 and 5.18, we see that the slopes of the regression lines give exactly this information. The slope (defined as an increment in the ordinate value over an increment in the corresponding abscissa value) is the increment in oxygen consumption (liters O_2 kg^{-1} h^{-1}) over the increment in speed (km h^{-1}). The units cancel out to give the slope in liters O_2 kg^{-1} km^{-1}, precisely the units used to define the cost of running at the beginning of this paragraph.

We can now compare directly the cost of running 1 km per kilogram body mass for the animals in Figure 5.18. The result is plotted in Figure 5.19. It is immediately apparent that the cost of running decreases quite regularly as the body size increases. The significance of this relationship is that, for a running mammal, it is metabolically relatively less expensive, and thus an advantage, to be of large body size.

The metabolic cost of locomotion can be seen in a different light. The preceding discussion considered what it costs to run, above the cost of not running. However, an animal while it moves does in fact have to supply energy for its total metabolism. Thus, a sheep that grazes must eat enough food to cover its total metabolic rate, not merely the excess attributable to the act of locomotion per se. If we consider the total metabolic cost while moving, we obtain higher figures because the non-moving metabolic rate is added in.

FIGURE 5.16 The oxygen cost for a horse of moving one meter. In each gait the minimum cost of moving a given distance was nearly the same. The speeds at which the horse preferred to move when free to choose its speed is shown by the histograms on the abscissa. In each case, the preferred speed coincided with the speed of minimum cost. The horse had been trained to extend the speed in each gait beyond the normal range, and at these artificial speeds the cost of moving increased greatly (Hoyt and Taylor, 1981).

FIGURE 5.17 The oxygen consumption of running white rats increases linearly with increasing speed. [Taylor et al. 1970]

FIGURE 5.18 The oxygen consumption of a variety of running mammals increases linearly with speed. The increase per unit body weight is smaller the larger the body size of the animal. [Taylor et al. 1970]

Swimming and flying

We can examine the cost of swimming or of flying in terms similar to those we have used for running, and we again find that the cost of moving is related to the body size.

Although the oxygen consumption during running increases regularly and linearly with increasing speed (Figure 5.17), the situation is different for flying birds. Determinations on budgerigars (parakeets) have shown that there is an optimum speed of flight at which the bird's oxygen consumption is at a minimum (Figure 5.20). Flying slower or faster increases the oxygen consumption,

so that the metabolic cost of flight relative to speed yields a U-shaped curve.

Although flying horizontally at 35 km h^{-1} gives the lowest metabolic rate for the flying budgerigar, the speed that enables the bird to fly a certain distance most economically is higher, in the case of the budgerigar about 40 km h^{-1}. This is because the bird that flies faster covers the same distance in less time.

The most economical flight speed can be calculated, but it can also be found directly from the graph by taking the tangent to the curve from the origin. The point of contact gives the lowest pos-

FIGURE 5.19 The cost of running for mammals of various body sizes. The net cost designates the cost of moving 1 kg body mass over a distance of 1 km, calculated from the increase in metabolism caused by running (and obtained from slopes of regression lines). The total cost includes the total metabolism while running and is therefore somewhat higher. [Schmidt-Nielsen 1972a]

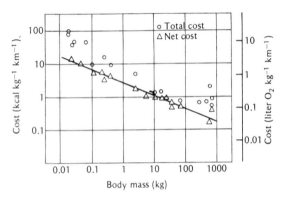

FIGURE 5.20 Oxygen consumption of a flying budgerigar (parakeet) in relation to flying speed. The lowest oxygen consumption occurs at about 35 km per hour, but the lowest cost of transport is at about 40 km per hour (see text). [Schmidt-Nielsen 1972b; data from Tucker 1968]

sible slope for the oxygen consumption relative to flight speed, and this is the cost of transport as previously defined, expressed in liters O_2 km^{-1}. (The cost of transport derived by using the tangent to the curve in Figure 5.20 gives the total cost of transport, rather than the net transport as we calculated for the mammals. For birds this is relatively unimportant, because their metabolic rate during flying is about 10 times as high as the resting metabolic rate, and subtracting the latter makes only a minor difference in the result. For mammals the difference between net and total cost of locomotion is greater.)

Information about the cost of locomotion for swimming, flying, and running has been put together in Figure 5.21. It is interesting that for a given body size, flying is a far less expensive way to move to a distant point than running. At first glance this is surprising, for we have the intuitive feeling that flying must require a great deal of effort just to stay up in the air. We know, however, that migrating birds fly nonstop for more than 1000 km, and it would be difficult to imagine a small

FIGURE 5.21 Comparison of the energy cost of moving one unit body weight over 1 km for running, flying, and swimming. [Schmidt-Nielsen 1972a]

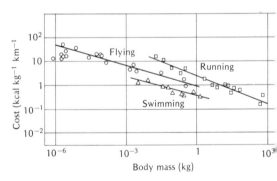

mammal such as a mouse that could run that far without stopping to eat and drink.

Swimming fish use even less energy than flying birds. This may seem surprising, for if we consider the high viscosity of water, we would expect considerable resistance to moving a body through water. There are two factors to take into account, how-

OXYGEN CONSUMPTION DURING FLIGHT The budgerigar, or parakeet (*Melopsittacus undulatus*), photographed during flight in a wind tunnel. To determine the rate of oxygen consumption in flight, this trained bird is equipped with a plastic face mask that, through the attached tube, permits the collection of all exhaled air. [Courtesy of V. A. Tucker, Duke University]

ever. One is that fish on the whole do not move very fast, and their streamlined bodies are highly adapted to moving in a medium of relatively high density and viscosity. The other factor is that fish are in near-neural buoyancy; the body is fully supported by the medium, and no effort goes into keeping from sinking.

For animals less well adapted to swimming the situation is different. If we calculate the cost of transportation for a swimming duck, it turns out to be similar to the cost of running on land for a mammal of the same size. For a person swimming is even more expensive. Compared with a fish he moves through water with difficulty; his body shape and his appendages make him an ineffective swimmer, and the cost of moving through water is 5 or 10 times as high as moving the same distance on land. He is, in fact, so unsuited to swimming that the cost of transportation puts a swimming person well above the range represented by the regression line for running in Figure 5.21.

PHYSIOLOGICAL TIME

Small animals live at a much faster pace than large animals. They move their legs faster, they breathe more quickly, and their hearts beat faster. Everything is speeded up, and it seems that clock time cannot have the same meaning for small and large animals.

The heart of a shrew beats 1000 times each minute and an elephant's heart perhaps 30 times. It requires 0.5 hour for the elephant's heart to beat 1000 times, a number that the heart of a shrew rushes through in only 1 minute. Obviously, a time unit on the clock must have a very different meaning for the two animals.

Time and frequency

The high frequency of the small heart means that the duration of each heartbeat is shorter. Frequency and time are inversely related; that is, frequency and time are reciprocals:

$$\text{frequency} = \frac{1}{\text{time}}$$

The empirical equation for the heart frequency of mammals in Table 5.9 was:

$$\text{heart rate (min}^{-1}) = 241\, M_b^{-0.25}$$

The duration of each heartbeat is the reciprocal value of this equation:

$$\text{duration (min)} = \frac{1}{241}\, M_b^{0.25}$$

This equation merely says that the duration of the heartbeat of mammals increases in proportion to the body mass raised to the power 0.25. We shall return to the inverse relationship between time and frequency.

Metabolic time

The most important measure of how fast an animal lives is its metabolic rate. Consider the specific metabolic rate, or the metabolic rate per unit body mass, which decreases with increasing body size according to the equation:

$$\text{metabolic rate} \propto M_b^{-0.25}$$

Metabolic time is the reciprocal of metabolic rate, and physiological time thus changes with body mass according to:

$$\text{metabolic time} \propto M_b^{0.25}$$

The exponent in this expression is the same as for the duration of the heartbeat, and the relationship holds for many other metabolic rate processes as well. The principle that physiological time, relative to clock time, increases with increasing body size applies to all sorts of rate processes.

The life span of small animals is scaled to their fast pace. They don't live as long as large animals. The life span of mammals increases with body size in roughly the same proportion as the duration of the heartbeat. The surprising result is that the small and the large animal have roughly the same number of heartbeats allotted for their normal life spans.

A mouse, whose heart ticks away at 600 beats per minute, has an allotment of very roughly 800 million heartbeats during its 2- or 3-year life span. The elephant, with its heart beating 30 times per minute, generates a similar number of heartbeats during its 50- or 60-year life span, 20 times the life span of the mouse.

If this has led you to use your own heart rate of 70 or so to calculate your life expectancy, you have arrived at a disappointing 20 or 25 years. Luckily, the rule is not very precise; humans live several times as long as our body size would suggest.

Real time

Animals live in the real world and cannot escape from real time, such as the day-and-night cycle and the seasons of the year. These cycles must have a very different physiological meaning for large and small animals.

Consider the high specific metabolic rate of the small animal, the correspondingly high requirements for food, and the amount of energy available to an animal that is deprived of food. There may be some food in the digestive tract, and in addition there is the energy stored in the body, mostly as fat. The total available energy, and the rate at which it is used, determine how long the reserves will last; this we can call the *endurance time*. The endurance time is directly related to the available energy reserves and inversely related to the rate of use:

$$\text{endurance time} = \frac{\text{available energy}}{\text{rate of use}}$$

If the relative amounts of stored fat are similar in large and small animals, and available evidence indicates that this is so, the endurance time should increase as the rate of use (the metabolic rate) decreases in the larger animal. The endurance time should therefore increase in the inverse ratio of metabolic rate:

$$\text{endurance time} \propto M_b^{0.25}$$

This relation states that the endurance time increases with the size of the animal. For a small animal the endurance time is short, and for the large animal it is long. The smallest mammals and birds must eat almost continuously, and we shall later see that a hummingbird solves the dilemma of getting through the night without feeding by dropping its metabolic rate. Many of the smallest warm-blooded vertebrates do the same; when they are unable to feed they go torpid, their body temperature drops, their metabolic rate decreases, and the endurance time increases.

Now consider an entire unfavorable season, such as the winter, which makes the situation much more difficult. The options are few but well known. The animal can migrate to a better climate, as many birds do. Small mammals cannot easily migrate over long distances, but they can stockpile food. However, a better solution is to store body fat and combine this with a decreased metabolic rate: that is, go torpid and hibernate. Hibernation makes it possible for the small animal to endure the long unfavorable season and survive the winter.

Large animals can more easily bridge seasonal barriers because their endurance time is much longer. Bears sleep through the winter without any great decrease in body temperature and metabolic rate, and their fat stores are still sufficient to carry them through months without food.

We have now seen that real time, whether we speak about a day-and-night cycle or the changes of seasons, must have a very different physiological meaning for small and large animals.

EFFECT OF HIGH ALTITUDE

Most of what we know about the responses to high altitude and low air pressure comes from studies on humans. As described in Chapter 1, the percentage composition of the atmosphere is unchanged at high altitudes, but the decrease in barometric pressure has severe physiological effects.

Mountain climbers unaccustomed to high altitude begin to feel decreased physical performance at some 3000 to 4000 m altitude, but after a few days they become acclimatized* and perform better. The highest known permanent human habitation is a mining camp in the central Andes in South America at an altitude of nearly 5000 m.

A very few highly trained and selected mountain climbers have barely managed to reach the summit of Mount Everest at 8848 m (about 29 000 ft) without supplementary oxygen, but this record performance is most exceptional. Severe or even lethal altitude sickness often develops at much lower than the record altitudes.

It has been established on many mountain-climbing expeditions that acclimatization is progressive and increases with altitude, but at heights above 6000 m there is also an underlying process of deterioration that is more rapid and severe the greater the altitude. This eventually forces the climber to descend (Pugh 1964).

The symptoms of low oxygen in humans are known as *mountain sickness*. The severity varies from person to person, but even at moderate altitude there is a feeling of weakness and disinclination to move about and perform work. This is followed by headache, nausea, sometimes vomiting, as well as a deterioration of mental functions, judgment, ability to perform simple arithmetic, and so on. Sleep and rest become difficult, which contributes to the problems.

Except for the danger of developing bends during rapid ascent in an unpressurized airplane or balloon, all the known symptoms of high altitude are related to the low partial pressure of oxygen and the inability to supply sufficient oxygen at the tissue level. Thus, breathing pure oxygen at an altitude of about 12 000 m, where the barometric

---

*It is now agreed that the word *acclimatize* is used when an organism adjusts to a certain climate, say the seasonal change between summer and winter. For example, a fish caught in winter in cold water is acclimatized to cold water. The word *acclimate* is used when an organism adjusts to an artificially imposed condition. For example, a goldfish kept in the laboratory for 2 months in water at 10 °C is said to be acclimated to 10 °C.

pressure is about 150 mm, removes all symptoms of mountain sickness. At this altitude humans could survive for no more than a few minutes if they were to breathe atmospheric air.

The adequate supply of oxygen to the tissues depends on a sequence of the following five steps:

1. Ventilation of the lungs
2. Diffusion from lung air to blood
3. Transport of oxygen in the blood
4. Diffusion of oxygen from blood to tissues
5. Tissue utilization of oxygen

Each of these steps is affected during acclimatization to high altitude.

The immediate response to an inadequate supply of oxygen is an increased ventilation of the lungs, due to increased tidal volume as well as frequency. This increased ventilation leads to an excessive loss of carbon dioxide from the blood, and thus causes a shift in the pH of the blood in the alkaline direction, reducing the normal stimulation of the respiratory center. The result is a decrease in ventilation and oxygen supply. This decrease in oxygen supply in turn causes increased ventilation and further loss of carbon dioxide. This continues, leading to a cycling between increased and decreased respiration.

During acclimatization to altitude there is a change in the response of the respiratory center and an adjustment of the bicarbonate concentration in the blood that permits the maintenance of acid–base balance and a normal blood pH at a lower P_{CO_2}. As these changes progress, the increased lung ventilation can be maintained with less difficulty.

The diffusion capacity of the lungs of the lowland natives remains unchanged at high altitudes. Highland natives, however, have larger lung volumes than lowland natives, and because pulmonary diffusing capacity is related to lung surface area, highland natives have greater pulmonary diffusing capacity. The difference results from an increased growth of the lungs during childhood. Similar changes occur in young rats that have been exposed to low pressure during growth. Thus, the difference in diffusion capacity seems related to a lifetime acclimatization.

Transport of oxygen in the blood is also affected by altitude acclimatization, in regard to both oxygen capacity and oxygen affinity of the blood. In humans the number of red blood cells is increased from a normal of about 5 million per cubic millimeter to as many as 8 million per cubic millimeter at altitudes about 4000 m. This change is found both in permanent residents and in acclimatized lowland natives when they reside at high altitudes.

The oxygen affinity of the blood was discussed in Chapter 2. Llamas and vicuñas, animals native to high altitude, have blood with a higher oxygen affinity than other mammals, and this facilitates the uptake of oxygen in the lung. Humans, at altitude, on the other hand, display a slight shift in the opposite direction, with a decrease in the oxygen affinity of the blood. This is of no aid to the uptake of oxygen in the lung, but slightly facilitates the delivery of oxygen to the tissues.

Diffusion of oxygen from blood to tissues can be enhanced by decreasing the diffusion pathway and by increasing the myoglobin concentration in the tissues (facilitated diffusion). Both responses have been observed, with an increased number of capillaries per unit volume of tissues (reducing the diffusion distance) and with an increased muscle myoglobin concentration (increasing the rate of diffusion).

Utilization of oxygen at the tissue level is the last step. Adaptations in the enzymatic pathways and their importance in high altitude acclimatization is a more poorly understood subject.

The importance of these steps in the oxygen supply from air to tissue in humans is summarized in Figure 5.22. This graph compares native residents of Lima, which is at sea level, with native residents of Morococha, a mining town located at 4540 m altitude. The average barometric pressure at Morococha is 446 mm Hg, compared with the normal 760 mm at sea level.

FIGURE 5.22 Mean oxygen partial pressures observed in native residents of Lima (sea level) and Morococha, Peru (4540 m altitude). [Hurtado 1964]

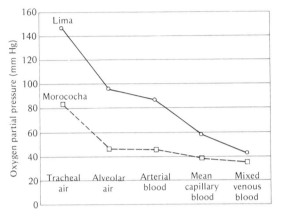

TABLE 5.10 Oxygen partial pressures (mm Hg) observed in native residents of Lima (sea level) and Morococha, Peru (4540 m). Mean of observations on eight healthy subjects in each group. Same data as in Figure 5.22. [Hurtado 1964]

	Tracheal air	Mixed venous blood	Drop
Lima	147.2	42.1	105.1
Morococha	83.4	34.8	48.6
Difference	63.8	7.3	56.5

The observations on the oxygen partial pressure (P_{O_2}) in the two groups of permanent residents are summarized in Table 5.10.

The drop in P_{O_2} from tracheal air (essentially humidified outside air) to venous blood was more than twice as great at sea level as at high altitude. Nevertheless, the P_{O_2} of mixed venous blood was only 7.3 mm Hg lower at high altitude; this indicates that the oxygen supply to the tissues in the acclimatized natives was nearly as adequate as at sea level (Hurtado 1964).

Birds are conspicuously more tolerant of high altitude than mammals, and mountain climbers have often reported seeing birds flying overhead at altitudes where they themselves suffered severely from oxygen deficiency.

There are several reasons for this difference. One is the greater effectiveness of the respiratory system of birds compared with mammals. In birds the air moves through the lungs during both inhalation and exhalation. Because of the cross-current flow between air and blood, the oxygenated blood reaches a higher oxygen partial pressure than that in the exhaled air. This is impossible in the mammalian lung, where the oxygen partial pressure of the blood can approach, but never exceed, that in the exhaled air.

A second reason is the exceptional tolerance of birds to alkalosis. When increased ventilation causes excessive loss of carbon dioxide from the blood, the birds do not suffer unduly. The resulting shift in the pH of the blood in the alkaline direction causes a left-shift of the oxygen dissociation curve of the blood (the Bohr effect). This signifies an increased oxygen affinity, which increases the amount of oxygen taken up by the blood at low oxygen partial pressures.

A third factor is that birds, as opposed to mammals, maintain a normal blood flow to the brain at low blood P_{CO_2}. In mammals, in contrast, a low blood P_{CO_2} causes reduced blood flow to the brain and thus worsens the already poor oxygen supply. The maintenance of the blood supply to the brain of birds frees them from the severe effects of brain anoxia observed in mammals at high altitude (Grubb et al. 1977).

The three factors listed here are the main differences that distinguish birds from mammals in their exceptional tolerance to high altitude and low oxygen.

To summarize the preceding discussions, we can see the close connection between the energy metabolism of animals and not only their supply of food, but also the essential roles of oxygen and respiration. When we move on to temperature as another environmental variable, we will again find that we must consider many of the principles discussed in the preceding chapters and that there are close connections among the functions of all physiological systems.

REFERENCES

Adolph, E. F. (1949) Quantitative relations in the physiological constitutions of mammals. *Science* 109:579–585.

Bartholomew, G. A., and MacMillen, R. E. (1961) Oxygen consumption, estivation, and hibernation in the kangaroo mouse, *Microdipodops pallidus*. *Physiol. Zool. 34*:177–183.

Black, E. C., Manning, G. T., and Hayashi, L. (1966) Changes in levels of hemoglobin, oxygen, carbon dioxide, pyruvate, and lactate in venous blood of rainbow trout (*Salmo gairdneri*) during and following severe muscular activity. *J. Fish. Res. Bd. Canada* 23:783–795.

Blazka, P. (1958) The anaerobic metabolism of fish. *Physiol. Zool. 21*:117–128.

Brody, S. (1945) *Bioenergetics and Growth: With Special Reference to the Efficiency Complex in Domestic Animals.* New York: Reinhold. 1023 pp. Reprinted (1964), Hafner, New York.

Butler, P. J. (1979) The use of radiotelemetry in the studies of diving and flying of birds. In *A Handbook on Biotelemetry and Radio Tracking* (C. J. Amlaner, Jr., and D. W. MacDonald, eds.), pp. 569–577. Oxford: Pergamon Press.

Castellini, M. A., and Somero, G. N. (1981) Buffering capacity of vertebrate muscle: Correlations with potentials for anaerobic function. *J. Comp. Physiol. 143*:191–198.

Childress, J. J. (1975) The respiratory rates of midwater crustaceans as a function of depth of occurrence and relation to the oxygen minimum layer off southern California. *Comp. Biochem. Physiol. 50A*:787–799.

Dawson, T. J., and Hulbert, A. J. (1970) Standard metabolism, body temperature, and surface areas of Australian marsupials. *Am. J. Physiol. 218*:1233–1238.

Douglas, E. L., Friedl, W. A., and Pickwell, G. V. (1976) Fishes in oxygen-minimum zones: Blood oxygenation characteristics. *Science 191*:957–959.

Drorbaugh, J. E. (1960) Pulmonary function in different animals. *J. Appl. Physiol. 15*:1069–1072.

Elsner, R. (1965) Heart rate response in forced versus trained experimental dives in pinnipeds. *Hvalrådets Skrifter,* no. 48, pp. 24–29.

Felger, R. S., Cliffton, K., and Regal, P. J. (1976) Winter dormancy in sea turtles: Independent discovery and exploitation in the Gulf of California by two local cultures. *Science 191*:283–285.

Fry, F. E. J., and Hart, J. S. (1948) The relation of temperature to oxygen consumption in the goldfish. *Biol. Bull. 94*:66–77.

Gaunt, A. S., and Gans, C. (1969) Diving bradycardia and withdrawal bradycardia in *Caiman crocodilus. Nature, Lond. 223*:207–208.

Gordon, M. S., Boëtius, I., Evans, D. H., McCarthy, R., and Oglesby, L. C. (1969) Aspects of the physiology of terrestrial life in amphibious fishes. 1. The mudskipper, *Periophthalmus sobrinus. J. Exp. Biol. 50*:141–149.

Graham, J. B. (1974) Aquatic respiration in the sea snake *Pelamis platurus. Respir. Physiol. 21*:1–7.

Grubb, B., Mills, C. D., Colacino, J. M., and Schmidt-Nielsen, K. (1977) Effect of arterial carbon dioxide on cerebral blood flow in ducks. *Am. J. Physiol. 232*:H596–H601.

Hawkins, A. E., Jewell, P. A., and Tomlinson, G. (1960) The metabolism of some British shrews. *Proc. Zool. Soc. Lond. 135*:99–103.

Heatwole, H., Seymour, R. S., and Webster, M. E. D. (1979) Heart rates of sea snakes diving in the sea. *Comp. Biochem. Physiol. 62A*:453–456.

Hemmingsen, A. M. (1960) Energy metabolism as related to body size and respiratory surfaces, and its evolution. *Rep. Steno Hosp. 9*(2):1–110.

Hochachka, P. W. (1980) *Living Without Oxygen: Closed and Open Systems in Hypoxia Tolerance.* Cambridge, Mass.: Harvard University Press. 182 pp.

Hochachka, P. W., Fields, J., and Mustafa, T. (1973) Animal life without oxygen: Basic biochemical mechanisms. *Am. Zool. 13*:543–555.

Hoyt, D. F. and Taylor, C. R. (1981) Gait and the energetics of locomotion in horses. *Nature, Lond. 292*:239–240.

Hudson, J. W. (1962) The role of water in the biology of the antelope ground squirrel *Citellus leucurus. Univ. Calif. Publ. Zool. 64*:1–56.

Hurtado, A. (1964) Animals in high altitudes: Resident man. In *Handbook of Physiology*, sec. 4, *Adaptation to the Environment* (D. B. Dill, E. F. Adolph, and C. G. Wilber, eds.), pp. 843–860. Washington, D.C.: American Physiological Society.

Jackson, D. C., and Schmidt-Nielsen, K. (1966) Heat production during diving in the fresh water turtle, *Pseudemys scripta. J. Cell. Physiol. 67*:225–232.

Jones, D. R., West, N. H., Bamford, O. S., Drummond, P. C., and Lord, R. A. (1982) The effect of the stress of forcible submergence on the diving response in muskrats (*Ondatra zibethica*). *Can. J. Zool. 60*:187–193.

Kanwisher, J. W., Gabrielsen, G., and Kanwisher, N. (1981) Free and forced diving in birds. *Science 211*:717–719.

King, J. R. (1957) Coments on the theory of indirect calorimetry as applied to birds. *Northwest Sci. 31*:155–169.

Kleiber, M. (1961) *The Fire of Life: An Introduction to Animal Energetics*. New York: Wiley. 454 pp.

Kooyman, G. L., Wahrenbrock, E. A., Castellini, M. A., Davis, R. W., and Sinnett, E. E. (1980) Aerobic and anaerobic metabolism during voluntary diving in Weddell seals: Evidence of preferred pathways from blood chemistry and behavior. *J. Comp. Physiol. 138*:335–346.

Kylstra, J. A. (1968) Experiments in water-breathing. *Sci. Am. 219*:66–74.

Lasiewski, R. C., and Dawson, W. R. (1967) A reexamination of the relation between standard metabolic rate and body weight in birds. *Condor 69*:13–23.

Lusk, G. (1931) *The Elements of the Science of Nutrition*, 4th ed. Philadelphia: Saunders. 844 pp.

MacDonald, A. G. (1975) *Physiological Aspects of Deep Sea Biology*. London: Cambridge University Press. 450 pp.

Mangum, D. C. (1980) Sea anemone neuromuscular responses in anaerobic conditions. *Science 208*:1177–1178.

Margaria, R., Cerretelli, P., Aghemo, P., and Sassi, G. (1963) Energy cost of running. *J. Appl. Physiol. 18*:367–370.

Olsson, K. E., and Saltin, B. (1968) Variation in total body water with muscle glycogen changes in man. In *Biochemistry of Exercise*, Medicine and Sports Series, vol. III, pp. 159–162. Basel: Karger.

Paulev, P. (1965) Decompression sickness following repeated breath-hold dives. *J. Appl. Physiol. 20*:1028–1031.

Pearson, O. P. (1960) The oxygen consumption and bioenergetics of harvest mice. *Physiol. Zool. 33*:152–160.

Pugh, L. G. C. E. (1964) Animals in high altitudes: Man above 5,000 meters – mountain exploration. In *Handbook of Physiology*, sect. 4, *Adaptation to the Environment* (D. B. Dill, E. F. Adolph, and C. G. Wilber, eds.), pp. 861–868. Washington, D.C.: American Physiological Society.

Schmidt-Nielsen, K. (1972a) Locomotion: Energy cost of swimming, flying, and running. *Science 177*:222–228.

Schmidt-Nielsen, K. (1972b) *How Animals Work.* Cambridge: Cambridge University Press. 114 pp.

Schmidt-Nielsen, K. (1984) *Scaling: Why is animal size so important?* Cambridge: Cambridge University Press. 241 pp.

Scholander, P. F. (1940) Experimental investigations on the respiratory function in diving mammals and birds. *Hvalrådets Skrifter*, no. 22, pp. 1–131.

Shepard, M. P. (1955) Resistance and tolerance of young speckled trout (*Salvelinus fontinalis*) to oxygen lack, with special reference to low oxygen acclimation. *J. Fish. Res. Bd. Can. 12*:387–446.

Shoubridge, E. A., and Hochachka, P. W. (1980) Ethanol: Novel end product of vertebrate anaerobic metabolism. *Science 209*:308–309.

Smith, E. N., Allison, R. D., and Crowder, W. E. (1974) Bradycardia in a free-ranging American alligator. *Copeia 1974*:770–772.

Snyder, G. K. (1983) Respiratory adaptations in diving mammals. *Respir. Physiol. 54*:269–294.

Somero, G. N., and Hochachka, P. W. (1976) Biochemical adaptations to pressure. In *Adaptation to Environment: Essays on the Physiology of Marine Animals* (R. C. Newell, ed.), pp. 480–510. London: Butterworth.

Stahl, W. R. (1967) Scaling of respiratory variables in mammals. *J. Appl. Physiol. 22*:453–460.

Taylor, C. R., Schmidt-Nielsen, K., and Raab, J. L. (1970) Scaling of the energetic cost of running to body size in mammals. *Am. J. Physiol. 219*:1104–1107.

Thomas, H. J. (1954) The oxygen uptake of the lobster (*Homarus vulgaris* Edw.). *J. Exp. Biol. 31*:228–251.

Tucker, V. A. (1968) Respiratory exchange and evaporative water loss in the flying budgerigar. *J. Exp. Biol. 48:*67–87.

Whalen, W. J. (1966) Intracellular P_{O_2}: A limiting factor in cell respiration. *Am. J. Physiol. 211:*862–868.

Whalen, W. J., Buerk, D., and Thuning, C. A. (1973) Blood-flow-limited oxygen consumption in resting cat skeletal muscle. *Am. J. Physiol. 224:*763–768.

Zapol, W. M., Liggins, G. C., Schneider, R. C., Qvist, J., Snider, M. T., Creasy, R. K., and Hochachka, P. W. (1979) Regional blood flow during simulated diving in the conscious Weddell seal. *J. Appl. Physiol. 47:*968–973.

ADDITIONAL READING

Bouverot, P. (1985) *Adaptation to Altitude-Hypoxia in Vertebrates.* Berlin: Springer-Verlag. 176 pp.

Brody, S. (1964) *Bioenergetics and Growth: With Special References to the Efficiency Complex in Domestic Animals,* reprint ed. New York: Hafner. 1023 pp.

Butler, P. J., and Jones, D. R. (1982) The comparative physiology of diving in vertebrates. *Adv. Comp. Physiol. Biochem. 8:*179–364.

Calder, W. A., III (1984) *Size, Function, and Life History.* Cambridge, Mass.: Harvard University Press. 431 pp.

Dejours, P. (1988) *Respiration in Water and Air: Adaptations – Regulation – Evolution.* Amsterdam: Elsevier. 179 pp.

Elsner, R., and Gooden, B. (1983) *Diving and Asphyxia: A Comparative Study of Animals and Man.* Cambridge: Cambridge University Press. 168 pp.

Fields, J. H. A., (1988) Fermentative adaptations to the lack of oxygen. *Can. J. Zool. 66:* 1036–1040.

Heath, J. E., and Heath, M. S. (1982) Energetics of locomotion in endothermic insects. *Annu. Rev. Physiol. 44:*133–143.

Heatwole, H., and Seymour, R. (1975) Diving physiology. In *The Biology of Sea Snakes* (W. A. Dunson, ed.), pp. 289–327. Baltimore: University Park Press.

Hochachka, P. W. (1980) *Living Without Oxygen: Closed and Open Systems in Hypoxia Tolerance.* Cambridge, Mass.: Harvard University Press. 182 pp.

Hochachka, P. W., and Guppy, M. (1987) *Metabolic Arrest and the Control of Biological Time.* Cambridge, Mass.: Harvard University Press. 227 pp.

Kleiber, M. (1961) *The Fire of Life: An Introduction to Animal Energetics.* New York: Wiley. 454 pp.

Kooyman, G. L. (1981) *Weddell Seal: Consummate Diver.* New York: Cambridge University Press. 135 pp.

Kooyman, G. L., Castellini, M. A., and Davis, R. W. (1981) Physiology of diving in marine mammals. *Annu. Rev. Physiol. 43:*343–356.

Kylstra, J. A. (1975) Liquid breathing and artificial gills. In *The Physiology and Medicine of Diving and Compressed Air Work,* 2d ed. (P. B. Bennett and D. H. Elliott, eds.), pp. 155–165. London: Bailliere Tindall.

Rayner, J. M. V. (1982) Avian flight energetics. *Annu. Rev. Physiol. 44:*109–119.

Schmidt-Nielsen, K. (1972) *How Animals Work.* Cambridge: Cambridge University Press. 114 pp.

Schmidt-Nielsen, K. (1984) *Scaling: Why Is Animal Size so Important?* Cambridge: Cambridge University Press. 241 pp.

Sleigh, M. A., and MacDonald, A. G. (eds.) (1972) *The Effects of Pressure on Organisms.* Symposium, Society for Experimental Biology, no. 26. Cambridge: Cambridge University Press. 516 pp.

Stevens, E. D., and Dizon, A. E. (1982) Energetics of locomotion in warm-bodied fish. *Annu. Rev. Physiol. 44:*121–131.

Taylor, C. R., and Heglund, N. C. (1982) Energetics of terrestrial locomotion. *Annu. Rev. Physiol. 44:*97–107.

West, J. B. (1984) Human physiology at extreme altitudes on Mount Everest. *Science 223:*784–788.

West, J. B. (1985) *Everest: The Testing Place.* New York: McGraw-Hill, 187 pp.

TEMPERATURE

Temperature effects

In the preceding chapter we discussed the energy metabolism of living animals and the influence of variables such as oxygen concentration, body size, and activity. In this chapter we shall discuss the profound influence that temperature has on living organisms and their metabolic processes.

Active animal life is limited to a narrow range of temperatures, from a few degrees below the freezing point of pure water (0 °C) to approximately +50 °C. We are now concerned with the temperature of the organism itself, not the surroundings. For example, a human maintains a body temperature of about 37 °C whether he goes out in subfreezing weather or enters a sauna with an air temperature near the boiling point of water.

The temperature range for animal life is very narrow compared to cosmic temperatures, but nevertheless, suitable temperatures can be found throughout the oceans and over much of the surface of the earth, at least for part of the year.

In polar regions numerous fish and invertebrates live in water at −1.8 °C. At the other extreme, in hot springs, a few animals can live at about 50 °C. Some blue-green algae live above 70 °C, and a few thermophilic bacteria thrive above the boiling point of water.

Outside the temperature range that permits active life, many animals can survive in an inactive or torpid state. In fact, some can survive freezing at extremely low temperatures, such as that of liquid air (about −190 °C) or even liquid helium (−269 °C).

Most animals, among them all aquatic invertebrates, have nearly the same temperature as their surroundings. Birds and mammals, in contrast, usually maintain their body temperature nearly constant and independent of the environment. However, some other animals, both invertebrates and vertebrates, can at times maintain a substantial difference between their own temperature and that of the surroundings.

CONFUSING TERMINOLOGY

There is no simple and easy way to classify the various responses to the thermal environment. Birds and mammals usually maintain their body temperature above that of the environment and have traditionally been called warm-blooded. These terms are so well established that it is convenient to continue to use them. They are, however, inaccurate and may be quite misleading.

A cold-blooded animal is not necessarily cold; a tropical fish or a desert lizard or an insect sitting in the sun may have a higher body temperature than a mammal. Furthermore, quite a few mammals and some birds undergo periods of torpor or hibernation during which their temperature may decrease to near the freezing point of water, with no harm to the animal. In that state it seems odd to call them warm-blooded.

The corresponding scientific terms, *poikilothermic* for cold-blooded and *homeothermic* for warm-blooded, are also imprecise. The word poikilothermic (Greek *poikilos* = changeable) refers to the fact that the temperature of a cold-blooded animal fluctuates with that of its surroundings. A fish has the temperature of the water it swims in and an earthworm that of the soil it crawls in. However, a deep-sea fish that spends its entire life in water that has barely measurable temperature fluctuations is truly an animal with a virtually constant body temperature. It might be logical to call such a fish homeothermic,* but this term is used specifically to refer to birds and mammals, animals that in fact have body temperatures that normally fluctuate by several degrees and, in hibernation, may even drop to near 0 °C. Animals that at times have high and well-regulated body temperatures, but at other times are more like cold-blooded animals, are often called *heterothermic* (Greek *heteros* = different).

Homeothermic animals (birds and mammals)

---

*The terms *homothermic, homoiothermic,* and *homeothermic* are used interchangeably. They are derived from the Greek *homos* = like and *homoios* = similar. Homeothermic is an anglicized spelling form that is most commonly used.

usually maintain a high body temperature and remain active in cold as well as in warm surroundings. Most poikilothermic animals become more and more inactive as the temperature decreases. There are some remarkable exceptions. For example, by exposing itself to the sun, a lizard can maintain its body temperature far above that of the surrounding air. To distinguish the lizard from birds and mammals, which maintain their high body temperature by metabolic heat production, we have the terms *ectothermic* and *endothermic.*

Endothermic animals are able to maintain a high body temperature by internal heat production; ectothermic animals depend on external heat sources, primarily solar radiation.

These definitions also have their limitations; as we shall see later, a tuna fish may keep the temperature of some of its muscles 10 to 15 °C above that of the water. The heat is derived from muscular metabolism, but we do not consider the tuna an endothermic animal in the same sense as birds and mammals. Insects make this terminology even more difficult to apply. They may sit in the sun and warm up and at the same time produce additional heat by intense muscular contractions. Obviously, this case does not clearly fit one of the terms to the exclusion of the other. The same is true of some mammals and birds that, under appropriate circumstances, take advantage of solar heat to reduce internal heat production.

The choice of terminology is primarily a matter of convenience, and it may at times be difficult to find terms that are suitable in every situation. The question is not whether a certain terminology is right or wrong, but how useful it is for a given purpose. Nevertheless, the terms we use must always be accurately defined so that there is no uncertainty about their precise meaning.

EFFECTS OF TEMPERATURE CHANGE

Temperature change has striking effects on many physiological processes. Within limits, a temperature increase accelerates most processes. For example, consider the effect on the rate of oxygen consumption, which is a convenient expression for the overall metabolic activity of an animal.

Within the temperature range an animal can tolerate, the rate of oxygen consumption often increases in a fairly regular manner with increasing temperature. In general, a rise of 10 °C in temperature causes the rate of oxygen consumption to increase about twofold to threefold.

The increase in a rate caused by a 10 °C increase in temperature is called the Q_{10}. If the rate doubles, Q_{10} is 2; if the rate triples, Q_{10} is 3; and so on. This term is used not only for oxygen consumption but for all rate processes affected by temperature (see Appendix D.)

If an animal has a wide range of temperature tolerance, its rate of oxygen consumption may accelerate vastly as temperature increases. Thus, with a Q_{10} of 2, and a beginning temperature of 0 °C, the rate would double with a temperature increase to 10 °C; it would quadruple with an increase to 20 °C; and it would increase 8-fold with a temperature increase to 30 °C. With a Q_{10} of 3, the increases in oxygen consumption would be 3-fold, 9-fold, and 27-fold, respectively, for the same temperature intervals.

If we plot these numbers on graph paper (Figure 6.1), we obtain rapidly rising curves, known as exponential curves because mathematically they are described by exponential functions,* of the general form:

$$y = b \cdot a^x$$

If R_2 and R_1 are the rates at two temperatures T_2 and T_1, and we use the customary symbol Q_{10}, the equation will be:

$$R_2 = R_1 \cdot Q_{10}^{\frac{T_2 - T_1}{10}}$$

Many rate processes, such as acceleration, radioactive decay, and growth curves, are described by exponential equations. Calculations that in-

*See also Appendix C.

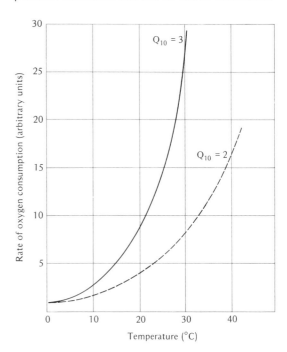

FIGURE 6.1 Many temperature-dependent rate processes proceed more and more rapidly as temperature increases. See text for further discussion.

volve exponential equations can conveniently be carried out if we use the logarithmic form:

$$\log y = \log b + x \cdot \log a$$

or

$$\log R_2 = \log R_1 + \log Q_{10} \cdot \frac{T_2 - T_1}{10}$$

We can now see that $\log R_2$ increases linearly with the temperature change $(T_2 - T_1)$. This means that if we plot log rate against temperature, we get a straight line. This is commonly done by using graph paper with a logarithmic ordinate and a linear abscissa (so-called semilog graph paper). A graph representing this form of the equation is given in Figure 6.2.

Obviously it is not necessary to determine two rates exactly 10 °C apart in order to calculate the

FIGURE 6.2 A temperature-dependent rate process of the kind shown in Figure 6.1 will, when plotted on logarithmic ordinate, give a straight line. Note that the abscissa is linear and only the ordinate is logarithmic. For further explanation, see text.

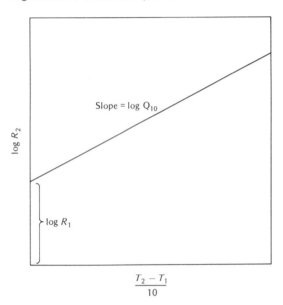

TABLE 6.1 Rates of oxygen consumption of the Colorado potato beetle (*Leptinotarsa decemlineata*) at various temperatures between 7 and 30 °C. Before the experiment the animals had been maintained at 8 °C. [Data from Marzusch 1952]

Temperature (°C)	O₂ consumed (μl g^{-1} h^{-1})	Temperature interval (°C)	Q_{10}
7	61		
10	81	7–10	2.57
15	126	10–15	2.41
20	200	15–20	2.52
25	290	20–25	2.10
30	362	25–30	1.56

Q_{10}. Any two temperatures can be used, provided they are sufficiently far apart to give reliable information about the temperature effect. To solve for Q_{10} when the rates at two different temperatures have been observed, we can rearrange the equation and use one of the two following forms:

$$\log Q_{10} = (\log R_2 - \log R_1) \cdot \frac{10}{T_2 - T_1}$$

or

$$Q_{10} = \left(\frac{R_2}{R_1}\right)^{\frac{10}{T_2 - T_1}}$$

We can then calculate the Q_{10} for the interval between T_1 and T_2. However, most often Q_{10} does not remain the same throughout the range of temperatures an animal can tolerate, and it is necessary to specify accurately the conditions under which the observations were made.

Let us now turn to a real example. The rates of oxygen consumption in the Colorado potato beetle between 7 and 30 °C are given in Table 6.1. Throughout this range the oxygen consumption increases with temperature, the Q_{10} for the entire range being 2.17. However, if we calculate the Q_{10} for each temperature interval for which observations were made, we find that up to 20 °C the Q_{10} remains rather constant at nearly 2.5, but at higher temperatures the Q_{10} falls off (Table 6.1, last column). If the same data are plotted in graph form, we can more clearly see that the observations, when the temperature exceeds about 20 °C, deviate from a regular exponential function (Figure 6.3).

It is a bit cumbersome to calculate the Q_{10} for each set of observations, and it is also inconvenient to construct an exponential curve on arithmetic coordinates for comparison with the observed data. If, instead, we plot the numbers on a logarithmic ordinate, we obtain a straight line for that interval in which the temperature effect remains constant and the Q_{10} does not change (Figure 6.4). The straight line in this figure corresponds to a Q_{10} of 2.50, and we can see that the observations up to 20 °C follow this line completely. Above 20 °C, however, the rate of oxygen consumption deviates increasingly from the straight line (i.e., although

FIGURE 6.3 The rate of oxygen consumption of the Colorado potato beetle increases with temperature (solid line). The broken line shows the expected curve, had the Q_{10} remained constant at 2.5. Same data as in Table 6.1.

FIGURE 6.4 When plotted on a logarithmic ordinate, the rate of oxygen consumption of the Colorado potato beetle shows a linear dependence on temperature up to 20°C. Above this temperature the rate does not increase as rapidly as expected from a constant Q_{10} of 2.5 (broken line). Same data as in Table 6.1 and Figure 6.3.

there is still an increase in rate with temperature, the extent of the temperature effect is diminished). At even higher temperatures the drop in oxygen consumption becomes more pronounced, and eventually the lethal temperature limit is reached.

EXTREME TEMPERATURES: LIMITS TO LIFE

Animals differ in the range of temperatures they can tolerate. Some have a very narrow tolerance range; others have a wide tolerance range. Furthermore, temperature tolerance may change with time, and a certain degree of adaptation is possible so that continued exposure to a temperature close to the limit of tolerance often extends the limit. Some organisms are more sensitive to extreme temperatures during certain periods of their lives, particularly during the early stages of development.

When discussing tolerance to extreme tempera-

tures, we must distinguish between temperatures an organism can survive and those at which it can carry out its entire life cycle. We should also realize that we cannot determine an exact temperature that is lethal for a given organism because the duration of exposure is of great importance. A certain high temperature that can be tolerated for several minutes may be lethal if continued for several hours.

One further matter is important: We are concerned with the temperature of the organism itself, not that of its environment. For virtually all aquatic organisms the two are nearly equal, but for terrestrial organisms, as we have indicated, they may differ a great deal. A lizard sitting in the sun may, because of solar radiation, acquire a body temperature 10° or 20° above the surrounding air temperature. Among the so-called warm-blooded animals, mammals and birds, most can tolerate only a rather narrow range of body temperatures, and this must not be confused with the wide range of temperatures of the environment in which they live – from the Arctic to the hottest deserts.

TOLERANCE TO HIGH TEMPERATURE

No animal is known to live and carry out its complete life cycle at a temperature over 50 °C. Multicellular plants also have an upper temperature limit of about 50 °C, but some unicellular plants are more heat-tolerant. For example, the unicellular blue green alga *Synechococcus* is found in hot springs at temperatures as high as 73 to 75 °C. This seems to be the upper limit for photosynthetic life. Thermophilic bacteria are more heat-tolerant, some live and grow in the hot springs of Yellowstone National Park, where the temperature is about 92 °C, which is the boiling point of water at the altitude of Yellowstone. Other thermophilic bacteria thrive at temperatures well above 100 °C (Brock, 1985).

As indicated at the beginning of the chapter, an animal in a resting stage may be extremely tolerant to high temperatures. For example, a fly larva (*Polypedilum,* from Nigeria and Uganda) can tolerate dehydration, and in the dehydrated state it can survive a temperature of 102 °C for 1 minute and afterward grow and metamorphose successfully (Hinton 1960). Another example of extreme tolerance is provided by the eggs of a fresh-water crustacean (*Triops,* from Sudan); these eggs survive through winter and early summer in dry mud, where they may be exposed to temperatures up to 80 °C. In the laboratory they withstand temperatures within 1 °C of boiling. With the boiling point of water increased by subjecting the vessel to pressure, they have withstood 103 °C ± 1 °C for 16 hours, although they are killed within 15 minutes at 106 °C (Carlisle 1968).

From these examples it should be clear that the upper temperature limit for life cannot be accurately defined.

FIGURE 6.5 To determine the temperature that is lethal to 50% of a group of organisms (the T_{L50}), experiments are carried out at a range of temperatures, and after the data are plotted, the temperature for 50% survival is read from the graph.

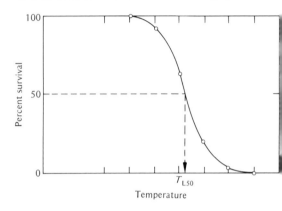

Determinations of lethal temperatures

When several individuals of an animal species are exposed to a temperature close to the limit of their tolerance, some may die and others survive. What is the exact lethal temperature for this species?

The *lethal temperature* is commonly defined as that temperature at which 50% of the animals die and 50% survive, often written as T_{L50}. To find this exact temperature by trial and error would require a great deal of experimentation before one experiment ends with exactly half the animals dead. Instead, after the approximate lethal temperature is established, the procedure indicated in Figure 6.5 is followed. Several groups (say, four) are exposed for the same time period (say, 2 hours) to a series of temperatures, and the percent survival at each temperature is plotted. The T_{L50} can now be readily found from the graph.

This procedure gives the lethal temperature only for the exposure time used in the experiment; shorter exposure gives higher survival and longer exposure usually lower survival. The effect of exposure time can be determined by carrying out tests similar to that in Figure 6.5 for different periods of

FIGURE 6.6 The effect of exposure time on temperature tolerance becomes evident if the T_{L50} is plotted against the duration of each experiment in which the T_{L50} was determined.

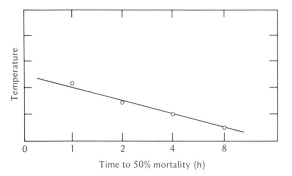

exposure – say 1, 2, 4, and 8 hours. If the various T_{L50} values obtained are plotted against exposure time, the points will often fall on a straight line on a log scale abscissa (Figure 6.6). This is the reason the time intervals in this example were chosen to correspond directly to a logarithmic scale. We now have a single graph that gives a quite complete picture of the temperature tolerance of the organism in question.

Lethal temperatures and cause of heat death

Although active animal life has an upper limit at about 50 °C, many animals die when they are exposed to much lower temperatures. This is particularly true of aquatic animals, and especially marine animals, which usually are not exposed to very high temperatures. Even in tropical seas the temperature is rarely as high as 30 °C, although it may be higher in small enclosed bays and pools.

For intertidal animals, however, the situation is different. At low tide, intertidal animals are exposed to warm air and to solar radiation, and their temperatures may increase appreciably. To some extent they remain cool by evaporation of water, but while in air their restricted water reserves limit this way of cooling.

TABLE 6.2 The heat resistance of snails found in the upper part of the tidal zone is almost 10 °C higher than the resistance of snails from near the low-water mark. The temperature given is the highest from which snails can recover after 1 hour's exposure. [Fraenkel 1968]

Occurrence	Snail	Temperature (°C)
Spray zone, upper part of intertidal zone	*Tectarius vilis*	48.5
	Planaxis sulcatus	48
	Nodilittorina granularis	47
	Littorina brevicola	47
Shallow splash pools	*Paesiella raepstorttiana*	47
Middle part of intertidal zone	*Nerita japonica*	46
	Nerita albicilla	44
Lower part of intertidal zone in sheltered or shadowy places or under stones	*Lunella coronata*	43
	Drupa granulatus	42
	Purpura clavigerus	42
	Monodonta labis	42
Lower part of intertidal zone at water's edge	*Tegula lischkei*	39

Intertidal snails show a clear correlation between heat resistance and their location in the tidal zone (Table 6.2). Those snails that are located high in the tidal zone, and are thus out of water and exposed for the longest time, have substantially higher heat resistance than those located in the lower part of the tidal zone, close to the low-water mark.

The most heat-tolerant fish of any are probably the desert pupfish that live in warm springs in California and Nevada. The species *Cyprinodon diabolis* lives in a spring known as Devil's Hole, in which the temperature (33.9 °C) probably has remained almost unchanged for at least 30 000 years (Brown and Feldmeth 1971). The upper lethal limit for this tiny fish, which as an adult weighs less than 200 mg, is about 43 °C – as high as is known for any species of fish.

In contrast, some Arctic and Antarctic animals

have a surprisingly low heat tolerance. Antarctic fish of the genus *Trematomus* are particularly heat-sensitive and have an upper lethal temperature of about 6 °C. In nature these fish live in water with an average temperature of -1.9 °C that through the year varies by only about 0.1 °C (Somero and DeVries 1967).

The extremely low thermal tolerance of these animals is important to the understanding of the mechanism of heat death. Some factors that have been suggested as contributing to heat death are:

1. Denaturation of proteins, thermal coagulation.
2. Thermal inactivation of enzymes at rates that exceed rates of formation
3. Inadequate oxygen supply
4. Different temperature effects (Q_{10}) on interdependent metabolic reactions
5. Temperature effects on membrane structure

Let us consider heat denaturation of proteins. It is true that many animals die at temperatures where thermal denaturation of protein takes place, for many proteins are denatured at temperatures above 45 to 55 °C. It is difficult to imagine, however, that thermal denaturation of any protein could take place at $+6$ °C, the lethal temperature for *Trematomus*.

Another suggestion for the cause of thermal death is thermal inactivation of certain particularly temperature-sensitive enzyme systems. Again, it is difficult to imagine enzymes so thermolabile that they become inactivated at $+6$ °C. Some enzyme systems from *Trematomus* that have been tested in this regard actually show increased activity up to about 30 °C (Somero and DeVries 1967).

The third possibility, that thermal death is caused by inadequate oxygen supply as the increased temperature accelerates the demand for oxygen, can be excluded in many cases. For example, supplying insects with pure oxygen instead of air does not enable them to survive higher temperature. Likewise, the trout, a cold-water fish, dies at the same temperature in warm water even if the oxygen content of the water is increased several-fold by aeration with pure oxygen.

The fourth possibility is consistent with much of the information at hand. If various processes in the intermediary metabolism are influenced differently by temperature (have different Q_{10} values), it may lead to the depletion or accumulation of certain intermediary metabolic products. Consider the following scheme:

$$A \rightarrow B \rightarrow C \rightarrow D$$
$$\downarrow$$
$$E$$

If in this scheme the process $C \rightarrow D$ is accelerated by increased temperature more than the process $B \rightarrow C$, the intermediary C will be depleted at high temperature. If intermediary C is also needed for some other metabolic purpose, E, it will no longer be available in sufficient quantity. Different temperature sensitivities of the several hundred metabolic enzymes that participate in intermediary metabolism may therefore easily lead to a derangement of the normal biochemical balance of the organism.

Although the possibility of different thermal sensitivities of enzyme systems gives a plausible explanation of heat death under some circumstances (e.g., the Antarctic fish that die at $+6$ °C), it is not necessarily the only factor in heat death. Any of the other possibilities mentioned above may be contributory, and for organisms that have a very high lethal temperature, it is more likely that several of the listed factors play a role.

The last possibility, changes in membrane structure, is very important and covers a broad range of subjects. The generally accepted view is that a membrane consists of a lipid bilayer with a variety of functional proteins embedded or attached, and that any change in interactions among these can be expected to change the functional properties of the membrane. Temperature (i.e., changes in kinetic energy) has profound effects on such interactions by affecting higher orders of protein structure, protein–lipid interactions, lipid–lipid interactions, and so on: in short, molecular structures that depend on weak interactions that are easily changed by temperature. Such disturbances in the integrity of membrane function appear to be a primary factor in heat damage to organisms.

TOLERANCE TO COLD AND FREEZING

The effects of low temperature are at least as perplexing as those of high temperature. Some organisms can tolerate extensive freezing, but most animals cannot. Some can survive submersion in liquid nitrogen at −196 °C or even liquid helium at −269 °C (about 4 K), but some animals are so sensitive to cold that they die at temperatures far above freezing. People who keep tropical fish in home aquaria know this. If the heater is off, the fish may be found dead on a cold morning.

Guppies (*Lebistes reticulatus*), for example, that have been kept at 23 °C room temperature or above will die if they are cooled to about 10 °C (Pitkow 1960). For this fish death apparently is caused by cold depression of the respiratory center, followed by damage from anoxia. The basis for this conclusion is that an increase in the oxygen content of the water improves survival, and a decrease in the oxygen content increases susceptibility to cold.

Cold hardiness

Animals that live in temperate and cold regions are often exposed to long periods of winter temperatures that are far below the freezing point of water. Survival of cold-blooded animals at such subzero temperatures depends on physiological and biochemical characteristics that can be described as *cold hardiness.*

Two general strategies for cold hardiness are possible. An animal can be *freeze-tolerant,* which means that it survives extensive freezing and ice formation in the body. In contrast, an animal can be *freeze-intolerant* and die if there is any internal ice formation; in cold areas such animals must be able to avoid formation of ice even if exposed to temperatures as low as −40 °C or −50 °C.

Freezing and ice formation

To appreciate that animals can survive freezing temperatures, and especially that freeze-intolerant animals survive without ice formation, it is necessary to understand the physical conditions under which ice is formed when water or aquatic solutions are cooled to subzero temperatures.

If water is cooled to below 0 °C, the freezing point of water, freezing does not necessarily take place immediately. A sample of pure water can be cooled far below 0 °C without formation of ice, and we say then that the water sample is *supercooled.* The probability that such supercooled water will freeze depends on three important variables: temperature, the presence of nuclei for ice formation, and time. In the absence of foreign nucleating materials, a sample of very pure water is readily supercooled to −20 °C before it freezes, and if extraordinary precautions are taken, water can be supercooled to near −40 °C. The moment an initial ice nucleus is formed, freezing progresses rapidly throughout the sample.

If we start with a completely frozen sample of pure water, adding heat will initially raise its temperature to 0 °C, and some of the ice will then begin to melt. Adding more heat turns more ice into water, but the temperature remains at 0 °C as long as any ice remains in the sample. Removing heat makes the crystals of ice grow in size, but as long as there is any liquid water left, the temperature remains at 0 °C. (These procedures necessitate heating and cooling at sufficiently slow rates for the temperature to remain uniform throughout the sample.) Once any minute crystal of ice is present, the *freezing point* and the *melting point* for pure water is the same: 0 °C. The *supercooling point,* the temperature where ice formation first takes place when a sample is being cooled, can be much lower.

We should now consider the presence of dissolved substances in the water. Solutes lower the freezing point and usually also the extent of supercooling that can take place before freezing occurs. Assume that we have a solution at a given temperature below 0 °C which is in thermal equilibrium with a few ice crystals present. If heat is removed, more water turns to ice and the crystals grow; if heat is added, some of the ice melts. When cooling turns some water into ice, the solute concentration in the remaining liquid increases, thus gradually lowering the temperature at which more ice can form. This process can be continued until the entire sample is frozen throughout and all water has turned to ice, but this point is difficult to determine and is only rarely considered.

What we need to know is the temperature at which a minute amount of ice is present in a liquid sample that is in thermodynamic equilibrium at that temperature. This is often referred to as the "freezing point," but it could equally well be called the "melting point." It is most often determined by freezing the entire sample, observing it during slow heating, and reading the temperature when the last minute crystal of ice is about to disappear. This, then, is the freezing point as well as the melting point: In water and simple solutions containing salts, sugars, and the like, the two are identical. In biological systems, however, freezing and melting points are not necessarily the same.

Supercooling and freeze-intolerant animals

An animal may be exposed to a temperature considerably below the point where its body fluids might freeze but nevertheless remain supercooled unless ice formation begins through nucleation. Such supercooling is in fact of great importance to the survival of many animals. During an occasional cold night, for example, supercooling may be essential for freeze-intolerant animals that are unable to escape to a warmer hiding place. Reptiles and amphibians, whose body fluids if seeded would begin to freeze at -0.6 °C, have been supercooled to as low as -8 °C without freezing (Lowe et al. 1971). This may mean the difference between life and death for an animal that is unexpectedly exposed to a frosty night before it has found a frost-free winter hideout.

A compound particularly effective in lowering both the freezing point and also the supercooling

point is glycerol. Glycerol often occurs in high concentration in overwintering insects and is very effective in increasing cold tolerance. An extreme case is the willow gallfly (*Rhabdophaga strobiloides*), which survives the winter in Alaska by extreme supercooling, down to temperatures of about -60 °C, as cold as the lowest temperatures it would be exposed to (Miller 1982). The overwintering immature willow gallfly, a dipteran, contains as much as 50% glycerol and thereby avoids freezing.

We shall later return to glycerol because it not only lowers freezing and supercooling points, it also improves the tolerance to freezing in animals that tolerate ice formation.

Antifreeze in fish

Teleost fish have an osmotic concentration in their body fluids of 300 to 400 mOsm, which corresponds to a freezing point of about -0.6 to -0.8 °C. Fish that live in the polar regions swim in water that is regularly at about -1.8 °C. This is the temperature at which sea water freezes, and in areas where sea ice is formed, the sea water remains at this temperature throughout.*

Why are fish that swim in water at -1.8 °C not killed? They cannot remain supercooled in the ordinary sense because they continuously encounter ice crystals in the water that would serve as nuclei for ice formation. However, they are not freeze-tolerant, for any formation of ice leads to destruction of the frozen areas, and if freezing extends to vital organs the fish rapidly die.

The explanation for this anomalous situation is that the blood and tissue fluids of the fish contain antifreeze compounds that lower the freezing point

---

*The salt content of sea water corresponds to a concentration of nearly 1.0 osmole per liter. The molar freezing point depression for water is 1.86 °C, and sea water therefore does not freeze until its temperature is this low. When ice begins to form, it floats at the surface, both in fresh water and sea water. Fresh water has a point of highest density at $+4$ °C, and when ice forms at the surface the bottom water in a lake may still be at $+4$ °C. Sea water, however, does not exhibit this peculiarity; the temperature of the polar seas, when there is ice at the surface, is about -1.8 °C throughout.

without affecting the melting point. This requires a more detailed explanation.

To determine the freezing point, a tiny sample of blood plasma or tissue fluid is sealed inside a glass capillary that is placed in a refrigerated alcohol bath where it can be observed through a microscope. To avoid supercooling of the sample, a tiny ice crystal is produced inside the capillary by local freezing with a coolant. The temperature of the bath is then raised very slowly until the crystal begins to disappear. The temperature of the bath at the moment the crystal disappears is the melting point of the sample.

A new tiny crystal of ice is then produced in the sample by local cooling, and the temperature of the bath is slowly lowered until the crystal begins to grow. This temperature is taken as the freezing point. In a normal solution the freezing point is virtually identical to the melting point. However, the freezing point of the body fluids of the polar fishes is a degree or two below the melting point, and this explains why they do not freeze in the cold water. The difference between melting and freezing points is known as *thermal hysteresis.*

The substance responsible for this lowering of the freezing point has become known as an "antifreeze" substance. It was originally identified and its composition clarified in the blood of an Antarctic fish, *Trematomus* (DeVries 1970). It is a glycoprotein that acts by preventing the addition of water molecules to the crystal lattice of ice and, therefore, the growth of ice crystals.

The way the antifreeze acts is suggested in Figure 6.7. By attaching itself to the surface of an ice crystal, the antifreeze prevents the addition of water molecules to the crystal and thus the further growth of the crystal (DeVries, 1982).

The effectiveness of the antifreeze is shown in Figure 6.8, which gives the freezing point depression caused by various concentrations of glucose, sodium chloride, and the antifreeze. For both sodium chloride and glucose the freezing point depression increases linearly with the concentration, but sodium chloride, because of its dissociation into sodium and chloride ions, has twice the freezing point depression of glucose. In contrast, the antifreeze substance prevents the formation of ice in water several hundred times as effectively as other solutes.

The antifreeze isolated from Antarctic fish is made up of long chains of amino acids with the major structure consisting of two molecules of the amino acid alanine alternating with one molecule of the amino acid threonine, with a disaccharide derived from galactose attached to each threonine unit, as follows:

FIGURE 6.7 A diagram of how an antifreeze substance acts to prevent growth of ice crystals. The antifreeze molecules are absorbed to the crystal and force water molecules, represented by cubes, to join the crystal lattice between the absorbed molecules. This results in growth of the crystal in areas that have a high radius of curvature so that a lower temperature is required for growth. [DeVries 1982]

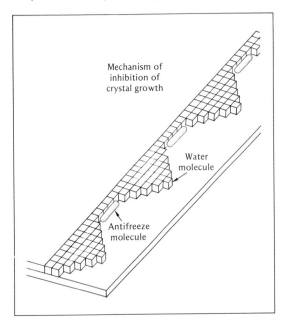

—ALA—ALA—THR—ALA—ALA—THR—ALA—ALA—
                    | |
                disacch   disacch

The glycoproteins range in molecular weight from 2600 to 32 000 and occur in several discrete size groups. In mixtures of the various groups the activities are additive. There are minor variations in the amino acid composition, but the characteristic for all is that the disaccharide is attached to each threonyl residue.

Other antifreeze substances have been found in Arctic fish, but those from the winter flounder (*Pseudopleuronectes americanus*) and the Alaskan sculpin (*Myoxocephalus verrucosus*) are

FIGURE 6.8 The antifreeze substance from the Antarctic fish *Trematomus* is abnormally effective in preventing formation of ice in an aqueous solution. On a molar basis it is several hundred times as effective in this regard as expected. The abscissa gives the molal concentration of the substance; the ordinate gives the apparent osmolal concentration as determined from freezing point depression. [DeVries 1970]

carbohydrate-free polypeptides that contain 60% alanine residues (DeVries 1984). However, glycoproteins similar to those from Antarctic fish have been found in several fish of the cod family; they contain alanine, threonine, and galactosamine similar to the antifreeze of the Antarctic fishes (Fletcher et al. 1982).

All in all, antifreezes have been found in fish that belong to at least 11 different families of distant phylogenetic relationships. The structures of the glycoproteins and the carbohydrate-free polypeptides are so different that it can safely be assumed that the ability to produce antifreeze substances has evolved independently in different fish families.

Freeze-tolerant animals

Marine invertebrates

Fully aquatic marine invertebrates have no problem with freezing. Their osmotic concentration equals that of the surrounding water, and their freezing point therefore is the same. Even contact with a crystal of ice suspended in the water has no effect in causing ice formation.

The situation is different for invertebrate animals that live in the intertidal area. At high tide, when they are covered with water, they are safe from freezing; but when the tide recedes they are exposed to air that in winter is far below the freezing point. In winter they must tolerate the regular alternation between freezing and thawing twice a day: When they are in air their body water rapidly freezes to ice, and as the water rises again they thaw out.

An intertidal animal may be exposed to air temperatures as low as −30 °C for 6 hours or more, and as its internal temperature rapidly approaches that of the air, its body water freezes. At −30 °C more than 90% of the body water is frozen, and the remaining fluid therefore contains solutes in extremely high concentrations (Kanwisher 1959). This means that the cells, in addition to losing water to the ice crystals, must be able to tolerate an exceptional increase in osmotic concentration.

Examination of intertidal animals shows that they indeed freeze rather than become supercooled and that the ice causes a remarkable distortion of muscles and internal organs. The ice crystals are located outside the cells, which are shrunken and probably have no ice formed within them. Within a few seconds of thawing, the tissues again assume a normal appearance and the organism functions normally.

The intertidal animals obviously are freeze-tolerant in the sense that they survive extensive ice formation within their bodies. Many other animals can also survive in spite of extensive ice formation. Midge larvae (*Chironomus*) from Alaska can be frozen and thawed repeatedly without injury.

It may seem difficult to decide whether these tiny animals are supercooled or frozen to ice. However, the amount of water frozen to ice can be accurately determined without injury to the animal by an ingenious method that takes advantage of the change

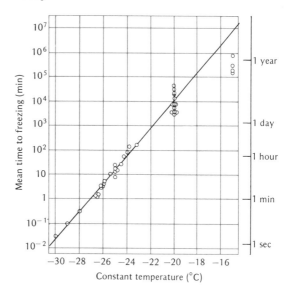

FIGURE 6.9 The relation between maintenance temperature and mean time to freezing for larvae of the wheat-stem sawfly. Each point shows the time required for 50% of a group of larvae to freeze. [Salt 1966]

in volume as water freezes to ice.* By determining the density of the frozen larva, it was shown that 70% of the body water was frozen at −5 °C, and 90% was frozen at −15 °C. As no harm is done to the larvae by these determinations, the very same specimen in which the extent of ice formation was known could be thawed out and tested for survival (Scholander et al. 1953a).

In addition to the degree of cooling, the duration of exposure to a given low temperature determines whether freezing takes place. The probability of ice formation in a supercooled liquid is increased by the passage of time. This time dependence is shown in Figure 6.9 for hibernating larvae of the wheat-stem sawfly, a plant-feeding wasp (*Cephus cinctus*). When these larvae are kept at constant subfreezing temperature, no freezing takes place at

---

*The density of water at 0 °C is 1.000 and the density of ice is 0.917.

temperatures above $-15\ ^{\circ}C$. As the temperature is lowered further, freezing takes place more and more rapidly so that at $-30\ ^{\circ}C$, freezing occurs in about 1.2 seconds, whereas at $-17\ ^{\circ}C$ it takes more than a year for 50% of the larvae in a large sample to freeze (Salt 1966).

These and many other insects contain a high concentration of glycerol in their body fluids. It has long been known that glycerol protects red blood cells and mammalian spermatozoa from injury caused by freezing. Glycerol is widely used for this purpose, and samples of human or bull sperm can be kept frozen and remain viable for several years if glycerol is added in a suitable concentration before freezing. Without such treatment, freezing is lethal to spermatozoans.

Because of its well-known protective action, the natural occurrence of high concentrations of glycerol in insects seems to explain their ability to survive very low temperatures. Glycerol could affect the cold resistance of insects in two ways: (1) by lowering the supercooling point, thereby increasing the probability that an insect will completely avoid ice formation, and (2) by its protective action against freezing damage once ice formation takes place. It appears that both effects are of importance and that all those insects that tolerate ice formation contain high concentrations of glycerol and other polyhydric alcohols such as sorbitol.

In many insects the concentration of these substances increases before winter in response to exposure to falling temperatures. The gallfly of goldenrod (*Eurosta solidaginis*) usually has a high concentration of glycerol, and when the insects are exposed to low temperatures, sorbitol is synthesized at the expense of stored glycogen, adding to the protective action of glycerol (Figure 6.10).

In the parasitic wasp *Bracon cephi* the concentration of glycerol may reach 30% in winter and decrease again in spring. This large amount of glycerol depresses the freezing point of the blood to $-17.5\ ^{\circ}C$, with the supercooling point lowered even more, so that the larvae of *Bracon* can be supercooled without ice formation to as low as $-47\ ^{\circ}C$.

FIGURE 6.10 When larvae of the gallfly *Eurosta solidaginis* are acclimated to low temperatures, their glycogen content is decreased and the concentrations of sorbitol and, to a lesser extent, glycerol are increased. Glycogen is expressed as glucose units after enzymatic hydrolysis. [Storey et al. 1981]

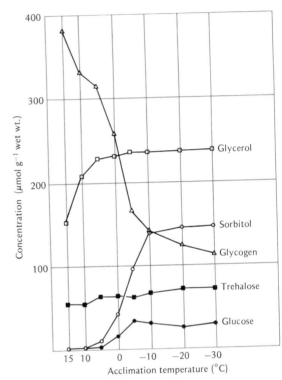

Glycerol, however, is not always a protection against injury from cold and freezing. Of 11 insect species collected in Alberta, Canada, 8 species, most of which contained more than 15% glycerol, were killed by freezing. Evidently, the glycerol could serve these animals by lowering the freezing point and permitting substantial supercooling, but it could not protect them against damage once ice formation occurred (Sømme 1964).

Ice nucleating agents

Although supercooling may protect an organism from ice formation, it is not an advantage if freezng eventually occurs. When ice begins to form in an animal that has been supercooled, ice crystals grow very rapidly and cause extensive damage. When such ice crystals puncture cell membranes and disrupt subcellular structures, it is always lethal.

Natural freeze-tolerance involves mechanisms that restrict ice formation to the extracellular fluid. When ice begins to form in the extracellular compartment, the still unfrozen fraction of extracellular fluid becomes osmotically more concentrated and withdraws water from the cells. This leaves the intracellular fluid more concentrated so that it remains unfrozen.

Supercooling and subsequent damage caused by freezing can be avoided by the presence of *nucleating agents* that actually aid in ice formation. Among a large number of beetles, all the freeze-tolerant species had nucleating agents in the hemolymph, and nucleating agents were lacking in those beetles that were freeze-intolerant (Zachariassen 1980). Nucleating agents have been found in a large number of freeze-tolerant insects.

A nucleating agent purified from the hemolymph of the bald-faced hornet, *Vespula maculata,* turned out to be a protein with a molecular weight of 74 000. This protein is highly hydrophilic, with glutamic acid and/or glutamine making up 20% of the amino acid residues, a strikingly high fraction. It is likely that the strong hydrophilic character of the protein is essential for its nucleating function (Duman et al. 1984).

Freeze-tolerant vertebrates

Although a natural tolerance to freezing and ice formation is the key to winter survival of many insects, only a few vertebrates tolerate extensive ice formation. Fish seem unable to tolerate freezing, and most higher vertebrates are freeze-intolerant under natural conditions.

Exceptions are found among some amphibians that overwinter on land. The treefrog, *Hyla versicolor,* can tolerate ice formation and in winter has 2% glycerol in its body fluids. In contrast, frogs that are freeze-intolerant, such as the common

leopard frog, *Rana pipiens,* have no glycerol in their body fluids. This frog overwinters in an aquatic habitat, while the treefrog overwinters on land, where temperatures readily fall to below freezing. Other freeze-tolerant frogs, such as the wood frog, *Rana sylvatica,* rapidly increase the blood glucose level in response to beginning ice formation, and this seems to increase their tolerance (Storey and Storey 1985).

Recently it has been found that some turtles may be freeze-tolerant. Hatchlings of the painted turtle, *Chrysemys picta,* remain in the nest where they have been hatched, and when the temperature falls in winter, they may survive with more than 50% of the extracellular water frozen to ice (Storey et al. 1988). Birds and mammals, however, are not known to tolerate freezing.

TEMPERATURE ADAPTATION

The limits of temperature tolerance for a given animal are not fixed. Exposure to a near-lethal temperature often leads to a certain degree of adaptation so that a previously lethal temperature is tolerated.

Frequently the range of thermal tolerance is different for the same species in summer and in winter. A winter animal often tolerates and even is active at a temperature so low that it is lethal to a summer animal; conversely, the winter animal is less tolerant to high temperature than a summer animal. Such changes in the temperature tolerance with climatic changes are called *acclimatization.*

Similar effects can be simulated in laboratory experiments by keeping animals for some time at given temperatures. To distinguish the adaptations or adjustments that take place in laboratory experiments from natural acclimatization, the response to experimental conditions is often described by the term *acclimation* (Hart 1952; Prosser 1973).

Acclimatization and acclimation can take place in response to many environmental factors, such as temperature, oxygen tension, nature of the food, moisture. Here we shall be concerned primarily with responses to temperature, a subject about which we

FIGURE 6.11 The relation between temperature tolerance and breeding time in four species of North American frogs. The horizontal lines show the temperature range for normal development, with the average water temperature at the time of egg-laying indicated by a circle. The crosses indicate lethal temperatures. [Moore 1939]

FIGURE 6.12 The bullhead, *Ictalurus* (*Ameiurus nebulosus*, has an upper lethal temperature limit that changes with the seasons. Each point indicates the temperature that is fatal to 50% of the fish after an exposure of 12 hours. [Fry 1947]

have more information than about the responses to other environmental factors.

Geographical differences and seasonal adjustments

It is self-evident that many Arctic animals live, breathe, and move about at temperatures close to freezing and in a range that would completely inactivate or kill similar animals from the tropics. Arctic fish and crustaceans are active in ice-cold water and die when the temperature exceeds 10 to 20 °C. Comparable tropical forms die below 15 to 20 °C. There are obvious differences in the temperature range that can be tolerated by animals from different geographical areas (Scholander et al. 1953b).

Closely related animals, such as various species of frogs of the genus *Rana*, have a geographical distribution that is closely related to their temperature tolerance. Figure 6.11 shows that, of four species of North American frogs, the species that has the northernmost geographical limit is also the most cold-tolerant.

The correlation between distribution and temperature tolerance extends to the temperature of the water at the time of breeding and to the lower as well as the upper limiting temperatures for embryonic development. Frogs that breed at relatively low

environmental temperatures have low minimal and maximal temperatures for breeding and for embryonic development (Table 6.3).

The geographically most widely distributed frog of these four species is *R. pipiens*. The population of this species from the northern part of the range differs from southern populations in the same way that the northern species differs from southern species. Thus, *R. pipiens* exists in races that show physiological differences in their responses to temperature. This is clearly related to the unusually wide geographical range over which this species is distributed (Moore 1949).

We should now examine more closely some natural and experimentally induced changes in temperature tolerance. In large parts of the world the seasonal temperature fluctuations are appreciable and during one season animals tolerate temperature extremes that in other seasons are fatal. For example, the bullhead, a catfish (*Ictalurus* [*Ameiurus*] *nebulosus*) has an upper lethal temperature of nearly 36 °C in summer, but in winter this fish will die if the temperature of the water exceeds 28 °C (Figure 6.12). These changes in temperature tolerance occur in response to the natural climatic change between summer and winter and, therefore, present an example of natural acclimatization.

The lower lethal temperature also changes with the seasons. As an example, summer individuals

TABLE 6.3 Comparison of breeding and development of four species of the frog *Rana*. [Moore 1939]

	R. sylvatica	*R. pipiens*	*R. palustris*	*R. clamitans*
Order of breeding	1	2	3	4
Water temperature at time of breeding (°C)	10	12	15	25
Northernmost record (°N)	67	60	55	50
Lower limiting embryonic temperature (°C)	2.5	6	7	11
Upper limiting embryonic temperature (°C)	24	28	30	35
Time between stages 3 and 20 at 18.5 °C (h)	87	116	126	138

of the beetle *Pterosticus brevicornis*, from Alaska, invariably die if they are frozen (which happens at −6.6 °C. In winter, however, they tolerate temperatures below −35 °C and complete freezing as well (Miller 1969).

We have now seen that there are phylogenetic differences among animals that are reflected in their geographical distribution and that there are seasonal changes in their thermal tolerance (acclimatization), changes that can be mimicked by laboratory exposure to appropriate thermal conditions (acclimation) while the phylogenetic limits remain unchanged.

Maximum range of tolerance

Since the limits of temperature tolerance are modified by the previous thermal history of an animal, how far can the limit be pushed by slow and gradual acclimation? Is there an ultimate limit that cannot be exceeded? To answer this question, groups of animals are kept at various constant temperatures for sufficient periods of time to ensure that they are fully acclimated at these temperatures. They can then be tested in order to establish the upper and lower limits of thermal tolerance, and when this has been done for a series of different temperatures, the results can be plotted in a single graph. Figure 6.13 is such a plot for the goldfish, showing its complete thermal tolerance.

The goldfish has an exceptionally wide temperature tolerance, and the area enclosed by the tolerance curve is therefore large. The tolerance curve

FIGURE 6.13 The complete range of thermal tolerance of goldfish is shown by the solid line. If a goldfish is kept at some temperature on the diagonal broken line until it is fully acclimated, its tolerance to high temperature can be read on the upper solid line, and its tolerance to cold on the lower solid line. For example, a fish kept at 30 °C has an upper lethal limit of 38 °C and a lower lethal limit of 9 °C. [Fry et al. 1942]

for a fish with a narrower range, such as the chum salmon (*Oncorhynchus keta*), covers a much smaller area.

Like other salmonids the chum salmon is a typ-

FIGURE 6.14 The ultimate thermal tolerance of the chum salmon, a cold-water fish, covers a much smaller range than that of the bullhead. The latter has a wide temperature tolerance similar to that of the goldfish (Figure 6.13). [Brett 1956]

TABLE 6.4 Thermal tolerance of various fresh-water and salt-water fish, expressed in units of degrees squared (see text for explanation). The American lobster is included for comparison with the fish.

Fish	Thermal tolerance ($°C)^2$
Goldfish (*Carassius auratus*)	1220[a]
Bullhead (*Ictalurus nebulosus*)	1162[a]
Greenfish (*Girella nigricans*)	800[b]
Silverside (*Menidia menidia*)	715[c]
Winter flounder (*Pseudopleuronectes americanus*)	685[c]
Speckled trout (*Salvelinus fontinalis*)	625[d]
Puffer (*Spheroides maculatus*)	550[c]
Chum salmon (*Oncorhynchus keta*)	468[a]
Lobster (*Homarus americanus*)	830[e]

[a] Brett (1956).
[b] Doudoroff (cited in McLeese 1956).
[c] Hoff and Westman (1966).
[d] Fry et al. (1946).
[e] McLeese (1956).

ical cold-water fish, and its upper lethal temperature cannot be extended beyond 24 °C. This is shown in Figure 6.14, which gives the thermal tolerance range for the chum salmon and for the brown bullhead, *Ictalurus nebulosus,* a fish with a large tolerance range, similar to the goldfish.

The total area enclosed by the tolerance curve for the bullhead is more than twice as large as that for the chum salmon. To quantify the area, we can take as our unit of measurement a small square that corresponds to the division of 1 degree on each coordinate. The area enclosed by the bullhead curve is 1162 such squares, which properly have the dimension degrees squared, $(°C)^2$; the curve for chum salmon enclosed 468 squares.

Most other fish whose thermal tolerance has been established fall between these two extremes. Both the bullhead and the chum salmon tolerate water at freezing temperatures, but many warm-water spe-

cies have curves that do not extend down to 0 °C, but are closed at some higher temperature. However, the complete thermal tolerance is known for only a relatively small number of fish, and it seems likely that, on the whole, fresh-water fish may have a wider tolerance than marine fish.

The tolerance of fresh-water fish ranges from 450 $(°C)^2$ for salmon to 1220 for goldfish (Brett 1956). Of three salt-water fish that have been examined, the silverside (*Menidia menidia*) has the largest range, 715 $(°C)^2$; a puffer (swellfish, *Spheroides maculatus*) has the smallest, 550 $(°C)^2$; and the winter flounder (*Pseudopleuronectes americanus*) has a range intermediate between the other two (Hoff and Westman 1966).

For easy comparison these data are combined with information about other fish in Table 6.4. The much greater thermal range of bullheads and goldfish is attributable to their better tolerance of high water temperatures, a fact obviously related to their habitat, which is shallow fresh water in which the tem-

FIGURE 6.15 The thermal acclimation of the bullhead takes place within about 24 hours when the water contains adequate oxygen, but proceeds very slowly at low oxygen concentration. [Fry 1947]

TABLE 6.5 Brown bullheads (*Ictalurus nebulosus*) acclimated to different temperatures between 7 and 32 °C initially show different thermal preferences, but during 24 hours all approach the same final thermal preference of about 29 °C. [Crawshaw 1975]

Acclimation temperature (°C)	Preference temperature (°C)		
	Immediate	10 h	24 h
7	16	25	~29
15	21	25	~29
24	26	29	~29
32	31	30	~29

perature can rise more readily than in the cool streams where the salmon is at home. The American lobster is added at the bottom of the list as a representative of an economically important invertebrate; if the lobster were a fish it would be closer to the top of the list.

More information about the thermal tolerance of fish is much needed, for the limits to thermal adaptation will become increasingly important in connection with problems of thermal pollution from industrial activity, especially power-generating plants and atomic reactors.

Rate of acclimation

The time needed for a fish to increase its ability to tolerate high temperatures is relatively short. It takes less than 24 hours for a bullhead to adjust fully when it is moved from 20 °C to 28 °C (Figure 6.15). The process of temperature acclimation depends on an adequate supply of oxygen.

Although the bullhead is tolerant to low oxygen and survives quite well, the process of thermal acclimation seems to be completely impeded in the absence of sufficient oxygen. In general, the acclimation goes faster the higher the temperature. Therefore it seems to be related to metabolic processes. However, there is no clear understanding

of the mechanisms responsible for the modification of the lethal limit, although changes in metabolic enzyme systems have an important role (Hochachka and Somero 1973).

If bullheads are given the opportunity to select their own preferred water temperature by presenting them with a temperature gradient, the water temperature they select depends on their previous thermal history (Table 6.5). Individuals acclimated to 7 °C initially select water at 16 °C, but they soon drift upward in their preference. After 10 hours they prefer 25 °C, and after 24 hours they select the same range of 29 to 30 °C that the fish warm-acclimated at 32 °C also select after 24 hours.

When a fish is moved from a higher to a lower temperature, the loss in tolerance to high temperatures and the gain in resistance to low temperatures are slower. In a minnow (*Pimephales promelas*) the complete adaptation takes between 10 and 20 days when the fish is moved from 24 °C to 16 °C (Brett 1956).

We have now seen that the thermal tolerance of a fish depends on the thermal history of the individual and that oxygen plays a role in the speed with which adaptation to a new temperature regimen takes place. However, other factors are important, such as the age and size of the fish and the quality of the water. Of particular interest is the finding that goldfish, even if they are kept in the laboratory at completely constant conditions of

FIGURE 6.16 Hypothetical change in metabolic rate (oxygen consumption, on logarithmic ordinate) in a fish acclimated to 5 °C water (a). If the fish compensates fully, acclimation to 25 °C water will make its metabolic rate follow the line marked b. If it does not compensate at all, its metabolic rate will follow the broken line marked (a').

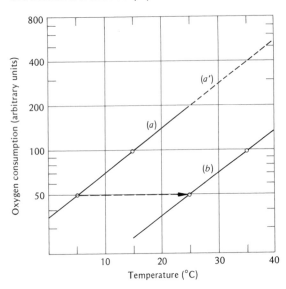

temperature and diet, show a marked seasonal change in thermal tolerance, with a greater cold resistance in the winter. This difference has been ascribed to the photoperiod (Hoar 1956). In general, a long photoperiod, as in summer, increases resistance to heat, and a short photoperiod, as in winter, causes an increase in tolerance to cold (Roberts 1964).

Thermal acclimation and metabolic rate

Until now we have been concerned with lethal temperatures and adaptation as a result of the thermal history of an animal. However, adaptation is not restricted to a change in lethal temperature limits; there are other compensating mechanisms as well, and these usually tend to counteract the acute effects of temperature change. Anybody who has caught fish in winter has noticed that the fish are

not appreciably slower and more sluggish than in summer. Laboratory studies confirm the impression that fish, if given time, to a great extent can compensate for temperature changes by appropriate alterations in metabolic rate. This is true not only for fish, but for many other poikilothermic animals as well.

Let us assume that a fish, adapted to a winter temperature of 5 °C, has a temperature tolerance range of 0 to 25 °C and that the Q_{10} throughout this range is 2.0. The metabolic rate of this fish follows the straight line marked a in Figure 6.16. Assume that we acclimate this fish to a summer temperature of 25 °C and that its upper lethal limit is extended to 40 °C. If no metabolic compensation takes place, the fish will have a metabolic rate extending along the broken line marked a'. If, on the other hand, the fish compensates fully for the change in temperature, the summer fish after adaptation to 25 °C will have the metabolic rate it previously had at 5 °C. If the Q_{10} remains the same (2.0), the oxygen consumption of the fish will now follow the line marked b. Thus, the fish will have compensated fully for the change in temperature and will have returned to its previous metabolic rate.

In addition to the two possibilities just mentioned, no compensation or complete compensation, there are several other long-term types or patterns of adaptation. The various patterns can be generalized as indicated in Figure 6.17.

If an animal shows no compensation and continues to consume oxygen at the same rate as caused by an acute effect of a temperature change, its metabolic rate will follow line a (the same as a' in Figure 6.16). Complete compensation, on the other hand, involves a return to the original oxygen consumption after the animal has had time to adapt to a new temperature, and the oxygen consumption will therefore follow line b, showing complete compensation. Any response between these two lines indicates partial compensation for the temperature change. The possibility also exists that the animal may overcompensate, as indicated by line 1. A few cases have been reported when a temper

FIGURE 6.17 The compensation for a temperature change, from T_1 to T_2, may take any of several courses, as indicated by the various curves. For details, see text. [Precht et al. 1955]

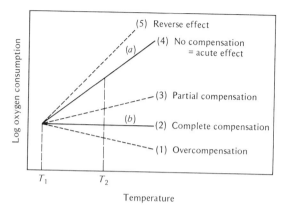

FIGURE 6.18 Metabolic response to temperature change of crayfish and trout. Initially the animals were acclimated to 15 °C. When the water temperature was dropped by 10 °C, the trout rapidly achieved a new plateau; the crayfish reached a plateau more slowly. When the temperature was again raised to 15 °C, the trout overcompensated before returning to the initial level; the crayfish only slowly responded to the temperature rise. [Bullock 1955]

ature change has a reverse effect, as indicated by line 5, at least within part of the temperature range.

In reality, animals do not often follow these idealized patterns; their responses are frequently far more complex. As an example, we can examine the responses of two aquatic animals, the trout and the European crayfish, to a 10 °C drop in temperature, followed by a 10 °C increase in temperature after full compensation has taken place (Figure 6.18).

An animal that shows little or no compensation is the beach flea (*Talorchestia megalophthalma*) from the area of Woods Hole, Massachusetts. The oxygen consumption of this animal in winter is the same as in summer, when measured at the same temperature. Thus, there is virtually no acclimatization or compensation for the seasonal change in temperature. Accordingly, the winter animals have such a low metabolic rate that they remain inactive, and they can be found immobile in small burrows, hibernating beneath the sand. If they are disturbed in winter, they respond very sluggishly, but when warmed they soon become as active as in summer (Edwards and Irving 1943b).

In contrast, another crustacean from the same

habitat, the sand crab (*Emerita talpoida*), compensates for low winter temperatures by increasing its metabolic rate. At 3 °C, winter animals consume oxygen at a rate four times as high as that of summer animals at the same temperature. These animals can remain active in cold weather, but because the sandy beaches of Woods Hole are most inhospitable during winter, the animals move from the intertidal zone out to deeper water (Edwards and Irving 1943a). Thus the sand crab adjusts to the seasonal changes in temperature so that it can remain active at the low winter temperatures, instead of becoming inactive and hibernating. Its lethal temperature also changes with the seasons. Temperatures above 27 °C are lethal in winter, but for summer animals death occurs above 37 °C.

We have now discussed a variety of important physiological effects of temperature, not only the lethal effects of very high or very low temperatures, but also the way organisms react and adjust within the range that is normal to them. We shall now turn to the subject of how animals make themselves independent of the thermal whims and vagaries of their environment.

REFERENCES

Brett, J. R. (1956) Some principles in the thermal requirements of fishes. *Q. Rev. Biol. 31:*75–87.

Brock, T. D. (1985) Life at high temperatures. *Science 230:*132–138.

Brown, J. H., and Feldmeth, C. R. (1971) Evolution in constant and fluctuating environments: Thermal tolerances of desert pupfish (*Cyprinodon*). *Evolution 25:*390–398.

Bullock, T. H. (1955) Compensation for temperature in the metabolism and activity of poikilotherms. *Biol. Rev. 30:*311–342.

Carlisle, D. B. (1968). *Triops* (Entomostraca) eggs killed only by boiling. *Science 161:*279–280.

Crawshaw, L. I. (1975) Attainment of the final thermal preferendum in brown bullheads, acclimated to different temperatures. *Comp. Biochem. Physiol. 52A:*171–173.

DeVries, A. L. (1970) Freezing resistance in antarctic fishes. In *Antarctic Ecology,* vol. 1 (M. Holdgate, ed.), pp. 320–328. New York: Academic Press.

DeVries, A. L. (1982) Biological antifreeze agents in coldwater fishes. *Comp. Biochem. Physiol. 73A:*627–640.

DeVries, A. L. (1984) Role of glycopeptides and peptides in inhibition of crystallization of water in polar fishes. *Phil. Trans. R. Soc. Lond. B304:*575–588.

Duman, J. G., Morris, J. P., and Castellino, F. J. (1984) Purification and composition of an ice nucleating protein from queens of the hornet, *Vespula maculata. J. Comp. Physiol. B. 154:*79–83.

Edwards, G. A., and Irving, L. (1943a) The influence of temperature and season upon the oxygen consumption of the sand crab, *Emerita talpoida* Say. *J. Cell. Comp. Physiol. 21:*169–182.

Edwards, G. A., and Irving, L. (1943b) The influence of season and temperature upon the oxygen consumption of the beach flea, *Talorchestia megalophthalma. J. Cell. Comp. Physiol. 21:*183–189.

Fletcher, G. L., Hew, C. L., and Joshi, S. B. (1982) Isolation and characterization of antifreeze glycoproteins from the frostfish, *Microgadus tomcod. Can. J. Zool. 60:*348–355.

Fraenkel, G. (1968) The heat resistance of intertidal snails at Bimini, Bahamas; Ocean Springs, Mississippi; and Woods Hole, Massachusetts. *Physiol. Zool. 41:*1–13.

Fry, F. E. J. (1947) Effects of the environment on animal activity. University of Toronto Studies, Biological Services, no. 55. *Publ. Ontario Fish. Res. Lab. 68:*1–62.

Fry, F. E. J., Brett, J. R., and Clawson, G. H. (1942) Lethal limits of temperature for young speckled trout (*Salvelinus fontinalis*). University of Toronto Studies, Biological Series, no. 54. *Publ. Ontario Fish. Res. Lab. 66:*1–35.

Hart, J. S. (1952) Geographic variations of some physiological and morphological characters in certain freshwater fish. University of Toronto Studies, Biological Series, no. 60. *Publ. Ontario Fish. Res. Lab. 72:*1–79.

Hinton, H. E. (1960) A fly larva that tolerates dehydration and temperatures of −270 °C to +102 °C. *Nature, Lond. 188:*336–337.

Hoar, W. S. (1956) Photoperiodism and thermal resistance of goldfish. *Nature, Lond. 178:*364–365.

Hochachka, P. W., and Somero, G. N. (1973) *Strategies of Biochemical Adaptation.* Philadelphia: Saunders. 358 pp.

Hoff, J. G., and Westman, J. R. (1966) The temperature tolerances of three species of marine fishes. *J. Mar. Res. 24:*131–140.

Kanwisher, J. (1959) Histology and metabolism of frozen intertidal animals. *Biol. Bull. 116:*258–264.

Lowe, C. H., Lardner, P. J., and Halpern, E. A. (1971) Supercooling in reptiles and other vertebrates. *Comp. Biochem. Physiol. 39A:*125–135.

McLeese, D. W. (1956) Effects of temperature, salinity, and oxygen on the survival of the American Lobster. *J. Fish. Res. Bd. Can. 13:*247–272.

Marzusch, K. (1952) Untersuchungen über die Temperaturabhängigkeit von Lebensprozessen bei Insekten unter besonderer Berücksichtigung winterschlafender Kartoffelkäfer. *Z. Vergl. Physiol. 34:*75–92.

Miller, L. K. (1969) Freezing tolerance in an adult insect. *Science 166:*105–106.

Miller, K. (1982) Cold-hardiness strategies of some

adult and immature insects overwintering in interior Alaska. *Comp. Biochem. Physiol. 73A*:595–604.

Moore, J. A. (1939) Temperature tolerance and rates of development in the eggs of Amphibia. *Ecology 20*:459–478.

Moore, J. A. (1949) Geographic variation of adaptive characters in *Rana pipiens* Schreber. *Evolution 3*:1–24.

Pitkow, R. B. (1960) Cold death in the guppy. *Biol. Bull. 119*:231–245.

Precht, H., Christophersen, J., and Hensel, J. (1955) *Temperatur und Leben*. Berlin: Springer-Verlag, 514 pp.

Prosser, C. L. (1973) *Comparative Animal Physiology*, 3d ed. Philadelphia: Saunders. 966 pp.

Roberts, J. L. (1964) Metabolic responses of freshwater sunfish to seasonal photoperiods and temperatures. *Helgol. Wiss. Meeresunters. 9*:459–473.

Salt, R. W. (1966) Relation between time of freezing and temperature in supercooled larvae of *Cephus cinctus* Nort. *Can. J. Zool. 44*:947–952.

Scholander, P. F., Flagg, W., Hock, R. J., and Irving, L. (1953a) Studies on the physiology of frozen plants and animals in the arctic. *J. Cell. Comp. Physiol. 42*:(Suppl. 1)1–56.

Scholander, P. F., Flagg, W., Walters, V., and Irving, L. (1953b) Climatic adaptation in arctic and tropical poikilotherms. *Physiol. Zool. 26*:67–92.

Somero, G. N., and DeVries, A. L. (1967) Temperature tolerance of some antarctic fishes. *Science 156*:257–258.

Sømme, L. (1964) Effects of glycerol on cold-hardiness in insects. *Can. J. Zool. 42*:87–101.

Storey, J. M., and Storey, K. B. (1985) Triggering of cryoprotectant synthesis by the initiation of ice nucleation in the freeze tolerant frog, *Rana sylvatica. J. Comp. Physiol. B 156*:191–195.

Storey, K. B., Baust, J. G., and Storey, J. M. (1981) Intermediary metabolism during low temperature acclimation in the overwintering gall fly larva, *Eurosta solidaginis. J. Comp. Physiol. 144*:183–190.

Storey, K. B., Storey, J. M., Brooks, S. P. J., Churchill, T. A., and Brooks, R. J. (1988) Hatchling turtles survive freezing during winter hibernation. *Proc. Natl. Acad. Sci. U.S.A. 85*:8350–8354.

Zachariassen, K. E. (1980) The role of polyols and nucleating agents in cold-hardy beetles. *J. Comp. Physiol. 140*:227–234.

ADDITIONAL READING

See also reference list at end of Chapter 8

Asahina, E. (1969) Frost resistance in insects. *Adv. Insect Physiol. 6*:1–49.

Baust, J. G. (1981) Biochemical correlates to cold hardening in insects. *Cryobiology 18*:186–198.

Bowler, K., and Fuller, B. J. (eds.) (1987) *Temperature and Animal Cells*. Symposium, Society for Experimental Biology, no. 41. Cambridge: The Company of Biologists Ltd. 460 pp.

Brock, T. D. (1985) Life at high temperatures. *Science 230*:132–138.

DeVries, A. L. (1983) Antifreeze peptides and glycopeptides in cold-water fishes. *Annu. Rev. Physiol. 45*:245–260.

Duman, J., and Horwath, K. (1983) The role of hemolymph proteins in the cold tolerance of insects. *Annu. Rev. Physiol. 45*:261–270.

Hochachka, P. W., and Somero, G. N. (1973) *Strategies of Biochemical Adaptation*. Philadelphia: Saunders. 358 pp.

Levitt, J. (1980) *Responses of Plants to Environmental Stresses*, 2d ed.: vol. 1, *Chilling, Freezing, and High Temperature Stresses*. New York: Academic Press. 512 pp.

Mazur, P. (1984) Freezing of living cells: mechanisms and implications. *Am. J. Physiol. 247*:C125–C142.

Miller, K. (1982) Cold-hardiness strategies of some adult and immature insects overwintering in interior Alaska. *Comp. Biochem. Physiol. 73A*:595–604.

Precht, H., Christophersen, J., Hensel, H., and Larcher, W. (1973) *Temperature and Life*. New York: Springer-Verlag. 779 pp.

Prosser, C. L., and Nelson, D. O. (1981) The role of nervous systems in temperature adaptation of poikilotherms. *Annu. Rev. Physiol. 43*:281–300.

Somero, G. N., and Hochachka, P. W. ((1976) Biochemical adaptations to temperature. In *Adaptation to Environment: Essays on the Physiology of Marine Animals* (R. C. Newell, ed.), pp. 125–190. London: Butterworth.

Storey, K. B. (1987) What contributes to freeze tolerance? *News Physiol. Sci. 2*:157–160.

Storey, K. B., and Storey, J. M. (1988) Freeze tolerance in animals. *Physiol. Rev. 68*:27–84.

Zachariassen, K. E. (1985) Physiology of cold tolerance in insects. *Physiol. Rev. 65*:799–832.

CHAPTER SEVEN

Temperature Regulation

The temperature of most animals follows passively that of the surroundings. It would seem advantageous if they could free themselves from the vagaries of this environmental variable, and in this chapter we shall deal with animals that keep their body temperature more or less independent of the environmental temperature.

Birds and mammals live through most of their lives with body temperatures that fluctuate no more than a few degrees. We shall consider these animals first, and then discuss what other animals have achieved in keeping their temperature independent of the environment.

First, we should clarify what we mean by *body temperature,* by no means a simple concept. Second, to maintain a constant temperature, the heat gained and the heat lost from the organism must be equal. To understand these two processes, we must be familiar with the simple physics of heat transfer.

In a *cold environment* a high temperature can be maintained by reducing the heat loss and/or increasing the heat gain (heat production). Most birds and mammals do this very well. However, some mammals and a few birds seem to give up the fight against cold and permit their temperature to drop precipitously; they go into torpor or hibernation. Nevertheless, temperature regulation is not abandoned; on the contrary, hibernation is a well-regulated physiological state.

In a *hot environment* the problems of maintaining the body temperature are reversed: The animal must keep the body temperature from rising and is often compelled to cool itself by evaporation of water.

Not only birds and mammals, but certain other animals are amazingly adept at keeping their temperature above that of the environment. This applies, for example, to lizards that bask in the sun, to many insects, and amazingly, even to some fish that maintain parts of their body at temperatures approaching those of ''warm-blooded'' animals.

TABLE 7.1 Heat production in the major organs of a man at rest (body mass, 65 kg; heat production, 1872 kcal per day = 78 kcal per hour = 90.65 W). The main internal organs weigh about 5 kg but account for 72% of the total heat production. [Aschoff et al. 1971]

Organ	Organ mass				Heat production at rest			
	kg		% of body mass		kcal h^{-1}		% of total	
Kidneys	0.29 ⎤		0.45 ⎤		6.0		7.7 ⎤	
Heart	0.29 ⎥		0.45 ⎥		8.4		10.7 ⎥	
Lungs	0.60 ⎬ 5.03		0.9 ⎬ 7.7		3.4		4.4 ⎬ 72.4	
Brain	1.35 ⎥		2.1 ⎥		12.5		16.0 ⎥	
Splanchnic organs[a]	2.50 ⎦		3.8 ⎦		26.2		33.6 ⎦	
Skin	5.00 ⎤		7.8 ⎤		1.5		1.9 ⎤	
Muscle	27.00 ⎬ 59.97		41.5 ⎬ 92.3		12.2		15.7 ⎬ 27.6	
Other	27.97 ⎦		43.0 ⎦		7.8		10.0 ⎦	
Total	65.00		100.0		78.0		100.0	

[a] Abdominal organs, not including kidneys.

BODY TEMPERATURE OF BIRDS AND MAMMALS

What is body temperature?

The heat produced by an animal must be transported to the surface before it can be transferred to the environment. Therefore, the surface of the organism must be at a lower temperature than the inner parts, for if the temperature were the same throughout, no heat could be transferred. The conclusion is that the temperature of an organism of necessity cannot be uniform throughout.

If we examine where in the mammalian body heat production takes place, we find that some parts produce more heat than others. In humans the organs of the chest and the abdomen, although they make up less than 6% of the body mass, produce 56% of the total heat (Table 7.1). If we include the brain, which in humans is large and has a high heat production, we have accounted for 72%, or more than two-thirds of the total heat production, in less than 8% of the body mass.

We can therefore consider that the body consists of a core where most of the heat is produced, and a much larger shell that includes skin and muscles

and produces only a small fraction of the total body heat.

During exercise, the situation is different, for the total metabolic rate may increase 10-fold or more. Most of this increase occurs in the muscles (including the diaphragm and other respiratory muscles). During exercise, then, for the internal temperature to remain constant, more than 10 times as much heat as was produced at rest must be transported to the surface of the organism.

Temperature distribution in the body

The inner, or core, temperature remains reasonably constant, but this does not mean that the temperature throughout the core is uniform. Organs that have a high rate of heat production may be warmer than others, but they are cooled by the blood (i.e., the venous blood that leaves these organs is warmer than the arterial blood). The temperature differences in the core may be as much as 0.5 °C from one site to another. We therefore cannot speak about a single core temperature, but for practical purposes the deep rectal temperature is often used as a representative measure.

The surface temperature of a person who is in

FIGURE 7.1 Temperature distribution in the body of a person at room temperatures of 20 °C (left) and at 35 °C (right). The isotherms, which indicate sites of equal temperature, show that, at 35 °C room temperature, a core temperature of 37 °C (shaded area) extends into the legs and arms. At 20 °C room temperature, the temperature gradients in the shell extend throughout the legs and arms, and the core temperature is restricted to the trunk and head. [Aschoff and Wever 1958]

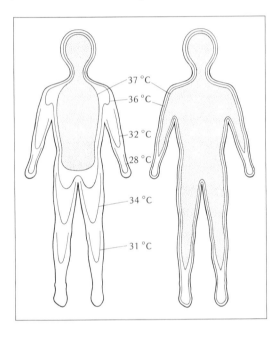

heat balance is always lower than the core temperature. This means that the arterial blood that flows to the shell loses heat and returns as colder venous blood. This is, of course, how most of the heat produced in the core is brought to the surface or, in other words, how the core organs are cooled. Depending on the circumstances (external temperatures and the need for heat loss), the surface temperature varies a great deal. Also, the underlying tissues, including a large part of the muscle mass, can take on temperatures considerably below the core temperature (Figure 7.1).

What is the mean temperature of the body? We

can calculate a mean body temperature from multiple measurements at various sites, provided we have a weighting procedure to apply to the various measurements. However, mean body temperature calculated from multiple measurements is physiologically relatively meaningless.

We have already seen that the core temperature does not represent the heat status of the whole body, that the temperature of the shell can vary widely, and that the depth in the body to which the shell temperature extends can change drastically. A change in the temperature of the shell means that the total amount of heat in the body changes, although the core temperature may remain constant. If a person moves from a room temperature of 35 °C to 20 °C, the drop in shell temperature may involve a heat loss of 200 kcal (>800 kJ) from the shell. The magnitude of this much heat becomes clear when we realize that it corresponds roughly to 3 hours' resting metabolism of the person.

"Normal" body temperature of birds and mammals

Daily fluctuations in core temperature

The core temperature of humans and of other mammals and birds undergoes regular daily fluctuations. Within 24 hours these fluctuations are usually between 1 and 2 °C. Diurnal animals show a temperature peak during the day and a minimum at night; nocturnal animals show the reverse pattern. However, these daily cycles are not caused directly by the alternating periods of activity and rest, for they continue even if the organism is at complete rest.

The daily body temperature pattern of many mammals and birds consistently follows the light cycle. When the towhee (*Pipilo aberti*) is kept in 12 hours of darkness and 12 hours of light, its body temperature follows a cycle synchronized with the light cycle. At a room temperature of 23 °C, the core temperature at night is about 39 °C and when the light comes on, it rapidly rises to nearly 42 °C

FIGURE 7.2 When the towhee (a finch, *Pipilo aberti*) is kept at a constant room temperature of 23 °C, its body temperature varies with the light cycle. When the lights come on at 0600 hours, the body temperature rises by nearly 3 °C, to drop again when the lights go off at 1800 hours. If the room temperature is reduced to 5 °C, the body temperature cycle is similar, but at a slightly higher level. [Dawson 1954]

external cues, it must be inherent in the organism; it must be a truly endogenous cycle.

The cycle is self-sustained in the sense that, in the absence of external timing cues, it continues for days, weeks, and even months. It remains extremely regular, but its duration usually is not exactly 24 hours. As a result, the timing of a free-running cycle in the absence of an external cue gradually drifts away from the environmental 24-hour day-and-night cycle.

We have now seen that the temperature of the body core and shell may differ greatly and that there are regular variations of several degrees that make measurements of the core temperature meaningful only if we also have information about activity, the time of day, and the usual temperature cycle of the animal.

Temperature differences among animal groups

If we disregard variations that amount to a couple of degrees, we see that the usual or ''normal'' core temperature is almost uniform within each of the major groups of warm-blooded vertebrates, but that there are characteristic differences among the groups (Table 7.2).

Allowing for the fact that it is difficult to establish what is normal body temperature for a given animal, and disregarding variations caused by external conditions and activity, we can as a rule of thumb say that most birds maintain their body temperature at 40 ± 2 °C, eutherian mammals at 38 ± 2 °C, marsupials at 36 ± 2 °C, and monotremes at 31 ± 2 °C. Small birds may have somewhat higher body temperatures than large birds (McNab 1966), but in mammals there is no clear relationship between body size and body temperature (Morrison and Ryser 1952). For marsupials the available information is insufficient to show whether there is any relationship between body size and body temperature.

The fact that the so-called primitive groups – insectivores, marsupials, and especially monotremes – consistently have low body temperatures

(Figure 7.2). At a different room temperature (5 °C, for example) the core temperature follows the same day-and-night cycle, although it is, on the average, about 0.5 °C higher than the corresponding core temperature measured at 23 °C.

Similar temperature cycles have been recorded in a variety of other birds from several different families. Diurnal forms always have the highest temperature during the day and nocturnal species at night. One example of a nocturnal bird is the flightless kiwi (*Apteryx*) from New Zealand. Its core temperature is lower than in most other birds, with a daytime mean of 36.9 °C. At night, which is the normal period of activity for the kiwi, the temperature increases to a mean of 38.4 °C (i.e., at night the temperature is about 1.5 °C higher than in the daytime) (Farner 1956).

Both the temperature cycle and the corresponding cycle in metabolic rate can be reversed by reversing the periods of light and dark. This shows that the cycle is governed by the illumination. However, even if all variations in the light are removed and the animals are kept in continuous uniform light, their temperature variations continue on nearly the same timing. This is called the *free-running period*, and because it persists without any

TABLE 7.2 Approximate normal and lethal core temperatures of some major groups of mammals and birds. The lethal temperatures are based on observations made under a wide variety of conditions. There is rather consistently an approximately 6 °C interval between the normal and the lethal temperatures for the same animal.

Animal	Approximate normal core temperature (°C)	Approximate lethal core temperature (°C)
Monotreme (echidna)	30–31[a]	37[a]
Marsupials	35–36[b]	40–41[e]
Insectivore (hedgehog)	34–36	41[f]
Man	37	43
Eutherian mammals	36–38[c]	42–44[g]
Bird (kiwi)	38[d]	
Birds, nonpasserine	39–40[b]	46[h]
Birds, passerine	40–41[b]	47[i,j]

[a] Schmidt-Nielsen et al. (1966).
[b] Dawson and Hulbert (1970).
[c] Morrison and Ryser (1952).
[d] Farner (1956).
[e] Robinson and Morrison (1957).
[f] Shkolnik and Schmidt-Nielsen (1976).
[g] Adolph (1947).
[h] Robinson and Lee (1946).
[i] Calder (1964).
[j] Dawson (1954).

raises some interesting evolutionary problems. These groups are considered to be very ancient and presumably have had more time than more recent groups to evolve toward a high body temperature, if this is "desirable" or advantageous. Have they remained in their more "primitive" stage because they have lacked the capacity to evolve in this direction? They have certainly been successful as witnessed by their ability to survive for so long.

The fact is that we do not fully understand the advantages of any given body temperature. In any event, it would be a mistake to interpret a low body temperature as a sign of "primitive" and thus inadequate temperature regulation. It has been said that the egg-laying echidna is halfway to being a cold-blooded animal and is unable to regulate its body temperature adequately. In fact, the echidna is an excellent temperature regulator and can maintain its core temperature over a wide range of ambient temperatures down to freezing or below, although it has a poor tolerance to high temperature (Schmidt-Nielsen et al. 1966).

The approximate lethal body temperatures for the various groups of warm-blooded vertebrates are given in the last column of Table 7.2. It seems that the lethal temperature is regularly at roughly 6 °C above the normal core temperature. Thus, the echidna dies when its body temperature reaches 37 °C, which is a normal temperature for placental mammals and well below the normal temperature for birds. The margin of safety remains uniformly at about 6 °C from group to group, but we do not know why the lethal temperature is at such a constant level relative to the normal core temperature.

Temperature of cold-climate animals

Do Arctic birds and mammals maintain body temperatures within the same range as species from warmer climates?

To answer this question the body temperatures of a number of Alaskan birds and mammals were measured while the animals were exposed to a wide range of low air temperatures. The birds studied belonged to 30 different species and ranged in weight from 0.01 to 2 kg. In air temperatures from +20 °C down to −30 °C, their mean body temperature was 41.1 °C, which is within the normal range for birds from moderate or tropical climates.

Arctic mammals of 22 species, weighing from 0.1 to 1000 kg, were exposed to temperatures down to −50 °C, and one species (the white fox) even to −80 °C. The individuals of all species maintained their body temperatures within normal mammalian limits. The mean for all the observed species was 38.6 °C, which is about 0.5 °C higher than the previously reported mean temperature for a large number of mammals from temperate regions, but this difference is considered insignificant (Irving and Krog 1954).

We can therefore conclude that Arctic birds and

mammals maintain body temperatures characteristic of their groups, although they live in some of the coldest areas on earth.

Animals in natural conditions

In contrast to the large amount of information on body temperature in captive and laboratory animals, there are few long-term studies on animals under natural conditions. The development of methods to measure temperature with transducers that transmit information by radio (telemetry) has made it possible to obtain recordings from unrestrained and completely undisturbed animals.

Surgical implantation of such a telemetric unit has enabled the monitoring of the temperature of a single animal for as long as 1 year. When this technique was used on sheep grazing under field conditions, the daily range of the deep body temperature was less than 1 °C regardless of sun, rain, or storm. The total temperature range observed during a whole year from summer to the most severe winter was 1.9 °C, from 37.9 to 39.8 °C (Bligh et al. 1965).

Fever

Fever is an increase in body temperature that usually is associated with microbial or virus infections. Similar increases in body temperature can be produced by the injection of killed bacteria. These contain substances, *pyrogens,* that are the cause of the fever reaction. The organism behaves as if the set point of a thermostat has been increased by a few degrees and now regulates as if it wants to maintain the increased temperature.

Fever is definitely not a failure of temperature regulation. If a person with fever is challenged with cooling, increased heat production serves to maintain the higher temperature, and vice versa if an extra heat load is applied. The processes of temperature regulation respond as expected, except that the reactions revolve around the higher body temperature.

The concept that fever represents a well-regulated resetting of a "thermostat" is further supported by the following experiment. If a pyrogen is injected into a dog, the animal develops a fever. However, if heat is applied locally only to the hypothalamus, where the temperature regulation center is located, the fever in the body is suppressed. The body behaves as if the higher "set-point" has already been reached. This indicates that the bacterial pyrogen does not somehow damage the regulatory mechanism in the hypothalamus, but that a certain hypothalamic temperature is "desirable," and when this has been attained, no further heating is called for (Andersen et al. 1961).

Both mammals and birds develop fever after becoming infected with bacteria, but it is unclear whether the fever is beneficial or harmful. An interesting attempt at settling this much-argued question has been made by studying, not warm-blooded mammals or birds, but a cold-blooded animal, the lizard *Dipsosaurus dorsalis*. When *Dipsosaurus* is placed in an environment where there is a range of temperatures, it selects a preferred temperature where it tends to remain so that it maintains body temperature of about 38.5 °C. If the lizard is injected with an appropriate bacterial suspension, it seeks a somewhat warmer environment and maintains a higher body temperature, a "fever" of approximately 2 °C.

Amphibians and fish show similar reactions in response to bacterial pyrogens (Reynolds et al. 1976; Kluger 1977). They select a warmer environment and thus maintain higher preferred body temperatures, demonstrating that the "fever" response is present in a broad range of vertebrates and not only in the warm-blooded mammals and birds. Even cockroaches, when injected with bacteria and placed in a thermal gradient, prefer temperatures significantly elevated over what they normally do (Bronstein and Conner 1984).

Let us return to the question of whether fever is beneficial or harmful. An answer has been sought by performing the following experiment on *Dipsosaurus*. Five groups of lizards were infected with bacteria (*Aeromonas hydrophila*) and kept at uniform constant temperatures. One group was kept at a neutral 38 °C, two groups at lower temperatures (36 and 34 °C), and two at higher temperatures (40 and 42 °C). There was a striking correlation between survival and temperature; all the animals at the lowest temperature died in less than 4 days; survival increased with temperature and was highest in the animals at 42 °C. This is approximately the same temperature that injected animals voluntarily chose if they were allowed to select their temperature (Kluger et al. 1975). In these animals, then, a clear beneficial effect was associated with the higher body temperature.

The ability to maintain a constant body temperature, as birds and mammals do, and the ability to reset the temperature with full maintenance of regulation as occurs in fever, require extraordinary capacities of the physiological mechanisms for heat exchange. Before we discuss these mechanisms we will need an elementary knowledge of the physics of heat transfer.

TEMPERATURE, HEAT, AND HEAT TRANSFER

In the preceding section we were concerned with temperature, and the concept of heat was mentioned only incidentally. It is important to understand the difference between these two physical quantities and to realize that the measurement of temperature does not necessarily give any information about heat.

Temperature is usually measured in degrees Celsius (°C), although in physical chemistry and thermodynamics we use absolute temperature expressed in kelvins (K).*

In biology heat is usually measured in calories, and 1 calorie (cal) is defined as the amount of heat needed to raise the temperature of 1 g water by 1 °C. The calorie is not a part of the International System of Units (the SI System), but the term is so common that it will remain in use for some time to come. For conversion to SI units, 1 cal = 4.184 joules (J).

To heat 1 g water from room temperature (25 °C) to the boiling point (100 °C) requires 75 cal of heat. To raise the temperature of 100 g water by 75 °C requires 7500 cal (7.5 kcal). In both cases the initial and final temperatures are the same, and temperature alone therefore gives no information about the amount of heat added. If we know the amount of water, however, we can calculate the amount of heat from the temperature change, for by definition, the heat needed to raise the temperature of 1 g water by 1 °C is 1 cal. The amount of heat needed to warm 1 g of substance by 1° is known as the *specific heat capacity* of that substance.[†] The specific heat capacity of water is 1.0 cal g^{-1} $°C^{-1}$, which compared with that of other substances is very high. The specific heat capacity of rubber is 0.5, of wood 0.4, and of most metals 0.1 or less. The specific heat capacity of air is 0.24 cal g^{-1} $°C^{-1}$, and because the density of air is 1.2 g per liter (at 20 °C), the heat capacity of 1 liter of air is 0.3 cal $°C^{-1}$.

The amount of heat needed to increase the temperature of the animal body is slightly less than the amount needed to heat the same mass of water. The mean specific heat capacity of the mammalian body is about 0.8. Thus, to increase the temperature of a 1000-g mammal by 1 °C requires about 800 cal. The exact value for the specific heat capacity of the animal body varies somewhat. For example, the specific heat capacity of mouse bodies was found to vary between 0.78 and 0.85 with a mean of 0.824 (Hart 1951).

The bulk of the body is water, which has a specific heat capacity of 1.0, and the other components – proteins, bone, and fat – tend to reduce

---

*Absolute zero is at −273.15 °C. Celsius temperature (T_C) relates to absolute temperature (T_K) as follows: $T_K = T_C + 273.15$.

†It is recommended that the word *specific* before the name of a physical quantity be restricted to the meaning "divided by mass" (Council of the Royal Society 1975).

this value. Fat is particularly important, for its amount can vary within wide limits, and its specific heat capacity is only about 0.5.

For many purposes it is sufficient to use a mean value of 0.8 for the specific heat capacity of the animal body, for in order to determine a change in heat content, we must know the mean body temperature, and as we have seen, this variable is very difficult to determine with accuracy.

Physics of heat transfer

For a body to maintain a constant temperature, there is one absolute requirement: Heat loss must exactly equal heat gain. For an animal to maintain a constant temperature, heat must be lost from the body at the same rate it is produced by metabolic activity.

As we have seen, metabolic heat production can easily increase more than 10-fold with activity, and unless heat loss is increased in the same proportion, body temperature will rise rapidly. Furthermore, the conditions for heat loss vary tremendously with external factors such as air temperature and wind. An understanding of the physiological mechanisms involved in the regulation of heat production and heat loss requires an elementary knowledge of the physics of heat transfer.

Whenever physical materials are at different temperatures, heat flows from a region of higher temperature to one of lower temperature. This transfer of heat takes place by *conduction* and by *radiation*. A body cannot lose heat by conduction or radiation unless its environment, or some part of it, is at a lower temperature than the surface of the body. There is, however, a third way to remove heat: the *evaporation* of water. These three ways of heat transfer – conduction,* radiation, and

*Transfer of heat between a surface and a fluid (gas or liquid) in contact with it takes place by conduction. Mass movement in the fluid, termed *convection,* contributes to renewal of fluid in the boundary layer and thus complicates the conductive heat transfer (see later in this chapter).

FIGURE 7.3 Heat flow in a uniform conductor depends on its cross section, the temperature gradient, and the material from which it is made.

evaporation – are the only means available for the removal of the heat produced in the metabolic activity of living organisms.

Conduction

Conduction of heat takes place between physical bodies that are in contact with each other, whether they are solids, liquids, or gases. Conduction of heat consists of a direct transfer of the kinetic energy of molecular motion, and it always occurs from a region of higher temperature to one of lower temperature.

Assume that we have a uniform conductor in which we keep one end warm and the other cold (Figure 7.3). The rate of heat transfer by conduction $\dot{Q}$) can now be expressed as:

$$\dot{Q} = k\,A\,\frac{T_2 - T_1}{l}$$

where k is the thermal conductivity of the conductor, A is the area through which the heat flows

TABLE 7.3 Thermal conductivities (k) for a variety of common materials. [Hammel 1955; Hensel and Bock 1955; Weast 1969]

Material	k (cal s^{-1} cm^{-1} °C^{-1})
Silver	0.97
Copper	0.92
Aluminum	0.50
Steel	0.11
Glass	0.002 5
Soil, dry	0.000 8
Rubber	0.000 4
Wood	0.000 3
Water	0.001 4
Human muscle	0.001 1
Adipose tissue	0.000 51
Air	0.000 057
Animal fur	0.000 091

(normal to the direction of heat flow), and T_2 and T_1 are the temperatures at two points separated by the distance l. The fraction $(T_2 - T_1)/l$ is known as the *temperature gradient* and stands for the temperature difference per unit distance along the conductor.

This expression for heat flow in a conductor can be stated in simple intuitively obvious terms. Heat flow increases with the thermal conductivity (k) of the conducting material, with increasing cross-sectional area (A) of the conductor, and with an increased temperature difference between two points T_2 and T_1. Increasing the distance (l) between two given temperatures, T_2 and T_1 (if these remain unchanged), decreases the amount of heat flow.

The thermal conductivity coefficient (k), is an expression for how easily heat flows in a given material. Values for the thermal conductivity of some common materials are given in Table 7.3. As we know, metals are excellent conductors and have high conductivity coefficients. Glass and wood are poorer conductors; water and human tissues have slightly lower conductivities again, but of a similar order of magnitude. The similarity between the thermal conductivities of human tissue and water is, of

course, attributable to the fact that most tissues consist of roughly two-thirds to three-quarters water. Air and animal fur have very low thermal conductivities, which means that their insulation value (resistance to heat flow) is high. The main reason the thermal conductivity of fur is low is the large amount of air trapped between the hairs. Other materials that enclose a high proportion of air (e.g., felt, woolen fabrics, down) are also poor conductors or excellent insulators.

The simple equation for heat conduction given above unfortunately applies only when heat flows through a plane object such as a wall. Most animal surfaces are curved, and this makes the conduction equation considerably more complex. For practical purposes, we can consider the heat flow as dependent on temperature gradients and area, but if we wish to make a quantitatively satisfactory analysis of heat transfer in an animal, it is necessary to apply a more rigorous treatment of the physics of heat transfer.

Convection

The transfer of heat in fluids is almost invariably accelerated by the process of *convection,* which refers to mass movement of the fluid. Assume a cold fluid in contact with a warm solid surface. Heat flows into the fluid by conduction, and the fluid adjacent to the surface becomes warmer. If the fluid is in motion, the warm fluid adjacent to the solid surface is replaced by cold fluid, and the heat loss from the solid surface is therefore speeded up. Mass flow, or convection, in the fluid thus facilitates heat loss from the solid, although the transfer process between solid and fluid remains one of conduction.

Convection in a fluid may be caused by temperature differences or by external mechanical force. Heating or cooling of a fluid usually changes its density, and this in turn causes mass flow. For example, if a warm solid surface is in contact with a cold fluid, the heated fluid expands and therefore rises, being replaced by cool fluid. In this case the mass flow, or convection, is caused by the temperature difference and is called *free* or *natural con-*

vection. This term applies also if the wall is colder than the fluid and the fluid adjacent to the wall becomes denser and sinks.* Free convection, of course, can take place both in air and in water and contributes substantially to the rate of heat loss from living organisms.

Motion in the fluid can also be caused by external forces, such as wind, water currents, or an electric fan. Convection caused by external forces, as opposed to density changes, is referred to as *forced convection.*

Because convection depends on mass transfer in fluids, the process is governed by the rather complex laws of fluid dynamics, which include such variables as the viscosity and density of the fluids, in addition to their thermal conductivity. Convective heat loss does not depend only on the area of the exposed surface. Variables such as the curvature and the orientation of the surface give rise to rather complex mathematical expressions, which cause great difficulties in the analysis of the heat transfer from an animal. As we shall see below, however, we can find practical means for analyzing the heat transfer between an animal and the environment that circumvent the need for an exact analysis of the physical processes involved.

Radiation

Heat transfer by radiation takes place in the absence of direct contact between objects. All physical objects at a temperature above absolute zero emit electromagnetic radiation. The intensity and the wavelength of this radiation depend on the temperature of the radiating surface (and its emissivity, which we shall discuss below). All objects also receive radiation from their surroundings. Electromagnetic radiation passes freely through a vac-

---

*Water has a higher density at +4 °C than at the freezing point. This is of great importance in bodies of fresh water, but in most physiological situations the anomalous density of water is of no significance. Sea water does not exhibit the anomalous density properties at near-freezing temperature (see also page 226).

uum, and for our purposes atmospheric air can be regarded as fully transparent to radiation.

The *intensity* of radiation from an object is proportional to the fourth power of the absolute temperature of the surface. This is expressed by the *Stefan-Boltzmann law* for heat radiation flux:

$$\dot{Q}_R \propto \sigma T^4$$

where T is the absolute temperature of the radiating surface (in kelvin, K), and σ is Stefan-Boltzmann's constant (1.376×10^{-12} cal s^{-1} cm^{-2} K^{-4} or 5.67×10^{-8} W m^{-2} K^{-4}). Because the amount of heat or energy radiated increases with the fourth power of the absolute temperature, the emission increases very rapidly indeed with the surface temperature.

The *wavelength* of the emitted radiation depends on the surface temperature, and the hotter the surface, the shorter is the emitted radiation. As the surface temperature of a heated object increases, the radiation therefore includes shorter and shorter wavelengths.

The radiation from a heated piece of iron just barely begins to include visible red light when its temperature is about 1000 K. If heated further, it emits shorter wavelengths (i.e., more visible radiation is included). Therefore, as the temperature increases, the visible color shifts from red to yellow to white. The sun's radiation, which has its peak in the visible part of the spectrum, corresponds to a surface temperature of about 6000 K and includes an appreciable amount of radiation in the near ultraviolet as well.

Objects that are close to physiological temperatures emit most of the radiation in the middle infrared. For example, the infrared radiation from the living human skin ($T \approx 300$ K) has its peak at about 10 000 nm. As visible light is between 450 and 700 nm, the radiation from human skin includes no visible light. (The visible light from the skin that we perceive is, of course, only reflected light, and in darkness we see no light from the skin.)

Figure 7.4 shows how the wavelengths and intensity of emitted radiation change with the tem-

FIGURE 7.4 The thermal radiation from a body depends on its surface temperature, in regard to both the spectral distribution of the radiation and its intensity. The higher the surface temperature, the shorter is the wavelength and the higher is its intensity. This figure shows the spectral distribution of the thermal radiation from the sun (6000 K), a red-hot stove (1000 K), and the human body (300 K). [Hardy 1949]

perature of the radiating surface, as described in the preceding two paragraphs.

The next concept to consider is *emissivity*. The simplest way to approach this concept is to consider first the *absorptivity* of a surface for the radiation that falls on it. A black body, by definition, absorbs radiation completely in all wavelengths and reflects nothing. The absorptivity of a completely black body is therefore 100%. (Although we often think of "black" in relation to visible light, the physical concept applies to all wavelengths).

In contrast to a black body, the surface that reflects all radiation is a perfect reflector, and its absorptivity is zero. This condition is approached by a highly polished metallic surface (e.g., a silver mirror). Because incident radiation is either absorbed or reflected, absorptivity and reflectivity *must* add to 1 (or 100%). If 30% of incident radiation is reflected, 70% is absorbed, and so on.

Usually, the absorptivity (and thus reflectivity) is different in different wavelengths of incident radiation. In visible light we recognize this as the color of objects; an object we perceive as yellow reflects mostly yellow light and absorbs other components of the visible light. In the middle infrared, which is of the greatest interest in connection with heat radiation at physiological temperatures, most surfaces are black bodies. The human skin, for example, absorbs virtually 100% of incoming infrared radiation, and thus is a black body in these wavelengths, irrespective of whether it is light or dark in the visible light.

The concept of surface *absorptivity* should now be clear, and we can return to the *emissivity*. The two are numerically equal, a fact that can be intuitively understood from the following. Consider an object suspended in a vacuum within a hollow sphere of uniform wall temperature. The object within the sphere receives radiation from the wall of the sphere, part of which is absorbed and part of which is reflected. Likewise, the object radiates to the spherical surface, and part of this radiation is absorbed while part is reflected. (The reflected portion, in turn, is either intercepted by the suspended object or reaches another point of the sphere.)

When the system is in equilibrium, the object has attained the temperature of the surrounding spherical surface, and the absorbed and emitted radiations from the object are now precisely equal. If this were not so, the object would not be at the temperature of the sphere. This would be a physical impossibility, for otherwise we could tap energy from the system and have the makings of a *perpetuum mobile*. As a practical example, take a highly reflective metal coffee pot that has a reflectivity of 95% (i.e., an absorptivity of only 5%). Its emissivity, therefore, is also 5% (i.e., the polished coffee pot loses heat by radiation very slowly).

Let us next consider that the object within the sphere is nearly black and has an absorptivity of 99%. This object obviously will attain the same temperature as a highly reflective silver object similarly suspended, although the latter may have an absorptivity of only 1%. For each object at equilibrium the absorbed and the emitted radiations are precisely equal. The final temperature will be the same for both, but the highly reflective object will need longer to reach it.

To repeat, an object of high absorptivity also has high emissivity. A body that is a perfect emitter in a given wavelength is also a perfect black body in the same wavelength, and vice versa.

Human skin, animal fur, and all sorts of other nonmetallic surfaces have high absorptivities in the middle infrared range of the spectrum, between 5000 and 10 000 nm. For all practical purposes human skin is black in this range, and there is no difference between heavily pigmented and unpigmented skin. The difference we perceive is in the visible region of the spectrum, but because no radiation is emitted in this range, the difference in pigmentation does not influence radiation from human skin.

As a consequence of the high absorptivity in the middle infrared, emissivity is also close to 100% in this region. Consequently, heat losses by radiation from pigmented and unpigmented skins are about equal; both radiate as virtually black bodies. The same is true of radiation from animal fur, which likewise is independent of the color in the visible part of the spectrum.

Failure to appreciate this simple physical fact has led to some ill-conceived speculation in regard to animal coloration. For example, it has been suggested that black-colored animals lose heat by radiation faster than white-colored ones. Because the emission of radiation in the infrared has no relation to visible coloration, a difference in emissivity in the infrared can be established only by direct measurement in this range of the spectrum. No such differences have been found.

Skin and fur color may, however, be important to the heat absorbed from solar radiation, which has its peak intensity in the visible range. About half of the energy carried in solar radiation falls within what we call the visible light (Figure 7.4), and it is important to the heat balance whether this light is absorbed or reflected. When exposed to direct solar radiation, dark-colored skin or fur absorbs more of the incident energy than light-colored skin or fur.

Net heat transfer by radiation. If two surfaces are in radiation exchange, each emits radiation according to *Stefan-Boltzmann's law*, and the net radiation transfer ($\dot{Q}_R$) between them is:

$$\dot{Q}_R = \sigma \epsilon_1 \epsilon_2 (T_2^4 - T_1^4) A$$

in which σ is Stefan-Boltzmann's constant, ϵ_1 and ϵ_2 are emissivities of the two surfaces, T_1 and T_2 are their absolute temperatures, and A is an expression for the effective radiating area.* If the environment is a uniform sphere, A is a simple expression of the integrated "visible" surface in the direction of radiation.

If the environment is nonuniform, however, and in particular if it includes a point source of heat (such as the sun), the integration of the surfaces of exchange becomes more complex. In this regard, situations in nature are extremely complex, and to describe in exact terms the total heat transfer is very difficult. Nevertheless, once the elementary physics of radiation exchange is understood in principle, we can readily avoid some erroneous conclusions such as those relating to the role of surface pigmentation.

A practical simplification. Although radiation heat transfer changes with the fourth power of the absolute temperature, we can use a simplified expression, provided the temperature difference between the surfaces is not too great. Within a temperature range of about 20 °C, the error of not using the rigorous Stefan-Boltzmann equation can often be disregarded, and, as an approximation, we can regard the radiation heat exchange as being proportional to the difference in temperature between the two surfaces. For small temperature differences, the error is relatively insignificant, but it becomes increasingly important the greater the temperature difference.

---

*This equation for net radiation transfer assumes that one or both surfaces has an emissivity close or equal to unity.

We will not use this simplification and regard the radiation heat exchange as proportional to the temperature difference. The rate of heat loss from a warm-blooded animal in cool surroundings consists of conduction and radiation heat loss (for the moment disregarding evaporation). Because both can be considered proportional to the temperature difference $(T_2 - T_1)$, their sum will also be proportional to $(T_2 - T_1)$, or:

$$\dot{Q} = C (T_1 - T_1)$$

in which all the constants that enter into the heat-exchange equation have been combined to a simple proportionality factor, C.

We shall later return to the application of this simplified equation in the discussion of heat loss from warm-blooded animals in the cold.

Evaporation

The evaporation of water requires a great deal of heat. To transfer 1 g water at room temperature to water vapor at the same temperature requires 584 cal (2443 J). This is an amazingly large amount of heat, for when we consider that it takes 100 cal (418 J) to heat 1 g of water from the freezing point to the boiling point, we see that it takes more than five times again as much heat to change the liquid water into water vapor at the same temperature.

The amount of heat required to achieve the phase change from liquid water to vapor is known as the *heat of vaporization* (H_v). The heat of vaporization changes slightly with the temperature at which the evaporation takes place: At 0 °C the H_v is 595 cal per gram water; at 22 °C it is 584 cal per gram; and at the boiling point, 100 °C, it is 539 cal per gram.

In physiology it is customary to use the figure 580 cal per gram water, which is an approximation of the value for vaporization of water at the skin temperature of a sweating person, about 35 °C.

The measurement of heat loss by vaporization of water has one great convenience: It suffices to know the amount of water that has been vaporized. When a person is exposed to hot surroundings, he cools himself by evaporation of sweat from the general body surface, but a dog evaporates most of the water from the respiratory tract by panting. The amount of heat transferred per gram water is, of course, the same in both cases (i.e., we do not have to know the exact location or the area of the evaporating surface).

The respiratory air of mammals and other air-breathing vertebrates is exhaled saturated with water vapor, and therefore there is normally a considerable evaporation of water from the respiratory tract, even in the absence of heat stress. This evaporation must, of course, be included in any consideration of the total heat balance of an animal, which is the subject we shall now discuss.

HEAT BALANCE

We have repeatedly emphasized that for the body temperature to remain constant, heat loss must equal heat gain. The body temperature does not always remain constant, however. Assume that heat loss does not quite equal the metabolic heat production, but is slightly lower. The body temperature inevitably rises. This means that part of the metabolic heat remains in the body instead of being lost, and the increase in body temperature thus represents a storage of heat.

If the mean body temperature decreases, which happens when heat loss exceeds heat production, we can regard the excess heat loss as heat removed from storage. The amount of heat stored depends on the change in mean body temperature, the mass of the body, and the specific heat capacity of the tissues (which for mammals and birds usually is assumed to be 0.83).

The heat exchange between the body and the environment takes place by the three means described above: conduction (including convection), radiation, and evaporation. Usually each of these represents a heat loss from the body, but this is not always so. When the air temperature exceeds the body surface temperature, heat flow by conduction is to, not from, the body. When there is a strong radiation from external sources, the net radiation flux may also be toward the body. Evaporation is nearly always a negative entity, but under unusual circumstances it could be reversed; this happens, for example, when a cold body comes in contact with moist warm air.*

We can enter these variables into a simple equation:†

$$H_{tot} = \pm H_c \pm H_r \pm H_e \pm H_s$$

in which

H_{tot} = metabolic heat production
    (always positive)
H_c = conductive and convective heat exchange
    (+ for net loss)
H_r = net radiation heat exchange
    (+ for net loss)
H_e = evaporative heat loss (+ for net loss)
H_s = storage of heat in the body
    (+ for net heat gain by the body)

The three components of heat exchange – conduction, radiation, and evaporation – depend on external factors, among which the most important

---

* When a human with a skin temperature of, say, 35 °C, enters a Turkish steam bath in which the air is nearly saturated and above 40 °C, there is an immediate condensation of water on his skin. In this case the direction of evaporation heat exchange is the reverse of the usual. In the sauna bath, however, the air is usually dry, and the visible moisture seen on the skin shortly after entering the sauna is sweat.

† It is important to note that no work term appears in this equation. For heat balance, the equation is correct as it stands. However, if oxygen consumption is used to estimate metabolic rate and to calculate heat production, external work must be considered, for that fraction of the oxygen consumption that goes to perform external work does not appear as heat in the body. For example, a flying bird uses probably 20% of its metabolic rate to impart acceleration to the air through which it flies (ultimately this external work degenerates into heat in a trail of decaying air vortexes). For a flying bird, therefore, only 80% of the oxygen consumption represents heat released within the body, and only this part enters into the heat balance equation as metabolic heat (H_{tot}) to be dissipated through the means to the right of the equality sign. This also means that the caloric equivalent of oxygen (4.8 kcal formed per liter oxygen consumed) cannot be used to calculate the heat production of an animal performing external work.

single factor is temperature. It is obvious that heat losses increase when the external temperature falls. If, however, external temperature rises, the heat losses decrease, and if external temperature exceeds body surface temperature, both conduction and radiation heat exchange may be from the environment to the organism. The total heat gain is then the sum of heat gain from metabolism plus heat gain from the environment. This situation still permits the maintenance of a constant body temperature (storage = 0), provided evaporation is increased sufficiently to dissipate the entire heat load.

The physiological responses in cold and in heat differ in many respects, and it is convenient to treat the two conditions separately. We shall first deal with temperature regulation in cold surroundings, and afterward discuss temperature regulation in the heat.

Temperature regulation in the cold: keeping warm

To maintain a constant body temperature, an animal must satisfy the steady-state condition in which the rate of metabolic heat production ($\dot{H}$) equals the rate of heat loss ($\dot{Q}$). For the moment we will consider that the heat loss (in the cold) takes place only through conduction (including convection) and radiation, and that heat loss through evaporation can be disregarded.* To describe the heat loss we will use the simplified equation that relates heat loss ($\dot{Q}$) to ambient temperature as developed in the preceding section:

$$\dot{H} = \dot{Q} = C\,(T_b - T_a)$$

This equation simply says that the rate of metabolic heat production equals the rate of heat loss, which in turn is proportional to the temperature difference between the body and the environment ($T_b - T_a$). The term C is a conductance term that will be discussed later.

---

* At moderate to low temperature, evaporation takes place primarily from the respiratory tract, at a rate corresponding to a few percent of the metabolic heat production.

FIGURE 7.5 An insulated box is maintained at a given temperature (T_b) by a heater supplying heat at the rate $\dot{H}$. If the outside temperature (T_a) is lowered, T_b can be maintained constant by increasing the rate of heat input in proportion to the temperature difference ($T_b - T_a$). This is a model of thermoregulation in a mammal.

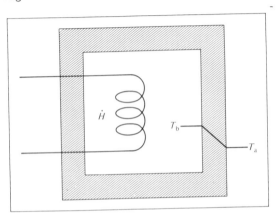

FIGURE 7.6 The oxygen consumption of the pygmy possum (*Cercaertus nanus*) at various ambient temperatures. Below a certain point, the lower critical temperature (T_{lc}), the oxygen consumption increases linearly with decreasing ambient temperature. [Bartholomew and Hudson 1962]

What can an animal do to maintain the steady state? Among the terms in the above equation, ambient temperature (T_a) is one an animal can do little about, short of moving to a different environment. To adjust to an unfavorable T_a, an animal can adjust only three remaining terms: heat production ($\dot{H}$), the conductance term (C), or body temperature (T_b). As we are concerned with the maintenance of the body temperature (T_b), we are left with adjustments to be made either in heat production ($\dot{H}$) or in the conductance term (C). (The third alternative, a change in body temperature, does occur in hibernation, a subject to be discussed later.)

Increase in heat production

Although heat production (metabolic rate) cannot be turned down below a certain minimum level, increased metabolic rate permits a wide range of adjustments. The major ways in which heat production is increased are through (1) muscular activity and exercise, (2) involuntary muscle contractions (shivering), and (3) so-called nonshivering

thermogenesis. The last term refers to an increased metabolic rate that takes place without noticeable muscle contractions.

To clarify the role of increased heat production, let us consider a simple physical model of an animal (Figure 7.5). An insulated box is equipped with an electric coil that delivers heat at the rate $\dot{H}$. The temperature inside the box (T_b) is therefore higher than the ambient temperature (T_a). If we lower the ambient temperature (T_a), the temperature within the box can be maintained only by increasing the rate of heat input ($\dot{H}$). The increase in $\dot{H}$ must be proportional to the increase in temperature difference as long as the insulation (or conductance) in the wall remains unchanged.

This is precisely the situation for an animal exposed to cold. The lower the ambient temperature, the greater the increase in metabolic rate needed to stay warm. In Figure 7.6 we see that below a certain ambient temperature, called the *lower critical temperature* (T_{lc}), the metabolic rate increases linearly with decreasing temperature. Above the critical temperature, heat production, which cannot be

FIGURE 7.7 The metabolic rates of various mammals in relation to air temperature. The normal resting metabolic rate for each animal, in the absence of cold stress, is given the value 100%. Any increase at lower temperature is expressed in relation to this normalized value, making it possible to compare widely differing animals. [Scholander et al. 1950a]

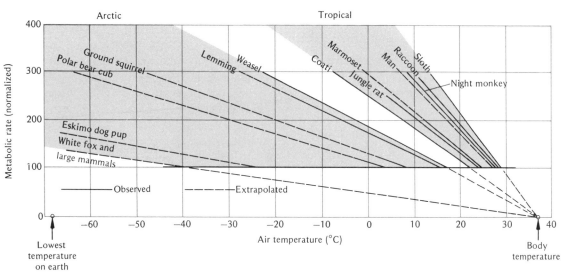

reduced lower than the resting metabolic rate, remains constant as temperature is increased.

If we imagine an animal that can reduce its metabolic rate further, less and less heat is needed to keep warm as the outside temperature (T_a) is increased. When T_a reaches T_b, no heat is lost from the animal, and no heat input is needed to maintain T_b. This is quite feasible for our model in Figure 7.5; the straight line that describes its heat loss relative to changing air temperature intersects with the abscissa when T_b equals T_a. The straight line that describes the heat production of the mammal in Figure 7.6 likewise extrapolates to the abscissa to intersect at the temperature of the body.

If we want to compare a variety of different animals, we can conveniently do so if we assign to the normal resting metabolic rate of each the value 100%. This has been done in Figure 7.7, where we can see that most tropical mammals have critical temperatures between +20 and +30 °C; below these temperatures their metabolic rates increase rapidly.

Arctic mammals, however, have much lower critical temperatures, and a well-insulated animal such as an Arctic fox does not increase its metabolic rate significantly until the air temperature is below −40 °C.

The slopes of the lines in Figure 7.7 can be regarded as an expression of the conductance term (C) in the equation we have been using.* As a rule, tropical mammals have high conductances, and Arctic mammals low conductances (high insulation values). As a consequence, a tropical mammal, because of its high conductance (low insulation), must increase its metabolism drastically for even a moderate temperature drop.

For example, a monkey with a body temperature of 38 °C and a lower critical temperature of 28 °C must double its metabolic rate for a further 10° drop

---

*When the data have been normalized by assigning to the resting heat production the value of 100%, the slope does not indicate conductance in absolute units.

FIGURE 7.8 If two animals have the same body temperature and the same conductance, but have different resting metabolic rates, they will also have different lower critical temperatures (top). If their metabolic rates are normalized to 100%, this procedure will suggest, incorrectly, that they have different conductance values (bottom).

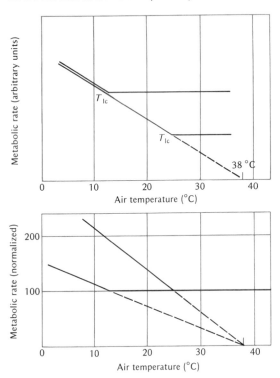

thermoneutral zone* is much greater in the Arctic animal and the metabolic response to cold is less pronounced than in tropical mammals. The way the graph is drawn, however, is in some ways misleading. Assume that we examine two mammals of equal size and equal conductance value, but that one has a metabolic rate in the thermoneutral zone that is twice as high as the other. If, instead of normalizing the metabolic rate at 100%, we plot their actual metabolic rates (Figure 7.8), the lower critical temperature will be different for the two animals, but below this temperature, the two animals must expend the same number of calories to keep warm.

It is therefore obvious that the location of the lower critical temperature (the width of the thermoneutral zone) by itself is not sufficient information about how well adapted an animal is to cold conditions. Information about the level of the resting metabolic rate, or better, the conductance value, is also necessary to evaluate the relationship between energy metabolism and heat regulation.

Conduction, insulation, and fur thickness

Conductance, in the sense it has been used above, is a measure of the heat flow from the animal to the surroundings. The term includes the flow of heat from deeper parts of the body to the skin surface and from the skin surface through the fur to the environment. When conductance is low, the insulation value is high. In fact, insulation is the reciprocal value of conduction.†

Our discussion is concerned with animals in the cold, where the fur is a major barrier to heat flow. The insulation values of the fur from various animals differ a great deal. Some data, plotted relative to the thickness of the fur, are shown in Figure 7.9.

in the temperature, to 18 °C. An Arctic ground squirrel, on the other hand, which has a lower critical temperature at about 0 °C, will not need to double its metabolic rate until the temperature has dropped to nearly −40 °C. The Arctic fox, with its lower critical temperature at −40 °C, should be able to sustain the lowest temperatures measured in Arctic climates (−70 °C) with less than a 50% increase in its metabolic rate.

The difference between Arctic and tropical animals is very clear in Figure 7.7. The width of the

---

*The *thermoneutral zone* is the temperature range within which the metabolic heat production is unaffected by a temperature change.

†*Thermal conductance* (heat flow per unit time per unit area per degree temperature difference) has the units W m^{-2} °C^{-1} (or cal s^{-1} cm^{-2} °C^{-1}). *Insulation* is the reciprocal of conductance, and its units are therefore W^{-1} m^2 °C (or cal^{-1} s cm^2 °C).

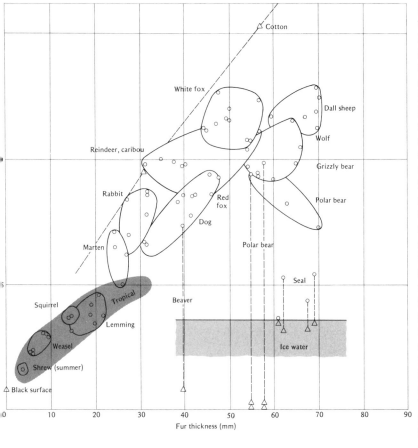

FIGURE 7.9 The insulation value of animal fur is related to fur thickness. Because small mammals of necessity carry relatively short fur, they have poor insulation compared with larger animals. Insulation values for fur in air are indicated by open circles; values for fur submerged in water, by triangles. Insulation values for fur (ordinate) are measured in air or water at 0 °C with the inside of the fur maintained at 37 °C. The sloping broken line represents the insulation values of cotton of various thicknesses. In this figure the insulation value of 4 cm of fur is about 1 W^{-1} m^2 °C, which equals 41 840 cal^{-1} s cm^2 °C. For 1 cm thickness of fur the insulation value will be one-fourth, or 10 460 cal^{-1} s cm^2 °C. The conductance value (the reciprocal of insulation) is therefore 0.000 096 cal s^{-1} cm^{-2} °C^{-1}, which is in good agreement with the value for fur listed in Table 7.3. [Scholander et al. 1950b]

As expected, the insulation value increases with the thickness of the fur and reaches a maximum for some of the larger animals that have thick fur, such as the white fox.

Among the smaller animals there is a clear correlation between fur thickness (and insulation) and the size of the animal. A small animal must have relatively short light fur or it could not move about. This is particularly true for the smallest mammals,

small rodents and shrews. Because of their relatively poor insulation, these animals must either take advantage of microclimates (e.g., by living in burrows) or hibernate to escape the problem of keeping warm.

The polar bear is interesting, for its open and coarse fur provides poor insulation relative to its thickness. More important, when polar bear skin is submerged in ice water, most of the insulation value of the fur is lost, and heat transfer is 20 to 25 times faster than in air. If the water is agitated (as it would be when the bear is swimming), the heat loss is increased even more, some 50 times. This is because water penetrates to the skin surface, dislodging all air from the fur. When in water, the polar bear is helped by subcutaneous blubber, which affords insulation (see later in this chapter).

The skin of the seal has a relatively thin fur, but a heavy layer of blubber affords considerable insulation. Therefore, the difference in insulation value for seal skin in air and in water is not very great. In air, seal skin with the blubber attached has only a slightly better insulation value than the skin of a lemming (i.e., 60 to 70 mm of mostly blubber insulates about as well as 20 mm of fur). When the seal skin is submerged in water, the insulation value is reduced, but not as drastically as for polar bear skin.

It is well known that the thickness of animal fur changes with the seasons and that winter fur is heavier and presumably affords better insulation than the thinner summer fur. The seasonal changes that have been measured are greatest in large animals and relatively insignificant in small rodents. The black bear, for example, in summer loses 52% of the insulation value of its winter fur (i.e., the summer pelt affords less than half the insulation of the winter pelt). The smallest seasonal change observed in a sub-Arctic mammal was a 12% reduction of the winter value in the muskrat (Hart 1956).

Conductance in birds

As we have seen, the conductance values for mammals in the cold require that the metabolic rates increase in proportion to the imposed cold stress. The straight lines describing the metabolic rates at low temperature extrapolate to body temperature, showing that mammals fit our assumed model (Figure 7.5).

For birds the situation is not so simple. Some birds conform to the mammalian pattern with metabolic curves that extrapolate to body temperature. Other birds, however, deviate from this pattern.

FIGURE 7.10 The oxygen consumption of the pigeon and of the roadrunner increases at low temperature, but the regression lines do not extrapolate to the body temperature of the animal as they commonly do in mammals. [Calder and Schmidt-Nielsen 1967]

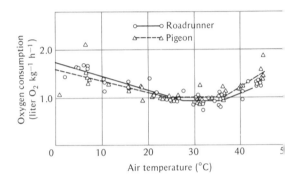

For example, the metabolic heat production of pigeons and roadrunners does not increase as much as expected at low ambient temperature (Figure 7.10). For these birds the metabolic line extrapolates, not to the body temperature of about 40 °C but to a much higher temperature, well above 50 °C.

There is only one possible explanation for such a slope: The bird does not adhere to the simple equation $\dot{H} = C\ (T_b - T_a)$, where C remains constant at low temperature.

One way to explain the observed regression line is to assume that the conductance value (C) gradually decreases with falling ambient temperature (T_a). The other way to achieve a change in slope is to permit the body temperature (T_b) to decrease. The birds represented in Figure 7.10 maintained normal body temperatures, and the question is then how can conductance change gradually as the ambient temperature decreases?

One way to decrease conductance is to raise the feathers and withdraw feet and head as far as possible into the feathers, and birds do this in the cold. Another way is to allow the peripheral tissues to undergo an appreciable temperature drop while the core temperature is maintained. The drop in shell temperature has several effects: Circulation in the

FIGURE 7.11 Body temperature of a willow tit (*Parus montanus*) during the night hours at three different ambient temperatures. The records are from midafternoon one day to the following morning. (Reinertsen and Haftorn 1986).

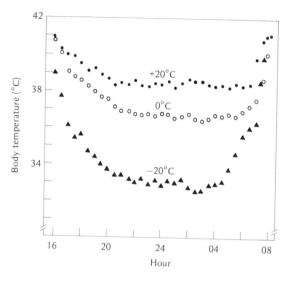

shell is decreased, shell thickness is increased, and the volume of the core is reduced. Each of these variables contributes to a change in conductance and permit the maintenance of core temperature.

A change in conductance can also be combined with a drop in core temperature, and this could achieve even greater savings. Small birds in northern climates have serious problems in winter. The ambient temperature is low, snow covers the ground, and the days are short with only a couple of hours of weak daylight. How do small birds manage through the winter?

Studies in Norway show that the willow tit (*Parus montanus*) uses both a decreased conductance and a drop in core temperature to survive from day to day in winter. The drop in core temperature is greater at low ambient temperature (Figure 7.11) and may be as great as 10 °C. Even at an ambient temperature as high as 20 °C the drop may be nearly 4 °C, which represents an energy saving of 35% compared to what it would cost to maintain day-

time temperature during the night (Reinertsen and Haftorn 1986).

The relative savings of dropping the core temperature are not as great at lower ambient temperatures, but in combination with other adjustments it enables the bird to remain in subarctic regions during the winter. The birds use food storage, they find sheltered roosting places, and they assume a sleeping posture that utilizes heat conservation to the fullest.

Huddling

Among the penguins of the Antarctic the most striking is the large emperor penguin (*Aptenodytes forsteri*), which lives under colder conditions than any other bird and has the unique characteristic of breeding during midwinter.

As winter approaches, emperor penguins leave the open water and walk on the sea ice for 50 or 100 km toward the permanent ice shelf to the rookeries. There the female lays a single egg, which is placed on the feet of the male, who remains at the rookery to incubate the egg while the female returns to the sea to feed.

While incubating, the male stands on the ice for over 2 months in air temperatures that may be as low as −30 to −40 °C with high wind velocities. The female usually returns to the rookery with a full stomach about the time the egg hatches; she feeds the chick while the male returns to the sea to feed.

Emperor penguins can feed only at sea, and to sustain himself during the long fast the male begins the period with large deposits of subcutaneous fat. The fast may exceed 100 days, and by the end the male may have lost up to 40% of his initial body weight.

How much energy is needed for the long walk to and from the rookery, and how much is needed to keep warm in the Antarctic cold? And do penguins keep warm? These questions have been the subject of detailed studies (LeMaho et al. 1976; Pinshow et al. 1976).

The penguins do keep warm; their body temper-

HUDDLING PENGUINS Chicks of the emperor penguin photographed on the ice-covered Antarctic breeding grounds. The chicks huddle together and thus reduce the surface area exposed to cold air. This behavior leads to a substantial reduction in the metabolic cost of keeping warm in the frigid climate. The group in this photo contains about 50 birds; a few have momentarily raised their heads to watch the photographer. [Courtesy of Yvon LeMaho, University of Lyon, France]

ature remains at the normal 38 °C during the entire breeding period. How much fuel do the birds need for this and how much for walking?

A large male penguin may weigh 35 kg when he leaves the sea and may have lost 15 kg at the end of the fast. The amount of energy needed for walking can be determined by training penguins to walk on a treadmill in the laboratory and determining their oxygen consumption. This shows that walking 200 km to the breeding grounds and back to the sea may require the use of about 1.5 kg of fat.

Is the remainder sufficient to keep the bird warm for several months of starvation while he incubates the egg?

One might expect penguins to have physiological characteristics that distinguish them from other birds, such as a thermoneutral zone extending to extremely low temperatures (i.e., a very low thermal conductance). This is not the case. Penguins have a thermal conductance to be expected for any bird of the same body size, and their lower critical temperature is at −10 °C, far above the usual Antarctic winter temperatures. The metabolic rates of penguins under laboratory conditions at temperatures similar to those in nature suggest that perhaps 25 kg of fat would be needed to keep warm during the winter, and this by far exceeds the reserves that even a large male carries.

The apparent contradiction is explained by the behavior of the penguins in the rookery, where they huddle together in large groups that may number several thousands. The metabolic rates of huddling penguins have not been determined, but weight loss, which presumably results from metabolism of fat, is a good indication. Single emperor penguins in the cold lose about 0.2 kg per day; huddling penguins lose only about half as much, about 0.1 kg. Huddling together helps, and it is easy to understand why. Instead of being exposed to the cold on all sides, a major part of each bird's surface is in contact with neighboring penguins. When two bodies have the same temperature, there is no heat flow between them. Huddling together obviously is a prerequisite for survival and for the success of breeding. Why the emperor penguins choose to breed in winter is another question that remains unanswered.

Other animals also huddle together to keep warm, although they may be less spectacular than the penguins. Huddling reduces the exposed surface, thus reducing the cold stress and the metabolic requirements for heat production. This is of particular importance for newborn mammals and bird nestlings that live together as a litter: They can remain warm in the absence of the parent for longer periods, and the young animal's decreased dependence on the use of energy for heat production permits faster growth.

Insulation in aquatic mammals

Many seals and whales live and swim in the near-freezing water of the Arctic and Antarctic seas. Not only are the temperatures low, but water has a high thermal conductance and a high heat capacity, and the thermal loss to water is therefore much higher than to air of the same temperature. The heat con-ductivity coefficient for water is about 25 times as high as for air, but because convection, both free and forced, plays a major role, the cooling power of water may under some circumstances be even greater, perhaps 50 or even as high as 100 times as great as for air.

What can seals and whales do about their heat balance in water with such cooling power? Evidently they manage quite well, for both seals and whales are far more numerous in cold water than in the tropics. As there is nothing they can do about the water temperature (short of moving to warmer seas), they have a limited choice. They could either (1) live with a lowered body temperature, (2) increase the metabolic rate to compensate for the heat loss, or (3) increase the body insulation to reduce the heat loss.

With regard to body temperature, seals and whales are similar to other warm-blooded animals; their usual temperature is around 36 to 38 °C (Irving 1969). We must therefore look at the other possibilities for an explanation of their ability to live in ice water.

The metabolic rate has been measured in several species of seals and in some dolphins (porpoises), but not in any of the large whales, which are rather unmanageable as experimental animals. Most of these animals have resting metabolic rates about twice as high as would be expected from their body size alone (Irving 1969). In the harp seal (*Phoca groenlandica*), an Arctic species, the metabolic rate remained the same in water all the way down to the freezing point; in other words, even the coldest water did not cause the heat loss to increase enough to require increased heat production. The lower critical temperature for the harp seal in water is thus below freezing and has not been determined (Irving and Hart 1957).

The third possibility, an effective insulation, is obviously the main solution to the problem. Both seals and whales have thick layers of subcutaneous blubber that afford the major insulation. Measurements of the skin temperature support this concept, for the temperature at the skin surface is nearly

FIGURE 7.12 The temperature of the skin surface of a living seal immersed in ice water is nearly identical to that of the water. Most of the insulation is provided by the thick layer of blubber. [Irving and Hart 1957]

FIGURE 7.13 The skin surface temperature of seals in air and water. Solid line indicates equality of skin and environmental temperature. [Hart and Irving 1959]

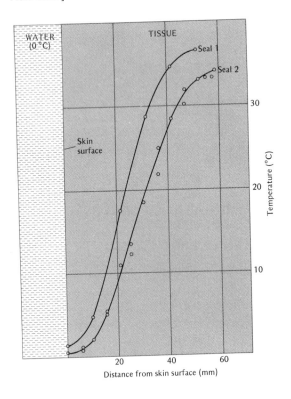

Distance from skin surface (mm)

identical to that of the water (Figure 7.12). If the surface temperature is nearly the same as that of the water, very little heat can be transferred to the water. The temperature gradient is sustained by the blubber, and at a depth of some 50 mm (roughly the thickness of the blubber), the temperature is nearly at body core temperature.

In Arctic land mammals the temperature gradients are different. The surface temperature of the body skin under the fur is regularly within a few degrees of core temperature (Irving and Krog 1955), and most of the insulation therefore resides outside the skin surface. The polar bear, whose fur loses most of its insulation value in water, also has a

substantial layer of blubber that seems to be of major importance when the polar bear swims in ice water. Without the layer of blubber, the semi-aquatic way of life would seem impossible (Øritsland 1970).

If seals and whales are so well insulated, how do they manage to avoid overheating when the water is warmer or when the metabolic rate is increased during fast swimming? If a seal is removed from water and placed in air, its skin temperature increases considerably (Figure 7.13). The higher skin temperature is necessary for dissipation of heat to the air, due to the reduced cooling power of air relative to water. The increase in skin temperature results from an increased blood flow through the blubber to the superficial layer of the skin, which is well supplied with blood vessels. This system of cutaneous blood vessels permits a precise regulation of the amount of heat that reaches the skin surface and thus is lost to the environment.

We can now see that the main difference in insulation between aquatic and terrestrial mammals is that the insulator of the aquatic mammals (the blubber) is located internally to the surface of heat dissipation. Therefore, the blood can bypass the insulator, and heat dissipation during heavy exer-

SEAL BLUBBER This cross section of a frozen seal shows the thick layer of blubber. Of the total area in the photo, 58% is blubber and the remaining 42% is muscle, bone, and visceral organs. The measuring stick is graduated in inches. [Courtesy of P. F. Scholander, University of California, San Diego]

tion or in warm water can be independent of the insulator. Terrestrial animals, in contrast, have the insulator located external to the skin surface. Therefore, they cannot modulate the heat loss from the general skin surface to any great extent and must seek other avenues of heat loss when they need to dissipate excess heat to the environment (Figure 7.14).

Distribution of insulating material

We must remember that furred Arctic animals cannot afford to insulate all parts of the body equally well; they also need surfaces from which heat can be dissipated, especially during exertion. Because the main part of the body is well insulated for max-

imum heat retention, they need thinly covered skin areas on the feet, face, and other peripheral locations, from which heat loss can take place when the demand for heat dissipation increases.

Estimates of the fraction of the body surface area that is covered with fur of different thicknesses suggests that some animals have considerable flexibility in regulating their conductance by changes in posture. Estimates on a male guanaco (a relative of the llama) are shown in Table 7.4 and in diagrammatic form in Figure 7.15 (Morrison 1966).

Because the thermal conductance through fur is inversely related to the thickness of the fur, the nearly bare areas can potentially transfer much more heat than the heavily furred areas. With the limbs

FIGURE 7.14 The insulation afforded by blubber can be bypassed when the need for heat dissipation increases. Fur, in contrast, is located outside the skin surface and its insulation value cannot be drastically changed by a bypass.

Blubber

Skin surface

Fur

TABLE 7.4 Partitioning of surface areas with fur of different thickness in the South American guanaco (*Lama guanicoe*). About 40% of the area is covered by heavy fur, but almost 20% is nearly bare and permits extensive modulation of heat dissipation. [Morrison 1966]

	Depth of fur (mm)	Percent of total skin area
Nearly bare	1	19
Short hair	4	20
Medium-length hair	15	20
Long hair	30	41

FIGURE 7.15 Distribution of fur of different thicknesses on the body of the guanaco. Areas on the chart are proportional to measured areas on the animal. [Morrison 1966]

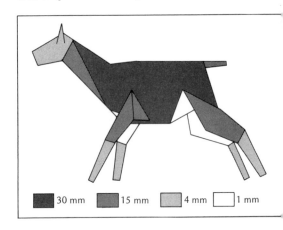

30 mm   15 mm   4 mm   1 mm

extended and their surfaces fully exposed, the bare areas can serve as "heat windows" that can account for more than half the total heat transfer.

A guanaco lying or standing in a normal position would have the bare areas on the inside of the legs opposed, and this would reduce surface heat transfer by one-half. If the animal were to curl up to cover as much as possible of the remaining short-haired areas, surface conductance would be at a minimum: only one-fifth of that for the stretched out animal. Other animals have similar short-haired areas whose role in heat transfer can be modulated by changing their exposure and by changing the blood flow to these areas, thus giving a great deal of flexibility in the regulation of heat loss.

Heat exchangers

Seals and whales have flippers and flukes that lack blubber and are poorly insulated. These appendages are well supplied with blood vessels and receive a rich blood supply. This means that these relatively thin structures with their large surfaces can lose substantial amounts of heat and aid in heat dissipation, but how is it possible to avoid excessive heat loss from the flippers when heat needs to be conserved? If the blood that returns to the body

FIGURE 7.16 In the flippers and fluke of a dolphin (porpoise) each artery is surrounded by several veins. This arrangement permits the venous blood to be warmed by heat transfer from the arterial blood before it reenters the body. [Schmidt-Nielsen 1970]

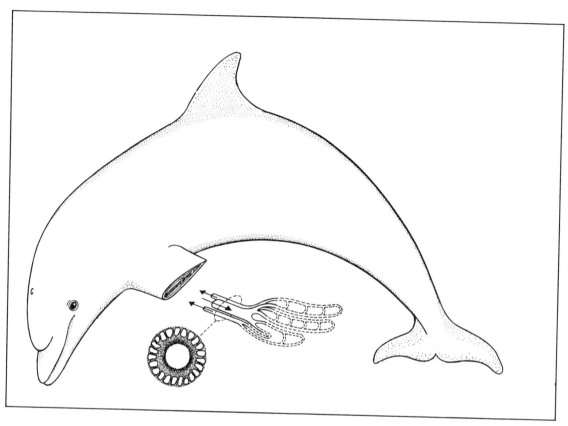

core from these appendages were at near-freezing temperature, the deeper parts of the body would rapidly be cooled.

Excessive heat loss from the blood in the flippers is prevented by the special structure of the blood vessels, which are arranged in such a way that they function as heat exchangers. In the whale flipper each artery is completely surrounded by veins (Figure 7.16), and as warm arterial blood flows into the flipper, it is cooled by the cold venous blood that surrounds it on all sides. The arterial blood therefore reaches the periphery precooled and loses little heat to the water. The heat has been transferred to the venous blood, which thus is pre-

warmed before it reenters the body. If the heat exchange is efficient, the venous blood can reach nearly arterial temperatures and thus contribute virtually no cooling to the core. This type of heat exchanger is known as a *countercurrent heat exchanger* because the blood flows in opposite directions in the two streams (Scholander and Schevill 1955).

A diagram of a countercurrent heat exchanger may help explain how it works (Figure 7.17). Assume that a copper pipe carries water at 40 °C into a coil suspended in an ice bath, and that the water returns in a second pipe located in contact with the first so that heat is easily conducted between them.

FIGURE 7.17 Model of a countercurrent heat exchanger. In this case heat is conducted from the incoming water to the outflowing water so that in the steady-state condition the outflowing water is prewarmed to within 1 °C of the incoming water. For explanation, see text.

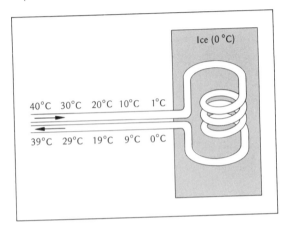

As the water leaves the coil, it is very nearly at the temperature of the ice bath, 0 °C. As this water flows adjacent to the warmer pipe leading into the coil, it picks up heat and thus cools the incoming water. After some time a steady state of temperatures is reached, which may be as indicated in the diagram, depending on the conditions for heat exchange between the two pipes and their length.

For animals swimming in warm water, heat dissipation presents a greater problem than heat conservation. The anatomical arrangement of the heat exchanger is such that an increased blood supply and an increase in arterial blood pressure cause the diameter of the central artery to increase, and this in turn causes the surrounding veins to be compressed and collapse. The venous blood must now return in alternate veins, which are located closer to the surface of the flipper. Because the heat exchanger is now bypassed, the arterial blood loses heat to the water, and the venous blood returns to the core without being rewarmed, thus cooling the core. In this way the circulatory system of the flipper can function both in heat conservation and in heat dissipation.

It is interesting that such countercurrent heat exchangers are found in a number of other animals. For example, sea cows, which live in tropical and subtropical waters, have heat exchangers in their appendages. This might seem unnecessary, for they live in warm water, but sea cows move slowly and have relatively low metabolic rates for their size, and thus would have difficulties unless they could reduce the heat loss from the appendages.

Heat exchangers in the limbs are not restricted to aquatic animals. Even in the limbs of humans some heat exchange takes place between the main arteries and the adjacent larger veins, located deep within the tissues. In cold surroundings most of the venous blood from the limbs returns in these deep veins, but in warm surroundings much of the venous blood returns in superficial veins under the skin and is thus diverted from heat exchange with the artery (Aschoff and Wever 1959).

In birds, heat exchange in the legs is very important, especially for birds that stand or swim in cold water. Unless the blood flowing to these thin-skinned peripheral surfaces went through a heat exchanger, the heat loss would be very great indeed. However, because of the heat exchanger the heat loss is minimal. A gull placed with its feet in ice water for 2 hours lost only 1.5% of its metabolic heat production from the feet, a quite insignificant heat loss (Scholander 1955).

Even some tropical animals have vascular heat exchangers in their limbs. In the sloth the artery to the foreleg splits into several dozen thin parallel arteries, which are intermingled with a similar number of veins. It may seem superfluous for a tropical animal to have such heat exchangers, but on a rainy and windy night the heat loss from an animal sleeping in the crown of a tree can be very great indeed.

The value of the heat exchanger in the sloth has been clearly demonstrated by immersing a limb in ice water. The blood temperature below the heat exchanger decreased drastically, but in the part of the foreleg above the heat exchanger the temperature of the venous blood was nearly at core temperature (Scholander and Krog 1957).

BLOOD VESSELS IN A BIRD LEG Cross section (left) and surface view of the blood vessels in the leg of a European rook (*Corvus frugilegus*), a crow-like bird. The thick-walled artery runs in the center and is surrounded by several thin-walled veins that branch and anastomose so that they virtually cover the artery (right). The diameter of the structure is 2 mm. [Courtesy of Uffe Midtgård, University of Copenhagen]

It has been suggested that the vascular bundles in the forelegs of the sloth and the loris have an entirely different function: that they are related to the ability of these climbing animals to maintain a strong grip (Buettner-Janusch 1966; Suckling et al. 1969). However, the vascular structure is located in the upper part of the foreleg, and the muscles responsible for the grasp are in the lower part. The fact that the blood runs through a bundle of arteries that join again into a single vessel cannot have any imaginable influence on the muscles in the lower part of the foreleg. The blood still contains the same oxygen and nutrients, and the only difference these muscles could detect would be a slightly decreased blood pressure, which certainly could not improve the strength of the grip.

An animal in the cold has two opposed problems: to conserve heat and to keep peripheral tissues such as the feet from freezing. At ambient temperatures down to freezing, there is no conflict between the two; reduction in blood flow and cooling of the arterial blood in a heat exchanger help to minimize the heat loss.

However, when the temperature falls much below freezing, it is necessary to add enough heat to the peripheral tissues to keep them from freezing. The bare feet of the duck and other birds are suitable examples. As long as there is no danger of freezing, the heat loss from the feet is minimal, but as the temperature is lowered below 0 °C, the heat loss increases in proportion to the decrease in temperature (Kilgore and Schmidt-Nielsen 1975). The increase in heat drain is reflected in an increased metabolic heat production, appearing as a distinct inflection point in the metabolic curve (Figure 7.18).

Temperature regulation in the heat: keeping cool

We have seen that mammals and birds must increase their heat production in order to stay warm at ambient temperatures below a certain point, the

FIGURE 7.18 At low temperature, below 24 °C, the heat loss from the feet of a duck is a small fraction of the metabolic heat production. Below the freez-ing point the heat transferred to the feet (and lost to the water) increases in proportion to the drop in temperature. [Kilgore and Schmidt-Nielsen 1975]

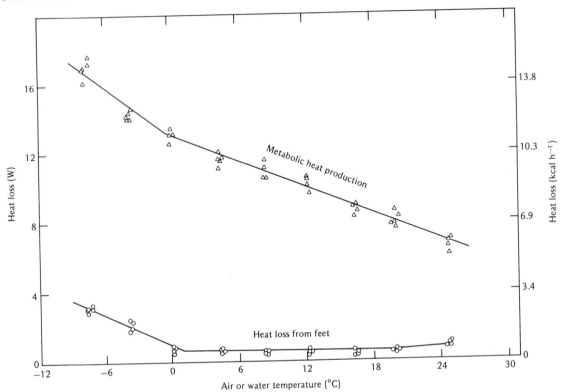

lower critical temperature. We will now be concerned with what happens above this critical temperature.

Above the critical temperature the metabolic heat production (resting metabolic rate) remains constant through a range usually known as the *thermoneutral range* (see Figure 7.6). As before, for the body temperature to remain constant, metabolic heat must be lost at the same rate it is produced. If we return to the now familiar equation for heat balance, $\dot{H} = C\,(T_b - T_a)$, we see that if $\dot{H}$ and T_b remain constant and T_a changes, the conductance term (C) must also change.

The conductance term refers to the total heat flow

from the organism, and it can be increased in several ways. One way is by increasing circulation to the skin so that heat from the core is moved more rapidly to the surface. Another way is by stretching out and exposing increased surface areas, especially naked or thinly furred areas.

Let us return to the basic equation for heat balance (page 253):

$$H_{tot} = \pm H_c \pm H_r \pm H_e \pm H_s$$

As the ambient temperature increases, the conditions for heat loss by conduction and convection (H_c) and radiation (H_r) become increasingly unfavorable. Because the metabolic heat production

FIGURE 7.19 The cooling by evaporation in jack-rabbits, expressed in percent of the simultaneous metabolic heat production. Evaporation increases as the difference between body temperature (T_b) and air temperature (T_a) decreases. As this difference approaches zero (body and air temperatures equal), the entire heat production (100%) is dissipated by evaporation. [Dawson and Schmidt-Nielsen 1966]

(H$_{tot}$) remains unchanged (or even increases slightly), the heat balance equation must increasingly emphasize the evaporation and storage terms (H_e and H_s).

In the discussions of heat regulation in the cold we disregarded heat removed by evaporation because it made up a fairly small fraction of the total heat exchange. At high temperature evaporation is the key item in the heat balance. Assume that we have increased the temperature of the environment to equal the body temperature of an animal. If the temperature of the environment equals the body temperature, there can be no loss of heat by conduction, and the net radiation flux approaches zero. If we now wish to maintain a constant body temperature (i.e., keep the storage term H_s at zero), the consequence is that the entire metabolic heat production (H_{tot}) must be removed by evaporation of water.

This prediction has been tested on a number of mammals and birds. When they are kept in air that equals their body temperature, those that are able to keep their body temperature from rising (not all species can) evaporate an amount of water that is equivalent to their metabolic heat production. An example is given in Figure 7.19, which shows that the desert jackrabbit (actually a hare) behaves according to the prediction.

The ambient temperature can, of course, increase further and exceed the body temperature, as frequently happens in deserts. In this event the body receives heat from the environment by conduction from the hot air, by radiation from the heated ground surface, and in particular by radiation from the sun.

Under these circumstances evaporation must be used to dissipate the sum of metabolic heat production and the heat gain from the environment. The possibility of storing heat is not a good solution, for as we saw before, animals have only limited tolerance to increased body temperature. Nevertheless, as we shall see later, even a moderate increase in body temperature can indeed be important for a desert animal.

To evaluate the importance of the heat load from the environment, consider how much water is used for evaporation (sweating) by a person in a hot desert. His total resting metabolic rate is about 70 kcal per hour, and this amount of heat, if dissipated entirely by evaporation, requires 120 ml, or 0.12 liter, water. We also know that a person exposed to the sun on a hot desert day may sweat at a rate of between 1 and 1.5 liters per hour. To make our calculations simple, let us say 1.32 liters per hour. If there is no change in his body temperature, and 0.12 liter is used to dissipate the metabolic heat, the excess of 1.2 liters used must be attributable to the heat gain from the environment. Thus the sum of conduction from the hot air, radiation from the hot ground, and radiation from the sun must require the evaporation of 1.2 liters of water. Therefore, in this example, the heat load from the environment is exactly 10 times as high as the metabolic heat production of the person at rest. If the body temperature were to increase by 1 °C, the storage of heat would be about 60 kcal, or the equivalent of 0.1 liter water evaporated – a fairly insignificant amount.

We now have a good idea of how important the environmental heat load can be when the temperature gradients are reversed, and it is worth considering how this affects animals. If a person were to continue evaporating at the indicated rate and had no water to drink, he would be near death from dehydration at the end of a single hot day in the desert. Yet the deserts of the world have a rich and varied animal life, although in most deserts there is no drinking water available. This requires further discussion.

The importance of body size

If we put a large block of ice outside in the hot sun, and a small piece next to it, the small piece will melt away long before the large block. If a big rock and a small pebble are placed on the ground in the sun, the pebble will be hot long before the big rock. The reason is that a small object has a much larger surface relative to its volume. If the big rock or block of ice is broken into many small pieces, numerous new surfaces, previously unexposed, are in contact with the warm air and receive solar radiation. Because of the many new exposed surfaces the total mass of rock heats up more rapidly and the crushed ice melts faster.

The relation between surface and volume of similar-shaped objects is simple. If a given cube is cut into smaller cubes with each side one-tenth that of the larger cube, the total surface of all the small cubes is 10 times that of the original cube. If the linear dimension of the small cube is one-hundredth of the larger one, the aggregate surface is 100 times the original, and so on. This rule holds for any other similar-shaped objects. Thus, any small body has a surface area that, relative to its volume, increases as the linear dimension decreases. Small and large mammals have sufficiently similar shapes that their body surface area is a regular function of the body volume, and the same rule applies.

Let us return to the animal in hot surroundings.

Its total heat gain consists of two components: heat gain from the environment and metabolic heat gain. The heat gain from the environment by conduction and convection, as well as radiation from ground and sun, are surface processes, and the total environmental heat load therefore is directly related to the surface area. In Chapter 5 we saw that the metabolic heat production of a mammal is not quite proportional to body surface, but sufficiently close that, as an approximation, we can assume this to be so. Therefore, the total heat load – the sum of metabolic and environmental heat gain – is roughly proportional to the body surface area. This puts the small animal, with its larger relative surface, in a far more unfavorable position with regard to heat load than a large animal.

If a person in the desert must sweat at a rate of 1 liter per hour (which equals about 0.60 liter per square meter body surface area per hour), we can use the surface relationship to estimate or predict how much water other animals should evaporate under similar desert conditions in order to dissipate the heat load. We thus obtain a theoretical curve that predicts the amount of water needed to keep cool (Figure 7.20). The amount is an exponential function of body mass, and on logarithmic coordinates the curve will be a straight line.

However, Figure 7.20 is plotted on a linear ordinate in order to emphasize the exponential increase in water required if a small animal were to depend on evaporation to keep cool in the desert. Many small rodents weigh between 10 and 100 g and would have to evaporate water at a rate of 15 to 30% of their body weight per hour. As a water loss of between 10 and 20% is fatal for mammals, such rates of water loss are impossible. Obviously, because of their small size alone, desert rodents must evade the heat, which they do by retreating to their underground burrows during the day.

Large body size: the camel. Figure 7.20 shows that large animals, such as camels, derive a substantial advantage simply from being big. The benefit, however, does not increase much as the body size increases further. Theoretically, if a camel in

FIGURE 7.20 For a mammal to maintain a constant body temperature under hot desert conditions, water must be evaporated in proportion to the heat load. Because of the larger relative surface area of a small animal, the heat load, and therefore the estimated evaporation in relation to the body size, increases rapidly in the small animal. The curve is calculated on the assumption that heat load is proportional to body surface. [Schmidt-Nielsen 1964]

FIGURE 7.21 The daily temperature fluctuation in a well-watered camel is about 2 °C. When the camel is deprived of drinking water, the daily fluctuation may increase to as much as 7 °C. This has a great influence on the use of water for temperature regulation (see text). [Schmidt-Nielsen 1963]

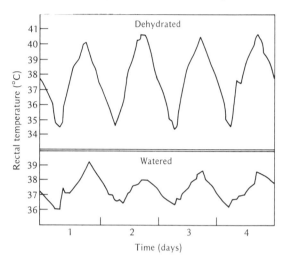

other respects were like a man, its body size would make it evaporate water at about half the rate of a man. The camel, however, is not a man and does not use this much water for heat regulation. As we shall see, the camel uses a combination of several approaches that help reduce the heat gain from the environment and thus the use of water.

First consider what a camel can do about two of the variables in the heat balance equation, evaporation (H_e) and storage of heat (H_s). Storage of heat is reflected in an increase in body temperature. In a normal camel that is watered every day and is fully hydrated, the temperature varies by about 2 °C, between about 36 and 38 °C (Figure 7.21). When the camel is deprived of drinking water, however, the daily temperature fluctuations become much greater (Schmidt-Nielsen et al. 1957). The morning temperature may be as low as 34 °C, and the highest temperature in the late afternoon may be nearly 41 °C. This large increase in body temper-

ature during the day constitutes a storage of heat. For a camel that weighs 500 kg, the amount of heat stored by a 7 °C temperature rise corresponds to 2900 kcal of heat, which equals a saving of 5 liters of water. In the cool night, the stored heat can be unloaded by conduction and radiation without use of water.

The high body temperature during the day has a further advantage, beyond that of storing heat. When the body temperature is increased, the temperature gradient from the hot environment to the body (which determines the amount of heat gain from the environment) is reduced. This reduction in environmental heat gain is at least as important as heat storage with regard to water savings and is reflected in the reduced amount of water used. This is illustrated in Figure 7.22, which shows the heat balance of a small adult camel studied in the Sahara Desert. In the fully hydrated animal the evaporation during the 10 hottest hours of the day was

FIGURE 7.22 The change in body temperature of a camel deprived of drinking water greatly affects the heat gain from the environment and in turn the amount of water used for heat regulation (by evaporation). [Schmidt-Nielsen et al. 1957]

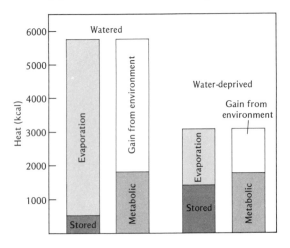

9.1 liters, corresponding to the dissipation of 5300 kcal heat. In the same camel when deprived of drinking water, the evaporation was reduced to 2.8 liters in 10 hours, corresponding to the dissipation of 1600 kcal heat. Thus, the use of water in the dehydrated camel was reduced to less than one-third that in the fully hydrated animal.

The main reason the camel uses substantially less water than a man (about 0.28 liter per hour in the camel compared with more than 1 liter per hour for a man who weighs about one-quarter as much) is the higher body temperature in the camel, which means both storage of heat and a reduced environmental heat gain. In contrast, man maintains his body temperature nearly constant at about 37°C, and thus does not store appreciable amounts of heat. A human also maintains much steeper gradients between the environment and his body surface, for the mean skin surface temperature of a sweating human is about 35°C.

One further important reason is that the camel has thick fur with a high insulation value. This im-

poses a heavy insulating layer between the body and the source of the heat and thus reduces the heat gain from the environment. The simplest way of testing the importance of this fur is to shear a camel. In one such experiment, the water expenditure of a camel, under otherwise similar conditions, was increased by about 50% when the animal was shorn.

We have now seen some of the major reasons for a lower-than-predicted (from body size alone) water loss in the camel: (1) heat is stored because of an increase in body temperature; (2) the increase in body temperature decreases the heat flow from the environment; and (3) the fur is a substantial barrier to heat gain from the environment.

The camel has one further advantage over humans: It can tolerate a greater degree of water depletion of the body. A human is close to the fatal limit when he has lost water amounting to between 10 and 12% of the body weight, but the camel can tolerate about twice as great a water depletion without apparent harm. As a result, the camel can go without drinking for perhaps 6 to 8 days under desert conditions that would be fatal in a single day to a human without water.

In the end, however, a camel must drink to replenish its body water, and when water is available, it may drink more than a third of its body weight. The immense capacity for drinking has given rise to the legend that a camel, before a long desert journey, fills its water reservoirs. In reality, the camel, like other mammals, drinks to replenish lost water and to restore the normal water content of the body, and there is no evidence that it overdrinks in anticipation of future needs.

Small body size: the ground squirrel. We saw earlier that because of their small size and large relative surface, small rodents should be unable to remain above ground and active during the day. Seemingly, the small ground squirrels of the North American deserts defy this conclusion. These small rodents, which weigh from less than one hundred to a few hundred grams, are often seen outside their burrows in the daytime. They move around quickly, dashing from one place to another; they frequently disappear in a hole, but soon appear again, often within minutes. During the hottest midsummer days they are less active in the middle of the day, but are often seen during the morning and late afternoon hours.

Usually, there is no free water available in their environment. How is it possible for ground squirrels to remain active during the day? Apparently, heat storage plays a great role, but not on a diurnal cycle, as for the camel.

When they are outside their burrows on a hot day, the ground squirrels heat up very rapidly. They cannot tolerate any higher body temperature than other mammals and die if heated to 43°C. However, a temperature of 42.4°C can be tolerated without apparent ill effects. As an active ground squirrel has become heated outside, it returns to the relatively cool burrow, were it rapidly cools off, helped by its large relative surface (Chappell and Bartholomew 1981). By moving in and out of the cool burrow the ground squirrel can rapidly pick up heat and rapidly unload it again and in this way avoid using water for evaporation. This permits it to move about and be active during the day.

Evaporation: sweating or panting

We know from our own experience that humans sweat to increase cooling by evaporation. Dogs, in contrast, have few sweat glands, and they cool primarily by panting – a very rapid, shallow breathing that increases evaporation from the upper respiratory tract. Some animals use a third method for increasing evaporation: They spread saliva over their fur and lick their limbs, thus achieving cooling by evaporation.

The amount of heat needed to evaporate 1 kg of water (580 kcal or 2426 kJ) is, of course, the same irrespective of the source of the water and where it is evaporated. Let us see if there are other differences that might make one method of evaporation more or less advantageous than another for a particular animal. We will therefore examine the characteristic features of each method.

In humans, sweating carries virtually the entire burden of increased evaporation. Evaporation from the body surface is an effective way of cooling, and in the absence of fur the water can evaporate readily. However, several furred mammals, in particular those of large body size, such as cattle, large antelopes, and camels, also depend mainly on sweating for evaporation. The camel in particular has a thick fur, but evaporation is not much impeded because in a desert atmosphere the air is very dry and the sweat evaporates so rapidly that some observers have reported that camels do not sweat at all. In contrast, a number of smaller ungulates – sheep, goats, and many small gazelles – evaporate mostly by panting. Those carnivores that have been studied also pant.

The third method, salivation and licking, is common in a fairly large number of Australian marsupials, including the large kangaroos, and can also be found in some rodents, including the ordinary laboratory rat. The salivation-and-licking method is not very effective and seems to be used primarily as an emergency measure when the body temperature threatens to approach a lethal level. The importance of salivation can be demonstrated in rats by removal of the major salivary glands. The rats are then much less tolerant of high air temperatures, and when they are exposed to a heat load their body temperature increases faster and to higher levels (Hainsworth 1967).

In addition to being less effective than sweating and panting, salivation and licking are less widespread, and we shall not discuss the method further. Instead, we shall compare sweating and panting and see what possible advantages and disadvantages each method may have.

Birds, in contrast to mammals, have no sweat glands. They increase evaporation either by panting or by a rapid oscillation of the thin floor of the mouth and upper part of the throat, a mechanism known as *gular flutter*. Either mechanism seems to be an effective means of cooling, for birds may use one or the other or often a combination of both (Bartholomew et al. 1968).

An obvious difference between panting and

sweating is that the panting animal provides its own air flow over the moist surfaces, thus facilitating evaporation. In this regard the sweating animal is not so well off. Another difference is that the sweat (at least in humans) contains considerable amounts of salt, and a heavily sweating human may lose enough salt in the sweat to become salt-deficient. This is the reason for the recommendation that we should increase the intake of sodium chloride when sweating is excessive in very hot situations. Panting animals, in contrast, do not lose any of the electrolytes secreted from glands in the nose and the mouth (unless the saliva actually drips to the ground) and are thus better off in this regard as well.

Panting has two obvious disadvantages. One is that increased ventilation easily causes an excessive loss of carbon dioxide from the lungs, which can result in severe alkalosis; the other is that increased ventilation requires muscular work, which in turn increases the heat production and thus adds to the heat load. The tendency to develop alkalosis can in part be counteracted by shifting to a more shallow respiration (smaller tidal volume) at an increased frequency, so that the increased ventilation takes place mostly in the dead space of the upper respiratory tract. Nevertheless, heavily panting animals regularly become severely alkalotic, and thus they do not utilize fully the possibility of restricting the ventilation to the dead space (Hales and Findlay 1968).

The increased work of breathing during panting would be a considerable disadvantage were it not for the interesting fact that the muscular work, and thus heat production, can be greatly reduced by taking advantage of the elastic properties of the respiratory system. When a dog begins to pant, its respiration tends to shift rather suddenly from a frequency of about 30 to 40 respirations per minute to a relatively constant high level of about 300 to 400. A dog subjected to a moderate heat load does not pant at intermediate frequencies; instead, it pants for brief periods at the high frequency, alternating with periods of normal slow respiration.

The meaning of this becomes clear when we realize that the entire respiratory system is elastic and has a natural frequency of oscillation, like other elastic bodies. That is, on inhalation, much of the muscular work goes into stretching elastic elements, which on exhalation bounce back again, like a tennis ball bouncing. To keep the respiratory system oscillating at its natural frequency (the *resonant frequency*) requires only a small muscular effort. As a consequence, the heat production of the respiratory muscles is small, adding only little to the heat load (Crawford 1962).

It has been estimated that if panting were to take place without the benefit of a resonant elastic system, the increased muscular effort of breathing at the high frequency of panting would generate more heat than the total heat that can be dissipated by panting. Apparently, birds can also pant at a resonant frequency, thus achieving the same benefits as mammals (Crawford and Kampe 1971).

Although we recognize some characteristic functional differences between panting and sweating, we do not have a clear picture of why one method or the other is used. Many animals use both methods, and this may indicate that both have valuable characteristics we do not yet fully appreciate.

Brain temperature. The greatest advantage of panting may be that an animal under sudden heat stress, such as a fast-running African gazelle pursued by a predator, can reach a high body core temperature and yet keep the brain, the most heat-sensitive organ, at a lower temperature. This may, at first glance, seem impossible, for the brain is supplied with arterial blood at a high flow rate. It is achieved as follows (see Figure 7.23).

In gazelles and other ungulates, most of the blood to the brain flows in the external carotid artery, which at the base of the skull divides into hundreds of small arteries, which then rejoin before passing into the brain. These small arteries lie in a large sinus of venous blood that comes from the walls of the nasal passages where it has been cooled. The blood that flows through these small arteries is

FIGURE 7.23 The brain of a gazelle can be kept at a lower temperature than the body core because the arterial blood, before it reaches the brain, passes in small arteries through a pool of cooler venous blood that drains from the nasal region, where evaporation takes place. [Taylor 1972]

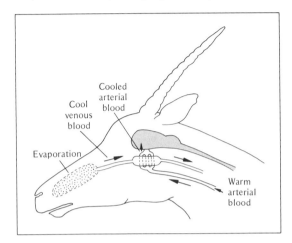

therefore cooled before it enters the skull, and as a result the brain temperature may be 2 or 3°C lower than the blood in the carotid artery and the body core.

Such temperature differences have been measured in the small East African gazelle, Thomson's gazelle, which weighs about 15 to 20 kg. When it ran for 5 minutes at a speed of 40 km per hour, a rapid buildup of body heat caused the arterial blood temperature to increase from a normal 39 to 44°C. The brain temperature did not even reach 41°C, which is a safe level (Taylor and Lyman 1972). It would be difficult to design a cooling system that can keep the entire body from heating up when an animal runs in hot surroundings at a speed that may require a 40-fold increase in metabolic rate, but by selective cooling of the blood to the brain the most serious hazard of overheating is avoided.

Similar cooling of blood to the brain has been observed in a number of domestic ungulates, and it may be a fairly common mechanism in animals that pant. We could surmise that the existence of a

brain-cooling mechanism is unlikely in humans because we do not pant, and sweating and cooling take place over the entire body surface. This conclusion is too hasty, however. It appears that selective cooling of the brain can result with the aid of venous blood returning from facial skin via the ophthalmic vein to the cavernous sinus, where a cooling of arterial blood ascending to the brain can take place (Cabanac 1986).

It has been known for many years that some reptiles, when they are exposed to a severe heat stress, increase their respiration frequencies and breathe with open mouth, a situation reminiscent of panting. It has been difficult to establish with certainty that this increased respiration, which does increase respiratory evaporation, plays any major role in heat regulation. The amount of water evaporated does not suffice to keep the animals substantially cooler than the environment, and the mechanism therefore seems to be no more than a rudiment of panting found in birds and mammals. However, in view of the ability of mammals to cool the brain selectively, the panting of lizards can better be understood.

In a desert lizard, the chuckawalla (*Sauromalus obesus*), panting is of marginal importance in the overall heat balance. When the ambient temperature is kept at 45°C, the body is barely cooler (44.1°C), but the brain remains at 42.3°C, or nearly 3°C below the ambient temperature. If a chuckawalla is moved from a room at 15°C to 45°C, the brain and cloacal temperatures initially increase rapidly (Figure 7.24), but as these temperatures reach 41 to 42°C, the brain temperature stabilizes at this level, while the cloacal temperature continues to rise until it nearly reaches the ambient temperature.

In the chuckawalla the carotid arteries run very close to the surface of the pharynx, so close that they are visible through the open mouth of the animal. As the arteries pass right under the moist surfaces where evaporation takes place, the blood is cooled before it enters the brain. This is again an example of an animal that is able to maintain dif-

FIGURE 7.24 When the lizard *Sauromalus* is moved from 15 °C to 45 °C, cloacal and brain temperatures increase rapidly. As these temperatures reach about 41 °C, they separate and the brain remains about 2 °C below the cloacal temperature and 3 °C below the air temperature. [Crawford 1972]

ferent parts of its body at different, well-regulated temperatures.

TORPOR AND HIBERNATION

Maintaining the body temperature in the cold at a cost of a several-fold increase in metabolic rate is expensive. Small animals have high metabolic rates to begin with, and a further increase may be too expensive when food is scarce or unavailable. The easy way out, and the only logical solution, is to give up the struggle to keep warm and let the body temperature drop. This not only eliminates the increased cost of keeping warm, but cold tissues use less fuel and the energy reserves last longer. This, in essence, is what hibernation is all about.*

Many mammals and a few birds hibernate regularly each winter. This means that the body tem-

*The word hibernation (from Latin *hiberna* = winter) is also used to designate an overwintering, inactive stage of poikilothermic animals such as insects or snails.

perature drops almost to the level of the surroundings. Metabolic rate, heart rate, respiration, and many other functions are greatly reduced; and the animal is torpid and shows little response to external stimuli such as noise or being touched. With active life virtually suspended, the animal can survive a long winter. Before entering into a period of hibernation, most hibernators also become very fat (i.e., they deposit large fuel reserves). Without the burden of keeping warm, the reserves can last for extended periods of unfavorable conditions.

Most animals that hibernate are small. This makes sense, for their high metabolic rates require a high food intake. Thus, many rodents – hamsters, pocket mice, dormice – hibernate. Insect eaters at high latitudes can find little food in winter, and bats and insectivores (e.g., hedgehogs) could not possibly manage without hibernation. Hibernators are also found among Australian marsupials, such as the pygmy possum (Bartholomew and Hudson 1962). Hibernating birds include hummingbirds, the smallest of all birds, the insect-eating swifts, and some mouse birds (*Colius,* an African genus).

It is not easy to give a completely satisfactory definition of the term *hibernation.* In physiology the word refers to a torpid condition with a substantial drop in metabolic rate. Thus, bears may sleep during much of the winter, but most physiologists say that they are not true hibernators. Their body temperature drops only a few degrees, they show only a moderate drop in metabolic rate and other physiological functions, and the females often give birth to the cubs during the winter. In other words, the bear does not conform to what a physiologist customarily considers true hibernation.

Another term, *estivation* (Latin *aestas* = summer), refers to inactivity during the summer and is even less well defined. It may be applied to snails that become dormant and inactive in response to drought, or it may refer to ground squirrels that during the hottest months disappear into their burrows and remain inactive. The Columbian ground squirrel begins to estivate in the hot month of August, but then it remains inactive throughout the autumn and winter and does not appear again until the following May. Does it estivate or hibernate, and when does one condition change into the other? The fact is that no clear physiological distinction can be made between the two states. Furthermore, many animals, such as bats and hummingbirds, undergo daily periods of torpor with decreased body temperature and metabolic rate. Their physiological state is then quite similar to hibernation, although it lasts only for hours instead of weeks or months.

Because we lack precise definitions and there are no sharp lines between these different states, we will treat torpor and hibernation as one coherent physiological phenomenon.

It was earlier believed that hibernation and torpor result from a failure of temperature regulation in the cold and express some kind of ''primitive'' condition or poor physiological control. It is quite clear, however, that hibernation is not attributable to inadequate temperature regulation; it is a well-regulated physiological state, and the superficial similarity between a hibernating mammal and a cold-blooded animal (say, a lizard or frog) is misleading. We shall find that hibernation can in no way be considered a physiological failure.

Body temperature and oxygen consumption

Bats are among the animals that may have daily periods of torpor and may also hibernate for longer periods. If exposed to low ambient temperature, bats may respond in one of two different characteristic ways. This is illustrated in Figure 7.25, which shows measured body temperatures of two North American bat species. At temperatures below about 30°C these bats can be either torpid and have body temperatures within a degree or two of the air temperature, or they can be metabolically active with normal body temperatures between 32 and 36°C. Even at air temperatures below 10°C, these bats may be either active or torpid.

The two bats are of similar size. The larger, the big brown bat (*Eptesicus fuscus*), which weighs about 16 g, ranges over most of the United States;

FIGURE 7.25 The body temperatures of two species of North American bats. At low air temperature these bats may be either active with normal body temperatures about 34 °C or torpid with body temperatures near the air temperature. [Herreid and Schmidt-Nielsen 1966]

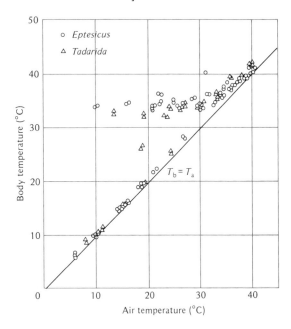

and at least in the northern part of its range it hibernates. The smaller species, the Mexican free-tailed bat (*Tadarida mexicana*), weighs about 10 g; it inhabits the hot, dry southwestern United States and Mexico, and in winter in migrates southward and is not known to hibernate.

The response to low temperature of these two species is amazingly similar, although in nature one hibernates and the other does not. The oxygen consumption of those individuals that maintain a high body temperature at low air temperature shows the typical homeothermic pattern. As ambient temperature decreases, the cost of keeping warm is reflected in the increased oxygen consumption, which is at a minimum at about 35°C (Figure 7.26). Those individuals that become torpid, however, have very low metabolic rates. At 15°C, for example, the dif-

FIGURE 7.26 The oxygen consumption of active bats increases with decreasing air temperature; the oxygen consumption of torpid bats drops to a small fraction of the active rate (same bat species as in Figure 7.25). [Herreid and Schmidt-Nielsen 1966]

FIGURE 7.27 Oxygen consumption over a 24-hour period of a pocket mouse kept on a restricted food ration of 1.5 g seeds per day. During 9 hours the animal was torpid and had a very low rate of oxygen consumption followed by a peak as it returned to the active state. Air temperature was 15 °C. [Tucker 1965a]

ference in oxygen consumption between an active and a torpid bat is about 40-fold. The torpid bat thus uses fuel at a rate only one-fortieth the rate of the active bat, and its fat reserves would last 40 times as long.

The amount of energy that can be saved by going into torpor has been more carefully analyzed by Tucker (1965a,b, 1966). The oxygen consumption during torpor is easily determined, but in addition, the energy cost required to rewarm the animal at the end of a period of torpor must be included in the cost. Tucker studied the California pocket mouse (*Perognathus californicus*) and found that for a short period of torpor the energy cost of reheating was a substantial part of the total cost.

The pocket mouse weighs about 20 g and readily becomes torpid at any ambient temperature between 15 and 32°C. Below 15°C the normal torpid state is disrupted and the animals usually cannot arouse on their own; in other words, if cooled below 15°C they are unable to produce enough metabolic heat to get the rewarming cycle started.

Pocket mice enter torpor much more readily if their food ration is restricted. With a gradual restriction of the food supply, longer and longer periods are spent in torpor. In this way a pocket mouse can maintain its body weight on a food ration that is about one-third of what it normally consumes when it remains active. A characteristic curve for the oxygen consumption of a pocket mouse on a restricted food ration is given in Figure 7.27. This particular individual received 1.5 g seed per day, and on this ration it was torpid for about 9 hours out of every 24. Although the food ration was less than half of what the animal would normally need if it remained active, it maintained a constant body weight.

When an animal goes into torpor, what happens first? Does heat loss first increase, the body temperature therefore drop, and the metabolic rate as a consequence decrease, thus accelerating the temperature drop? Or does the animal decrease its metabolic rate, the body temperature therefore begin to drop, the temperature drop causing a further decrease in heat production, and so on?

With information about heat production and heat loss during all stages of the torpor cycle, this question can be answered. The conclusion is that entry into torpor can result simply from a cessation of any thermoregulatory increase in metabolism at air temperatures below the lower critical temperature (about 32.5°C in the pocket mouse). At any temperature below this point, a decrease in metabolic rate to the thermoneutral resting level alone will, without change in conductance, lead to a decrease in body temperature. In other words, there is no need for a special mechanism to increase the heat loss. As the body temperature begins to fall, metabolism decreases and the pocket mouse slides into torpor with a further drop in heat production and body temperature.

Arousal can take place at any temperature above 15°C if the animal changes to maximum heat production at that temperature, which is about 10 to 15 times the minimum oxygen consumption at the same temperature. Arousal is thus an active process that requires a considerable expenditure of energy for a considerable period until the body temperature has reached normal.

If a pocket mouse goes into torpor at 15°C and immediately arouses again, does it save any energy, or does rewarming cost too much? The decline in body temperature takes about 2 hours, and the total oxygen consumption during this period is 0.7 ml O_2 per gram. If arousal follows immediately, normal body temperature is reached in 0.9 hour at a cost of 5.8 ml O_2 per gram. Entering into torpor and then immediately arousing therefore costs a total of 6.5 ml O_2 per gram in 2.9 hours. The cost of maintaining normal body temperature for the same period of time at 15°C ambient temperature would be 11.9 ml O_2 per gram. This period of torpor thus consumes only 55% of the energy required to keep warm for the same period of time (Tucker 1965b).

Remaining in torpor for longer periods, for example, 10 hours, is even more favorable for the

animal. We must now add to the previously calculated cost of entry and arousal the cost of 7.1 hours at 15°C, which is 1.2 ml O_2 per gram, giving a total for a 10-hour cycle, including entry and arousal, of 7.7 ml O_2 per gram. This is less than 20% of the cost of maintaining normal body temperature for 10 hours, which at 15°C would be over 40 ml O_2 per gram.

We can see that in all these cases the cost of arousal is the major part of the cost of the torpor cycle. For the 10-hour period of torpor the cost of arousal alone is 75% of the total energy expenditure for the period.

The widespread occurrence of torpidity as a response to unfavorable conditions suggests that this ability is a very fundamental trait. Animals that can enter torpor include the egg-laying echidna (the spiny anteater, *Tachyglossus aculeatus*), which is only remotely related to modern mammals. The echidna is normally an excellent temperature regulator in the cold and can maintain its normal body temperature at freezing temperatures (Schmidt-Nielsen et al. 1966). However, when it is without food and kept at 5°C, it readily becomes torpid. The heart rate decreases from about 70 to 7 beats per minute, and the body temperature remains about +5.5°C. In this state the oxygen consumption of the echidna is about 0.03 ml O_2 g^{-1} h^{-1}, roughly one-tenth its normal resting oxygen consumption (Augee and Ealey 1968).

Control mechanisms

Torpor and hibernation are under accurate physiological control. We have just seen how the duration of the daily torpor of a pocket mouse is adjusted to the food ration and the need for energy savings. Obviously, not only must entry into torpor be controlled, but the duration of the cycle and the arousal process must be accurately regulated.

In nature the beginning of the hibernation cycle is usually associated with the time of the year, but it is not necessarily induced by low temperature or lack of adequate food. The yearly cycle of hibernation is influenced by the duration of the daily light cycle and is associated with endocrine cycles. For example, it is often impossible to induce even a good hibernator to begin a cycle of torpor during early summer and especially during the reproductive period.

The torpid animal with its low body temperature seems quite passive; it cannot perform coordinated movements, and it hardly responds to sensory stimuli. Superficially it resembles a cold-blooded animal that has been chilled. All of us have experienced how our fingers become numb when the hands are cold; this is because nerve conduction ceases at temperatures below some 10 to 15°C, and we sense nothing. (The reason we can still move our cold fingers is that the appropriate muscles are located above the hand in the lower part of the arm.)

If the nerves of a hibernator became inoperative at lower temperatures, the nervous system could not remain coordinated at, say, 5°C. Yet respiration and many other functions continue in a well-coordinated fashion although at a lower rate. If the ambient temperature decreases toward or below freezing, some hibernators die. Others, however, respond in one of two ways, leaving no doubt about the integrity of the central nervous system. Either they arouse and return to the fully active condition, or they resist the decrease in body temperature by a regulated increase in heat production, keeping the body temperature at some low level, say +5°C.

Such a well-regulated heat production does indeed require a well-coordinated central nervous system. For example, the European hedgehog maintains its body temperature at +5 to +6°C as the ambient temperature decreases to below freezing. This prevents freezing damage and at the same time saves fuel, for it eliminates the need of going through an expensive complete arousal and the cost of then maintaining the high body temperature of the active state. For a hibernator that may be exposed to freezing temperatures several times during winter, such repeated arousals might prove too expensive; maintaining a low body temperature, just sufficient to keep from freezing, is far more economical.

FIGURE 7.28 Oxygen consumption of the tropical hummingbird *Eulampis*. In the awake *Eulampis* oxygen consumption increases linearly with decreasing temperature. In the torpid *Eulampis* oxygen consumption drops to a low level, but if the air temperature goes below 18 °C, the animal produces more heat and maintains its body temperature at 18 to 20 °C without arousing from torpor. [Hainsworth and Wolf 1970]

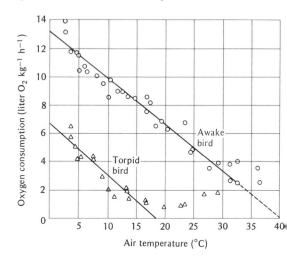

A similar well-regulated state of torpor has been observed in the West Indian hummingbird, *Eulampis jugularis,* which like other hummingbirds readily becomes torpid. The temperature of *Eulampis,* when torpid, approaches air temperature, but if the air temperature falls below 18°C, *Eulampis* resists a further fall and maintains its body temperature constant at 18 to 20°C. Below this point the heat production must therefore be increased (Figure 7.28), and the increase must be linearly related to the drop in the ambient temperature.

Eulampis is particularly interesting because it is a clear case of torpidity in a tropical warm-blooded animal, and this shows that torpidity is not restricted to animals from cold climates. Also, the well-regulated metabolic rate at two different levels of body temperature certainly shows that torpor in no way is a failure of the process of thermoregulation. Another interesting aspect is that the overall thermal conductance (the slope of the metabolic regression line) is the same in torpid and in normal nontorpid hummingbirds.

Arousal

The rewarming during arousal, as we have seen, is metabolically the most expensive part of the torpor cycle. The arousing animal displays violent shivering and muscle contractions and apparently uses fuel at a maximal rate. However, not only the muscles but also a particular kind of tissue, known as *brown fat*, is important in the rewarming process.

Brown fat is found in smaller and larger patches along the neck and especially between the shoulders. The cells are filled with fat, but in contrast to ordinary white fat cells, they are packed with large mitochondria that give the brown fat its characteristic color. The tissue has a rich blood supply and is innervated by sympathetic nerves.

In the cells of brown fat the mitochondrial respiration is uncoupled from the mechanism of ATP synthesis so that the energy of oxidation is dissipated as heat instead of being used for ATP synthesis. A special protein with a molecular weight of 32 000, known as *thermogenin,* is responsible for the uncoupling. As a result fat is oxidized and heat produced at an extremely high rate. The cells use the stored fat, but can also utilize substrates supplied with the blood (Himms-Hagen 1985). The heat production is turned on by noradrenalin or by nervous stimulation.

Brown fat is found in all hibernating mammals, but it also occurs in many newborn mammals, including humans, in whom its importance is to produce heat and maintain body temperature. In most adult mammals the brown fat has been lost, but in hibernators it remains prominent. During arousal large amounts of heat are produced in the brown fat, and because the venous blood returns directly to the heart, this essential organ is heated rapidly.

It is characteristic of an arousing hibernator that the temperature is not the same throughout the body;

FIGURE 7.29 The temperature increase in a hamster arousing from hibernation proceeds more rapidly in the anterior part of the body (cheek pouch temperature) and more slowly in the posterior part (rectal temperature). [Lyman 1948]

the rewarming is far from uniform. Figure 7.29 shows temperatures recorded separately in the cheek pouch and in the rectum of an arousing hamster. The anterior part of the body, which contains vital organs such as heart and brain, warms much more rapidly than the posterior part. Rewarming of the heart is not only essential but must be an initial step, for the heart is needed to provide circulation and oxygen for all the other organs. The major masses of brown fat are also located in the anterior part of the body. The records show that not until the anterior portion of the body has reached near-normal temperatures does the posterior part enter into the reheating process. At this moment, when the entire body rapidly attains normal temperature, there is a peak in the oxygen consumption, which afterward declines and settles at a normal level.

The uneven temperature increase in the various parts of the body leads to an inevitable conclusion: During the early stages of arousal the blood flow is directed almost exclusively to the vital organs in the anterior part of the body, and not until these have been rewarmed does circulation increase substantially in the posterior parts as well.

The blood flow to the various organs can be followed by means of radioactive tracers, and such studies have shown that the skeletal muscles in the anterior part of arousing animals receive more than 16 times as much blood as in an awake, nonhibernating individual. This confirms that the muscles are highly involved in the increased heat production.

During early arousal the muscles in the hind part of the animal receive only one-tenth as much blood as in the front part. The brown fat, however, receives even more blood than the most active muscles, implicating this tissue in heat production during arousal. As could be expected, the digestive tract, in particular the small intestine, is among the tissues that receive the least blood during arousal (Johansen 1961).

It is perplexing that brown adipose tissue, which is so important for mammalian hibernators, appears to be absent from a number of birds that regularly hibernate, such as hummingbirds, swifts, and nighthawks (Johnston 1971).

BODY TEMPERATURE IN "COLD-BLOODED" ANIMALS

Birds and mammals are traditionally known as warm-blooded, the term cold-blooded being used for all other animals. However poor the term is, it is well established and convenient, and we will use it, for no better single word is available that is both complete and accurate.

Most cold-blooded animals are more or less at the mercy of their environment in regard to the body temperatures they attain. However, some so-called cold-blooded animals can and indeed do stay warmer than the medium in which they live, whether it be air or water. What they can achieve in this regard is governed by simple physical principles, although some of the solutions seem quite ingenious.

Let us return to the heat balance equation on page 253 and rearrange the terms with only the term for heat storage (H_s) on the left side. On the right side are the total heat production (H_{tot}) and heat

exchange by radiation (H_r), by conduction and convection (H_c), and by evaporation (H_e). If the body temperature is higher than the surrounding medium, the terms for conduction (H_c) and evaporation (H_e) will signify heat losses and be negative. The equation can therefore be written:

$$H_s = H_{tot} \pm H_r - H_c - H_e$$

If the goal is to increase the body temperature, we must see what the animal can do to maximize heat storage (H_s).

Aquatic animals

For an aquatic animal the situation is simple, for in water there is no evaporation and no significant radiation source (infrared radiation is rapidly absorbed in water). This leaves the equation: $H_s = H_{tot} - H_c$. Only two parameters can be manipulated in order to increase heat storage: Either total heat production must be increased, or conductive heat loss must be minimized.

Because of the high thermal conductivity and high heat capacity of water, a small animal loses heat rapidly and has no chance of attaining a body temperature very different from the medium. Even if it has a high level of heat production (metabolic rate), it needs oxygen to sustain this heat production.

Here is where the heat problem comes in. A high rate of oxygen uptake requires a large gill surface and blood for oxygen transport. As the blood flows through the gills, it is inevitably cooled to water temperature. The gill membrane, which must be thin enough to permit passage of oxygen, provides virtually no barrier to heat loss. The blood is therefore cooled to water temperature, and it is impossible for the animal to attain a high temperature unless a heat exchanger is placed between the gills and the tissues.

This solution is used by some large, fast-swimming fish (e.g., tunas and sharks) to achieve independent control of the temperature in limited parts of the body. They have heat exchangers that permit them to maintain high temperatures in their

HEAT EXCHANGER Cross section of a 2-kg skip-jack tuna (*Katsuwonis pelamis*) shows how this powerful swimming machine consists mostly of muscle (left). The red muscle, which is maintained at high temperature, appears nearly black in the photo. In some tunas the heat exchanger is located laterally, but in the skipjack the main heat exchanger is located just below the vertebral column, al-most exactly in dead center of the photo. A cross section of the vascular heat exchanger (right) shows a roughly equal number of arteries and veins. The arteries (smaller and thick-walled) are interspersed with veins (larger and thin-walled). The diameter of the arteries is about 0.04 mm and of the veins 0.08 mm, and their length is about 10 mm. [Courtesy of E. D. Stevens, University of Guelph, Ontario]

swimming muscles, independent of the water in which they swim.

The heat exchangers that supply the swimming muscles of the tuna are, in principle, similar to the countercurrent heat exchangers in the whale flipper, but anatomically they are somewhat differently arranged. In the ordinary fish the swimming muscles are supplied with blood from the large dorsal aorta that runs along the vertebral column and sends branches out to the periphery. In the tuna the pattern is different.

The blood vessels that supply the dark red muscles (which the tuna uses for steady cruising) run along the side of the fish just under the skin. From these major vessels come many parallel fine blood vessels that form a slab in which arteries are densely interspersed between veins running in the opposite direction (Figure 7.30). This puts the cold end of the heat exchanger at the surface of the fish and the warm end deep in the muscles.

The arterial blood from the gills is at water temperature, and as this blood runs in the fine arteries between the veins, it picks up heat from the venous blood that comes from the muscles. Heat exchange is facilitated by the small diameter of the vessels, about 0.1 mm. When the venous blood has reached the larger veins under the skin, it has lost its heat, which is returned to the muscle via the arterial blood.

Rete
mirabile

Vertebra

Dorsal
aorta

Dark
muscle

Cutaneous
artery

Vein

Cutaneous
vein

FIGURE 7.30 The blood supply to the swimming muscle of the tuna comes through a heat exchanger, which is arranged so that these muscles retain a high temperature, although the arterial blood is at water temperature. A network of blood vessels arranged in this way is known as a *rete mirabile*, meaning "wonderful net." [Carey and Teal 1966]

As a result, the tuna can maintain muscle temperatures as much as 14°C warmer than the water in which it swims.

The advantage of keeping the swimming muscles warm is that high temperature increases their power output.* A high power output gives the tuna a high swimming speed relatively independent of water temperature, and this in turn enables the tuna to pursue successfully prey that otherwise swims too fast to be caught, such as pelagic fish (e.g., mackerel) or squid. The greatest advantage is probably that it makes the tuna relatively independent of abrupt changes in water temperature as it moves rapidly between the surface and deeper cold water.

Several of the large sharks, but by no means all of them, have similar heat exchangers that permit their muscles to be kept considerably warmer than the water in which they swim (Carey et al. 1971).

The tuna enjoys an additional advantage from the substantial temperature differences between various parts of the body. Separate heat exchangers permit the digestive organs and the liver to be kept at a high temperature. The high power output of the muscles requires a high rate of fuel supply, and this puts a premium on a rapid rate of digestion, which is most readily achieved by a high temperature in the digestive tract (Carey et al. 1984).

---

*The force exerted by a contracting muscle is relatively independent of temperature, and the work performed in a single contraction (work = force × distance) is therefore also temperature-independent. However, at higher temperature the muscle contracts faster, and because the number of contractions per unit time increases, the power output (work per unit time) increases accordingly.

Swordfish and the closely related marlins and sailfish have the remarkable ability to keep eyes and brain warm, again by using countercurrent heat exchangers to prevent the loss of heat. These predatory fish have some of the largest eyes of any animal. A large swordfish has eyes the size of grapefruit, which undoubtedly is helpful when it pursues prey in deep cold water where the light is minimal.

In these fish it is not the eye and brain that produce the large amounts of heat; the heat is produced in a special organ that is derived from one of the eye muscles. The organ has lost all contractile characteristics but has extensive systems of membranes that correspond to what in muscle cells is known as the sarcoplastic reticulum (Block 1986). The uncoupling of the ATP synthesis seems to be associated with an increased flux of Ca^{++} ions across these membranes (Block 1987).

Similar eye and brain heater organs are found also in some of the large sharks, the porbeagle and the mako sharks (Block and Carey 1985).

Let us now compare the warm-bodied fish with marine mammals and consider the main differences and similarities between them. Seals and whales have heavily insulated surfaces (blubber); heat loss from the extremities is reduced by heat exchangers; and most important, they are air breathers whose blood escapes being cooled to water temperature at the respiratory surface. The warm-bodied fish, in contrast, would gain no advantage from a better surface insulation, for all the arterial blood coming from the gills is already at water temperature. They do achieve independent temperature control in limited parts of the body by strategic location of efficient heat exchangers that help retain locally produced heat.

Terrestrial animals

The heat balance of terrestrial animals includes all the terms listed in the equation on page 253. To raise the body temperature (i.e., to increase heat storage), it is important to reduce evaporation and heat loss by conduction and to maximize heat gain by radiation and metabolic heat production. The only commonly available radiation source is the sun, and in the absence of solar radiation, metabolic heat production is the only way to increase the heat gain. Of course, external and internal sources of heat (i.e., sun and metabolism), can be used simultaneously, but it is more economical to use an external source rather than body fuel.

Solar radiation is used especially by insects and reptiles. To increase the amount of radiant energy absorbed, these animals depend both on their color and on their orientation relative to the sun. Many reptiles can change their color by dispersion or contraction of black pigment cells in their skin. Because about half the solar radiation energy is in the visible light, a dark skin substantially increases the amount of solar energy that is absorbed rather than reflected. The absorption in the near infrared is not much affected by color change, for the animal surface is already close to being "black" in this range of the spectrum.

The other way of increasing heat gain from solar radiation is to increase the exposed area. This is done by orienting the body at right angles to the direction of the sun's rays and by spreading the legs and flattening the body. In this way a lizard can attain a temperature much higher than the surrounding air. When a suitable temperature has been reached, further heating is avoided by lightening the skin color and changing the orientation to a posture more parallel to the sun's rays, and eventually by moving into shade. The temperature of the substratum is also important, for a cool lizard can place itself in close contact with a warm rock and thereby increase its heat gain.

The temperature a lizard can attain by warming itself in the sun can be spectacular. In the mountains of Peru, at an altitude of 4000 m where the air temperature is low even in summer, the lizard *Liolaemus* remains active and comes out at temperatures below freezing. One lizard that weighed 108 g was tethered in the sun in the morning while the air temperature was still at −2°C. Within the next hour the air temperature increased to +1.5°C,

FIGURE 7.31 The Peruvian mountain lizard *Liolaemus* comes out in the morning while the air temperature is still below freezing and, by exposing itself to the sun, rapidly heats up to a body temperature that permits full activity. [Data from Pearson 1954]

but the cloacal temperature of the lizard had increased from +2.5°C to 33°C (Figure 7.31). The lizard's body temperature was 30°C higher than the air temperature, allowing it to be active in the near-freezing surroundings.

Once the lizard is warm, its body temperature remains more or less constant around 35°C, while the air temperature slowly continues to rise. This is achieved because the lizard reduces the amount of absorbed solar energy by changing posture, so that it maintains its *preferred body temperature*, also known as the *eccritic temperature*.

It is, of course, important for the animal to heat up rapidly and become fully active as fast as possible. However, when the animal moves out of the sun, retarding the inevitable cooling would be an advantage. Can the lizard do anything to increase the rate of heating and decrease the rate of cooling? The heating of the body core can be accelerated by increasing the circulation to the heated skin area, but what about retarding cooling?

This question has been studied in several species of lizards. It is of particular importance to the Galápagos marine iguana, which feeds in water where

FIGURE 7.32 Heating and cooling rates of the Galápagos marine iguana (body weight 652 g) when immersed in water (open triangles) or in air (open circles). The initial temperature difference between animal and medium (ΔT) was, in all cases, 20 °C. [Bartholomew and Lasiewski 1965]

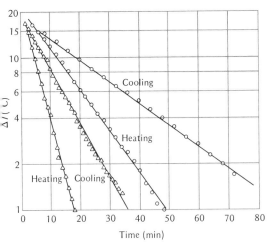

FIGURE 7.33 The heart rate of a soft-shelled turtle weighing 650 g. The heart rate, measured at any given body temperature, was much higher if the turtle was being heated than if it was undergoing cooling. The animal was in water during both heating and cooling. [Smith et al. 1981]

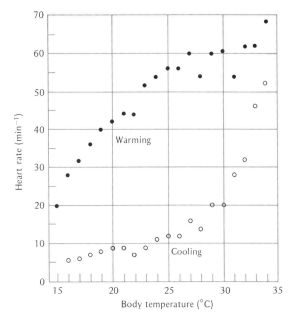

the temperature is 22 to 27°C. When it is out of water, the animal sits on the rocks near the surf where it can warm itself in the sun. When it enters the water to feed on the seaweed, its temperature drops toward that of the water, which is below the eccritic temperature. It would be advantageous for the marine iguana to reduce its rate of cooling while in water, and thus extend the period during which it can feed actively and move fast enough to escape predators. Conversely, after leaving the sea, it would be advantageous to heat up as rapidly as possible.

A comparison of the rates of heating and cooling for the marine iguana shows that heating rates are about twice as high as cooling rates, both in air and in water (Figure 7.32). When a cold animal is immersed in water that is 20°C above body temperature, the heating is twice as fast as the cooling when a warm animal is immersed in water 20°C below its body temperature. In air both rates are lower, but again, heating is about twice as fast as cooling at the same temperature differential.

The different rates of heating and cooling can best be explained by changes in circulation in the skin. During heating, the heart rate is high and increases with body temperature, but in an iguana being cooled the heart rate drops to half of what it otherwise is at the same body temperature.

For blood flow to be effective in control of heat exchange, it must occur in a region of the body where the resistance to heat flow is relatively high, and in an environment where the external resistance to heat exchange is relatively low. This means that control of heat exchange should be especially effective in water (Turner 1987).

A soft-shelled turtle (*Trionyx spinifer*) when kept in water showed very different responses of the heart rate to heating and to cooling (Figure 7.33). During heating the heart rate increased with increasing

FIGURE 7.34 A brooding female python while incubating a clutch of eggs keeps her body temperature above that of the surroundings by strong muscular contractions reminiscent of shivering in mammals. The diagonal line indicates equality between the snake's temperature and air temperature. [Vinegar et al. 1970]

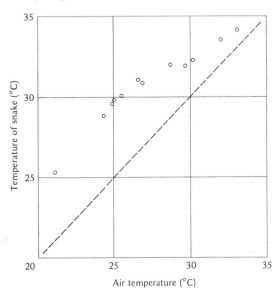

body temperature, and during cooling the heart rate dropped rapidly to very low values, thus reducing circulation and retarding heat loss (Smith et al. 1981).

Many reptiles habitually bask in the sun. Both tortoises and snakes, and even the most aquatic reptiles, crocodiles, can often be seen sunning themselves at the edge of the water.

Internal heat production (i.e., metabolic rate) can also be used to attain a high body temperature. It has been reported that some female snakes maintain a high body temperature while incubating their eggs. This has been observed in the New York Zoological Park, where a 2.7-m long female python started incubating a clutch of 23 eggs. She coiled herself around the eggs, and temperatures were measured with thermocouples placed between the

tightly appressed coils of the snake. As the air temperature was lowered, the animal increased her rate of oxygen consumption and kept her body temperature at 4 to 5°C above the air (Figure 7.34).

The increase in oxygen consumption of the brooding snake seemed to be caused by strong contractions of the body musculature, reminiscent of the shivering of mammals. At 25.5°C the oxygen consumption was 9.3 times higher than during the nonbrooding period. When the temperature was decreased further to 21°C, the animal seemed unable to increase her metabolic rate further, for her temperature fell and there was a drop in the oxygen consumption. After 30 days of incubation the animal had decreased in weight from 14.3 kg to 10.3 kg, a decrease of nearly 30%, supposedly attributable to the fuel consumption needed to keep warm (Hutchison et al. 1966).

Flying insects

Most insects become increasingly sluggish and are unable to fly at low temperatures. Some insects, however, can warm up their flight muscles and be active in quite cold air. The flight muscles are located in the thorax and can produce large amounts of heat by a process similar to shivering in vertebrates. Heating of the flight muscles before takeoff occurs mainly in large insects such as locusts, large moths, butterflies, and bumblebees, and also in wasps and bees that are strong and rapid fliers.

A bumblebee must have a thoracic temperature of at least 29 to 30°C before it is able to fly (Heinrich 1972). If the flight muscles are at lower temperature, their speed of contraction is too slow to support flight, which requires a wing-beat frequency of 130 Hz.

The maintenance of a high thoracic temperature enables the bumblebee to forage for nectar at temperatures as low as 5°C. However, remaining warm on a continuous basis is not feasible unless the bee can find food at a rate at least equal to the rate at which fuel is consumed. A bumblebee weighing 0.5 g may have an oxygen consumption of 50 ml

FIGURE 7.35 The rate of oxygen consumption of two bumblebee queens (*Bombus vosnesenskii*) during periods of uninterrupted incubation. The weight of the thorax, which contains the muscles that produce essentially all the heat, is about 0.2 g; the total body weight of a queen may be around 0.5 g, but varies with the contents of the honey stomach. [Heinrich 1974]

FIGURE 7.36 The increase in thoracic temperature of the sphinx moth (*Manduca sexta*) proceeds more rapidly when the initial temperature of the animal is higher. [Heinrich and Bartholomew 1971]

O_2 per hour (100 liter O_2 kg^{-1} h^{-1}), which corresponds to the use of 60 mg sugar. This investment in fuel may be worthwhile, for at low temperature the competition for the available food supply is reduced because many other nectar-feeding insects are inactive.

The advantage of a small body size, relative to mammals, is not only that warmup can be achieved rapidly, but also that cooling is rapid. Therefore, the body temperature can easily be adjusted to the energy supplies, and the insect can choose to keep warm only when the energy supplies warrant the expense; in other words, it can be utterly opportunistic about being "warm-blooded."

The thermoregulation in bumblebees extends to the incubation of their brood. An incubating queen keeps her abdomen in close contact with the brood clump and regulates her abdominal temperature at 31 to 36°C, which helps maintain the brood at a relatively high temperature. Even at an air temperature as low as 5°C, the brood can be maintained nearly 20 °C higher by the incubating queen. The oxygen consumption of the incubating queen increases linearly at lower temperature, so that her

metabolic curve (Figure 7.35) resembles those we saw for mammals (e.g., Figure 7.7).

Another insect that has been carefully studied in regard to thermoregulation is the sphinx moth, *Manduca sexta* (whose larva is the tobacco hornworm). Sphinx moths are fairly large and weigh between 2 and 3 g. They feed on nectar while hovering in front of a flower, much like a hummingbird. Because they are nocturnal, they often fly at quite low air temperatures, and flight would be impossible unless they could heat themselves.

Hovering flight requires a muscle temperature of at least 35 °C; nevertheless, sphinx moths can fly and feed when the air temperature is as low as 10 °C. This is possible through preflight heating of the thorax. The lower the initial temperature, the lower the rate of heating (Figure 7.36). During the warmup the animal seems to shiver as the wings vibrate lightly.

The heat is generated by contraction of the wing muscles, and the reason that the wings do not flap is that the muscles for upstroke and downstroke contract simultaneously, rather than alternately as in flight. We might expect that the warmup should progress faster and faster as the temperature of the moth increases, for more heat can be produced at higher temperatures. However, as the insect be-

FIGURE 7.37 The flight muscles of honeybees and bumblebees are supplied with blood from the heart, which is located in the abdomen. A countercurrent arrangement in the petiole serves to retain heat in the flight muscles. When, at high air temperature, the bee needs to dissipate excess heat, the countercurrent arrangement is circumvented by alternating the flows in the artery and in the returning venous channel, thus eliminating countercurrent exchange. [Heinrich 1976]

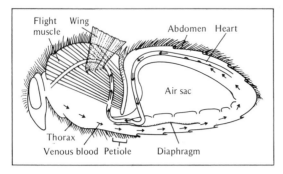

comes warmer, the rate of heat loss to the air also increases, and for this reason the temperature curves in Figure 7.36 appear as more or less straight lines.

The thorax of the moth is covered by long, furry scales. This helps retain the generated heat in the thorax during the warmup. In free flight the sphinx moth maintains its thoracic temperature at about 40 to 41 °C over a wide range of air temperatures, but because of the high rate of heat production in the flight muscles, it must be able to increase the rate of heat loss to keep the temperature from rising even higher. This is achieved through an increased circulation of blood between the thorax and the abdomen.

Heat loss from the abdomen is facilitated by its larger surface and a smaller amount of surface insulation. If we tie off the large dorsal blood vessel leading from the heart (located in the abdomen) so that the flight muscles cannot be cooled, the moth rapidly overheats and cannot continue to fly even at an air temperature of 23 °C. The cause could, of course, be that the blood supply to the muscles is necessary for supplying fuel. However, a simple experiment can answer this objection. If the scales are removed from the thorax of a moth with ligated circulation, the heat loss is sufficiently improved to enable the animal to fly again (Heinrich 1970).

Bumblebees also depend on cooling from the abdomen when they need to dump excess heat from the flight muscles in the thorax. At low air temperatures, however, the problem is to retain heat, and this is aided by a countercurrent exchange system (Figure 7.37). The main blood vessel to the thorax passes from the heart (in the abdomen) through the thin stem, the petiole, that connects to the thorax. Blood that returns from the flight muscles to the abdomen flows through the petiole in the opposite direction in a closely apposed space. This arrangement forms a typical countercurrent heat exchanger, which serves to retain heat in the thorax.

At higher air temperatures, heat must be dissipated rather than retained. The thin petiole has no alternate route for the returning blood to bypass the exchanger; the blood must run in the one available pathway. The only way to block heat exchange is therefore to separate the two flows in time, so that the flow alternates between arterial and venous blood.

The respiratory movements help in the following way. When the abdomen distends during inhalation, venous blood is pulled into the abdomen and when the bee exhales, the abdomen is compressed and arterial blood is pushed into the thorax. The increased arterial pressure causes the artery in the petiole to expand, flattening and blocking the venous channel while the arterial blood flows, thus blocking the heat exchanger (Heinrich 1976).

Honeybees regulate not only their own individual temperature, but that of the entire colony as well. In summer the temperature of a beehive is maintained at about 35 °C and fluctuates very little. This provides the colony with an optimum temperature for development of the brood. Rapid reproduction permits a fast buildup of the population during the most favorable season, and an optimal temperature for the brood is therefore ecologically highly desirable.

If the beehive temperature rises above the regulated level, the bees spread droplets of water over the comb, and evaporation is aided by air currents created by a large number of bees fanning their wings. Below 30 °C bees do not usually consume any water, but at higher temperatures their water intake increases enormously as water is used for cooling (Free and Spencer-Booth 1958).

At low outside temperatures the bees cluster together and maintain a temperature within the cluster of about 20 to 30 °C. The temperature at the center of the cluster is the highest, and it appears that the bees at the periphery force their way into the center where it is warmer, thus exposing others to the outside. The core temperature of the cluster is regulated between 18 and 32 °C at outside air temperatures ranging as low as −17 °C. In fact, there is a tendency for the cluster temperature to be higher at the very lowest ambient temperatures. For each 5 °C drop in the outside air temperature, the cluster temperature increases by about 1 °C (Southwick and Mugaas 1971). In this way a cluster of bees behaves much like a single large organism that maintains a well-regulated high temperature at external temperature conditions that would rapidly kill individual bees.

At low air temperatures the oxygen consumption and heat exchange in clusters of bees change with the size of the cluster, similar to what we have seen about body size in mammals. In a study of honey bee clusters between 5 and 1200 g, the specific rate of oxygen consumption (ml O_2 g^{-1} h^{-1}) was related to "body" (cluster) size by the relation:

$$\dot{V}_{O2} = 22.7 \, M_b^{-0.48}$$

For mammals the specific rate of oxygen consumption is related to body size by the exponent −0.27. Thus, the advantage for bees to congregate in large clusters in the cold is very substantial indeed, and for very small clusters the demand on

fuel is too great to be feasible during winter (Southwick 1985).

We have now discussed the temperature relations of animals, and in this last chapter we have seen how important water can be in maintaining proper temperature conditions. We shall now move on to a more detailed treatment of this important substance and its physiological role.

REFERENCES

Adolph, E. F. (1947) Tolerance to heat and dehydration in several species of mammals. *Am. J. Physiol.* 151:564–575.

Andersen, H. T., Hammel, H. T., and Hardy, J. D. (1961) Modifications of the febrile response to pyrogen by hypothalamic heating and cooling in the unanesthetized dog. *Acta Physiol. Scand.* 53:247–254.

Aschoff, J., Günther, B., and Kramer, K. (1971) *Energiehaushalt und Temperaturregulation*. Munich: Urban & Schwarzenberg. 196 pp.

Aschoff, J., and Wever, R. (1958) Kern und Schale im Wärmehaushalt des Menschen. *Naturwissenschaften* 45:477–485.

Aschoff, J., and Wever, R. (1959) Wärmeaustausch mit Hilfe des Kreislaufes. *Dtsch. Med. Wochenschr.* 84:1509–1517.

Augee, M. L., and Ealey, E. H. M. (1968) Torpor in the echidna, *Tachyglossus aculeatus*. *J. Mammal.* 49:446–454.

Bartholomew, G. A., and Hudson, J. W. (1962) Hibernation, estivation, temperature regulation, evaporative water loss, and heart rate of the pigmy possum, *Cercaertus nanus*. *Physiol. Zool.* 35:94–107.

Bartholomew, G. A., and Lasiewski, R. C. (1965) Heating and cooling rates, heart rate and simulated diving in the Galápagos marine iguana. *Comp. Biochem. Physiol.* 16:575–582.

Bartholomew, G. A., Lasiewski, R. C., and Crawford, E. C., Jr. (1968) Patterns of panting and gular flutter in cormorants, pelicans, owls, and doves. *Condor* 70:31–34.

Bligh, J., Ingram, D. L., Keynes, R. D., and Robinson, S. G. (1965) The deep body temperature of an unrestrained Welsh mountain sheep recorded by a radio-telemetric technique during a 12-month period. *J. Physiol. 176*:136–144.

Block, B. A. (1986) Structure of the brain and eye heater tissue in marlins, sailfish, and spearfishes. *J. Morphol. 190*:169–189.

Block, B. A. (1987) Billfish brain and eye heater: A new look at the nonshivering heat production. *News Physiol. Sci. 2*:208–213.

Block, B. A., and Carey, F. G. (1985) Warm brain and eye temperatures in sharks. *J Comp. Physiol. B. 156*:229–236.

Bronstein, S. M. and Conner, W. E. (1984) Endotoxin-induced behavioral fever in Madagascar cockroach, *Gromphadorhina portentosa. J. Insect Physiol. 30*:327–330.

Buettner-Janusch, J. (1966) *Origins of Man.* New York: Wiley. 674 pp.

Cabanac, M. (1986) Keeping a cool head. *News Physiol. Sci. 1*:41–44.

Calder, W. A. (1964) Gaseous metabolism and water relations of the zebra finch, *Taeniopygia castanotis. Physiol. Zool. 37*:400–413.

Calder, W. A., and Schmidt-Nielsen, K. (1967) Temperature regulation and evaporation in the pigeon and the roadrunner. *Am. J. Physiol. 213*:883–889.

Carey, F. G., Teal, J. M. (1966) Heat conservation in tuna fish muscle. *Proc. Natl. Acad. Sci. U.S.A. 56*:1464–1469.

Carey, F. G., Kanwisher, J. W., and Stevens, D. E. (1984) Bluefin tuna warm their viscera during digestion. *J. Exp. Biol. 109*:1–20.

Carey, F. G., Teal, J. M., Kanwisher, J. W., and Lawson, K. D. (1971) Warm-bodied fish. *Am. Zool. 11*:137–145.

Chappell, M. A., and Bartholomew, G. A. (1981) Activity and thermoregulation of the antelope ground squirrel *Ammospermophilus leucurus* in winter and summer. *Physiol. Zool. 54*:215–223.

Council of the Royal Society (1975) *Quantities, Units, and Symbols*, 2d ed. London: The Royal Society. 55 pp.

Crawford, E. C., Jr. (1962) Mechanical aspects of panting in dogs. *J. Appl. Physiol. 17*:249–251.

Crawford, E. C., Jr. (1972) Brain and body temperatures in a panting lizard. *Science 177*:431–433.

Crawford, E. C., Jr., and Kampe, G. (1971) Resonant panting in pigeons. *Comp. Biochem. Physiol. 40A*:549–552.

Dawson, T. J., and Hulbert, A. J. (1970) Standard metabolism, body temperature, and surface areas of Australian marsupials. *Am. J. Physiol. 218*:1233–1238.

Dawson, T., and Schmidt-Nielsen, K. (1966) Effect of thermal conductance on water economy in the antelope jack rabbit, *Lepus alleni. J. Cell. Physiol. 67*:463–472.

Dawson, W. R. (1954) Temperature regulation and water requirements of the brown and abert towhees, *Pipilo fuscus* and *Pipilo aberti. Univ. Calif. Publ. Zool. 49*:81–124.

Farner, D. S. (1956) The body temperatures of the North Island kiwis. *Emu 56*:199–206.

Free, J. B., and Spencer-Booth, Y. (1958) Observations on the temperature regulation and food consumption of honeybees (*Apis mellifera*). *J. Exp. Biol. 35*:930–937.

Hainsworth, F. R. (1967) Saliva spreading, activity, and body temperature regulation in the rat. *Am. J. Physiol. 212*:1288–1292.

Hainsworth, F. R., and Wolf, L. L. (1970) Regulation of oxygen consumption and body temperature during torpor in a hummingbird, *Eulampis jugularis. Science 168*:368–369.

Hales, J. R. S., and Findlay, J. D. (1968) Respiration of the ox: Normal values and the effects of exposure to hot environments. *Respir. Physiol. 4*:333–352.

Hammel, H. T. (1955) Thermal properties of fur. *Am. J. Physiol. 182*:369–376.

Hardy, J. D. (1949) Heat transfer. In *Physiology of Heat Regulation and the Science of Clothing* (L. H. Newburgh, ed.), pp. 78–108. Philadelphia: Saunders.

Hart, J. S. (1951) Calorimetric determination of average body temperature of small mammals and its variation with environmental conditions. *Can. J. Zool. 29*:224–233.

Hart, J. S. (1956) Seasonal changes in insulation of the fur. *Can. J. Zool. 34*:53–57.

Hart, J. S., and Irving, L. (1959) The energies of harbor seals in air and in water, with special consideration of seasonal changes. *Can. J. Zool. 37*:447–457.

Heinrich, B. (1970) Thoracic temperature stabilization by blood circulation in a free-flying moth. *Science 168*:480–482.

Heinrich, B. (1972) Temperature regulation in the bumblebee *Bombus vagans:* A field study. *Science 175*:185–187.

Heinrich, B. (1974) Thermoregulation in bumblebees. 1. Brood incubation by *Bombus vosnesenskii* queens. *J. Comp. Physiol. 88:*129–140.

Heinrich, B. (1976) Heat exchange in relation to blood flow between thorax and abdomen in bumblebees. *J. Exp. Biol. 64:*561–585.

Heinrich, B., and Bartholomew, G. A. (1971) An analysis of pre-flight warm-up in the sphinx moth, *Manduca sexta. J. Exp. Biol. 55:*223–239.

Hensel, H., and Bock, K. D. (1955) Durchblutung und Wärmeleitfähigkeit des menschlichen Muskels. *Pflügers Arch. 260:*361–367.

Herreid, C. F., II, and Schmidt-Nielsen, K. (1966) Oxygen consumption, temperature, and water loss in bats from different environments. *Am. J. Physiol. 211:*1108–1112.

Himms-Hagen, J. (1985) Brown adipose tissue metabolism and thermogenesis. *Annu. Rev. Nutr. 5:*69–94.

Hutchison, V. H., Dowling, H. G., and Vinegar, A. (1966) Thermoregulation in a brooding female Indian python, *Python molurus bivittatus. Science 151:*694–696.

Irving, L. (1969) Temperature regulation in marine mammals. In *The Biology of Marine Mammals* (H. T. Andersen, ed.), pp. 147–174. New York: Academic Press.

Irving, L., and Hart, J. S. (1957) The metabolism and insulation of seals as bare-skinned mammals in cold water. *Can. J. Zool. 35:*497–511.

Irving, L., and Krog, J. (1954) Body temperatures of arctic and subarctic birds and mammals. *J. Appl. Physiol. 6:*667–680.

Irving, L., and Krog, J. (1955) Temperature of skin in the arctic as a regulator of heat. *J. Appl. Physiol. 7:*355–364.

Johansen, K. (1961) Distribution of blood in the arousing hibernator. *Acta Physiol. Scand. 52:*379–386.

Johnston, D. W. (1971) The absence of brown adipose tissue in birds. *Comp. Biochem. Physiol. 40A:*1107–1108.

Kilgore, D. L., Jr., and Schmidt-Nielsen, K. (1975) Heat loss from ducks' feet immersed in cold water. *Condor 77:*475–478.

Kluger, M. J. (1977) Fever in the frog *Hyla cinerea. J. Thermal Biol. 2:*79–81.

Kluger, M. J., Ringler, D. H., and Anver, M. R. (1975) Fever and survival. *Science 188:*166–168.

LeMaho, Y., Delclitte, P., and Chatonnet, J. (1976)

Thermoregulation in fasting emperor penguins under natural conditions. *Am. J. Physiol. 231:*913–922.

Lyman, C. P. (1948) The oxygen consumption and temperature regulation of hibernating hamsters. *J. Exp. Zool. 109:*55–78.

McNab, B. K. (1966) An analysis of the body temperatures of birds. *Condor 68:*47–55.

Morrison, P. (1966) Insulative flexibility in the guanaco. *J. Mammal. 47:*18–23.

Morrison, P. R., and Ryser, F. A. (1952) Weight and body temperature in mammals. *Science 116:*231–232.

Øritsland, N. A. (1970) Temperature regulation of the polar bear (*Thalarctos maritimus*). *Comp. Biochem. Physiol. 37:*225–233.

Pearson, O. P. (1954) Habits of the lizard *Liolaemus multiformis multiformis* at high altitudes in Southern Peru. *Copeia 1954* (2):111–116.

Pinshow, B., Fedak, M. A., Battles, D. R., and Schmidt-Nielsen, K. (1976) Energy expenditure for thermoregulation and locomotion in emperor penguins. *Am. J. Physiol. 231:*903–912.

Reinertsen, R. E., and Haftorn, S. (1986) Different metabolic strategies of northern birds for nocturnal survival. *J. Comp. Physiol. B. 156:*655–663.

Reynolds, W. W., Casterlin, M.E., and Covert, J. B. (1976) Behavioural fever in teleost fishes. *Nature 259:*41–42.

Robinson, K. W., and Lee, D. H. K. (1946) Animal behaviour and heat regulation in hot atmospheres. *Univ. Queensland Papers, Dept. Physiol. 1:*1–8.

Robinson, K. W., and Morrison, P. R. (1957) The reaction to hot atmospheres of various species of Australian marsupial and placental animals. *J. Cell. Comp. Physiol. 49:*455–478.

Schmidt-Nielsen, K. (1963) Osmotic regulation in higher vertebrates. *Harvey Lect. 58:*53–93.

Schmidt-Nielsen, K. (1964) *Desert Animals: Physiological Problems of Heat and Water.* New York: Oxford University Press. 277 pp. Reprinted 1979, Dover Publications, New York.

Schmidt-Nielsen, K. (1970) *Animal Physiology,* 3d ed. Englewood Cliffs, N.J.: Prentice-Hall.

Schmidt-Nielsen, K., Dawson, T. J., and Crawford, E. C., Jr. (1966) Temperature regulation in the echidna (*Tachyglossus aculeatus*). *J. Cell. Physiol. 67:*63–72.

Schmidt-Nielsen, K., Schmidt-Nielsen, B., Jarnum,

S. A., and Houpt, T. R. (1957) Body temperature of the camel and its relation to water economy. *Am. J. Physiol. 188:*103–112.

Scholander, P. F. (1955) Evolution of climatic adaptation in homeotherms. *Evolution 9:*15–26.

Scholander, P. F., Hock, R., Walters, V., Johnson, F., and Irving, L. (1950a) Heat regulation in some arctic and tropical mammals and birds. *Biol. Bull. 99:*237–258.

Scholander, P. F., and Krog, J. (1957) Countercurrent heat exchange and vascular bundles in sloths. *J. Appl. Physiol. 10:*405–411.

Scholander, P. F., and Schevill, W. E. (1955) Countercurrent vascular heat exchange in the fins of whales. *J. Appl. Physiol. 8:*279–282.

Scholander, P. F., Walters, V., Hock, R., and Irving, L. (1950b) Body insulation of some arctic and tropical mammals and birds. *Biol. Bull. 99:*225–236.

Shkolnik, A., and Schmidt-Nielsen, K. (1976) Temperature regulation in hedgehogs from temperate and desert environments. *Physiol. Zool. 49:*56–64.

Smith, E. N., Robertson, S. L., and Adams, S. R. (1981) Thermoregulation of the spiny soft-shelled turtle *Trionyx spinifer. Physiol. Zool. 54:*74–80.

Southwick, E. E. (1985) Allometric relations, metabolism and heat conductance in clusters of honeybees at cool temperatures. *J. Comp. Physiol. B. 156:*143–149.

Southwick, E. E., and Mugaas, J. N. (1971) A hypothetical homeotherm: The honeybee hive. *Comp. Biochem. Physiol. 40A:*935–944.

Suckling, J. A., Suckling, E. E., and Walker, A. (1969) Suggested function of the vascular bundles in the limbs of *Perodicticus potto. Nature, Lond. 221:*379–380.

Taylor, C. R. (1972) The desert gazelle: A paradox resolved. In *Comparative Physiology of Desert Animals* (G. M. O. Maloiy, ed.), Symposium, Zoological Society of London, no. 31, pp. 215–227. London: Academic Press.

Taylor, C.R., and Lyman, C. P. (1972) Heat storage in running antelopes: Independence of brain and body temperatures. *Am. J. Physiol. 222:*114–117.

Tucker, V. A. (1965a) Oxygen consumption, thermal conductance, and torpor in the California pocket mouse *Perognathus californicus. J. Cell. Comp. Physiol. 65:*393–403.

Tucker, V. A. (1965b) The relation between the torpor cycle and heat exchange in the California pocket mouse

*Perognathus californicus. J. Cell. Comp. Physiol. 65:*405–414.

Tucker, V. A. (1966) Diurnal torpor and its relation to food consumption and weight changes in the California pocket mouse *Perognathus californicus. Ecology 47:*245–252.

Turner, J. S. (1987) The cardiovascular control of heat exchange: Consequences of body size. *Amer. Zool. 27:*69–79.

Vinegar, A., Hutchison, V. H., and Dowling, H. G. (1970) Metabolism, energetics, and thermoregulation during brooding of snakes of the genus *Python* (Reptilia, Boidae). *Zoologica 55:*19–50.

Weast, R. C. (1969) *Handbook of Chemistry and Physics,* 50th ed. Cleveland: Chemical Rubber.

ADDITIONAL READING

American Society of Zoologists, Symposium (1987) Cardiovascular adaptation in reptiles. *Am. Zool. 27:*3–131.

Aschoff, J. (1982) The circadian rhythm of body temperature as a function of body size. In *A Companion to Animal Physiology* (C. R. Taylor, K. Johansen, and L. Bolis, eds.), pp. 173–188, Cambridge: Cambridge University Press.

Baker, M. A. (1982) Brain cooling in endotherms in heat and exercise. *Annu. Rev. Physiol. 44:*85–96.

Bernstein, M. H. (1982) Temperature regulation in exercising birds. In *A Companion to Animal Physiology* (C. R. Taylor, K. Johansen, and L. Bolis, eds.), pp. 189–197, Cambridge: Cambridge University Press.

Bligh, J. (1973) *Temperature Regulation in Mammals and Other Vertebrates.* Amsterdam: North-Holland. 436 pp.

Blix, A. S., and Steen, J. B. (1979) Temperature regulation in newborn polar homeotherms. *Physiol. Rev. 59:*285–304.

Crawshaw, L. I. (1980) Temperature regulation in vertebrates. *Annu. Rev. Physiol. 42:*473–491.

Fischer, K. C., Dawe, A. R., Lyman, C. P., Schönbaum, E., and South, F. E., Jr. (eds.) (1967) *Mammalian Hibernation,* vol. III. Edinburgh: Oliver & Boyd. 535 pp.

Folk, G. E., Jr. (1980) Protein and fat metabolism during mammalian hypophagia and hibernation. *Fed. Proc. 39:*2953–2979.

Gates, D. M. (1962) *Energy Exchange in the Biosphere*. New York: Harper & Row. 151 pp.

Gates, D. M., and Schmerl, R. B. (eds.) (1975) *Perspectives of Biophysical Ecology*. New York: Springer-Verlag. 609 pp.

Hardy, J. D., Gagge, A. P., and Stolwijk, J. A. J. (eds.) (1970) *Physiological and Behavioral Temperature Regulation*. Springfield, Ill.: Thomas. 944 pp.

Heinrich, B. (1974) Thermoregulation in endothermic insects. *Science 185:*747–756.

Heinrich, B. (1979) *Bumblebee Economics*. Cambridge, Mass.: Harvard University Press. 245 pp.

Heinrich, B. (ed.) (1981) *Insect Thermoregulation*. New York: Wiley. 328 pp.

Heller, H. C., and Glotzbach, S. F. (1977) Thermoregulation during sleep and hibernation. *Int. Rev. Physiol. 15:*147–188.

Himms-Hagen, J. (1985) Brown adipose tissue metabolism and thermogenesis. *Annu. Rev. Nutr. 5:*69–94.

Kluger, M. J. (1979) *Fever: Its Biology, Evolution, and Function*. Princeton: Princeton University Press. 195 pp.

Lyman, C. P., Willis, J. S., Malan, A., and Wang, L. C. H. (1982) *Hibernation and Torpor in Mammals and Birds*. New York: Academic Press. 328 pp.

Maloiy, G. M. O. (ed.) (1972) *Comparative Physiology of Desert Animals*. Symposium, Zoological Society of London, no. 31. London: Academic Press. 413 pp.

Monteith, J. L. (1973) *Principles of Environmental Physics*. London: Edward Arnold. 241 pp.

Newburgh, L. H. (ed.) (1968) *Physiology of Heat Regulation and the Science of Clothing*. New York: Hafner. 457 pp. Reprinted.

Precht, H., Christophersen, J., Hensel, H., and Larcher, W. (1973) *Temperature and Life*. New York: Springer-Verlag. 779 pp.

Prosser, C. L., and Nelson, D. O. (1981) The role of nervous systems in temperature adaptation of poikilotherms. *Annu. Rev. Physiol. 43:*281–300.

Schmidt-Nielsen, K. (1964) *Desert Animals: Physiological Problems of Heat and Water*. New York: Oxford University Press. 277 pp.

Stevens, E. D., and Dizon, A. E. (1982) Energetics of locomotion in warm-bodied fish. *Annu. Rev. Physiol. 44:*121–131.

Thomas, R. D. K., and Olson, E. C. (eds.) (1980) *A Cold Look at the Warm-Blooded Dinosaurs*. Boulder: Westview Press. 514 pp.

Whittow, G. C. (ed.) (1970, 1971, 1973) *Comparative Physiology of Thermoregulation:* vol. 1, *Invertebrates and Nonmammalian Vertebrates* (1970), 333 pp.; vol. 2, *Mammals* (1971), 410 pp.; vol. 3, *Special Aspects of Thermoregulation* (1973), 278 pp. New York: Academic Press.

WATER

Water and Osmotic Regulation

In a very crude way, the living organism can be described as an aqueous solution contained within a membrane, the body surface. Both the *volume of the organism* and the *concentration of solutes* should be maintained within rather narrow limits. The reason is that optimal function of an animal requires a well-defined, relatively constant composition of its body fluids, and substantial deviations are usually incompatible with life.

The problem is that the proper concentrations of animal body fluids invariably differ from those of the environment. Animals must maintain appropriate concentrations, but the concentration differences tend to decline, upsetting the steady state of the internal conditions. Animals can minimize the difficulties by decreasing (1) *their permeabilities*, and (2) *the concentration gradients* between the body fluids and the environment. Both strategies are used.

Even a greatly reduced permeability doesn't solve all problems, for there will always be some diffusive leak. Steady-state internal conditions can then be maintained only if the organism generates a counterflow that exactly equals the diffusive leak. Such a counterflow requires input of energy.

The problems of keeping water and solute concentrations constant vary with the environment and are entirely different in *sea water*, in *fresh water*, and on *land*. It is therefore helpful to treat these environments separately. We shall analyze the main physiological problems in each and see how different animals have solved their problems. We shall first deal with aquatic animals and then move on to terrestrial animals.

THE AQUATIC ENVIRONMENT

Before we discuss the physiological problems peculiar to an environment, it is helpful to be familiar with its most important physical and chemical characteristics.

More than two-thirds (71%) of the earth's surface is covered with water. Most of this is ocean; the total fresh water in lakes and rivers makes up less than 1% of the area and 0.01% of the volume of sea water (Sverdrup et al. 1942; Hutchinson 1967). On land, life exists in a thin film on, just below, and just above the surface; in water, organisms not only live along the solid bottom but extend throughout the water masses to the greatest depths of the oceans, in excess of 10,000 m.

All water contains dissolved substances – salts, gases, minor amounts of organic compounds, and various pollutants. In addition to the temperature of the water these are of the greatest physiological importance. In this chapter we are concerned primarily with dissolved salts; the role of dissolved gases was discussed in Chapter 1 and the effects of temperature in Chapters 6 and 7. A brief summary of the properties of solutions, osmotic pressure, and related matters is given in Appendix E.

Sea water contains about 3.5% salts (i.e., 1 liter sea water contains 35 g salts).* The major ions are sodium and chloride, with magnesium, sulfate, and calcium present in substantial amounts (Table 8.1). The total salt concentration varies somewhat with the geographical location. The Mediterranean, for example, has a salt content of nearly 4% because the high evaporation is not balanced by an equal inflow of fresh water from rivers. In other areas, particularly in coastal regions, the salt content is somewhat lower than in the open ocean, but the relative amounts of the dissolved ions remain very

---

* It is not feasible to determine the exact amount of dissolved salts by evaporating the water and weighing the residue. The reason is that the drying conditions affect the amounts of water included as water of crystallization as well as the weight of the bicarbonate and carbonate salts.

TABLE 8.1 Composition of sea water. In addition to the ions listed, sea water contains small amounts of virtually all elements found on earth. [Potts and Parry 1964]

Ion	Amount per 1 liter sea water		Amount per 1 kg water[a]	
	mmol	g	mmol	g
Sodium	470.2	10.813	475.4	10.933
Magnesium	53.57	1.303	54.17	1.317
Calcium	10.23	0.410	10.34	0.414
Potassium	9.96	0.389	10.07	0.394
Chloride	548.3	19.440	554.4	19.658
Sulfate	28.25	2.713	28.56	2.744
Bicarbonate	2.34	0.143	2.37	0.145

[a] In thermodynamics concentrations are calculated per kilogram water and are referred to as the molality of the solution. A brief synopsis of the physics of solutions is found in Appendix E.

nearly constant in the proportions given in Table 8.1.

Fresh water, in contrast to sea water, has a highly variable content of solutes. Minute amounts of salts are present already in rainwater, but its composition is greatly modified as it runs over and through the surface of the earth. The salts in rainwater are derived from the sea; droplets of ocean spray evaporate and salt particles are carried by air currents, often far inland, and deposited with the rain.

If the water runs over hard, insoluble rock such as granite, it dissolves little additional material and is called *soft*. If, on the other hand, it percolates over and through porous limestone, it can dissolve relatively large amounts of calcium salts and is called *hard*.

The total salt content in fresh water may vary from less than 0.1 mmol per liter to more than 10 mmol per liter, and the relative amounts of different ions can vary over a tremendous range. This is of physiological significance, especially if magnesium and sulfate are the major ions. The composition of various types of hard and soft waters is given in Table 8.2.

TABLE 8.2 Typical composition of soft water, hard water, and inland saline water, given in millimoles per kilogram water and listed in the same order as in Table 8.1. [Recalculated from Livingstone 1963 (a, b, c, and d) and from Steinhorn et al. 1979 (e)]

Ion	Soft lake water[a]	River water[b]	Hard river water[c]	Saline water[d]	Dead Sea[e]
Sodium	0.17	0.39	6.13	640	1955
Magnesium	0.15	0.21	0.66	6	2028
Calcium	0.22	0.52	5.01	32	481
Potassium	—	0.04	0.11	16	219
Chloride	0.03	0.23	13.44	630	7112
Sulfate	0.09	0.21	1.40	54	5.3
Bicarbonate	0.43	1.11	1.39	3	3.7

[a] Lake Nipissing, Ontario.   [b] Mean composition of North American rivers.   [c] Tuscarawas River, Ohio.   [d] Bad Water, Death Valley, California.   [e] Dead Sea, Israel. This water also contains 74 mmol per kg H_2O of bromide.

Acid rain

Natural rain is slightly acid. This is because carbon dioxide from the atmospheric air dissolves in the rainwater, which then can be expected to attain a pH of about 5.6. However, much of the rain that falls in the northeastern United States has a pH as low as 4. It has been suggested that this is because sulfur dioxide and nitrous oxides from the combustion of fossil fuels form strong acids that lower the pH of the precipitation (Galloway et al. 1976).

The high acidity in rain is not unique to the United States. Large areas of Scandinavia downwind of the dense industrial areas of Central Europe receive acid rains that cause noticeable problems for animal life and in some areas have severe effects on the fish population. These effects have been particularly well documented in Norway, where the yield of the salmon fisheries in the southern parts of the country has declined precipitously, although there has been no corresponding decrease in more distant areas. Concurrently, many lakes in southern Norway have totally lost their population of brown trout, and the number of barren lakes is rapidly increasing (Overrein et al. 1980).

Why is Norway particularly prone to damage from acid rain, when areas closer to the industries do not report similar effects? The explanation is that large parts of southern Norway are underlain by granitic bedrock that is highly resistant to weathering. As a result, rivers and lakes have extremely soft waters, with virtually no mineral content to buffer the acids.

Similar problems are causing increasing concern in the northeastern United States and adjoining parts of Canada. These areas are located downwind from some of the most heavily industrialized regions of the United States, where large amounts of sulfur dioxide and nitrous oxides are released into the atmosphere. In 1980 approximately 33 million metric tons of sulfur dioxide and 21 million metric tons of nitrogen oxides were produced in the United States and Canada, and about one-half of this was produced upwind of the New England states. The areas that are most heavily affected are those characterized by granitic rocks and soils. In addition to the effect of acids, there is also an increased atmospheric fallout of such potentially toxic trace metals as lead, zinc, manganese, arsenic, vanadium, antimony, selenium, and chromium (Bormann 1982).

The degree of acidity that can be tolerated by a given fish is not related to pH alone; the amount of dissolved salts is extremely important. The effect of acid is probably an inhibition of the active uptake of sodium, for sodium uptake is inhibited when the water acidity is high (Maetz et al. 1976). This is in accord with the observation that the fish disappear first from lakes with extremely small amounts of dissolved salts and that fish kills have been particularly severe during the snow melt in spring when the runoff of melting snow carries a sudden surge of acid in virtually unbuffered water (Leivestad and Muniz 1976).

Saline waters

Some inland water has a very high salt content. The Great Salt Lake in Utah is saturated with sodium chloride, which crystallizes out on the shore. The Dead Sea in Israel is likewise saturated, but the predominant ions are magnesium and chloride, with calcium sulfate crystallizing out.

In the Great Salt Lake no fish can live, but some animals thrive: the brine shrimp (*Artemia*), for example. The saturated salt solution of the Dead Sea has a different composition. It harbors no higher animal or plant life; only microorganisms survive. Some springs have high and unusual salt contents, but such habitats, although interesting, are of minor importance compared with the larger bodies of fresh and sea water.

Brackish water occurs in coastal regions where sea water is mixed with fresh water. At the mouth of a large river fresh water dilutes the ocean water for a considerable distance, and if the tides are large the estuary at the mouth may extend far up the river. In this area the salinity varies rapidly with the tides, often from nearly fresh to nearly undiluted sea water. In a larger enclosed area, such as the Baltic Sea, the situation is different. The Baltic's salinity at the west coast of Sweden is some 3% and gradually declines to less than 0.5% in the northernmost part. In this large area there is a relatively stable geographical gradient in the salinity with an almost imperceptible change from brackish to fresh water.

It is difficult to say exactly where sea water becomes brackish and where a very dilute brackish water for practical purposes is fresh. As a commonly accepted definition we can say that brackish water refers to salinities between 3.0 and 0.05%.

Brackish water is physiologically extremely important. It forms a barrier to the distribution of many marine animals on one hand as well as fresh-water animals on the other, and it also forms an interesting transition between marine and fresh-water habitats. In geographical extent, however, brackish water covers less than 1% of the earth's surface.

Definitions

Most *marine invertebrates* have body fluids with the same osmotic pressure as the sea water; they are *iso-osmotic* or *isosmotic* with the medium in which they live (Greek *isos* = equal).

When there is a change in the concentration of the medium, an animal may respond in one of two ways. One way is to change the osmotic concentration of its body fluids to conform with that of the medium, thus remaining isosmotic with the medium; such an animal is an *osmoconformer*. The other way is to maintain or regulate its osmotic concentration in spite of external concentration changes; such an animal is called an *osmoregulator*. For example, a marine crab that maintains a high concentration of salts in its body fluids after it has been moved to dilute brackish water is a typical osmoregulator.

Fresh-water animals have body fluids that are osmotically more concentrated than the medium; these animals are *hyperosmotic*. If an animal has a lower osmotic concentration than the medium, as does a marine teleost fish, it is said to be *hypo-osmotic* or *hyposmotic*.

The concentrations of the various individual solutes in the body fluids of an animal usually differ substantially from those in the medium, even if the animal is isosmotic with the medium. The differences are usually carefully regulated, a subject dealt with under the term *ionic regulation*. Some degree of ionic regulation seems to occur in all living organisms, both in osmoregulators and in osmoconformers.

Some aquatic animals can tolerate wide variations in the salt concentration of the water in which they live; they are called *euryhaline* animals (Greek *eurys* = wide, broad; *halos* = salt). Other animals have a limited tolerance to variations in the concentration of the medium; they are called *stenohaline* (Greek *stenos* = narrow, close).

A marine animal that can penetrate into brackish water and survive is euryhaline. An extremely euryhaline animal may even be able to tolerate shorter or longer periods in fresh water. The term euryhaline is used also for fresh-water animals that can withstand considerable increases in the salt content of the water. A stenohaline organism, whether marine or fresh-water, can tolerate only a narrow range of salt concentrations of the water in which it lives.

There is no sharp separation between euryhaline and stenohaline animals, and there is no commonly accepted definition that places a given animal in one or the other group.

Units

The concentration of a dissolved substance is usually expressed in units of molarity: moles per liter solution. In a biological context, it is often convenient to use the unit millimole (mmol). For example, a solution of 0.5 mol per liter is equal to 500 mmol per liter.

The osmotic concentration of a solution can be expressed as the *osmolarity* (osmoles per liter). The osmolarity of a solution depends on the number of dissolved particles and can be stated without knowledge of which specific solutes are present. The osmolarity of a solution of a nonelectrolyte (e.g., sucrose or urea) equals the molar concentration.

A solution of an electrolyte (e.g., sodium chloride, which in solution dissociates into Na^+ and Cl^-) has a higher osmotic concentration than is expressed by its molarity. However, it does not have exactly twice the osmotic concentration because the salt is not fully dissociated (see Appendix E). The degree of dissociation depends on the concentration of sodium chloride and on interactions with other ions. However, the total osmotic concentration can readily be determined. In biological work this is most commonly done by measuring the freezing point depression or the vapor pressure of the solution. It is the osmotic concentration, rather than a detailed list of all solutes, that is important in most considerations of osmotic regulation in animals.

Ordinary sea water, which contains about 470 mmol sodium and about 550 mmol chloride per liter plus substantial amounts of divalent ions (magnesium and sulfate), has an osmotic concentration of about 1000 milliosmoles (mOsm) per liter.

The term *isotonic* has a meaning quite different from isosmotic. We say that a living cell is isotonic with a given solution if the cell neither swells nor shrinks in that solution. For example, if mammalian red blood cells are suspended in a sodium chloride solution of 150 mmol per liter (about 0.9%), the cells retain their size, shape, and volume. If, however, they are suspended in an isosmotic solution of urea (0.3 mol per liter), they rapidly swell and burst. The urea solution, although isosmotic, is not isotonic because urea rapidly penetrates the red cell membrane so that the urea concentrations inside and outside the cell are equal. The electrolytes do not move out of the cell, which now behaves as if it were suspended in distilled water. Because there are salts inside and none outside, water flows into the cell, which swells and bursts.

Isosmotic is defined in terms of physical chemistry; *isotonic* is a descriptive word based on the behavior of cells in a given solution.

AQUATIC INVERTEBRATES

Marine animals

Most marine invertebrates have an osmotic concentration in their body fluids equal to that of the surrounding sea water; they are *osmoconformers*.

TABLE 8.3 Concentrations of common ions (in millimoles per kilogram water) in sea water and in the body fluids of some marine animals. [Potts and Parry 1964]

	Na	Mg	Ca	K	Cl	SO$_4$	Protein (g liter^{-1})
Sea water	478.3	54.5	10.5	10.1	558.4	28.8	—
Jellyfish (*Aurelia*)	474	53.0	10.0	10.7	580	15.8	0.7
Polychaete (*Aphrodite*)	476	54.6	10.5	10.5	557	26.5	0.2
Sea urchin (*Echinus*)	474	53.5	10.6	10.1	557	28.7	0.3
Mussel (*Mytilus*)	474	52.6	11.9	12.0	553	28.9	1.6
Squid (*Loligo*)	456	55.4	10.6	22.2	578	8.1	150
Isopod (*Ligia*)	566	20.2	34.9	13.3	629	4.0	—
Crab (*Maia*)	488	44.1	13.6	12.4	554	14.5	—
Shore crab (*Carcinus*)	531	19.5	13.3	12.3	557	16.5	60
Norwegian lobster (*Nephrops*)	541	9.3	11.9	7.8	552	19.8	33
Hagfish (*Myxine*)	537	18.0	5.9	9.1	542	6.3	67

From the osmotic viewpoint, this eliminates one major physiological difficulty: They do not have to cope with the problem of osmotic movement of water.

Although as a rule marine invertebrates are osmoconformers, this does not mean that their body fluids have the same *solute* composition as sea water. On the contrary, they maintain concentrations of salt that are out of equilibrium with the environment, and this requires extensive regulation.

Ionic concentration in body fluids

The concentrations of the most important ions in the blood of some invertebrates are given in Table 8.3. In some animals the concentrations are similar to those in sea water, but in others they differ substantially. For example, several invertebrates have magnesium present in the same concentration as in sea water, but others have much lower concentrations of this ion. The same holds true for sulfate.

Such differences can be maintained only if the body surface, including the thin surface membrane of the gills, is relatively impermeable to the ion in question. Some amounts of these ions will enter anyway, for no animal is completely impermeable

and all ingested food contains some solutes. Therefore, the animals must have a mechanism for eliminating some ions while maintaining others at a higher level than in the water. The regulated elimination of solutes is a major function of excretory organs, such as the kidney.

Several of the animals listed in Table 8.3 regulate their concentrations of sulfate at less than half that in sea water. Such great differences obviously indicate both exclusion and elimination of this ion – in other words, active ion regulation. If the concentration in the animal differs only slightly from that in sea water, it is less clear whether the difference is attributable to regulation. In particular, proteins have a considerable influence on the distribution of ions across a semipermeable membrane (known as the Donnan effect, see Appendix E). A difference in ion concentrations, therefore, does not necessarily indicate active regulation of the ion in question.

The difficulty of evaluating accurately the role of proteins and the Donnan effect on the various ions was circumvented by the British investigator J. D. Robertson (1957) in a simple way. He put the fluid sample in a semipermeable cellophane bag,

TABLE 8.4 Ionic regulation in some marine invertebrates. Concentrations in plasma or coelomic fluid expressed as percentage of concentration in body fluid dialyzed against sea water. [Robertson 1957]

Animal	Na	Mg	Ca	K	Cl	SO₄
Coelenterates						
Aurelia aurita	99	97	96	106	104	47
Echinoderms						
Marthasterias glacialis	100	98	101	111	101	100
Tunicates						
Salpa maxima	100	95	96	113	102	65
Annelids						
Arenicola marina	100	100	100	104	100	92
Sipunculids						
Phascolosoma vulgare	104	69	104	110	99	91
Crustaceans						
Maia squinado	100	81	122	125	102	66
Dromia vulgaris	97	99	84	120	103	53
Carcinus maenas	110	34	108	118	104	61
Pachygrapsus marmoratus[a]	94	24	92	95	87	46
Nephrops norvegicus	113	17	124	77	99	69
Molluscs						
Pecten maximus	100	97	103	130	100	97
Neptunea antiqua	101	101	102	114	101	98
Sepia officinalis	93	98	91	205	105	22

[a] This grapsoid crab is the only animal listed that is hyposmotic to sea water (ionic concentration 86% that of sea water).

which he then placed in sea water. Because salts and water can pass through the cellophane but proteins cannot, the ion concentrations inside the bag will, at equilibrium, differ somewhat from those in the sea water, the differences being caused by the Donnan effect of the proteins. This procedure is called *dialysis*.

The concentrations inside the bag serve as a baseline, and the concentration of each ion as found in the living animal is expressed as a percentage of this baseline. An appreciable difference between the observed value and the concentration reached passively by dialysis must be the result of active regulation of this particular ion. Some results of such experiments are listed in Table 8.4.

Echinoderms show no significant regulation of any ion. The coelenterate *Aurelia,* a jellyfish, reg-

ulates only sulfate, which is kept considerably below the concentration in sea water. In this animal the low sulfate is directly related to a problem of buoyancy: Exclusion of the heavy sulfate ion reduces the density of the jellyfish, which thus is kept from sinking (See Chapter 10).

The crustaceans listed in Table 8.4 demonstrate the interesting feature that the magnesium level in the plasma is low in those that have the ability to move quickly. The crab *Pachygrapsus* and the lobster-like *Nephrops* are active and fast-moving animals; the spider crab *Maia,* on the other hand, is slow moving and has a high magnesium concentration. Magnesium is an anesthetic that depresses neuromuscular transmission, and one might hastily conclude that a high magnesium concentration is related to a low level of activity of these crustaceans. However, the further observation that the cuttlefish *Sepia,* which moves quickly and swims well, has a magnesium concentration as high as the scallop (*Pecten*) makes us suspicious of any cause-and-effect correlation between activity and magnesium concentration.

Intracellular concentrations and volume regulation

So far we have discussed the relationships between the surrounding water and the solute concentrations in blood and other body fluids. The concentrations of ions *inside* the cells are usually very different from those outside; for example, sodium and chloride concentrations are usually low inside the cells and high outside, and potassium is usually high inside and low outside. However, the cells are isosmotic with the surrounding tissue fluid and blood, although the concentrations of individual solutes differ.

To illustrate the problems of volume regulation, we can consider the cell membrane as permeable to water but impermeable to solutes. A change in the extracellular concentration will cause a change in cell volume. If the extracellular concentration is lowered, the cell takes up water and swells; if the extracellular concentration is increased, water is withdrawn and the cell shrinks. In reality, in most animals the cell volume is regulated so that the cell after an initial disturbance returns to its original volume, which then remains relatively constant.

It is now well established that intracellular concentration of free amino acids is an important factor in the control of cell volume in salinity-stressed environments. As the salinity in the water rises or falls, the amino acids in the cells increase or decrease so that the cells remain isotonic with the surroundings.

It has been suggested that the decrease in amino acid concentration could be achieved by the synthesis of protein and the concentration restored by degradation of protein. Another suggestion is that when an animal is stressed by low salinity the intracellular amino acids cross the initially distended

cell membranes, accompanied by osmotically obligated water, and the cell volume thus is restored. According to this view, which is supported by studies of molluscan heart muscle, the volume regulation in the cells depends on the permeability characteristics of the cell membrane, which in turn are influenced by the extracellular osmotic concentration (Pierce and Greenberg 1973).

The changes in vertebrate cells during adjustment to changing salinities resemble those of the invertebrates. As the salinity of the medium increases or decreases, corresponding changes occur in the intracellular amino acid concentrations that are in accord with the maintenance of constant cell volume (Fugelli and Zachariassen 1976). In principle, therefore, cell volume regulation seems to be accomplished by similar mechanisms in vertebrates and invertebrates.

Why is intracellular regulation based on organic molecules when the increase in blood concentrations is caused mainly by inorganic ions?

Intracellular ion concentrations are always very different from those outside, but this is not the entire story. It turns out that the common cations, Na and K, have a considerable perturbing effect on the action of metabolic enzymes, and substantial changes in their concentrations could be very upsetting to cellular metabolism (Figure 8.1). The Michaelis constant (which indicates the substrate concentration at which an enzymatic reaction proceeds at one-half of its maximum velocity) changes several-fold when the concentration of sodium or potassium ions is changed from 0.1 to 0.5 mol per liter. However, the Michaelis constant is virtually unaffected by changes in several of the amino acids, up to concentrations as high as 1 mol per liter. These are the amino acids that commonly occur in high concentrations in marine invertebrates: glycine, serine, beta-alanine, and several others (Bowlus and Somero 1979).

Other amino acids, in contrast, have a strongly perturbing effect on enzyme reactions. This applies, for example, to arginine and lysine, which are among those that do not change in response to

FIGURE 8.1 The activity of the enzyme phospho-enolpyruvate from the crab *Pachygrapsus* is strongly affected by the sodium and potassium ion concentrations, but changes in several common amino acids have little or no effect on the enzyme. Two other amino acids (lysine and arginine) have even more strongly perturbing effects than the Na$^+$ and K$^+$ ions. K_m is the substrate concentration at which the reaction proceeds at one-half its maximum velocity. [Bowlus and Somero 1979]

osmoregulatory changes in intracellular amino acid concentrations.

With the knowledge that sodium and potassium ions greatly affect enzymatic reactions, it is easier to understand why amino acids are so important as intracellular solutes in balancing intracellular and extracellular osmotic concentrations. For example, in muscle fibers of barnacles, 70% of the intracellular osmotic concentration is made up of amino acids, with glycine alone making up more than one-half (Clark and Hinke 1981).

Animals in fresh and brackish water.

If we transfer a variety of marine animals to somewhat dilute sea water, say 80% of the usual strength, most of them are likely to survive. When, after some time, we examine their body fluids, we find that they have adjusted to the dilution and have

established new and lower concentrations of ions in their body fluids. The osmoconformers (e.g., starfish and oysters) will have the same osmotic concentration as the dilute medium, although the concentrations of individual ions still differ from those in the diluted sea water. The osmoregulators, on the other hand, resist the dilution more or less successfully and remain hyperosmotic.

Thus, marine animals that penetrate into brackish water are of two types: passive osmoconformers and active osmoregulators. An osmoconformer, such as the oyster, can tolerate considerable dilution; it can also to some extent resist the effects of periodic dilution of the water in an estuary by keeping its shells closed. In the long run, however, the active regulators can better resist the fluctuations in the environment.

Let us see how the active regulators manage in dilute sea water. The relationship between the concentrations in the body fluids and the water is shown in Figure 8.2 for a number of osmoregulators. Even the good regulators meet certain limitations. An example is the European shore crab, *Carcinus*, which cannot survive in brackish water more dilute than about one-third of normal sea water.

The limit of dilution tolerated by the shore crab, and by many other brackish-water animals, varies with the geographical locale. A shore crab from the North Sea is far less tolerant to dilute water than a shore crab from the Baltic Sea, where the salt concentration normally is much lower than in the open sea.

Another crab, the Chinese mitten crab *Eriocheir*, can tolerate far greater dilutions and, in fact, is able to penetrate into fresh water. To reproduce, the mitten crab must return to the sea, and it therefore cannot establish itself permanently and carry out its complete life cycle in fresh water.

In principle, fresh-water animals behave osmotically in a way similar to the successful osmoregulators in brackish water, but there are great differences in the concentrations at which they maintain their body fluids (Figure 8.3). A crustacean such as the crayfish *Potamobius* maintains an osmotic

FIGURE 8.2 Relation between the concentrations of body fluids and of medium in various brackish-water animals. Full-strength sea water is indicated by an arrow. Diagonal line indicates equal concentrations in body fluid and medium. [Beadle 1943]

FIGURE 8.3 Relation between the concentrations of body fluids and of medium in various fresh-water animals. Full-strength sea water is indicated by an arrow. Diagonal line indicates equal concentrations in body fluid and medium. [Beadle 1943]

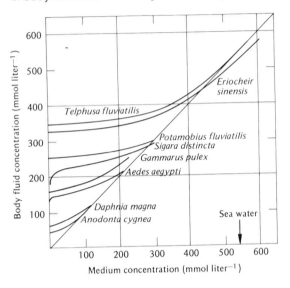

concentration of about 500 mOsm per liter, but the fresh-water clam *Anodonta* maintains less than one-tenth of this, only about 50 mOsm per liter. *Anodonta* still is hyperosmotic, and no fresh-water animal is known that permits its concentration to decrease to the level of the fresh water in which it lives.

Most of the major animal phyla have representatives both in the sea and in fresh water, although the number of species in the sea is much greater. No echinoderms are found in fresh water, and among molluscs the entire class Cephalopoda (octopus and squid) is absent in fresh water.

Mechanism of osmoregulation

When an animal is hyperosmotic to the surrounding medium, it encounters two physiological problems: (1) water tends to flow into the animal because of the higher solute concentration inside, and (2) solutes tend to be lost because the inside

concentration is higher and also because the water that enters must be excreted and carries some solutes with it. The magnitude of these problems could be reduced by making all surfaces highly impermeable, but no completely impermeable animal is known. At the least, the respiratory surfaces must be thin and large enough to permit adequate diffusion of gases, and the respiratory surfaces are usually the site of the greatest loss of solutes and gain of water.

Active transport. How can the animal compensate for the solute loss? It might be possible to obtain the necessary ions in food, and this might suffice for a highly impermeable animal. However, many fresh-water animals are not particularly impermeable, and yet, even when starved they maintain their normal high concentrations. This is achieved by a direct uptake of ions from the medium.

The simplest way to demonstrate such uptake is

to keep an animal, for example a crayfish, in running or frequently changed distilled water. This will gradually deplete the salts in the crayfish so that its concentration is reduced from, say, 500 mOsm per liter to 450, which it tolerates without difficulty. If the crayfish is now returned to ordinary fresh water, its blood concentration increases again, although the fresh water is 100 times more dilute than the blood (e.g., 5 mOsm per liter).

Because, in this experiment, the crayfish has removed ions from a dilute solution and moved them to a much higher concentration in the blood, the ions have been moved against a concentration gradient. Such transport is called an *active transport.**

The organs responsible for active ion uptake are not always known. It has been assumed that the general body surface of some animals is the organ of uptake, but this is a conclusion based mostly on lack of any obvious recognizable organs that could be tested for this role. In other animals, notably crustaceans and insects, the general body covering does not participate in the active transport. There is conclusive evidence that the gills of crustaceans are organs of active ion transport, and that certain appendages of aquatic insect larvae, notably the "anal gills," are organs for ion uptake. There is no conflict between osmoregulation and respiration; the same organ can serve both functions, as do the gills of fresh-water crabs and crayfish. But the "anal gills" of insect larvae probably have no respiratory function and are exclusively organs of osmoregulation.

Active transport requires *energy,* and it would be interesting to know the increase in energy requirement as an animal moves to more dilute water. If the shore crab *Carcinus,* which is quite tolerant of brackish water, is moved from sea water to gradually more diluted solutions, its rate of oxygen consumption increases appreciably. When placed in sea water of one-quarter full strength, the oxygen consumption of *Carcinus* is increased by about 50%, and it could easily be concluded that this increase is attributable to the work of active ion transport necessary for the crab to maintain itself in the dilute water.

In contrast, the mitten crab *Eriocheir* has the same metabolic rate in sea water, in brackish water, and in fresh water. The mitten crab therefore does not seem to spend measurable extra energy to maintain itself – not even in fresh water. This difference could be explained if the mitten crab were more impermeable than the shore crab, but there is no striking difference in the permeabilities of the two animals. Determining the change in oxygen consumption therefore does not give reliable information about the energy cost of osmoregulation (Potts and Parry 1964).

The energy required for osmoregulation can, however, be calculated from thermodynamic considerations. If we know how much solute is lost in a given time, we know how much must be taken up from the medium for the animal to remain in steady state. The osmotic work (W_{osm}) required depends on the solute concentration in the medium (C_{med}) and in the blood (C_{bl}), and the work to move 1 mol is given by the following equation, in which R is the universal gas constant and T is the absolute temperature:

$$W_{osm} = R\,T\,\ln\frac{C_{bl}}{C_{med}}$$

In our model we will assume that the surface of the animal is permeable to water but impermeable to solutes. In a hypotonic medium, water enters by osmosis, and the same volume of water must be removed as urine. The permeability of the animal determines the water influx, and the urine volume is therefore a measure of the permeability. Some solutes are lost with the urine, and the total amount of solutes lost is the product of the volume of the urine and its concentration, $V_u \times C_u$.

We now can see that the osmotic work for ion

---

*Active transport is defined as transport against an electrochemical gradient and as such is an energy-requiring process. It is often referred to as an "uphill" transport, as opposed to passive or "downhill" diffusion along the concentration gradient.

TABLE 8.5 Estimated metabolic cost of osmotic regulation for three animals, each weighing 60 g, in fresh water with a solute concentration of 6 mmol per liter. Observations on blood and urine concentrations (columns a and b) and urine volume (column c) are used to calculate the minimum theoretical cost of osmotic regulation (column d). In the last column the calculated osmotic work is compared with the observed metabolic rates of the animals. [Potts 1954]

Animal	(a) Blood concentration (mmol liter^{-1})	(b) Urine concentration (mmol liter^{-1})	(c) Urine volume (ml h^{-1})	(d) Osmotic work (cal h^{-1})	(e) Metabolic rate (cal h^{-1})	(f) Osmotic work (% of metabolic rate)
Mitten crab	320	320	0.1	0.076	14	0.5
Crayfish	420	124	0.1	0.037	10	0.3
Clam	42	24	0.5	0.015	1.2	1.3

uptake depends on the permeability of the animal and the solute loss, according to:

$$W_{osm} = R\,T\,V_u\,C_u\,\ln\frac{C_{bl}}{C_{med}}$$

Some animals produce urine isotonic with the blood. Many others reabsorb some solutes from the urine and produce a hypotonic urine. In the latter case, less solute is lost, but work is required for solute reabsorption from the urine to make it hypotonic. This complicates the calculation of the minimum osmotic work required, but the result is that the reabsorption of solutes from the urine constitutes a considerable saving.

Estimates of the minimum osmotic work required for osmoregulation in fresh water for three animal species are compiled in Table 8.5. The information about blood concentration, urine concentration, and urine volume is used to calculate the osmotic work.

The first animal, the mitten crab *Eriocheir sinensis* is a marine animal that readily penetrates into fresh water. Its urine concentration equals the blood concentration, which is common for marine invertebrates. The osmotic work required to recover lost solutes from the dilute medium is quite small: 0.5% of the metabolic rate of the animal.

The crayfish *Potamobius fluviatilis* is a fresh-water animal. It produces a dilute urine at about one-third of the blood concentration. The urine volume is the same as in the mitten crab, and because the blood concentration of the crayfish is higher, its permeability must be somewhat lower. The greatest saving, however, is derived from the recovery of solutes from the urine, so that the total osmotic work is about one-half that for the mitten crab. It is only 0.3% of the metabolic rate of the crayfish.

The third animal, the fresh-water clam *Anodonta cygnea,* has a very low blood concentration, only one-tenth that of the crayfish. It also produces a dilute urine, at about one-half the blood concentration. However, the urine volume is high, which indicates a high permeability due to the large surface area of the soft tissues exposed to the medium. The high permeability makes it virtually mandatory for the clam to have a very low internal concentration; if it were to have a concentration similar to that of the crayfish, the osmotic inflow of water and thus the urine volume would be 10 times as high. This would give a corresponding increase in the amount of solutes to be recovered from the medium. The increase in the required work, however, would be much greater, because the uptake would be against the 10-fold higher blood concentration. For an animal with high permeability and low metabolic rate it is therefore essential that the blood concentration be as low as possible.

Although the animals described in Table 8.5 differ in their approach to osmoregulation, for all of them the minimum energy required for osmoregulation is a small fraction of the total metabolic rate. In reality the cost is higher, for the process does not have an efficiency of 100%. Nevertheless, we must conclude that the large increase in metabolic rate observed in some animals when they are moved to dilute media does not directly measure the increased energy requirements for osmoregulation.

Animals in saline habitats: hyporegulation

The shrimp *Palaemonetes* and a close relative, *Leander*, differ in one important way from all the other animals described in Figure 8.2. In normal, full-strength sea water they are hypotonic (i.e., their body fluids are osmotically more dilute than the medium), and this must require active osmoregulation. Hyporegulation is very uncommon for a marine invertebrate, and it is generally assumed that these two shrimps belong to a group that originally was at home in fresh water, and that they have secondarily invaded the sea while maintaining a lower concentration than is common in sea water.

Some saline waters are far more concentrated than sea water, and in such extreme environments hyporegulation is of great importance. The best known example is probably the brine shrimp *Artemia*, which is found in tremendous numbers in salt lakes and in coastal evaporation ponds where salt is obtained for commercial use by evaporation of sea water.

Although the brine shrimp cannot survive in fresh water, it can adapt to media that vary from about one-tenth sea water to crystallizing brine, which contains about 300 g salt per liter. In dilute sea water, *Artemia* is hypertonic to its medium and behaves like a brackish-water organism. At higher concentrations, *Artemia* is an excellent hyporegulator; in concentrated brine it still maintains its body fluid at an osmotic pressure not much more than one-tenth that of the medium (Figure 8.4).

The brine shrimp maintains its low osmotic concentration, not by being exquisitely impermeable to water and ions, but by active regulation. It continuously swallows the medium, and the osmotic

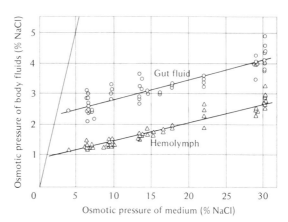

FIGURE 8.4 Osmotic pressure of body fluids of brine shrimp in highly concentrated NaCl solutions. The steep line running through the origin indicates equal concentrations in body fluid and medium. [Croghan 1958a]

pressure of the gut fluid is appreciably greater than that of the hemolymph (blood). However, although the osmotic pressure of the gut fluid remains greater than the hemolymph's, the concentrations of sodium and chloride in the gut fluid are considerably below those in the hemolymph (Croghan 1958b). Sodium and chloride must therefore be removed from the gut by active uptake, and to eliminate these ions from the body, so that their blood concentration is kept low, excretion must take place elsewhere. It is probable that the epithelium of the gills has the major role in this process.

The larva of *Artemia* seems to use a different site for the active secretion of sodium, the "neck organ," a specialized structure located on the dorsal surface of the animal (Conte et al. 1972).

Several species of mosquito larvae can thrive equally well in fresh water and saline water that is several times as concentrated as their hemolymph and can even tolerate salinities three times as high as sea water. Even more remarkable, the larvae of the mosquito *Aedes campestris* can thrive in alkaline salt lakes in which the dominant salt is sodium bicarbonate and the pH exceeds 10.

FIGURE 8.5 Ion and water exchange per day in an 8-mg larva of *Aedes campestris* acclimated to hy per-osmotic alkaline water. [Phillips et al. 1977]

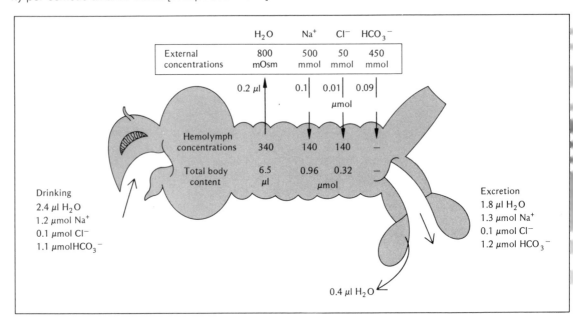

In low-salinity water the larvae of *Aedes* are hyperosmotic to the medium, but at higher concentrations they are hyposmotic. These larvae can adapt to concentrations that extend over a 500-fold range without changing the concentrations of the major ions in the hemolymph more than twofold.

The larvae of *Aedes* respond to rising external salinities by increasing their drinking rate several-fold. This seems to be the only way of obtaining water to compensate for the loss to the concentrated medium, but drinking also means ingestion of a heavy load of dissolved ions. The excess salt load is excreted with the aid of the Malpighian tubules and the rectum (to be discussed in Chapter 9). The so-called anal papillae, which in dilute media are the major sites of active ion uptake, seem to be insignificant for the elimination of excess salt.

The total ion balance in an *Aedes* larva in alkaline water is shown in Figure 8.5. In one day an 8-mg larva drinks 2.4 μl water, which is more than one-third of its total body water content (6.5 μl). The amount of sodium ingested in a day (1.2 μmol) is tremendous – more than the total sodium content of the animal (0.96 μmol). All this ingested sodium, together with a small amount that enters through the body surface (0.1 μmol), is excreted. The amount of water needed for this excretion (1.8 μl) is less than the amount ingested (2.4μl), leaving enough to cover what is lost by diffusion to the concentrated medium through the body surface (0.2 μl) and from the anal papillae (0.4 μl). Thus, the hypotonic larva is able to osmoregulate and remain in balance in the highly concentrated solution.

Hyporegulation is, on the whole, the exception rather than the rule among invertebrates. As we now turn to the vertebrates we shall find that os-

TABLE 8.6 Concentrations of major solutes (in millimoles per liter) in sea water and in the blood plasma of some aquatic vertebrates.

| | Habitat | Solute | | | Osmotic concentration (mOsm liter^{-1}) |
		Na	K	Urea[a]	
Sea water		~450	10	0	~1000
Cyclostomes					
Hagfish (*Myxine*)[b]	Marine	549	11		1152
Lamprey (*Petromyzon*)[c]	Marine				317
Lamprey (*Lampetra*)[b]	Fresh water	120	3	<1	270
Elasmobranchs					
Ray (*Raja*)[b]	Marine	289	4	444	1050
Dogfish (*Squalus*)[b]	Marine	287	5	354	1000
Fresh-water ray (*Potamotrygon*)[d]	Fresh water	150	6	<1	308
Coelacanth (Latimeria)[e]	Marine	197	7	350	954
Teleosts					
Goldfish (*Carassius*)[b]	Fresh water	115	4		259
Toadfish (*Opsanus*)[b]	Marine	160	5		392
Eel (*Anguilla*)[b]	Fresh water	155	3		323
	Marine	177	3		371
Salmon (*Salmo*)[b]	Fresh water	181	2		340
	Marine	212	3		400
Amphibians					
Frog (*Rana*)[f]	Fresh water	92	3	~1	200
Crab-eating frog (*R. cancrivora*)[g]	Marine	252	14	350	830[h]

[a] When no value is listed for urea, the concentration is of the order of 1 mmol per liter and osmotically insignificant. Values for ray, dogfish, and coelacanth include trimethylamine oxide.
[b] Bentley (1971). [c] Robertson (1954). [d] Thorson et al. (1967). [e] Locket (1980).
[f] Mayer (1969). [g] Gordon et al. (1961).
[h] Values for frogs kept in a medium of about 800 mOsm per liter, or four-fifths of normal sea water.

motic hyporegulation is much more widespread, although still not universally present, for there are other ways to solve osmotic problems.

AQUATIC VERTEBRATES

This section will deal primarily with fish and amphibians. Nobody would deny that whales and sea turtles are aquatic vertebrates, but we will treat them in a different context. They are descended from terrestrial ancestors and are air breathers, and it is more convenient to discuss them as terrestrial

animals living in an environment where no fresh water is available.

The major strategies used by aquatic vertebrates are evident from an examination of Table 8.6. The table lists examples of both marine and fresh-water vertebrates.

The marine representatives fall into two distinct groups: those whose osmotic concentrations are the same as or slightly above sea water (hagfish, elasmobranchs, *Latimeria,* and crab-eating frogs) and those whose osmotic concentrations are about one-third that of sea water (lampreys, teleosts). The former group has no major problem of water balance, for if the inside and outside concentrations are equal, there is no osmotic water flow. In contrast, those that remain distinctly hyposmotic live in constant danger of losing water to the osmotically more concentrated medium. The osmotic problems and the means to solve them thus differ drastically among marine vertebrates.

Fresh-water vertebrates, on the other hand, uniformly have concentrations of about one-quarter to one-third that of sea water; thus they are hyperosmotic to the medium and in principle similar to fresh-water invertebrates.

Cyclostomes

The cyclostomes are eel-shaped fish and considered the most primitive of all living vertebrates. They lack a bony skeleton, paired fins, and jaws (they are grouped in the class Agnatha, jawless vertebrates).

There are two groups of cyclostomes: lampreys and hagfish. Lampreys live both in the sea and in fresh water. Hagfish are strictly marine and stenohaline. Interestingly, lampreys and hagfish have employed two different solutions to the problem of life in the sea. Hagfish are the only true vertebrates whose body fluids have salt concentrations similar to that of sea water; in fact, the normal sodium concentration in hagfish blood slightly exceeds that in the medium. Nevertheless, hagfish have pronounced ionic regulation, but in being isosmotic and having high salt concentrations, they behave osmotically like invertebrates.

With the exception of hagfish, all marine vertebrates maintain salt concentrations in their body fluids at a fraction of the level in the medium. This situation has been cited in favor of the argument that the vertebrates originally evolved in fresh water and only later invaded the sea. Cyclostomes show many similarities to the ancestral forms of modern vertebrates, and knowledge of their anatomy has been of great importance to the interpretation of the fossil records of vertebrates and to the understanding of their early evolution.

The fact that hagfish deviate from the general vertebrate pattern of low salt concentrations means that the evolutionary theory of a fresh-water origin of all vertebrates is not supported by physiological evidence: A low salt concentration is not a universal vertebrate trait. However, present-day physiological characteristics cannot be used in evolutionary arguments, for, on the whole, physiological adaptation takes place more easily than morphological change. Anatomical structure and fossil records therefore remain more important to evolutionary hypotheses than physiological evidence.

The other group of cyclostomes, the lampreys, live both in fresh water and in the sea, but even the sea lamprey (*Petromyzon marinus*) is *anadromic** and ascends rivers to breed in fresh water.

The lampreys, whether fresh-water or marine, have osmotic concentrations about one-quarter to one-third the concentration of sea water. Their main osmotic problem is similar to that of teleost fish, whether marine or fresh-water species. These problems will be discussed in detail later in this chapter.

Marine elasmobranchs

The elasmobranchs – sharks and rays – are almost without exception marine. They have solved the osmotic problem of life in the sea in a very interesting way. Like most vertebrates they maintain salt concentrations in their body fluids at roughly one-third the level in sea water, but they still maintain osmotic equilibrium. This is achieved by adding to the body fluids large amounts of organic compounds, primarily urea, so that the total osmotic concentration of their blood equals or slightly exceeds that of sea water (Table 8.6).

The blood urea concentration in marine elasmo-

---

* A fish that ascends from the sea to spawn in fresh water is called *anadromic* (Greek *ana* = up; *dramein* = to run). Shad and salmon are well-known examples. *Catadromic* (Greek *kata* = down) refers to living in fresh water and descending to the sea. The common eel is catadromic; it grows to adult size in fresh water and descends to the sea to breed.

branchs is more than 100 times as high as in mammals, and such concentrations could not be tolerated by other vertebrates. In the elasmobranchs urea is a normal component of all body fluids, and the tissues cannot function normally in the absence of a high urea concentration. The isolated heart of a shark can continue to contract normally for hours when perfused with a saline solution of ionic composition similar to the blood, provided urea is also present in a high concentration. If the urea is removed, however, the heart rapidly deteriorates and stops beating.

In addition to urea, an osmotically important organic compound in elasmobranch blood is trimethylamine oxide (TMAO).

$$(CH_3)_3\ N{-}CH_2C{=}O \qquad\qquad CH_3$$
$$\big|\qquad\quad O\quad\big|\qquad\qquad\big|$$
$$\qquad\qquad\qquad\qquad\quad N{-}CH_2COOH$$
$$\qquad\qquad\qquad\qquad\quad \big|$$
$$\qquad\qquad\qquad\qquad\quad H$$

Betaine   Sarcosine

$$\qquad\qquad\qquad CH_3$$
$$\qquad\qquad\qquad \big|$$
$$NH_2 \qquad\qquad H_3C{-}N{=}O$$
$$O{=}C\qquad\qquad\qquad \big|$$
$$NH_2 \qquad\qquad\qquad CH_3$$

Urea   TMAO

Urea is the end product of protein metabolism in mammals and some other vertebrates; it is excreted by the mammalian kidney, but the shark kidney actively reabsorbs this compound so that it is retained in the blood. TMAO is found in many marine organisms, but its origin and metabolism are still poorly understood.

To use urea to obtain osmotic equilibrium is a rather elegant solution for a marine animal that must maintain salt concentrations much lower than those in the sea water. There is a disturbing aspect to urea, however; urea is known to destabilize many proteins and especially enzymes. Elasmobranch proteins could perhaps have evolved urea adaptation, but paradoxically, this does not seem to be the case.

The dilemma was clarified when it was discovered that the inhibitory effect of urea on enzymes is counteracted by TMAO. Similar counteracting effects are produced also by two other compounds that are methylated amines, betaine and sarcosine.

The counteracting effects of these methylated compounds on urea are maximal when the ratio urea/methylamine is about 2:1, which is the ratio that is commonly found in those vertebrates that use organic nitrogen compounds as a solution to the osmotic problem of life in sea water (Yancey and Somero 1980, Somero 1986).

Although the elasmobranchs have solved the osmotic problem of life in the sea by being isosmotic, they still are capable of extensive ionic regulation. The sodium concentration, for example, is maintained at about half that in sea water. This means that sodium tends to diffuse from the medium into the shark, primarily through the thin gill epithelium; furthermore, some sodium is ingested with the food. Because the sodium concentration tends to increase but must be kept down, excess sodium must be eliminated.

Part of the sodium excretion is handled by the kidney, but a special gland, the *rectal gland,* is probably more important. This small gland opens via a duct into the posterior part of the intestine, the rectum. The gland secretes a fluid with high sodium and chloride concentrations – in fact, somewhat higher than in sea water. The rectal gland has provided excellent material for the clarification of the cellular processes that are involved in the active transport of monovalent ions (Greger et al. 1986)

The function of the rectal gland does not fully explain the salt elimination in elasmobranchs. If the rectal gland is surgically removed in the spiny dogfish (*Squalus acanthias*), this shark is still able to maintain its plasma ion concentration at the usual level of about half that in sea water. As the gills are slightly permeable to salts, the blood concentrations should gradually increase unless some other means of excretion is available. It is likely that the kidney plays a major role in this excretion, but whether the gills are also sites for active transport of ions from elasmobranch blood is not known.

The fact that elasmobranchs are nearly in osmotic equilibrium with sea water eliminates the problem of a severe osmotic water loss (which for marine teleost fish is very important). Elasmobranchs do not need to drink, and thus they avoid the high sodium intake associated with drinking sea water.

It is an interesting fact, however, that elasmobranch blood is usually slightly more concentrated than sea water. This higher concentration inside causes a slight osmotic inflow of water via the gills. In this way, the elasmobranch slowly gains water osmotically, and this water is used for the formation of urine and for the secretion from the rectal gland. Because the excess osmotic concentration is attributable to urea, the retention of urea is a solution to the otherwise difficult osmotic problem of maintaining a low salt concentration while living in the sea.

Fresh-water elasmobranchs

The overwhelming number of elasmobranchs belong in the sea, but a few enter rivers and lakes, and some belong permanently in fresh water. Even those elasmobranchs that are thought of as typically marine include some species with a remarkable tolerance to low salinity in the external medium.

In several parts of the world both sharks and rays enter rivers and apparently thrive in fresh water. A well-known case is the existence in Lake Nicaragua of the shark *Carcharhinus leucas*. It was previously assumed that this shark was land-locked, but the shark in Lake Nicaragua is morphologically indistinguishable from the marine form and is able to move in free communication with the sea (Thorson et al. 1966).

Four elasmobranch species found in the Perak River in Malaysia probably do not live permanently in fresh water, but enter regularly from the sea. Their blood concentrations are lower than those

TABLE 8.7 Concentrations of solutes in the blood serum of the Amazon sting ray are similar to those of a teleost fish. Although the ray is an elasmobranch, urea is virtually absent from its body fluids. [Thorson et al. 1967]

Solute	Amount per liter
Sodium	150 mmol
Potassium	5.9 mmol
Calcium	3.6 mmol
Magnesium	1.8 mmol
Chloride	149 mmol
Urea	0.5 mmol
TMAO	0 mmol
Osmolality	308 mOsm
Protein	18 g

of strictly marine forms; in particular, the urea is reduced to less than a third the value for marine sharks, although it remains far above the normal level for other vertebrates.

The low level of solutes in the blood reduces the problems of osmotic regulation, for the osmotic inflow of water is diminished and lower salt concentrations are easier to maintain. The reduced osmotic inflow of water gives less water to be eliminated by the kidney. Because the urine inevitably contains some solutes, a low urine flow in turn reduces the urinary salt losses. It is, of course, difficult to say whether the lowered blood concentration is a primary adjustment or merely a passive result of the increased inflow of water and concomitant urinary losses.

One elasmobranch, the Amazon sting ray *Potamotrygon,* is permanently established in fresh water. It is common in the Amazon and Orinoco drainage systems up to more than 4000 km from the sea. It does not survive in sea water, even when the transfer takes place through a gradual concentration increase, but it can be adapted to one-half of sea water concentration (Bittner and Lang 1980). The average composition of its blood (Table 8.7) shows complete adaptation to fresh water, with a low blood urea concentration similar to that of fresh-water teleosts.

The most striking feature is the low urea concentration – even lower than in most mammals. Clearly, the retention of urea is not a universal requirement for elasmobranchs – an interesting physiological observation, showing again that physiological function is more changeable than most anatomical structures and that evolutionary arguments cannot be reliably based on physiological similarities or differences.

The coelacanth

Until 1938 it was believed that the group of fishes known as the Crossopterygii had been extinct for more than 75 million years, for they disappeared completely from the later fossil record. Their evolutionary position is far away from modern fish, close to the lungfish, and in the ancestry of amphibians. In 1938 a living specimen of the coelacanth *Latimeria* was caught off the coast of South-East Africa, causing a worldwide scientific sensation. It was a large specimen, more than 1.5 m long, and weighed over 50 kg, but it was poorly preserved and no detailed information on its anatomy was obtained.

After intensive search, several additional specimens of *Latimeria* have been caught near Madagascar, and although no living specimen has been kept long enough to allow physiological experimentation, it is known that the coelacanth has solved its problems of osmoregulation in the same way as elasmobranchs (Locket 1980).

Teleost fish

Teleost fish maintain their osmotic concentrations at about one-quarter to one-third the level in sea water (Table 8.6). On the whole, marine and fresh-water fish are within the same range, although marine fish tend to have somewhat higher blood concentrations. Some fish can tolerate a wide range of salinities and move between sea, brackish, and fresh water.

Such moves are often associated with the life cycle; the salmon, for example, reproduces in fresh water, migrates to the sea, and after reaching maturity returns to fresh water to spawn. The life cycle of the common eel is the reverse: the eel breeds in the sea, the larvae drift with the currents and reach coastal areas where they ascend into fresh water, and when maturity approaches the eel returns to the sea to reproduce. The change from one environment to the other requires profound changes in the osmoregulatory process.

Marine teleosts

Marine fish are hyposmotic and in constant danger of losing body water to the more concentrated sea water, for their body surfaces, in particular the large gill surfaces, are somewhat permeable to water. They must somehow compensate for the inevitable osmotic loss of water, and this they do by drinking sea water.

Although drinking restores the water content, large amounts of salts are also ingested and are absorbed from the intestinal tract together with the water. The salt concentration in the body increases, and the problem becomes one of eliminating the excess salt. To achieve a net gain in water from the ingestion of sea water, the salts must be excreted in a higher concentration than in the water taken in. The teleost kidney cannot serve this purpose, for it cannot produce a urine that is more concentrated than the blood.

Some other organ must therefore eliminate the excess salt. This is done by the gills, which thus have the dual function of participating in both osmotic regulation and gas exchange. The secretion of salt across the gill epithelium must be an active transport, for it takes place from a lower concentration in the blood to a higher concentration in the surrounding medium.

The main aspects of osmotic regulation in marine teleosts are summarized in Figure 8.6. The top diagram shows the movements of water. Water is lost osmotically across the gill membrane and in the urine. To compensate for the losses, the fish drinks sea water, absorbing both water and salts from the intestine.

The lower diagram in Figure 8.6 shows the movement of salts, which are taken in by mouth as the fish drinks from the surrounding sea water. The

FIGURE 8.6 A marine teleost is osmotically more dilute than the water in which it lives. Because of the higher osmotic concentration in the medium, the fish constantly loses water (top diagram), primarily across the thin gill membranes. Additional water is lost in the urine. To compensate for the water loss, the marine teleost drinks substantial amounts of sea water. Of the ingested salts, sodium and chloride are absorbed in the intestine and eliminated via the gills by active transport (double arrow, bottom diagram); magnesium and sulfate are excreted by the kidney.

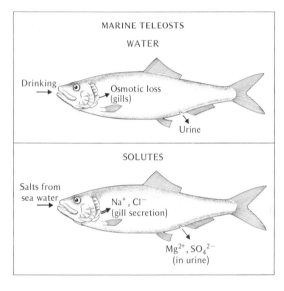

drinking; the larger part of the sodium influx takes place elsewhere, presumably because the gills are somewhat permeable. Whether the general body surface or the gills are responsible, it is certain that fish adapted to sea water are relatively permeable to ions, and those adapted to fresh water are relatively impermeable (Motais and Maetz 1965).

The killifish (*Fundulus heteroclitus*), which readily adapts to both fresh and sea water, has been used to study the changes in permeability to sodium and chloride that take place during adaptation to various concentrations. The permeability decreases within a few minutes of transfer to fresh water, but the increase in permeability on return to sea water takes many hours (Potts and Evans 1967).

The advantage of a low permeability to ions in fresh water is obvious, but it is difficult to see the advantage of a higher permeability in sea water. Marine fish must expend work to maintain their osmotic steady state in sea water, and it seems that a low permeability would reduce this work. It takes the fish several hours to return to the higher permeability in sea water, and we can only wonder why they do not permanently retain the low permeability that seems to be within their physiological capacity.

The ion transport is carried out, not by the general epithelial cells of the gills, but specifically by some large cells known as *chloride cells*. The cells were named chloride cells without definite knowledge of their function (Keys and Willmer 1932), and for many years it was uncertain whether they actively transport chloride ions with sodium following passively, or whether sodium is actively transported with chloride following passively.

Studies of the chloride cells became much easier when it was discovered that they are found also on the opercular cover of fish. This meant that a flat piece of epithelium could be obtained and used for experiments (Karnaky and Kinter 1977). During active transport a potential difference is established across the transporting surface, indicating that it is indeed chloride that is being actively transported (Karnaky 1986).

double arrow at the gills indicates the elimination, by active transport, of sodium and chloride. The excretion of sodium and chloride in the urine is of minor importance because teleost urine is usually more dilute than the body fluids. However, the kidney plays a major role in the excretion of divalent ions, magnesium and sulfate, which make up roughly one-tenth of the salts in sea water. These ions are not eliminated by the gills, which seem to transport only sodium and chloride.

Although marine fish drink sea water, measurements of the drinking rates have shown that only a small part of the total sodium intake comes from

FIGURE 8.7 A fresh-water teleost is osmotically more concentrated than the medium and therefore suffers a steady osmotic influx of water, mainly through the gills (top diagram). The excess water is eliminated as urine. Loss of solutes through the gills and in the urine is compensated for primarily through active uptake in the gills (double arrow, bottom diagram).

FIGURE 8.8 By placing a fish in a chamber where the front and hind parts are separated by a rubber membrane it can be shown that the skin does not participate in the active uptake of ions, which is carried out by the gills. [Krogh 1937]

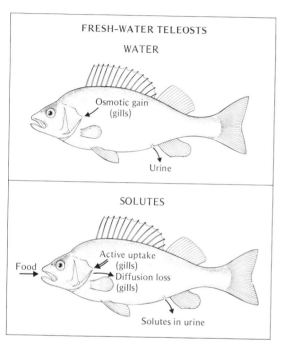

The final demonstration that the chloride cell is responsible for active chloride transport was made by inserting a microprobe directly into a single cell. By measuring the current flow, this showed that the chloride cells are the only significant element in the active ion transport in the opercular membrane of the tilapia (*Sarotherodon mossambicus*), a fish that can be adapted to any concentration between fresh water and double-strength sea water (Foskett and Scheffey 1982).

Fresh-water teleosts

The osmotic conditions for fish in fresh water resemble those for fresh-water invertebrates. The osmotic concentration in the blood, roughly in the range of 300 mOsm per liter, is much higher than in the surrounding fresh water.

The main events in the osmoregulation of fresh-water teleosts are outlined in Figure 8.7. The major problem is the osmotic water inflow. The gills are important in this gain because of their large surface and relatively high permeability; the skin is less important. Excess water is excreted as urine, which is very dilute and may be produced in quantities up to one-third of the body weight per day. Although the urine contains no more than perhaps 2 to 10 mmol per liter of solutes, the large urine volume nevertheless causes a substantial loss of solutes that must be replaced. The gills are also slightly permeable to ions, and this loss must likewise be covered by ion uptake.

Some solutes are taken in with the food, but the main intake is by active transport in the gills. This has been shown by placing a fish in a divided chamber, in which the head and the remainder of the body can be studied separately (Figure 8.8). In such experiments active uptake of ions occurred only in the front chamber. It was therefore concluded that the skin plays only a minor, if any, role in active absorption.

Catadromic and anadromic fish

Most teleost fish have a limited ability to move between fresh water and the sea; they are relatively stenohaline. We have already mentioned, however, that lampreys, salmon, and eels move between fresh and salt water as part of their normal life cycles. Such moves between fresh and sea water expose the fish to cataclysmic changes in the demands on their osmoregulatory mechanisms.

If an eel is moved from fresh to sea water, the osmotic loss of water reaches 4% of the body weight in 10 hours (Keys 1933). If the eel is prevented from drinking sea water by placing an inflated balloon in the esophagus, it continues to lose water and dies from dehydration within a few days. If it can drink, however, it soon begins to swallow sea water, the weight loss subsides, and a steady state is reached within 1 or 2 days. If the eel is transferred in the opposite direction, from sea water to fresh water, there is an initial gain in weight, but as urine formation increases, a steady state is reached, again within 1 or 2 days.

When the eel is moved between fresh and sea water, not only does the osmotic flow of water change direction, but to achieve a steady state and compensate for the solute gain or loss, the active ion transport in the gills must change direction. How this change takes place is unknown, although we assume that endocrine mechanisms are involved. Neither is it known whether different cell populations are responsible for the transport in the two directions, one or the other being activated as need arises. The other possibility is that the polarity of the transport mechanism in all participating cells can be reversed on demand. These questions remain unanswered.

Present knowledge indicates that it is unlikely that the direction of transport in the individual cell can be reversed. Of the many organs and cell types

that participate in active transport of some sort, none is known that with certainty can carry out such reversals. The frog skin, which is analogous to the fish gill in performing active uptake from the dilute solutions of fresh water, does not seem able to reverse the direction of the transport in the one species known to live in sea water, the crab-eating frog (discussed in the next section).

Amphibians

Most amphibians are aquatic or semiaquatic. The eggs are laid in water, and the larvae are gill-breathing aquatic animals. At the time of metamorphosis, many amphibians (but not all) change to respiration by lungs. Some salamanders retain the gills and remain completely aquatic as adults; most frogs, on the other hand, become more terrestrial, although usually tied to the vicinity of water or moist habitats.

Some atypical frogs, at home in very dry habitats and highly resistant to water loss by evaporation, have recently been studied in Africa and South America. Their unusual physiological characteristics are described later in this chapter.

Fresh-water amphibians

With regard to osmotic regulation, amphibians are quite similar to teleost fish. Virtually all amphibians are fresh-water animals, and in the adult the skin serves as the main organ of osmoregulation. When the animal is in water, there is an osmotic inflow of water, which is excreted again as a highly dilute urine. There is, however, a certain loss of solutes, both in the urine and through the skin. This loss is balanced by active uptake of salt from the highly dilute medium. The transport mechanism is located in the skin of the adult, and the frog skin has become a well-known model for studies of active ion transport.

Pieces of frog skin can be readily removed and used as a membrane separating two chambers, which can be filled with fluids of various concentrations. By analyzing the changes in the two chambers, the transport function by the skin can be studied (Fig-

FIGURE 8.9 Apparatus used for determining the sodium transport in an isolated piece of frog skin. The skin separates two chambers containing Ringer's solution, and sodium transport in the skin will produce a potential or voltage across the skin. If an external current is now applied in a direction opposite to the potential produced by sodium transport, this current will, when it is adjusted to give a zero voltage across the skin, be a direct measure of the sodium transport in the skin. A and A' indicate agar bridges to connect solutions to calomel electrodes; B and B' are agar bridges to provide electrical connections to the outside voltage source. [Ussing and Zerahn 1951]

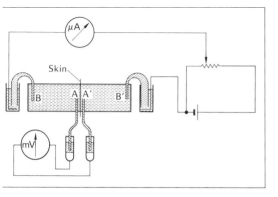

re 8.9). Such isolated pieces of skin survive for many hours. This apparatus for the study of active transport processes was originally devised by the Danish physiologist Hans Ussing and is known as an Ussing cell.

When frog skin in the Ussing cell separates two salt solutions* of the same composition, a potential difference of about 50 mV is rapidly established between the inside and the outside of the skin. The inside is positive, and it is therefore postulated that the potential is caused by an inward active trans-

*The salt solution should have the same pH and osmotic concentration as the blood and have approximately the same concentrations of the major ions, Na^+, K^+, Ca^{2+}, and Cl^-. Such a balanced salt solution is called Ringer's solution, named after a British physiologist who discovered that survival of the isolated frog heart depended on balanced proportions of these ions.

port of the positive sodium ion. When the potential difference has been established, the chloride ion passes through the skin by diffusion, accelerated by the electric field. An enormous amount of evidence has accumulated to indicate that this interpretation is correct, and the fact that the transport is active is clear from the developed potential and from the fact that metabolic inhibitors (e.g., cyanide) prevent the formation of a potential and inhibit the transport.

By applying an external potential across the skin, equal to the sodium potential but in the opposite direction, the potential across the skin can be reduced to zero. The current needed to maintain the potential at zero must equal the current generated by sodium transport in the skin. The current, called the short-circuit current, is therefore a direct measure of the inward transport of sodium. This method has, in similar form, become an extremely valuable tool in measuring active transport of ions in many other systems. It was also used to demonstrate that the ion transport in the fish gill, as mentioned earlier, depends on an active chloride transport.

A salt-water frog

Frogs and salamanders are usually restricted to fresh water and die within a few hours if placed in sea water. There is one exception: the crab-eating frog (*Rana cancrivora*) of Southeast Asia. This small, ordinary-looking frog, lives in coastal mangrove swamps, where it swims and seeks its food in full-strength sea water.

If a frog in sea water is to maintain the characteristic relatively low salt concentrations of vertebrates, it has two possible avenues for solving the problems. One strategy (the one used by marine teleosts) is to counteract the osmotic water loss through the skin and compensate for the inward diffusion of salt through the skin. The other strategy (that of marine elasmobranchs) is to retain urea and have body fluids in osmotic equilibrium with the medium, thus eliminating the problem of osmotic water loss. The salt-water frog has employed the elasmobranch strategy of adding large amounts of urea to the body fluids, which may contain as much as 480 mmol urea per liter (Gordon et al. 1961).

It seems reasonable that the salt-water frog uses this solution. Amphibian skin is relatively permeable to water, and it is therefore simpler to prevent osmotic water loss by maintaining the same osmotic concentration as that of the medium. To eliminate water loss solely by increasing the internal salt concentrations, the frog would need a salt tolerance unique among vertebrates (except the hagfish). If it were to use the teleost strategy and remain hyposmotic, the salt balance would be further impaired by the need to drink from the medium.

A crab-eating frog placed in sea water is not completely isosmotic with the medium; like sharks, it remains slightly hyperosmotic. The result is a slow osmotic influx of water, which is desirable because it provides the water required for formation of urine. This is certainly more advantageous than obtaining water by drinking sea water, which inevitably increases the salt intake.

In the crab-eating frog, as in the elasmobranch, urea is an important osmotic constituent and not merely an excretory product. In addition to its osmotic importance, urea is necessary for normal muscle contraction, which rapidly deteriorates in the absence of urea (Thesleff and Schmidt-Nielsen 1962).

Because urea is essential for the normal life of the animal, it should be retained and not be excreted in the urine. In sharks, urea is retained by active reabsorption in the kidney tubules (see Chapter 9). In the crab-eating frog, however, urea retention is achieved primarily by a reduction in urine volume when the frog is in sea water. It seems that urea is not actively reabsorbed, for the urea concentration in the urine consistently remains slightly above that in the plasma (Schmidt-Nielsen and Lee 1962).

The tadpoles of the crab-eating frog have an even greater tolerance for high salinities than the adults.

Their pattern of osmotic regulation, however, is similar to that of teleost fish and thus differs from the elasmobranch pattern adopted by adult frog (Gordon and Tucker 1965).

Although both the tadpole and the adult crab-eating frog are highly tolerant to sea water, this frog is not independent of fresh water, for both fertilization of the eggs and metamorphosis to the adult depend on a relatively low salt concentration in the water. Because of frequent torrential rains in the tropics, temporary fresh-water pools readily form near the shore, and spawning can therefore take place in dilute water. The tadpole is highly tolerant to salt, but metamorphosis is delayed for as long as the salinity remains high, and the frog goes through this critical stage only after the water has been diluted by heavy rain.

Although the crab-eating frog depends on fresh water for reproduction, its tolerance to sea water permits the exploitation of a rich tropical coastal environment that is closed to all other amphibians.

THE TERRESTRIAL ENVIRONMENT

The greatest physiological advantage of terrestrial life is the easy access to oxygen; the greatest physiological threat to life on land is the danger of dehydration. Successful large-scale evolution of terrestrial life has taken place in only two animal phyla, arthropods and vertebrates, which live and thrive in some of the driest and hottest habitats found anywhere. In addition, some snails thrive on land and are truly terrestrial; some even live in deserts.

The threat to dehydration is evidently a serious barrier to terrestrial life, for many ''terrestrial'' animals, except in the phyla just mentioned, depend on the selection of a suitably moist habitat and are terrestrial only in the technical sense of the word. An earthworm, for example, is highly dependent on a moist environment and soon succumbs to desiccation if exposed to the open atmosphere for any length of time. It has little resistance to water loss, depends on behavior, and seeks out microhabitats where air and surroundings are humid. Other ex-

FIGURE 8.10 The water vapor pressure over a free water surface increases rapidly with temperature (open circles, fully drawn line). At the body temperature of mammals (38 °C) it is roughly twice as high as at room temperature (25 °C), and at higher temperature the rise is increasingly steep.

as at a room temperature of 25 °C (24 mm Hg) and more than 10 times as high as the vapor pressure at the freezing point (4.6 mm Hg).

If the air already contains some water vapor, the difference between the vapor pressure at the surface of liquid water and in the air will be less, and the driving force for water vapor to diffuse from the saturated boundary layer at the water surface into the air will be correspondingly reduced.

As a rough approximation to the measure of this driving force for evaporation, many investigators have used the *saturation deficit* or *vapor pressure difference*. The vapor pressure difference is expressed as the difference between the vapor pressure over a free water surface at the temperature in question and the water vapor pressure in the air. If the relative humidity of the air is 50%, the vapor pressure difference increases with temperature, as shown by the difference between the two curves in Figure 8.10.

The vapor pressure difference has been useful in many ecological studies of both plants and animals, but it is theoretically inadequate, for the vapor pressure difference is not the only physical factor that determines the rate of evaporation (Ramsay 1935; Edney 1957). For one thing, air movement caused by free or forced convection drastically alters the rate of evaporation.

Another factor that is important under some circumstances is the loss of heat from and the cooling of the surface from which evaporation takes place. The evaporating surface may therefore be cooler than other parts of the system, the temperature differences depending on the rate of evaporation as well as the transfer of heat from other parts, a parameter that is often difficult to measure. Furthermore, the rate of diffusion of water vapor into air increases with increasing temperature. And finally, the diffusion rate increases with decreasing barometric pressure. Although each variable is a complex function that in turn influences all the others, mathematical modeling can provide a successful analysis of water exchanges in a terrestrial environment (Tracy 1976).

amples are frogs and snails, which we can group together as moist-skinned animals that have a high rate of evaporation.

However, before we discuss animals, we must know something about the physical factors that influence evaporation. That subject covered, we will discuss first some moist-skinned terrestrial animals that depend heavily on water in their environment and then turn to arthropods and terrestrial vertebrates.

Evaporation

We know that evaporation from a free water surface increases with temperature and that evaporation is faster in a dry atmosphere than under humid conditions. In order to understand evaporation from animal surfaces, it is necessary to have a more precise concept of the physical laws that govern the rate of transfer of water from the liquid to the gas phase.

The water vapor pressure over a free water surface increases rapidly with temperature (Figure 8.10). As a rule of thumb we can remember that at mammalian body temperature (38 °C) the water vapor pressure (50 mm Hg) is about twice as high

For animals under natural conditions, the extent of air movement is one of the most important variables. If there is wind (and in nature the air is virtually never completely still), the air layer close to the surface is rapidly renewed, and evaporation increases. This we feel as a cooling effect when moist skin is exposed to the moving air from wind or a fan.

The cooling of a surface caused by evaporation also causes a change in the density of the adjoining air, producing convection currents. The extent of such convection differs between a horizontal and a vertical surface, and as a consequence, under otherwise identical physical conditions the evaporation from a moist surface varies with the orientation of the surface. Finally, evaporation also depends on the curvature of a surface.

Because of these complexities, it is difficult to describe accurately the variables that govern the rate of evaporation. The vapor pressure difference remains the most practical approximation we have, and expressing a certain rate of evaporation relative to the vapor pressure difference gives a reasonably good basis for comparisons. But we should realize that results on animals of various sizes and shapes obtained under seemingly similar conditions do not give an accurate basis for comparison and, in particular, cannot be directly compared with the evaporation from the horizontal surface of free liquid water.

MOIST-SKINNED ANIMALS

Consider the two extreme cases. From a moist animal surface the evaporation is high, and the rate of water loss is determined primarily by the transfer of water vapor into the surrounding air (diffusion aided by convection). In the opposite case (e.g., the dry insect cuticle), the greatest resistance to evaporation is in the surface itself, and any change in the permeability of the barrier greatly affects the rate of evaporation. In this case changes in the vapor pressure difference, convection, and so on are of minor importance and in some instances may

TABLE 8.8 Evaporation of water from the body surface of various animals at room temperature. The data indicate orders of magnitude; exact figures vary with experimental conditions. All data refer to micrograms of water evaporated per hour from 1 cm^2 body surface at a vapor pressure difference of 1 mm Hg (0.13 kPa). [Schmidt-Nielsen 1969]

Earthworm	400
Frog	300
Salamander	600
Garden snail, active	870
Garden snail, inactive	39
Man (not sweating)	48
Rat	46
Iguana lizard	10
Mealworm	6

even be disregarded. The two cases can be described as a *vapor-limited system* where the resistance to evaporation lies in water transport in the air, and a *membrane-limited system* in which the membrane is the major barrier to water evaporation (Beament 1961).

Earthworms

An earthworm kept in dry air rapidly loses weight because of evaporation. The rate of water loss (Table 8.8) is quite high, and in dry air the earthworm soon dies. If a partially dehydrated worm is placed in a U-shaped tube, covered with water but with the mouth and anus above the water surface, water is absorbed by the worm. Its skin is readily permeable to water in both directions.

A more careful study will show that the earthworm behaves much like a typical fresh-water animal. If it is placed in salt solutions of various concentrations, its body fluids remain hypertonic to the medium (Figure 8.11). The urine, however, remains hypotonic to the body fluids, and also in this regard the earthworm resembles a typical fresh-water animal (Ramsay 1949).

These osmotic relations show that, with regard to water balance, the earthworm is really more of a fresh-water animal than a truly terrestrial animal.

FIGURE 8.11 The earthworm, when palced in sodium chloride solutions of various concentrations,
behaves like a typical fresh-water osmoregulator. It
remains hyperosmotic to the medium at all concentrations, and more so at the lower concentrations.
[Ramsay 1949]

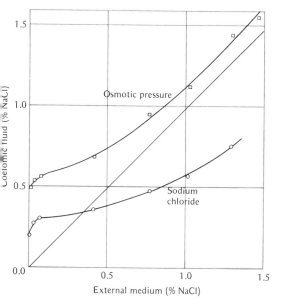

Accordingly, those adult amphibians that are terrestrial and air-breathing usually live near water and in humid habitats where evaporation is low. When they enter water, they behave osmotically like the typical fresh-water animal.

Knowing that amphibians are moist-skinned and tied to the vicinity of free water makes it seem a paradox that some live successfully in deserts. Several species of frogs are found in the dry arid interior of Australia. These animals retreat to burrows several feet deep in the ground, where they estivate during the long periods of drought. Because reproduction takes place in water, they can breed only after a rainfall, which gives an abundant but temporary supply of water.

When it rains these frogs reappear from their burrows, restore their water content, and deposit their eggs, which develop at exceptional speed into tadpoles that metamorphose before the pools dry up. When the frogs enter estivation again, their urinary bladders are filled with dilute urine, which is of considerable importance in their water balance. Some can store as much as 30% of their gross body weight as urine in the urinary bladder. The urine is extremely dilute and has an osmotic concentration corresponding to less than 0.1% NaCl. This urine is the main water reserve and is gradually depleted during estivation (Ruibal 1962).

It has been reported that the Australian aborigines use the desert frog (*Chiroleptes*) as a source of drinking water because the frog is so distended with dilute urine as to ''resemble a knobbly tennis ball.'' Once the animal has used up the water reserve in the bladder, further dehydration results in increased body concentrations, and the animal then undergoes progressive dehydration of blood and tissues.

The situation for frogs that remain deep in their burrows during dry periods is similar to that described for lungfish, which also estivate in the ground during drought. When the lake dries out, the lungfish wiggles into the mud, where it remains in a dry cocoon with a breathing channel to the rock-hard surface. It may remain alive in this con-

t tunnels in the soil where the air is fully saturated, and it is in contact with soil particles covered with a thin film of free water. An earthworm is unable o live in completely dry soil where it rapidly becomes dehydrated and dies.

Frogs and other amphibians

A frog, even when picked up on land, has a moist and cool skin. The rate of evaporation from the frog's skin is similar to that from the earthworm's (Table 8.8), and the twofold difference between frog and salamander is, for reasons explained above, not a meaningful difference. For both, the rate of evaporation is of the same magnitude as evaporation from a free water surface, and the amphibian skin does not seem to present any significant barrier to evaporation.

FIGURE 8.12 When kept at 25 °C and 20 to 30% relative humidity, the frog *Chiromantis* loses water by evaporation at a very low rate, similar to that of a reptile (*Chamaeleo*). After 6 days in air its weight loss is only a few grams. Other amphibians lose weight rapidly. A toad (*Bufo*) loses over 40% of its body weight and dies within 2 days; an ordinary frog (*Rana*) dies in a day, and the aquatic *Xenopus* in less than 10 hours. [Loveridge 1970]

be minor and not sufficient to explain the great variations in habitat. It has been claimed that frogs lose water more rapidly than toads, which have thicker and more cornified skins, but this does not appear to be true (Bentley 1966). Therefore, resistance to cutaneous evaporation does not seem to be a major factor in the adaptation of amphibians to terrestrial life.

There are, however, some striking exceptions. The South African frog *Chiromantis* does not depend on a humid environment to prevent rapid dehydration. When placed in dry air it loses water by evaporation at a rate that is a small fraction of the rate in other frogs; in fact, its rate of water loss is similar to that of reptiles (Figure 8.12).

Several frogs of the South American genus *Phyllomedusa* also exhibit rates of evaporation that are only about one-twentieth of those typical of other frogs (Shoemaker and McClanahan 1975). *Phyllomedusa* has specialized glands in the skin that secrete a waxy substance that apparently waterproofs the skin. This wax makes the skin surface hydrophobic so that droplets of water, instead of spreading on the skin, form beads (Shoemaker 1988). *Chiromantis* lacks glands that secrete wax, and its skin must be rendered impermeable by some other mechanism.

These unusual frogs have another physiological characteristic completely different from other frogs. Instead of excreting urea, as all other amphibians do, they excrete uric acid, which is characteristic of reptiles. This is important for their water balance and will be discussed further in Chapter 9, in connection with nitrogen excretion.

The finding of reptilian physiological characteristics in amphibians illustrates the point we have made before. Physiological characteristics as a rule are more plastic and adaptable to the environment than morphological characters and are therefore not suitable for establishing evolutionary relationships.

Snails

Snails and slugs are moist-skinned, and evaporation from their surfaces is high (Table 8.8). Na-

dition for several years, and to keep its minimal metabolic rate going, it gradually uses up its body proteins. No water is available for the excretion of the urea formed in protein metabolism, and the concentration of urea in its body fluids may rise as high as 500 mmol per liter (Smith 1959).

Is the rate of evaporation from the skin of a desert amphibian less than in other amphibians? When determinations are made under similar conditions, including temperature and atmospheric humidity, the differences in the rates of evaporation seem to ed slugs depend on the humidity of their habitat, and they are active mostly after rain and at night when the relative humidity is high. Otherwise they withdraw to and remain in microhabitats where the humidity is favorably high. Snails, in contrast, carry with them a water-impermeable shell into which they can withdraw.

It has been suggested that the *mucus* that covers a snail forms a barrier to evaporation, but this is difficult to understand. When snail mucus is removed from the animal, water evaporates from the isolated mucus at practically the same rate as from a free water surface, until the mucus is completely dried out. Nevertheless, this situation is not entirely clear, for if the surface of an otherwise undisturbed snail is lightly touched, fresh mucus is extruded from mucus glands in the skin and the rate of evaporation increases (Machin 1964).

A terrestrial snail that is inactive or estivates has an excellent barrier to water loss. It withdraws into its shell and covers the opening with a membrane, an epiphragm, which consists of dry mucus; in some cases it also contains a large amount of crystalline calcium carbonate.

The extremely low rate of water loss from an inactive snail in the shell and covered with an epiphragm permits some species to survive in hot, dry desert areas. One such snail, *Sphincterochila,* is found in the deserts of the Near East. Withdrawn into the shell and dormant, this snail can be found on the desert surface in midsummer, fully exposed to sun and heat. The snails become active after winter rains, from November to March, when they feed and reproduce.

During the long dry season the snails remain dormant. The water loss from a dormant *Sphincterochila* is less than 0.5 mg per day, and since one animal contains about 1.5 g water, it can survive for several years without becoming severely dehydrated. In fact, the snail itself (not including the calcareous shell), contains over 80% water. This high water content remains the same, even during the hottest and driest part of the year. This shows that these snails do not gradually deplete their water resources, but have solved the water problem by

having a uniquely low water loss while remaining dormant (Schmidt-Nielsen et al. 1971).

ARTHROPODS

The most successful terrestrial animals are found among the arthropods, which comprise insects, crustaceans, and several other animals characterized by a jointed, rigid exoskeleton. Insects are by far the most numerous animals in regard to number of species, for nearly 1 million different species have been described.

Insects and arachnids (spiders, ticks, mites, scorpions, etc.) are primarily terrestrial, and they are highly adapted to life on land and respiration in air. Only a small number have secondarily invaded fresh water, and almost no marine forms are found in these two classes. Two additional classes of arthropods are terrestrial: the millipedes (Diplopoda) and the centipedes (Chilopoda). In contrast, most crustaceans are aquatic, as are the Xiphosura, represented by the living fossil, the horseshoe crab (*Limulus*).

Crustaceans

Although most crustaceans are aquatic, appreciable numbers, including some decapods (crabs and crayfish), are terrestrial, even to the extent that they drown if kept submerged. However, they are mostly restricted to moist habitats and the vicinity of water. On the other hand, some isopods, notably sowbugs (or pillbugs) and woodlice, are completely terrestrial, and a few even live in hot, dry deserts.

Crabs are common in the intertidal zone, and in the tropics many live permanently above the high-tide mark. They depend on a moist habitat, however, and often have burrows where a pool of water remains at the bottom. In the more terrestrial species emancipation from water is nearly complete, but the crabs do return to water for spawning.

The semiterrestrial and terrestrial crabs are generally excellent osmoregulators. Depending on conditions and need, they can usually carry out effectively both hyperregulation and hyporegulation. An interesting aspect of some land crabs (*Cardiosoma* and *Gecarcinus*) is their ability to take up water from damp sand or a moist substratum, even in the absence of visible amounts of free water. In this way they can survive for many months without direct access to water (Gross et al. 1966; Bliss 1966).

The ghost crab, *Ocypode quadrata,* which is often seen scurrying over sandy beaches, requires frequent immersion in sea water. Similarly, the well-known fiddler crabs, *Uca,* which are found along the coasts of all tropical seas and into temperate areas, dig deep burrows into the beach near or above high-tide mark, but depend on frequent returns to water.

The common terrestrial isopods known as pill-bugs, sowbugs, or woodlice are mostly found in humid habitats, well away from exposure. They remain hidden during the day and move around at night when the relative humidity is higher. These crustaceans are completely independent of free water for reproduction and are thus truly terrestrial.

Although woodlice mostly live in microhabitats with high humidity, they are sometimes exposed to drier air. Then they lose water by evaporation, and this water must be replaced. Normally, woodlice feed on moist decaying plant material, which is probably their normal source of water. It has been found, however, that several isopods can take up free water by drinking and also through the anus. When kept on a moist porous slab of plaster of Paris, they can even absorb water from this surface (Edney 1954).

Compared with that of an insect, the cuticle of an isopod is relatively permeable to water. The main reason for this difference seems to be that insect cuticle is covered by a thin layer of wax, which greatly reduces water loss. Attempts at demonstrating a wax layer in isopods have been unsuccessful (Edney 1954).

It may seem strange to find a land crustacean in a desert, but a woodlouse, *Hemilepistus,* is quite common in desert areas. These animals still breathe by gills, and they survive by digging narrow vertical holes about 30 cm deep, where they spend the hot part of the day. Temperature and humidity measurements show that their hideouts remain much cooler than the desert surface, and most importantly, the relative humidity is high, around 95%. The cuticle of *Hemilepistus* is somewhat less permeable to water than that of other woodlice, but the animal's major adaptation to life in the desert seems to be a question of behavior (Edney 1956).

Insects and arachnids

If numbers indicate success, insects are undoubtedly the most successful animals on earth. There are more species of insects than of all other animals combined, and furthermore, many insect species are extremely numerous. They live in almost every conceivable habitat on land and in fresh water; only the polar regions and the ocean seem to be relatively free from them. Dryness, desert, and absence of free water seem to be no barrier to these animals.

Clothesmoths that thrive on a woolen garment, or flour beetles that complete their life cycle from egg to adult in dry flour, live, grow, and thrive virtually without any free water in the environment. Nevertheless, as the adult insect consists of more than two-thirds water, these animals must have extraordinary capacities to retain water and reduce losses.

Because there are considerable similarities between insects and arachnids, these two classes will be discussed together.

Water balance: gains and losses

For an organism to remain in water balance, all loss of water must, over a period of time, be balanced by an equal gain of water. The components in the water balance are:

Water loss
  Evaporation
    from body surface
    from respiratory organs
  Feces
  Urine
  Other (specialized secretions)

Water gain
   Drinking
   Uptake via body surface
      from water
      from air
   Water in food
   Oxidation water (metabolic water)

The problem of maintaining balance may work either way. There may be an excess of water or a shortage of water, and the physiological mechanisms must therefore be able to cope with both situations.

For example, a fresh-water insect may swallow large amounts of water with its food and also be subject to osmotic inflow of water. The excess water must be eliminated, a task normally carried out by the renal organ or kidney, which then must produce a large volume of dilute urine. A terrestrial insect living in a dry habitat, on the other hand, has a very limited water intake, and all losses must be reduced to such a level that their sum in the long run does not exceed the total gain.

Any organism can tolerate a certain variation in its water content, and some are more tolerant than others. Mammals, for example, can usually withstand losing 10% of their body water, although they are then in rather poor condition; a loss of 15 or 20% is probably fatal to most. Many lower organisms can withstand greater losses. Some frogs, for example, can withstand the loss of 40% of their body water, but very few animals can tolerate losing half their body water.*

*Two methods are available for the determination of total body water. The water content can be obtained by killing the animal, drying the body completely (e.g., in a drying oven), and determining the weight loss. If such a destructive method is employed, some individuals are used to establish a normal or baseline value, and others are used for experimentation. A nondestructive method is the isotope dilution technique. A small volume of water labeled with an isotope (deuterium or tritium) is injected into the animal. After the labeled water is evenly distributed in all body water, a sample of blood or any other body fluid is withdrawn and the concentration of the label determined. It is then simple to calculate the water volume into which the isotope label was distributed. The method permits repeated determinations on the same individual animal.

When the water content of an animal is expressed as a percent of its body weight, the results are not always easy to interpret. For example, the fat content of animals varies a great deal, and fatty tissues have a low water content, about 10% or less. An animal with a large fat storage therefore will have a lower overall percent water content in the body than a lean animal in which muscles and internal organs have exactly the same degree of hydration. If the water content is expressed in relation to the fat-free tissue mass, the figures are usually much less variable. The disadvantage is that this method requires an estimate of fat content and therefore becomes much more complicated.

The custom of expressing water content as a percent of the body weight tends to be misleading in another way that is best explained by an example.

Consider an insect that initially contains 75% water. If it loses exactly half of all water in its body, it will contain 60% water. The result may seem surprising, but an examination of Figure 8.13 shows that arithmetic to be correct.

At first glance, a reduction in water content from 75% to 60% does not seem excessive, although in reality half of all body water was lost. The sometimes deceptive effect of percentage values can be avoided by expressing results in absolute units. If the insect in the above example weighed 100 mg, it would initially contain 75 mg water and after dehydration 37.5 mg water, and it is immediately apparent that half its water was lost.

Water loss: evaporation

Evaporation takes place both from the respiratory organs and from the general body surface, but it is often experimentally difficult to determine these two variables separately.

The respiratory organs of insects and vertebrates differ profoundly in structure and function. The lining of the vertebrate lung is always moist, the respiratory air is saturated with water vapor, and the water loss from the respiratory tract is substantial. The respiratory organs of insects, however, consist of tubes lined with chitin, and only the finest branches of these are relatively permeable to

FIGURE 8.13 Diagram to illustrate that it can be deceptively misleading to express body water content in percent of body weight. If an animal originally has a water content of 75% of its body weight (a) and loses exactly half its body water, the water content of the dehydrated animal will be 60% of the body weight (b). Thus a water loss that in reality is very serious appears not very great.

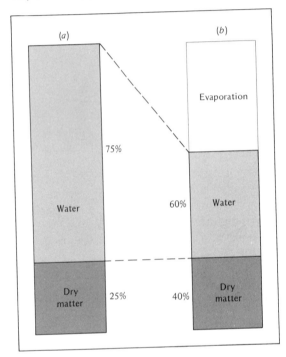

TABLE 8.9 Evaporation of water from the body surface of insects and arachnids. Water loss from respiratory organs is not included. All data refer to micrograms of water evaporated per hour from 1 cm² body surface at a vapor pressure difference of 1 mm Hg (0.13 kPa). [Schmidt-Nielsen 1969]

Marsh fly larvae (*Bibio*)	900
Cockroach (*Periplaneta*)	49
Desert locust (*Schistocerca*)	22
Tsetse fly (*Glossina*)	13
Mealworm (*Tenebrio*)	6
Flour mite (*Acarus*)	2
Tick (*Dermacentor*)	0.8

marsh fly, when in air, is of the same order of magnitude as evaporation from a moist-skinned animal (see Table 8.8). This means that the cuticle essentially is no barrier to evaporation. On the other hand, in insects that live in very dry habitats the permeability of the cuticle can be lower than in any other animals about which we have information.

The insect cuticle is a highly complex organ, which consists of several layers. The hard chitin itself is not particularly impermeable to water; the resistance to evaporation resides in a thin covering *layer of wax*. If this layer is scratched by abrasives (e.g., alumina particles) the evaporation increases greatly and insects that can otherwise live in very dry material, such as stored grain, die of dehydration (Wigglesworth 1945). This is of some practical interest, for abrasives have been used as nontoxic insecticides that can later be separated from the grain with relative ease because of differences in density. Not only sharp particles, but also materials such as dry clay dust have a similar effect, presumably because they too disturb the structural arrangement of the wax layer.

The evaporation from insect cuticle increases with rising temperatures. The increase is far greater than the increase in water vapor pressure and seems related to physical changes in the wax layer.

For arthropods a wax layer is not a universal prerequisite for successful terrestrial life. It was

water. Nevertheless, for insects the water loss from the respiratory system is important, and the possible means to reduce this loss were discussed under insect respiration in Chapter 1.

The general body surface of an insect is covered by a hard, dry cuticle, and it is deceptively easy to assume that this cuticle is completely impermeable to water. It is not. There are enormous differences in evaporation between species that normally live in moist environments and others that live in completely dry surroundings (Table 8.9).

The evaporation from the aquatic larvae of the

mentioned above that many isopods are adapted to terrestrial life and that some even live in deserts. No wax has been found on their cuticle, and their water loss from evaporation is directly related to the vapor pressure difference in the atmosphere and is otherwise temperature-independent. In insects, in contrast, evaporation changes with temperature even if the vapor pressure difference is kept constant.

Water loss in feces and urine

Insects eliminate their feces and urine through the same opening, the anus. The urine is formed by the Malpighian tubules, which open into the posterior part of the gut. Liquid urine, as well as fecal material from the intestine, enter the rectum, where water reabsorption takes place. This matter will be discussed in greater detail in Chapter 9.

Storage excretion

If excretory products, instead of being eliminated in the urine, are withheld in the body, no water is expended for their excretion. Because uric acid is a highly insoluble compound, its retention in the body is in fact a feasible approach to the problem of ''excretion,'' and the deposition of uric acid in various parts of the insect body seems to be a regular feature. For example, in several species of cockroaches, as much as 10% of the total dry weight of the body may be uric acid.

Many insects store uric acid in the fat body; others store it in the cuticle. Uric acid deposited in the cuticle is probably never mobilized again. In the fat body, however, the situation may be different. It is possible that this uric acid represents a storage or depot from which nitrogen can be mobilized during periods of nitrogen deprivation and used for metabolic purposes. So far evidence for such a function is inadequate.

''Active'' transport of water

An insect that lives in very dry surroundings (e.g., a mealworm) withdraws water from the rectal contents until the fecal pellets are extremely dry. This

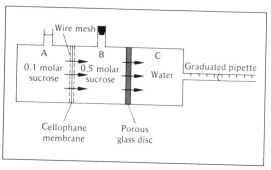

FIGURE 8.14 The Curran model to demonstrate net transfer of water against an osmotic gradient between chambers A and C. During experiments the exit tube from chamber B is closed. For details, see text. [Curran and MacIntosh 1962]

withdrawal of water has the appearance of an active transport of water, for water is moved from a high osmotic concentration in the rectum to a lower concentration in the blood.

However, because of the complexity of the rectum and its surrounding structures, alternate hypotheses are possible. In general, most uphill transport of water in animal systems can be explained by a primary transport of a solute, with water following passively due to osmotic forces, acting as explained by the *three-compartment theory* proposed by Curran (1960).

The Curran hypothesis can be understood by reference to the artificial model in Figure 8.14, which shows a cylinder divided into three compartments. Compartment A and B are separated by a cellophane membrane, which is readily permeable to water but not to sucrose. Compartments B and C are separated by a porous glass disc (a sheet of filter paper would do if given suitable mechanical support).

Let us initially fill compartment A with 0.1 molar sucrose solution, compartment B with 0.5 molar sucrose, and compartment C with distilled water. The concentrated solution in B will immediately draw water osmotically from A and C, but if the liquid in B is kept from expanding, it will flow

FIGURE 8.15 Three-compartment model for water transport across an epithelial cell. Heavy arrows indicate active Na⁺ transport. Dotted area corresponds to compartment B in Figure 8.14.

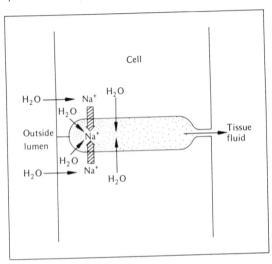

through the porous disc into compartment C. The overall result is therefore a net movement of water from A to C, although the osmotic concentration in A is higher than in C. The movement of water against the overall osmotic gradient obviously cannot continue forever, for the concentration differences will gradually decline and the system run down. However, if somehow we can maintain a high solute concentration in compartment B, the water movement will continue.

Movement of water through a variety of biological membranes can be explained according to the Curran model. The diagram in Figure 8.15 represents a typical epithelial cell, which has deep infoldings of the membrane on the side facing the tissue fluid. Active transport of sodium establishes a high osmotic concentration within the lumen of the infolding. Then, because of the osmotic forces, water diffuses into the lumen, and the increased hydrostatic pressure causes a bulk flow of liquid through the opening into the tissue fluid. The result is an overall movement of water from the outside lumen to the tissue fluid. This water transport, which can be against an overall osmotic gradient from the outside of the cell to the tissue fluid, depends on the active transport of sodium as the primary driving force.

Transport as explained by this mechanism may be a common way of moving water. Long, narrow intercellular or intracellular spaces are characteristic of many epithelia, and available experimental evidence supports the hypothesis (Diamond and Bossert 1967).

Water reabsorption from the insect rectum has been studied in cockroaches, and the results conform to the three-compartment hypothesis. Samples of fluid, as small as 0.1 nl (0.0001 mm³) have been withdrawn from the rectal pads. This fluid is more concentrated than the fluid in the rectal lumen, while the blood is less concentrated. The fluid thus corresponds to compartment B in the Curran model (Wall and Oschman 1970).

Water gain: drinking

The most obvious form of water intake is drinking free water. Such water is, of course, available to all fresh-water insects and to insects under many other circumstances, such as when there is dew or rain. For most insects, however, free water is available only intermittently and at irregular intervals, and many live in dry habitats where no free water is available at any time. These must obtain water from elsewhere.

Water in food

Insects that eat plants may obtain large amounts of water in the food, for fresh vegetable material has a high water content. Succulent fruits, leaves, and so on may contain over 90% water, but even the driest plant material contains some free water. Dry grains and seeds, flour, wool, and other seemingly completely dry substances on which insects are able to subsist, grow, and reproduce may contain about 5 to 10% free water.

When water is plentiful in the food, the problem is to eliminate the excess. This is the normal func-

TABLE 8.10 Amounts of water formed in the oxidation of various foodstuffs. [King 1957; Schmidt-Nielsen 1964]

Foodstuff	Water formed (g H_2O g^{-1} food)	Metabolic energy value (kcal g^{-1})	Water formed (g H_2O kcal^{-1})
Starch	0.56	4.2	0.13
Fat	1.07	9.4	0.11
Protein (urea excretion)	0.39	4.3	0.09
Protein (uric acid excretion)	0.50	4.4	0.11

tion of the kidney, which then produces dilute urine in large quantities. Under dry conditions, in contrast, the goal is to eliminate the excretory products with a minimum of water and produce a urine as concentrated as the renal mechanism is able to achieve.

Oxidation water

For animals living under dry conditions, the most significant water gain is from water formed in the oxidation of organic materials. We can properly call it *oxidation water*, but the term *metabolic water* is also common.

Everybody knows that water is formed when organic materials burn; we can see water condense on the outside of a cold pot placed over a gas flame, or water dripping from the exhaust pipe of a car on a cold morning. For glucose the oxidation reaction is:

$$C_6H_{12}O_6 + 6\ O_2 \rightarrow 6\ CO_2 + 6\ H_2O$$
$$(180g)\ +\ (192g) \rightarrow (264g)\ +\ (108g)$$

In the overall reaction water can be regarded as formed by oxidation of the hydrogen. In the oxidation of food, the amount of water depends on the amount of hydrogen present in the foodstuff in question. From the above equation it can be readily calculated that 1 g glucose yields 0.60 g water, and no amount of physiological maneuvering can obtain more water than what is indicated by the equation.

In the metabolism of other carbohydrates, such as polysaccharides and starch, slightly less water is formed because of the lower hydrogen content of these materials. The oxidation of starch, for example, is as follows:

$$(C_6H_{10}O_5)_n + n\ O_2 \rightarrow 6n\ CO_2 + 5n\ H_2O$$

In this case, the amount of water formed in oxidation is 0.56 g water per gram starch metabolized (Table 8.10).

Oxidation of fat gives more oxidation water than is obtained from carbohydrates, about 1.07 g water per gram fat, a figure that varies slightly with the composition of the fat and the degree of saturation of the fatty acids. Thus, about twice as much water is formed in fat oxidation as in the oxidation of starch. In some respects, however, this figure is misleading, for fat also gives more energy per gram (9.4 kcal against 4.2 for starch).

For a given metabolic rate, an animal therefore uses less than half the amount of fat, with a corresponding reduction in the yield of oxidation water. The amount of oxidation water formed, relative to a given metabolic rate, is therefore slightly more favorable for starch than for fat. This is shown in the last column of Table 8.10, which lists the amount of water formed in relation to the energy value of the food.

Protein metabolism is somewhat more complex, because the nitrogen contained in the protein yields excretory products that contain some hydrogen; this hydrogen is excreted and not oxidized to water. The amount of oxidation water depends on the nature of the end product of protein metabolism. If it is urea, the amount of oxidation water formed is 0.39 g water per gram protein; if the end product is uric acid, which is common in insects, the amount of oxidation water formed is higher.

Urea (CH_4ON_2) contains two hydrogen atoms per nitrogen atom; uric acid ($C_5H_4O_3N_4$) contains only one hydrogen per nitrogen atom, that is, only half as much. This increases the yield of oxidation water to 0.50 g water per gram protein when uric acid is the end product. To give an accurate account of the amount of oxidation water, we must therefore know not only the exact amounts and composition of the foodstuffs oxidized, but also the nature of the metabolic end product of protein metabolism. This is important only if a very accurate account is needed, for the differences in oxidation water formed for the various foodstuffs are relatively small (Table 8.10, last column).

Water uptake via body surface

In aquatic insects the uptake of water through the body surface is the same as in other fresh-water animals: The higher osmotic concentration of solutes in the body fluids causes an osmotic inflow of water. For aquatic insects the problem is to eliminate excess water, and, as mentioned above, this is usually done by the kidney or equivalent excretory organ.

In terrestrial insects, however, the situation is different; they are often in water shortage. A very interesting phenomenon has been observed in some terrestrial insects and arachnids: They are able to absorb water vapor directly from atmospheric air.

When first reported, this observation was met with considerable doubt because it is extremely difficult to suggest any plausible mechanism for water uptake directly from the air. The early reports have repeatedly been confirmed, and some well-documented examples are listed in Table 8.11. The table also contains information about the lowest relative humidity at which each animal can carry out such absorption of water vapor.

The characteristics of the uptake mechanism can

TABLE 8.11 Arthropods that can absorb water from atmospheric air. The right-hand column gives the limiting relative humidity below which a net gain in water is no longer possible.

Animal	Limiting relative humidity (%)
Tick (*Ornithodorus*)	94
Tick (*Ixodes*)	92
Mealworm (*Tenebrio*)	90
Mite (*Echinolaelaps*)	90
Desert roach (*Arenivaga*)	83
Grasshopper (*Chortophaga*)	82
Flour mite (*Acarus*)	70
Flea (*Xenopsylla*)	50
Firebrat (*Thermobia*)	45

best be described with the aid of an example, and we will use the desert roach *Arenivaga*. *Arenivaga* takes up water from the atmosphere only after it has been partly dehydrated. If it has been placed in a very dry atmosphere until, say, 10% of the body weight has been lost, uptake of water begins when the animal is moved to any relative humidity above 83%. The uptake continues until the animal is fully rehydrated and then ceases (i.e., the absorption process is accurately regulated according to the need for water).

Figure 8.16 shows the weight loss due to water loss of a desert roach that was dehydrated by being kept in dry air. This was followed by a subsequent weight gain when the animal was transferred to a higher humidity and took up water from air.

We saw that the atmospheric humidity below which uptake cannot take place differs from species to species (Table 8.11). It is curious that this limit is unaffected by temperature, although the rate of uptake changes (Figure 8.17). Therefore the limit to uptake is determined by the relative humidity rather than the vapor pressure difference, i.e., the process appears to be similar to a hygroscopic effect. This point is difficult to understand, and any hypothesis ought to explain it adequately.

One hypothesis for the uptake of water vapor is

FIGURE 8.16 Relative proportions of water and dry material in the desert roach *Arenivaga* before and after dehydration in dry air, and after subsequent rehydration by absorption of water vapor from air at 95% relative humidity. The animals, which initially weighed about 250 mg, were not fed during the experiment and therefore lost dry substance. As a result the percentage of water was higher at the end of the experiment (73.2%) than initially (67.2%). The ordinate refers all measurements to the initial weight, normalized to 100%. [Edney 1966]

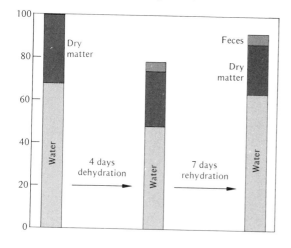

FIGURE 8.17 Desert roaches that have been partially dehydrated take up water from the air, provided the relative humidity exceeds about 83%. The rate of uptake is faster at higher temperatures. However, the lower limit for uptake, 83%, remains the same at different temperatures. [Edney 1966]

that temperature microgradients within the animal lead to condensation of water vapor. To achieve condensation from an atmosphere of 90% relative humidity requires a gradient in excess of 2 °C, which is rather unthinkable when experiments are done under constant temperature conditions. At 50% relative humidity, roughly the lowest humidity from which uptake has been reported in any species, the air temperature would have to be lowered from 25 °C to 14 °C to achieve condensation. No known mechanism could establish and maintain such temperature differences in a tiny insect.

The anatomical location for the uptake of water vapor is difficult to establish. It might seem simple to determine whether the respiratory system is involved, for it is relatively easy to seal the openings of the tracheae with wax. In some insects such

sealing stops water absorption, but in the tick *Ixodes* uptake still continues after the spiracles have been sealed. In those animals in which uptake ceases, this could be explained as a secondary effect of sealing the respiratory system, for this prevents oxygen uptake, and anoxia as such also stops water vapor uptake.

The many uncertainties are in part attributable to the very small size of some of the experimental animals (the house dust mite weighs 0.01 mg), but present information indicates that there is not one uniform mechanism in all those animals that are able to absorb water vapor. An oral site for absorption is likely for ticks and mites; the rectum almost certainly is the site of uptake in the firebrat and the mealworm (Noble-Nesbitt 1977).

In the desert roach *Arenivaga* the mechanism is associated with two well-defined bubblelike structures in the hypopharynx that are extruded and visible during periods of active water absorption. These bubbles or bladders have a hygroscopic surface that

HOUSE MITE The house mite (*Dermatophagoides farinae*) can absorb water vapor from air. In very dry air it becomes dehydrated, but if the relative humidity exceeds 70% it can take up water directly from the atmosphere. The top photo shows a dehydrated mite that weighs 8 μg (0.008 mg); the bottom photo shows the mite after it has increased its weight to 13 μg by absorption of water vapor from the air. The length of the animal is about 0.4 to 0.5 mm. [Courtesy of L. G. Arlian, Wright State University, Dayton, Ohio]

absorbs water from the atmosphere. A fluid with solute concentrations similar to the blood is then secreted onto the bladders. The salt contained in the fluid reduces the affinity of the cuticle for water.

This frees water from the cuticle, and the dilute fluid is then sucked from the bladders into the mouth and ingested. The bladder surface is now free to absorb more water; the cycle of fluid secretion is repeated, freeing the absorbed water; the fluid is ingested; and so on (O'Donnell 1982). A similar sequence of events seems to explain the mechanism of water vapor uptake in booklice (*Psyllipsocus*) (Rudolph 1982).

TERRESTRIAL VERTEBRATES

Reptiles

There are four major orders of living reptiles. Of these, the crocodilians are always associated with water. The other three – snakes, lizards, and tortoises – are considered well adapted to dry habitats, but they also have some representatives that are aquatic or semiaquatic. All the aquatic reptiles have lungs, are air breathers, and are obviously descended from terrestrial stock.

The skin of a reptile is dry and scaly and has been assumed to be impermeable to water. Let us therefore examine evaporation from the skin in a number of reptiles (Table 8.12). To make a comparison easier, we will use the same units as we used before (Tables 8.8 and 8.9). We can see that cutaneous evaporation in a dry-habitat reptile is only a small fraction of that in an aquatic reptile, and that even in the aquatic reptile (when kept in air) evaporation is one magnitude lower than in moist-skinned animals such as frogs (see Table 8.8).

We might expect that in a reptile evaporation from the moist respiratory tract would be much greater than evaporation from the dry skin, but this is not so. The contribution made by the skin to the total evaporation always exceeds the respiratory evaporation by a factor of two or more (Figure 8.18). When the water snake *Natrix* is kept in air, the skin contributes nearly 90% of the total evaporation. The relationships are similar for turtles and lizards: The skin remains more important than the respiratory tract in the water loss. Even in the chuckawalla, a

TABLE 8.12 Evaporation of water from the body surface of reptiles at 23 to 25°C. All data refer to micrograms of water evaporated per hour from 1 cm² body surface at a vapor pressure difference of 1 mm Hg (0.13 kPa). [Schmidt-Nielsen 1969]

Caiman (*Caiman*)	65
Water snake (*Natrix*)	41
Pond turtle (*Pseudemys*)	24
Box turtle (*Terrapene*)	11
Iguana (*Iguana*)	10
Gopher snake (*Pituophis*)	9
Chuckawalla (*Sauromalus*)	3
Desert tortoise (*Gopherus*)	3

FIGURE 8.18 In reptiles, evaporation from the dry skin exceeds evaporation from the respiratory tract, ranging in these animals from 66 to 88% of the total evaporation. The figure in parentheses next to each animal's name is the total evaporation per day (skin and respiratory tract combined) from that animal in grams water per 100 g body mass. The total evaporation is habitat-related and is more than 10 times as high in a water snake as in a desert tortoise. All observations in dry air at 23 °C. [Schmidt-Nielsen 1969]

desert lizard, two-thirds of the total evaporation is from the skin and only one-third from the respiratory tract.

There is a close correlation between evaporation and habitat. The drier the normal habitat, the lower the rate of evaporation. The total evaporation, combining body surface and respiratory evaporation, from a desert rattlesnake is less than 0.5% of its body weight per day, and the snake could probably survive for 2 to 3 months at this rate. If it remained in an underground burrow or tunnel where the humidity is higher, it could undoubtedly last even longer.

In addition to water lost through evaporation, water is also needed for urine formation. Reptiles excrete mainly uric acid as an end product of protein metabolism, and as this compound is highly insoluble it requires only small amounts of water for excretion. The relationship between nitrogen excretion and water metabolism will be discussed in greater detail in Chapter 9, which deals with excretion and excretory organs.

Marine reptiles have a special problem because the water they live in and much of their food contain large amounts of salt. How this affects excretion will be discussed later in this chapter.

The gain in water that is necessary to balance the losses must be derived from the same sources as for other animals, drinking, food, and oxidation

KANGAROO RAT This rodent (*Dipodomys spectabilis*) is common in the deserts of North America. It does not drink and subsists mostly on seeds and other dry plant material. Through economical use of water for urine formation and evaporation from the respiratory tract, it manages, using the water formed in metabolic oxidation processes as its main water source. [K. Schmidt-Nielsen, Duke University]

water. There is no indication that reptiles have the ability to obtain water by absorption from the atmosphere, as some insects can. Therefore, when drinking water is unavailable, their total water intake must be derived from food and from oxidation water.

Birds and mammals

Until now we have discussed water balance without considering problems of temperature regulation, but some animals, in particular birds and mammals, use water to keep cool in hot surroundings. Humans and some other mammals sweat; dogs and many other mammals and birds pant; and the increased evaporation cools the animal. This aspect of water expenditure was discussed in Chapter 7, and we shall not discuss it further here.

Mammals the size of rodents are convenient for a discussion of the basic aspects of water balance, for they do not pant and they lack skin glands in sufficient numbers to be of importance in heat regulation. This permits us to examine the basic, un-avoidable components of their water expenditure – evaporation and the losses in urine and feces – without regard for a large and highly variable amount used for heat regulation. We can then see whether the available sources of water cover the needs.

In most deserts free water is available only on the rare occasions of rain, yet birds and small rodents live in many deserts where rain falls once a year or even less often. These animals must obtain all their water from the food, for dew seems to play a minimal role, at least for higher vertebrates. Many animals obtain more than sufficient water from green leaves, stems, fruits, roots, and tubers; carnivorous animals obtain much water from the body fluids of their prey, which may contain from 50 to 80% water.

Kangaroo rats

Some desert rodents depend on such moist food, but others live primarily on dry seeds and other dry plant material, and their intake of free water is minimal. The kangaroo rats and pocket mice that

TABLE 8.13 Overall water metabolism, balancing gains and losses, during a period in which a kangaroo rat consumes and metabolizes 100 g barley (usually about 4 weeks). Air temperature, 25 °C; relative humidity, 20%. [Schmidt-Nielsen 1964]

Water gains	ml	Water losses	ml
Oxidation water	54.0	Urine	13.5
Absorbed water	6.0	Feces	2.6
		Evaporation	43.9
Total water gain	60.0	Total water loss	60.0

are abundant in the North American deserts are well-known examples of this type. They can live indefinitely on dry food, yet never drink. An examination of their water balance will therefore explain the mechanisms through which a mammal can live and remain in water balance, seemingly without any water intake.

A kangaroo rat is not an exceptionally "dry" animal; its body contains as much water as other mammals (about 66%). Even when a kangaroo rat has lived on a diet of only dry barley or oats for weeks or months, its water content remains the same. It maintains its body weight or may even gain in weight. This shows that the animal remains in water balance on the dry food and does not depend on some kind of water storage; in other words, water loss does not exceed water gain.

An account of the overall water metabolism of a kangaroo rat is given in Table 8.13. The account refers to a period during which a kangaroo rat would eat and metabolize 100 g barley. For a kangaroo rat that weighs about 35 g, this amount of food might be consumed in about 1 month. The exact time is of no consequence, however, for the balance would look exactly the same if the animal had a higher metabolic rate and the same amount of food were consumed in, say, 2 weeks.

On the gain side we find that 54 g water is formed in the oxidation of the food. This figure can be calculated from the composition of the grain. By using the figures for oxidation water formed in the metabolism of starch, protein, and fat (as given in

Table 8.10) we arrive at the listed figure. The grain also contains a small amount of free water, the exact amount depending on the humidity of the air. At 20% relative humidity there are 6 g water in 100 g of barley, giving a total amount on the gain side of 60 g. This has to suffice for all the animal's needs.

On the loss side we find that nearly one-quarter of the available water goes for urine formation. In quantity, the most important excretory product is urea, formed from the protein in the grain. This excretory product must be eliminated (there is no evidence for the kind of storage excretion described for insects), and the more concentrated the urine, the less water is used. The kangaroo rat has a remarkable renal concentrating ability, far better than most nondesert mammals (see Table 9.2). If we know the amount of protein in 100 g barley, we can calculate how much urea must be excreted and, in turn, how much water is needed. The amount of water lost in the feces can be determined by collecting the feces and measuring their water content.

The most important avenue for water loss is evaporation, occurring mostly from the respiratory tract. In Table 8.13, water gains and losses just balance. If the atmospheric humidity is lower, the water gain is slightly reduced because less free water is absorbed in the grain, and at the same time, in the drier air more water is evaporated from the respiratory tract. In a very dry atmosphere, therefore, the kangaroo rat cannot maintain water balance.

In more humid air, above 20% relative humidity, less water evaporates from the respiratory tract because the inhaled air contains more water and the kangaroo rat now readily maintains water balance. In nature, kangaroo rats spend much time in their underground burrows, where the air humidity is somewhat higher than in the outside desert atmosphere, and this aids in their water balance.

Evaporation from respiratory tract. The amount of evaporation from the respiratory tract depends on how much air is brought into the lungs (the ventilation volume) and on the fact that the exhaled air

FIGURE 8.19 Model of heat exchange in the nasal passages. Ambient air is 28 °C and saturated; body temperature is 38 °C. As inhaled air flows through the passages (left), it gains heat and water vapor and is saturated and at 38 °C before it reaches the lungs. On exhalation (right), the air flows over the cool walls and gives up heat, and water recondenses. As heat and water exchange approaches completion, exhaled air approaches 28 °C, saturated. [Schmidt-Nielsen 1972]

evaporation is achieved by exhalation of air at a lower temperature than the body core. Although the lung air is at core temperature and saturated, it is cooled as it passes out through the nose. The mechanism is simple (Figure 8.19): During inhalation the walls of the passages lose heat to the air flowing over them; the wall temperature therefore decreases, and because of evaporation it may fall below the temperature of the inhaled air. On exhalation, as warm air from the lungs passes over the cool surfaces, the air is cooled and water condenses on the walls. The extent of cooling (i.e., the final temperature of the exhaled air) varies with the temperature and humidity of the inhaled air.

How important is the cooling for the water balance? That varies, of course, with the conditions of the air the kangaroo rat breathes, but let us use a reasonable example with the air at 30 °C and 25% relative humidity. This air when inhaled into the lungs becomes heated to body temperature (38 °C), and is saturated with water vapor so that it contains 46 mg H_2O per liter. On exhalation, this air is cooled to 27 °C, and although still saturated, it contains only half as much water (i.e., half of the water is retained in the nasal passages and is used to humidify the next inhaled breath). But note that the exhaled air is saturated and always contains more water vapor than inhaled air; in other words, the nose is not a system to wring water out of the air, and there is always a net loss of water from the respiratory tract.

Heat exchange in the nasal passages is not unique for kangaroo rats; it occurs in all animals. It is an inevitable result of heat exchange between air and the nasal walls, and this heat exchange is more complete when the passages are narrow and the surface is large, as in small rodents. In birds, which have shorter and wider nasal passages, the cooling is less effective (Figure 8.20).

Kangaroo rats, in dry air, exhale air at temperatures below the inhaled air; birds, although they do cool the air appreciably, have higher exhaled-air temperatures. Humans, with their much wider nasal passages, have very incomplete heat exchange,

is saturated with water vapor. The amount of water already present in inhaled air determines how much additional water is needed to saturate the air.

Ventilation volume is determined by the rate of oxygen consumption (which in turn determines how much oxidation water is available). Mammals in general remove about 5 ml O_2 from 100 ml alveolar air before it is exhaled. If the oxygen extraction could be increased, the volume of respired air could be reduced (for a given oxygen consumption), and the amount of water evaporated would be correspondingly reduced.

This avenue for reducing the respiratory evaporation apparently has not been used by any mammal. To extract more oxygen in the lungs the hemoglobin of the blood would have to have a higher affinity for oxygen, but kangaroo rat blood does not differ from the blood of other rodents in this regard. Increased oxygen extraction would also lead to an increase in carbon dioxide in the blood, which would be reflected in the acid–base balance, but the kangaroo rat shows no such deviation from the usual mammalian pattern.

However, an important reduction in respiratory

FIGURE 8.20 Temperature of exhaled air in seven species of birds, measured in dry air. The body temperatures of the birds were between 40 and 41 °C. Two of the birds, cactus wren and budgerigar (parakeet), are desert species, but there is no difference in the extent of cooling (and thus water recovery) between these and birds from other habitats. The less effective cooling of exhaled air in the duck comes with the larger size of its respiratory passages. [Schmidt-Nielsen et al. 1970]

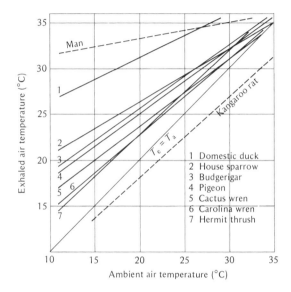

decreased to approximately 75% relative humidity. The combination of cooling and desaturation of the exhaled air provided a saving of water of 60% relative to the exhalation of saturated air at body temperature, a substantial saving indeed (Schmidt-Nielsen et al. 1981).

The mechanism responsible for desaturation of the air depends on the hygroscopic properties of the nasal surfaces. When a camel is dehydrated, the nasal membranes dry out and are covered by a layer of dried mucus, cellular debris, and salts. These surfaces give off water vapor during inhalation of the dry desert air, and take up water again from the moist respiratory air during exhalation. The absorbed water is evaporated again during the next inhalation, restoring the hygroscopic properties of the surfaces, so that water is again absorbed during the subsequent exhalation. The system is an entirely passive alternation between evaporation and absorption of water, and it functions only when the nasal passageways dry out in dehydrated animals.

and the exhaled air is commonly only a few degrees below body temperature. In humans, therefore, heat exchange in the nose and the associated water conservation are of lesser importance, and larger amounts of water are lost in the exhaled air.

The water savings we have discussed relate to the cooling of exhaled air that nevertheless remains saturated. However, severely dehydrated camels can further reduce the water content of exhaled air. They not only decrease the temperature of the exhaled air as described above, they also remove additional water vapor from the air.

In camels kept under desert conditions and deprived of drinking water, exhaled air in the daytime was fully saturated, but at night its humidity

MARINE AIR-BREATHING VERTEBRATES

The higher vertebrates – reptiles, birds, and mammals – are typically terrestrial, and, as we have seen, some are at home in the most arid deserts of the world. However, several lines of terrestrial vertebrates have secondarily invaded the sea but have remained air breathers.

In regard to problems of water and salt, they are essentially terrestrial animals, and, compared with fish, they are physiologically isolated from the surrounding sea water. In contrast to fish, which have gills that are relatively permeable to water, the higher marine vertebrates have lungs and thus escape the osmotic problem of an intimate contact with the sea water over a large gill surface. The higher marine vertebrates differ physiologically from their terrestrial relatives primarily in that they have only sea water to drink and that much of their food has a high salt content.

The sea contains enough water; the problem is

that all this water contains about 35 g salt per liter and has an osmotic concentration of some 1000 mOsm. Much of the food also has a high salt content, although in ionic composition it often differs substantially from sea water.

Plants and invertebrate animals, when eaten, present roughly the same osmotic problem as the drinking of sea water. Marine teleost fish, as we have seen, contain much less salt, and animals that subsist primarily on fish have a less severe salt problem than those that feed on plants or invertebrates. In any event, all of them must excrete the end products of protein metabolism, which in mammals is mainly urea and in birds and reptiles is uric acid.

If a vertebrate drinks sea water, the salts are absorbed and the concentration of salt in the body fluids increases. Unless the salts are eliminated with a smaller volume of water than that which was taken in, there can be no net gain of water. In other words, the salts must be excreted in a solution at least as concentrated as sea water, otherwise the body will become more and more dehydrated.

The reptilian kidney cannot produce urine that is more concentrated than the body fluids, and the bird kidney can usually produce urine no more than twice as concentrated as the blood. Because the concentration of their body fluids is about 300 to 400 mOsm per liter, the urine cannot reach the concentration of sea water (1000 mOsm per liter). The kidney therefore does not have a sufficient concentrating ability to permit these animals to drink sea water or eat food with a high salt content, and if they do, they must have other mechanisms for salt excretion.

Marine reptiles

Three orders of reptiles – turtles, lizards, and snakes – have marine representatives. The fourth living order of reptiles, the crocodiles, has no truly marine representative. The salt-water crocodile, *Crocodylus porosus,* is primarily estuarine in its habits, but it could probably survive indefinitely in a truly marine environment.*

Some of the sea snakes are completely independent of land, even in their reproduction, for they bear live young, never leave the sea, and would in fact be quite helpless on land. Sea turtles spend most of their lives in the open ocean, but they return to sandy tropical beaches for reproduction. Only the female turtles go on land to lay their eggs; the males never set foot on land after they enter the sea as hatchlings.

The marine lizards are more tied to land. An example is the Galápagos marine iguana, *Amblyrhynchus cristatus.* It lives in the surf of the Galápagos Islands, where it climbs on the rocks and feeds exclusively on seaweed.

The excretion of excess salt, which the reptilian kidney is unable to handle, is carried out by glands in the head called *salt-excreting glands* or simply *salt glands.* The salt glands produce a highly concentrated fluid that contains primarily sodium and chloride in concentrations substantially higher than in sea water.

The salt glands do not function continually, as the kidney does; they secrete only intermittently in response to a salt load that increases the plasma salt concentrations. Similar salt-secreting glands are found in marine birds, in which they have been studied in greater detail.

In the marine lizard the salt glands empty their secretion into the anterior portion of the nasal cavity, and a ridge keeps the fluid from draining back and being swallowed. A sudden exhalation occasionally forces the liquid as a fine spray of droplets out through the nostrils. The Galápagos iguana eats seaweed that has a salt concentration similar to that of sea water. It is therefore a necessity for this animal to have a mechanism for excretion of salts in high concentration (Schmidt-Nielsen and Fänge 1958).

---

*The tuatars *(Sphenodon)* from New Zealand is the only living representative of the fifth order, the Rhynchocephalia.

A marine turtle, whether plant eating or carnivorous, has a large salt-excreting gland in the orbit of each eye. The duct from the gland opens into the posterior corner of the orbit, and the turtle that has been salt-loaded cries salty tears. (The tears of humans, which everybody knows have a salty taste, are isosmotic with the blood plasma. Human tear glands therefore play no particular role in salt excretion.)

Sea snakes, which also excrete a salty fluid in response to a salt load, have salt glands that open into the oral cavity, from which the secreted liquid is expelled (Dunson 1968). The sea snakes are close relatives of the cobras and are highly venomous, an aspect that confers some excitement on their handling and tends to slow down their study.

The estuarine or salt-water crocodiles were believed to lack salt glands and to manage their salt balance by living in estuaries, subsisting on a diet of fish and having a highly impermeable skin. However, both the Australian salt-water crocodile, *Crocodylus porosus,* and the North American *Crocodylus acutus* possess salt glands that are distributed over the surface of the tongue (Taplin and Grigg 1981, Taplin et al. 1982)

We can now see that the four groups of marine reptiles have used four different anatomical systems for the elimination of excess salt. The lizards use a nasal gland, the turtles an orbital gland, the sea snakes a sublingual gland, and the crocodiles a large number of small glands on the surface of the tongue. They can all secrete a highly concentrated fluid, but the question of whether many of these animals normally drink sea water in appreciable quantities remains unresolved.

Marine birds

Many birds are marine, but most of them live on and above, rather than in, the ocean. Many are coastal, but some are truly pelagic. The young albatross that is hatched on a Pacific island spends 3 or 4 years over the open ocean before it returns to the breeding grounds.

Penguins, the most highly adapted marine birds, have lost their power of flight. They are excellent swimmers and are well advanced in their evolution toward a fully aquatic life. Nevertheless, in the physiological sense they have remained essentially terrestrial air-breathing animals that reproduce on land. The saying that the emperor penguin does not even breed on land is a play on words, for it hatches its egg while standing on the ice during the long, dark Antarctic winter.

All marine birds have paired nasal salt glands that, through a duct, connect with the nasal cavity (Fänge et al. 1958). Nasal glands are found also in most terrestrial birds, although in these they are very small; in marine birds the glands are without exception large. Most frequently they are located on top of the skull, above the orbit of each eye, in shallow depressions in the bone (Figure 8.21). In birds that are fed a diet high in salt or receive salt solutions to drink, the glands increase in size and become even larger than normal (Schmidt-Nielsen and Kim 1964).

The salt glands usually remain inactive and start secreting only in response to an osmotic stress (e.g., when sea water or salty food is ingested). Otherwise, the glands remain at rest, and in this respect they differ from the kidneys, which produce urine continuously. The secreted fluid has a simple composition; it contains mostly sodium and chloride in rather constant concentrations. Also in this regard the salt gland differs drastically from the kidney, which changes the concentrations and the relative proportions of the secreted components over a very wide range. Urine also contains a variety of organic compounds, whereas the salt gland secretion has no more than a trace amount of nonelectrolytes (Schmidt-Nielsen 1960).

Although the salt concentration in the secreted fluid is consistently high, there are characteristic species differences related to the normal ecology of the birds and their feeding habits. For example, the cormorant, which is a coastal bird and a fish eater, secretes a fluid with a relatively low salt con-

FIGURE 8.21 The salt-secreting glands of the sea gull are located on top of the head, above each eye, in shallow depressions in the bone. [Schmidt-Nielsen 1963]

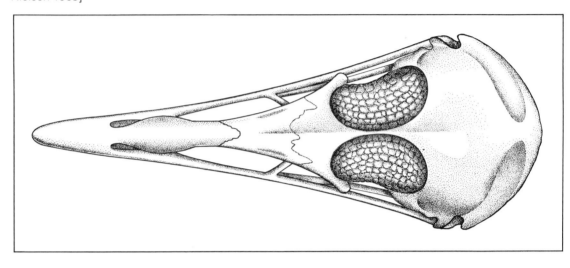

TABLE 8.14 Concentration of sodium in the nasal secretion of different species of birds. The chloride concentration in a given sample is nearly identical to the sodium concentration, and other ions are found only in small amounts. [Schmidt-Nielsen 1960]

Species	Sodium concentration (mmol liter^{-1})
Duck, mallard	400–600
Cormorant	500–600
Skimmer, black	550–700
Pelican, brown	600–750
Gull, herring	600–800
Gull, black-backed	700–900
Penguin, Humboldt's	725–850
Guillemot	750–850
Albatross, blackfooted	800–900
Petrel, Leach's	900–1100

tent, about 500 to 600 mmol Na per liter (Table 8.14). The herring gull, which eats more invertebrate food and consequently ingests more salt, has a concentration of 600 to 800 mmol Na per liter.

The petrel, a highly oceanic bird that feeds on planktonic crustaceans, has sodium concentrations in the nasal fluid up to 1100 mmol per liter.

The salt gland has an extraordinary capacity to excrete salt. In an experiment in which a gull received nearly one-tenth of its body weight of sea water (which would correspond to 7 liters for a human), the entire salt load was eliminated in about 3 hours. The details of this experiment are recorded in Table 8.15. The bird weighed 1420 g and was given 134 ml sea water by stomach tube. During the following 3 hours the combined volume of excreted liquid, 131.5 ml, equaled that ingested, with less than half (56.3 ml) coming from the salt gland and the remainder (75.2 ml) from the cloaca. The amount of sodium excreted by the nasal gland was, however, about 10 times as high as the amount appearing in the cloacal discharge (urine mixed with some fecal material).

Examination of Table 8.15 shows that the volume of nasal secretion was highest during the second hour, after which it gradually tapered off. The

TABLE 8.15 Nasal and cloacal excretion by a black-backed gull during 175 minutes following the ingestion of 134 ml of sea water, an amount nearly one-tenth of its body weight. [Schmidt-Nielsen 1960]

Time (min)	Nasal excretion			Cloacal excretion		
	Volume (ml)	Sodium concentration (mmol liter^{-1})	Sodium amount (mmol)	Volume (ml)	Sodium concentration (mmol liter^{-1})	Sodium amount (mmol)
15	2.2	798	1.7	5.8	38	0.28
40	10.9	756	8.2	14.6	71	1.04
70	14.2	780	11.1	25.0	80	2.00
100	16.1	776	12.5	12.5	61	0.76
130	6.8	799	5.4	6.2	33	0.21
160	4.1	800	3.3	7.3	10	0.07
175	2.0	780	1.5	3.8	12	0.05
Total	56.3		43.7	75.2		4.41

concentration of the nasal secretion was quite constant, in spite of changes in flow rate. The volume of cloacal discharge (mostly urine) exceeded the volume of nasal secretion, but its salt concentration was quite low.

The bird kidney can produce urine with a maximum sodium concentration of about 300 mmol per liter, but in this experiment the urine concentration was only a small fraction of this. Also, the urine sodium concentration varied about eightfold during the experiment, by no means unusual for the kidney, but in sharp contrast to the very constant concentration in the nasal secretion.

The volume of fluid secreted by the salt gland is remarkably high, particularly in view of the exceptional osmotic work performed in producing the concentrated salt solution. The flow rate from the salt glands, estimated per kilogram body weight, was about twice as high as the urine production of a human in maximum water diuresis (Table 8.16). Calculated per gram gland, the difference is even more striking. One gram of the salt gland of a gull can produce 36 ml fluid per hour, whereas the kidney of human in maximum water diuresis produces only about one-twentieth as much. For the human

TABLE 8.16 Secretion rates of the salt gland of the herring gull and of the human kidney during maximum water diuresis and antidiuresis.

Secretion rate	Salt gland	Kidney, diuresis	Kidney, antidiuresis
Per kilogram body weight (ml h^{-1})	30	15	0.15
Per gram gland (ml h^{-1})	36	0.2	0.002

kidney, when it produces a concentrated urine (which is still only half as concentrated as the secretion from the gull's salt gland), the flow rate may decline to only 1% of that in water diuresis. Clearly, the salt gland is one of the most effective ion transport systems known.

The fluid secreted from the salt gland of a marine bird always contains mostly sodium and very little potassium. The ratio between these two ions remains at about 30:1, and if the amount of potassium in the diet of a gull is increased, the Na/K ratio does not change much (Schmidt-Nielsen 1965).

Sea turtles and sea snakes have similarly high

TABLE 8.17 The concentrations of sodium and potassium (in millimoles per liter) in sea water and in salt gland secretion. The Na^+/K^+ ratio is high in the marine species and low in terrestrial plant eaters.

	Na^+	K^+	Ratio: Na^+/K^+
Sea water	470	10	47
Herring gull (*Larus*)[a]	718	24	30
Sea turtle (*Lepidochelys*)[b]	713	29	25
Sea snake (*Pelamis*)[b]	607	28	24
Estuarine crocodile (*Crocodylus*)[c]	663	21	32
Marine iguana (*Amblyrhynchus*)[b]	1434	235	6.7
False iguana (*Ctenosaura*)[d]	78	527	0.15
Chuckawalla (*Sauromalus*)[d]	121	379	0.32

[a] Schmidt-Nielsen (1960). [b] Dunson (1969). [c] Taplin and Grigg (1981). [d] Templeton (1967).

Na/K ratios, but in the marine iguana we find a lower ratio and the relative amount of potassium excreted by the salt gland is higher (Table 8.17). This is easy to understand; the diet of the marine iguana is primarily marine algae, and in general plants contain large quantities of potassium. The marine iguana therefore lives on a diet with a relatively high potassium content, and its need for potassium excretion is correspondingly high.

A typical land lizard from a dry habitat, the false iguana (*Ctenosaura*) has a nasal secretion with a very high potassium concentration and a low sodium concentration, giving a Na/K ratio of 0.15.

Another way in which the salt gland secretion of a land reptile differs from that of its marine counterpart is that the composition of the secreted fluid changes with the nature of the load. If a land reptile is given a sodium load, the secretion changes in favor of this ion; if it is given a potassium load, the secretion contains mainly potassium. In terrestrial reptiles the composition of the secreted fluid is regulated according to need, while the salt glands of marine birds and reptiles appear to be organs highly specialized specifically for the excretion of sodium and chloride.

Marine mammals

Three orders of mammals – seals, whales, and sea cows – are exclusively marine in the sense that they spend practically their entire lives in the sea. Seals return briefly to land to bear and nurse their young, but whales and sea cows (manatee and dugong) even bear their young in water.

The food these mammals eat varies greatly in salt content. Seals and whales are carnivorous and feed on fish, certain large invertebrates, and marine plankton organisms. Those that feed on fish obtain food with a rather low salt content (less than 1%), but with a relatively high protein content. The crab-eating seal of the Antarctic and the walrus (which feeds on clams and other bottom organisms) live on food organisms that are isosmotic with sea water. The baleen whales feed on crustacean plankton organisms with the high salt content characteristic of marine invertebrates. If sea water is incidentally ingested with the food, this adds further to the salt load. Dugongs and manatees are herbivorous and feed on plants that are in osmotic equilibrium with sea water; their salt intake is therefore high.

Do marine mammals have some physiological

TABLE 8.18 Effect on the water balance of ingesting 1 liter of sea water in a human and in a whale.

	Sea water consumed		Urine produced		
	Volume (ml)	Cl concentration (mmol liter^{-1})	Volume (ml)	Cl concentration (mmol liter^{-1})	Water balance: gain or loss (ml)
Human	1000	535	1350	400	−350
Whale	1000	535	650	820	+350

mechanism that corresponds to the salt-secreting glands of birds and reptiles? This should not be necessary, for the kidneys of whales and seals can produce urine more concentrated than sea water. A chloride concentration of 820 mmol per liter has been reported for whale urine (Krogh 1939). This is well above the concentration of sea water (about 535 mmol Cl per liter), and a whale that takes food with a high salt content, or even sea water, should be able to eliminate the salts without difficulty.

It is well known that sea water is toxic to humans and that a castaway at sea only hastens the dehydration processes if he drinks sea water. Table 8.18 shows the effect of drinking sea water on the water balance of a human and of a whale. A whale can drink 1 liter of sea water and have a net gain of about one-third liter of pure water after the salts are excreted. The kidney of a human is less powerful; the maximum urine salt concentration is below that of sea water, and if he drinks 1 liter of sea water, he inevitably ends up with a net water loss of about one-third liter and is worse off than if he had not drunk at all. His dehydration is further aggravated by the large amounts of magnesium and sulfate in sea water, which act as a laxative and cause diarrhea, thus increasing the water loss.

There is still inadequate information about whether seals and whales ingest any appreciable amounts of sea water, either by drinking or incidentally with the feed. There is good evidence, however, that seals do not have to resort to drinking sea water. California sea lions (*Zalophus californianus*) have been kept in captivity and given nothing but fish to eat. Even after 45 days without access to water, these animals were perfectly normal and in positive water balance (Pilson 1970). It is interesting to note that if sea lions eat invertebrate animals, such as squid, they probably manage equally well.

Although the blood of the squid is in osmotic equilibrium with sea water and has a high salt content, the osmotically active substances within the cells of the squid are only in part salts, for organic compounds are important tissue constituents and make up a substantial fraction of the total osmotic concentration. Thus, squid and many other invertebrates actually impose a lower salt load than we would expect merely from the osmotic concentrations of their body fluids.

Mammals have in their water balance an item that does not apply to birds and reptiles: The female nurses her young, and large quantities of water are required for production of milk. One way of reducing this loss of water would be to produce a more concentrated milk. It has long been known that seal and whale milk has a very high fat content and a higher protein content than cow's milk (Table 8.19). This has usually been interpreted as necessary for the rapid growth of the young, and particularly as a means of transferring a large amount of fat to be deposited as blubber and serve as insulation.

The high fat content of seal milk can also be viewed in light of the limited water resources of the mother. In the Weddell seal, which has been better studied than other seals, the fat content of the milk gradually increases during the lactation period, while the water content decreases corre-

TABLE 8.19 Composition and energy value of mammalian milk.

| Animal | Composition (g per 100 g milk) | | | | Energy value (kcal per g H_2O) |
	H_2O	Fat	Protein	Carbohydrate	
Cow[a]	87.3	3.7	3.3	4.8	0.8
Human[a]	87.6	3.8	1.2	7.0	0.8
Harp seal[b]	45.3	42.7	10.5	0	9.9
Weddell seal					
Mean value[c]	43.6	42.2	(14.1)[d]	—	10.5
Extreme value[c]	27.2	57.9	(19.5)[d]	—	23

[a]Kon and Cowie (1961).   [b]Sivertsen (1935).   [c]Kooyman and Drabek (1968).
[d]Figures refer to total nonfat solids, but as seal milk is virtually carbohydrate-free, most of it is protein.

spondingly. The highest fat content found in Weddell seal milk is 57.9% (nearly twice as much fat as is contained in whipping cream), with a water content in the same sample of only 27.2% (which contrasts to a water content in ordinary lean meat of about 65%).

Indeed, seals provide nutrients for their young with a minimal expenditure of water. For each gram of water used, seals transfer nourishment more than 10 times as effectively as land mammals (Table 8.19, last column). The observation that female Weddell seals in captivity do not suffer dehydration during the lactation period is in accord with these findings (Kooyman and Drabek 1968).

This neat way for a lactating mammal to save water leads to the question of whether desert mammals economize their limited water supply in the same way when they nurse their young. This seems to be the case, for kangaroo rat milk has a water content of only about 50% and contains up to 30% fat (Kooyman 1963), obviously a great help in the precarious water balance of the female kangaroo rat. But what about camels, whose milk has a fat content of about 4%, similar to cow's milk?

A moment's thought tells us that the camel's situation is entirely different from the kangaroo rat's. The kangaroo rat pup in the burrow uses no water for heat regulation. The camel's calf is born in late

winter and nurses through the hot summer, and at this time the calf needs water for heat regulation as much as or more than the mother. The female kangaroo rat transfers nutrients with minimal loss of water, but the female camel must also supply water to cover the needs of the calf.

In our discussions of osmotic regulation we have over and over again seen how important the organs of excretion are for the maintenance of the relative constancy, or steady state, of internal concentrations and water content of living organisms. We shall now move on to a more detailed discussion of excretory organs and their function.

REFERENCES

Beadle, L. C. (1943) Osmotic regulation and the faunas of inland waters. *Biol. Rev. 18:*172–183.
Beament, J. W. L. (1961) The water relations of insect cuticle. *Biol. Rev. 36:*281–320.
Bentley, P. J. (1966) Adaptations of amphibia to arid environments. *Science 152:*619–623.
Bentley, P. J. (1971) *Endocrines and Osmoregulation: A Comparative Account of the Regulation of Water and Salt in Vertebrates.* New York: Springer-Verlag. 300 pp.
Bittner, A., and Lang, S. (1980) Some aspects of the osmoregulation of Amazonian freshwater stingrays

(*Potamotrygon hystrix*). 1. Serum osmolality, sodium and chloride content, water content, hematocrit and urea level. *Comp. Biochem. Physiol. 67A:*9–13.

Bliss, D. E. (1966) Water balance in the land crab, *Gecarcinus lateralis,* during the intermolt cycle. *Am. Zool. 6:*197–212.

Bormann, F. H. (1982) The New England landscape: Air pollution stress and energy policy. *Ambio 11*(4):188–194.

Bowlus, R. D., and Somero, G. N. (1979) Solute compatibility with enzyme function and structure: Rationales for the selection of osmotic agents and end-products of anaerobic metabolism in marine invertebrates. *J. Exp. Zool. 208:*137–152.

Burger, J. W., and Hess, W. N. (1960) Function of the rectal gland in the spiny dogfish. *Science 131:*670–671.

Clark, M. E., and Hinke, J. A. M. (1981) Studies on water in barnacle muscle fibres. 1. The dry weight components of fresh fibres. *J. Exp. Biol. 90:*33–41.

Conte, F. P., Hootman, S. R., and Harris P. J. (1972) Neck organ of *Artemia salina* nauplii: A larval salt gland. *J. Comp. Physiol. 80:*239–246.

Croghan, P. C. (1958a) The osmotic and ionic regulation of *Artemia salina* (L.). *J. Exp. Biol. 35:*219–33.

Croghan, P. C. (1958b) The mechanism of osmotic regulation in *Artemia salina* (L.): The physiology of the gut. *J. Exp. Biol. 35:*243–249.

Curran, P. F. (1960) Na, Cl, and water transport by rat ileum *in vitro. J. Gen. Physiol. 43:*1137–1148.

Curran, P. F., and MacIntosh, J. R. (1962) A model system for biological water transport. *Nature, Lond. 193:*347–348.

Diamond, J. M., and Bossert, W. H. (1967) Standing-gradient osmotic flow. *J. Gen. Physiol. 50:*2061–2083.

Dunson, W. A. (1968) Salt gland secretion in the pelagic sea snake *Pelamis. Am. J. Physiol. 215:*1512–1517.

Dunson, W. A. (1969) Electrolyte excretion by the salt gland of the Galápagos marine iguana. *Am. J. Physiol. 216:*995–1002.

Edney, E. B. (1954) Woodlice and the land habitat. *Biol. Rev. 29:*185–219.

Edney, E. B. (1956) The micro-climate in which woodlice live. *Proc. 10th Int. Congr. Entomol. 2:*709–712.

Edney, E. B. *(1957) The Water Relations of Terrestrial Arthropods.* Cambridge: Cambridge University Press. 109 pp.

Edney, E. B. (1966) Absorption of water vapour from unsaturated air by *Arenivaga* sp. (Polyphagidae, Dictyoptera). *Comp. Biochem. Physiol. 19:*387–408.

Fänge, R., Schmidt-Nielsen, K., and Osaki, H. (1958) The salt gland of the herring gull. *Biol. Bull. 115:*162–171.

Foskett, J. K., and Scheffey, C. (1982) The chloride cell: Definitive identification as the salt-secretory cell in teleosts. *Science 215:*164–166.

Fugelli, K., and Zachariassen, K. E. (1976) The distribution of taurine, gamma-aminobutyric acid, and inorganic ions between plasma and erythrocytes in flounder (*Platichthys flesus*) at different plasma osmolalities. *Comp. Biochem. Physiol. 55A:*173–177.

Galloway, J. N., Likens, G. E., and Edgerton, E. S. (1976) Acid precipitation in the northeastern United States: pH and acidity. *Science 194:*722–724.

Gordon, M. S., Schmidt-Nielsen, K., and Kelly, H. M. (1961) Osmotic regulation in the crab-eating frog (*Rana cancrivora*). *J. Exp. Biol. 38:*659–678.

Gordon, M. S., and Tucker, V. A. (1965) Osmotic regulation in the tadpoles of the crab-eating frog (*Rana cancrivora*). *J. Exp. Biol. 42:*437–445.

Greger, R., Schlatter, E., and Gögelein, H. (1986) Sodium chloride secretion in rectal gland of dogfish, *Squalus acanthias. News Physiol. Sci. 1:*134–136.

Gross, W. J., Lasiewski, R. C., Dennis, M., and Rudy, P., Jr. (1966) Salt and water balance in selected crabs of Madagascar. *Comp. Biochem. Physiol. 17:*641–660.

Hutchinson, G. E. (1967) *A Treatise on Limnology,* vol. II, *Introduction to Lake Biology and the Limnoplankton.* New York: Wiley. 1115 pp.

Karnaky, K. J. (1986) Structure and function of the chloride cell of *Fundulus heteroclitus* and other teleosts. *Am. Zool. 26:*209–224.

Karnaky, K. J., Jr., and Kinter, W. B. (1977) Killifish opercular skin: A flat epithelium with a high density of chloride cells. *J. Exp. Zool. 199:*355–364.

Keys, A. B. (1933) The mechanism of adaptation to varying salinity in the common eel and the general problem of osmotic regulation in fishes. *Proc. R. Soc. Lond. B. 112:*184–199.

Keys, A. B., and Willmer, E. N. (1932) "Chloride

secreting cells'' in the gills of fishes, with special reference to the common eel. *J. Physiol. 76:*368–378.

King, J. R. (1957) Comments on the theory of indirect calorimetry as applied to birds. *Northwest Sci. 31:*155–169.

Kon, S. K., and Cowie, A. T. (1961) *Milk: The Mammary Gland and Its Secretion,* vol. II. New York: Academic Press. 423 pp.

Kooyman, G. L. (1963) Milk analysis of the kangaroo rat, *Dipodomys merriami. Science 142:*1467–1468.

Kooyman, G. L., and Drabek, C. M. (1968) Observations on milk, blood, and urine constituents of the Weddell seal. *Physiol. Zool. 41:*187–194.

Krogh, A. (1937) Osmotic regulation in fresh water fishes by active absorption of chloride ions. *Z. Vergl. Physiol. 24:*656–666.

Krogh, A. (1939) *Osmotic Regulation in Aquatic Animals.* Cambridge: Cambridge University Press. 242 pp. Reprinted by Dover Publications, New York, 1965.

Leivestad, H., and Muniz, I. P. (1976) Fish kill at low pH in a Norwegian River. *Nature, Lond. 259:*391–392.

Livingstone, D. A. (1963) Chemical composition of rivers and lakes. In *Data of Geochemistry,* 6th ed. (M. Fleischer, ed.). Geological Survey Professional Paper 440, Chapter G. Washington, D.C.: GPO. 64 pp.

Locket, N. A. (1980) Some advances in coelacanth biology: Review lecture. *Proc. R. Soc. Lond. B. 208:*265–307.

Loveridge, J. P. (1970) Observations on nitrogenous excretion and water relations of *Chiromantis xerampelina* (Amphibia, Anura). *Arnoldia 5:*1–6.

Machin, J. (1964) The evaporation of water from *Helix aspersa.* 1. The nature of the evaporating surface. *J. Exp. Biol. 41:*759–769.

Maetz, J., Payan, P., and de Renzis, G. ((1976) Controversial aspects of ionic uptake in freshwater animals. In *Perspectives in Experimental Biology,* vol. 1, *Zoology* (P. S. Davies, ed.), pp. 77–92. Oxford: Pergamon Press.

Mayer, N. (1969) Adaptation de *Rana esculenta* à des milieux variés: Etude speciale de l'excrétion rénale de l'eau et des électrolytes au cours des changements de milieux. *Comp. Biochem. Physiol. 29:*27–50.

Motais, R., and Maetz, J. (1965) Comparaison des échanges de sodium chez un téléostéen euryhalin (le flet) et un téléostéen sténohalin (le serran) en eau de mer: Importance rélative du tube digestif et de la branchie dans ces échanges. *C. R. Acad. Sci. Paris 261:*532–535.

Noble-Nesbitt, J. (1977) Absorption of water vapour by *Thermobia domestica* and other insects. In *Comparative Physiology: Water, Ions and Fluid Mechanics* (K. Schmidt-Nielsen, L. Bolis, and S. H. P. Maddrell, eds.), pp. 53–66. London: Cambridge University Press.

O'Donnell, M. J. (1982) Hydrophilic cuticle: The basis for water vapour absorption by the desert burrowing cockroach, *Arenivaga investigata. J. Exp. Biol. 99:*43–60.

Overrein, L. N., Seip, H. M., and Tollan, A. (1980) *Acid Precipitation: Effects on Forest and Fish.* Final report of the SNSF Project 1972–1980. Oslo, Norway: SNSF Project. 175 pp.

Phillips, J. E., Bradley, T. J., and Maddrell, S. H. P. (1977) Mechanisms of ionic and osmotic regulation in saline-water mosquito larvae. In *Comparative Physiology: Water, Ions and Fluid Mechanics* (K. Schmidt-Nielsen, L. Bolis, and S. H. P. Maddrell, eds.), pp. 151–171. London: Cambridge University Press.

Pierce, S. K., Jr., and Greenberg, M. J. (1973) The initiation and control of free amino acid regulation of cell volume in salinity-stressed marine bivalves. *J. Exp. Biol. 59:*435–440.

Pilson, M. E. Q. (1970) Water balance in sea lions. *Physiol. Zool. 43:*257–269.

Potts, W. T. W. (1954) The energetics of osmotic regulation in brackish- and fresh-water animals. *J. Exp. Biol. 31:*618–630.

Potts, W. T. W., and Evans, D. H. (1967) Sodium and chloride balance in the killifish *Fundulus heteroclitus. Biol. Bull. 133:*411–425.

Potts, W. T. W., and Parry, G. (1964) *Osmotic and Ionic Regulation in Animals.* Oxford: Pergamon Press. 423 pp.

Ramsay, J. A. (1935) Methods of measuring the evaporation of water from animals. *J. Exp. Biol. 12:*355–372.

Ramsay, J. A. (1949) The osmotic relations of the earthworm. *J. Exp. Biol. 26:*46–56.

Robertson, J. D. (1954) The chemical composition of the blood of some aquatic chordates, including members of the Tunicata, Cyclostomata and Osteichthyes. *J. Exp. Biol. 31:*424–442.

Robertson, J. D. (1957) Osmotic and ionic regulation

n aquatic invertebrates. In *Recent Advances in Inverte-brate Physiology* (B. T. Scheer, ed.), pp. 229–246. Eugene: University of Oregon Press.

Rudolph, D. (1982) Site, process and mechanism of active uptake of water vapour from the atmosphere in the Psocoptera. *J. Insect Physiol. 28*:205–212.

Ruibal, R. (1962) The adaptive value of bladder water in the toad, *Bufo cognatus. Physiol. Zool. 35*:218–223.

Schmidt-Nielsen, K. (1960) The salt-secreting gland of marine birds. *Circulation 21*:955–967.

Schmidt-Nielsen, K. (1963) Osmotic regulation in higher vertebrates. *Harvey Lect. 58*:53–95.

Schmidt-Nielsen, K. (1964) *Desert Animals: Physio-logical Problems of Heat and Water.* Oxford: Claren-don Press. 277 pp. Reprinted by Dover Publications, New York, 1979.

Schmidt-Nielsen, K. (1965) Physiology of salt glands. In *Funktionelle und morphologische Organisation der Zelle: Sekretion und Exkretion*, pp. 269–288. Berlin: Springer-Verlag.

Schmidt-Nielsen, K. (1969) The neglected interface: The biology of water as a liquid gas system. *Q. Rev. Biophys. 2*:283–304.

Schmidt-Nielsen, K. (1972) *How Animals Work.* Cambridge: Cambridge University Press. 114 pp.

Schmidt-Nielsen, K., and Fänge, R. (1958) Salt glands in marine reptiles. *Nature, Lond. 182*:783–785.

Schmidt-Nielsen, K., Hainsworth, F. R., and Mur-rish, D. E. (1970) Countercurrent heat exchange in the respiratory passages: Effect on water and heat balance. *Respir. Physiol. 9*:263–276.

Schmidt-Nielsen, K., and Kim, Y. T. (1964) The ef-fect of salt intake on the size and function of the salt gland of ducks. *Auk 81*:160–172.

Schmidt-Nielsen, K., and Lee, P. (1962) Kidney function in the crab-eating frog (*Rana cancrivora*). *J. Exp. Biol. 39*:167–177.

Schmidt-Nielsen, K., Schroter, R. C., and Shkol-nik, A. (1981) Desaturation of exhaled air in camels. *Proc. R. Soc. Lond. B 211*:305–319.

Schmidt-Nielsen, K., Taylor, C. R., and Shkolnik, A. (1971) Desert snails: Problems of heat, water and food. *J. Exp. Biol. 55*:385–398.

Shoemaker, V. H. (1988) Physiological ecology of amphibians in arid environments. *J. Arid Environ. 14*:145–153.

Shoemaker, V. H., and McClanahan, L. L., Jr.

(1975) Evaporative water loss, nitrogen excretion and osmoregulation in Phyllomedusine frogs. *J. Comp. Physiol. 100*:331–345.

Siversten, E. (1935) Ueber die chemische Zusammen-setzung von Robbenmilch. *Magazin for Naturvidenska-berne 75*:183–185.

Smith, H. W. (1959) *From Fish to Philosopher.* Sum-mit, N.J.: CIBA. 304 pp. Reprinted by Doubleday, Garden City, N.Y., 1961.

Somero, G. N. (1986) From dogfish to dogs: Trime-thylamines protect proteins from urea. *News Physiol. Sci. 1*:9–12.

Steinhorn, I., Assaf, G., Gat, J. R., Nishry, A., Nissenbaum, A., Stiller, M., Beyth, M., Neev, D., Garber, R., Friedman, G. M., and Weiss, W. (1979) The Dead Sea: Deepening of the mixolimnion signifies the overture to overturn of the water column. *Science 206*:55–57.

Sverdrup, H. U., Johnson, M. W., and Fleming, R. H. (1942) *The Oceans: Their Physics, Chemistry, and General Biology.* Englewood Cliffs, N.J.: Pren-tice-Hall. 1087 pp.

Taplin, L. E., and Grigg, G. C. (1981) Salt glands in the tongue of the estuarine crocodile *Crocodylus poro-sus. Science 212*:1045–1057.

Taplin, L. E., Grigg, G. C., Harlow, P., Ellis, T. M., and Dunson, W. A. (1982) Lingual salt glands in *Crocodylus acutus* and *C. johnstoni* and their ab-sence from *Alligator mississipiensis* and *Caiman cro-codilus. J. Comp. Physiol. 149*:43–47.

Templeton, J. R. (1967) Nasal salt gland excretion and adjustment to sodium loading in the lizard, *Cteno-saura pectinata. Copeia 1967*:136–140.

Thesleff, S., and Schmidt-Nielsen, K. (1962) Os-motic tolerance of the muscles of the crab-eating frog *Rana cancrivora. J. Cell. Comp. Physiol. 59*:31–34.

Thorson, T. B., Cowan, C. M., and Watson, D. E. (1967) *Potamotrygon* spp.: Elasmobranchs with low urea content. *Science 158*:375–377.

Thorson, T. B., Watson, D. E., and Cowan, C. M. (1966) The status of the freshwater shark of Lake Nica-ragua. *Copeia 1966*:385–402.

Tracy, C. R. (1976) A model of the dynamic ex-changes of water and energy between a terrestrial am-phibian and its environment. *Ecol. Mongr. 46*:293–326.

Ussing, H. H., and Zerahn, K. (1951) Active trans-port of sodium as the source of electric current in the short-circuited isolated frog skin. *Acta Physiol. Scand.* 23:110–127.

Wall, B. J., and Oschman, J. L. (1970) Water and solute uptake by rectal pads of *Periplaneta americana*. *Am. J. Physiol. 218*:1208–1215.

Wigglesworth, V. B. (1945) Transpiration through the cuticle of insects. *J. Exp. Biol. 21:* 97–114.

Yancey, P. H., and Somero, G. N. (1980) Methylamine osmoregulatory solutes of elasmobranch fishes counteract urea inhibition of enzymes. *J. Exp. Zool.* 212:205–213.

ADDITIONAL READING

American Society of Zoologists (1984) Cellular mechanisms of ion regulation in arthropods: Symposium. *Am. Zool.* 24:137–284.

Bentley, P. J. (1976) Osmoregulation. In *Biology of the Reptilia*, vol. 5, *Physiology A.* (C. Gans and W. R. Dawson, eds.), pp. 365–412. London: Academic Press.

Burton, R. F. (1973) The significance of ionic concentrations in the internal media of animals. *Biol. Rev.* 48:195–231.

Committee of Atmospheric Transport and Chemical Transformation, Environmental Studies Board (1983) *Acid Deposition: Atmospheric Processes in Eastern North America*. Washington, D.C.: National Academy Press. 375 pp.

Dunson, W. A. (1975) Salt and water balance in sea snakes. In *The Biology of Sea Snakes* (W. A. Dunson, ed.), pp. 329–353. Baltimore: University Park Press.

Edney, E. B. (1977) *Water Balance in Land Arthropods*. Berlin: Springer-Verlag. 282 pp.

Gupta, B. L., Moreton, R. B., Oschman, J. L., and Wall, B. J. (eds.) (1977) *Transport of Ions and Water in Animals*. London: Academic Press. 817 pp.

Hadley, N. F. (1982) Cuticle ultrastructure with respect to the lipid waterproofing barrier. *J. Exp. Zool.* 222:239–248.

Keynes, R. C. (1971) A discussion on active transport of salts and water in living tissues. *Philos. Trans. R. Soc. Lond. (Biol.)* 262:83–342.

Krogh, A. (1965) *Osmotic Regulation in Aquatic Animals*. New York: Dover. 242 pp. Republication of 1st ed. (1939).

Lahlou, B. (ed.) (1980) *Epithelial Transport in the Lower Vertebrates*. New York: Cambridge University Press. 366 pp.

Machin, J., O'Donnell, M. J., and Coutchié, P. A. (1982) Mechanisms of water vapor absorption in insects. *J. Exp. Zool.* 222:309–320.

Peaker, M., and Linzell, J. L. (1975) *Salt Glands in Birds and Reptiles*. Cambridge: Cambridge University Press. 307 pp.

Peaker, M., Vernon, R. G., and Knight, C. H. (eds.) (1984) *Physiological Strategies in Lactation*. Symposia of the Zoological Society of London, no. 51. London: Academic Press. 432 pp.

Potts, W. T. W., and Parry, G. (1964) *Osmotic and Ionic Regulation in Animals*. Oxford: Pergamon Press. 423 pp.

Rankin, J. C., and Davenport, J. (1981) *Animal Osmoregulation*. New York: Wiley. 202 pp.

Rudolph, D., and Knülle W. (1982) Novel uptake systems for atmospheric water vapor among insects. *J. Exp. Zool.* 222:321–333.

Shoemaker, V. H., and Nagy, K. A. (1977) Osmoregulation in amphibians and reptiles. *Annu. Rev. Physiol.* 39:449–471.

Skadhauge, E. (1981) *Osmoregulation in Birds*. Berlin: Springer-Verlag. 203 pp.

Smith, H. W. (1961) *From Fish to Philosopher*. Garden City, N.Y.: Doubleday.

Somero, G. N. (1986) Protons, osmolytes, and fitness of internal milieu for protein function. *Am. J. Physiol.* 251:R197–R213.

Yancey, P. H., Clark, M. E., Hand, S. C., Bowlus, R. D., and Somero, G. N. (1982) Living with water stress: Evolution of osmolyte systems. *Science* 217:1214–1222.

CHAPTER NINE

Excretion

In the preceding chapter we saw that excretory organs have important roles in osmoregulation, that is, in the maintenance of the desired water and solute concentrations in the organism. This chapter will discuss how excretory organs work.

The functions of an excretory organ are all related to one basic principle: To maintain a constant internal environment, any material an organism takes in must be balanced by an equal amount removed. This, in turn, requires that the excretory functions must have a variable capacity that can be adjusted to remove judiciously controlled amounts of each of a tremendous variety of different substances.

The major functions of excretory systems are:

1. Maintenance of proper *concentrations of solutes*
2. Maintenance of proper *body volume* (water content)
3. Removal of *metabolic end products*
4. Removal of *foreign substances* or their metabolic products

The first two functions are necessary for osmotic regulation and were repeatedly brought up in Chapter 8. One major metabolic end product, carbon dioxide, is removed primarily by the respiratory organs; most other metabolic end products are removed by the excretory organs. Foreign substances may be removed either unchanged or after modifications that renders them harmless (detoxification) or more easily excreted.

There is a great variety of excretory organs. In spite of the morphological variety, the basic principles of the excretory processes can be reduced to two major functional domains, *ultrafiltration* and *active transport*.

ORGANS OF EXCRETION

Basic processes

There is a wide variety of excretory organs, but there are, in principle, only two basic processes responsible for the formation of the excreted fluid: ultrafiltration and active transport.

In *ultrafiltration,* pressure forces a fluid through a semipermeable membrane that withholds protein and similar large molecules but allows water and small molecular solutes, such as salts, sugars, and amino acids, to pass (see Appendix E).

Active transport is the movement of solute against its electrochemical gradient by processes requiring the expenditure of metabolic energy. If active transport is directed from the animal into the lumen of the excretory organ or organelle, we call it an *active secretion.* If the active transport is in the opposite direction, from the lumen back into the animal, we speak about an *active reabsorption.*

In those excretory organs where the initial fluid is formed by ultrafiltration, this fluid is modified as it passes through the canals of the excretory organ, certain substances being removed from the filtrate by active reabsorption and others added to the ultrafiltrate by active secretion. Thus, active transport is often, if not always, superimposed on the filtration system.

Functional types

In spite of the great variety of morphological structure and anatomical location, excretory organs can be classified into a relatively small number of functional types. Some excretory organs are generalized, or unspecialized, and can, in a general sense, be regarded as kidneys and their excretory product as urine. Other excretory organs have more specialized roles in that they carry out one particular function. Some examples will help:

Generalized excretory organs
  Contractile vacuoles of protozoans
  Invertebrate nephridial organs
  Malpighian tubules of insects
  Vertebrate kidneys

Specialized excretory organs
  Gills (crustaceans, fish)
  Rectal glands (elasmobranchs)
  Salt glands (reptiles, birds)
  Liver (vertebrates)
  Intestine (insects)

The vertebrate kidney is the best known of the generalized excretory organs. It has been studied extensively, and its function is well understood. The contractile vacuole is only a part of a cell (i.e., an organelle) and is so small that study of the fluid it produces has been extremely difficult; nevertheless, a few facts about its function are known. The Malpighian tubules of insects have been carefully studied in a few species and are quite well understood.

The excretory organs of other invertebrates, however, are less well understood, and we have only approximate ideas about how some of them work. For some we do not even know whether a secretory process or ultrafiltration is the initial step in the formation of the excreted fluid.

Several specialized excretory organs were discussed in the preceding chapter on osmotic regulation. However, no mention was made of the vertebrate liver. One of its excretory functions is as follows. Red blood cells that have reached their normal life span (in humans ca. 100 days) are sequestered from circulation and broken down; the porphyrin of the hemoglobin molecule is transformed by the liver to compounds known as bile pigments. These are excreted in the bile, discharged into the intestine, and finally eliminated with the feces. Thus, in addition to its many other functions, the liver is an excretory organ specialized for porphyrin excretion. It also plays a major role in the metabolism and detoxification of a wide variety of foreign substances and may have other minor roles in excretion.

Classification

The function of various excretory organs is easier to discuss if we first classify the organs into groups or types.* The major types and their distribution in the animal kingdom are approximately as follows:

None demonstrated
    Coelenterates
    Echinoderms

Contractile vacuoles
    Protozoans
    Sponges

Nephridial organs
    Protonephridium, closed
        Platyhelminthes
        Aschelminthes
    Metanephridium, open-ended
        Annelids
    Nephridium
        Molluscs

Antennal gland (green gland)
    Crustaceans

Malpighian tubules
    Insects

Kidneys
    Vertebrates

No specific excretory organs have been identified in coelenterates and echinoderms. This is cu-

---

*This functional grouping of the renal organs is widely used, but is not in accord with the morphological grouping of excretory organs of Goodrich (1945), whose classification is based on whether the excretory organ is derived from a tube that develops from the outside and inward (a *nephridium*) or from the inside (the coelom) and outward (a *coelomoduct*). The renal organs of molluscs, arthropods, and vertebrates are derived from coelomoducts; those of other invertebrates (except protozoans, sponges, coelenterates, and echinoderms) are nephridia. Goodrich also recognizes a complex variety of mixed renal organs. The renal organs of insects, the *Malpighian tubules*, arise from the posterior part of the gut and do not fit into the Goodrich classification.

rious, for fresh-water coelenterates are distinctly hypertonic to the medium, and they undoubtedly gain water by osmotic influx. How the excess water is eliminated is not known. Echinoderms, on the other hand, have no problem of osmoregulation, for echinoderms do not occur in fresh water and the marine forms are always isosmotic with sea water.

CONTRACTILE VACUOLES

Two animal groups have contractile vacuoles: protozoans and sponges. It seems that all fresh-water protozoans have contractile vacuoles. Whether contractile vacuoles are present in all marine forms is more uncertain, but they are found in at least some marine ciliates. The occurrence of contractile vacuoles in fresh-water sponges, although previously doubted, has been confirmed beyond doubt.

Because a fresh-water form is always hyperosmotic to the medium in which it lives, and its surface is permeable to water, it will tend to swell as a result of osmotic inflow of water. Consequently, it must continually bail out the water. It must not only eliminate excess water, but also replace lost solutes, presumably by active uptake of salts from the external medium. Estimates of the water permeability of the large ameba *Chaos chaos* indicate that the calculated osmotic influx of water is in good agreement with the observed volume of fluid eliminated by the contractile vacuole. This confirms the widely held opinion that the primary function of the contractile vacuole is in osmotic and volume regulation.

Microscopic observation of the contractile vacuole in fresh-water protozoans reveals continuous cyclic changes. It collects fluid and gradually increases in volume until it reaches a critical size. It then suddenly expels its contents to the outside and decreases in size, whereupon it begins to enlarge again to repeat the cycle.

The lumen of the contractile vacuole of an ameba is surrounded by a single thin membrane. Sur-

FIGURE 9.1 The contractile vacuole of *Amoeba proteus* is enclosed in a membrane and surrounded by a layer of tiny vesicles that are filled with fluid and appear to empty into the vacuole. Around this structure is a layer of mitochondria, which presumably provide the energy for the secretory process. [Based on Patterson 1980]

FIGURE 9.2 Rate of fluid elimination by the contractile vacuole of *Amoeba lacerata* in relation to the concentration of the medium. The amebas were tested in the solution in which they were grown. [Hopkins 1946]

rounding this membrane is a thick layer, the *spongiome* (0.5 to 2μm thick) densely packed with small vesicles, each some 0.02 to 0.2 μm in diameter. Around this layer of small vesicles is a layer of mitochondria, which presumably provide the energy required for the osmotic work of producing the hypotonic contents of the vacuole. In electron micrographs it appears that the small vesicles empty into the contractile vacuole by fusion of their membranes (Figure 9.1).

The role of the contractile vacuole complex in osmotic regulation has been well demonstrated in the euryhaline ameba, *Amoeba lacerata*. This ameba is originally a fresh-water organism, but it has a high salt tolerance and can eventually be adapted to 50% sea water. The rate of emptying of the contractile vacuole of this ameba, when it is adapted to various salt concentrations, varies inversely with the osmotic concentration of the medium (Figure 9.2).

Apparently, the contractile vacuole complex works to eliminate water as fast as it enters by osmotic inflow, for as the concentration of the medium increases, the amount of entering water decreases. In a marine habitat, where the inside and outside osmotic concentrations must be assumed to be nearly equal, contractile vacuoles (in those organisms where they have been observed) empty at a very low rate. In these cases we must assume that they are not primarily organelles of osmoregulation, but carry out other excretory functions.

If the primary function of the contractile vacuole of fresh-water protozoans is to remove water, the contents of the vacuole should be hypotonic to the remainder of the cell. This is indeed the case. Minute samples of fluid withdrawn from the contractile

TABLE 9.1 Solute concentrations in cytoplasm and in the contractile vacuole of a fresh-water ameba. Mean volume of the vacuole was about 0.2 nl. [Riddick 1968]

Concentration	Medium	Cytoplasm	Vacuole	Ratio: vacuole/cytoplasm
Osmotic (mOsm liter^{-1})	<2	117	51	0.44
Na$^+$ (mmol liter^{-1})	0.2	0.60	19.9	33
K$^+$ (mmol liter^{-1})	0.1	31	4.6	0.15

vacuole have an osmotic concentration less than half that of the cytoplasm, yet several times as high as the surrounding medium's (B. Schmidt-Nielsen and Schrauger 1963).

The contractile vacuole can eliminate a hypotonic fluid and serve to remove water. Nevertheless, because the eliminated fluid has a higher osmotic concentration than the medium, solutes are continuously lost. It is therefore necessary to postulate that the ameba can take up the needed solutes, presumably by active transport directly from the medium.

How can the vacuole increase in volume and yet contain a fluid less concentrated than the cytoplasm? This could be explained in several ways. One is active water transport into the vacuole. However, for a number of reasons this is an unlikely hypothesis. Another alternative is that the vacuole originally contains an isotonic fluid from which osmotically active substances are withdrawn before the fluid is discharged to the surface. This proposal is contrary to the observation that the fluid concentration is hypotonic and relatively constant in composition throughout the growth of the vacuole.

Information about the composition of the vacuolar fluid permits us to propose a third mechanism. Table 9.1 shows that the osmotic concentration of the vacuolar fluid is about half that of the cytoplasm, but more than 25 times as high as the surrounding medium's. The sodium concentration in the vacuolar fluid is relatively high – in fact, 33

times as high as the cytoplasmic concentration of sodium. The potassium concentration in the vacuolar fluid, on the other hand, is relatively low, and substantially below the potassium concentration in the cytoplasm. The sum of sodium and potassium in the vacuolar fluid is nearly 25 mmol per liter, and if the anion is chloride, almost the total osmotic concentration of the fluid (51 mOsm per liter) would be accounted for.

The most likely mechanism for formation of the contractile vacuole is as follows. The small vesicles that surround the contractile vacuole are initially filled with a fluid isotonic with the cytoplasm. The vesicles then pump sodium into this fluid by active transport and remove potassium, also by active transport, in such a manner that removal of potassium exceeds sodium accumulation. The membrane of the vesicles must be relatively impermeable to water to permit the formation of a vesicle fluid hypotonic to the cytoplasm.

If the hypotonic vesicles now fuse and open into the contractile vacuole, as the electron micrographs indicate, the contractile vacuole becomes a receptacle for fluid produced by the vesicles, the necessary energy for the osmotic work being provided by the layer of mitochondria adjacent to the vesicles. Because sodium is continuously lost through the activity of the contractile vacuole, we must assume that it is replaced by active uptake at the cell surface (Riddick 1968).

We can summarize the evidence that contractile vacuoles eliminate water to offset osmotic water

FIGURE 9.3 The excretory system of a planarian is a widely branched system. The excreted fluid is initially formed at the solenocytes or flame cells, and then passed down the nephridial ducts and discharged through excretory pores.

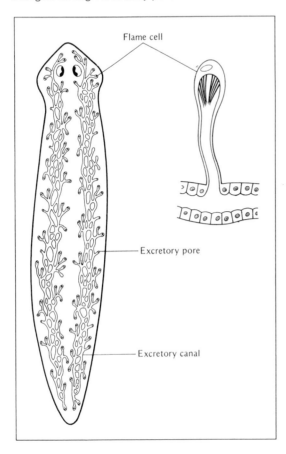

influx as follows: (1) contractile vacuoles are widespread in fresh-water organisms and are often absent from marine forms; (2) contractile vacuoles occur mainly in organisms that lack rigid external cell walls and therefore swell because of water influx; (3) vacuolar output decreases when the osmotic concentration of the medium is increased, thus reducing water influx; (4) contractile vacuole

activity is reduced by metabolic inhibitors. In the absence of evidence to the contrary, we can safely maintain this interpretation.

INVERTEBRATE EXCRETORY ORGANS

Protonephridia and metanephridia

True organs of excretion are found only in those animal phyla that show bilateral symmetry. The most common type, widely distributed among invertebrates, is a simple or branching tube that opens to the outside through a pore (nephridial pore). There are two major types: the *protonephridium,* whose internal end is closed and terminates blindly, and the *metanephridium,* which connects to the body cavity through a funnel-like structure called a nephridiostome or nephrostome.

Protonephridia occur mainly in animals that lack a true body cavity (coelom). An animal may have two or more protonephridia, which are often extensively branched. The closed ends terminate in enlarged bulblike structures, each with a hollow lumen and one or several long cilia inside (Figure 9.3). If there is a single cilium, the terminal cell is called a *solenocyte;* if there are numerous (often several dozen) cilia projecting into a lumen, the structure is called a *flame cell* because the tuft of cilia, as it undulates, has some resemblance to the flickering flame of a candle. No information is available about functional differences between flame cells and solenocytes.

The metanephridia are characteristically unbranched, and the inner ends open through funnels into the coelomic cavity. Metanephridia are found only in animals with a coelom, but the reverse is not true; some animals with a coelom have protonephridia, which otherwise are characteristic of acoelomate and pseudocoelomate animals (Figure 9.4).

The function of the metanephridium (often called simply nephridium) is reasonably clear, primarily

FIGURE 9.4 Diagram of the front end of a primitive annelid in longitudinal section, showing relations of nephridia and coelomoducts to segmental coelomic cavities (C). On left side, protonephridia (Pn) with solenocytes (Sol); on right side, metanephridia (Mn) with funnel-shaped nephrostomes (Nst). The gonadal products (G) empty through the coelomoducts (Cd) at the nephrostome. [Goodrich 1945]

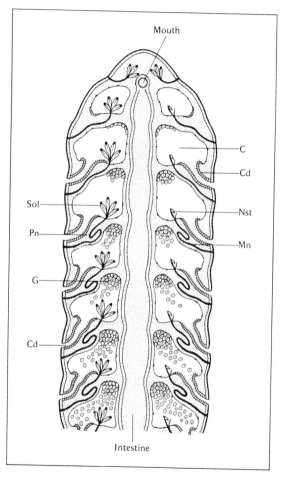

thanks to studies by J. A. Ramsay, who succeeded in removing minute samples of fluid from various parts of the earthworm nephridium. The results corroborate the following view.

Fluid from the coelom drains into the nephridium through the funnel-shaped nephrostome, and as it passes down through the extensively looped duct, its composition is modified. When originally entering the nephridium the fluid is isotonic, but salt is withdrawn in the terminal parts of the organ, and a dilute urine is discharged (Ramsay 1949). (The importance of a dilute urine for the osmoregulation of the earthworm was discussed in Chapter 8). The metanephridium therefore functions as a filtration–reabsorption kidney, in which an initial fluid is formed by ultrafiltration and is modified as it passes through a uriniferous tubule.

How the protonephridium functions is more uncertain, for it has been assumed that the closed end is not a suitable structure for ultrafiltration. Furthermore, the closed end is located in tissue between the body cells, rather than in contact with coelomic fluid, as in the case of metanephridia. The function of solenocytes and flame cells is obscure, although it has been suggested that their constant beating could produce sufficient negative pressure to cause ultrafiltration. Direct evidence for this hypothesis may be difficult to obtain because of the small size of these structures.

There is evidence, however, that the protonephridium of a rotifer, *Asplanchna*, functions on the basis of filtration and reabsorption. *Asplanchna* has body fluids hypertonic to the medium and produces a dilute urine; if the animal is moved to a more dilute medium, more dilute urine is formed. This shows at least that the protonephridium is involved in osmoregulation and the excretion of water (Braun et al. 1966).

The most important question concerning the protonephridium is whether ultrafiltration takes place. The generally accepted method for demonstrating ultrafiltration is to inject the substance inulin into the body and observe whether it appears in the urine. Inulin (which will be further discussed in connection with the vertebrate kidney) is a soluble polysaccharide with a molecular weight of about 5000. Inulin is not metabolized in the body,

FLAME CELL The flame cell of the excretory organ of a larval liver fluke (*Fasciola hepatica*). The photograph (*a*) shows, in cross section, that the "flame" (F) is a densely packed bundle of cilia. As also indicated in the drawings (*b* and *c*), the flame is surrounded by a thin wall (W) of pillarlike rods. The initial urine is presumably formed by ultrafiltration through this wall. [Courtesy of G. Kümmel, Freie Universität Berlin]

and it can appear in the urine only if filtration takes place. It is never excreted by cellular transport or secretion, for it is inert to all known active transport processes.

When injected into *Asplanchna*, inulin does appear in the urine, and this suggests a filtration process. Furthermore, the concentration of inulin in the urine of *Asplanchna* is higher than in the body, indicating that water was reabsorbed before the urine was discharged to the exterior. The wall of the flame cell of *Asplanchna* is extremely thin and seems suitable for ultrafiltration (Pontin 1964).

Obviously, other protonephridia might function differently but since in *Asplanchna* there is an ini-

tial filtration, the need to distinguish functionally among the several types of nephridia seems less important.

The molluscan kidney

The major groups of molluscs are cephalopods (octopus and squid), bivalves (clams, etc.), and gastropods (snails). Octopus and squid are strictly marine, but clams and snails occur in both sea and fresh water, and snails include terrestrial species as well.

Renal function has been studied in representatives of all these groups. In all, an initial fluid is formed by ultrafiltration of the blood. This fluid contains the same solutes as are present in the blood, except the proteins, and in virtually identical concentrations.* The ultrafiltrate therefore contains not only substances to be excreted, but also valuable substances such as glucose and amino acids. It would be very wasteful to discharge the ultrafiltrate to the outside, were not these valuable substances reabsorbed before the urinary fluid is eliminated.

In addition to ultrafiltration followed by selective reabsorption, there is also active secretion of certain substances, which are added to the urinary fluid in specific portions of the kidney. Two compounds that are actively secreted in a wide variety of kidneys (including vertebrate kidneys), are *paraaminohippuric acid* (often abbreviated *PAH*) and the dye *phenol red* (phenolsulfonphthalein). Phenol red has the advantage that, because of its color, its presence is readily observed and its concentration is easily measured. Phenol red and PAH will be mentioned again in the discussion of the vertebrate kidney because of their importance in demonstrating secretory processes.

One aspect of molluscan excretion seems rather peculiar: The function of the two kidneys is not always identical. For example, the following information concerns the abalone, *Haliotis*. Both kidneys have an initial ultrafiltration, for inulin appears in the urine from both kidneys, and its concentration is the same in blood, in pericardial fluid, and in the final urine from both kidneys. This indicates that the volume of urine is the same as the filtered volume and that water is not reabsorbed to any appreciable extent.

A marine animal, which is isotonic with sea water, has no major problem of water regulation. It is therefore reasonable that it does not reabsorb water in its kidney. It must, however, regulate its ionic composition and excrete metabolic products, and in this regard the abalone's two kidneys differ

in function, for PAH and phenol red are actively secreted primarily by the right kidney, while glucose reabsorption seems to take place primarily in the left kidney (Harrison 1962).

The antennal gland of crustaceans

The renal organ of crustaceans is the *antennal gland* or *green gland*. The paired glands are located in the head. Each consists of an initial sac, a long coiled excretory tubule, and a bladder. The opening is an excretory pore near the base of the antennae. Hence the name, antennal gland.

Urine is formed in the antennal gland by filtration and reabsorption, with tubular secretion added. Ultrafiltration can be demonstrated by injection of inulin, which then appears in the urine. The lobster, a typical marine animal, produces urine with the same inulin concentration as the blood (i.e., the urine/blood concentration ratio is 1.0), which shows that water is not reabsorbed. Injected PAH and phenol red, however, are found in higher concentrations in the urine than in the blood. This shows that they are secreted into the urine, and that, in addition to the filtration–reabsorption process, secretion is involved in urine formation (Burger 1957).

In contrast to the lobster, the common shore crab *Carcinus* reabsorbs water from the ultrafiltrate. *Carcinus* is a good osmoregulator and can penetrate into brackish water. The inulin concentration in its urine may exceed that in the blood by several-fold and this must be the result of water reabsorption from the initial filtrate. The most likely explanation for the water movement is active reabsorption of sodium followed by passive reabsorption of water. The result is a urine with a lower sodium concentration than the blood, and this is of obvious value to an animal that penetrates into a dilute environment where it will have difficulty maintaining high blood concentrations (Riegel and Lockwood 1961).

In marine crustaceans the antennal gland serves to retain potassium and calcium and to eliminate magnesium and sulfate. The urine magnesium concentration varies a great deal, but tends to increase

---

*The small concentration differences between blood and ultrafiltrate are attributable to the Donnan effect of the blood proteins (see Appendix E).

FIGURE 9.5 The relation between blood, urine, and medium concentrations in two species of amphipod. Top: the fresh-water species, *Gammarus pulex,* can form a highly dilute urine. Bottom: the closely related brackish-water species, *G. duebeni,* is unable to produce urine more dilute than the medium. Triangles = blood concentration. Circles = urine concentration. [Lockwood 1961]

substantially when crabs are moved to more concentrated sea water. It appears that magnesium is actively transported into the urine while the urine is in the bladder and that its concentration increases with the time that urine is held in the bladder. This may explain the large variations in the magnesium concentration in the urine and the difficulties in showing a close relationship to the concentration of the external medium (Gross and Capen 1966).

A comparison of the fresh-water amphipod *Gammarus pulex* and a close relative, the brackish-water *Gammarus duebeni,* reveals an interesting difference. The fresh-water *G. pulex* produces urine that is very dilute and may even be more dilute than the medium (Figure 9.5). This means that the animal can readily eliminate excess water that enters because of osmotic water inflow. The brackish-water species, *G. duebeni,* can also tolerate fresh water, but its urine concentration is closer to the blood concentration. This animal is unable to form a very dilute urine, which means that additional energy must be expended to compensate for the urinary loss of solutes. In fresh water this species survives, but is unable to compete successfully with *G. pulex.*

The Malpighian tubules of insects

For many invertebrate excretory organs, the question of whether the initial urine is formed by ultrafiltration is unresolved. For the excretory organs of insects, however, the answer seems to be unequivocally negative. This conclusion is supported by the fact that inulin, considered the most reliable indicator of ultrafiltration, when injected into an insect does not appear in the urine, thus indicating the absence of ultrafiltration (Ramsay and Riegel 1961). How, then, is insect urine formed?

The excretory system of insects consists of tubules, known as *Malpighian tubules,* which may number anywhere from two to several hundred. Each tubule opens into the intestine between the midgut and the hindgut; the other end is blind and, in most insects, lies in the hemocoele (Figure 9.6). Some insects, however, notably beetles that feed on dry substances (e.g., the mealworm, *Tenebrio*), have a particular arrangement associated with a remarkable ability to withdraw water from the excrement. In these, the blind end of the tubule lies closely associated with the rectum, the entire structure being surrounded by a membrane (the *perirectal membrane*). The space formed by this membrane is filled with fluid (*perirectal fluid*), which surrounds both the Malpighian tubule and the rectal epithelium, but is separated from the general hemolymph.

The Malpighian tubule functions as follows. Po-

FIGURE 9.6 The excretory system of the bug *Rhodnius*. Only one of the four Malpighian tubules is shown in full. The Malpighian tubules have two distinct parts: upper and lower. They empty into the rectum at its junction with the midgut. [Wigglesworth 1931]

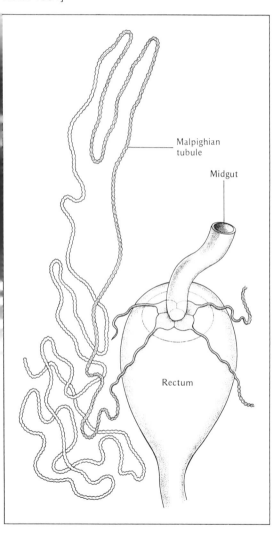

Malpighian tubule

Midgut

Rectum

tassium is actively secreted into the lumen of the tubule and water follows passively, due to osmotic forces. As a result, a copious amount of a potassium-rich fluid is formed in the tubule. This fluid is isotonic with blood, but has a very different composition. This is in striking contrast to the mammalian kidney, where the initial urinary fluid is an ultrafiltrate of the blood plasma.

Within the Malpighian tubule the fluid is changed by secretory as well as reabsorptive processes. The fluid then enters the hindgut. In the hindgut solutes and much of the water are reabsorbed, the uric acid (which entered the fluid as water-soluble potassium urate) precipitates. This facilitates the further withdrawal of water, for the precipitated uric acid does not contribute to the osmotic activity of the rectal contents. What remains in the rectum is eventually expelled as mixed urine and feces.

An insect that lives on fresh vegetation eats food with a high water content and excretes copious amounts of liquid urine. An insect that lives on dry food can produce very dry excreta and therefore may lose virtually no water in feces and urine. The mealworm, the larva of the flour beetle (*Tenebrio molitor*), is a good example. It spends its entire life cycle in dry flour and yet has all the water it needs.

The mealworm has the characteristic perirectal membrane described above, and it can produce excreta so dry that they absorb water from air with 90% relative humidity (Ramsay 1964). Let us see how the mechanism of water withdrawal can be explained. The osmotic concentration of the fluid in the perirectal space can be very high, its freezing point depression ($\triangle T$) reaching as much as 8 °C. This is very much greater than the concentration in the hemolymph ($\triangle T = 0.7$ to 1.4 °C). The difference is more pronounced if the mealworm has been kept at particularly low humidity.

The rectal complex of the mealworm could function in either of two ways: (1) active transport of water from the lumen of the rectal complex into the hemolymph, or (2) active transport of a solute (presumably potassium chloride) from the hemolymph into the perirectal space, a high concentration in this space in turn accounting for the osmotic withdrawal of water from the rectal lumen.

The second of these alternatives for water reabsorption is, in principle, the same as Curran's three-

FIGURE 9.7 Model of fluid movements in the rectal wall of the roach *Periplaneta americana*. Movement of solute is shown by black arrows; of water by open arrows. They are shown separately for reasons of clarity, although both occur in the same intracellular sinus. Solute is pumped actively into the intracellular sinus, both from the rectal lumen and from the fluid flowing into the subepithelial sinus. The high osmotic concentration in the intracellular sinus causes water to enter osmotically, and as a result there is a bulk flow of water and solutes toward the subepithelial sinus. The system is similar to Curran's three-compartment model (Figure 8.14). [Oschman and Wall 1969]

compartment theory for fluid transport (Chapter 8). The study of cockroaches has given further support for this hypothesis. When cockroaches are provided with water, they excrete a dilute urine; if deprived of water, they produce dry fecal pellets and their rectal contents are highly hyperosmotic to the hemolymph. A detailed analysis of osmotic concentrations in the rectal complex suggested the mechanism for water reabsorption shown in diagrammatic form in Figure 9.7.

In conclusion, it appears (1) that the excretory system of insects operates without any initial ultrafiltration; (2) that it is based on a primary secretion of potassium into the Malpighian tubules, followed by passive movement of water; and (3) that water as well as solutes are withdrawn in the hindgut and the rectal complex. There is ample evidence that the movement of water is based on a primary solute transport.

VERTEBRATE KIDNEYS

The kidneys of all vertebrates – fish, amphibians, reptiles, birds, and mammals – are similar in that they function on the *filtration–reabsorption principle* with tubular secretion added. A few teleost fish differ from this general pattern; they lack the ultrafiltration mechanism and depend entirely on a secretory-type kidney.

What are the advantages and disadvantages of a filtration mechanism? The initial ultrafiltrate contains all the compounds present in the blood, except substances of large molecular size, such as proteins. Many of the filtered compounds are valuable and should not be lost. Reabsorption mechanisms must therefore be present so that compounds such as glucose, amino acids, and vitamins are conserved.

A filtration–reabsorption kidney can process large volumes of fluid, and often more than 99% of the filtered volume is reabsorbed and less than 1% is excreted as urine. It might seem more advantageous to design a kidney that functions by tubular secretion only, for this would save a great deal of the work of reabsorption. We know that this solution in the evolutionary sense is feasible, for some teleost fish manage this way, and, as we have seen, the insect kidney also lacks ultrafiltration.

Moreover, there is one unique and important consequence of ultrafiltration: Any substance that has been filtered remains in the urine unless it is reabsorbed. An organism that encounters ''new'' substances to be excreted can, with the aid of a filtration kidney, eliminate these without having to develop a specialized secretory mechanism for each particular new substance it may meet. This gives an organism much greater freedom in exploring new environments, changing food habits, and so on. A secretory kidney is in this sense much more restrictive. It is in accord with this viewpoint that, among vertebrates, secretory-type kidneys are found primarily in a few marine fish that live in a stable and ''conservative'' environment.

All vertebrates can produce urine that is isotonic with or hypotonic to the blood, but only birds and mammals can produce urine more concentrated than the body fluids. In fresh water a dilute urine serves to bail out excess water while withholding solutes. Animals in sea water cannot use a dilute or isotonic urine to eliminate excess salt, and they have accessory organs for salt excretion, e.g., gills, rectal glands, salt glands.

Marine mammals have kidneys with exceptional concentrating ability, and they manage their salt problems by renal excretion. For land mammals the ability to produce a concentrated urine is of great importance in their water balance. The fact that birds and reptiles excrete uric acid (rather than urea) enables them to produce a semisolid urine that puts only moderate demands on the use of water for excretion (see under Nitrogen Excretion, later in this chapter).

Structure

To understand how the kidney functions, we must know its structure (Figure 9.8). All vertebrate kidneys consist of a large number of units, *nephrons.* A small fish may have only a few dozen nephrons in its kidneys; a large mammal may have several million in each kidney. Each nephron begins with a *Malpighian body,* in which ultrafiltration of the blood plasma takes place. A small artery leads into each Malpighian body, where it splits up into a bundle or tuft of capillaries, the *glomerulus.* It is through the walls of the glomerular capillaries that fluid is forced out by the blood pressure; this fluid, the glomerular filtrate, enters the tubule that leads from the Malpighian body. Within the tubule the fluid is modified, both by tubular reabsorption and by tubular secretion, to form the final urine.

The tubule can be divided into two parts: the first or *proximal tubule* in which many solutes such as salt and glucose as well as water are reabsorbed, and a *distal tubule* that continues the process of changing the tubular fluid into urine. The distal tubules join to form collecting ducts, and these empty the urine into the renal pelvis. From here, the urine passes through the ureters to the urinary bladder, from which, at intervals, it is discharged to the exterior.

In the nephron of mammals the proximal and distal segments of the tubule are separated by a characteristic thin segment that forms a loop, like a hairpin, known as *Henle's loop.* A similar structure, although not as well developed, is found in the bird kidney. The loop is the particular structure in the nephron that is responsible for the formation of a urine that is more concentrated than blood plasma. The loop is missing in fish, amphibians, and reptiles, and these can produce urine no more concentrated than the blood plasma.

FIGURE 9.8 Diagram of a mammalian kidney. The kidney contains a large number, up to several million, of single nephrons. Only one nephron is indicated in this diagram and is shown enlarged to the right. The outer layer of the kidney, the *cortex*, contains the *Malpighian bodies* and the proximal and the distal convoluted tubules. The capillary network within the Malpighian body is known as the *glomer-* *ulus*. The inner portion, the *medulla*, contains *Henle's loops* and *collecting ducts*.

The urine is initially formed by ultrafiltration in the Malpighian bodies. The filtered fluid is modified and greatly reduced in volume as it passes down the renal tubule and into the collecting ducts. These empty the urine into the renal *pelvis*, from where it is conveyed via the *ureter* to the bladder.

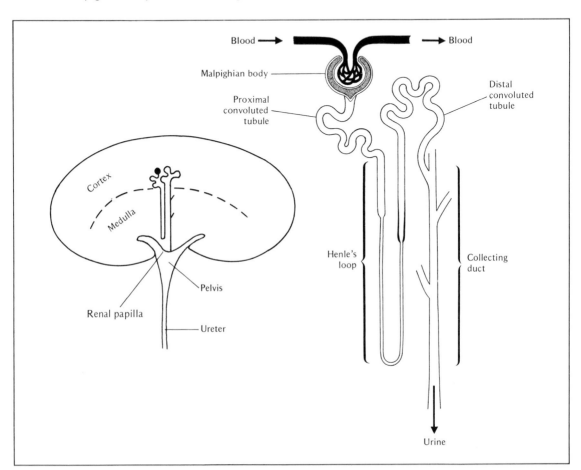

How renal function is studied

The methods used to study renal function can give a surprising amount of information without any direct experimentation on the kidney itself. From the intact animal, we can obtain information about the amount of fluid formed by ultrafiltration, the amount of fluid reabsorbed in the renal tubule, the process of tubular secretion, and even the rate of blood flow to the kidney. The methods are routinely used, without undue hazard or discomfort, to study kidney function in humans. We should therefore understand the principles involved.

Much information about renal function has also been obtained by direct experimentation on the kidney. This is especially true of studies of the mechanism responsible for formation of a concentrated urine. Through so-called micropuncture methods, in which fine pipettes are inserted into various segments of the nephron to withdraw minute samples of fluid, information has been gathered about the function of various parts of the nephron. Such methods require extensive surgery and are not suitable for use on humans.

Filtration

Ultrafiltration, which takes place in the Malpighian body, is caused by the blood pressure, which forces fluid out through the thin walls of the capillaries. Because the wall does not permit proteins to pass, the plasma proteins are withheld, and only substances of low molecular weight are filtered with the water.

For ultrafiltration to take place, the blood pressure must exceed the osmotic pressure of the blood proteins (the colloidal osmotic pressure). If the artery to a dog kidney is gradually constricted with a clamp so that the blood pressure in the kidney is decreased, urine production ceases when the blood pressure in the kidney falls to equal the colloidal osmotic pressure. This was shown many years ago by Starling. When the clamp is gradually loosened again so the blood pressure in the capillaries is increased, filtration begins again when the blood pressure just exceeds the colloidal osmotic pressure, and the higher the excess pressure, the higher the filtration rate (Starling 1899).

The Malpighian bodies of the mammalian kidney are located deep within the organ, but in the amphibian kidney they are visible at the surface. The distinguished American renal physiologist A. N. Richards succeeded in inserting micropipettes into the Malpighian bodies of amphibian kidneys, and he showed that the concentrations of chloride, glucose, urea, and phosphate, as well as

the total osmotic pressure in the filtrate, were the same as in the blood, except for the minor differences caused by the Donnan effect of the proteins. These experiments gave the essential proof that the initial urine is indeed an ultrafiltrate of the blood (Richards 1935).

The total amount of fluid formed by ultrafiltration can be determined after an injection of the polysaccharide inulin, without exposing the kidney or performing any other drastic manipulation. Inulin has a molecular weight of about 5000 and therefore passes through the capillary wall, which only withholds molecules larger than about 70 000 molecular weight. Inulin is not metabolized and is neither reabsorbed nor secreted by the renal tubule; all the inulin found in the final urine is therefore derived only from the filtration process.

Let us write this fact as a simple equation:

inulin in filtrate = inulin in urine

The amount of inulin in the filtrate equals the volume of filtrate ($\dot{V}_{\text{filtr}}$) times the inulin concentration in the filtrate (C_{filtr}); likewise, the amount of inulin in the urine equals the urine volume ($\dot{V}_{\text{urine}}$) times the inulin concentration in the urine (C_{urine}). The two amounts are equal, and we therefore have:

$$\dot{V}_{\text{filtr}} \times C_{\text{filtr}} = \dot{V}_{\text{urine}} \times C_{\text{urine}}$$

or

$$\dot{V}_{\text{filtr}} = \frac{\dot{V}_{\text{urine}} \times C_{\text{urine}}}{C_{\text{filtr}}}$$

The three terms on the right side can readily be determined. $\dot{V}_{\text{urine}}$ and C_{urine} are obtained by measuring the urine volume in a given period of time and analyzing the urine for its inulin content. C_{filtr} is equal to the inulin concentration in the blood plasma and can be determined from a blood sample. We thus have all the information necessary to calculate the glomerular filtration rate ($\dot{V}_{\text{filtr}}$).

Let us try an example. Say that we measure a urine flow of 1.3 ml per minute in a person and that the inulin concentration in the urine is 2% and the plasma inulin concentration is 0.02%.

$$\dot{V}_{filtr} = \frac{1.3 \text{ ml min}^{-1} \times 2\%}{0.02\%} = 130 \text{ ml min}^{-1}$$

This calculation gives a glomerular filtration rate of 130 ml per minute, which is a normal filtration rate for humans.

We can now see that in order to determine the filtration rate in an intact animal we need only (1) inject a suitable amount of inulin,* (2) collect the urine during a known period of time and measure its volume, and (3) determine the inulin concentration in the urine and in the plasma. It turns out that in humans the filtration rate is rather constant, about 130 ml per minute, and variations in urine volume are attributable primarily to variations in the amount of water reabsorbed. In many other animals, particularly in the lower vertebrates, the filtration rate can vary considerably, and in the frog the filtration rate can, without harm, cease completely for long periods.

A filtration rate in humans of 130 ml per minute, or 0.13 liter per minute means that in 1 hour 7.8 liters of filtrate is formed. This is roughly twice the total volume of blood plasma in humans. If glucose, amino acids, and other important substances in this large volume of filtrate were lost, the filtration process would soon deplete the organism of these valuable compounds. The saving process is tubular reabsorption.

Tubular reabsorption

One important substance that is reabsorbed in the proximal tubule is glucose, and it can serve as an example to illustrate the general principle of tubular reabsorption. Reabsorption of glucose is an active transport, and normally all filtered glucose is reabsorbed.

If the amount of glucose in the plasma is increased above the normal level, the reabsorption mechanism is presented with larger than usual

*Usually, inulin is infused continuously throughout an experiment. In this way the plasma concentration of inulin is maintained constant, instead of declining continuously during the observation period.

amounts of glucose, and if the amount exceeds the capacity of the transport mechanism, not all the glucose is reabsorbed and some remains in the urine. This is what happens to persons suffering from diabetes mellitus; their blood glucose readily increases to higher than normal levels, and when the filtered amount exceeds the capacity for reabsorption, glucose appears in the urine.

Let us examine the situation by reference to Figure 9.9. Glucose is continually filtered, and because the filtration rate remains constant, the filtered amount increases linearly with the plasma concentration of glucose. At low plasma glucose levels, all filtered glucose is reabsorbed, and none appears in the urine. This is the normal situation, for the plasma glucose value usually remains about 1 mg per milliliter. If plasma glucose increases, all filtered glucose is still reabsorbed up to the threshold value, about 2.3 mg per milliliter, and none is found in the urine.

Should the plasma glucose exceed the threshold value, however, reabsorption is incomplete and some glucose remains in the urine. In other words, the maximum capacity of the reabsorption mechanism has been reached. This amount is often designated the *tubular maximum* (T_m) for glucose. Above the threshold value, larger and larger amounts of glucose are found in the urine (Figure 9.9). The line for filtered glucose and the line for glucose in the urine run parallel; this means that the tubular maximum remains constant, as indicated by the constant distance between these two lines.

The tubular maximum for any other reabsorbed substance can be determined in a similar way by increasing the plasma concentration to above the threshold value for that substance, so that the reabsorption mechanism is saturated and the substance begins to appear in the urine.

Tubular secretion

In addition to filtration and reabsorption, a third process, tubular secretion, is important in urine formation. It was mentioned above that phenol red and PAH (para-aminohippuric acid) are secreted by

FIGURE 9.9 Excretion of glucose by the dog. Below a certain limit (2.3 mg glucose per milliliter plasma) no glucose appears in the urine; this is because all the filtered glucose has been removed by tubular reabsorption. Above this limit the reabsorption mechanism is fully saturated, and although it continues to work at full capacity, glucose appears in the urine in amounts that increase with further increases in the plasma glucose concentration. The amount reabsorbed remains constant above the saturation point and thus represents the tubular maximum (T_m) for glucose reabsorption. [Shannon and Fisher 1938]

FIGURE 9.10 Excretion of phenol red by the bullfrog. The amount filtered increases in proportion to the plasma concentration. In addition, phenol red is added to the urine by active tubular transport, increasing the amount in the urine. The amount added by tubular secretion remains constant at plasma concentrations above 0.05 mg per milliliter; this indicates that the tubular maximum for phenol red has been reached. [Forster 1940]

the renal tubules. Because these substances are also filtered, the amount appearing in the urine is the sum of the filtered and the secreted amounts.

If we inject one of these substances into the bloodstream, the total amount secreted will depend on the plasma concentration, as shown in Figure 9.10. The amount filtered increases linearly with the plasma concentration. At low plasma concentrations the amount added by tubular secretion exceeds the filtered amount. However, at a given plasma concentration, the transport mechanism becomes saturated. The total excretion now increases parallel to the filtered amount, and the difference between the total excretion and the filtered amount (the distance between the two parallel lines) gives

the amount added by tubular secretion. In other words, this amount is the tubular maximum for phenol red excretion.

Other substances that are foreign to the body are also eliminated by tubular secretion. Often this is the case with phenolic compounds and their detoxification products. One well-known substance rapidly eliminated by tubular secretion is penicillin. This disadvantage is counteracted by injection of sufficient amounts of penicillin to maintain an effective blood concentration.

Renal blood flow

The process of tubular secretion can, in a very interesting way, be used to determine how much blood flows to the kidney (i.e., to determine the renal blood flow). For this we can use a substance that is excreted by tubular secretion so readily that it is completely removed from the blood that passes through the kidney.

One such substance is para-aminohippuric acid

(PAH), which is removed so completely that none is present in the blood that leaves the kidney (unless its tubular maximum is exceeded). The method depends on the fact that the renal artery is the only blood supply to the mammalian kidney; all the blood first flows through the glomerular capillaries, and then on to the capillaries that surround the renal tubules.

We inject some PAH and determine the amount of PAH that appears in the urine in a given time. From the concentration of PAH in the blood plasma, we can calculate the volume of plasma that has flowed through the kidney in that period. In humans the renal plasma flow is normally about 0.7 liter per minute. If the hematocrit is 45%, this corresponds to a blood flow to the kidneys of 1.25 liters per minute. This is between one-quarter and one-fifth of the total cardiac output at rest (5 to 6 liters per minute). Thus, the kidneys of mammals receive a surprisingly large blood supply.

The same method cannot be used to determine arterial blood flow to the kidneys of lower vertebrates, because the kidneys in these animals have a different type of circulation and receive a double blood supply. In addition to receiving blood from the renal artery, these kidneys also receive venous blood from the posterior part of the body (called the renal portal system).

The importance of the renal portal circulation is not clear. It may be related to the fact that when a renal portal circulation is present urine production can continue even if glomerular filtration ceases. In the frog, for example, filtration may stop, but phenol red that arrives in the kidney via the renal portal system is still secreted by tubular secretory activity.

Excretion in fish

The kidneys of most teleosts consist of nephrons with typical vertebrate characteristics. Fresh-water fish are hypertonic to the medium, and excess water that enters the body due to the osmotic gradient is eliminated as a hypotonic urine.

Marine teleosts, which suffer a water shortage because they are hypotonic to the medium, have a low rate of urine production. They cannot produce a concentrated urine, and excess salts are excreted by the gills. To make up for the osmotic water loss, they drink sea water, and one major function of the kidneys is to excrete the divalent ions found in sea water, magnesium and sulfate.

Some marine teleosts have kidneys that lack glomeruli. This holds true for the goosefish (*Lophius piscatorius*), the toadfish (*Opsanus tau*), and some pipefish (*Syngnathus*). Aglomerular marine teleosts have been very important in the study of renal function, for they permit the analysis of secretory processes in a kidney where no filtration can take place. If these fish are injected with inulin, no inulin appears in the urine. The absence of inulin from the urine is an excellent functional test for the absence of a filtration mechanism.

Among the fish that lack glomeruli, and thus the initial filtration process in urine formation, are several Antarctic genera. For these fish it makes a great deal of sense to discard the filtration process. We saw in Chapter 6 that many Antarctic fish avoid freezing because they possess glycoproteins with antifreeze properties. The molecular weight of these glycoproteins ranges from a few thousand up to about 30 000. Because the gomerulus withholds proteins of about 70 000 molecular size, but permits smaller molecules to pass, the glycoproteins would rapidly be lost in a filtration kidney unless a suitable recovery mechanism were present. In any event, a mechanism for reabsorbing the glycoproteins would involve a considerable energy expenditure, which is avoided by never permitting these substances to enter the tubular fluid. It therefore appears that an aglomerular kidney is a great bonus to those fish that use antifreeze substances as a protection against damage from ice formation (Dobbs et al. 1974; Eastman et al. 1979).

It is peculiar that a few fresh-water pipefish also seem to lack glomeruli, but these fish and their renal function have not been studied in detail. A fresh-water fish should have a considerable osmotic inflow of water, and it is difficult to explain how a nonfiltering kidney can eliminate excess water. Obviously, these fish deserve further study.

The elasmobranchs have glomerular kidneys, and as we saw earlier, urea is essential for elasmobranchs to maintain their osmotic balance. Urea, being a small molecule (molecular weight = 60), is filtered in the glomerulus, but it is reabsorbed by the tubules. This means that there is, in the elasmobranch kidney, a specific mechanism for active transport of urea. Details of this mechanism are discussed later in this chapter.

Excretion in amphibians

The first unequivocal evidence for an initial ultrafiltration process in the kidney was obtained on frogs. This is because many of the glomeruli in the amphibian kidney are rather large and located so close to the surface of the kidney that it is possible to insert a fine micropipette and withdraw minute samples of fluid for analysis. Another advantage of experimentation on cold-blooded animals is that the kidney functions quite well at room temperature, as opposed to the mammalian kidney, which must be elaborately maintained at body temperature.

Most amphibians live at or near fresh water. When they are in water there is an osmotic influx of water, and to eliminate the water they produce a large volume of highly dilute urine. The inevitable loss of sodium is compensated for by cutaneous active uptake of sodium from the dilute medium.

In the frog kidney urea is eliminated, not only by glomerular filtration, but by tubular secretion as well (Marshall 1933). This has certain advantages, for the following reason. When a frog is exposed to dry air, it reduces its urine output (water loss) by reducing the glomerular filtration rate and increasing the tubular reabsorption of water. However, urea excretion may still remain high. In fact, of the total urea excreted by the kidney, as little as one-seventh may be filtered and the remainder added to the urine by tubular secretion (B. Schmidt-Nielsen and Forster 1954). This means that the excretion of urea can continue at a high rate even when glomerular filtration has almost ceased. Thus, in contrast to the situation in the elasmobranch, where urea is actively reabsorbed, in the amphibian kidney urea is actively secreted.

In Chapter 8, on osmoregulation, we saw that one amphibian, the crab-eating frog of Southeast Asia (*Rana cancrivora*), lives and seeks its food in full-strength sea water. Physiologically this frog resembles elasmobranchs in that it carries a high level of urea in the blood, thus achieving osmotic equilibrium with the surrounding sea water and yet retaining the relatively low salt concentrations characteristic of vertebrates in general. However, the crab-eating frog differs from sharks in the way its kidney handles urea.

The kidney of the crab-eating frog does not reabsorb urea, but when the animal is in sea water, very little water enters its body and a very low rate of urine flow combined with a cessation of urea secretion reduces the loss of urea. If the crab-eating frog is transferred to more dilute water, the osmotic influx of water increases, urine volume goes up, and more urea is lost. This reduces the osmotic difference between the frog and the medium, which in turn reduces the osmotic influx of water (Schmidt-Nielsen and Lee 1962).

Excretion in reptiles

The reptilian kidney conforms to the typical vertebrate pattern. It can produce dilute or isotonic urine, but it cannot produce urine more concentrated than the blood plasma. For aquatic fresh-water reptiles (e.g., crocodiles and aquatic turtles) this is adequate, for a dilute urine serves to eliminate excess water.

Marine reptiles have the opposite problem: water shortage and an overabundance of salt. Their kidneys cannot handle excess salts, which are eliminated instead by salt glands that secrete highly concentrated solutions of sodium chloride.

Terrestrial reptiles that live in dry habitats and have limited water supplies excrete their nitrogenous waste in the form of uric acid, which is highly insoluble and therefore precipitates in the urine. As

TABLE 9.2 The maximum concentrating ability of the kidney of various mammals is correlated with the normal habit of the animal, desert animals having the highest urine concentrations and fresh-water animals the lowest.

Animal	Urine maximum osmotic concentration (Osm liter^{-1})	Urine/plasma concentration ratio
Beaver[a]	0.52	2
Pig[a]	1.1	3
Man[b]	1.4	4
White rat[b]	2.9	9
Cat[b]	3.1	10
Kangaroo rat[b]	5.5	14
Sand rat[b]	6.3	17
Hopping mouse[c]	9.4	25

[a] B. Schmidt-Nielsen and O'Dell (1961).
[b] K. Schmidt-Nielsen (1964).
[c] MacMillen and Lee (1967).

a result, urine can be eliminated as a paste or as a semisolid pellet, requiring very little water.

The excretion of uric acid provides an extra dividend. Not only uric acid, but also its sodium and potassium salts, have very low solubilities. By incorporating Na^+ and K^+ in the precipitated fraction of the urine, these important ions can be eliminated without additional expense of water, as is required when they are excreted by the mammalian kidney or by the avian nasal salt gland (McNabb and McNabb 1975).

Many terrestrial reptiles do have nasal salt glands that eliminate excess sodium and potassium, but in desert reptiles that lack salt glands (e.g., *Amphibolurus*) and in terrestrial snakes, the uric acid excretion provides a mechanism for eliminating salts with a minimal loss of water (Minnich 1972).

Urine concentration in birds and mammals

Uniquely among vertebrates, birds and mammals can produce urine hyperosmotic to the blood plasma. In birds this ability is not very pro-

nounced; the maximum urine osmotic concentration can reach about two times the plasma concentration. The mammalian kidney, in contrast, can produce urine up to perhaps 25 times the plasma concentration. The highest urine concentrations are found in mammals from desert habitats; animals such as beavers, which always have a superabundance of water, have only a moderate concentrating ability (Table 9.2).

The mechanism responsible for formation of a concentrated urine in birds and mammals is based on an ingenious geometric arrangement of the renal tubule. It would seem that the production of a concentrated urine would require either reabsorption of water from the tubular fluid or active secretion of solute into the tubule. Neither is the case. The mechanism responsible for forming a concentrated fluid in the renal tubule is based on the reabsorption of sodium and chloride from the tubular fluid, which, paradoxically, in the end serves to produce a highly concentrated urine. The mechanism is found in birds and is developed to the highest perfection in those mammals that form the most concentrated urine.

The concentrating ability of the mammalian kidney is closely related to the length of Henle's loop, the hairpin loop formed by the thin segment interposed between the proximal and the distal tubules. Most mammalian kidneys have two types of nephrons (Figure 9.11). Some nephrons have long loops; others have short ones. Those animals that produce the most highly concentrated urine have only long-looped nephrons with loops so long that they extend into the renal papilla. Others, such as the beaver and the pig, have only short loops. These are the animals that cannot produce urine more than about twice as concentrated as the plasma.

The role of the loop structure in the formation of a concentrated urine has been clarified by the Swiss investigator Wirz and his collaborators. Investigators in Europe and the United States have added further evidence. The hypothesis proposed by Wirz was based on an initial active transport of sodium out of the renal tubule, but recent evidence indi-

MAMMALIAN KIDNEY The vascular system of the kidney of a desert rodent, the jird (*Meriones*). The arteries and capillaries have been injected with silicone rubber, and the glomeruli can be seen as numerous tiny, beadlike dots in the outer layer (the renal cortex). The renal papilla in the jird is longer than in rodents from less arid habitats; this enhances the concentrating ability of its kidney. [Courtesy of Lise Bankir, Institut National de la Santé et de la Recherche Médicale, Paris]

cates that the primary pump may be a chloride pump. For the understanding of the concentrating mechanism, this makes little difference: Whichever ion is transported actively is followed by the appropriate counter-ion, so we can speak about an initial active transport of NaCl.

The mechanism for the formation of a concentrated urine can best be understood by reference to Figure 9.12. The site of active reabsorption of sodium chloride is the ascending limb of the loop of Henle, specifically in the thick segment of the ascending loop. In this segment chloride is transported into the interstitium, followed by sodium as the counter-ion.

The next site to examine is the descending loop of Henle, which is permeable to water but impermeable to solutes. As the chloride pump has made the interstitium more concentrated, water is withdrawn from the descending limb of the loop, making its contents more concentrated. The tubular fluid that makes the hairpin turn has thus become more concentrated and is now presented to the pump; more chloride is pumped out; additional water is withdrawn; and so on. In this way the single effect of the ion pump is multiplied as more

water is withdrawn and more and more concentrated fluid is presented to the ascending limb.

The next question is: What happens to the water? The important structures in this regard are the capillary loops that run parallel to the loop of Henle. These capillaries, known as the *vasa recta,* are freely permeable to water and solutes. Their contents therefore display the same increased solute concentrations toward the hairpin turn as do the tubular fluid and the interstitium. However, the capillary wall is not permeable to protein, and the colloidal osmotic pressure of the plasma proteins therefore causes the withdrawal of water into the bloodstream, as it does in other capillaries. From there excess water is carried into the venous system. Consider the overall picture: If the kidney is to remove a concentrated fluid (i.e., urine) from the blood, the outgoing venous blood must, of necessity, end up more dilute than the incoming arterial blood.

The next area to consider is the distal portion of the tubule. The fluid that flows from the thick ascending segment into the distal tubule has had sodium chloride removed from it, and its concentration is less than that in the plasma. The sodium

DOG KIDNEY Close-up view of glomeruli from a dog kidney after arterial injection of silicone rubber. The glomerulus in the center of the photo shows the slightly thicker vessel leading into the glomerulus and the somewhat thinner vessel leaving it. The diameter of these vessels is about 15 to 20 μm; the diameter of the glomerulus is about 150 μm. [Courtesy of A. Clifford Barger, Harvard University]

FIGURE 9.11 Diagram showing two typical nephrons, one "long-looped" (located close to the border of the medulla and therefore called juxtamedullary) and one "short-looped" (or cortical) nephron. The long-looped nephron is paralleled by a loop formed by the blood capillary. The short-looped nephron is surrounded by a capillary network. Most mammalian kidneys contain a mixture of the two types of nephrons, but some species have only one or the other kind. [Smith 1951]

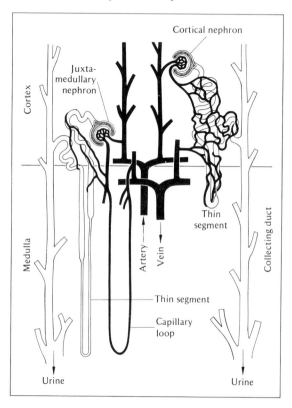

chloride that was removed in the thick segment remains around the hairpin loop, where it accumulates in high concentration.

As the dilute tubular fluid enters the collecting duct, it is low in salt but carries urea, which is an important factor in the system. The descending collecting duct is permeable to water and to urea. Because of the high osmotic concentration in the interstitium, water leaves the collecting duct, and as urea becomes more concentrated, it also diffuses out. This urea contributes significantly to the interstitial osmotic concentration. As the tubular fluid that flows in the descending collecting duct achieves osmotic equilibrium with the high concentration of the medullary interstitium, it becomes urine with a high solute concentration. The high concentration in the medulla results from urea derived from the collecting ducts and salt derived

FIGURE 9.12 Diagram of the concentrating mechanism in the loop of Henle in the mammalian kidney during the formation of concentrated urine. Active transport of chloride ion is indicated by heavy arrows; passive flux of water and urea by light arrows. [Kokko and Tisher 1976]

from the loops of Henle, and passive diffusion of water out of the collecting duct provides the mechanism for formation of a concentrated urine.

The high urea concentration in the tissue of the renal papilla raises the same question that we touched on in the discussion of elasmobranch osmoregulation – how can the cells in the papilla tolerate urea concentrations so high that they seriously perturb the structure of enzymes and other proteins? The answer is similar; in the cell of the papilla methylamines accumulate in a concentration of about one-half that of the urea concentration. Presumably this offsets fully the perturbing effect of urea on the intracellular proteins (Somero 1986).

Our model for the formation of a concentrated urine depends on highly selected permeabilities of the various segments of the nephron. In spite of the minute size of the tubules, it has been possible to carry out microperfusion of isolated fragments of loops of Henle, and the observed permeabilities are in accord with the requirements of the model (Kokko 1974).

The final question to be answered is: How can a dilute urine be formed? In water diuresis the tubular fluid that enters the collecting ducts from the distal convoluted tubule remains dilute. This results from a decreased permeability to water in the collecting ducts so that water is not withdrawn. The change in permeability to water is controlled by the antidiuretic hormone (ADH). In the absence of ADH the collecting ducts are impermeable to water, and a dilute urine is formed; in its presence the collecting ducts are water-permeable, water is withdrawn, and a concentrated urine is formed.

Countercurrent multiplier

The fluids in the descending and the ascending limbs flow in opposite directions; the flow is a countercurrent flow. In any one place along the loop the concentration difference between the ascending limb and the descending limb is moderate, but along the length of the loop this moderate concentration difference is additive (i.e., there is a multiplication effect). The entire system is therefore called a *countercurrent multiplier system.*

The final concentration that can be reached in this system should depend on the length of the loop; the longer the loop, the higher the final concentration. We now run into a paradox – a small rodent that can produce very high urine concentrations has loops that in absolute length are much shorter than the loops of a large animal that produces far lower concentrations. This apparent contradiction can be resolved if we consider the much higher mass-specific metabolic rates of small animals. Because the cells of the small animal are capable of more intense active transport, they can produce higher final concentrations for a given length of the loops (Greenwald and Stetson 1988).

The bird kidney differs from the mammalian kidney in that some nephrons have loops but others do not. It appears that the maximum concentrating ability of the bird kidney is related to the proportion of looped nephrons to nephrons without loops, rather than to the maximum length of the loops, as in mammals (Poulson 1965).

Many birds live under relatively dry conditions, eat dry food such as seeds, and have limited access to water. It is reasonable to ask why their kidneys cannot match the concentrating ability of mammalian kidneys. Perhaps the answer is to be found in the fact that birds excrete uric acid as an end product of their nitrogen metabolism.

If the bird kidney withdrew water to a much greater extent than it does, the highly insoluble uric acid would precipitate in the tubules and clog them, unless sufficient liquid remained to carry the precipitate along. Although bird urine contains precipitated uric acid, it must remain relatively liquid in the nephrons and in the ureter until it enters the cloaca, which is the very last portion of the intestine. Here urine and fecal material accumulate until, at intervals, they are emptied through the single exterior opening. After the urine has entered the cloaca, further water can be withdrawn before the cloacal contents are voided to the exterior.

Role of the cloaca

As the urine of a bird or a reptile enters the cloaca, it is liquid or semiliquid. If an animal needs to conserve water, however, the urine is dropped as a semisolid paste or pellet (i.e., water has been withdrawn). This reabsorption of water may, at least in part, take place outside the cloaca, for urine from the cloaca may migrate into the lower part of the intestine, where both salt and water are reabsorbed (Nechay et al. 1968; Skadhauge 1976, 1981).

What is the nature of the process of water withdrawal from the cloacal contents? It can be a solute-linked water transport, and, at least in lizards, it can be a passive process, resulting from the colloidal osmotic pressure of the plasma proteins.

In the lizard cloaca, water is absorbed until the force necessary to withdraw more water corresponds to the colloidal osmotic pressure of the plasma proteins; then further water absorption ceases. If the osmotic pressure of the plasma proteins is counterbalanced by placing within the cloaca a protein solution with the same colloidal osmotic concentration as the plasma, water reabsorption is prevented. Thus, it seems that in lizards a passive osmotic removal of water suffices for the production of a semidry urine within the cloaca. Nevertheless, it remains possible that an active transport of sodium or other solutes also takes place in the lizard cloaca, causing solute-linked water withdrawal (Murrish and Schmidt-Nielsen 1970).

When salt is withdrawn from the cloaca together with water, the animal will have difficulty eliminating excess salt. However, lizards (and many birds) can excrete excess salt via the nasal glands. Thus, lizards that live in a dry habitat can use the kidney primarily for the excretion of uric acid and lose a minimal amount of water, and use the nasal glands for the excretion of salts in a highly concentrated solution, again with only a small loss of water (Schmidt-Nielsen et al. 1963).

Regulation of urine concentration and volume

If an animal is in caloric balance, its body weight, and thus volume, remains remarkably constant. A major factor in this constancy is the accurate regulation of the rate of urine production. The total volume of urine depends on the need for water excretion. If there is an excess intake of water, a large volume of dilute urine is excreted; if there is a shortage of water in the organism, the urine volume decreases and its concentration increases toward the maximum possible.

In humans, the urine flow may vary more than 100-fold. When a man is deprived of water, urine production may be as low as 10 ml per hour, but it will not decrease further until the situation becomes critical and renal failure sets in. In the opposite case, when the water intake is high, urine flow may exceed 1 liter per hour.

One might think that by drinking more, urine

production could be increased further, but this is not possible. The maximum urine volume is determined by the amount of fluid filtered, which, in humans, remains quite constant and does not increase when we drink more. In the proximal tubule about 85% of the filtered volume of water is invariably withdrawn, leaving 15% of the filtered water as the maximum volume of urine that can be excreted.

The reason for the reabsorption of water in the proximal tubule is that major solutes such as glucose and sodium chloride are reabsorbed here. As solutes are removed, the tubular content becomes hyposmotic, and due to the osmotic effect, water follows passively through the tubular wall. Thus, in the proximal tubule the reabsorption of water is a direct and inevitable result of the reabsorption of glucose and sodium. This is called the *obligatory water reabsorption.*

The distal mechanism (Henle's loop and the collecting duct) is, as described above, the structure responsible for formation of a concentrated urine. The concentrating process is controlled by the *antidiuretic hormone (ADH)*, now usually called *vasopressin.* In the absence of this hormone, which is secreted from the neurohypophysis (posterior pituitary, see Chapter 12), a large volume of highly dilute urine is produced. Under the influence of vasopressin, more water is reabsorbed, urine volume decreases, and urine concentration increases.

Vasopressin acts by influencing the permeability of the collecting ducts. In its absence, the permeability of the collecting ducts to water is low. Thus, the dilute fluid that enters the collecting ducts from the distal tubules undergoes no further loss of water, even if the osmotic concentration in the surrounding tissues is high. With increasing amounts of vasopressin, the permeability of the collecting ducts increases. Thus, because of the high osmotic concentration in the tissue of the papilla (produced by the countercurrent multiplier mechanism), water is osmotically withdrawn from the fluid within the collecting ducts, leaving a small volume of concentrated urine.

This view of the role of the collecting duct has been verified by introducing filament-thin polyethylene tubes through the tip of the papilla in rats and withdrawing samples from various levels of the collecting duct (Hilger et al. 1958). When vasopressin is present in the blood, the collecting duct urine is in osmotic equilibrium with the fluid in the loop, but in the absence of vasopressin such equilibrium does not exist because of the lowered permeability of the collecting duct (Gottschalk 1961).

It may seem paradoxical that a hormone that causes the formation of a concentrated urine acts by increasing the permeability of the collecting ducts. Another paradoxical aspect is that the transport mechanism that leads to the formation of a concentrated urine is based on the withdrawal of sodium and chloride from the tubular fluid, rather than their addition. In the end, however, it is precisely this reabsorption which, in combination with the high permeability of the collecting ducts under the influence of vasopressin, causes the final urine to reach a maximum concentration.

It is interesting that the effect of vasopressin in increasing duct permeability is similar to its effect on frog skin, a much used model in studies of ion transport. In the frog skin the sodium transport takes place in the same direction as the sodium pump in the kidney tubule, by returning sodium to the organism. In both systems vasopressin increases the permeability to water. In the mammalian kidney it has a water-saving effect; in the frog skin the increased permeability leads to an increased osmotic influx of water.

NITROGEN EXCRETION

Most of the food that animals eat contains three major nutrient components: carbohydrates, fats, and proteins (plus smaller amounts of nucleic acids). When carbohydrates and fats are metabolized, they give rise to carbon dioxide and water as the only end products of oxidation. Proteins and nucleic acids also yield carbon dioxide and water, but in addition, the chemically bound nitrogen in these food-stuffs gives rise to the formation of some relatively simple nitrogen-containing excretory products. The three compounds of greatest interest are *ammonia, urea,* and *uric acid.*

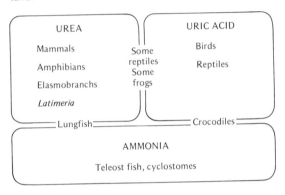

When amino acids are metabolized, the amino group ($-NH_2$) is removed by the process of deamination and forms ammonia (NH_3). The ammonia is excreted unchanged by many animals, in particular by aquatic invertebrates, but by some animals it is synthesized to urea and by others to uric acid before being excreted.

Animals are often grouped according to their main excretory products, as follows. Those that excrete mainly ammonia as the end product of protein metabolism are called *ammonotelic;* those that excrete mostly urea are *ureotelic;* and those that form mainly uric acid are *uricotelic.*

There is no clear correlation between the phylogenetic relationship of vertebrates and their major excretory products. The grouping in Figure 9.13 gives an overall picture, but there are many modifications. Crocodiles excrete mainly ammonia, as teleost fish do; frog tadpoles excrete ammonia (like fish), but adult frogs excrete urea. Some frogs, however, excrete uric acid (like reptiles). Some turtles excrete uric acid, and others excrete mainly urea or even ammonia, like teleost fish, which excrete mostly ammonia but also some urea. Mammals and birds seem to be clearly ureotelic and uricotelic, respectively, but other groups of vertebrates seem to have more exceptions than firm rules, and we shall soon examine some of them.

FIGURE 9.13 Different groups of vertebrates use different compounds as their major nitrogenous excretory product. There are many exceptions to the general pattern indicated in this diagram, most of them related to environmental factors rather than to phylogenetic relationships. See text for further details.

There are some other nitrogen-containing compounds that also need discussion. The nitrogen compounds contained in nucleic acids are either purines or pyrimidines, and they may be excreted as any of the degradation products listed in Table 9.3. The bulk of the excreted nitrogen compounds is derived from protein (amino acids) by deamination, and we shall discuss these first.

Other nitrogen-containing products are also excreted, some derived from the diet, others arising in metabolic pathways. For example, the urine of vertebrates normally contains a small amount of creatinine. This compound is normally formed by the muscles and is of particular interest because it can be used in studies of renal function in a manner similar to inulin. It is formed in the body at a relatively constant rate, and monitoring its excretion allows study of renal function without injection of a foreign substance. Creatinine is apparently derived from the metabolism of the high-energy phosphate compound, phosphocreatine.

Another group of nitrogen compounds frequently found in urine comes from detoxification

TABLE 9.3 Metabolic end products of the major groups of foodstuffs. Ammonia from protein metabolism may be excreted as such or may be synthesized into other N-containing excretory products; purines from nucleic acids may be excreted as such or as any of a number of degradation products, including ammonia.

Foodstuff	End product
Carbohydrate	$\rightarrow$ $CO_2 + H_2O$
Fat	$\rightarrow$ $CO_2 + H_2O$
Protein	$\rightarrow$ NH_3 $\longrightarrow$ Urea / $\longrightarrow$ Uric acid
Nucleic acids	$\rightarrow$ Purines + Pyrimidines
	Uric acid ↓ β-Amino acids ↓
	Allantoin ↓ NH₃
	Allantoic acid ↓
	Urea ↓
	NH₃

processes. For example, benzoic acid (present in certain plant foods), when combined with the amino acid glycine, forms the conjugation product hippuric acid (named after the Greek *hippos* = horse, because it was originally discovered in horse urine). Other phenolic compounds are detoxified in a similar manner and are then excreted in the urine. Hippuric acid and the closely related compound, para-aminohippuric acid (PAH), are important in studies of renal physiology because they are readily secreted by tubular activity and can be used to determine renal blood flow, as described earlier in this chapter.

One important aspect of nitrogen metabolism is the assumption that nitrogen gas never enters into the metabolism of animals. It has been universally accepted that only nitrogen-fixing bacteria can assimilate molecular nitrogen (N_2) into chemical compounds, and that only denitrifying bacteria have

metabolic pathways that lead to the formation of molecular nitrogen gas.*

Whether the nitrogen from protein metabolism is excreted as ammonia, urea, or uric acid is closely related to the normal habitat of the animal and the availability of water (Table 9.4). Ammonia is a highly toxic compound, even in low concentrations, and must therefore be removed rapidly, either to the exterior or by synthesis into less toxic compounds (urea or uric acid).

Ammonia

Most aquatic invertebrates excrete ammonia as the end product of protein metabolism. Because of its high solubility and small molecular size, ammonia diffuses extremely rapidly. It can be lost to a great extent through any surface in contact with water and need not be excreted by the kidney. In teleost fish most of the nitrogen is lost as ammonia from the gills: In the carp and goldfish the gills excrete 6 to 10 times as much nitrogen as the kidneys do, and only 10% of this is urea; the remaining 90% is ammonia (Smith 1929).

Urea

Urea is easily soluble in water and has a moderately low toxicity. The synthesis of urea in higher animals has been clarified by the British biochemist Hans Krebs, the same man whose name is attached to the Krebs cycle of oxidative energy metabolism (the tricarboxylic acid cycle).

The synthesis of urea begins with a condensation of ammonia and carbon dioxide with phosphate to form carbamyl phosphate, which enters a synthetic pathway to form citrulline, as shown in Figure 9.14. A second ammonia is added from the amino acid aspartic acid, leading to the formation

---

*There is convincing evidence that this is not true under all circumstances. Formation of gaseous nitrogen in mammals on a high-protein diet has been reported by several investigators (e.g., Costa et al. 1968, 1974). These findings are bound to upset some of our present ideas about protein metabolism and nitrogenous end products.

TABLE 9.4 Major nitrogen excretory products in various animal groups.

Animal	Major end product of protein metabolism	Adult habitat	Embryonic environment
Aquatic invertebrates	Ammonia	Aquatic	Aquatic
Teleost fish	Ammonia, some urea	Aquatic	Aquatic
Elasmobranchs	Urea	Aquatic	Aquatic
Crocodiles	Ammonia, some uric acid	Semiaquatic	Cleidoic egg[a]
Amphibians, larval	Ammonia	Aquatic	Aquatic
Amphibians, adult	Urea	Semiaquatic	Aquatic
Mammals	Urea	Terrestrial	Aquatic
Turtles	Urea and uric acid	Terrestrial	Cleidoic egg
Insects	Uric acid	Terrestrial	Cleidoic egg
Land gastropods	Uric acid	Terrestrial	Cleidoic egg
Lizards	Uric acid	Terrestrial	Cleidoic egg
Snakes	Uric acid	Terrestrial	Cleidoic egg
Birds	Uric acid	Terrestrial	Cleidoic egg

[a] The role of cleidoic eggs is discussed later in this chapter.

FIGURE 9.14 Urea is synthesized from ammonia and carbon dioxide by condensation with the amino acid ornithine. Through several more steps arginine is formed, which, with the aid of the enzyme arginase, splits off urea, forming ornithine, which can reenter the cycle.

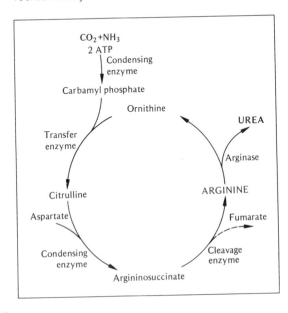

of the amino acid arginine. In the presence of the enzyme arginase, arginine is decomposed into urea and ornithine. This frees the ornithine for renewed synthesis of citrulline, repeating the entire cycle; the total pathway is therefore known as the *ornithine cycle* for urea synthesis.

The presence of arginase in an animal shows that the animal has the ability to produce urea and often indicates that urea is the major nitrogenous excretory product. This is not invariably so, however, for arginase can be present with the remaining pathway absent.

Urea in vertebrates

The vertebrates that excrete mainly urea and possess the ornithine cycle enzymes for urea synthesis were indicated in Figure 9.13. Some urea is excreted by teleost fish, and in elasmobranchs, amphibians, and mammals it is the main nitrogenous excretory product.

In elasmobranchs (sharks and rays), as well as in the crab-eating frog and in the coelacanth *Latimeria*, urea is retained and serves a major role in osmotic regulation. It is therefore a valuable met-

TABLE 9.5 Urea is actively transported by the kidney tubule of both shark (active reabsorption) and frog (active secretion). Three closely related compounds, however, are not transported alike by the two kinds of animals. This provides evidence that the cellular transport mechanism is not identical in the two kidneys. [B. Schmidt-Nielsen and Rabinowitz 1964]

Compound	Shark	Frog
Urea	Active	Active
Methylurea	Active	No
Thiourea	No	Active
Acetamide	Active	No

abolic product. Urea is filtered in the glomerulus of the elasmobranch kidney, but, because of its importance in osmoregulation, must not be lost in the urine. Accordingly, in elasmobranchs urea is recovered by active tubular reabsorption. In amphibians the situation is different. Urea is filtered, but in addition, a substantial amount is added to the urine by active tubular secretion.

Thus, both elasmobranchs and amphibians have active tubular transport of urea, but the transport is in the opposite direction in the two groups. Apparently, the mechanism of the pump is not metabolically identical, for some closely related urea derivatives are treated differently by the two animal groups (Table 9.5). This serves as an excellent example that a physiological function can arise independently in two groups and that it does not necessarily have to use an identical mechanism to achieve the same end (in this case the active transport of urea).

The crab-eating frog, which also retains urea for osmoregulation, does not show evidence of active tubular reabsorption of urea (Schmidt-Nielsen and Lee 1962). The crab-eating frog has a low rate of urine production, and the renal tubules are highly permeable to urea. Urea therefore diffuses from the tubular fluid back into the blood and appears in the urine in approximately the same concentration as in the blood. Because the urine volume is low when the frogs are in salt water, only small amounts of urea are lost in the urine.

If ordinary frogs have active tubular secretion of urea, why has the crab-eating frog not utilized this pump by simply reversing its direction? The question is not easily answered, but it appears that in general the direction of an active transport mechanism is a conservative physiological function not readily reversed. As we have seen, the frog skin and the mammalian kidney both retain the direction of active transport of sodium chloride from the exterior into the organism. In the mammalian kidney, however, the inward transport of sodium chloride from the tubular fluid back into the organism has been utilized in the countercurrent multiplier system in such a way that the end result nevertheless is the formation of a concentrated urine.

The common concept of urea excretion in the mammalian kidney is that urea is filtered in the glomerulus and then treated passively by the tubules, although some urea, because of its high diffusibility, reenters the blood by passive diffusion. There is, however, convincing evidence that urea is an important element in the function of the countercurrent multiplier system and that the excretory pattern for urea is an essential element in the function of the mammalian kidney.

Urea and amphibian metamorphosis

The tadpoles of frogs and toads excrete mostly ammonia; the adults excrete urea. At metamorphosis there is a well-defined changeover from ammonia to urea excretion in frog (*Rana temporaria*), toad (*Bufo bufo*), newt (*Triturus vulgaris*), and other amphibians. The South African frog *Xenopus,* however, which remains aquatic during adult life, continues to excrete ammonia as an adult.

The change to urea excretion at the onset of metamorphosis in the semiterrestrial amphibians is associated with a marked increase in the activities of all the liver enzymes of the ornithine cycle (Brown et al. 1959).

It is interesting that adult specimens of the aquatic *Xenopus,* if kept out of water for several weeks, accumulate urea in the blood and tissues. When *Xenopus* adults were kept out of water but in moist moss to prevent dehydration, the blood urea concentration increased between 10- and 20-fold, rising to nearly 100 mmol per liter. When the animals were returned to water, the excess urea was excreted (Balinsky et al. 1961). A group of *Xenopus* that was found naturally estivating in the mud near a dried pool had urea concentrations raised similarly by a factor of 15 to 20. Of the enzymes involved in urea synthesis, the enzyme carbamyl phosphate synthetase, responsible for the first synthetic step in Figure 9.14, was increased about sixfold, but the activities of the other enzymes of the cycle were unchanged. The synthesis of carbamyl phosphate may be the rate-limiting step in urea synthesis, and an increase in this enzyme is probably responsible for keeping the plasma ammonia low in animals out of water (Balinsky et al. 1967).

We shall return a little later to another striking variation in the nitrogen excretion in frogs. Some species that live in South America and southern Africa excrete mainly uric acid, which otherwise is the predominant and characteristic excretory product of reptiles and birds. This is again an example of plasticity and adaptability of physiological processes in response to the demands of the environment as animals invade new habitats.

Urea in the lungfish

The changes in the African lungfish, *Protopterus,* are analogous to those in amphibians. Normally, when the lungfish lives in water, it excretes mostly ammonia (and some urea), but when it estivates in a cocoon within the dried mud, it channels the entire waste nitrogen into urea, which accumulates in the blood and may reach concentrations as high as 3% (500 mmol per liter) at the end of 3 years of estivation (Smith 1959).

All five enzymes of the ornithine cycle are present in the liver of the African lungfish (Janssens and Cohen 1966). The levels of the two enzymes that are rate-limiting in urea synthesis are similar in the lungfish and in the tadpole of the frog *Rana*

catesbeiana, and considerably lower than the levels reported for adult frogs. This is consistent with the predominant excretion of ammonia by the lungfish when it is in water. It has been calculated, however, that the amounts of the ornithine cycle enzymes present in the liver of a nonestivating lungfish are sufficient to account for the accumulation of urea actually observed in estivating lungfish (Forster and Goldstein 1966).

In the Australian lungfish, *Neoceratodus,* the concentrations of the ornithine cycle enzymes are low. This is in accord with the life habits of the Australian lungfish, which uses its lung only as an accessory respiratory organ and cannot survive in air for any length of time (see Chapter 1). The rate of synthesis of urea by liver slices of the Australian lungfish is only one-hundredth of the rate in the African lungfish. This again is consistent with the completely aquatic nature of the Australian lungfish (Goldstein et al. 1967).

Uric acid

Excretion of uric acid prevails in insects, land snails, most reptiles, and birds. All these are typical terrestrial animals, and the formation of uric acid can be considered a successful adaptation to water conservation in a terrestrial habitat. Because uric acid and its salts are only slightly soluble in water (the solubility of uric acid is about 6 mg per liter water), the withdrawal of water from the urine causes uric acid and its salts to precipitate.

Uric acid in birds and insects

The semisolid white portion of bird droppings is urine and consists mostly of uric acid; very little water is used for excretion of the nitrogenous excretory products in these animals. Some insects have carried the reduction in urine water loss so far that they do not excrete the uric acid at all, but deposit it in various parts of the body, mainly in the fat body. In these forms, therefore, no water whatsoever is required for the elimination of nitrogenous end products.

It has been suggested that birds gain a further

advantage by using uric acid as their main excretory product. Because little water is needed for urine formation, uric acid excretion has been said to save weight for flying birds. This argument, however, is not convincing, for birds that have access to water, both fresh-water and marine species, often eliminate large quantities of liquid urine.

The cleidoic egg

It has been suggested by Joseph Needham that the difference between those higher vertebrates that form urea (mammals and amphibians) and those that form uric acid (reptiles and birds) is primarily correlated with their mode of reproduction. The amphibian egg develops in water, and the mammalian embryo develops in the liquid environment of the uterus, where waste products are transferred to the blood of the mother.

The embryonic development in reptiles and birds, on the other hand, takes place in a closed egg, a so-called *cleidoic** egg, where only gases are exchanged with the environment and all excretory products remain within the eggshell. In the cleidoic egg the embryo has a very limited water supply, and ammonia is, of course, too toxic to be tolerated in large quantities. If urea were produced, it would remain inside the egg and accumulate in solution. Uric acid, however, can be precipitated and thus essentially eliminated; this happens when it is deposited as crystals in the allantois, which thus serves as an embryonic urinary bladder.

Uric acid in reptiles

Lizards and snakes excrete mostly uric acid; many turtles excrete a mixture of uric acid and urea; and crocodiles excrete mainly ammonia (Cragg et al. 1961). This fits the generalization that nitrogen excretion is closely related to the availability of water in the environment.

Crocodiles and alligators excrete ammonia in the urine, where the principal cation is NH_4^+ and the

*Greek *kleistos* = closed, from *kleis* = key.

principal anion is HCO_3^- (Coulson and Hernandez 1955). It is possible that the presence of these ions in the urine permits an improved retention of sodium and chloride in these fresh-water animals, which, incidentally, also lose very little sodium and chloride in the feces.

There is little doubt that a close correlation exists between the habitats of turtles and their nitrogen excretion. Table 9.6 shows the composition of urine samples obtained from eight species of turtles in the London Zoo. The most aquatic species excrete considerable amounts of ammonia and urea and only traces of uric acid; the most terrestrial species excrete over half their nitrogen as uric acid.

There have been conflicting reports about whether turtles excrete mostly urea or uric acid. The reason is that not only do differences exist among species, but also different individuals of one species may excrete mainly uric acid, mainly urea, or a mixture of both (Khalil and Haggag 1955). One individual may even in the course of time change from one compound to the other. The fact that some precipitated uric acid may remain in the cloaca while the liquid portion of the urine is voided to the exterior makes it unreliable to determine the amount of uric acid formed by analyzing a single or a few urine samples. An incomplete emptying of the cloaca may give an entirely too low amount of uric acid, and an evacuation that includes precipitate accumulated over a period of time will give too large an amount.

The cause of the shift between urea and uric acid in the excretion of the tortoise *Testudo mauritanica* seems to be a direct function of temperature and hydration of the animal. Uric acid excretion increases when the water balance is unfavorable, but the mechanism that controls the shift in biochemical activity from urea to uric acid synthesis is not understood.

Exceptional frogs

It was mentioned in Chapter 8 that the African frog *Chiromantis xerampelina* loses water from the skin very slowly, at rates similar to those in rep-

TABLE 9.6 Partition of nitrogen in the urine of turtles (in percent of total nitrogen excretion). The most aquatic species excrete almost no uric acid, whereas this compound dominates in the most terrestrial species. [Moyle 1949]

Species	Habitat	Urine component[a]				
		Uric acid	Ammonia	Urea	Amino acids	Unaccounted for
Kinosternon subrubrum	Almost wholly aquatic	0.7	24.0	22.9	10.0	40.3
Pelusios derbianus	Almost wholly aquatic	4.5	18.5	24.4	20.6	27.2
Emys orbicularis	Semiaquatic; feeds on land in marshes	2.5	14.4	47.1	19.7	14.8
Kinixys erosa	Damp places; frequently enters water	4.2	6.1	61.0	13.7	15.2
Kinixys youngii	Drier than above	5.5	6.0	44.0	15.2	26.4
Testudo denticulata	Damp, swampy ground	6.7	6.0	29.1	15.6	32.1
Testudo graeca	Very dry, almost desert conditions	51.9	4.1	22.3	6.6	4.0
Testudo elegans	Very dry, almost desert conditions	56.1	6.2	8.5	13.1	12.0

[a] There were small amounts of allantoin, guanine, xanthine, and creatinine, and a variable amount not accounted for.

tiles. *Chiromantis* resembles reptiles also in that it excretes mainly uric acid, rather than urea as adult amphibians normally do. This is a sensational finding, for it challenges our commonly accepted view of nitrogen excretion in amphibians. There is no doubt about the accuracy of the report, for uric acid was determined in *Chiromantis* urine by means of an enzymatic method specific for uric acid, and uric acid was found to make up 60 to 75% of the dry weight of the urine (Loveridge 1970).

A South American frog *Phyllomedusa sauvagii* shows similar reptile-like traits. Its cutaneous water loss is of the same magnitude as from the dry skin of reptiles, and its urine contains large amounts of a semisolid urate precipitate (Shoemaker et al. 1972). The fraction of total nitrogen excreted as urate was found to be 80% in *Phyllomedusa*, and increased water intake did not change the rate of urate production. Thus, this species continues to produce mainly uric acid, even when water is freely available.

When the frog is in need of water conservation, the excretion of uric acid, rather than urea, becomes very important. It has been estimated that

this frog, if urea were its excretory product, would require about 60 ml water per day per kilogram body weight for urine formation. Instead, *P. sauvaggi*, because it excretes uric acid, uses as little as 3.8 ml water per day per kilogram body weight for urine formation (Shoemaker and McClanahan 1975).

Other aspects

Ammonia and urinary pH

The preceding discussion makes it appear that ammonia is excreted mainly by aquatic animals; but this is not completely true. Ammonia is normally found in the urine of terrestrial animals, where it serves to regulate the pH of the urine. If urine becomes acid due to the excretion of acid metabolic products, ammonia is added to neutralize the excess acid.

Excess acid is normally formed in protein metabolism, for the end product of oxidation of the sulfur-containing amino acid cysteine is sulfuric acid. The more acid the urine is, the more ammonia is added. The ammonia used to neutralize an

TABLE 9.7 Nitrogenous end products of purine metabolism in various animals. [Keilin 1959]

Animal	End product
Birds	Uric acid
Reptiles, terrestrial	Uric acid
Insects	Uric acid
Man, apes, dalmatian dog	Uric acid
Mammals (except those above)	Allantoin
Gastropod molluscs	Allantoin
Diptera	Allantoin
Some teleost fish	Allantoic acid
Amphibians	Urea
Teleosts	Urea
Elasmobranchs	Urea
Lamellibranch molluscs (fresh-water)	Urea
Lamellibranch molluscs (marine)	Ammonia
Most other aquatic invertebrates	Ammonia

acid urine is produced in the kidney and is derived from the amino acid glutamine. Glutaminase is found in the kidney, and its presence there serves the particular purpose of producing ammonia. The ammonia in mammalian urine, therefore, has no direct connection with the ammonia produced in the liver by the deamination of amino acids and in this sense should not be considered a normal end product of protein metabolism.

Nucleic acids and nitrogen excretion

Nucleic acids contain two groups of nitrogen compounds: *purines* (adenine and guanine) and *pyrimidines* (cytosine and thymine). In some animals purines are excreted as uric acid (which itself is a purine); in other animals the purine structure is degraded to a number of intermediaries or to ammonia, any one of which may be excreted.

The structure of the purines adenine and guanine is similar to that of uric acid: each contains one six-membered and one five-membered ring. The pyrimidines (cytosine and thymine), however, are single six-membered rings that contain two nitrogen atoms. In higher vertebrates the pyrimidines are degraded by opening this ring, forming one molecule of ammonia and one molecule of a β-

amino acid. These are then metabolized in the normal metabolic scheme by further deamination.

The metabolic degradation and excretion of purines have not been as carefully studied as the metabolism of protein nitrogen. The main features, however, are listed in Table 9.7. Birds, terrestrial reptiles, and insects degrade purines to uric acid and excrete this compound. These are the same animals that synthesize uric acid from amino nitrogen; it would obviously be meaningless for an animal to synthesize uric acid and also to have mechanisms for its degradation. We therefore cannot expect to find further breakdown of purines in animals in which uric acid is the end product of protein metabolism.

Among mammals, humans, apes, and the dalmatian dog form a special group: they excrete uric acid, whereas mammals in general excrete allantoin. Allantoin is formed from uric acid in a single step in the presence of the enzyme uricase. This enzyme is absent in man and apes. Because of its low solubility, uric acid is at times deposited in the human organism, causing swelling of the joints and the very painful disease gout. If humans had retained the enzyme uricase, gout would be unknown.

Although the dalmatian dog excretes uric acid in amounts much higher than other dogs, this is not the result of a metabolic defect. The dog's liver contains uricase and some allantoin is formed. However, the dalmatian has a kidney defect that prevents tubular reabsorption of uric acid (as occurs in other mammals, including humans); uric acid is therefore lost in the dog's urine faster than it can be converted to allantoin by the liver (Yü et al. 1960). There is considerable evidence that uric acid in the dalmatian dog, in addition to being filtered in the glomerulus, is also excreted by active tubular transport (Keilin 1959).

The most striking feature of nucleic acid metabolism is that the ''higher'' animals listed at the top of Table 9.7 completely lack the enzymes necessary to degrade the purines. Among the ''lower'' animals we find an increasing complexity in the biochemical and enzymatic systems for further degradation of the purines. Thus the "lowest" animals in this case possess the most complete enzyme systems.

Guanine

In spiders *guanine* is the major excretory product. It appears to be synthesized from amino nitrogen, although the complete pathway is unknown. Some spiders, notably the bird-eating tarantulas, after a meal excrete more than 90% of the total nitrogen as guanine (Peschen 1939). In the common garden spider, *Araneus diadematus,* the identification of guanine has been confirmed by a highly specific enzymatic method.

Guanine is also found rather widely in a variety of other animals. For example, the silvery sheen in fish scales is caused by a deposit of guanine crystals.

The swimbladder of fish, which contains gases that sometimes are under very high pressures (depending on the depth at which the fish lives), is made highly impermeable to diffusive loss of gases by guanine. The guanine is found as extremely thin crystals in multiple layers, each layer only about 0.02 μm thick. The low gas conductance of the swimbladder wall reduces the metabolic energy that otherwise would be needed to replace the gas lost by diffusion (see Chapter 10).

The garden snail *Helix* excretes guanine, but only to the extent of about 20% of the total purine excretion, the remainder being uric acid. It is possible that this fraction is derived from nucleic acid metabolism, while the remainder of the nitrogen is derived from protein metabolism.

Amino acids are not major end products of nitrogen metabolism, but they are found in small amounts in the urine of many animals. It would seem more economical for an animal to deaminate the amino acid, excrete the ammonia in the usual way, and use the resulting organic acid in energy metabolism. Because of the minor role excretion of amino acids plays, it will not be discussed further here.

The recapitulation theory

The nitrogen excretion of the developing chick embryo is said to change with time and go through a series of peaks, with first ammonia as the main end product, then urea, and finally uric acid. This development is said to *recapitulate* the evolutionary events that terminate in uric acid excretion in birds. Ammonia production in the chick embryo is reported to culminate at 4 days, urea at 9 days, and uric acid at 11 days of incubation (Baldwin 1949).

A more recent study claims that the nitrogen excretion in the chick embryo differs sharply from these reported peaks (Clark and Fischer 1957). All the three major excretory products – ammonia, urea, and uric acid – are present and are formed from the beginning of embryological development. Toward the end of the incubation period, uric acid is by far the dominant excretory product. However, the amounts of both urea and ammonia continue to increase throughout the incubation, and at the time of hatching both are present in approximately equal quantities. At the end of incubation the chick has excreted 40 mg nitrogen, of which 23% is divided equally between urea and ammonia, and the remainder is present as uric acid (Figure 9.15).

What is the reason for the conflicting reports? The older results may have been less accurate because of primitive analytical techniques, but this could hardly explain the observation of distinct peaks. The main reason is simply that the earlier results were expressed as the amount of each excretory product relative to the weight of the embryo. Because the embryo continuously increases in size, and at a rising rate, an artificial peak is created by dividing the amount of each excretory product by the weight of the embryo.

In fact, all three excretory products are present from the beginning and increase in amount throughout embryological development, but ammonia does not increase appreciably until after the tenth day of incubation. The urea produced by the chick embryo is not synthesized from amino acid nitrogen by the ornithine cycle, but from the action

FIGURE 9.15 The accumulation of nitrogenous products in the chick embryo during incubation of the egg. Urea and uric acid are formed from the beginning of incubation; uric acid, however, dominates later. Ammonia, also present from the beginning, remains constant for about 10 days and then increases further throughout the incubation period. [Clark and Fischer 1957]

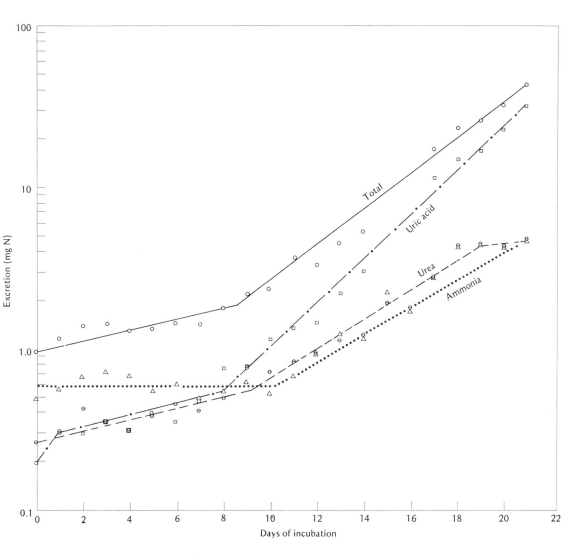

of arginase upon arginine (Eakin and Fisher 1958). Thus, neither ammonia production nor urea production in the chick embryo supports the claim that embryonic biochemical development recapitulates the evolutionary history of nitrogen excretion.

We have now discussed a wide range of excretory organs and have seen the characteristics they have in common. Excretory organs eliminate metabolic waste products, aid in the maintenance of proper concentrations of salts and other solutes, and regulate the water content of the organism by a judiciously regulated balance between conserving a scarce commodity and disposing of an excess.

REFERENCES

Baldwin, E. (1949) *An Introduction to Comparative Biochemistry*. Cambridge: Cambridge University Press. 164 pp.

Balinsky, J. B., Choritz, E. L., Coe, C. G. L., and Van Der Schans, G. S. (1967) Amino acid metabolism and urea synthesis in naturally aestivating *Xenopus laevis. Comp. Biochem. Physiol. 22:*59–68.

Balinsky, J. B., Cragg, M. M., and Baldwin, E. (1961) The adaptation of amphibian waste nitrogen excretion to dehydration. *Comp. Biochem. Physiol. 3:*236–244.

Braun, G., Kümmel, G., and Mangos, J. A. (1966) Studies on the ultrastructure and function of a primitive excretory organ, the protonephridium of the Rotifer *Asplanchna priodonta. Pflügers Arch. 289:*141–154.

Brown, G. W., Jr., Brown, W. R., and Cohen, P. P. (1959) Comparative biochemistry of urea synthesis. 2. Levels of urea cycle enzymes in metamorphosing *Rana catesbeiana* tadpoles. *J. Biol. Chem. 234:*1775–1780.

Burger, J. W. (1957) The general form of excretion in the lobster, *Homarus. Biol. Bull. 113:*207–223.

Clark, H., and Fischer, D. (1957) A reconsideration of nitrogen excretion by the chick embryo. *J. Exp. Zool. 136:*1–15.

Costa, G., Kerins, M. E., Kantor, F., Griffith, K., and Cummings, W. B. (1974) Conversion of protein nitrogen into gaseous catabolites by the chick embryo. *Proc. Natl. Acad. Sci. U.S.A. 71:*451–454.

Costa, G., Ullrich, L., Kantor, F., and Holland, J. F. (1968) Production of elemental nitrogen by certain mammals including man. *Nature, Lond. 218:*546–551.

Coulson, R. A., and Hernandez, T. (1955) Renal excretion of carbon dioxide and ammonia by the alligator. *Proc. Soc. Exp. Biol. Med. 88:*682–687.

Cragg, M. M., Balinsky, J. B., and Baldwin, E. (1961) A comparative study of nitrogen excretion in some amphibia and reptiles. *Comp. Biochem. Physiol. 3:*227–235.

Dobbs, G. H., III, Lin, Y., and DeVries, A. L.

(1974) Aglomerularism in antarctic fish. *Science 185:*793–794.

Eakin, R. E., and Fisher, J. R. (1958) Patterns of nitrogen excretion in developing chick embryos. In *The Chemical Basis of Development* (W. D. McElroy and B. Glass, eds.), pp. 514–522. Baltimore: Johns Hopkins Press.

Eastman, J. T., DeVries, A. L., Coalson, R. E., Nordquist, R. E., and Boyd, R. B. (1979) Renal conservation of antifreeze peptide in Antarctic eelpout, *Rhigophila dearborni. Nature, Lond. 282:*217–218.

Forster, R. P. (1940) A renal clearance analysis of phenol red elimination in the frog. *J. Cell Comp. Physiol. 16:*113–122.

Forster, R. P., and Goldstein, L. (1966) Urea synthesis in the lungfish: Relative importance of purine and ornithine cycle pathways. *Science 153:*1650–1652.

Goldstein, L., Janssens, P. A., and Forster, R. P. (1967) Lungfish *Neoceratodus forsteri:* Activities of ornithine-urea cycle and enzymes. *Science 157:*316–317.

Goodrich, E. S. (1945) The study of nephridia and genital ducts since 1895. *Q. J. Microsc. Sci. 86:*113–392.

Gottschalk, C. W. (1961) Micropuncture studies of tubular function in the mammalian kidney. *Physiologist 4:*35–55.

Greenwald, L. and Stetson, D. (1988) Urine concentration and the length of the renal papilla. *News Physiol. Sci. 3:*46–49.

Gross, W. J., and Capen, R. L. (1966) Some functions of the urinary bladder in a crab. *Biol. Bull. 131:*272–291.

Harrison, F. M. (1962) Some excretory processes in the abalone, *Haliotis rufescens. J. Exp. Biol. 39:*179–192.

Hilger, H. H., Klümper, J. D., and Ullrich, K. J. (1958) Wasserrückresorption und Ionentransport durch die Sammelrohrzellen der Säugetierniere (Mikroanalytische Untersuchungen). *Pflügers Arch. 267:*217–237.

Hopkins, D. L. (1946) The contractile vacuole and the adjustment to changing concentration in freshwater amoebae. *Biol. Bull. 90:*158–176.

Janssens, P. A., and Cohen, P. P. (1966) Ornithine-urea cycle enzymes in the African lungfish, *Protopterus aethiopicus. Science 152:*358–359.

Keilin, J. (1959) The biological significance of uric acid and guanine excretion. *Biol. Rev. 34:*265–296.

Khalil, F., and Haggag, G. (1955) Ureotelism and uricotelism in tortoises. *J. Exp. Zool. 130:*423–432.

Kokko, J. P. (1974) Membrane characteristics governing salt and water transport in the loop of Henle. *Fed. Proc. 33:*25–30.

Kokko, J. P., and Tisher, C. C. (1976) Water movement across nephron segments involved with the countercurrent multiplication system. *Kidney Int. 10:*64–81.

Lockwood, A. P. M. (1961) The urine of *Gammarus duebeni* and *G. pulex. J. Exp. Biol. 38:*647–658.

Loveridge, J. P. (1970) Observations on nitrogenous excretion and water relations of *Chiromantis xerampelina* (Amphibia, Anura). *Arnoldia 5:*1–6.

MacMillen, R. E., and Lee, A. K. (1967) Australian desert mice: Independence of exogenous water. *Science 158:*383–385.

McNabb, R. A., and McNabb, F. M. A. (1975) Urate excretion by the avian kidney. *Comp. Biochem. Physiol. 51A:*253–258.

Marshall, E. K., Jr. (1933) The secretion of urea in the frog. *J. Cell. Comp. Physiol. 2:*349–353.

Minnich, J. E. (1972) Excretion of urate salts by reptiles. *Comp. Biochem. Physiol. 41A:*535–549.

Moyle, V. (1949) Nitrogenous excretion in chelonian reptiles. *Biochem. J. 44:*581–584.

Murrish, D. E., and Schmidt-Nielsen, K. (1970) Water transport in the cloaca of lizards: Active or passive? *Science 170:*324–326.

Nechay, B. R., Boyarsky, S., and Catacutan-Labay, P. (1968) Rapid migration of urine into intestine of chickens. *Comp. Biochem. Physiol. 26:*369–370.

Needham, J. (1931) *Chemical Embryology,* vol. I, pp. 1–614; vol. II, pp. 615–1254; vol. III, pp. 1255–1724. Cambridge: Cambridge University Press.

Oschman, J. L., and Wall, B. J. (1969) The structure of the rectal pads of *Periplaneta americana* L. with regard to fluid transport. *J. Morphol. 127:*475–510.

Patterson, D. J. (1980) Contractile vacuoles and associated structures: Their organization and function. *Biol. Rev. 55:*1–46.

Peschen, K. E. (1939) Untersuchungen über das Vorkommen und den Stoffwechsel des Guanins im Tierreich. *Zool. Jahrb. 59:*429–462.

Pontin, R. M. (1964) A comparative account of the protonephridia of *Asplanchna* (Rotifera) with special reference to the flame bulbs. *Proc. Zool. Soc. Lond. 142:*511–525.

Poulson, T. L. (1965) Countercurrent multipliers in avian kidneys. *Science 148:*389–391.

Ramsay, J. A. (1949) The site of formation of hypotonic urine in the nephridium. *J. Exp. Biol. 26:*65–75.

Ramsay, J. A. (1964) The rectal complex of the mealworm *Tenebrio molitor,* L. (Coleoptera, Tenebrionidae). *Philos. Trans. R. Soc. Lond. B. 248:*279–314.

Ramsay, J. A., and Riegel, J. A. (1961) Excretion of inulin by Malpighian tubules. *Nature, Lond. 191:*1115.

Richards, A. N. (1935) Urine formation in the amphibian kidney. *Harvey Lect. 30:*93–118.

Riddick, D. H. (1968) Contractile vacuole in the amoeba, *Pelomyxa carolinensis. Am. J. Physiol. 215:*736–740.

Riegel, J. A., and Lockwood, A. P. M. (1961) The role of the antennal gland in the osmotic and ionic regulation of *Carcinus maenas. J. Exp. Biol. 38:*491–499.

Schmidt-Nielsen, B., and Forster, R. P. (1954) The effect of dehydration and low temperature on renal function in the bullfrog. *J. Cell. Comp. Physiol. 44:*233–246.

Schmidt-Nielsen, B., and O'Dell, R. (1961) Structure and concentrating mechanism in the mammalian kidney. *Am. J. Physiol. 200:*1119–1124.

Schmidt-Nielsen, B., and Rabinowitz, L. (1964) Methylurea and acetamide: Active reabsorption by elasmobranch renal tubules. *Science 146:*1587–1588.

Schmidt-Nielsen, B., and Schrauger, C. R. (1963) *Amoeba proteus:* Studying the contractile vacuole by micropuncture. *Science 139:*606–607.

Schmidt-Nielsen, K. (1964) *Desert Animals: Physiological Problems of Heat and Water.* Oxford: Clarendon Press. 277 pp. Reprinted by Dover Publications, New York, 1979.

Schmidt-Nielsen, K., Borut, A., Lee, P., and Crawford, E. C., Jr. (1963) Nasal salt excretion and the possible function of the cloaca in water conservation. *Science 142:*1300–1301.

Schmidt-Nielsen, K., and Lee. P. (1962) Kidney function in the crab-eating frog (*Rana cancrivora*). *J. Exp. Biol. 39:*167–177.

Shannon, J. A., and Fisher, S. (1938) The renal tubular reabsorption of glucose in the normal dog. *Am. J. Physiol. 122:*766–774.

Shoemaker, V. H., Balding, D., and Ruibal, R. (1972) Uricotelism and low evaporative water loss in a South American frog. *Science 175:*1018–1020.

Shoemaker, V. H., and McClanahan, L. L., Jr.

(1975) Evaporative water loss, nitrogen excretion and osmoregulation in Phyllomedusine frogs. *J. Comp. Physiol. 100:*331–345.

Skadhauge, E. (1976) Cloacal absorption of urine in birds. *Comp. Biochem. Physiol. 55A:*93–98.

Skadhauge, E. (1981) *Osmoregulation in Birds.* Berlin: Springer-Verlag. 203 pp.

Smith, H. W. (1929) The excretion of ammonia and urea by the gills of fish. *J. Biol. Chem. 81:*727–742.

Smith, H. W. (1951) *The Kidney: Structure and Function in Health and Disease.* New York: Oxford University Press. 1049 pp.

Smith, H. W. (1959) *From Fish to Philosopher.* Summit, N.J.: CIBA. 304 pp. Reprinted by Doubleday, Garden City, N.Y., 1961.

Somero, G. N. (1986) From dogfish to dogs: Trimethylamines protect proteins from urea. *News Physiol. Sci. 1:*9–12.

Starling, E. H. (1899) The glomerular functions of the kidney. *J. Physiol. 24:*317–330.

Wigglesworth, V. B. (1931) The physiology of excretion in a blood-sucking insect, *Rhodnius prolixus* (Hemiptera, Reduviidae). 2. Anatomy and histology of the excretory system. *J. Exp. Biol. 8:*428–442.

Yü, T. F., Berger, L., Kupfer, S., and Gutman, A. B. (1960) Tubular secretion of urate in the dog. *Am. J. Physiol. 199:*1199–1204.

ADDITIONAL READING

Campbell, J. W. (1970) *Comparative Biochemistry of Nitrogen Metabolism,* vol. 1, *The Invertebrates.* New York: Academic Press. 493 pp.

Cochran, D. G. (1975) Excretion in insects. In *Insect Biochemistry and Function* (D. J. Candy and B. A. Kilby, eds.), pp. 179–281. New York: Wiley.

Dantzler, W. H. (1976) Renal function (with special emphasis on nitrogen excretion). In *Biology of the Reptilia,* vol. 5, *Physiology A* (C. Gans and W. R. Dawson, eds.), pp. 447–503. London: Academic Press.

Dantzler, W. H., and Braun, E. J. (1980) Comparative nephron function in reptiles, birds, and mammals. *Am. J. Physiol. 239:*R197–R213.

Denton, D. (1982) *The Hunger for Salt. An Anthropological, Physiological, and Medical Analysis.* New York: Springer-Verlag. 650 pp.

Kirschner, L. B. (1967) Comparative physiology: Invertebrate excretory organs. *Annu. Rev. Physiol. 29:*169–196.

Knox, F. G. (ed.) (1982) Comparative physiology of the control of renal function. *Fed. Proc. 41:*2347–2384.

McNabb, R. A., and McNabb, F. M. A. (1980) Physiological chemistry of uric acid: Solubility, colloid and ion-binding properties. *Comp. Biochem. Physiol. 67A:*27–34.

Maddrell, S. H. P. (1971) The mechanism of insect excretory systems. *Adv. Insect Physiol. 8:*199–331.

Orloff, J., Berliner, R. W., and Geiger, S. R. (eds.) (1973) *Handbook of Physiology,* sect. 8, *Renal Physiology.* Washington, D. C.: American Physiological Society. 1082 pp.

Patterson, D. J. (1980) Contractile vacuoles and associated structures: Their organization and function. *Biol. Rev. 55:*1–46.

Peaker, M., and Linzell, J. L. (1975) *Salt Glands in Birds and Reptiles.* Cambridge: Cambridge University Press. 307 pp.

Phillips, J. (1981) Comparative physiology of insect renal function. *Am. J. Physiol. 241:*R241–R257.

Riegel, J. A. (1972) *Comparative Physiology of Renal Excretion.* Edinburgh: Oliver & Boyd. 204 pp.

Skadhauge, E. (1981) *Osmoregulation in Birds.* Berlin: Springer-Verlag. 203 pp.

Smith, H. W. (1951) *The Kidney: Structure and Function in Health and Disease.* New York: Oxford University Press. 1049 pp.

Wall, B. J., and Oschman, J. L. (1975) Structure and function of the rectum in insects. *Fortschr. Zool. 23 (2/3):*193–222.

Wessing, A. (ed.) (1975) Excretion. *Fortschr. Zool. 23:*1–362.

Yancey, P. H., Clark, M. E., Hand, S. C., Bowlus, R. D., and Somero, G. N. (1982) Living with water stress: Evolution of osmolyte systems. *Science 217:*1214–1222.

MOVEMENT, INFORMATION, INTEGRATION

CHAPTER TEN

Movement, muscle, biomechanics

We may think of movements primarily in connection with locomotion, i.e., an animal moving from place to place. However, even animals that remain attached and never move about (corals and sponges, for example), show a great variety of movements. Also consider how animals move water over their gills, or food through the intestinal tract, or blood through the vascular system.

The number of mechanisms used to achieve motion is limited, although their uses vary greatly. We shall discuss the three basic mechanisms, *ameboid, ciliary,* and *muscular* movements.

Ameboid movement derives its name from the motion of the ameba, the unicellular organism described in every biology textbook. Ameboid locomotion involves extensive changes in cell shape, flow of cytoplasm, and pseudopodal activity.

Ciliary locomotion is the characteristic way in which ciliated protozoans such as paramecium move. However, cilia are found in all animal phyla and serve a variety of functions. For example, cilia set up the currents that move water over the gills of bivalves or move fluid in the water-vascular system of echinoderms. The respiratory passages of air-breathing vertebrates are lined with ciliated cells that slowly remove foreign particles that lodge on their surfaces. The sperm of most animals move with the aid of a tail, which in principle acts like a cilium.

Muscular movement is the fundamental mechanism used in an overwhelming majority of different movements. It depends on the use of muscle, which throughout the animal kingdom has one universal characteristic – the ability to exert a force by shortening.

In this chapter we shall be concerned mostly with muscle and its use in animal locomotion. One important aspect will be how animals can achieve the most economical use of muscle energy.

AMEBOID, CILIARY, AND FLAGELLAR LOCOMOTION

Ameboid movement

What we designate as ameboid movement is characteristic of some protozoans, slime molds, and vertebrate white blood cells. The movement of these cells is connected with cytoplasmic streaming, change in cell shape, and extension of pseudopodia. These changes are easily observed in a microscope, but the mechanisms involved in achieving the movement are not well understood.

When an ameba moves, its cytoplasm flows into newly formed armlike extensions of the cell (*pseudopodia*), which gradually extend and enlarge so that the entire cell occupies the space where previously only a small pseudopodium began to form. As the cell moves, new pseudopodia are formed in the direction of movement, while the posterior parts are withdrawn.

In an ameba the outermost layer, the *ectoplasm*, is a somewhat stiffer gel-like layer. As a pseudopodium is formed, the more liquid *endoplasm* streams into it, and new ectoplasm is formed on the surface. In the rear part of the advancing cell the ectoplasmic gel should then be converted to a more liquid endoplasmic sol by a gel-sol transformation.

What is the driving force in the streaming of the endoplasm? One suggestion is that contraction of material at the posterior end of an ameba drives endoplasm forward and forces it into the extending pseudopodium. A lower viscosity in certain parts of the endoplasm would facilitate streaming, and the driving force would be minute pressure differences.

The hypothesis that hydrostatic pressure plays a major role in pseudopodium extension is contradicted by experimental evidence. If one pseudopodium of an ameba is sucked into a capillary connected to a reduced pressure, the extension of other pseudopodia is not prevented, as would be expected if the pressure hypothesis were correct (Allen et al. 1971). If the endoplasmic streaming could

be attributed to pressure gradients along the length of the stream, streaming should be reversed under an applied pressure gradient of opposite sign.

Ameboid movement has obvious similarities to cytoplasmic streaming (*cyclosis*), a commonly observed phenomenon in all sorts of cells, both plant and animal, which plays an important role in intracellular transport. Ideas about how intracellular forces arise have been revolutionized during the last few years because two important force-generating proteins in muscle, actin and myosin, have been found in all eukaryotic cells (i.e., all cells – animal, plant, and protozoan – except bacteria and blue-green algae). The actins from the most varied sources are remarkably similar in structure and occur as thin filaments within the cells.

It thus appears that both cytoplasmic streaming and the formation of pseudopodia may depend on the interaction between myosin and actin filaments, in other words, the same fundamental mechanism as muscle contraction. The myosins from various sources are more diverse than the actins, but all bind reversibly to actin filaments and all catalyze the hydrolysis of adenosine triphosphate (ATP), the universal energy source for movement (Pollard 1977).

The existence of a force-generating system in the ameba helps us understand how the ameboid movement can be generated. However, further experimental work will be difficult because we are unable to measure directly the mechanical forces involved, partly because of the small size of amebas and partly because of the absence of mechanical structures that can be subjected to direct manipulation and measurement.

Movement by cilia and flagella

Cilia and flagella have a similar internal structure; the difference between them relates to their different beating patterns. A flagellum, like the tail of a sperm, beats with a symmetrical undulation that is propagated as a wave along the flagellum. A cilium, in contrast, beats asymmetrically with a fast or dashlike stroke in one direction, followed

FIGURE 10.1 Diagram showing the timing of the ciliary beat of the ctenophore *Pleurobrachia,* drawn from a film taken at 17 °C. The position of one cilium is shown at 5-ms intervals. The power stroke takes 10 ms, and the complete return stroke takes 50 ms. The cilium then remains in the initial position until the next stroke begins at 65 ms. [Sleigh 1968]

FIGURE 10.2 The typial beat of a flagellum (left) propels water parallel to the main axis of the flagellum (arrow). The beating of a cilium (right) propels water parallel to the surface to which the cilium is attached. [Sleigh 1974]

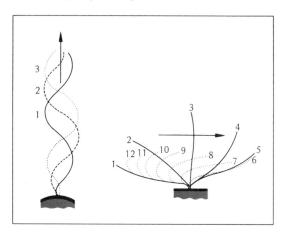

by a slower recovery motion in which the bending cilium returns to its original position (Figure 10.1). In flagellar motion water is propelled parallel to the long axis of the flagellum; in ciliary motion water is propelled parallel to the surface that bears the cilia (Figure 10.2).

A flagellated cell usually carries only one or a few flagella; a ciliated cell such as a paramecium may have several thousand cilia distributed on the surface. It is, however, difficult to maintain a sharp distinction between flagella and cilia: Their internal structures are identical, the motions are generated within the organelles themselves, and intermediate patterns of movement are common.

The flagella of bacteria are quite different. They are thinner (about 0.02 μm in diameter, against 0.25 μm for true flagella and cilia), short and relatively rigid, and are rotated by forces at the base where they are attached to the cell (Berg and Anderson 1973).

Cilia and flagella are found in many protozoans and are of primary importance in locomotion. The sperm of a vast number of animals swim by means of flagella. Ciliated gills and tentacles, common in many invertebrates, serve two main functions: respiratory exchange and filtering of water to obtain food particles.

Cilia are common also in more highly organized animals where they serve, for example, in moving fluid through tubes, such as reproductive and excretory systems (e.g., the nephridia of annelids). In mammals ciliated epithelia aid in the transport of material on internal surfaces, such as moving mucus in the respiratory tract and the egg within the oviduct.

Cilia are found in all animal phyla. Insects were formerly considered an exception, for they do not

CILIATED PROTOZOAN The coordinated beating of the cilia of *Paramecium caudatum* gives the appearance of waves passing over its surface. The length of the entire animal is about 150 μm. [Courtesy of Robert L. Hammersmith, Indiana University]

FLAGELLA This electron micrograph shows cross sections of the flagella of the protozoan *Trichonympha*. The characteristic two-plus-nine arrangement of the filaments is clearly visible in each flagellum. [Courtesy of Dr. A. V. Grimstone, Cambridge University]

have functional cilia. However, modified ciliary structures, recognized by the characteristic nine-plus-two arrangement of the internal filaments, occur in insect eyes, as well as in a majority of other sense organs in most animal phyla. Main exceptions are the eyes of some invertebrates and the taste receptors of vertebrates. In many cases where a ciliary structure is recognized in a sense organ, it is difficult to determine the importance of the ciliary structure in sensory reception, whether of light or of chemical molecules. These problems are difficult to solve, and only studies on a broad comparative basis are likely to yield acceptable generalizations.

Examination in the electron microscope has revealed that cilia and flagella have a common structure: a central pair of filaments surrounded by an additional nine thin filaments. This arrangement of two plus nine filaments is striking; it occurs from protozoans to vertebrates and in the overwhelming majority of animal sperm. The particular combination of nine plus two is almost universal, although not essential, for a few sperm are known that have three central filaments, one central filament, or none at all (Blum and Lubliner 1973).

Cilia used for movement can function only in an

FIGURE 10.3 The flagellate *Ochromonas* swims with the single flagellum etending in the direction of movement. The undular motion of the flagellum consists of a wave that travels from its base to its tip (i.e., in the same direction as the direction of swimming). This paradoxical situation is attributable to the fixed protrusions extending at a 90° angle from the flagellum, causing a movement of water as shown by the curved arrows. The organism therefore is propelled in the opposite direction. [Jahn et al. 1964]

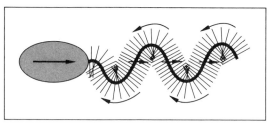

aqueous medium and are therefore found only on cell surfaces that are covered by liquid or a film of mucus.

We intuitively expect an organism that swims with the aid of a single flagellum, such as a sperm, to move in the opposite direction from the wave that travels down the tail. This is correct for a smooth, undulating filament, which propels the organism away from the filament or "tail." Therefore, it seems peculiar that some flagellates move with the long flagellum beating anteriorly, in the direction of movement, and that the waves move out from the body to the anterior tip of the flagellum (Jahn et al. 1964).

This seemingly paradoxical situation, that wave propagation and locomotion are in the same direction, may occur if the filament is rough or covered with projections. The flagella of species that move in this way have tiny appendages in the form of thin lateral projections that are responsible for the unexpected direction of movement (Figure 10.3).

The mechanism of movement of cilia and flagella has long been a matter of speculation. Three types of mechanisms have been suggested: (1) that

the flagellum is moved passively, much like a whip, by forces exerted at its base; (2) that elements along the inner curvature of a propagating wave contract while the opposite side does not; and (3) that the thin filaments inside the cilium slide relative to each other due to forces between them of a nature similar to the sliding filaments of muscle contraction.

The idea that the flagellum is a passive element driven from its base is incompatible with the form of the waves (Machin 1958). In the sperm tail the bending waves pass along without decrease in amplitude, and in some cases with increasing amplitude (Rikmenspoel 1965). It is therefore necessary to assume that there are active elements within the flagellum and that the energy is generated locally.

Assuming that the filaments are responsible for the movement of the cilium, two mechanisms present themselves as possible explanations for the motion: (1) bending of the cilium could be achieved by contraction of the filaments on one side or (2) movement could be achieved by a longitudinal sliding between the filaments. The first hypothesis is contradicted by the fact that the ciliary tubules maintain constant length during the bending cycle (Satir 1968, 1974), and present evidence is entirely in accord with the second alternative.

The characteristic fine structure of a flagellum or cilium is shown in Figure 10.4. The doublet tubules are made up of a protein, *tubulin*, with a molecular weight of about 55 000. Each A-tubule has attached to it a doublet of "arms," consisting of the protein *dynein*. This protein shows little resemblance to any of the contractile proteins of muscle, except for its ability to split ATP enzymatically, a property it shares with the important muscle protein myosin (Gibbons 1977).

Bending of the flagellum occurs when the extending dynein arms attach to the neighboring B-tubule, inducing active sliding movements at the expense of ATP. This view was recently confirmed by direct observation of the flagella of sea urchin sperm. Photomicrographs were made of minute gold particles attached to adjoining filaments, and careful measurements showed that the distance be-

FIGURE 10.4 Diagram showing the internal structure of a cilium or flagellum in cross section. The doublet tubules consist of the protein tubulin. Attached to the A-tubules are "arms" containing the protein dynein. [Brokaw and Gibbons 1975]

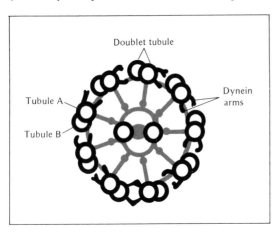

FIGURE 10.5 The range of body lengths of organisms that use flagella, cilia, or muscles for swimming. There is considerable overlap; some organisms propelled by muscles are smaller than the largest that use flagella, and ciliated organisms overlap both these groups. Comb plates are the comblike groups of cilia that ctenophores (comb jellies) use for swimming. [Sleigh 1977]

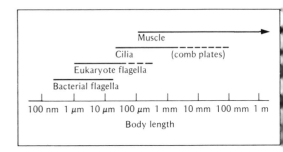

tween particles changed in accord with the bending of the flagellum (Brokaw 1989). The process is similar to the sliding of the filaments in muscle, but the exact molecular movements in the flagellar movement have not been as well clarified.

It is possible to prepare an experimental model of the ciliary apparatus of a paramecium that can be used to study the control of the ciliary movement. The cell is extracted with a detergent that leaves the ciliary apparatus functional, but the cell membrane disrupted. Such an extracted paramecium can be reactivated to swim in a solution of ATP containing magnesium ions. It can then be shown that the concentration of calcium ions determines the direction in which the cilia beat (Naitoh and Kaneko 1972).

When a living paramecium meets an obstacle, it retreats and swims backward because the ciliary beat is reversed; the initial process in this reversal is an increase in the permeability of the cell to calcium ions (Eckert 1972). As we shall see later, calcium ions are important in the control of a variety of processes, including muscle contraction.

The overall efficiency of conversion of metabolic energy to mechanical energy in the swimming sperm has been estimated to be at least 19%, and probably closer to 25% (Rikmenspoel et al. 1969). This value is strikingly similar to the efficiency of muscles in performing external work.

Body size and method of propulsion

Small organisms use cilia or flagella for swimming; larger organisms use muscle (Figure 10.5).

There are some very good reasons for this distribution of propelling mechanisms. One is a matter of simple geometry. The small organism has a much larger surface area relative to its volume and obtains forward thrust by moving the cilia against the surrounding surface layer of water, using the viscosity of the water to obtain the necessary thrust. Imagine a whale with a ciliated surface depending on the viscosity in a layer of water, only a fraction of a millimeter thick, along its surface to push against. In relation to its bulk, the thrust it could obtain would be infinitesimal.

Another reason is found in the laws of fluid dynamics. For a very small organism that swims at low speed, the movements are completely dominated by viscous forces – a situation analogous to a person swimming in syrup. For a large animal, by contrast, swimming depends on inertial forces – the thrust obtained by accelerating a mass of water in the opposite direction. For this the animal uses suitably large surfaces, such as fins and flukes, and muscle to provide the power.

We shall now turn to the function of muscle as it is used not only in locomotion, but also in achieving a variety of movements that all are based on the fact that muscle exerts a force by contraction.

MUSCLE AND MOVEMENT

Muscle, What It Is: Structure

As far as we know, the biochemical mechanism of muscle contraction is the same in all muscles. Two proteins, *actin* and *myosin,* are part of the machinery, and ATP is the immediate energy source for the contraction. The detailed arrangement, however, varies a great deal, and it is convenient to classify various kinds of muscles accordingly. The classification is based primarily on vertebrate muscles for the simple reason that we know and understand more about these than about invertebrate muscles.

The broadest classification is based on the presence or absence in the muscle of regular cross striations that can be seen in an ordinary light microscope. Vertebrate skeletal muscles and heart muscle are *striated;* muscles of the internal organs – in the walls of the bladder, intestine, blood vessels, uterus, and so on – are *unstriated* (also called *smooth*). The heart muscle, although striated, is often considered as a separate type because it differs from skeletal muscle in characteristic ways, the functionally most important being that a beginning contraction of the heart muscle spreads to the entire organ.

Skeletal muscles are usually called *voluntary,* for the muscles of limb and trunk are under control of the will. This does not mean that we are always aware of or decide about our movements; on the contrary, locomotion, breathing, and so on take place without conscious knowledge of the muscles involved. Voluntary muscle is always striated, but the term is very poor and, at best, useful only for humans.

Vertebrate smooth muscle almost always occurs in the walls of hollow internal organs. It is not under control of the conscious mind, and is called *involuntary.* The contractions are usually much slower than those of striated muscle, and normally we are completely unaware of the state of contraction of the smooth muscle in our blood vessels, stomach, intestine, and so on.

Striated muscle. The organization of striated muscle is shown in diagram form in Figure 10.6. The muscle is made up of a large number of parallel *fibers,* which are between 0.1 and 0.01 mm in diameter, but may be several centimeters long. The fibers in turn are made up of thinner *fibrils.* These fibrils have characteristic cross striations, the so-called Z-*lines* or Z-*bands,* which are repeated at completely regular intervals.

The region between two Z-lines is called a *sarcomere.* In vertebrate muscle the distance between the Z-lines, the sarcomere length, tends to be very uniform, about 2.5 μm in a relaxed muscle. Many invertebrate muscles have similar sarcomere length, but some have much longer sarcomeres, 10 or 15 μm or even longer.

From the narrow Z-line very thin *filaments* extend in both directions, and in the center these thin filaments are interspersed with somewhat thicker filaments. The result is a number of less conspicuous bands located between the Z-lines; the appearance of these bands changes with the state of contraction of the muscle.

In vertebrates the thin filaments are about 0.005 μm in diameter and the thick filaments are about twice that, about 0.01 μm in diameter. The length of the thick filaments is about 1.5 μm. The thin filaments in different muscles differ somewhat more

FIGURE 10.6 Schematic diagram of a vertebrate striated muscle. The whole muscle is composed of fibers, which in the light microscope appear cross-striated. The fibers consist of fibrils, which have lighter and darker bands. The electron microscope reveals that these bands result from a repeating pattern in the regular arrangement of thick and thin filaments.

in length. They are often between 2 and 2.6 μm from tip to tip (or about half this length measured from the Z-band to the tip). The arrangement of the filaments is extremely regular and well ordered.

The thick filaments consist of *myosin* and the thin filaments of *actin*. The thick and thin filaments are linked together by a system of molecular cross-linkages, and when the muscle contracts and shortens, these cross-linkages are rearranged so that the thick filaments slide in between the thin filaments, reducing the distance between the Z-bands (Huxley 1969).

This model of muscle structure is well supported by electron micrographs of the filaments in various states of contraction. Neither the thick filaments nor the thin filaments change in length during contraction, but it can clearly be seen that they move relative to each other. What is even more important in support of the sliding filament model is that the force developed by a muscle is related to the amount of overlap of the filaments (page 403).

Cardiac muscle. Cardiac muscle is cross-striated, like skeletal muscle, but its functional properties differ in two important respects. One is that, when a contraction starts in one area of the heart muscle, it rapidly spreads throughout the muscle mass. Another important property of cardiac muscle is that a contraction is immediately followed by a relaxation period during which the muscle cannot be stimulated to contract again. As a result, a long-lasting contraction, like the sustained contraction of a skeletal muscle, does not occur. These two properties are essential for the normal rhythmic contraction of the heart, which we shall return to shortly.

Smooth muscle. Smooth muscle lacks the cross-striations characteristic of skeletal muscle, but the contraction depends on the same proteins as in

MUSCLE FIBER This electron micrograph of a fibril from the flight muscle of a giant water bug (*Lethocerus cordofanus*) shows the very regular arrangement of thick and thin muscle filaments. The two dark bands crossing the fibril are Z-lines; the distance between them is about $3\mu m$ and shows the length of a single sarcomere. The darker bodies to the upper left and lower right of the fiber are mitochondria. [Courtesy of Barbara M. Luke, Oxford University]

striated muscle, actin and myosin, and on a supply of energy from ATP.

Smooth muscle has not been as extensively studied as striated muscle. There are several reasons for this. One is that smooth muscle is often interspersed with connective tissue fibers. Another is that smooth muscle fibers do not form neat parallel bundles that can readily be isolated and studied.

Also, smooth muscle consists of much smaller cells; the fibers are often only a fraction of a millimeter long.

Substances for energy storage

The immediate source of energy for a muscle contraction is *adenosine triphosphate* (*ATP*), a compound that is the primary energy source for nearly every energy-requiring process in the body. ATP is the only substance the muscle proteins can use directly. When the terminal phosphate group of this compound is split off, the high energy of the bond is available for the energy of muscle contraction.

In spite of its great importance, ATP is present in muscle in very small amounts. The total quantity may be sufficient for no more than 10 rapid contractions; it follows, therefore, that ATP must be restored again very rapidly, for otherwise the muscles would soon be exhausted. The source is another organic phosphate compound, creatine phosphate, which is present in larger amounts. Its phosphate group is transferred to adenosine diphosphate (ADP) and thus the supply of ATP is restored.

The creatine phosphate must, however, eventually be replenished, and the ultimate energy source is the oxidation of carbohydrates or fatty acids. Carbohydrate is stored in the muscle in the form of glycogen, which is often present in an amount of between 0.5 and 2% of the wet weight of the muscle and provides an amount of energy perhaps 100 times as great as the total quantity of creatine phosphate. In the absence of sufficient oxygen, the glycogen can still yield energy by being split into lactic acid, but the amount of energy is then only a small fraction, about 7%, of that available from complete oxidation (see Chapter 5).

Creatine phosphate has been found in the muscles of all vertebrates examined, but it is conspicuously absent from the muscles of some invertebrates. In these animals we find other organic phosphates. One such compound is arginine phosphate, which is never found in vertebrate muscle.

These various organic phosphate compounds all provide energy through the splitting off of a phosphate group, the high bond energy being available for energy-requiring processes. The fact that creatine phosphate is found in only some invertebrate groups but is universally present in vertebrates has led to the theory that the biochemical similarity between these invertebrate groups and the vertebrates indicates which invertebrate phylum gave rise to the vertebrates.

Present knowledge suggests, however, that the distribution of arginine and creatine phosphate among invertebrates is very irregular and not related to the classification of these animals. In particular, there seems to be no pattern in the occurrence of creatine phosphate among invertebrates that can give a clue to the phylogeny of vertebrates.

Muscle, how it works: contraction

In muscle physiology *contraction* refers to a state of mechanical activity. It may involve a shortening of the muscle, but if the muscle is kept from shortening by having its ends solidly attached, we still use the word *contraction* to describe the active state. In the latter case, contraction of the muscle causes a force to be exerted on the points of attachment; because no change in length takes place, this is called an *isometric contraction*. If, on the other hand, we attach to one end of the muscle a weight it can lift, the muscle shortens during contraction; because the load remains the same throughout the contraction, this is called an *isotonic contraction*.

Virtually no movements of muscles in the body are purely isometric or purely isotonic, for usually both the length of a muscle and the load change during contraction. For example, when an arm lifts a certain weight, both the length of the contracting muscles and the load on them change continuously, for the leverage and the angle of attack change throughout the movement.

Although pure isometric or pure isotonic contractions are unimportant for muscle function in the body, it is convenient in the study of isolated muscles and their function to distinguish between isometric and isotonic contractions. From isometric contractions we obtain a great deal of information about the forces a muscle can exert; in the study of isotonic contractions the amount of work performed by the muscle can easily be calculated. In the laboratory we therefore usually choose one

FIGURE 10.7 The force developed in isometric contraction by the gastrocnemius muscle from a cat's leg: (*a*) after a single stimulus; (*b–d*) after double stimuli applied at shorter and shorter intervals. Time marks are 20 ms apart. Temperature 34.5 °C. [Cooper and Eccles 1930]

FIGURE 10.8 Force developed in isometric contraction by an eye muscle of a cat in response to repeated stimulation: (*a*) at a rate of 125 per second (8 ms interval); (*b*) at a rate of 210 per second. The latter gives almost complete fusion to a tetanus. Temperature 36 °C. [Cooper and Eccles 1930]

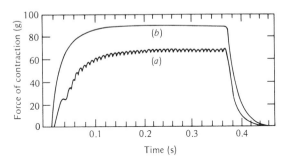

method or the other, according to the purpose of our study.

Force

A muscle can be stimulated to contract by applying a single brief electric pulse, and if the muscle is kept from shortening (isometric contraction), the force it produces gives a record as in Figure 10.7*a*. Immediately after the stimulus is applied, there is a very short period (the *latent period*) of a few milliseconds before contraction starts. Then the force exerted by the muscle rises rapidly to a maximum and declines again somewhat more slowly. Such a single contraction in response to a single stimulus is called a *twitch*.

The time required to reach maximum tension (the *contraction time*) varies from muscle to muscle. The contraction time for the major muscles of locomotion of a cat may be as long as 100 ms; the fastest muscles, such as the eye muscles, may have contraction times of less than 10 ms.

If a second stimulus is applied before a contraction is ended, a new contraction is superimposed

on the first twitch (Figure 10.7*b*). The second contraction reaches a higher peak of force than the single twitch. If we continue to apply two stimuli spaced more closely, the two contractions become more and more fused (Figure 10.7*c, d*).

Repeated stimuli applied at suitably long intervals give a series of separate twitches, but if we space the stimuli more and more closely, the twitches become fused and eventually yield a smooth, sustained contraction (Figure 10.8*b*). Such a smooth, sustained contraction is called a *tetanus*.

The maximum force developed in the tetanus depends on the initial length of the muscle. If a muscle is made to contract isometrically (i.e., with its ends fixed), the force is reduced if the distance between the ends is less than the resting length of the muscle, and the force is also reduced if the length is greater than the resting length. The maximum force is obtained when the muscle is kept at its resting length.

This is closely related to the position of the filaments relative to each other (Figure 10.9). If the muscle is initially stretched so that there is no overlap between thick and thin filaments (position 1), no force is developed. This is consistent with the

FIGURE 10.9 The force developed by a contracting muscle (ordinate) in relation to its initial length (abscissa). The force is expressed in percent of the maximal isometric force the muscle can develop. The amount of overlap between thick and thin filaments is shown at the top by diagrams located above the corresponding place on the force curve. [Gordon et al. 1966]

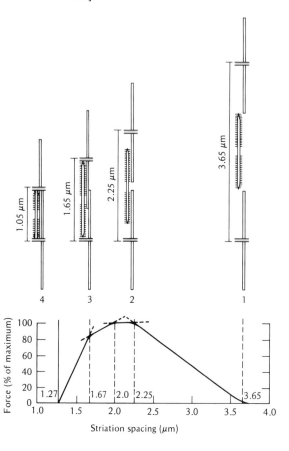

FIGURE 10.10 The thin filament (above) consists of a double helix of the globular protein, actin. In the groove between the actin strands are located two thin strands of the protein tropomyosin and, for every seven actin molecules, one molecule of troponin. The thick filament (below) consists of the rod-shaped protein, myosin, with the larger "heads" protruding toward the actin strands.

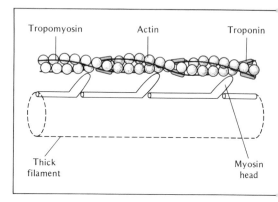

If the muscle is even shorter before contraction begins, the force is again diminished and rapidly decreases to zero as the ends of the thick filaments make contact with the Z-lines (positions 3 and 4).

What makes the filaments slide?

The molecular events during muscle contraction are better known than most other processes in the living organism. To describe these events, we must refer to the structural composition of the thick and thin filaments.

The thick filaments consist of the protein *myosin*. Each myosin molecule resembles a thin rod with a globular "head." One thick filament contains several hundred myosin molecules, lined up as shown in Figure 10.10 with the heads facing in both directions away from the center of the filament. This leaves a bare zone in the middle and the heads protruding all along the remainder of the thick filament.

The thin filaments are more complex, with three important proteins. The backbone of the filament

complete inability of the filaments to form crossbridges in this position. If the muscle is permitted to be shorter, however, the force increases as the filament overlap is greater. The maximum force is developed when the spacing is such that all the crossbridges along the full length of the thick filaments are in contact with the thin filaments (position 2).

is the protein *actin,* a relatively small globular protein, arranged in the filament as a twisted double strand of beads. Each actin molecule is slightly asymmetrical, and in each filament all the actin is lined up facing in the same direction. This directionality is important for muscle contraction.

The next important protein in the thin filament is *tropomyosin,* long and thin molecules attached to each other end to end, forming a very thin threadlike structure that lies in the grooves between the double helix of the actin molecules. The length of one tropomyosin molecule is such that it extends over seven actin molecules. The single thin filament is about 1 μm long and contains about 400 actin molecules and 60 tropomyosin molecules.

A third protein molecule, *troponin,* is attached to each tropomyosin molecule. Troponin, a calcium-binding protein, is a key to the contraction process. When troponin binds calcium ions, it undergoes a conformational change that is essential for the interaction between the myosin heads of the thick filaments and the actin of the thin filaments (Squire 1975).

The interaction between thick and thin filaments consists of a cyclic attachment and detachment of cross-bridges between the two filaments. The myosin heads attach to the thin filaments at a certain angle; they undergo a conformational change that makes the bridge swivel to a different angle, pulling the thin filament past the thick (Huxley 1973). The cross-bridges on opposite sides of the bare zone in the middle of the thick filament swivel in opposite directions, pulling at the opposing ends of the thin filaments, thereby shortening the distance between the Z-lines. This decrease in the distance between the Z-lines causes the shortening of the muscle.

For any significant shortening to occur, each cross-bridge must be attached, swivel, detach, and then attach again at a point further along the thin filament. Each cycle of attachment requires the expenditure of energy in the form of ATP. The ATP molecule binds temporarily to the myosin head, forming an active complex that will attach to the actin strand, provided the appropriate site is available and not blocked by the thin tropomyosin strand (Weber and Murray 1973).

The myosin head will not detach again from the actin strand until it is combined with a new ATP molecule. The head is then "charged" and ready to attach anew, the energy for the "swivel" action coming from the splitting of ATP into ADP. The sequence can be repeated, each detachment requiring ATP. The rigid condition of muscles after death (*rigor mortis*) sets in as the cross-bridges cannot be detached because the ATP supply in the muscle is depleted.

The role of calcium. The tropomyosin is located in the grooves of the actin filaments, where it blocks the interaction of the myosin head with the actin filament. The controlling protein, troponin, has a high affinity for calcium ion. Immediately after stimulation of the muscle, the calcium ion concentration within the muscle fiber rises abruptly; the calcium ion binds to troponin, which in turn undergoes a conformational change that allows tropomyosin to move out of its blocking position on the actin strands.

A direct demonstration of the increase of the calcium ion concentration immediately after stimulation has been made with the aid of the protein *aequorin.* This protein, which can be isolated from a luminescent jellyfish, emits light in the presence of calcium ions. Aequorin injected into the muscle fibers of the giant barnacle *Balanus nubilis* yields a faint glow of light when the fibers are stimulated electrically, indicating an increase in calcium ion concentration within the muscle fibers. Measurement of this light emission is a sensitive indicator of very rapid and transient changes in the intracellular calcium concentration (Ridgway and Ashley 1967).

How is contraction triggered?

Normally, muscle contraction is initiated when a nerve impulse arrives at the neuromuscular junction (the *motor end-plate*). The impulse spreads rapidly as an electric depolarization over the surface of the muscle fiber (the *sarcolemma*), momentarily abolishing the normal surface potential of about -60 mV. This depolarization is almost instantaneously communicated to the interior of the muscle cell, causing contraction to take place.

As we shall see later, the message from the motor end-plate to a vertebrate muscle fiber consists in the diffusion of a messenger chemical, *acetylcholine,* across the minute distance between the motor end-plate and the sarcolemma. However, the distance from the surface to the interior of the muscle fiber is so great that, if the communication were based on the diffusion of a chemical, the impulse could not be communicated to all the filaments simultaneously. One of the classic problems in muscle physiology has been to explain how communication throughout the muscle fiber could be rapid enough to achieve simultaneous contraction of all the filaments.

The answer can be found in the ultrastructure of the muscle fiber. The sarcolemma (the muscle cell membrane) connects with a complex system of *transverse tubules* that run across the muscle cells, typically near the Z-lines, known as the *T-system* (Figure 10.11). When an impulse arrives at the motor end-plate, it causes depolarization of the sarcolemma; the depolarization continues within the T-system, and this triggers the further events of contraction.

The next important structure to consider is a system of flattened vesicles, the *sarcoplasmic reticulum,* which surrounds the muscle fiber like a sleeve (Figure 10.11).

We now have the structural basis for the rapid communication throughout the muscle fibers. As an impulse spreads over the sarcolemma and the T-system, the change is transmitted to the membrane of the sarcoplasmic reticulum, causing an increase in its permeability to calcium. As a result, calcium ions that were sequestered inside the sarcoplasmic reticulum of the resting muscle can escape (Figure 10.12).

The sudden appearance of calcium triggers the

SARCOPLASMIC RETICULUM Sarcoplasmic reticulum (SR) surrounding the contractile filaments in the semitendinosus muscle of the European green frog (*Rana temporaria*). M = mitochondria, T = transverse tubule, Z = Z-lines. The distance from one Z-line to the next is about 2.4 μm. [Courtesy of Joachim R. Sommer, Duke University]

change in the configuration of the troponin, which permits the interaction between myosin and actin that is the basis of contraction. The calcium concentration necessary to activate the contractile events is no more than about 0.01 mmol per liter, roughly one-thousandth of the concentration in blood plasma.

FIGURE 10.11 The system of T-tubules and the sarcoplasmic reticulum that surrounds the fibers of striated muscle. [Peachey 1965. Courtesy of Lee D. Peachey and The Rockefeller University Press]

FIGURE 10.12 The sliding between the thick and thin filaments of muscle is triggered by the release of Ca^{2+} ions (indicated by dots) from the sarcoplasmic reticulum.

The calcium that was released is immediately taken up again by the sarcoplasmic reticulum, where it is held until the next impulse arrives; in the relaxed skeletal muscle the calcium concentration is believed to be as low as 0.0002 mmol per liter.

We are now in a position to list the major events in the contraction of vertebrate muscle (Table 10.1). The detailed events are not as well known for other types of muscles, but it can be assumed that the main outline is similar, that ATP is the ultimate energy source, and that calcium ions play a key role.

Force, work, and power

If a muscle is free to contract when it is stimulated, it may shorten by, say, one-third. However, if a load is attached to the muscle, the shortening is smaller. If the load is gradually increased, the muscle will be able to lift the load over a shorter and shorter distance until the load is so heavy that the muscle is unable to lift it at all. We thus obtain a curve for shortening versus force, as shown in Figure 10.13. In this particular case, a load of just below 100 g completely prevented shortening.

Force. We must now discuss the strength of muscles and ask whether the muscles of various animals differ much in strength. When we see an ant carry in its jaws a seed that weighs more than

TABLE 10.1 Sequence of events in stimulation and contraction of muscle.

Stimulation
1 Sarcolemma depolarized
2 T-system depolarized
3 Ca^{2+} released from sarcoplasmic reticulum
4 Ca^{2+} diffuses to thin filament

Contraction
5 Ca^{2+} binds to troponin
6 Troponin–Ca^{2+} complex removes tropomyosin blockage of actin sites
7 Heads of thick filament (containing preexisting myosin–ATP complex) form cross-bridges to actin strand
8 Hydrolysis of ATP induces conformational changes in heads that cause cross-bridges to swivel

Relaxation
9 Ca^{2+} sequestered from thin filament by sarcoplasmic reticulum
10 Ca^{2+} diffuses from thin filament toward sarcoplasmic reticulum
11 Ca^{2+} released from troponin–Ca^{2+} complex
12 Troponin permits tropomyosin return to blocking position
13 Myosin–actin cross-bridges break
14 ATP–myosin complex re-formed in heads of thick filament

FIGURE 10.13 Isotonic contraction of the sartorius muscle of the toad *Bufo marinus*. The shortening of the muscle decreases to zero as the load is increased to 97 g. The work performed in the contraction increases to a maximum at a load of about 40% of the maximum, and decreases again as the load is increased. [Courtesy of Frans Jöbsis]

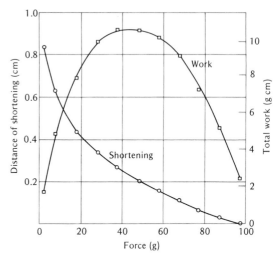

the animal itself, we gain the impression that insect muscles must be inordinately strong. An objective comparison of muscle strength shows that this is not so. In order to compare different muscles we must, of course, take into consideration their different sizes.

The force that a muscle of a given kind can exert is directly related to its cross-sectional area, but not to its length. Let us consider the length. Imagine that we suspend a muscle that is barely able to lift from the table a 100-g load if maximally stimulated. Next, let us join two such muscles end to end and attach a 100-g load. The joined muscles will each just be able to contract under this load, but if we increased the load, neither muscle would be able to contract. Thus, differences in length do not increase the force that muscles can exert. The cross-sectional area of the muscle, on the other hand, is a different matter.

Consider vertebrate muscles, where the sarcomere length is quite constant, about 2.5 μm, and the thick and thin filaments are of the usual dimensions. A given cross section of such muscles contains a given number of filaments, and along each filament there is a given maximal number of crossbridges between actin and myosin. Each cross-bridge is based on the same molecular mechanism, and the force it can exert is therefore constant. As a result of the constant number of cross-bridges, the maximal force per cross-sectional area of the muscle is constant. In vertebrate muscle this maximum force is around 4 to 5 kg force per square centimeter (kgf cm^{-2}) cross-sectional area. Many muscles are less strong, but the maximal limit appears to be roughly the same for all vertebrate and many invertebrate muscles.

Could this maximal muscle force somehow be increased? All muscle contraction we know about is based on sliding filaments of actin and myosin, and if we could pack more filaments into a given cross-sectional area, the force would be increased. This is most unlikely because the diameter of the filaments is determined by the size of the protein molecules that make up the filaments, and their size is probably determined by the requirements of the molecular mechanism.

A second possible way of increasing the force would be to increase the number of cross-bridges along the overlapping filaments, and this can be achieved by increasing the length of the filaments. This has indeed been achieved in some invertebrate muscles. For example, the anterior byssus retractor muscle of the mussel *Mytilus* has a filament length of 25 μm and a maximum force of 10 to 14 kgf cm^{-2}. In this case the thick filaments are thicker so that the force borne per cross-sectional area of the filament remains acceptable (Lowy et al. 1964). These thicker filaments have a different composition and contain a protein, *paramyosin,* which has not been found in vertebrate muscle but is detected in muscles of several other invertebrates, both arthropods and molluscs (Levine et al. 1976).

Let us for a moment consider two animals of widely different body size. An elephant is obviously much stronger than a mouse, but what about elephant and mouse muscle? Both muscles have sarcomeres of the same length and are made up of the same kinds of filaments. A microscopist would have difficulty telling which muscle came from which animal because they are so similar. For the mechanical reasons just discussed, both muscles can exert the same force per unit cross-sectional area, and the length is irrelevant to the maximum force.

What about the ant's muscles? When the ant with seeming ease carries an object heavier than its own body weight, its muscles appear much stronger than mammalian muscle. However, measurements on insect muscles show that these are not stronger; they exert the same force per cross-sectional area as vertebrate muscles.

What then makes an ant seem inordinately strong? The answer is related to its size. With decreasing size of an animal, its volume, or mass, decreases in proportion to the third power of its length measurement. The cross-sectional area of its muscles (which determines the muscle force) decreases only as the square of the length measurement. Thus, the force exerted by the muscles (of similarly shaped animals), relative to the body mass, *increases* in proportion to the *decrease* in the length measurement.

Work. When it comes to external work performed by a muscle, the situation looks quite different. When a muscle lifts a weight, the external work performed is the product of the load and the distance over which it is moved.

When the muscle is free to contract when stimulated, it performs no external work because it contracted, or shortened, without exerting any force. If a load is attached to the muscle, the muscle does perform external work, which equals the product of the force and the shortening distance. If the load is gradually increased, the muscle will be able to lift the load over a shorter and shorter distance until the load is so heavy that the muscle is unable to lift it at all and the external work is again zero. This is shown by the arched curve in Figure 10.13. Between the two extremes of zero external work we obtain a curve that describes the work performed. As the load is increased from zero, the external work first increases rapidly and reaches a peak at a load about 40% of the maximum the muscle can lift. As the load is increased further, the external work drops again and eventually reaches zero.

The work that can be performed by a muscle is obviously related to its size. Consider two muscles that have the same dimensions and can develop the same maximum force per unit cross-sectional area. If they shorten to the same extent, the work they perform must be the same. Now consider two muscles of the same cross section but different lengths, one twice as long as the other, that can both contract by the same fraction of their resting length, say, by one-third. The maximum work each can perform is directly related to its initial length (i.e., twice as high in one as in the other), and because the force of contraction is the same (same cross-sectional area) the longer muscle performs twice as much work.

We can extend this argument by saying that because work is the product of force and distance, and the volume of the muscle is proportional to the product of cross section and length, the work a muscle can perform must be directly related to its volume. This, of course, is a generalization from which there are exceptions, but as a general rule it is a useful fact to keep in mind.

Power. For power output the situation is different again. Power is the work per unit time, and because a fast muscle contracts in a short period of time, the power produced is greater than in a slow muscle.

Consequently, if we compare the muscles of elephants and mice, we can expect (1) that the muscles have approximately the same contractile force per square centimeter of cross-sectional area, (2) that they can shorten to approximately the same fraction of their resting length, (3) that the work performed during the contraction of 1 g of muscle will be similar in the two animals, and (4) that the power output of 1 g mouse muscle will be much higher.

The higher power output is because the contraction of a mouse muscle takes place much faster (i.e., in a much shorter time), and its power output is correspondingly higher. This is precisely what is reflected in the metabolic rates of the animals (see Chapter 5), which are usually expressed in power units (work or energy per unit time) and are much higher in the small animal.

Muscle, how it is used

The way muscle is used by various animals differs a great deal (depending primarily on the function of the particular muscle). The demands on the flight muscles of an insect, which contract several hundred times per second, and on the muscle that closes the shells of a clam and remains contracted perhaps for several hours, are very different indeed. The best way to describe how muscle can serve different purposes is to examine some characteristic types of muscles and how they are modified to meet specific demands.

Vertebrate fast and slow muscles

Vertebrate striated muscle is composed of muscle fibers that fall into two (or more) distinct classes that are frequently referred to as fast and slow fibers. This terminology can easily lead to confusion, and it is now common to designate the fast fibers as *twitch fibers* and the slower as *tonic fibers*. Usually there is a difference in the amount of myoglobin present in the two kinds of fibers. The twitch fibers, which have a lower myoglobin content, have been known as pale or white; the tonic fibers, with a higher myoglobin content, as red. Any one muscle may consist of only twitch fibers, only tonic fibers, or a mixture of both.

The distinction between twitch (fast) and tonic (slow) muscle fibers was first made for frog muscles. The main functional difference is that twitch fibers are used for rapid movements, and tonic fibers are used to maintain low-force prolonged contractions. The twitch system is associated with large nerve fibers (10 to 20 μm diameter) with conduction velocities about 8 to 40 m s^{-1}, which lead to quick contractile responses. This twitch system is used, for example, in jumping.

The tonic (slow) system has small nerve fibers (about 5 μm diameter), which conduct at 2 to 8 m s^{-1} and lead to slower, graded muscular contractions accompanied by nonpropagated muscle potentials of small amplitude and long duration. This is the tonic system used, for example, to maintain the posture of the body (Kuffler and Williams 1953a).

The sartorius muscle of the frog, which runs along the thigh and is used primarily in jumping, consists entirely of twitch fibers. The nerve to the muscle contains large, fast-conducting nerve fibers, and each makes contact with the muscle fibers through a single terminal (the end-plate). The twitch fibers respond according to the all-or-none rule, which means that when a stimulus exceeds a certain minimum (the *threshold value*), the fibers respond with a complete contraction. A further increase in stimulus intensity gives no further increase in response. Thus, the muscle fiber either does not respond or it responds maximally. This is known as an *all-or-none response*.

Tonic muscle is different. The muscle fibers, whose behavior in many respects bears little resemblance to that of the twitch fibers, are innervated by nerve fibers, all of small diameter, whose terminal contacts are distributed along the length of the muscle fiber. In contrast to the fast twitch-producing muscle fibers, the response of the tonic fibers does not necessarily spread throughout the fiber to give a complete all-or-none response, and there is a summation on repetitive stimulation.

The response to a single stimulus is small, and the tension rises with repeated stimulus frequency. Thus, repeated stimulation of small nerves is necessary to cause a significant tension rise in tonic fibers. Increased frequencies speed the rate of tension rise. Relaxation after slow muscle fiber contraction is at least 50 to 100 times slower than after twitch fiber action. This type of response plays an important role in the general postural activity of the frog (Kuffler and Williams 1953b).

When we wish to study separately the function of the different types of muscle fibers, we do not search for muscles that contain only one type. The observed differences between a muscle consisting of only twitch fibers and one with only tonic fibers could just as well result from the different locations and uses of the two muscles. It is therefore better to examine different fiber types as they are found in a single muscle.

The gastrocnemius muscle of the cat (which runs from the knee to the heel tendon and is used for stretching the foot) can be used for this purpose. It has not only two, but three, distinct functional types of muscle fibers. This is shown in the following way. A single motor neuron in the appropriate region of the spinal cord is stimulated with the aid of an inserted microelectrode. This activates only one motor unit, which consists of many muscle fibers jointly innervated by a single nerve axon. Stimulation of another motor neuron will activate another motor unit. The different motor units may give one of three different types of contraction, depending on which particular neuron is stimulated.

Figure 10.14 shows the response of the three types of fibers: (1) a fast-contracting and fast-fatiguing type, (2) a fast-contracting and fatigue-resistant type, and (3) a slow-contracting type, which does not fatigue even during prolonged stimulation. The slow-contracting (tonic) fibers have contraction times more than twice as long as the contraction time for the fast-contracting, fast-fatiguing (twitch) fibers (Table 10.2). The difference in the force developed by the muscle is even more striking: The fast-contracting fibers develop a force more than 10 times as high as the slow-contracting fibers. This is evidently paid for in the much faster fatigue of these fibers.

Whether a muscle is twitch or tonic is not necessarily a constant inherent characteristic; somehow it seems to depend on how the muscle is used. By changing the position of insertion of the tendon of muscles in the hindlimb of a rabbit, it has been possible to make tonic muscle acquire characteristics more nearly those of twitch muscle.

In the newborn cat all muscles seem to be equally slow, and the differentiation takes place as the muscles are used. By severing the nerves to two muscles and then reconnecting each nerve to the other muscle, it has been shown that when a nerve from a fast (twitch) motor neuron is connected to a slow (tonic) muscle, the muscle is transformed into a fast (twitch) muscle. Likewise, slow (tonic) motor neurons make twitch muscles change toward tonic (Buller et al. 1960). It has become apparent, however, that cross innervation does not com-

FIGURE 10.14 The gastrocnemius muscle of a cat contains three types of fibers that respond to stimulation (40 per second) in three different ways. They may be fast-contracting and fast-fatiguing (top); fast-contracting and fatigue-resistance (center); or slow-contracting and nonfatiguing (bottom). [Burke et al. 1971]

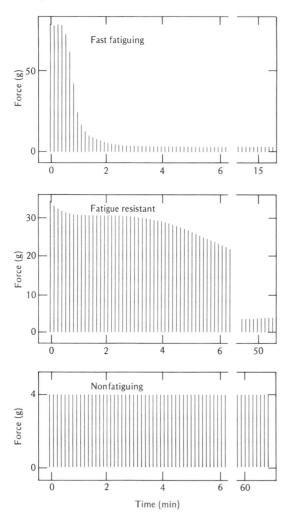

TABLE 10.2 Contraction time and maximum tetanic force developed by three different functional types of muscle fibers in the gastrocnemius muscle of a cat. [Burke et al. 1971]

	Fiber type		
	Fast-contracting, fast-fatiguing	Fast-contracting, fatigue-resistant	Slow-contracting
Twitch contraction time (ms)	34	40	73
Tetanic force (gram force)	60	20	5

swim continuously at relatively low speeds, have the two types of fibers separated in different muscle masses, which have strikingly different appearances. The tonic muscle is deep red because of a high concentration of myoglobin. It is located along the sideline and stretches in toward the vertebral column. Swimming during cruising is entirely executed by this red muscle; the large mass of white muscle (twitch type) represents a reserve of power for short bursts of high-speed activity (Figure 10.15).

The swimming muscles of sharks are also divided into similar types, with less than 20% of the muscle mass made up of tonic fibers. Again, the tonic fibers are for cruising and the twitch fibers are used during vigorous swimming such as in the pursuit of prey. The twitch fibers are capable of a high level of anaerobic power output, as would be expected during maximal activity, and accordingly have a high glycogen content. After vigorous activity the glycogen level in the fast fibers falls markedly. During prolonged periods of slow swimming, however, there is no change in their glycogen level; this indicates that they are not used at all during cruising (Bone 1966).

The separation into red and white swimming muscle has an interesting counterpart in an invertebrate. In squid the muscles of the mantle show a clear separation between layers that have a high

pletely convert one type of muscle to the other, but to some mixed form (Buller and Lewis 1964).

The muscles of fish also have twitch and tonic fibers. Pelagic fish such as mackerel and tuna, which

FIGURE 10.15 In the skipjack tuna (*a*) the red (tonic) swimming muscle, which is used during steady cruising, is located along the sideline and in toward the vertebral column (stippled in *b*). The remaining large mass of white (twitch) muscle is used in bursts of high-speed swimming. [Rayner and Kennan 1967]

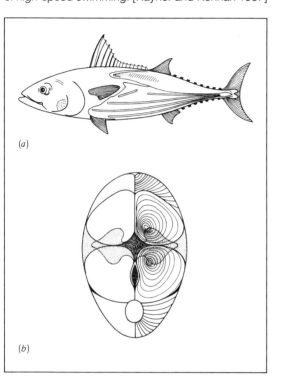

(*a*)

(*b*)

density of mitochondria and oxidative enzymes and a layer low in mitochondria but high in glycolytic enzymes. This metabolic differentiation is analogous to the red muscle of fish that supports steady swimming and the white muscle that supports burst-type swimming (Mommsen et al. 1981).

Cardiac muscle

The muscle of the heart, like skeletal muscle, contains actin and myosin filaments and cross striations identical to those in skeletal muscle, but the mitochondria are much more abundant. This is understandable in view of the constant demand on cardiac muscle for continued work throughout the lifetime of the organism.

One important contrast to skeletal muscle is that, when a contraction starts in the heart, it rapidly spreads to the entire muscle. This is a result of the characteristic structure of cardiac muscle, which in several respects differs from skeletal muscle (Figure 10.16). The heart muscle consists of a large number of branching cells that are connected end to end by specialized *intercalated discs*. As the fibers branch, they connect to adjacent fibers and form a complex three-dimensional network. The intercalated discs serve to maintain cell-to-cell cohesion; at irregular intervals along the disc there are *gap junctions*,* which are important because they form areas of low electric resistance that permit the rapid spread of excitation from one fiber to the next throughout the heart.

The other important aspect of cardiac muscle is related to the electric phenomena that occur during muscle contraction. During contraction the cell membrane of the cardiac muscle fibers, like other muscle fibers, undergoes electric changes, known as action potentials (see later in this chapter). The peculiarity of the heart muscle is that the cell membrane, after completion of an action potential, remains in a refractory state for a given period of time, long enough to allow the muscle to relax after each action potential. Because of this refractory period, the cardiac muscle cannot go into a sustained contraction (a tetanus, see p. 403); the refractory period is thus essential for the rhythmic alternation between contraction and relaxation of the heart.

Because of the spread of contraction throughout the heart, the whole heart is ''driven'' by the cells where contraction begins. In mammals the initial contraction begins spontaneously where the vena cava enters the heart. Cells located here are known as the *pacemaker* cells, and the excitation spreads

*Gap junctions are areas where the membranes of adjacent cells are brought close together, leaving a gap that is much narrower than elsewhere. The gap junctions presumably are responsible for the low electric resistance.

FIGURE 10.16 The main structural characteristics of the three major types of vertebrate muscle.

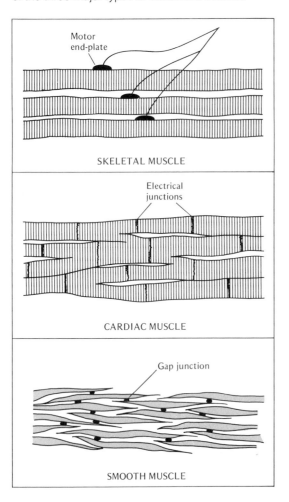

SKELETAL MUSCLE

CARDIAC MUSCLE

SMOOTH MUSCLE

from these throughout the atrium and then to the ventricle. The muscle of the atrium is separated from the ventricle, except through a thin bundle of fibers that have a smaller diameter. This results in a slowed-down conduction between atrium and ventricle, ensuring that there is a delay between the contraction of atrium and of the ventricle. As a consequence, the relaxed ventricle is filled with blood during the contraction of the atrium, where-upon the ventricle contracts and expels the blood into the arterial system.

We have seen that the contraction of the heart originates in the pacemaker cells. These are specialized muscle cells, and because the repetitive contractions originate in muscular tissue, we speak about the vertebrate heart as being *myogenic*. The hearts of some invertebrates, in contrast, are called *neurogenic* because the contractions are initiated by impulses from the nervous system. The myogenic nature of the vertebrate heart does not mean, however, that there is no influence from the nervous system. On the contrary, both rate and amplitude can be modified by nerves from the autonomic nervous system. One is the vagus nerve, which contains parasympathetic nerve fibers that release acetylcholine, causing slowing of the heart rate. Stimulation from sympathetic nerves causes the release of noradrenaline and an increase in rate and more powerful contractions of the heart muscle.

Smooth muscle

In contrast to striated muscle (in vertebrates the skeletal and heart muscle), the muscles of internal organs are smooth (also called unstriated, nonstriated, plain, or involuntary muscle). In vertebrates this type of muscle occurs in the stomach, intestine, bladder, ureters, uterus, bronchi, blood vessels, and so on. Smooth muscle also occurs in a vast number of invertebrates, but it does not form coherent functional groups in these.

Contraction of smooth muscle depends on filaments of the same proteins as in striated muscle, actin and myosin, and a supply of energy from ATP. Smooth muscle lacks the cross striations characteristic of skeletal muscle because the arrangement of the actin and myosin filaments is less orderly.

Smooth muscle has not been as extensively studied as striated muscle. There are several reasons for this. One is that smooth muscle consists of much smaller cells than striated muscle (see Figure 10.16). The fibers are only a fraction of a millimeter long and often run in many different directions. Bundles of the cells are often covered by thin sheets of connective tissue and these bundles are frequently interconnected. This makes it difficult to isolate neat preparations of smooth muscle fibers for direct study.

Smooth muscle need not be stimulated through the nerves in order to contract. It shows *spontaneous rhythmic contractions* that can vary greatly in both frequency and intensity. This is particularly true of muscles in the intestinal wall, which serve to move the intestinal contents as peristaltic waves of contraction pass along the intestine. The spontaneous activity of smooth muscle can be modified not only by nerves, but also by hormones such as adrenaline and noradrenaline.

If a strip of smooth muscle is isolated, contraction can be initiated by electrical stimulation, and contraction is propagated along the bundles of muscle cells. This is because adjacent cells and bundles are connected through gap junctions through which the action potentials are propagated. Contraction can also be initiated in strips of smooth muscle by stretching. Finally, the rate of contraction is usually much, much slower than in striated muscle; in fact, virtually all smooth muscle is incapable of producing rapid contractions.

Vertebrate smooth muscle is innervated from the autonomic nervous system by two sets of nerves. One set is stimulatory and the other is inhibitory, that is, they act in an antagonistic manner. Smooth muscle differs from voluntary muscle in that the muscle layers, mostly located in the walls of hollow organs, contain numerous nerve fibers and nerve cells. This has impeded experimentation with isolated smooth muscle, for when a stimulus is applied, it is difficult to distinguish direct effects on the muscle cells from effects on the nerve cells.

The spontaneous activity of smooth muscle can be modified, not only by nerves, but also by hormones such as adrenaline and noradrenaline.

Sudden stretching of smooth muscle causes an immediate contraction. Distention of a hollow organ is therefore frequently followed by contraction. An isolated piece of smooth muscle behaves somewhat differently, however. If a piece is loaded slightly, it stretches slowly and may, without change in the load, adopt different lengths at different times. Thus, smooth muscle has *no particular resting length*.

A distinctive feature of vertebrate smooth muscle is the slowness of response. Its most characteristic property is that it can maintain contraction for prolonged periods with very little energy expenditure (Murphy 1988).

The ability to maintain a slight contraction or *tonus* at various different lengths is most important to the economy of maintaining such tension, for in general the cost of maintaining tension is inversely related to the speed of contraction (Rüegg 1971).

Many invertebrate muscles are smooth. One special kind is the slow closing muscle of bivalves, which not only pulls the shells together, but can maintain the contraction for hours or days.

Molluscan catch muscle

Clams and mussels protect themselves by closing the shells. One or more adductor muscles pull the shells together and keep them closed against the springy action of the elastic hinge. Species that live in the intertidal zone must keep their shells closed when the water recedes, and the muscle is kept contracted for hours with the shells tightly closed under conditions that are probably completely anaerobic.

If a starfish attacks a mussel, it attaches its tube feet to the two shells and tries to pull them apart. During the ensuring endurance test, the closing muscle of the clam remains in a contracted state, known as *catch,* which is quite different from an ordinary tetanic contraction. A muscle in catch seems to maintain contraction without fatigue and with very little or no increase in oxygen consumption.

In some species, but not all, the closing muscles are divided into two portions, one of smooth and one of striated fibers. Functionally, the striated

FIGURE 10.17 (a) The ABRM responds to nerve stimulation by a brief phasic contraction. (b) If a serotonin antagonist is present, nerve stimulation causes the muscle to go into catch. (c) If an acetylcholine inhibitor is present, nerve stimulation causes relaxation of catch. [Twarog 1979]

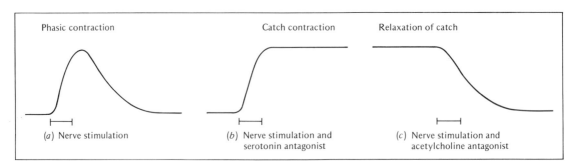

Phasic contraction   Catch contraction   Relaxation of catch

(a) Nerve stimulation   (b) Nerve stimulation and   (c) Nerve stimulation and
                                serotonin antagonist   acetylcholine antagonist

portion contracts quickly and is referred to as the fast or phasic (twitch) portion; the nonstriated is much slower and is known as the slow or tonic portion. The long-maintained contractions that keep the shells closed for extended periods are produced primarily by the slow portion of the closing muscle (Hoyle 1964).

The most studied muscle of the catch type is the anterior byssus retractor muscle (ABRM) of the blue mussel (*Mytilus edulis*), which contains only smooth fibers. It can be made to contract by electric stimuli, but, depending on the nature of the stimulus, relaxation may be complete within a matter of seconds or contraction may be maintained for minutes or hours (Johnson and Twarog 1960).

The nerve to the ABRM of *Mytilus* is a mixed nerve; that is, it contains both excitatory and relaxing nerve fibers. The excitatory fibers presumably act through the release of acetylcholine and the relaxing fibers by the release of another transmitter substance, serotonin. Direct application of serotonin to a muscle in catch causes immediate relaxation. In the absence of an activation of the relaxing system (i.e., in the absence of serotonin), probably all excitatory stimuli cause catch.

The unique feature of catch is that stiffness of the muscle (resistance to stretch) is maintained after excitation has terminated. There is strong evidence that this is because the cross-bridges remain attached and that these detach only in response to relaxing nerve impulses (or serotonin). This implies that the closing muscles of bivalves, once they have shortened and entered into catch, maintain their length without further expenditure of energy. Maintaining a state of tension without use of energy is advantageous for animals under anaerobic conditions, especially if the tension is to be maintained for prolonged periods of time.

These concepts of the control of the catch mechanism are summarized in Figure 10.17, which shows three experimental results. First, nerve stimulation alone causes a brief, phasic contraction. Second, if the relaxing transmitter serotonin is prevented from acting by an appropriate inhibitor, the muscle goes into catch. Third, if the stimulating transmitter acetylcholine is blocked, nerve stimulation of a muscle already in catch causes relaxation, evidently in response to the release of serotonin from the relaxing nerve fibers (Twarog 1979).

Crustacean muscle

Crustacean muscles show in pure form some characteristics that to varying degrees are present in the muscles of a number of other animals. We shall discuss two points: (1) the gross arrangement of the muscle fibers into parallel-fibered or pinnate

FIGURE 10.18 The muscles in the claw of a crab or lobster are arranged so that a mechanical advantage is gained. This pinnate arrangement also permits the muscle to thicken during contraction, which otherwise would not be possible within the confined space of the claw. [Alexander 1983]

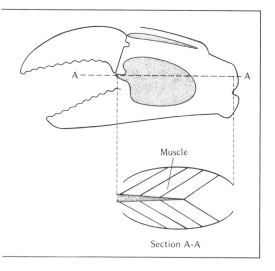

Muscle

Section A-A

muscles, and (2) the multiple innervation of a single muscle fiber and its responses to the different nerves.

Nearly everybody who has handled live crabs or lobsters has made painful contact with the impressive force these animals can exert with their claws. This is not because the strength of the muscle is extraordinarily high, but rather because of the anatomical arrangement of the fibers. In the claw of the crab the muscle is *pinnate;* this means that the fibers, instead of being parallel to the direction of the pull, are arranged at an angle, which greatly increases their mechanical advantage (Figure 10.18).

In the enclosed space of the claw, a pinnate muscle of a given volume can have more and shorter fibers than a parallel-fibered muscle. The increased force, attributable to this mechanical arrangement, is gained at the expense of the distance over which the attachment can be moved. In the crab claw, the closing muscle can exert, because of the pinnate arrangement, about twice as much force as it could if it were parallel-fibered (Alexander 1983).

Another advantage of the pinnate muscle in the crab claw is related to the confined space in which it works. As a parallel-fibered muscle contracts, its cross section increases, and in the rigid enclosure of the claw this would cause difficulties. Because of the arrangement of the pinnate muscle, however, the cavity of the crab claw can be almost completely filled with muscle, for as the fibers shorten, their angle is changed so that there is space for the thickening. This is impossible with a parallel-fibered muscle in a limited space.

A pinnate muscle can perform only the same amount of work as a parallel-fibered muscle of equal volume because the increased force of contraction is offset by a corresponding decrease in shortening distance. It should be noted that pinnate muscles are not unique to crustaceans; they also occur quite commonly in vertebrates.

The most interesting aspect of crustacean muscle is the *multiple innervation:* Individual muscle fibers may be innervated by two or more nerve fibers. In addition to multiple nerves that stimulate the muscle to contract, many arthropod muscles also have inhibitory nerves whose stimulation causes the muscle to relax if it is already in a state of contraction.

Another crustacean characteristic is that a whole muscle is often innervated by only a few or a single axon. This makes the whole muscle act much as a single unit. Gradation of response results from variations in the nerve impulses in combination with the balance between excitatory and inhibitory impulses. One muscle may, therefore, depending on the impulses it receives, act as a fast or as a slow muscle.

For example, the jumping muscle of the hindleg of a locust is innervated by three axons: one fast excitatory, one slow excitatory, and one inhibitory. The response to the fast axon is a rapid twitch; the response to an impulse in the slow fiber is barely

FIGURE 10.19 Top: a crustacean muscle is innervated from a few neurons, and the nerve fibers branch extensively and supply the entire muscle. Vertebrate muscle is innervated from a large number of neurons in the spinal cord, each one supplying a relatively small number of muscle fibers. Bottom: a crustacean muscle fiber supplied by three different nerve fibers, one for fast contraction, one for slow contraction, and one for inhibition.

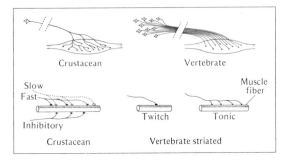

detectable, but repetitive impulses produce increasing intensity in the response. Some of those muscle fibers that are innervated by the slow axon are also innervated by an inhibitory axon. Impulses in this axon reduce the response to the slow axon and induce muscle relaxation (Usherwood 1967).

The difference in the innervation of vertebrate and crustacean muscle is shown in Figure 10.19. As stated above, the most striking characteristic of crustacean muscle is that it is supplied by two or more different nerve fibers. Often, there is a fast excitatory fiber, a slow excitatory fiber, and finally an inhibitory nerve fiber. In many muscles the majority of fibers is innervated by the fast nerve, while a smaller number is innervated by the slow nerve. In combination with the inhibitory fibers this permits a wide range in the gradation of the contractions.

The other characteristic of crustacean muscle that differs from vertebrate muscle is that a whole muscle is usually innervated from a very few neurons in the central nervous system and is reached by a correspondingly small number of axons that branch and connect to the fibers of the entire muscle.

Vertebrate muscle, in contrast, is innervated from a large number of neurons in the central nervous system, and each nerve fiber reaches only a small number of the fibers in the entire muscle. In the fast (twitch) vertebrate muscle a gradation in contraction of a muscle depends on the number of axons that are carrying excitatory impulses to their respective small groups of fibers. Inhibition does not occur in vertebrate muscle; in vertebrates, inhibition is achieved at the level of the neurons of the central nervous system and not at the level of the muscle itself (Hoyle 1957; Wiersma 1961).

Insect flight muscle

Many insects, such as dragonflies, moths, butterflies, and grasshoppers, have relatively low wingbeat frequencies, and each muscle contraction occurs in response to a nerve impulse. Because the muscle contraction is synchronized with the nerve impulse, this type is known as *synchronous muscle*.

Many small insects, such as bees, flies, and mosquitoes, beat their wings at frequencies from 100 to more than 1000 beats per second (Sotavalta 1953). This is much too fast for one nerve impulse to arrive at the muscle for each contraction, exert its effect, and decay before the next impulse arrives. It follows that for the muscles to contract at the high rates, they must be stimulated by some other mechanism. These fast muscles do have nerves, but the nerve impulses arrive at a lower frequency than the contractions, and the muscles are therefore known as *asynchronous muscles*. Because the myofibrils are very distinct in this type of muscle, it is also known as *fibrillar muscle*. It is found in four insect orders: flies and mosquitoes (Diptera), wasps and bees (Hymenoptera), beetles (Coleoptera), and true bugs (Hemiptera).

The first surprise is that the muscles inside the thorax of such a fast-beating insect are not attached to the wings at all, but rather to the wall of the thorax. There are two sets of muscles, which for convenience we can call vertical and horizontal. The primary effect of muscle contraction is to dis

FIGURE 10.20 In some insects the flight muscles together with the elastic thorax form an oscillating system that permits the wingbeat to occur at a much higher frequency than the frequency of nerve impulses. For details, see text.

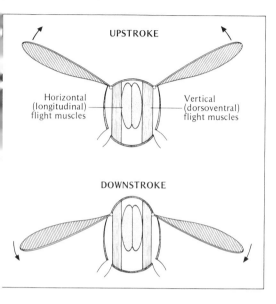

UPSTROKE

Horizontal (longitudinal) flight muscles

Vertical (dorsoventral) flight muscles

DOWNSTROKE

tort the thorax, and because the amount of shortening is very small, only a few percent of the muscle length, it can take place very fast. The most important property of asynchronous muscle is, however, that it contracts in response to being stretched.

When the vertical muscles contract, the thorax is distorted and springs into a new position with a click, so that the tension is suddenly removed from the contracting vertical muscles (Figure 10.20). However, the sudden change in shape of the thorax causes the horizontal set of flight muscles to be stretched, and this in turn acts as a stimulus for these to contract. This contraction now distorts the thorax in the opposite direction, and it suddenly clicks back to the initial position, releasing the tension on the horizontal muscles. This stretches the vertical muscles, which respond with a new contraction, and so on.

The elastic thorax with its muscles constitutes a mechanical oscillator, which by the energy input from the muscles is kept oscillating without being damped out by frictional forces. The most important damping is, of course, the drag on the wings as they propel the insect, and most of the muscle work is used to overcome this drag. The flight continues although the nerve impulses arrive at intervals of several muscle contractions. In some cases as many as 40 wingbeats have been recorded for each nerve impulse. Nerve impulses are needed, however, both for the initiation of flight and for maintenance of continued flight activity (Pringle 1949, 1957).

SKELETONS

To exert their force muscles must be connected to some mechanical structure; they cannot do much for an animal unless they have something to pull on. The appropriate mechanical structure for transmission of force is the skeleton. It usually consists of a rigid, jointed structure, although a fluid can also be used to transmit force, the same way the hydraulic system is used in an automobile.

We have the following major categories to consider:

Rigid skeleton
  Endoskeleton: vertebrates
  Exoskeleton: insects, crustaceans

Hydraulic skeleton
  Fluid + soft walls: worms, octopus, starfish
  Fluid + rigid elements: spider legs

A vertebrate has a rigid *internal skeleton,* the bones; an insect or a crustacean uses its stiff cuticle as an *external skeleton.* Some animals (e.g., earthworms) use their internal fluids for transmission of force; they have *hydraulic skeletons.*

Rigid skeletons

A vertebrate has most of the soft tissues of the body draped around the skeleton. Is this the best arrangement? An arthropod has its skeleton on the outside, and all the soft parts are well protected. Is this a better arrangement?

Let us compare the two kinds of structures. For reasons of simple mechanics, a hollow cylindrical tube can support a much greater weight without buckling than can a solid cylindrical rod made from the same amount of material. Therefore, if an animal can afford only a certain weight of skeleton, it is advantageous to use the material for a tube. This is precisely what arthropods have done, and they seem to have an advantage over vertebrates in this regard.

Why, then, do not vertebrates utilize a similar arrangement? First, for a constant weight of material, the rigidity of a hollow cylinder decreases very rapidly as its radius is increased, and eventually the wall will be so thin that the entire structure collapses. To prevent buckling of the hypothetical exoskeleton of a very large animal, it would be necessary to increase the thickness and thus the weight.

An exoskeleton has another mechanical disadvantage: Even if it is strong enough to provide the necessary support, it will be very sensitive to impact, with the risk of buckling as well as being punctured. This danger, which is of little significance for small animals such as insects, increases rapidly with increasing body size.

Another difficulty with an exoskeleton is growth. Most adult insects never grow; they retain the same rigid shape and size throughout their lifetime, and the larval growing phase is equipped with a softer cuticle that can be stretched. Even so, the larval skin is shed at intervals to permit adequate size increase.

Aquatic arthropods that molt periodically go through a period of increase in size when the old exoskeleton has been shed and the new exoskeleton is still soft and can be stretched. Aside from the lack of protection and the inability to move during this ''soft'' period, the mechanical disadvantage of a periodic loss of rigidity is not too inconvenient for an aquatic animal, for its body weight is supported by the water. For a land animal, the problem would be enormous: Any sizable animal would collapse under its own weight if it were to go through a molting process.

Nevertheless, some of the protective advantages of an exoskeleton are, in fact, utilized by vertebrates: The brain, certainly a vital organ, is enclosed in a rigid bone structure. The braincase is not part of the locomotory skeleton, however.

One advantage of an exoskeleton stems from its elastic properties. Elastic deformation of skeletal elements constitutes a storage of energy that is utilized for purposes of locomotion; some examples are insect flight (discussed above) and the jump of fleas (to be examined later).

Hydraulic skeletons

There is one important difference between rigid and hydraulic skeletons. A rigid skeleton consists of elements that usually act around a pivot point (a joint). The muscles are attached at both ends, and the force they exert is transmitted through the rigid element and acts somewhere else. For the hydraulic skeleton to work, a fluid must be enclosed in a limited space, and the muscle force must be used to produce pressure in this fluid (i.e., the muscle must usually enclose the fluid space in circular layers). In this case, the muscle does not have a point of attachment; a ring of muscle pulls only on itself.

Earthworms

The role of the hydraulic skeleton in the locomotion of worms has been studied in detail. The common earthworm is a familiar example. Its body wall has layers of muscles that run in two distinct directions, either in circular sheets around the body or in the longitudinal direction.

The movement of the earthworm when it crawls forward is approximately as follows. A wave of contraction of the circular muscles begins at the anterior end. This causes the body to become thinner and elongated, pushing the anterior end forward. The circular contraction passes down the body as a wave, followed by a wave of contraction of the longitudinal muscles.

FIGURE 10.21 The crawling earthworm uses its body fluids as an internal hydraulic skeleton. Contraction of circular muscles pushes the front end forward and is followed by a contraction in longitudinal muscles that thickens the body. The thickened segments remain in place relative to the ground as the other parts move forward. The track of individual points on the worm's body and their movements relative to each other are shown by the lines running obliquely forward from left to right of the diagram. Diagram was prepared from a movie film. [Gray and Lissmann 1938]

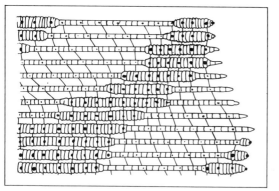

Tiny bristles on the side of the body prevent backsliding, and the anterior end pulls the adjacent part of the worm in its direction. In this way a portion of the worm is pulled forward, the motion passing as a wave backward on the heels of the wave of circular contraction. The cycle is repeated with a new elongation at the anterior end, the worm moving forward as the waves of contraction move backward (Figure 10.21).

When the earthworm moves, the muscles exert their force on the fluid contents of the animal; the circular muscles build up a pressure that causes the longitudinal muscles to be stretched, and when the longitudinal muscles in turn contract, the pressure stretches the circular muscles and the worm thickens.

The body of the earthworm is separated into compartments by septa. This helps a great deal in locomotion, for it makes the various regions of the worm relatively independent of the rest of the worm.

Many worms, the lugworm (*Arenicola*), for example, lack such septa (Seymour 1971). One disadvantage of the lack of septa is that if the worm is wounded and body fluid is lost, it is virtually unable to move. In contrast, as nearly everybody knows, if an earthworm is cut in two, each part can move about more or less like an intact worm.

Spider legs

Spiders cannot extend their legs with the aid of muscles because they have no extensor muscles. To flex the legs they use muscles, but how are the legs extended? There are two possibilities: the use of elastic forces or of hydraulic forces.

Elastic hinge joints in the legs could be used if they were arranged to extend the legs whenever the flexing muscles relax. This would be similar to the hinge of bivalves, in which the shells gape when the closing muscle relaxes. We can exclude this possibility by a simple observation: The joints of a detached spider leg remain neutral over a wide angle.

The other possibility, using hydraulic pressure, is what spiders do. This can be shown in various ways, most simply by detaching a leg and pinching the open end with a pair of forceps. This closes the opening and raises the fluid pressure inside, and as a result the leg extends. Furthermore, there is a direct relationship between the internal pressure applied to the leg and the joint angles of the leg. Also, if a spider is wounded so that blood fluid is lost, it is completely unable to extend the legs, which remain flexed in a disorderly fashion up against the body. The blood pressure in a spider is surprisingly high – up to 400 mm Hg, or about 0.5 atm (ca. 50 kPa). This differs very much from the blood pressure of other arthropods, which is usually only a few millimeters of mercury (Parry and Brown 1959a).

Some small spiders, known as jumping spiders, can leap on their prey at distances of more than 10 cm. For jumping they use a sudden extension of one pair of legs. Can the spiders use hydraulic pressure for this very rapid motion?

The jump itself is produced entirely by the sud-

JUMPING SPIDER The jumping spider (*Sitticus pubescens*), like other spiders, lacks muscles for extension of its legs. When this spider leaps on its prey, it uses the hind pair of legs, which are extended hydraulically by blood pressure. [Courtesy of G. A. Parry, Cambridge University]

den straightening of the last (fourth) pair of legs. The extension torque necessary for the jump can be estimated from the length of the jump, and the required hydraulic forces, which are difficult to measure directly in the jumping spider (it weighs only about 10 mg), fall within the range of blood pressures commonly measured in spiders.

High-speed photographs of jumping spiders support the view that blood pressure is involved, for at the moment of the jump the minute spines on the legs become erect precisely in the way they do when there is an increase in fluid pressure in the leg. The exact release mechanism is not fully understood, however (Parry and Brown 1959b).

LOCOMOTION: BIOMECHANICS

The skeletal muscles are used primarily to provide mechanical energy for animals to move about. Animals can walk and run and jump, fly in the air, and swim in the water. With the exception of ameboid, ciliary, and flagellar locomotion, and a few more unusual means of moving about, muscles are universally used in animal locomotion. In Chapter 5 we discussed the energy requirements for moving about; in the following pages we shall consider some mechanical aspects of these problems.

Running

An animal that runs on a horizontal surface uses power in three major ways:

1. Moving through the air requires power to overcome the frictional drag of the air.
2. As the animal runs, its center of mass is constantly lifted and falls again. Work is done each time it is raised, increasing its potential energy, and most of this is lost again as the center of mass is lowered.
3. As the animal moves, the limbs are constantly accelerated and decelerated again. Work is required to accelerate a limb and provide kinetic energy; to decelerate the limb requires additional work.

The mechanical power to overcome friction in joints is minimal, but some goes into increased breathing movements and pumping activity of the heart. The amount, however, is fairly small, compared with the power required for the mechanical movements of the body.

The air resistance can be calculated from fluid dynamics considerations and from direct measurements of wind pressure on shapes of similar area to the running animal. It can also be measured from the increase in oxygen consumption during running in winds of various speeds.

The results show that for humans the energy cost of overcoming air resistance in track running at middle distance speed may be 7.5% of the total energy cost. The air resistance can be expected to increase approximately with the square of the speed, and at sprint speed the cost may be 13% of the total energy cost of running. Running at low speed or walking involves negligible air resistance. It has been found that running 1 m behind another runner virtually eliminates air resistance (Pugh 1971). For other mammals, except those that run at extremely high speeds, air resistance is of minor importance in the total cost.

The friction against a firm substratum is likewise negligible (most of us know from experience that walking in loose sand is an entirely different matter). Thus, the work done on the environment (air resistance plus frictional resistance against the ground) is a small fraction of the total output, and as a result, we must conclude that most of the energy used in running is dissipated internally as heat. (It was mentioned in Chapter 5 that the situation is different for a flying bird, where perhaps 25% of the power output goes into work done on the environment.)

Running uphill

Additional power is required to run uphill. A body that moves uphill acquires potential energy, the product of its weight and the vertical distance through which it has moved. We can therefore predict that the energy required to move one unit of body weight for one unit of vertical distance should be the same for animals of widely different body sizes.

To lift 1 kg of body weight 1 m vertically increases its potential energy by 1 kgf m (which equals 9.8 J or 2.34 cal and corresponds to an oxygen consumption of 0.49 ml O_2). If the efficiency of the muscles of large and small animals is similar, say 20%, they would all use five times this amount. This expected similarity in the relative metabolic cost of moving vertically has been amply documented by many investigators for animals ranging in size from mice to horses.

This similarity in the cost of moving vertically has an interesting consequence: It must be much easier for a small animal to run uphill than for a large animal. Why?

The answer is simple. The normal specific met-

FIGURE 10.22 Running uphill on a 15° incline causes an insignificant increase in the oxygen consumption of a small animal, such as a 30-g mouse. For larger animals, in this case a chimpanzee, running uphill greatly increases the oxygen consumption. Similarly, running downhill makes little difference for the mouse, but it makes running much cheaper for the chimpanzee. [Taylor et al. 1972]

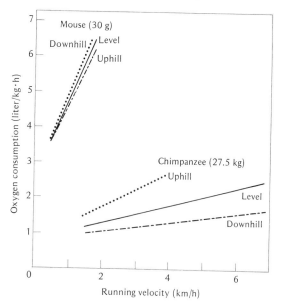

abolic rate of a 30-g mouse is about 15 times higher than for a 1000-kg horse. Because the vertical component of moving one unit body weight is the same for the two animals, the relative increase in metabolic rate attributable to the vertical component should be only one-fifteenth as great in the mouse as in the horse.

This surprising conclusion has been tested experimentally. Mice were trained to run on a treadmill at a 15° incline, which indeed is a rather steep hill. The results showed no statistically significant difference in the oxygen consumption of mice running uphill, on the level, or downhill. For chimpanzees (17 kg body weight) the oxygen consumption nearly doubled on the uphill slope, relative to running horizontally (Figure 10.22).

Taylor concluded that moving vertically at 2 km

per hour requires an increase in oxygen consumption of 23% for a mouse (hardly noticeable), of 189% for a chimpanzee (nearly doubling), and of 630% for a horse (heavy work). This explains the ease with which a squirrel moves up and down a tree trunk, apparently without effort. For an animal of such small size it makes little difference whether it runs up or down.

Energy saved by elasticity

A tennis ball bouncing on a hard surface bounces repeatedly, but the height of the bounce gradually decays. When the ball hits the surface, its kinetic energy is taken up in elastic forces and recovered again as the ball bounces off. As the ball rises, its kinetic energy is transformed into potential energy, only to be changed into kinetic energy on the way down again. Energy is gradually lost because of friction in the air as well as friction during elastic deformation of the tennis ball. However, a small input of energy, just enough to overcome the frictional losses, can keep the ball bouncing.

Does this have any similarity to animal locomotion? We saw that a great deal of energy goes into accelerating and decelerating the limbs and in moving the center of mass in the gravitational field. If elastic elements could act as springs and take up some of the kinetic energy, a great deal of energy could be saved.

The elastic elements we must consider are the tendons and the muscles. At first thought it may seem surprising that muscles can be considered as elastic, for their main function is to contract and exert a force. However, consider a running human and the moment his foot hits the ground. If the leg muscles were flaccid and limp, the person would collapse; obviously, the muscles must resist stretching and thus already be contracted in advance of the precise moment we consider. The kinetic energy of the body provides the force to stretch the muscles, and these resist, which requires work. Can some of the kinetic energy be taken up in elastic elements and be recovered?

The most significant studies of these problems have been made by a group of Italian investigators in Milan, among them Rodolfo Margaria and Giovanni Cavagna, and their results are most helpful in understanding the energetics of locomotion.

The fluctuations in potential and kinetic energy during running and measurements of the forces exerted on the ground can be analyzed with the use of a *force platform,* an instrument that in principle is quite simple. The subject under study, whether animal or human, runs over a solid platform of known mass, suspended on very stiff springs, while force transducers measure the minute movements of the platform in three dimensions. The recorded movements can then be used to calculate the forces and their direction during the entire time a foot is in contact with the force platform. From the information, it is possible to compute the total mechanical work needed for a single step. It turns out that the work performed per kilometer is independent of speed, the total mechanical work for a running person being about 0.40 to 0.50 kcal kg^{-1} km^{-1} (Cavagna et al. 1964).

Simultaneous measurement of the oxygen consumption shows how much metabolic energy the person uses. A calculation of the efficiency (the ratio of work output to energy input) gives an apparent efficiency of approximately 50%. This value seems quite unreasonable, for all available evidence shows that the mechanical efficiency of muscles in transforming chemical energy to external work cannot exceed 25%. This discrepancy, however, could vanish if one-half of the mechanical work recorded by the force platform were obtained from stored kinetic energy recovered by the elastic recoil from the stretched leg muscles and tendons. The conclusion that elastic storage is important in human and animal locomotion has been amply confirmed by later investigations.

Anyone can easily convince himself that elastic recoil is important in saving energy. When doing deep knee bends from a standing position, it is much easier to raise the body if the legs are stretched immediately, than if one waits a second or so in the low position before standing straight again.

TABLE 10.3 Oxygen consumption and mechanical efficiency of doing knee bends. In one case, the subjects made a short pause in the squatting position; in the other case, they took advantage of elastic rebound by straightening immediately. [Thys et al. 1972]

	Oxygen consumption (liter O_2 min^{-1})	Efficiency
No rebound	1.89	0.19
Rebound	1.49	0.26
Difference (%)	−22%	+37%

During the bending movement of the knees, the extensor muscles are kept tensed, controlling the downward movement. If they shorten again immediately, elastic recoil helps, but if the tension in the muscles is allowed to decay before raising the body, there is no elastic rebound. Is the apparent difference in effort real?

An objective way to finding out is to measure the oxygen consumption. When the knee-bending exercise is carried out on a force platform, the mechanical work performed in the stretching can be measured accurately, assuring that there is no difference in the two exercises. A difference could hardly be expected, because a subject lifts the same mass (his body) to the same height each time he stretches, and the measurements confirm this conclusion.

In a series of experiments the subjects performed exactly 20 knee bends per minute, in one case stretching immediately after flexing the knees, and in the other stretching after an interval of 1.5 seconds in the squatting position; the total number of knee bends per minute being the same. The results confirm the subjective impression (Table 10.3). When the exercise was performed with the advantage of elastic recoil, the oxygen consumption decreased by 22%. The mechanical work performed was measured with the force platform, and it was therefore possible to calculate the efficiency of the work, which increased by an amazing 37%.

A similar exercise is to stand on tiptoes with stiff knees and perform repetitive small jumps on both feet. In this exercise elastic energy is stored in the contracted muscles of the calf as these are forcibly stretched. Thus, only part of the positive work done to lift the body is derived from energy-yielding chemical processes in the muscles; the other part is derived from elastic energy stored in the contracted muscles. The latter accounts for as much as 60% of the total, with only 40% coming from metabolic processes (Thys et al. 1975).

The fact that in exercise and locomotion muscles often shorten actively immediately after being stretched by an external force is extremely important. Experiments on isolated muscles show that the force developed when the muscle shortens after being stretched is greater than that developed, at the same speed and length, when the muscle shortens starting from a state of isometric contraction. There is evidence that in addition to the elastic characteristics of the muscle, there may also be some change in the contractile process, for the force-velocity characteristics change during contraction after stretching (Cavagna and Citterio 1974).

A different approach has been used in studying animal performance. A trained German shepherd performed a single running jump, or long jump, clearing about 3 m over a series of hurdles. In these experiments a force platform was placed at the spot of takeoff, giving information about direction and forces throughout the jump. The muscles we are concerned with are the combined gastrocnemius and plantaris, the only muscles exerting a force on the ankle (Figure 10.23). These muscles must exert a force throughout the period of contact of the hindpaws with the ground, and from the force on the platform and the moment arm of the bone at the ankle joint, the total force exerted by the two muscles can be computed.

The change in force during landing and takeoff is shown in Figure 10.24. In this graph the force exerted by the combined gastrocnemius and plantaris is plotted against the change in length of the muscles, calculated from films taken of the jump. The numbers on the graph correspond to the frames

FIGURE 10.23 Outline tracings of a dog making a running long jump, taking off from a force platform. The numbers correspond to film frames, taken at 56 frames per second. The arrow superimposed on each outline represents the direction and the mag-nitude of the resultant force in that frame. Dots mark joints used for calculation of forces. To the right is an outline of the skeleton of the hindlimb at the moment of takeoff, showing the location of the gastrocnemius and plantaris muscles. [Alexander 1974]

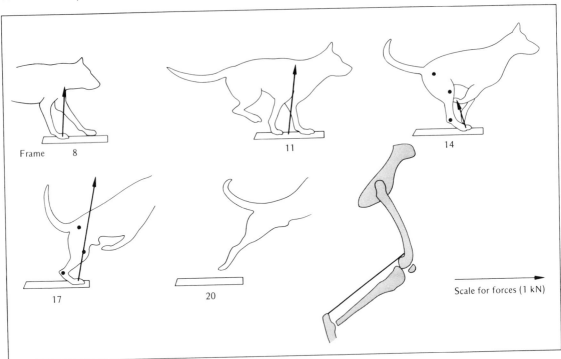

on the film, beginning at an instant after the hindlegs hit the platform. The next frames show the increase in force as the muscles are being stretched (15 and 16); then the muscles shorten again (17 through 20), and the force reaches zero as the hindlegs leave the platform (20).

The record is extremely interesting. During the extension of the muscles, work is being done on the muscles, the work being force times distance. During the takeoff phase, this work is being returned, and because it follows exactly the same path, the work (force × distance) during the recoil must be exactly the same as the work done on the muscle during the stretching. Thus, the gastrocnemius and plantaris muscles with their tendons have be-haved as passive elastic bodies. From the amount of stretching of the tendons and their elastic properties (the force being known from the records), it can be calculated that most of the elastic energy probably is stored in the tendons.

If no work is done by these muscles, what brings the dog over the hurdles? Only a part is attributable to kinetic energy accumulated during the run and recovered elastically. The extensors of the hip shorten and perform work throughout the period of contact of the hindfeet with the ground. The mass of these muscles is substantial, and they contribute a major part of the work that enters into the long jump (Alexander 1974).

Kangaroos jump more than most animals; what

FIGURE 10.24 Force exerted by the combined gastrocnemius and plantaris muscles during the jump of a 35-kg dog, plotted against the changes in muscle length. Numbers correspond to film frames, as in Figure 10.23. A force of 1 kN equals 102 kgf; thus the maximum force of about 2 kN (frame 16) equals about 200 kgf. [Alexander 1974]

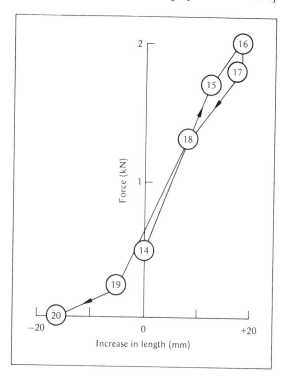

not change much in spite of a threefold increase in speed; the increase in speed is almost entirely the result of an increase in the length of each hop.

What does this signify? If the duration of a single hop does not increase, the vertical height of the hop must be unchanged, and only the length is increased. That is, no increase in vertical force is necessary, and only the forward vector of the force is increased to lengthen the hop. The longer hop signifies an increase in the kinetic energy of the kangaroo as it hits the ground, and if a major fraction of this kinetic energy can be taken up in elastic elements, it will constitute a great savings in energy required for the increased speed.

Measurements from force platforms and calculations of the forces in the long Achilles tendon of a kangaroo suggest that the savings attributable to elastic storage in the tendon amounts to 40% of the positive work the leg muscles must do during contact of the hindfeet with the ground (Alexander and Vernon 1975).

More about jumping

Fleas and grasshoppers can make jumps that are perhaps 50 times the length of their bodies. This has given rise to the popular pastime of calculating how far a human should be able to jump, could he jump in the same proportion. In North America he would jump to the top of the Empire State Building, and in France he would clear the top of the Eiffel Tower. Let us instead examine this question realistically.

Assume that a grasshopper weighs 1 g and can make a standing jump of 50 cm, both figures being quite realistic. Assume a larger animal of isometric build with all linear dimensions increased exactly 10-fold. What is the effect of scaling on the expected height of a jump? The large animal will have a mass 1000 times that of the small one, but the cross section of its jumping muscles (which determines the force the muscle can exert) will be increased only 100-fold. Relative to its larger mass, the large animal therefore has only one-tenth the force available to accelerate the body.

can we learn from their ways? When kangaroos move slowly with the use of all four feet and the tail, their oxygen consumption increases steeply with increasing speed (Figure 10.25). However, at higher speeds, when they jump on the hindlegs only, there is no further increase in the rate of oxygen consumption as the speed increases from 8 km per hour to 22 km per hour, almost a three-fold increase (Dawson and Taylor 1973).

In the context of our discussion of elasticity, it is interesting to examine the kangaroo's frequency of jumping. The number of jumps per minute does

FIGURE 10.25 Records from two kangaroos (18 and 28 kg) moving on a treadmill: (a) oxygen consumption; (b) stride frequency; (c) stride length. The left part of each graph refers to movement on all four legs and supported by the tail ("pentapedal" movement); the right part refers to hopping on the powerful hindlegs with the heavy tail used as a balancing device. The dashed line in (a) represents oxygen consumption for an 18-kg mammal running quadrupedally. [Dawson and Taylor 1973]

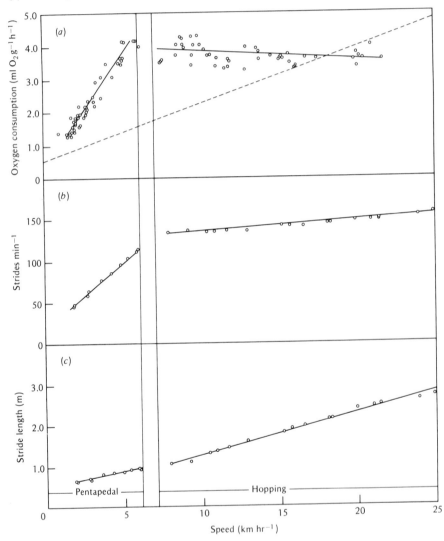

Since acceleration equals force/mass, the acceleration during takeoff is also one-tenth that in the small animal. However, as all linear dimensions are increased 10 times, acceleration continues throughout a takeoff distance that is 10 times the takeoff distance of the small animal. The net result is therefore that the takeoff speed is the same for the two animals.

TABLE 10.4 Mechanical performance of four animals in standing jumps. The height of the jumps are all within the same order of magnitude, although the body masses differ by a factor of more than 100 million. (The flea is so small that air resistance is significant, and to reach the indicated height it must in fact have a higher than calculated initial acceleration, estimated at 320 g.)

	Flea (Pulex)[a]	Click beetle (Athous)[b]	Locust (Schistocerca)[c]	Man (Homo)[d]
Body mass, m	0.49 mg	40 mg	3 g	70 kg
Height of jump, h	20 cm	30 cm	59 cm	60 cm
Acceleration distance, s	0.075 cm	0.077 cm	4 cm	40 cm
Takeoff speed, $v = \sqrt{h \cdot 2g}$	190 cm s^{-1}	240 cm s^{-1}	340 cm s^{-1}	343 cm s^{-1}
Time of acceleration, $t = 2s/v$ (takeoff time)	0.00079 s	0.00064 s	0.00235 s	0.233 s
Acceleration, $a = v/t$	241 000 cm s^{-2}	374 000 cm s^{-2}	14 500 cm s^{-2}	1471 cm s^{-2}
Acceleration relative to gravitational acceleration (g)	245 g	382 g	15 g	1.5 g

[a] Bennet-Clark and Lucey (1967). [b] Evans (1972, 1973). [c] Hoyle (1955), Alexander (1968).
[d] Estimate based on center of mass lowered to 60 cm above ground at beginning of jump, accelerated over 40 cm to 100 cm at takeoff, and lifted to 160 cm. World record for standing jump is about 165 cm above ground.

After takeoff both animals are slowed by the effect of the force of gravity. Because both the kinetic energy at takeoff and the retarding force of gravity are proportional to the body mass, the deceleration attributable to gravity permits the two animals to rise to equal heights. (Here we have disregarded the additional loss of speed caused by air resistance, which for a very small animal such as a flea becomes important.)

We can conclude that if the muscles of the small and large animals exert the same force per cross-sectional area, a small and a large animal of isometric build should be able to jump to exactly the same height.

Consider now the actual jumping performance of a variety of animals (Table 10.4). (We should consider only standing jumps, for a running jump utilizes the kinetic energy of the animal to increase the height of the jump. We will therefore disregard the world record high jump for man or a running horse, which both are about 2 m.) In a standing jump a man can clear about 1.6 m. His center of mass, however, is not lifted over this distance; it is lifted over a shorter distance, for it is not at ground level when he jumps but rather at about 1 m (see note to Table 10.4). Thus, man and grasshopper raise their center of mass by roughly the same amount, although the difference in their body mass is between 10 000- and 100 000-fold.

Of course, in reality animals are not isometric. Nevertheless, it is amazing how similar are the jumping records for a variety of animals. It is a matter of simple physics that man and other mammals do not jump in some proportion to their body lengths. Such jumps are indeed impossible.

The conclusion that similar animals irrespective of body size should be able to jump to equal height can also be stated as follows. Assume that the jumping muscles make up the same fraction of body mass. The muscle force is proportional to the cross-sectional area, and the shortening is proportional to the initial length of the muscle. The cross section times the length is the volume of the muscle, and the energy output of a single contraction is the product of force and distance. Energy output is therefore proportional to the muscle mass, and, in turn, to body mass. Because a jump uses only a single contraction of the jumping muscles, the work performed during takeoff and used for acceleration is the same relative to body mass. The conclusion is that isometrically built animals of different mass should all jump to the same height, provided their muscles contract with the same force.

The galago. How is it possible, then, for some animals to jump much higher? The record for a standing jump probably belongs to the lesser galago, a small tropical primate that weighs about 250 g. Under well-controlled conditions a galago has jumped 2.25 m vertically (Hall-Craggs 1965). This is more than three times the height of a standing jump for man. Unless the muscles can produce more force per square centimeter cross section (unlikely), the galago's superior performance can be explained only by the animal's having a larger muscle mass (more energy from the takeoff contraction) and possibly a more favorable mechanical structure of the limbs.

The galago indeed has large jumping muscles, nearly 10% of the body mass (Alexander 1968) or about twice as much as man. If we assume that all combined mechanical advantages of a highly specialized jumping animal could account for a 50% increase in performance, this in combination with a doubling in the mass of the jumping muscles, could account for a three-fold increase in jumping performance – by no means an unreasonable approximation. As a standing jump this is more impressive than the jump of the locust, but we must remember that the galago is a warm-blooded animal highly adapted to making long jumps in its natural jungle habitat.

The flea. Aside from the fact that animals are not completely isometric, we meet another difficulty. The smaller the animal, the shorter the takeoff distance, and since the takeoff speed should be the same for all, the smaller animal must accelerate its body mass much faster. Because much less time is available for the takeoff, the power output of the muscle must be increased accordingly (i.e., the muscle must contract very fast).

For an animal the size of a flea, the takeoff time is less than 1 ms, and the distance over which ac-

FIGURE 10.26 A click beetle (*Athous haemorroidalis*) lying on its back, ready for a jump, with the peg set on the lip of the pit in the hind part of the body [Evans 1972]

celeration takes place is only 0.75 mm. The average acceleration during takeoff exceeds 2000 m s^{-2}, or roughly the equivalent of 200 g (Rothschild et al. 1972).

Muscles just cannot contract this fast, so how can the flea jump at all? The flea uses the principle of the catapult and stores energy in a piece of elastic material (*resilin*), at the base of the hindlegs. Resilin is a protein with properties very similar to rubber (Weis-Fogh 1960; Andersen and Weis-Fogh 1964). The relatively slow muscles are used to compress the resilin, which returns the total energy again with close to 100% efficiency when a release mechanism is tripped. In this way the elastic recoil works much the same way as a slingshot and imparts the necessary high acceleration to the flea.

The click beetle. A click beetle placed upside down on a smooth, hard surface struggles for a moment to right itself, but when its legs find nothing to hold on to, the beetle stops moving. After a pause, the beetle springs into the air with an audible click. If it falls again upside down, it continues clicking until it does land right side up. This legless mechanism for jumping takes the beetle to heights as much as 30 cm, and the animal may reach a takeoff acceleration of nearly 400 g.

The mechanism is simple. The first segment of the thorax has a peg that points backward and fits part way into a pit in the hind section of the body (Figure 10.26). When the peg rests on the lip of the pit, it prevents the large jump muscle in the prothorax from swinging the front half of the body upward. The contracting muscle therefore builds up tension and acts like a spring; when the peg finally slips, the beetle jackknifes and the energy stored in the muscle provides the instant force needed for takeoff. Careful measurements of the jumping action show that during the click the center of mass of the beetle is raised by about 0.6 to 0.7 mm in just over 0.5 ms.

The click beetle, like the flea, is too small for the muscles to provide the necessary acceleration during the short acceleration time. Half a millisecond is not sufficient for muscle contraction, and the best way of obtaining the energy at the necessary speed is to store it in an elastic structure.

Flying and swimming

Air and water are fluid media, and animals moving through them have similar problems. They do not have a solid substratum to support their weight, and the forces needed to make them move are exerted against a medium that virtually slips away. Air is much less dense and viscous than water, a difference that has substantial effects on animal locomotion in the two media, but there are many fluid dynamics principles in common.

Flying

Judging by numbers, flying is extremely successful. Of the roughly 1 million living animal species, more than three-quarters are flying insects. The ability to fly has evolved independently at least four times: in insects, in the extinct large flying reptiles (Pterosaurs), in birds, and in mammals (bats). The size of actively flying animals ranges from about 1 μg for some tiny insects to more than 10 kg for the largest flying birds.

The power an animal needs to propel itself through the air in forward flight has two components: It must provide lift equal to its weight, and it must provide forward thrust equal to the frictional drag on the animal as it moves through the air. In flying animals lift and thrust are provided by the same surfaces. As we shall see later, this is not always true for animals swimming in water.

The main power for bird flight is provided by the large pectoral muscles, which in all birds make up about 15% of the body mass, irrespective of the size of the bird. The muscles responsible for the upstroke of the wing constitute only about one-tenth of the total mass of the flight muscles and thus cannot provide any major fraction of the power required for forward flight. There are, of course, minor deviations, but as a general rule, in birds the engine makes up the same fraction of the aircraft, irrespective of its size (Greenewalt 1975).

As a group, hummingbirds are different; their flight muscles make up 25 to 30% of the body mass. This is in accord with the greater power needed for hovering than for flapping forward flight (Weis-Fogh 1973). Furthermore, the muscles responsible for the upstroke constitute about one-third of the total mass of the flight muscles. This indicates that for hovering flight, both upstroke and downstroke are important in providing power, and because the bird remains in the same spot, all the power goes to provide lift.

The conclusion that lift is provided during both upstroke and downstroke in hovering hummingbirds is confirmed by analyses of high-speed movies of the wing movements (Figure 10.27). During hovering flight lift is provided, as could be expected, during the downstroke. In the upstroke the wing is then twisted so that lift is now provided by an airfoil virtually turned upside down (Stolpe and Zimmer 1939). Several other birds – gulls, pigeons, and doves – hover in a similar way with lift produced both in the downstroke and in the upstroke. Other birds, such as crows and starlings, use their wings differently and hover so that lift is produced in the downstroke but not in the upstroke (Dathe and Oehme 1978).

In spite of the intricate fluid dynamics theory of flight, both bird flight and insect flight are amazingly well understood (Pennycuick 1969; Tucker 1973; Weis-Fogh 1973, 1976; Lighthill 1974). The power required for bird flight is an extremely com-

FIGURE 10.27 The hovering hummingbird obtains lift during both upstroke and downstroke of the wings. The downstroke is actually a forward stroke (left), and in the return stroke (right) the wing is twisted, reversing the airfoil, thus providing lift in both strokes. The horizontal vector is reversed so that the hovering bird remains in place. [Stolpe and Zimmer 1939]

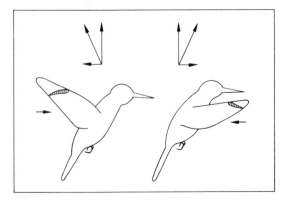

plex function of the size and shape of the bird, the speed, the size of the wings, and the shape of the wings. The exact analysis is further complicated by the constantly changing speed and shape of the wing during the wing stroke.

Aerodynamic theory (Lighthill 1974) suggests that the power required for bird flight should increase in proportion to the body weight to the power 1.17. Tucker's analysis, which includes consideration of empirical measurements, gave results which, in a simplified form, suggest that required power input should be proportional to the bird's weight (Tucker 1973). For the following argument it makes little difference whether we use the exponent 1.17 or 1.0.

We know that the resting metabolic rate of a bird increases in proportion to the body mass to the power 0.75 (page 197). Although we do not know for sure, we can assume that the maximum sustainable power output is similarly related to body size. This simply means that available power does not increase as rapidly with body size as the power required for

FIGURE 10.28 Power required for bird flight compared with maximum available power. The familiar relationship between resting metabolic rate and body mass of birds is drawn as a regression line with the slope 0.75. A regression line for an assumed maximum power output of 15 times the resting rate is drawn as a parallel line at a level 15 times higher. The estimated power required for flight is directly proportional to body mass, and a regression line with the slope 1.0 is drawn to intersect maximum power output at 10 kg body mass, the approximate size of the largest flying birds.

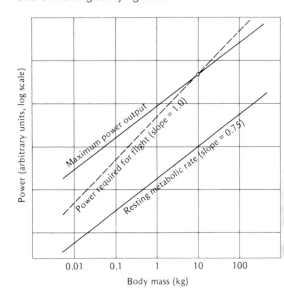

flight. This is shown in Figure 10.28, where the curve for power required for flight is drawn to cross the maximum available power at a body mass of 10 kg, the approximate size of the largest flying birds. It means that a very large bird, such as the ostrich (ca. 100 kg), even if it were anatomically designed to fly, could not be expected to muster enough power for steady-state forward flapping flight.

It is interesting to compare this information with the details of human-powered flight in which the famous Kremer Prize was won by a well-trained athlete weighing 62 kg who flew a plane weighing

32 kg. By cranking the propeller with a bicycle-type drive chain, he flew the plane successfully over a designated course of nearly 2000 m in about 7 minutes (Hirst 1977). This was a peak performance by a well-trained man under ideal conditions. Although a similar plane later crossed the Channel between England and France, much longer, sustained man-powered flights by anybody but top athletes are unlikely.

As a result of the advanced state of aerodynamic science, it has been possible for biologists to bring their understanding of animal flight in good agreement with the theoretical power requirements for propeller-driven flight. In fact, a propeller is simply an airfoil in which the surface providing thrust has been separated from the surface providing lift.

Swimming

Swimming is in principle not too different from flying through the air, except that the higher density of water reduces the power needed for providing lift. In fact, some swimming animals have the same density as the water, and no power is used for obtaining lift; all power used for swimming can be put into thrust. The higher density and viscosity of water are important; fish do not swim as fast as birds fly.

For larger animals, such as fish, the drag increases roughly in proportion to the square of the speed. The work needed to overcome the drag (a force) is drag times distance. The power needed is the work per unit time, which equals the drag times the speed. This makes the power used for swimming increase with the third power of the speed.

As the power requirement increases with the third power of the speed, it puts severe limits on the maximum attainable speed (Lighthill 1974). We have already seen how some of the largest and fastest predatory fish have developed means of keeping their power plant, the muscle mass, at a high temperature so that maximum power output is possible.

The problem of attaining the same density as water, and thus achieving neutral buoyancy and

avoiding the expenditure of energy to provide lift, has a wide variety of solutions. We shall now examine these.

BUOYANCY

If an animal that swims is heavier than the water, part of its energy expenditure goes to keep from sinking, and only part is available for forward locomotion. If the animal could have the same density as the water (i.e., have neutral buoyancy), it would have more energy available for moving.

The problem of buoyancy is more important for large than for very small animals. Small organisms have a large relative surface and often have extensions of various sorts that reduce the sinking rate in water; the smallest can remain suspended much like dust particles in air and live through their entire life cycle in this way. Surface extensions large enough to act as brakes on sinking become mechanically impossible for large organisms, which must therefore use other means.

How to reduce density

A reduction in the density or specific gravity of an organism is the only solution to staying afloat that does not involve some amount of swimming action (i.e., the expenditure of energy). We can list five possible avenues for reducing the tendency to sink and shifting the organism toward neutral buoyancy:

1. Reduction in the amount of heavy substances (e.g., calcium carbonate or calcium phosphate)
2. Replacement of heavy ions (e.g., Mg^{2+} and SO_4^{2-} by Na^+, Cl^-, or even lighter ions, H^+ and NH_4^+)
3. Removal of ions without replacement (i.e., making the organism more dilute)
4. Increasing the amount of substances lighter than water, primarily fats and oils
5. Using gas floats, such as the swimbladder of fish

We shall discuss these five possibilities: their advantages and disadvantages and the limitations of

TABLE 10.5 Specific gravity of sea water, fresh water, and substances commonly found as skeletal material in aquatic organisms. [Denton 1961]

Substance	Specific gravity
Sea water	1.026
Fresh water	1.000
Calcium carbonate	
Aragonite	2.9
Calcite	2.7
Calcium phosphate	
Apatite	3.2
Chitin	1.2

each. Because sea water is heavier (specific gravity = 1.026) than fresh water (specific gravity = 1.000), it is easier for an organism to achieve neutral buoyancy in the sea; in fact, some ways of producing neutral buoyancy in the sea can never attain this goal in fresh water.

Reduction in heavy substances

Many skeletons consist of very heavy substances, often calcium salts, either calcium carbonate or calcium phosphate. Calcium carbonate is common among invertebrates (the shells of clams and snails, corals, etc.); calcium phosphate is the major component of vertebrate bones, where it is found in the form of the mineral apatite. Some organisms (e.g., some sponges) have skeletons made of silica; others (many arthropods) use chitinous substances, either alone or together with calcium carbonate. The relative contribution of these substances to the weight of an animal in water is evident from their specific gravities (Table 10.5).

It is easy to see that reducing the amount of calcium salts can go far toward reducing the weight of an organism. Accordingly, pelagic nudibranch snails lack shells, whereas opistobranch snails, which are bottom organisms, have well-developed shells. Swimming coelenterates (e.g., jellyfish) completely lack the heavy calcium carbonate skeletons characteristic of many sessile coelenterates such as corals.

It is interesting to note that one group of pelagic organisms, the swimming crabs, has not gone very far toward reducing its heavy exoskeleton (mostly calcium carbonate). This holds for the common blue crab (*Callinectes*), which swims continuously to keep from sinking.

There is, of course, a considerable structural disadvantage to the removal of mechanical support. Therefore, this solution is more common in relatively small organisms. An alternative is replacement of a heavy skeleton by a skeleton built of lighter substances. For example, the ordinary squid has dispensed with the calcium carbonate skeleton found in its relative, the cuttlefish, and has replaced it with a lighter structure made of chitin, the so-called pen of the squid.

Replacement of heavy ions by lighter ions

The best examples of this solution are perhaps found among plants. Two multicellular algae, *Valonia* and *Halicystis,* show what can be achieved by manipulating the ionic composition of the organism. Both are almost neutrally buoyant: The former barely sinks in sea water, and the latter has a slight positive buoyancy. Both have abundant cell sap, which is isotonic with sea water but differs in composition (Table 10.6). Both exclude the heavy sulfate ion almost completely. *Valonia* likewise excludes magnesium, which is present in *Halicystis* in a concentration less than one-third that in sea water. Calcium, another important divalent ion, is also excluded by both, relative to its concentration in sea water.

Of the monovalent ions, K^+ is heavier than Na^+. *Valonia,* which is slightly heavier than sea water, accumulates primarily K^+ in the cell sap, as plants in general do. *Halicystis,* however, which is slightly lighter than sea water, has excluded most of the potassium and uses the lighter sodium as its most important cation.

A very common organism that causes luminescence in the sea is a dinoflagellate, *Noctiluca miliaris.* It is a little less than 1 mm in diameter and tends to accumulate at the surface because it is def-

TABLE 10.6 Ionic composition of sea water and of the sap of two multinuclear algae that have nearly neutral buoyancy. Ionic molar concentrations are expressed relative to chloride ($Cl^- = 100$). [Gross and Zeuthen 1948]

	Sea water	Sap of *Valonia*	Sap of *Halicystis*
Cl^-	100	100	100
Na^+	85.87	15.08	92.4
K^+	2.15	86.24	1.01
Ca^{2+}	2.05	0.285	1.33
Mg^{2+}	9.74	Trace	2.77
SO_4^{2-}	6.26	Trace	Trace
Specific gravity	1.0277	1.0290	1.0250
Artificial solution[a]	1.0285	1.0285	1.0252

[a] Containing the ion concentrations listed above.

TABLE 10.7 Ionic composition of sea water and of the cell sap of the dinoflagellate *Noctiluca*, a common source of phosphorescence in the sea. It achieves positive buoyancy by excluding heavy ions and replacing them primarily with NH_4^+. [Kesseler 1966]

	Sap (mEq liter^{-1})	Sea water (mEq liter^{-1})
Na^+	414	418
K^+	34	8.8
NH_4^+	58.5	0
Ca^{2+}	9.5	18.5
Mg^{2+}	15	95
Cl^-	496.5	498.9
SO_4^{2-}	Trace	46.6
$H_2PO_4^-$	13	Trace
pH	4.35	8.2

initely lighter than sea water. When it is mechanically stimulated, it emits light flashes that are visible as phosphorescence at night. Its light emission has been the subject of extensive studies, but we shall discuss only its buoyancy problems.

The composition of the cell sap of *Noctiluca* is given in Table 10.7. It is isotonic with sea water, and it achieves its positive buoyancy in part by reducing Ca^{2+}, Mg^{2+}, and especially the heavy SO_4^{2-}. Furthermore, it accumulates ammonium ion, which is lighter than the two commonest monovalent cations, Na^+ and K^+.

The hydrogen ion (H^+) is much lighter than any of the other cations. It has been suggested that positive buoyancy could be achieved by increasing the hydrogen ion concentration at the expense of other cations, but this is not feasible. Although the cell sap of *Noctiluca* is rather acid (i.e., has a high hydrogen ion concentration), it is easy to see that the effect on the specific gravity must be negligible.

If we take the pH of the *Noctiluca* sap to be 4.0, the hydrogen ion concentration is 10^{-4} mol per liter. In other words, the hydrogen ion concentration is 0.1 mmol per liter, and thus can replace no more than 1/5000 of the total cation content in *Nocti-*

luca. To replace a substantial amount of cation the pH value would have to be substantially less than 2, which corresponds to a hydrogen ion concentration in excess of 10 mmol per liter. Such high acidity is not known to exist in any cells; it occurs only in a few acid secretions such as the gastric juices of vertebrates and the saliva of some marine snails, which use acid to drill into the shells of their prey.

The principle of replacing heavier ions by light ones can be used just as effectively by large organisms. A striking example is the group of deep-sea squids known as cranchid squids. In these animals the fluid-filled coelomic cavity is very large and makes up about two-thirds of the entire animal. If it is opened and the fluid drained out, the animal loses its neutral buoyancy and sinks. The fluid has a specific gravity of about 1.010 (sea water is 1.026), it is osmotically isotonic with sea water, and the pH is 5.2. The concentration of NH_4^+ is about 480 mEq per liter and of sodium 80 mEq per liter. The anion is almost exclusively chloride; the heavy SO_4^{2-} ion is excluded (Denton et al. 1969).

The explanation of the high ammonium concentration is simple. The coelomic fluid is slightly acid, and ammonia, which is the normal metabolic end product of protein metabolism in most aquatic animals, diffuses into the fluid and becomes trapped there.

A mechanism that could produce the low pH would be a sodium pump that removes Na^+ from the coelomic fluid, for this would leave an excess of anions and thereby increase the hydrogen ion concentration, thus increasing the acidity. The acid fluid would then serve as a trap to sequester ammonia. It has been calculated that if ammonia is the only end product of protein metabolism in a cranchid squid, about 40% of all the ammonia produced throughout the lifetime of the animal is retained in order to maintain neutral buoyancy (Denton 1971).

The method used by the cranchid squids is quite effective in achieving neutral buoyancy. It has some structural disadvantages, however, for two-thirds of the bulk of the animal is fluid (i.e., the volume of fluid carried around amounts to twice the volume of the living animal itself).

Hypotonicity

Removal of some ions without replacing them by other ions would seem a possible avenue for reducing weight. There would, of course, be osmotic problems, for a dilute solution would be hypotonic to the sea water. In general, invertebrate organisms are in osmotic equilibrium with sea water, and it seems that they never use hypotonicity as a buoyancy mechanism.

Teleost fish, however, have ion concentrations much lower than those of sea water, and this contributes toward reducing their weight in water. This is only of minor importance, however, for teleost fish have skeletons and muscles that contribute sizable amounts of relatively heavy substances. In many deep-sea forms, which have greatly reduced skeleton and muscle masses, however, hypotonicity is of considerable help in balancing the relatively small amounts of heavy substances.

Fats and oils

Many planktonic organisms contain substantial amounts of fats. Fat, of course, is a common form of energy storage, both in animals and in many plants, but the contribution of fats and oils toward buoyancy is also important. Planktonic plants frequently deposit fat, rather than the heavier starch that is a common storage compound in plants in general. All diatoms, for example, deposit only fat.

Elasmobranch fish, as opposed to teleosts, never have swimbladders. But they do have large livers, especially those sharks that are good swimmers. The liver may constitute about one-fifth of the body weight; in a teleost fish it usually is about 1 to 2% of the body weight.

A survey of the black spiny shark (*Etmopterus spinax*) showed that on the average 17% of its body weight was liver, and 75% of the liver was oil (S. Schmidt-Nielsen et al. 1934). (In contrast, mammalian liver contains some 5% fat.) Half of the oil in the shark liver was squalene, an interesting unsaturated hydrocarbon ($C_{30}H_{70}$) that derives its name from *Squalus,* or shark.

Most fats and oils have specific gravities of about 0.90 to 0.92, but squalene has a lower specific gravity, 0.86. The difference is not great, but in sea water the buoyancy effect of a given volume of squalene is about 50% higher than the effect of a similar volume of fat or oil. Whether this is a reason for the tremendous accumulation of squalene in elasmobranch livers is difficult to say; nevertheless, its presence definitely aids in achieving nearneutral buoyancy for the fish. This is particularly effective because elasmobranchs, as opposed to teleosts, have relatively light cartilaginous skeletons that are not weighted down by heavy calcium phosphates.

Interestingly, rays and skates, which are bottomliving animals, have smaller livers with a lower fat content. Five species of rays (*Raja*) averaged a liver size of 7.53% with less than 50% fat in the liver. This makes sense, for rays and skates live at the bottom and are not very good swimmers.

It is worth noting that the relatively low ion concentrations in the body fluids of elasmobranchs, relative to sea water, do not contribute substantially to their buoyancy. The elasmobranchs are isosmotic with sea water, the difference in salt concentration is made up primarily by urea, and urea solutions give only very slight lift over sea water. Trimethylamine oxide (TMAO), which also is found in their body fluids, is appreciably less dense than urea and contributes a minor amount toward giving lift.

Several deep-sea teleosts have swimbladders filled with fat, rather than with air. This is true of some species of *Gonostoma* and *Cyclothone* (Marshall 1960), but since the fatty swimbladders of other fish have a substantial fraction of the fat as cholesterol (up to 49%), which has a higher specific gravity (1.067) than sea water (1.026), the role of the fat-invested swimbladder in buoyancy control is dubious (Phleger 1971).

Gas floats

Compared with an equal volume of water, gas has a very low density, and even a relatively small volume is an excellent solution to the problem of buoyancy. This solution, however, has some limitations and disadvantages. The situation differs depending on whether the gas space is enclosed by soft tissue or by rigid walls.

Soft-walled gas floats

Assume a gas-filled rubber balloon submerged below the surface. The further down it is brought, the greater is the pressure of water on its outside, and as the pressure of gas inside matches the water pressure outside, the balloon becomes increasingly compressed.

The water pressure increases by about 1 atm for each 10 m depth, and for an organism living at considerable depth this causes problems. Consider an organism that lives at 1000 m depth, where the water pressure is about 100 atm. To produce and maintain a volume of gas at that pressure entails three difficulties: (1) the gas must be secreted into the space against a pressure of 100 atm, (2) the *amount* of gas needed to fill the same volume at that pressure is 100 times as great as at the surface, and (3) it is a problem to keep any gas at such a

PORTUGUESE MAN-OF-WAR The gas float of this colonial coelenterate (*Physalia pelagica*) is filled with a gas that contains a large proportion of carbon monoxide. The length of the float may exceed 20 cm, and the tentacles, which give extremely painful stings, may extend for more than 10 m. [Courtesy of Charles E. Lane, University of Miami]

high pressure from diffusing out. The gas tensions in the water at any depth in the ocean are usually not very different from those at the surface, where there is equilibrium with 0.2 atm oxygen and 0.8 atm nitrogen. The gases for the float are obtained from the water, secreted into the high-pressure space of the float, and must then be kept from diffusing out.

A gas float with soft walls entails another difficulty: Its volume changes if the organism moves up or down in the water. Say that an organism has enough gas in its float to be in neutral buoyancy at a given depth. If it moves down, the gas is compressed; the animal gets heavier and loses buoyancy. If it moves up, the gas expands and positive buoyancy increases. The latter can be dangerous, for unless the animal actively swims down to the neutral level, the gas will expand faster and faster as the animal rises at increasing speed. Unless it has a mechanism for releasing gas, an organism that has lost control of its rise may even rupture as the gas expands.

If the gas space has rigid walls, the limitations on vertical movements are less strict, but the walls must have sufficient mechanical strength to support changes in pressure. If the walls are made stronger, they also become heavier, and this defeats the purpose of the gas chamber: to make the organism lighter. As we shall see later, the mechanical strength of the float is a limitation on the vertical distribution of the cuttlefish (*Sepia*), which uses rigid gas spaces for flotation.

The swimbladder of fish is undoubtedly the best known type of gas float, but the principle is also used by quite a few other organisms, in particular the colonial siphonophores (Coelenterata). One of these is the conspicuous Portuguese man-of-war (*Physalia*), commonly found along the Atlantic coast of North America. The colony is carried by a "sail," a balloon-like gas chamber, often more than 10 cm long. It floats on the surface of the ocean and carries the colony along as it drifts in the wind.

Because the gas chamber of *Physalia* floats on the surface rather than being submerged, the inside pressure is virtually equal to the atmospheric pressure. This simplifies the problem of filling it with gas. However, the composition of the gas is quite surprising, for it contains up to some 15% carbon monoxide. The oxygen concentration is less than in the atmosphere, the carbon dioxide concentration is negligible, and nitrogen makes up the balance. The gas is produced by a gas gland, which uses the amino acid serine as a substrate ($CH_2OHCHNH_2COOH$) for the formation of carbon monoxide (Wittenberg 1960).

Other colonial siphonophores also carry floats that contain carbon monoxide. One, *Nanomia bijuga*, has very small floats with a volume of 0.14 to 0.46 μl; in these the carbon monoxide concentration is from 77 to 93% (Pickwell et al. 1964). Carbon monoxide is, of course, highly toxic to man and many other animals, and usually occurs in nature only in trace amounts. It is found in the hollow stems of some marine algae, where it may constitute several percent of the total gas.

Carbon monoxide is found in minute amounts in the exhaled air of humans, even nonsmokers. It is not formed by bacteria in the intestine, but from metabolic breakdown of hemoglobin in the liver (Ludwig et al. 1957). The exhaled carbon monoxide does not originate externally from air pollution, for the Antarctic Weddell seal (which presumably breathes entirely unpolluted air) also exhales a measurable amount of carbon monoxide (Pugh 1959).

The presence of gas floats in plankton organisms may explain a rather mysterious phenomenon known as the *deep scattering layer*. Ships that use echo sounding for depth determination often find that sound waves are reflected not only by the bottom, but also by a layer in mid-water, which thus gives the impression of a false bottom. This extra layer, which often changes in depth from day to night, is the deep scattering layer. For many years the cause of this sound reflection was unknown; but it is now almost certain that the cause is schools of fish with swimbladders or pelagic siphonophores with gas floats.

Siphonophores have gas floats with an opening so that gas can escape when they rise in the water. A simple experiment has shown that gas can be both given off and secreted by these organisms. Pickwell et al. (1964) put them in a 10-ml syringe and exposed them to vacuum by pulling at the plunger. After gas had been pulled out, the floats were heavier than water, but within 1 hour they were again afloat. They were now filled primarily with carbon monoxide, presumably formed in a manner analogous to that of *Physalia*.

At night the deep scattering layer frequently migrates up toward the surface of the ocean. As the floats of siphonophores permit gas to escape, they presumably release bubbles as they ascend. This can explain the observation that the intensity of the sound reflection of the layer frequently increases as it rises.

If organisms release gas on the way up, we would expect that a similar amount of gas is secreted as they descend again. However, they could avoid repeatedly secreting and releasing gas by keeping a constant amount of gas and putting up with the volume changes during vertical migration. This might be feasible if the animal is in neutral buoyancy at the higher level, but it would hardly be possible if the neutral level is at the deeper level.

Let us consider an example. Say that a fish has a swimbladder with a volume (at neutral buoyancy) of 5% of the body volume. Descending from a night level of 50 m depth to 300 m would reduce the gas volume to one-fifth, or 1% of the body volume. The fish would now be negatively buoyant, but could keep at the new level by a moderate amount of swimming. However, if the fish were in neutral buoyancy at 300 m depth with the swimbladder constituting 5% of the body volume, rising to 50 m would increase the swimbladder volume five-fold unless gas were released.

A fivefold expansion of the swimbladder volume would involve serious danger of losing control and rising all the way to the surface where, at 1 atm pressure, the swimbladder would be a catastrophic 30 times its original volume. If organisms

of the deep scattering layer move vertically without releasing gas, we must therefore expect that they are in neutral buoyancy at the higher, rather than the lower, level.

Information about volume changes can to some extent be obtained without direct observation. The deep scattering layer reflects sound particularly well at frequencies that correspond to the resonant frequency of the gas bubbles. This in turn depends on the volume of the gas bubbles as well as on their pressure. If the pressure is known (the depth can be read directly from the echo-sounding record) and the resonant frequency is assumed to be the one that is best reflected, the size of the bubbles can be calculated. Now, when the layer moves up, one of two situations may prevail: Either gas is given off and the bubbles within the animals maintain constant volume, or the gas is retained and the bubbles expand with the pressure change.

If a bubble is kept at constant volume, its resonant frequency should vary with $P^{1/2}$ (P is hydrostatic pressure); if the volume changes with pressure while the mass of gas is kept constant, the resonant frequency should vary with $P^{5/6}$. The sound records are complex and often difficult to interpret, but it appears that organisms of the deep scattering layer may conform to either of these patterns (Hersey et al. 1962).

Rigid-walled gas floats

The problems connected with changes in volume, and thus in buoyancy with changing depth, are eliminated if the gas is enclosed in a rigid space. This solution is used by the cuttlefish (*Sepia officinalis*), a relative of squid and octopus that is common in the upper reaches of coastal water masses. Nearly one-tenth of its body volume is made up of a cuttlebone, a calcified internal structure well known by bird fanciers, who use it to supply calcium to their birds (Figure 10.29). The role of the cuttlebone has been studied by the British investigator Denton (Denton and Gilpin-Brown 1959, 1961).

The volume of the cuttlebone is 9.3% of the body

FIGURE 10.29 Diagram of a cuttlefish, a marine cephalopod that uses a gas-filled rigid structure called the cuttlebone to achieve neutral buoyancy in water. The posterior part of the cuttlebone (shown in black) is filled with liquid. In sea water the cuttlebone gives a net lift of 4% of the animal's weight in air and thus balances the excess weight of the rest of the animal. [Denton and Gilpin-Brown 1961]

CUTTLEBONE The buoyancy mechanism of the cuttlefish (a relative of the squid and octopus) depends on a gas-filled structure, the cuttlebone, made of calcium carbonate. The cuttlebone consists of thin layers of calcium carbonate, spaced about 0.66 mm apart, and supported by pillars of the same material. [Courtesy of M. L. Blankenship, Duke University]

volume, and its specific gravity when freshly removed is in the range of 0.57 to 0.64. This provides sufficient lift to give the animal neutral buoyancy in sea water.

The cuttlebone has a peculiar laminar structure. It is made up of thin layers of calcium carbonate, reinforced by chitin and held about 0.66 mm apart by a large number of ''pillars'' of the same material. In a fully grown animal the number of parallel layers may be about 100, and most of the space between the layers is filled with gas.

If a cuttlebone is removed from an animal and punctured under water, no gas escapes. This shows that the gas it contains is not under pressure. If the cuttlebone is crushed, gas escapes and can be collected for analysis. The gas is nitrogen under 0.8 atm pressure, plus a small amount of oxygen. This

indicates that the gas is in diffusion equilibrium with the body fluids.

What is the mechanism that brings gas into the cuttlebone? If it is possible to withdraw water from the spaces in the cuttlebone, gas will passively diffuse in and reach the same partial pressure as in the body fluids. The question is then whether fluid can be withdrawn.

Along the caudal surface of the cuttlebone, some

liquid is always present between the calcified layers. This liquid can be collected under paraffin oil by exposing the cuttlebone to a vacuum. It contains mostly sodium chloride and is hyposmotic to sea water and the body fluids (which are isosmotic with sea water). This requires the presence of a mechanism for active ion transport to reduce the concentration in the cuttlebone fluid. The higher osmotic pressure outside the cuttlebone tends to withdraw water from the cuttlebone. The osmotic withdrawal of water is opposed by the hydrostatic pressure, which tends to press water back into the cuttlebone.

The maximum force of water withdrawal is limited by the concentration difference. If all ions were removed from the fluid within the cuttlebone, the concentration difference between the cuttlebone fluid and blood would be about 1.1 osmolar. This corresponds to 24 atm pressure; that is, if the outside hydrostatic pressure exceeds 24 atm, fluid will be pressed into the cuttlebone in spite of the osmotic forces acting in the opposite direction. At a depth of about 240 m, the pressure is 24 atm, and below this depth it is impossible to withdraw fluid osmotically from the cuttlebone. However, the cuttlefish live in the upper reaches of the water masses and probably do not descend below about 200 m. Therefore, the osmotic forces suffice to keep water from being forced into the cuttlebone.

The gas pressure inside the rigid walls of the cuttlebone is about 0.8 atm, and the cuttlebone must be sufficiently strong to support the hydrostatic pressure of the surrounding water. This is indeed the case, for if a cuttlebone is covered with a thin plastic bag (so that sea water cannot penetrate into it) and is exposed to high pressures, it withstands more than 25 atm before it is crushed by the pressure.

The advantage of a rigid system is, of course, that the buoyancy is virtually unaffected by changes in depth. This gives the cuttlefish a great deal of freedom in moving up and down without making adjustments in the buoyancy mechanism. In addition to making the cuttlefish relatively independent

FIGURE 10.30 The shell of the chambered nautilus. Section of the shell of *Nautilus macromphalus*, oriented in its natural position. The chambers are gas-filled, and new chambers are added as the animal grows. A new chamber is initially filled with fluid, which is withdrawn osmotically and replaced by gases. A small amount of liquid remains in the most recently formed chambers. [Denton and Gilpin-Brown 1966]

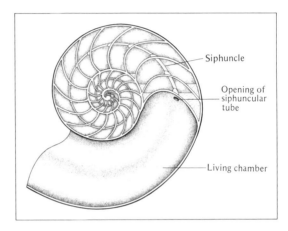

of depth, the cuttlebone has the further advantage that it serves as a skeleton.

The mechanism responsible for filling the shell of the pearly nautilus, or chambered nautilus, is in principle the same as in the cuttlefish (Figure 10.30). As the nautilus grows, it adds new chambers to its shell, one by one. A newly formed chamber (next to the animal) is initially filled with liquid in which sodium chloride is the major solute.

The wall of the chamber is impermeable to water and gases, and all exchange must take place via the siphuncle. The epithelium that covers the siphuncle contains extensive mitochondria-lined infoldings where NaCl is actively pumped from the fluid within the newly formed chamber. As the fluid becomes less concentrated, water is withdrawn osmotically in the same way as in the cuttlefish, and gas diffuses into the chamber to replace the water (Greenwald et al. 1982).

There is one difficulty in the proposed scheme:

PEARLY NAUTILUS An x-ray photograph of a living pearly nautilus. The last chamber (arrow), recently formed, is still full of liquid. In the next chamber the emptying process is nearly complete, but a small quantity of fluid remains in the lowest part of the chamber. All the other chambers are completely filled with gas. [Courtesy of P. Ward, University of California at Davis and L. Greenwald, Ohio State University]

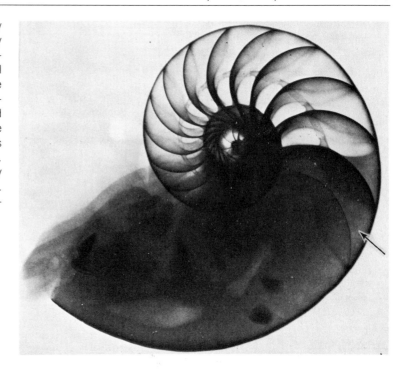

Nautilus are frequently found below the 240-m depth that should be the absolute limit for the removal of water by a simple osmotic mechanism. One workable mechanism would be to enhance the blood concentration within the siphuncle, possibly attained with the aid of a countercurrent multiplier mechanism. Some evidence in favor of such a mechanism has been obtained, but further studies are much needed (Greenwald et al. 1980).

By diffusion nitrogen reaches 0.8 atm, the same tension as in the water and the animal's body fluids. Oxygen, however, is lower and does not reach the concentration in the atmosphere, for the animal uses oxygen and its body fluids therefore have lower than atmospheric oxygen tension. The gas pressure in the shell is always about 0.9 atm, irrespective of the depth at which the animal is caught. Like the cuttlefish, nautilus has a freedom of vertical movement that is limited only by the pressure its rigid shell can support.

We have now discussed examples of the differ-ent mechanisms animals can use to improve their buoyancy. Each has advantages and disadvantages, summarized in Table 10.8. It is evident that gas is one of the most favorable solutions to the problem of providing lift and has other advantages as well. But if contained in a soft structure, it has one major disadvantage: It is extremely sensitive to pressure changes. Teleost fish use this mechanism in the form of an air-filled, soft-walled swimbladder, which is the subject we shall discuss next.

The swimbladder of fish

Many teleost fish have a *swimbladder* or *air bladder* that provides the necessary lift to give them neutral buoyancy. Whether a particular fish has a swimbladder is related more to its life habits than to its systematic position. Some bottom-living fish lack swimbladders, and it seems reasonable that neutral buoyancy has no particular advantage for a fish that is bound to the bottom. Many pelagic and surface forms do have swimbladders, but others lack

TABLE 10.8 Advantages and disadvantages of various mechanisms used by aquatic organisms to improve buoyancy.

Mechanism	Approx. volume needed (% of remaining body)	Buoyancy effectiveness		Pressure independence	Structural qualities	Energy requirement
		Sea water	Fresh water			
Reduction of heavy structures	0	Fair	Poor	Excellent	Poor	No maintenance
Replacement of heavy ions	200	Good	Poor	Excellent	Fair	Needs continuous maintenance
Fat	50	Good	Good	Excellent	Good	High initial cost, no maintenance
Squalene	35	Good	Good	Excellent	Good	High initial cost, no maintenance
Soft-walled gas float	5	Excellent	Excellent	Poor	Excellent	Needs continuous maintenance
Rigid-walled gas float	10	Excellent	Excellent	Good	Excellent	Probably needs maintenance

this structure. The Atlantic mackerel, for example, has no swimbladder. For a fast and powerful predator this is a definite advantage, for as we have seen, a soft-walled swimbladder places severe constraints on the freedom of moving up and down in the water masses. No elasmobranch (sharks and rays) has a swimbladder.

The swimbladder is a more or less oval, soft-walled sac, located in the abdominal cavity, just below the spinal column (Figure 10.31). The shape varies a great deal, but the volume is rather constant from species to species, mostly about 5% of the body volume in marine fish and 7% in freshwater species. The reason is that the specific gravity of a fish without the swimbladder is about 1.07. The volume of gas needed to give it neutral buoyancy in fresh water is therefore about 7% of the body volume, and in sea water (sp. gr. = 1.026) about 5%.

A swimbladder can provide a fish with perfect neutral buoyancy, and the fish can thus avoid expending energy to keep from sinking. There is, however, a serious disadvantage: The fish is in equilibrium or neutral buoyancy at only one particular depth. If it swims below this depth, the swimbladder is compressed by the increased water pressure, the buoyancy is decreased, and the fish must swim actively to keep from sinking further. This is not a very serious problem, for even if the swimbladder is almost completely compressed, the fish will not be much heavier than the water.

FIGURE 10.31 The swimbladder of a fish is located in the abdominal cavity, just below the vertebral column. The shape of the swimbladder can vary a great deal, although its volume is normally similar in different species of fish.

Swimbladder

TABLE 10.9 Composition of gas secreted by the gas gland of codfish. The gland was covered by a thin plastic film and stimulated to secrete gas. Bubbles ca. 0.5 μl in size were collected and analyzed. [Scholander 1956]

	O_2 (%)	CO_2 (%)	N_2 (%)
	81.1	8.6	10.3
	70.3	5.0	24.7
	76.6	15.8	7.6
	49.5	6.9	43.6
	61.1	6.8	32.1
	66.5	5.9	27.6
	30.1	13.9	56.0
	69.2	7.2	23.6
Mean	63.1	8.8	28.1

If the fish rises above the level of neutral buoyancy, however, the situation can become catastrophic. The pressure is reduced, the swimbladder expands, and the fish gains increased lift. It must now swim actively down, or it will continue to rise, and as the swimbladder expands more and more, the fish will eventually lose control.

Some fish have a connection from the swimbladder to the esophagus that permits gas to escape, but others have a completely closed swimbladder. If such a fish is brought up from deep water, it arrives at the surface with the expanded swimbladder everted through the mouth or even ruptured.

The swimbladder is embryologically formed as an evagination from the digestive tract. Those fish that maintain a connection between the swimbladder and the esophagus are known as *physostome* fish (Greek *physa* = bladder; *stoma* = mouth). These fish can fill their swimbladders by gulping air at the surface. For fish living at depth this would be impractical or impossible, for very large volumes of air would have to be taken into the bladder at the surface to achieve neutral buoyancy at depth where the pressure is many times greater.

In many fish the duct degenerates, and there is no connection from the swimbladder to the out-

side. In these fish, called *physoclist* fish (Greek *kleistos* = closed), the gases in the swimbladder must originate in the blood and be secreted into the swimbladder at a pressure equal to the depth at which the fish live. Fish with swimbladders have been caught below 4000 m depth. Because the swimbladder wall is a soft structure, the swimbladder gas at the depth must be under a pressure in excess of 400 atm. This poses the problem mentioned earlier in this chapter: How can gases be moved from the low tensions in sea water into the enormous pressures within the swimbladder?

The gases found in the swimbladder are the same as those in the atmosphere, oxygen, nitrogen, and carbon dioxide, but in different proportions. If we withdraw some gas from the swimbladder by puncturing it with a hypodermic needle, the fish is induced to secrete new gas to replace what was lost. In most fish the newly secreted gas is rich in oxygen (Table 10.9).

This is not true for all fish; the whitefish (*Coregonus albus*), for example, fills its swimbladder with virtually pure nitrogen. Because it lives in lakes at depths of 100 m or so, nitrogen must be secreted at a pressure of some 10 atm. Nitrogen gas is highly inert, and it has been difficult to understand how an inert gas can be secreted. Also, the truly inert gases (such as argon) are likewise enriched in the swimbladder gas in about the same proportion to nitrogen as they occur in the atmosphere (Wittenberg 1958). This indicates that the secretion mechanism cannot involve chemical reactions; it must be of a physical nature.

Even when the swimbladder gas is rich in oxygen, the nitrogen may still be more concentrated than in the surrounding water. Say, for example, that we obtain a fish from a depth of 500 m and that a sample of the swimbladder gas shows 80% O_2 and 20% N_2 (and a trace of carbon dioxide). At the depth where the fish was caught, the gases were under 50 atm total pressure, so that their partial pressures were 40 atm for oxygen and 10 atm for nitrogen. Relative to the gas tensions in the surrounding water, 0.2 atm O_2 and 0.8 atm N_2, oxy-

BUTTERFLY FISH An x-ray photograph of a long-nosed butterfly fish (*Chelmon rostratus*). The location of the swimbladder is clearly visible as the light area just below the vertebral column. The length of this specimen is about 10 cm. [Courtesy of John Lundberg, Duke University]

FIGURE 10.32 Diagram of the circulation to the swimbladder of a fish. Blood may reach the swimbladder either through a set of parallel capillaries, the rete, which supplies the gas gland, or through a vessel to the posterior portion where fine blood vessels that spread over the wall can serve to absorb gases rapidly. When gas is being secreted, the gas-absorption vessels are closed and carry no blood.

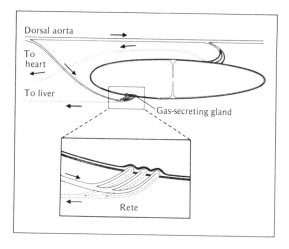

gen in the swimbladder gas was thus enriched 200-fold and nitrogen 12.5-fold. Before we discuss how gases can be secreted at these high pressures, we will examine the simpler question of how the gases are retained and kept from being lost by diffusion from the swimbladder.

How is gas kept in the swimbladder?

As always, to understand function we must know a bit about structure. In the wall of the swimbladder of most fish there is a gas gland, which stands out because of its bright red color. If the gas gland were supplied with ordinary arterial blood, which comes from the gills and therefore has gas tensions in equilibrium with the water, the blood would tend to dissolve the swimbladder gases and carry them away. How can this be prevented?

The gas gland is supplied with blood through a peculiar structure known as the *rete mirabile* (plural, *retia mirabilia*). The form of the rete and the gas gland varies from species to species, but in

principle it is as follows. Before the artery reaches the gas gland, it splits into an enormous number of straight, parallel capillaries, which may unite again into a single vessel before reaching the gas gland (Figure 10.32). The vein from the gland likewise splits into a similar number of parallel capillaries, which run interspersed between the arterial capillaries and then unite to form a single vein.

Let us see how this structure can help to keep gases inside the swimbladder, once they have been deposited there. The system is analogous to the countercurrent heat exchanger in the flipper of the whale, only it deals with dissolved gases. Assume that the swimbladder contains gas under a pressure of 100 atm. As venous blood leaves the swimbladder, it contains dissolved gases at this high pressure. When this blood enters the venous capillary, only the thin capillary wall separates it from blood in the arterial capillaries. Gases therefore diffuse across the wall to the arterial blood. As the blood runs along the venous capillaries, it loses more and more gas by diffusion to the arterial capillaries. At the end of the venous capillaries, as the blood is about to leave the rete, the surrounding capillaries contain arterial blood, which comes directly from the gills and is not yet enriched with additional gas. The venous blood can therefore lose gases until it is in diffusion equilibrium with the incoming arterial blood (i.e., the venous blood that leaves the rete contains no more gas than the incoming arterial blood).

In this way the rete serves as a trap to retain the gases in the swimbladder and avoid loss of gas to the circulating blood. The rete is a typical counter-current exchange system that depends on passive diffusion between two liquid streams that run in opposite directions. The exchange of gas is aided by a large surface (i.e., the large number of capillaries), by a short diffusion distance (i.e., the thin capillary wall, which is only a fraction of 1 μm thick), and by the length of the capillaries (which are exceptionally long).

Some measurements are informative in this regard (Krogh 1929). The rete from a common eel weighed about 65 mg, and had some 100 000 arterial capillaries and about the same number of venous capillaries. The capillaries were about 4 mm long, and this gives a total length of 400 m of each kind of capillary. Their diameter was, like most capillaries, some 7 to 10 μm, and simple arithmetic then gives their total wall surface area as being in excess of 100 cm^2 for each kind, all within the volume of a water drop. This large area is thus available for gas exchange between the arterial and venous blood.

The capillaries in the swimbladder rete are exceptionally long, often several millimeters. Muscle capillaries, which are otherwise among the longest anywhere, are only about 0.5 mm long. There is a clear correlation between the length of the rete capillaries and the gas pressure in the swimbladder. Those fish that live at the greatest depth and secrete gases under the highest pressures have the longest

capillaries, the record being held by a deep-sea fish, *Bassozetus taenia,* which has rete capillaries with the enormous length of 25 mm (Marshall 1960).

The effectiveness of the rete has been estimated by Scholander, who found that its dimensions would suffice for maintaining pressures in excess of 4000 atm in the swimbladder (Scholander 1954). Fish that have swimbladders almost certainly do not occur below 5000 m (500 atm), and we must therefore conclude that the rete is more than adequate as a device to keep the swimbladder gas from being carried away, dissolved in the blood. Because the rete depends entirely on diffusion and requires no energy, it provides a rather advantageous solution to the problem.

The other way gas could escape from the swimbladder is by diffusion through the swimbladder wall. Because the total pressure of gas in the swimbladder is nearly identical to the hydrostatic pressure in the water where the fish finds itself, the partial pressures are inevitably higher than in the surrounding water. Typically, the partial pressure of nitrogen in the water (and in the blood) is 0.8 atm. For oxygen the partial pressure in the water, and therefore in the blood, does not normally exceed 0.2 atm. In contrast, the partial pressure in the swimbladder may be several hundred atmospheres.

The water content of the swimbladder wall is as high as in other tissues, and we might therefore expect that oxygen would readily diffuse out. However, the diffusibility for oxygen is unexpectedly low. This is attributable to multiple layers of very thin platelets of crystalline guanine, each sheet having a thickness of only about 0.02 μm. These thin sheets form a barrier to diffusion of gases out of the swimbladder and thereby reduce the metabolic energy needed for secretion of gases into the swimbladder to maintain its volume (Lapennas and Schmidt-Nielsen 1977).

How is gas secreted into the swimbladder?

The problem of depositing gas into the swimbladder against a high pressure is more formidable

FIGURE 10.33 Diagram of the countercurrent multiplier system of the fish swimbladder. The numerous capillaries of the rete are represented as a single loop. As the gas gland produces lactic acid, the oxygen tension increases in the venous capillary, and gas diffuses across to the arterial capillary, thus remaining within the loop. The venous blood, as it exits, contains less oxygen than the incoming arterial blood.

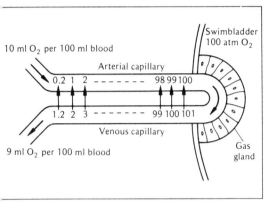

than the problem of keeping the gas in, once it is there.

We will first discuss oxygen, for this is the gas most commonly secreted into the swimbladder. Let us base this discussion on a swimbladder that already contains oxygen at a high pressure, say 100 atm (corresponding to 1000 m depth) and assume that additional gas is to be secreted against this enormous pressure. Because the oxygen tension in the arterial blood that comes from the gills at best equals that of the surrounding water, the arterial oxygen tension is no more than 0.2 atm.

Let us consider the oxygen *content* of the blood, rather than the gas tension. If oxygen is being removed from the blood and secreted into the swimbladder, it follows that the venous blood leaving the rete must contain less oxygen than the entering arterial blood. As an arbitrary example, say that the arterial blood contains 10 ml O_2 per 100 ml blood and that the venous blood contains 9 ml O_2

per 100 ml when leaving the rete. In Figure 10.33, in which for simplicity the entire rete structure is represented by a single loop, these two numbers are placed at the far left of the diagram. The difference between the arterial and the venous blood, 1 ml O_2 per 100 ml blood, is the oxygen that is being deposited into the swimbladder.

Assume now that the gas gland begins to produce lactic acid (which in reality it does when gas secretion takes place). The lactic acid enters the blood and reduces its affinity for oxygen. This effect of acid is much more pronounced in fish blood than in mammalian blood and is known as the *Root effect*. The lactic acid drives oxygen off the hemoglobin, and the oxygen tension in the blood leaving the swimbladder is therefore increased (Figure 10.34). The oxygen tension now is higher in the venous than in the arterial capillary, and oxygen diffuses across to the arterial capillary. This continues as long as lactic acid is added to the venous blood, which thus leaves the rete with less oxygen than is contained in the incoming arterial blood. Owing to the countercurrent flow in the rete, lactic acid also tends to remain in the loop, increasing the effect.

In this system it is important to distinguish clearly between *amount* and *tension*. The amount of oxygen in the venous blood that leaves the rete must be less than the incoming arterial blood, but for oxygen to diffuse across from the venous to the arterial capillary, the oxygen tension must be higher in the venous than in the arterial capillary. This higher tension is caused by the acid produced by the gas gland. In this way oxygen is constantly returned to the arterial capillary and gradually accumulates in the loop, where it can reach very high concentrations. How the gas gland itself functions is not known in detail, but the system described explains how the blood that enters the gland can have very high oxygen concentrations.

The mechanism described has received experimental verification. A Norwegian investigator, John Steen, has succeeded in taking minute blood samples from the artery and the vein of the rete in eels

FIGURE 10.34 Oxygen dissociation curves of fish blood demonstrate the pronounced effect of acid, known as the Root effect. The upper curve in each diagram represents untreated blood at 4 °C; in the lower curve of each diagram lactic acid has been added to give the pH stated for each curve. The solid bar on the abscissa gives the depth range at which the fish lives. Note that the abscissa gives the oxygen pressure in atmospheres. [Scholander and Van Dam 1954]

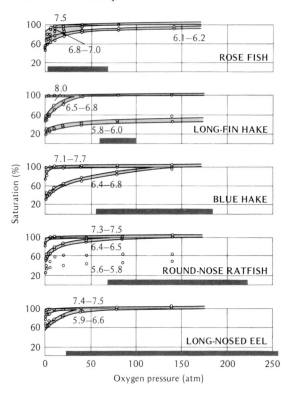

capillaries. Thus, the longer the capillaries, the higher the multiplication effect. Again, this is in accord with the much longer rete capillaries found in fish from greater depths. The multiplication requires energy, as does any process that transports a substance against a gradient; in this case the lactic acid released by the gas gland provides the necessary driving force.

The system described depends on the use of acid to decrease the affinity of hemoglobin for oxygen. However, acid is not the only substance that has an effect on the blood gases; any solute added to the blood decreases the solubility for gases. This we could call a *salting-out effect*. Thus, lactic acid, in addition to its Root effect, also has a salting-out effect on the blood gases – not only on oxygen, but also on the other gases, including nitrogen. It is possible that the concentration of sodium chloride may also play a role in the secretory mechanism because the salting-out effect of NaCl is much greater than that of lactic acid (Gerth and Hemmingsen 1982).

Because the addition of solute decreases the solubility for nitrogen, the rete acts as a multiplier for this gas as well as for oxygen, and we now can explain how inert gases can accumulate in the swimbladder. Until the salting-out effect was understood, it was very difficult to explain how any inert gas could be secreted. The possibility was pointed out as early as 1934 by Henri Koch, but received little attention until the late 1960s when sufficient experimental work had accumulated to yield a clear understanding of how the rete acts as a countercurrent multiplier system.

The salting-out effect, however, does not seem to be as effective as the Root effect. Careful estimates indicate that the salting-out effect of an increase in the blood salts of 0.02 mol per liter could result in a nitrogen concentration of 25 atm, but due to the Root effect a lactic acid concentration of only 0.005 mol per liter (as observed in actively secreting eels) could potentially produce some 3000 atm of oxygen (Kuhn et al. 1963).

One further finding seems to put the final piece

that were in the process of secreting oxygen. He did indeed find that the venous blood had a lowered oxygen *content*, but an increased oxygen *tension*, attributable to the production of lactic acid and carbon dioxide in the gas gland (Steen 1963).

We can now see that the rete, when it secretes, is a *countercurrent multiplier* system. A small difference in oxygen tension between venous and arterial blood is multiplied along the length of its of the puzzle in place. As acid is added to fish blood and oxygen is forced off the hemoglobin, the effect is spoken of as the *Root-off reaction*. The reverse, that oxygen is again bound to the hemoglobin when the pH is increased, is known as the *Root-on effect*. The interesting finding is that the reaction rates for the Root-off and Root-on effects are quite different. The half time for the off reaction is 50 ms (at 23 °C); for the on reaction it is in the order of 10 to 20 seconds. Blood in the arterial capillary of the rete, as acid diffuses into it, rapidly gives off its oxygen. But as the same blood runs in the venuous capillary and the lactic acid diffuses out and into the arterial capillary, the oxygen does not immediately enter into combination with hemoglobin; it remains as uncombined oxygen. In other words, a high oxygen tension persists. Thus, the slow Root-on effect permits a large fraction of the hemoglobin to leave the venous capillary in the deoxygenated state, although the tension of oxygen may still be quite high (Berg and Steen 1968).

Although the function of the rete as a countercurrent multiplier system is clear, the function of the secretory epithelium of the gas gland itself has not been well studied. One of the unsolved problems is how gas is moved from the capillary into the gas phase of the bladder itself. The role of the secretory epithelium in this process is not understood.

We have now seen how animal movement is achieved with the aid of various force-producing mechanisms. The underlying principle, the sliding between adjacent microfilaments, appears to be a universal mechanism for the production of force. In any but the smallest animals, locomotion depends on forces that act on skeletal elements, either a hydraulic skeleton or a rigid skeleton that can be internal (e.g., vertebrates) or external (e.g., arthropods). Animal adaptations to locomotion show many interesting principles of biomechanics, including how aquatic animals achieve neutral buoyancy and thus avoid the expenditure of energy to keep from sinking.

The subjects that come next concern the control and coordination of the many different physiological functions.

REFERENCES

Alexander, R. McN. (1974) The mechanics of jumping by a dog (*Canis familiaris.*) *J. Zool., Lond. 173*:549–573.

Alexander, R. McN. (1983) *Animal Mechanics,* 2d ed. Oxford: Blackwell Scientific Publ. 301 pp.

Alexander, R. McN. (1988) *Elastic Mechanisms in Animal Movement.* Cambridge: Cambridge University Press. 141 pp.

Alexander, R. McN., and Vernon, A. (1975) The mechanics of hopping by kangaroos (Macropodidae). *J. Zool., Lond. 177*:265–303.

Allen, R. D., Francis, D., and Zeh, R. (1971) Direct test of the positive pressure gradient theory of pseudopod extension and retraction in amoebae. *Science 174*:1237–1240.

Andersen, S. O., and Weis-Fogh, T. (1964) Resilin: A rubber-like protein in arthropod cuticle. *Adv. Insect Physiol. 2*:1–65.

Bennet-Clark, H. C., and Lucey, E. C. A. (1967) The jump of the flea: A study of the energetics and a model of the mechanism. *J. Exp. Biol. 47*:59–76.

Berg, H. C., and Anderson, R. A. (1973) Bacteria swim by rotating their flagellar filaments. *Nature, Lond. 245*:380–382.

Berg, T., and Steen, J. B. (1968) The mechanism of oxygen concentration in the swim-bladder of the eel. *J. Physiol. 195*:631–638.

Blum, J. J., and Lubliner, J. (1973) Biophysics of flagellar motility. *Annu. Rev. Biophys. Biomed. Engin. 2*:181–219.

Bone, Q. (1966) On the function of the two types of myotomal muscle fibre in elasmobranch fish. *J. Mar. Biol. Assoc. UK 46*:321–349.

Brokaw, C. J. (1989) Direct measurements of sliding between outer doublet microtubules in swimming sperm flagella. *Science 243*:1593–1596.

Brokaw, C. J., and Gibbons, I. R. (1975) Mechanisms of movement in flagella and cilia. In *Swimming and Flying in Nature,* vol. 1 (T. Y.-T. Wu, C. J. Brokaw, and C. Brennen, eds.), pp. 89–126. New York: Plenum Press.

Buller, A. J., Eccles, J. C., and Eccles, R. M.

(1960) Interactions between motoneurons and muscles in respect to the characteristic speeds of their responses. *J. Physiol. 150:*417–439.

Buller, A. J., and Lewis, D. M. (1964) The rate of rise of tension in isometric tetani of cross-innervated mammalian skeletal muscles. *J. Physiol. 170:*67P–68P.

Burke, R. E., Levine, D. N., Zajack, F. E., III, Tasiris, P., and Engel, W. K. (1971) Mammalian motor units: Physiological-histochemical correlation in three types in cat gastrocnemius. *Science 174:*709–712.

Cavagna, G. A., and Citterio, G. (1974) Effect of stretching on the elastic characteristics and the contractile component of frog striated muscle. *J. Physiol. 239:*1–14.

Cavagna, G. A., Saibene, F. P., and Margaria, R. (1964) Mechanical work in running. *J. Appl. Physiol. 19:*249–256.

Cooper, S., and Eccles, J. C. (1930) The isometric responses of mammalian muscles. *J. Physiol. 69:*377–385.

Dathe, H. H., and Oehme, H. (1978) Typen des Rüttelfluges der Vögel. *Biol. Zbl. 97:*299–306.

Dawson, T. J., and Taylor, C. R. (1973) Energetic cost of locomotion in kangaroos. *Nature, Lond. 246:*313–314.

Denton, E. J. (1961) The buoyancy of fish and cephalopods. *Prog. Biophys. Biophys. Chem. 11:*178–234.

Denton, E. J. (1971) Examples of the use of active transport of salts and water to give buoyancy in the sea. *Philos. Trans. R. Soc. Lond. B. 262:*277–287.

Denton, E. J., and Gilpin-Brown, J. B. (1959) On the buoyancy of the cuttlefish. *Nature, Lond. 184:*1330–1332.

Denton, E. J., and Gilpin-Brown, J. B. (1961) The distribution of gas and liquid within the cuttlebone. *J. Mar. Biol. Assoc. UK 41:*365–381.

Denton, E. J., and Gilpin-Brown, J. B. (1966) On the buoyancy of the pearly Nautilus. *J. Mar. Biol. Assoc. UK 46:*723–759.

Denton, E. J., Gilpin-Brown, J. B., and Shaw, T. L. (1969) A buoyancy mechanism found in cranchid squid. *Proc. R. Soc. Lond. B. 174:*271–279.

Eckert, R. (1972) Bioelectric control of ciliary activity. *Science 176:*473–481.

Evans, M. E. G. (1972) The jump of the click beetle (Coleoptera: Elateridae): A preliminary study. *J. Zool., Lond. 167:*319–336.

Evans, M. E. G. (1973) The jump of the click beetle (Coleoptera: Elateridae): Energetics and mechanics. *J. Zool., Lond. 169:*181–194.

Gerth, W. A., and Hemmingsen, E. A. (1982) Limits of gas secretion by the salting-out effect in the fish swimbladder rete. *J. Comp. Physiol. 146:*129–136.

Gibbons, I. R. (1977) Structure and function of flagellar microtubules. In *International Cell Biology* (B. R. Brinkley and K. R. Porter, eds.), pp. 348–357. New York: Rockefeller University Press.

Gordon, A. M., Huxley, A. F., and Julian, F. J. (1966) The variation in isometric tension with sarcomere length in vertebrate muscle fibres. *J. Physiol. 184:*170–192.

Gray, J., and Lissmann, H. W. (1938) Studies in animal locomotion. 7. Locomotory reflexes in the earthworm. *J. Exp. Biol. 15:*506–517.

Greenewalt, C. H. (1975) The flight of birds: The significant dimensions, their departure from the requirements for dimensional similarity, and the effect on flight aerodynamics of that departure. *Trans. Am. Philos. Soc.* (new series) 65(4):1–67.

Greenwald, L., Cook, C. B., and Ward, P. D. (1982) The structure of the chambered nautilus siphuncle: The siphuncular epithelium. *J. Morphol. 172:*5–22.

Greenwald, L., Ward, P. D., and Greenwald, O. E. (1980) Cameral liquid transport and buoyancy control in chambered nautilus (*Nautilus macromphalus*). *Nature, Lond. 286:*55–56.

Gross, F., and Zeuthen, E. (1948) The buoyancy of plankton diatoms: A problem of cell physiology. *Proc. R. Soc. Lond. B. 135:*382–389.

Hall-Craggs, E. C. B. (1965) An analysis of the jump of the lesser galago (*Galago senegalensis*). *J. Zool. 147:*20–29.

Hersey, J. B., Backus, R. H., and Hellwig, J. (1962) Sound-scattering spectra of deep scattering layers in the western North Atlantic Ocean. *Deep Sea Res. 8:*196–210.

Hirst, M. (1977) America's man-powered prizewinner. *Flight Int. 112:*1253–1256.

Hoyle, G. (1955) Neuromuscular mechanisms of a locust skeletal muscle. *Proc. R. Soc. Lond. B. 143:*343–367.

Hoyle, G. (1957) *Comparative Physiology of the Nervous Control of Muscular Contraction.* Cambridge: Cambridge University Press. 147 pp.

Hoyle, G. (1964) Muscle and neuromuscular physiology. In *Physiology of Mollusca,* vol. 1 (K. M. Wilbur and C. M. Yonge, eds.), pp. 313–351. New York: Academic Press.

Huxley, H. E. (1969) The mechanism of muscular contraction. *Science 164:*1356–1366.

Huxley, H. E. (1973) Muscular contraction and cell motility. *Nature, Lond. 243:*445–449.

Jahn, T. L., Landman, M. D., and Fonseca, J. R. (1964) The mechanism of locomotion of flagellates. 2. Function of the mastigonemes of *Ochromonas. J. Protozool. 11:*291–296.

Johnson, W. H., and Twarog, B. M. (1960) The basis for prolonged contractions in molluscan muscles. *J. Gen. Physiol. 43:*941–960.

Kesseler, H. (1966) Beitrag zur Kenntnis der chemischen und physikalischen Eigenschaften des Zellsaftes von *Noctiluca miliaris. Veröff. Inst. Meeresforsch. Bremerhaven 2:*357–368.

Koch, H. (1934) L'Emission de gaz dans la vésicule gazeuse des poissons. *Rev. Quest. Sci. 26:*385–409.

Krogh, A. (1929) *The Anatomy and Physiology of Capillaries,* 2d ed. New Haven: Yale University Press. 422 pp.

Kuffler, S. W., and Williams, E. M. V. (1953a) Small-nerve junctional potentials: The distribution of small motor nerves to frog skeletal muscle, and the membrane characteristics of the fibres they innervate. *J. Physiol. 121:*289–317.

Kuffler, S. W., and Williams, E. M. V. (1953b) Properties of the ''slow'' skeletal muscle fibres of the frog. *J. Physiol. 121:*318–340.

Kuhn, W., Ramel, A., Kuhn, H. J., and Marti, E. (1963) The filling mechanism of the swimbladder: Generation of high gas pressures through hairpin countercurrent multiplication. *Experientia 19:*497–511.

Lapennas, G. N., and Schmidt-Nielsen, K. (1977) Swimbladder permeability to oxygen. *J. Exp. Biol. 67:*175–196.

Levine, R. J. C., Elfvin, M., Dewey, M. M., and Walcott, B. (1976) Paramyosin in invertebrate muscles. 2. Content in relation to structure and function. *J. Cell. Biol. 71:*273–279.

Lighthill, J. (1974) *Aerodynamic Aspects of Animal Flight.* Fluid Science Lecture, British Hydromechanics Research Association. 30 pp.

Lowy, J., Millman, B. M., and Hanson, J. (1964) Structure and function in smooth tonic muscles of lamellibranch molluscs. *Proc. R. Soc. Lond. B. 160:*525–536.

Ludwig, G. D., Blakemore, W. S., and Drabkin, D. L. (1957) Production of carbon monoxide and bile pigment by haemin oxidation. *Biochem. J. 66:*38P.

Machin, K. E. (1958) Wave propagation along flagella. *J. Exp. Biol. 35:*796–806.

Marshall, N. B. (1960) Swimbladder structure of deep-sea fishes in relation to their systematics and biology. *Discovery Rep. 31:*1–122.

Mommsen, T. P., Ballantyne, J., MacDonald, D., Gosline, J., and Hochachka, P. W. (1981) Analogues of red and white muscle in squid mantle. *Proc. Natl. Acad. Sci. U.S.A. 78:*3274–3278.

Murphy, R. A. (1988) Muscle cells of hollow organs. *News Physiol. Sci. 3:*124–128.

Naitoh, Y., and Kaneko, H. (1972) Reactivated triton-extracted models of paramecium: Modification of ciliary movement by calcium ion. *Science 176:* 523–524.

Parry, D. A., and Brown, R. H. J. (1959a) The hydraulic mechanism of the spider leg. *J. Exp. Biol. 36:*423–433.

Parry, D. A., and Brown, R. H. J. (1959b) The jumping mechanism of salticid spiders. *J. Exp. Biol. 36:*654–664.

Peachey, L. D. (1965) The sarcoplasmic reticulum and transverse tubules of the frog's sartorius. *J. Cell Biol. 25:*209–231.

Pennycuick, C. J. (1969) The mechanics of bird migration. *Ibis. 111:*525–556.

Phleger, C. F. (1971) Pressure effects on cholesterol and lipid synthesis by the swimbladder of an abyssal *Coryphaenoides* species. *Am. Zool. 11:*559–570.

Pickwell, G. V., Barham, E. G., and Wilson, J. W. (1964) Carbon monoxide production by a bathypelagic siphonophore. *Science 144:*860–862.

Pollard, T. D. (1977) Cytoplasmic contractile proteins. In *International Cell Biology* (B. R. Brinkley and K. R. Porter, eds.), pp. 378–387. New York: Rockefeller University Press.

Pringle, J. W. S. (1949) The excitation and contraction of the flight muscles of insects. *J. Physiol. 108:*226–232.

Pringle, J. W. S. (1957) *Insect Flight.* Cambridge: Cambridge University Press. 132 pp.

Pugh, L. G. C. E. (1959) Carbon monoxide content of

the blood and other observations on Weddell seals. *Nature, Lond. 183:*74–76.

Pugh, L. G. C. E. (1971) The influence of wind resistance in running and walking and the mechanical efficiency of work against horizontal or vertical forces. *J. Physiol. 213:*255–276.

Rayner, M. D., and Kennan, M. J. (1967) Role of red and white muscles in the swimming of the skipjack tuna. *Nature, Lond. 214:*392–393.

Ridgway, E. B., and Ashley, C. C. (1967) Calcium transients in single muscle fibers. *Biochem. Biophys. Res. Commun. 29:*229–234.

Rikmenspoel, R. (1965) The tail movement of bull spermatozoa: Observations and model calculations. *Biophys. J. 5:*365–392.

Rikmenspoel, R., Sinton, S., and Janick, J. J. (1969) Energy conversion in bull sperm flagella. *J. Gen. Physiol. 54:*782–805.

Rothschild, M., Schlein, Y., Parker, K., and Sternberg, S. (1972) Jump of the oriental rat flea *Xenopsylla cheopis* (Roths) *Nature, Lond. 239:*45–48.

Rüegg, J. C. (1971) Smooth muscle tone. *Physiol. Rev. 51:*201–248.

Satir, P. (1968) Studies on cilia. 3. Further studies on the cilium tip and a ''sliding filament'' model of ciliary motility. *J. Cell. Biol. 39:*77–94.

Satir, P. (1974) The present status of the sliding microtubule model of ciliar motion. In *Cilia and Flagella* (M. A. Sleigh, ed.), pp. 131–142. London: Academic Press.

Schmidt-Nielsen, S., Flood, A., and Stene, J. (1934) On the size of the liver of some gristly fishes, their content of fat and vitamin A. *Kongelige Norske Videnskabers Selskab Forhandlinger 7:*47–50.

Scholander, P. F. (1954) Secretion of gases against high pressures in the swimbladder of deep sea fishes. 2. The rete mirabile. *Biol. Bull. 107:*260–277.

Scholander, P. F. (1956) Observations on the gas gland in living fish. *J. Cell. Comp. Physiol. 48:*523–528.

Scholander, P. F., and Van Dam, L. (1954) Secretion of gases against high pressures in the swimbladder of deep sea fishes. 1. Oxygen dissociation in blood. *Biol. Bull. 107:*247–259.

Seymour, M. K. (1971) Burrowing behaviour in the European lugworm *Arenicola marina* (Polychaeta: Arenicolidae). *J. Zool. 164:*93–132.

Sleigh, M. A. (1968) Patterns of ciliary beating. *Symp. Soc. Exp. Biol. 22:*131–150.

Sleigh, M. A. (1974) *Cilia and Flagella.* London: Academic Press. 500 pp.

Sleigh, M. A. (1977) Fluid propulsion by cilia and flagella. In *Comparative Physiology: Water, Ions and Fluid Mechanics* (K. Schmidt-Nielsen, L. Bolis, and S. H. P. Maddrell, eds.), pp. 255–265. Cambridge: Cambridge University Press.

Sotavalta, O. (1953) Recordings of high wing-stroke and thoracic vibration frequency in some midges. *Biol. Bull. 104:*439–444.

Squire, J. M. (1975) Muscle filament structure and muscle contraction. *Annu. Rev. Biophys. Biomed. Engin. 4:*137–163.

Steen, J. B. (1963) The physiology of the swimbladder in the eel *Anguilla vulgaris.* 3. The mechanism of gas secretion. *Acta. Physiol. Scand. 59:*221–241.

Stolpe, M., and Zimmer, K. (1939) Der Schwirrflug des Kolibri im Zeitlupenfilm. *J. Ornithol. 87:*136–155.

Taylor, C. R., Caldwell, S. L., and Rowntree, V. J. (1972) Running up and down hills: Some consequences of size. *Science 178:*1096–1097.

Thys, H., Cavagna, G. A., and Margaria, R. (1975) The role played by elasticity in an exercise involving movements of small amplitude. *Pflügers Arch. 354:*281–286.

Thys, H., Faraggiana, T., and Margaria, R. (1972) Utilization of muscle elasticity in exercise. *J. Appl. Physiol. 32:*491–494.

Tucker, V. A. (1973) Bird metabolism during flight: Evaluation of a theory. *J. Exp. Biol. 58:*689–709.

Twarog, B. M. (1979) The nature of catch and its control. In *Motility in Cell Function: Proceedings of the First John M. Marshall Symposium in Cellular Biology* (F. A. Pepe, J. W. Sanger, and V. T. Nachmias, eds.), pp. 213–241. New York: Academic Press.

Usherwood, P. N. R. (1967) Insect neuromuscular mechanisms. *Am. Zool. 7:*553–582.

Weber, A., and Murray, J. M. (1973) Molecular control mechanisms in muscle contraction. *Physiol. Rev. 53:*612–673.

Weis-Fogh, T. (1960) A rubber-like protein in insect cuticle. *J. Exp. Biol. 37:*889–907.

Weis-Fogh, T. (1973) Quick estimates of flight fitness in hovering animals, including novel mechanisms for lift production. *J. Exp. Biol. 59:*169–230.

Weis-Fogh, T. (1976) Energetics and aerodynamics of flapping flight: A synthesis. In *Insect Flight* (R. C. Rainey, ed.), pp. 48–72, New York: Wiley.

Wiersma, C. A. G. (1961) The neuromuscular system. In *The Physiology of Crustacea,* vol. II (T. H. Waterman, ed.), pp. 191–240. New York: Academic Press.

Wittenberg, J. B. (1958) The secretion of inert gas into the swimbladder of fish. *J. Gen. Physiol. 41:*783–804.

Wittenberg, J. B. (1960) The source of carbon monoxide in the float of the Portuguese man-of-war, *Physalia physalis.* L. *J. Exp. Biol. 37:*698–705.

ADDITIONAL READING

Alexander, R. McN. (1983) *Animal Mechanics,* 2d ed. Oxford: Blackwell Scientific Publ. 301 pp.

Alexander, R. McN. (1988) *Elastic Mechanisms in Animal Movement.* Cambridge: Cambridge University Press. 141 pp.

Alexander, R. McN., and Goldspink, G. (eds.) (1977) *Mechanics and Energetics of Animal Locomotion.* London: Chapman & Hall. 346 pp.

American Society of Zoologists (1978) Skeletal muscle tissue. *Am. Zool. 18:*97–166.

American Society of Zoologists (1984) Biomechanics: Symposium. *Am. Zool. 24:*1–134.

American Society of Zoologists (1987) Muscle fiber typing as a bioassay of nerve–muscle interaction: Comparison of arthropod and vertebrate systems. *Am. Zool. 27:*975–1098.

Blake, J. R., and Sleigh, M. A. (1974) Mechanics of ciliary locomotion. *Biol. Rev. 49:*85–125.

Blaxter, J. H. S., and Tytler, P. (1978) Physiology and function of the swimbladder. *Adv. Comp. Physiol. Biochem. 7:*311–367.

Bray, D., and White, J. G. (1988) Cortical flow in animal cells. *Science 239:*883–888.

Currey, J. D. (1984) *The Mechanical Adaptations of Bones.* New Jersey: Princeton University Press. 294 pp.

Day, M. H. (ed.) (1981) *Vertebrate Locomotion.* Symposium of the Zoological Society of London, No. 48. London: Academic Press. 471 pp.

Denton, E. J. (1974) On buoyancy and the lives of modern and fossil cephalopods. *Proc. R. Soc. London B. 185:*273–299.

Fänge, R. (1983) Gas exchange in fish swim bladder. *Rev. Physiol. Biochem. Pharmacol. 97:*111–158.

Hill, A. V. (1965) *Trails and Trials in Physiology: A Bibliography 1909–1964; with Reviews of Certain Topics and Methods and a Reconnaissance for Further Research.* London: Edward Arnold. 374 pp.

Hill, A. V. (1970) *First and Last Experiments in Muscle Mechanics.* Cambridge: Cambridge University Press. 141 pp.

Hoyle, G. (1983) *Muscles and Their Neural Control.* New York: Wiley. 689 pp.

Lighthill, M. J. (1969) Hydromechanics of aquatic animal propulsion. *Annu. Rev. Fluid Mech. 1:*413–446.

Margaria, R. (1976) *Biomechanics and Energetics of Muscular Exercise.* London: Oxford University Press. 146 pp.

Nachtigall, W. (1974) Locomotion: Mechanics and hydrodynamics of swimming in aquatic insects. In *The Physiology of Insecta,* 2d ed., vol. 3 (M. Rockstein, ed.), 381–432. New York: Academic Press.

Pedley, T. J. (ed.) (1977) *Scale Effects in Animal Locomotion.* London: Academic Press. 545 pp.

Rainey, R. C. (ed.) (1976) *Insect Flight.* New York: Wiley. 287 pp.

Sleigh, M. A. (ed.) (1974) *Cilia and Flagella.* London: Academic Press. 500 pp.

Treherne, J. E. (ed.) (1985) Design and performance of muscular systems. *J. Exp. Biol. 115:*1–412.

Twarog, B. M., Levine, R. J. C., and Dewey, M. M. (eds.) (1982) *Basic Biology of Muscles: A Comparative Approach.* Society of General Physiology Series, vol. 37. New York: Raven Press. 406 pp.

Vogel, S. (1981) *Life in Moving Fluids: The Physical Biology of Flow.* Boston: Willard Grant Press. 352 pp.

Vogel, S. (1988) *Life's Devices: The Physical World of Animals and Plants.* Princeton, N.J.: Princeton Univ. Press. 367 pp.

Wainwright, S. A., Biggs, W. D., Currey, J. D., and Gosline, J. M. (1976) *Mechanical Design in Organisms.* New York: Wiley. 423 pp.

Wu, T. Y.-T., Brokaw, C. J., and Brennen, C. (eds.) (1975) *Swimming and Flying in Nature,* vols. 1 and 2. New York: Plenum Press. 1005 pp.

CHAPTER ELEVEN

Control and Integration

It is obvious that physiological processes must be controlled and not allowed to run wild. In the preceding chapters we have often mentioned regulation and control, but we have not discussed the mechanisms of regulation.

To regulate means to adjust an amount, a concentration, a rate, or some other variable, usually in order to reach or keep it at some desired level. For example, we take it for granted that respiration must provide oxygen at the rate it is used by the organism. Similarly, all physiological processes must be controlled and integrated.

Integration means putting parts together. In physiology the word means to control all functional components so that they merge into a smoothly operating organism where no single process is permitted to proceed at its own independent rate.

Physiological functions can be controlled by *hormones* or by the *nervous system*, but two important differences stand out. One relates to the *speed of action;* the other to the *size of the target.*

The quick response of skeletal muscles depends on nerve impulses that move at speeds as high as 100 m per second; response time therefore is only milliseconds. In contrast, a hormone has no effect until it reaches the target, usually via the circulating blood. The minimum response time is therefore seconds or minutes or much longer.

The second important difference is the size of the target and the precision of control. An axon in a motor nerve affects only a single muscle or a fraction of it, without affecting other muscles. Hormones, in contrast, affect all sensitive cells they reach; whole organs or organ systems.

Nevertheless, there are much closer connections between the nervous and endocrine systems than was previously understood. The nervous system regulates many important endocrine functions and is itself an important producer of hormones.

Let us begin with what we mean by control.

CONTROL AND CONTROL THEORY

The control mechanisms responsible for the maintenance of steady states in living organisms constitute a major chapter in physiology, and it is helpful to examine their performance in the light of simple control theory. For example, birds and mammals maintain nearly constant body core temperatures in spite of wide variations in the external temperature and in internal heat production. This regulation of body temperature is carried out with the aid of a complex control system.

Theoretical considerations

In engineering, control systems are so important that control theory and design constitute an independent branch of engineering with its own terminology and theoretical approach. For physiology this is important in two respects: (1) many ill-defined and hazy old terms have been replaced by exact and well-defined concepts, and (2) control theory has resulted in a more precise definition of the components of physiological control mechanisms and a better understanding of the relations between the component parts.

Feedback

Let us examine a familiar control system: the thermostat that regulates the temperature in a house or a water bath. This and all other control systems operate with a *controlled variable* (in this case, temperature), which is maintained within a more or less narrow range around a desired value. A measurement of the value of the controlled variable is compared with the desired value, the *set point*. This is done by an *error detector,* which delivers a signal that in turn activates a control mechanism that results in the necessary correction.

If we want to keep a water bath at a "constant" temperature, we can use the arrangement shown in Figure 11.1, which is nearly self-explanatory. Person A can decide to add heat to the water bath by throwing an electric switch, thus causing the temperature to rise. A thermometer tells him the temperature of the water. Person B, who is in charge,

FIGURE 11.1 Diagram of a temperature-controlled water bath that includes a controlling system and a command. See text for details.

has already told A what temperature, or set point, he wants maintained in the water bath. If the water temperature falls below the set point, A can throw the switch to raise the water temperature. By looking at the thermometer, he can see when the temperature has reached the desired level, and then break the contact again.

We can, of course, replace A by an automatic mechanism, a thermostat, similar to that which controls the central heating system in a house. We can then represent the entire system by the block diagram shown in Figure 11.2*a*. Information about the output of the system, in this case water temperature, is fed back into the thermostat so that appropriate action is taken to correct any deviation in water temperature and keep it at the desired level. This is known as *feedback,* a term used when we compare the condition of the output of the control

FIGURE 11.2 Diagrams of (*a*) the control system represented in Figure 11.1 and (*b*) a control system that includes the terms most commonly used in control theory.

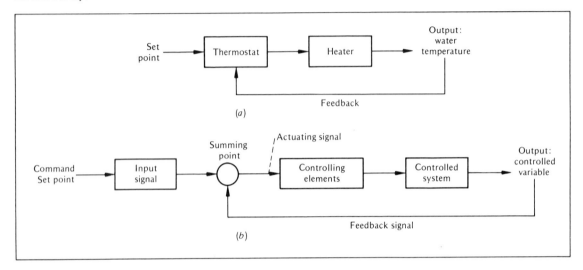

(*a*)

(*b*)

system with the set point. In this case, a rise in water temperature is corrected by a decrease in the heat input. This is *negative feedback,* a term used when a deviation is offset by a corrective action in the opposite direction.

For a further discussion it is useful to expand the diagram slightly and introduce standardized terms used by control system engineers (Figure 11.2*b*).

The typical negative feedback system is known as a *closed-loop control system,* a term whose meaning is obvious from the diagram. The signal from the controlled variable is fed back into the system, forming a closed loop.

For completeness we should also mention *open-loop systems,* although they are less important in physiological regulation. Assume that a house furnace has a variable fuel supply arranged so that a drop in outside temperature increases the flow of fuel to the furnace. We can carefully adjust the system so that a decrease in the outside temperature gives precisely the amount of fuel needed to keep the room temperature constant.

In this example the input is the outside temperature, and the output is the heat supplied by the furnace. However, if a disturbance enters into the carefully calibrated system (e.g., a strong wind), more heat is carried away from the house, but the furnace supplies no more heat and the room temperature falls. In this system there is no feedback; it is an *open-loop control system.*

Negative feedback serves to reduce the difference between the output and the desired value; a thermostat is a good example. What about *positive feedback?* Does it exist? Does it have any importance in biology?

The following situation has been suggested by my colleague, Steven Vogel, as an illustration of positive feedback. Assume that a husband and wife have separate electric blankets, each thermostatically controlled through negative feedback. Say that the husband prefers a rather cool blanket, and the wife a higher temperature. Now assume that the thermostats inadvertently get interchanged. The husband sets "his" thermostat at his preferred low emperature, and his wife, who now finds her blanket cooler than she likes, turns up ''her'' thermostat. The husband soon finds his blanket much too warm and turns ''his'' thermostat further down, whereupon the wife turns ''hers'' up even more. This is *positive feedback,* in which a deviation leads to an ever-increasing augmentation of the deviation.

Obviously, positive feedback is no good at all for control purposes, for the system will proceed to some extreme condition. However, positive feedback systems can be useful in some biological situations.

Negative feedback is used to maintain a steady state; positive feedback makes a system change more and more rapidly toward some extreme state.

A well-known example of positive feedback is the digestive enzyme trypsin, which is secreted from the pancreas as inactive trypsinogen. Trypsinogen is transformed into active trypsin by the intestinal enzyme enterokinase. However, trypsinogen is also activated by trypsin, and as more trypsin is formed, the activation of trypsinogen proceeds more and more rapidly. This is known as autocatalytic activation of the enzyme and is a typical case of positive feedback. A similar autocatalytic reaction applies to the activation of pepsin in the stomach, again a case of positive feedback.

Another example of positive feedback occurs in the generation of a nerve impulse. We shall later see that a slight decrease in the membrane potential of a nerve increases its permeability to sodium ions, and the consequent increased influx of sodium in turn further decreases the membrane potential, giving a positive feedback situation that results in a full action potential in the nerve. Positive feedback thus can serve to amplify a small signal and bring about a full response.

Positive feedback often is useful also in synchronizing events. A familiar example is the positive feedback during the process of mating. When suitable partners meet, progress toward mating is reinforced by the response to small signals. The responses lead to increased mutual reinforcement,

FIGURE 11.3 Diagram of a simple proportional control system that can adjust the water flow into a tank to match the outflow.

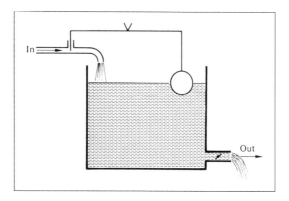

and continued positive feedback between the partners leads to increasing desire for copulation and eventually completion of mating.

On–off, proportional, derivative, and integral control

The thermostatic control of a house furnace is an *on–off control system.* The furnace is either on or off and therefore gives a discontinuous supply of heat. This inevitably leads to oscillations in temperature. In order to start the on cycle, a certain deviation from the set point is needed. Similarly, when the heater is on, there is usually an overshoot in temperature before the heater is cut off. The overshoot can be reduced by making the system more sensitive, but there is no way to eliminate oscillations entirely from an on–off system.

A more constant output of the controlled variable can be achieved with other types of control systems. One important system, known as *proportional control,* perhaps is best illustrated by a mechanical analogy (Figure 11.3). The water level in the tank is controlled with the aid of a float that gives continuous instead of on–off control. If the outflow from the tank for some reason increases, the water level sinks, thus opening the valve to allow increased inflow. The more the level sinks, the

greater the increase in the inflow. Should the outflow be impeded, the water level rises and causes
the inflow valve to close. This system has continuous control, and the degree of control action is
directly related to the deviation from the set point.
(The term *proportional* implies a continuous linear
relation between output and input with the origin
at zero, but the term is commonly used for systems
that do not meet this strict definition of proportionality.)

A proportional controller has an interesting
characteristic that is important. Assume a system
that is in steady state and controls exactly at the set
point. If a disturbance is introduced (e.g., a change
in outflow), the system readily reaches a new steady
state, but it is *not possible to attain the original set
point*. This may need an explanation.

Assume a given steady state, which is the original set point of the water level. Let us now decrease the outflow and examine the corrective action. The increased water level lifts the float and
decreases the inflow. The new steady state involves a higher water level and the float must remain at this higher level to keep the inflow equal
to the reduced outflow (i.e., the new level represents an error from the original set point). The corrective action cannot return the fluid level exactly
to the original set point (zero error), for the float
would then be in its original position and inflow
would not match the decreased outflow. Now assume that a different kind of disturbance is introduced by adding a second inflow. The fluid level
again rises and reduces the regulated inflow, but
the new steady state is again above the set point.

A *physiological system* that represents proportional control is the control of mammalian respiration (ventilation of the lungs) by the carbon dioxide level in the blood. When a mammal at rest has
an arterial P_{CO_2} of 40 mm Hg, we can consider this
as the set point for the arterial P_{CO_2}. If we increase
the carbon dioxide content of the inspired air, there
is an increase in arterial P_{CO_2} that results in an
increase in ventilation, and the amount of increase

is directly related to the increase in P_{CO_2} (Chapter
1).*

We can now ask: Does it matter how fast the
signal changes when an error appears? For example, if a gross error suddenly emerges, the signal changes rapidly. If the error is small, on the
other hand, the initial rate of change in the signal
will probably be small. Thus the rate of change in
the signal may give valuable information about the
expected size of the error and the magnitude of
corrective action needed.

If the rate of change is used to determine the
control action, this is known as *derivative control*
(or *rate control*) because the time derivative of a
signal, or the rate of change in the signal, is used
to anticipate the corrective action required. Because this control action responds only to a rate of
change, and not to the error signal itself, it is usually used in combination with other control action.
In physiology such combination is particularly
valuable, for it can be used to take appropriate action when there are temporary transients in a controlled variable.

We can ask an additional question: How can we
best handle a small error that creeps in very slowly?
We can integrate the error signal over time; that is,
the total error over time is used as an input to the
controller. This is called *integral control* because
the output is proportional to the time integral of the
input. If a disturbance in this system remains constant, the error in the output will with time tend
toward zero. Therefore, whatever the value of the
disturbance, if it remains constant, the integral ac-

*Respiratory control is in reality more complex. We know
that in exercise carbon dioxide production is increased; if there
were a simple proportional control system, the arterial P_{CO_2} should
be slightly increased in exercise to provide the actuating signal
for the increased ventilation. Actually, the situation is the opposite: During exercise arterial P_{CO_2} is slightly decreased. This
is often "explained" as a result of a "resetting" of the set point
in exercise. The real reason is that the respiratory center receives and integrates several different control inputs, including
nerve impulses from the working muscles and from the motor
cortex of the brain.

ion can achieve a steady state with zero error. As we saw before, a simple proportional control system cannot achieve a steady state with zero error, but integral control can do so.

Physiological mechanisms

Let us return to the regulation of body temperature. We used the analogy with the thermostatic control of a water bath through a negative feedback loop. Central heating in a house can be regulated in the same way, and if in summer the house tends to get too warm, we can add an air-conditioning (cooling) system. If the house temperature rises unduly, the thermostat starts the cooling cycle, an action that decreases the temperature toward the desired set point (i.e., there is negative feedback). Thus, regulation of both heating and cooling depends on negative feedback control.

The analogy to body temperature control is obvious. If body temperature tends to fall, heat production is increased, mainly through involuntary muscle contractions (shivering). If body temperature increases because of a heat load, whether external or internal, cooling is achieved through sweating or panting. Thus, both heating and cooling of the body depend on negative feedback control systems.

The temperature regulation center is located in the hypothalamus of the brain. This can be demonstrated in various ways. For example, if the blood in the carotid artery of a dog is heated, it causes the dog to pant. This shows that the regulation takes place in the head, because as the experimental animal is caused to pant, it loses too much heat and its core temperature drops. Conversely, if the carotid blood is cooled, the dog begins to shiver, and the body core temperature rises. The exact location of the heat regulation center can be pinpointed by heating or cooling small areas in the hypothalamus.

The heat regulation center can be regarded as the thermostat and the "normal" body temperature as the set point. The system is not so simple, however, for there is no constant set point and there are multiple inputs. To begin with, the body temperature fluctuates with a daily cycle, even if the external temperature and internal heat production remain constant. This means that the set point undergoes a diurnal cycle. Also, we learned that during exercise the core temperature is reset and regulated at a higher level than at rest. Of the many inputs to the heat regulation center, the warm and cold receptors in the skin provide important information, but many other inputs provide additional information, one of them being the temperature of the arterial blood that reaches the center.

We mentioned that regulation of body temperature depends on both heating and cooling (i.e., control action takes place in either direction). This is an important principle in physiological regulation and much more common than a superficial examination might indicate.

The heart is another example. Its rhythmic contraction is an inherent characteristic of the heart muscle. Contraction starts at the sinus node, spreads throughout the heart muscle, and is followed by relaxation, whereupon a new contraction begins. The rate of contraction is under the control of two nerves: the *accelerator nerve* speeds the heartbeat; the *decelerator nerve,* a branch of the vagus nerve, slows the rate. The heart rate is thus determined by the balance between two antagonistic nerves, one stimulating and the other inhibiting.

We shall see that a balance between stimulation and inhibition is common in physiological control systems. For example, several of the endocrine glands that are under the control of the hypophysis (pituitary) are subject to the balance between a stimulating and an inhibitory effect (in this case both transmitted as an endocrine signal).

Before we can deal with the two major interacting regulatory systems, the endocrine and nervous systems, we need an understanding of nerve cells and how they transmit excitatory and inhibitory impulses.

NERVES AND NERVOUS SYSTEMS

Nerve cells or *neurons* are the basic components of all nervous systems. We shall focus on two of their most important functional parts: the long fibrous extensions, the *axons,* and the connections between cells, the *synapses*. The single neuron has only one axon, but may have hundreds or even thousands of synaptic connections, and their role in integration is extremely important. The axons function as cables and the synapses serve as highly complex contact or switching devices.

To understand these functions, we must know how nerve cells operate. Because of the astounding uniformity inherent in the function of nerve cells, it is relatively unimportant what animal we deal with. A snail or a squid or a frog or a fish is equally relevant to the understanding of the principles on which nerves function.

After discussing the nerve cell we shall describe some important characteristics of axons and how synapses handle the information that impinges on the neuron. As an example of integration, we shall describe how simultaneous stimulatory and inhibitory control by the central nervous system is employed to control the muscles of a vertebrate limb used in locomotion.

Structure of nerve cells

All nervous systems consist of a large number of single nerve cells (*neurons*). Neurons can have a wide variety of shapes and sizes, but they have certain important features in common (Figure 11.4). There is a *cell body* that contains the nucleus and a large number of thin fibers extending from it. Each neuron usually has a single long fiber (the *axon*), which in a large animal may be several meters long, and a large number of shorter fibers (*dendrites*), which are heavily branched and for the most part are less than 1 mm long.

In vertebrates the cell body of the neuron is usually quite small, often less than 0.1 mm in diameter and the fibers less than 0.01 mm thick. The entire cell, including all the fibers, is surrounded by a thin membrane (the *nerve membrane*). A

FIGURE 11.4 Neurons can have various forms, but every neuron has a cell body and a number of long extensions. These schematic drawings represent (a) an arthropod motor neuron, (b) a mammalian spinal sensory neuron, (c) a bipolar neuron from the vertebrate retina, (d) a neuron from the nerve net of a coelenterate, and (e) a basket cell from the mammalian cerebellum. [Aidley 1971]

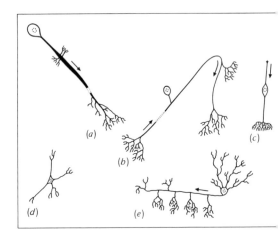

complex nervous system contains an immense number of neurons. In humans, for example, the brain alone contains around 10 000 000 000 cells.

The long fibers, the axons, are the main conduction lines in the body. What is commonly known as a nerve, or nerve trunk, consists of hundreds or thousands of axons, each originating from a different neuron. A nerve has no cell bodies in it; these are found in the central nervous system, in special aggregations known as *ganglia,* and in the sensory organs.

The points where nerve cells and their extensions make contact with other nerve cells are called *synapses*. A single nerve cell may, through the synapses, be connected to hundreds of other neurons. The most important feature of the synapse is that it functions as a one-way valve. Transmission of an impulse can take place in one direction only, from the axon to the next cell, and not in the opposite direction (for details, see later). As a result

FIGURE 11.5 The mantle of the common squid is supplied with nerves of exceptionally large diameter. These so-called giant axons are fast-conducting and serve to obtain a nearly simultaneous con- traction of the entire mantle musculature. [Keynes 1958. Copyright © by Scientific American, Inc. All rights reserved.]

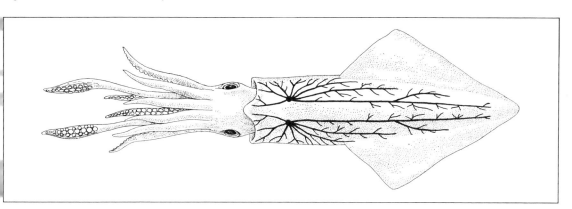

conduction in an axon anywhere in the nervous system takes place in one direction only.

An axon in itself is perfectly capable of conducting impulses in either direction, but in the integrated nervous system all conductance in any given axon is always in the same direction. In a nerve trunk, however, some axons may conduct in one direction, and others in the opposite direction. For example, in a nerve to a given muscle, the impulses stimulating the muscle to contract are transmitted from the central nervous system to the muscle, and at the same time sensory information from the muscle is carried in other axons back to the central nervous system.

HOW NERVE CELLS FUNCTION

It has been known for about two centuries that electric phenomena are associated with the transmission of nerve impulses and muscular contraction. However, the exact nature of the electric phenomena has been understood only for a few decades.

The single most important event was the discovery that certain nerve fibers in squid consist of a single axon of giant dimensions. Most axons mea-

sure from less than 1 μm to about 10 μm in diameter; the *giant axons* of squid may be nearly 1000 μm in diameter (Figure 11.5). Because of their exceptional size, giant axons can be used for many experiments that would be impossible on smaller axons. However, the general principles revealed in studies of giant axons apply to the function of other neurons and also to certain other tissues, such as muscle.

The normal neuron, including its axon, shows a potential difference between the inside and the outside of the cell membrane. In the inactive or resting neuron this membrane potential is known as the *resting potential*. In the giant axon of squid the resting membrane potential is about −60 mV (inside negative). During activity this can change to about +40 mV (inside positive), giving a total change, an *action potential* of about 100 mV.

Membrane resting potential

The most important information to consider in relation to the membrane potential is the concentration of ions inside and outside the axon and the permeability of the axon membrane to these ions. The inside axoplasm has a relatively high concen-

TABLE 11.1 Concentrations (in millimoles per liter) of major ions in axoplasm of giant axons from squid, in squid blood, and in sea water. [Data mainly from Hodgkin and Huxley 1952]

Ion	Axoplasm	Blood	Sea water
Potassium	400	20	10
Sodium	50	450	470
Chloride	40	570	550
Calcium[a]	0.3×10^{-3}	10	10
Magnesium	10	55	54

[a] The concentration of ionized calcium in axoplasm is uncertain, but extremely low (Baker et al. 1971).

$$E = \frac{RT}{F} \log_e \frac{[K]_o}{[K]_i}$$

In this equation R is the universal gas constant, T the absolute temperature, F the charge carried by 1 g equivalent of ions, and $[K]_o$ and $[K]_i$ the outside and inside potassium concentrations. At room temperature the values of R and F are such that the potential is given by:

$$E_K = 25 \log_e \frac{[K]_o}{[K]_i} \text{ mV} = 58 \log_{10} \frac{[K]_o}{[K]_i} \text{ mV}$$

The ratio of potassium concentrations in blood and axoplasm of the giant axon is 1:20, and the calculated resting potential should be -75 mV with the inside negative relative to the outside. The membrane potential can be measured by inserting a tiny microelectrode into the axon and reading the potential relative to the outside. The observed resting potential is close to the calculated potential, but usually several millivolts less, about -60 to -70 mV, a problem we shall return to in a moment.

The fundamental correctness of the conclusion that the membrane potential is caused by the potassium gradient can be verified by changing the ratio between the inside and outside potassium concentrations. This is most readily done by changing the potassium concentration in the outside medium. When the outside potassium concentration is high, the transmembrane potential responds to changes in the potassium concentration as predicted from the Nernst equation, but at low outside potassium concentration, the observed potential deviates from the prediction.

The experiment shown in Figure 11.6 shows the results of changing the external potassium concentration on the membrane potential of isolated frog muscle fibers. When the external potassium concentration was increased to equal the internal concentration, the membrane potential decreased to zero, exactly as predicted for a ratio of 1.0 between the outside and inside potassium concentra-

tration of potassium and a low concentration of sodium; the outside fluid, in contrast, is low in potassium and high in sodium (Table 11.1).

The nerve membrane, when at rest, is selectively permeable to potassium ions. The permeability to sodium and chloride ions is so low that for the moment we will assume the membrane to be impermeable to these ions. Because the potassium concentration inside is higher than outside, some potassium ions diffuse out, down the concentration gradient. Each potassium ion that leaves the axon without being accompanied by a chloride ion gives a positive charge to the outside of the membrane. The diffusion out of potassium ions therefore sets up a positive charge on the outside of the nerve membrane, which quickly reaches a level that opposes further outflux of potassium ions.

At equilibrium the membrane potential, positive on the outside, exactly balances the concentration gradient that drives potassium out. The magnitude of this potential depends on the potassium concentrations inside and outside and can be calculated from the laws of physical chemistry.

The potential difference (E, in millivolts) caused by the unequal distribution of ions inside and outside a permeable membrane is described by the Nernst equation (see Appendix E):

FIGURE 11.6 Changes in the potassium ion concentration in the external bath of isolated frog muscle fibers leads to predicted changes in the internal potential (straight line), except at the very lowest potassium concentrations. To avoid any effect of simultaneous changes in chloride ion concentration, the external anion was sulfate. [Hodgkin and Horowicz 1959]

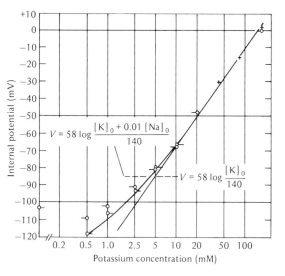

If the sodium permeability (P_{Na}) of the membrane to sodium, which we earlier assumed to be zero, is assumed to be 0.01 of the permeability to potassium, the Goldman equation predicts a membrane potential at low external potassium concentrations that corresponds fairly well to the actually observed points in Figure 11.6.

The most dramatic experiment to show that the nerve membrane is the essential element, and that the axoplasm plays no direct role, consists in removing the entire content of the axon and replacing it with an artificial salt solution (Figure 11.7). Such a perfused axon behaves amazingly like a normal axon; it can remain excitable for many hours and able to give several hundred thousand action potentials (Baker et al. 1961, 1962).

These elegant experiments, which are possible because of the large size of the giant axons, show two important things. First, the axoplasm does not contain any structural elements that are needed for the generation of membrane potentials; it can be replaced by an artificial salt solution. Second, if the internal potassium concentration is changed from normal, the membrane potential changes as predicted from theoretical equations, and if the internal potassium concentration equals the outside concentration, the resting potential is abolished. The membrane potential can even be reversed by making the inside potassium concentration lower than the outside.

tions. At low outside potassium concentrations, however, there was a systematic deviation from the straight line predicted by Nernst's equation, which suggests that the presence of other ions must also be considered.

A theoretical approach to this problem was developed by Goldman (1943). The resulting equation considers the permeabilities and concentrations of all the ions in the system, whether positively or negatively charged. The following equation, known as the Goldman equation, has been very valuable in examining the membrane potential of membranes permeable to more than one kind of ion:

$$E = \frac{RT}{F} \log_e \frac{P_K[K]_o + P_{Na}[Na]_o + P_{Cl}[Cl]_i}{P_K[K]_i + P_{Na}[Na]_i + P_{Cl}[Cl]_o}$$

Action potential

The next subject to consider is the action potential. When we discussed the resting potential, we found the permeability to sodium to be so low that it has a very minor effect on the normal resting membrane potential, which is caused by the potassium permeability that renders the inside negative relative to the outside. With a high sodium concentration outside, two forces tend to drive sodium in: both the negative potential inside and the high sodium concentration outside.

Let us for a moment disregard potassium and

FIGURE 11.7 The axoplasm can be squeezed out of a giant axon, and the axon can then be filled with an artificial solution of desired composition. An action potential recorded from the extruded axon filled with isotonic potassium sulfate (a) is virtually indistinguishable from an action potential recorded from an intact axon filled with its natural axoplasm (b). [Baker et al. 1962]

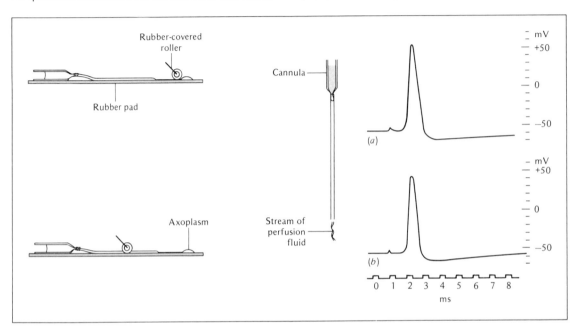

suppose that there is a momentary change in the membrane that for an instant makes it highly permeable to sodium ions. Both concentration gradient and membrane potential drive sodium in, making the inside positive until the potential (now positive) prevents further entry of sodium ion. The potential will, according to Nernst's equation, be +55 mV (inside positive). Thus, the membrane potential has changed from −70 mV (potassium potential) to +55 mV (sodium potential). In other words, the membrane potential of the axon can be changed by 125 mV, merely by altering the relative permeabilities to sodium and potassium ions.

If, an instant later, the membrane again becomes impermeable to sodium, the permeability to potassium reestablishes the resting potential. As we shall see below, the permeability to sodium ions does indeed change as outlined here.

The sodium pump.   During the brief change in permeability, some sodium has entered the cell, and this sodium must be removed to keep the system from running down. This is achieved through active transport of sodium out of the axon. The amount of sodium that entered during the action potential can be calculated from theoretical considerations and it can also be measured experimentally.

The experiment described in Figure 11.8 shows what happens when a giant axon is immersed in sea water that contains radioactive sodium ions. When the axon is stimulated repeatedly to produce action potentials, sodium ions enter the axon and the inside becomes radioactive. When stimulation ceases and the axon is at rest, the sodium is slowly removed from the inside. To show that this removal is an active transport, the sodium pump was poisoned by the inhibitor dinitrophenol (DNP); after

FIGURE 11.8 Entry of radioactive sodium into a squid axon placed in sea water containing radioactive sodium as a tracer. When the axon has been stimulated for 5 minutes to produce action potentials at a rate of 50 per second, a considerable amount of sodium has entered the axon. After stimulation ceases, the sodium concentration inside the axon slowly declines as active transport removes sodium to the outside. When the sodium pump is poisoned by the inhibitor DNP, the sodium content remains constant. Renewed stimulation of the axon causes more sodium to enter, and when the inhibitor DNP is removed, sodium is again extruded by active transport. [Hodgkin and Keynes 1955]

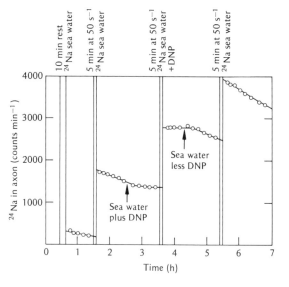

FIGURE 11.9 A giant axon from a squid was injected with radioactive sodium and then placed in artificial sea water. The rate at which sodium was leaving the axon could be followed by the appearance of radioactive sodium in the outside water. When the metabolic inhibitor DNP was added, the active extrusion of sodium decreased to a very low value, but when the inhibitor was again removed, the extrusion of sodium increased to the expected normal rate. [Hodgkin and Keynes 1955]

the pump was poisoned, sodium extrusion ceased and the inside concentration remained constant. After renewed stimulation of the axon, more sodium entered the axon, again showing that the action potentials and sodium entry are coupled.

Another experiment that gives conclusive evidence for an active transport of sodium out of the axon is the following. A solution of radioactive sodium is injected into a giant axon through a fine glass tube inserted into the axon. The axon is then placed in a known volume of sea water and the rate

at which the radioactive sodium leaves the axon is measured. If the axon is now poisoned with DNP, the extrusion rate of sodium drops drastically (Figure 11.9). This is because DNP interferes with the respiration processes that provide adenosine triphosphate (ATP) and the energy requirement of the sodium pump. If ATP is injected into the axon, there is a transient increase in the rate of sodium extrusion, confirming the view that sodium pumping is an energy-requiring process, moving sodium out of the axon against the high concentration in the surrounding sea water. The obvious conclusion is that the sodium pump is driven by the energy supplied by ATP (Caldwell et al. 1960).

The active extrusion of sodium from the axon is dependent on the external potassium concentration, and in potassium-free sea water the extrusion of sodium drops to a small fraction of its normal value. The implication is that the sodium pump is somehow coupled to the uptake of potassium. Experiments with simultaneous use of radioactive sodium and radioactive potassium have shown that there is indeed a coupling, and that for every three sodium ions that are extruded, two potassium ions are taken up.

Sodium pumps appear to be universally present in the cells of all animals, associated with an enzyme known as sodium-potassium ATPase (Na,K-ATPase). This enzyme was first isolated from crab nerve by the Danish physiologist Skou. Subsequently it has been found widely distributed in association with sites of active sodium transport. The enzyme is inhibited by a glycoside, ouabain, which attaches to the active site of the sodium pump. By using ouabain labeled with radioactive tritium, the number of sodium pumping sites in a membrane can be estimated. In the squid giant axon there are several thousand pumping sites per square micrometer of membrane.

The amount of sodium that enters the axon during a single action potential is extremely small. Giant axons in which the sodium pump has been inactivated with DNP or cyanide can continue to show normal excitability and action potentials for several hours. On one occasion such a giant axon continued to conduct impulses at a rate of 50 per second for a total of 70 minutes (i.e., a total of 210 000 impulses) before the system ran down due to inside accumulation of sodium (Hodgkin and Keynes 1955).

So far we have disregarded the chloride ion. Chloride ions are distributed according to the electrical potential to satisfy Nernst's equation, and for our purposes we need not consider their movements. Their effect is small, and, most importantly, chloride permeability does not change during the action potential.

We have now established that the membrane potential, whether the axon is at rest or active, is determined by the membrane permeability and that its magnitude can be calculated from basic principles. By selectively changing the permeabilities to sodium and potassium, the membrane potential can shift anywhere between -75 mV and $+55$ mV.

The permeability changes are extremely fast and take place in microseconds. The experimental difficulties involved in measuring the magnitude of the currents carried by the ions flowing through the membrane in such a short time have been solved by a technique known as the voltage clamp method. By maintaining the membrane potential constant ("clamped") at any particular desired value, one can determine the current flow carried across the membrane by different ions. From this one can estimate the magnitude of the permeability of the membrane to ions as a function of the membrane potential.

The most important results of such studies, primarily executed by the British investigators Hodgkin, Huxley, and Katz, is to establish the ion permeability of the membrane as a function of the potential across the membrane. A decrease in the resting potential (*depolarization*) increases sodium permeability of the membrane.

Depolarization of the membrane also causes the potassium permeability to change, but more slowly. If the permeability to both sodium and potassium were increased simultaneously, there would be no action potential, and the membrane potential would arrive at some intermediate value. However, because the potassium permeability changes after sodium permeability has declined, the effect is to restore the original state and drive the voltage back to its initial value, known as *repolarization* of the membrane (Figure 11.10).

Channels and gates

We have now discussed that, during the action potential, ions flow through the cell membrane because of downhill gradients. These movements take place with the aid of specific transport proteins, located in the lipid cell membrane, that form *channels* through which solutes of appropriate size and charge can move by simple diffusion.

Some channels are open continuously; others open only transiently and are said to be *gated channels*. If gated channels open in response to an outside agent, a ligand, they are said to be *ligand-gated*

FIGURE 11.10 Calculated changes in sodium and potassium conductances during a propagated action potential in a squid giant axon. The main increase in sodium conductance occurs during the rising phase of the action potential; the increase in potassium conductance is slower and serves to reestablish the resting potential after the sodium conductance has declined. [Hodgkin and Huxley 1952]

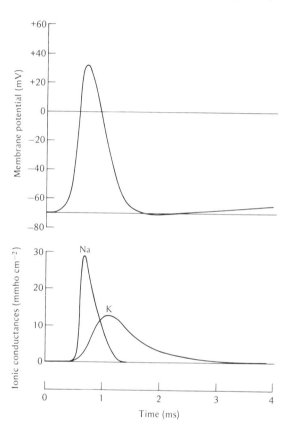

is still present. This is an essential characteristic of the voltage-gated channels in the axonal membrane.

When a voltage-gated channel opens, there is an abrupt change in the membrane conductance. By measuring the electric current that flows, we have a measure of the amount of an ion that flows through the membrane. We saw that during the action potential both sodium and potassium ions move across the membrane, but in opposite directions.

Do sodium and potassium ions flow through the same or different channels? In finding the answer we are aided by the discovery that a poison extracted from the puffer fish, the highly toxic *tetrodotoxin* (frequently abbreviated TTX), blocks the voltage-gated sodium channels while leaving the potassium channels unaffected. Another compound, *tetra methyl ammonium ion* (TEA), blocks the potassium channels while leaving the sodium channels unaffected.

We are now in the position to study experimentally the flow of sodium ions without being confused by the potassium ions moving in the opposite direction. We can measure the amount of sodium flowing through the membrane by measuring the electric current generated by the flow of ions.

When the membrane potential is at the normal resting value, there is very little sodium flowing into the axon, indicating that the sodium channels are closed. If the potential across the membrane is abruptly changed to a less negative value than the normal −60 or −70 mV, the voltage-gated sodium channels open and sodium rapidly flows into the axon, driven by the high concentration outside. This generates a sodium current that can be measured.

The sodium current appears within a fraction of a millisecond of the depolarization of the membrane, and it reaches a maximum within milliseconds. It then tapers off and returns to nearly zero, although the membrane is still depolarized. The gated channels have closed and remain closed and refractive to opening. Not until the membrane potential after several milliseconds has been restored

channels. Other channels open in response to a change in the membrane potential and are said to be *voltage-gated channels.* These are the channels we are interested in at the moment.

One characteristic of many voltage-gated channels is especially important for us: They are self-closing, that is, they close again shortly after opening, even though the stimulus that made them open to the original resting potential will the channels again open in response to membrane depolarization.

We can now see the role of the voltage-gated channels in the propagation of the action potential along the axon. Consider a certain spot on an axon that has been partly depolarized. The sodium gates open, and sodium ions flow in and change the membrane potential to a full action potential, say to +40 mV. This change causes the immediately adjoining area of the axonal membrane to be partly depolarized so that the channels open and sodium flows in and depolarizes the membrane fully. This affects the adjoining membrane area, causing the depolarization to continue spreading down the axon.

The time sequence of the gating is important. The sodium and potassium channels are separate and independently controlled. As we have seen, the potassium channels open more slowly. Furthermore, because channels are either closed or open (there is no halfway situation), we have the explanation for the all-or-none response of the axon. We have either a full action potential or none at all.

NERVE IMPULSES

Until now we have described only the local events at a given point of the nerve membrane. However, we are in possession of the information needed to understand how an action potential is propagated along the axon (i.e., how a nerve impulse is conducted along the axon).

We have seen that a local depolarization of the nerve membrane increases the inflow of sodium and through positive feedback leads to an action potential. The immediately adjacent area of the nerve membrane will be unable to retain its full resting potential and will be partly depolarized; its sodium conductance increases, causing further inflow of sodium, and so on, until it reaches a full action potential. The depolarization therefore spreads further and further and travels as a rapidly propagated action potential along the axon.

A very weak stimulus does not result in an ac-

tion potential. For this to happen, the stimulus must have a certain strength, referred to as the *threshold* strength. Below this threshold there is no action potential, but if the stimulus exceeds the threshold value, an action potential results. The size of the action potential, however, is not influenced by the magnitude of the stimulus. Let us assume that we double or triple the strength of the stimulus above the threshold; the resulting action potential remains the same. The reason, of course, is that the action potential is caused by the concentration of ions on the two sides of the membrane. Therefore, the stimulus causes either a full-strength action potential or none at all. In physiology, this type of response is called an *all-or-none response*.

Once the axon membrane has been depolarized and an action potential generated, it travels as a rapidly propagated action potential along the axon. Because the action potential results from the local concentrations of ions, it spreads along the axon without change in magnitude. This is called *conduction without decrement* and is a fundamental characteristic of the axon, different from conduction in, for example, an electric conductor.

The discussion of action potentials was focused on the giant axon of the squid because the size of this axon has made it a superior research tool. The general principle, that a rapid transient change in membrane permeability is accompanied by ion movements and a change in membrane potential, is universally applicable to nerve function. No regular variations in the size of the action potential are related to the size of the axon, the size of the animal, or whether it is an invertebrate or vertebrate animal (Table 11.2).

There is, however, an apparent exception to the principle that the action potential depends on a high outside sodium concentration and a rapid change in membrane permeability. Although most animals have an extracellular sodium concentration of about 100 mmol per liter or much higher, some plant-eating insects have extremely low sodium concentrations in their hemolymph, as shown in Table 11.3.

Because plants have a high potassium and very

TABLE 11.2 Resting and action potentials in neurons from a variety of animals. [Bullock and Horridge 1965]

Animal	Fiber or cell	Resting potential (mV)	Action potential peak (mV)	Spike duration (ms)
Squid (*Loligo*)	Giant axon	60	120	0.75
Earthworm (*Lumbricus*)	Median giant fiber	70	100	1.0
Crayfish (*Cambarus*)	Median giant fiber	90	145	2
Cockroach (*Periplaneta*)	Giant fiber	70	80–104	0.4
Shore crab (*Carcinus*)	30-μm leg axon	71–94	116–153	1.0
Frog (*Rana*)	Sciatic nerve axon	60–80	100–130	1.0
Sea slug (*Aplysia*)	Visceral ganglion	40–60	80–120	10
Land snail (*Onchidium*)	Visceral ganglion	60–70	80–100	9
Crayfish (*Cambarus*)	Stretch receptor cell	70–80	80–90	2.5
Puffer fish (*Sphaeroides*)	Supramedullary cell	50–80	80–110	3
Toad (*Bufo*)	Dorsal root ganglion	50–80	80–125	2.8
Toad (*Bufo*)	Spinal motor neuron	40–60	40–84	2
Rabbit (*Oryctolagus*)	Sympathetic cell	65–82	75–103	4–7
Cat (*Felis*)	Spinal motor neuron	55–80	80–110	1–1.5

TABLE 11.3 The concentrations of sodium and potassium (in millimoles per liter) in the hemolymph of some plant-eating insects, showing the low sodium concentration compared to a more omnivorous insect, the cockroach.

	Na$^+$	K$^-$
Cecropia moth, *Hyalophora cecropia*	2.5	54.0
Silk moth, *Bombyx mori*	9.0	41.3
Sphinx moth, *Sphinx ligustri*	3.6	49.8
Polyphemus moth, *Antheraea polyphemus*	2.5	54.1
Stick insect, *Carausius morosus*	15.0	18.0
Cockroach, *Periplaneta americana*	161.0	7.9

low sodium content, animals that feed exclusively on plant material have no readily available source for sodium. We know this well from domestic as well as wild grazing animals, which avidly seek out salt licks to satisfy their need for sodium.

In spite of the unusual ion concentrations in some insects, both the resting and the action potentials of their nerves are similar to those in other animals. How is this possible when the ion concentrations that set up these potentials are absent?

It is easier to understand this intriguing situation when we learn that the entire central nervous system of insects is surrounded by a nerve sheath that separates the nerve from immediate contact with the extracellular fluids. This sheath consists of an outer noncellular layer and an inner membrane, the *perineurium*, which is a single layer of specialized cells. The sheath acts as a barrier that separates the surface of the axon from the hemolymph and restricts the movement of materials between hemolymph and the fluid at the neuronal surfaces. The sodium concentration in the fluid bathing the neuronal surfaces may readily be an order of magnitude higher than in the hemolymph (Treherne and Pichon 1972).

The sheath explains why the function of the central nervous system is quite independent of substantial changes in the ionic composition of the hemolymph. Normal action potentials in the giant axon of a cockroach continue even if the outside potassium concentration is increased to as much as 150 mmol per liter. However, if the sheath is re-

TABLE 11.4 Conduction velocities (in meters per second) in nerves from various animals. [Data from Bullock and Horridge 1965]

Animal	Regular motor nerves	Giant axons
Vertebrates		
Cat	30–120	
Snake	10–35	
Frog	7–30	
Fish	3–36	
Invertebrates		
Cockroach	2	10
Squid	4	35
Earthworm	0.6	30
Crab	4	
Snail	0.8	
Sea anemone	0.1	

moved, the giant axon responds like the squid giant axon to changes in the outside ion concentrations.

The remaining question is how the sodium concentration is maintained at the axonal surface. It is most likely that the perineurium is involved, and that an active sodium pump helps maintain a high sodium concentration at the axonal surface, independently of changes in the hemolymph concentrations. In fact, this seems to be the only possible interpretation of the experimental results.

Conduction velocity

Although the size of the action potential is similar in a wide variety of animals, the speed of conduction is an entirely different matter; it varies tremendously from nerve to nerve and from animal to animal. A few typical values are listed in Table 11.4. Vertebrate motor nerves have a much higher speed of conduction than the regular motor nerves of invertebrates. This is not because the invertebrates live at lower temperatures, for the cold-blooded vertebrates also have relatively high conduction speeds. If the conduction velocities for the cold-blooded vertebrates were recalculated to the body temperature of mammals, using a Q_{10} of 1.8 (which is characteristic of nerve fibers), they would

all be within the mammalian range. For the invertebrates this would not be true. Some invertebrates, however, have a few fast-conducting axons, the *giant axons* we have discussed earlier, which are much larger than ordinary axons, up to nearly 1 mm in diameter. In these giant axons the speed of conduction is around 10 times as fast as in ordinary axons from the same animal.

The biological role of high-speed conduction is obvious. It is always related to a quick response mechanism the animal uses in locomotion, mostly to avoid predators. One of the fastest such responses is that of the cockroach, which within 25 ms reacts to an air puff on the tip of the abdomen (which is equipped with receptor hairs).

Another example is the squid, which has giant fibers that run the length of the mantle. Because of the rapid conduction in these giant axons, the entire mantle musculature can contract almost simultaneously, which is needed when the squid swims by jet propulsion. If conduction to the more distant parts of the mantle took appreciably longer, contraction would spread slowly and not give the necessary sudden forceful jet of water. In an earthworm, which on a moist morning reaches halfway out of its burrow, giant fibers permit almost instantaneous withdrawal in response to a mechanical disturbance, in sharp contrast to the relatively slow locomotion otherwise characteristic of this animal.

In general, there is a direct relationship between conduction velocity and axon diameter (Hodgkin 1954). It can be derived on theoretical grounds, based on cable theory, that the speed of conduction (u) should be proportional to the square root of the fiber diameter (d), or:

$$u = k\sqrt{d}$$

A great deal of observational material indicates that this relationship is approximately correct, but the value of the constant k varies from animal to animal. The equation therefore tells us that, for a given type of fiber, we can expect a 10-fold increase in conduction speed for a 100-fold increase in fiber diameter. The giant axons mentioned above may

MYELINATED NERVE A cross section of the optic nerve of a rat shows how each axon is surrounded by multiple layers of myelin. The optic nerve con- tains several hundred thousand such axons; each axon has a diameter of about $1\mu m$. [Courtesy of Alan Peters, Boston University Medical Center]

be 50 or even 100 times thicker than the ordinary axons in the same animals, and their high conduction velocities therefore fall within the expected range.

Vertebrate myelinated fibers

Vertebrates do not have giant axons; yet speed of conduction in their motor nerves is very high – the highest found anywhere in the animal kingdom. The reason mammalian axons conduct rapidly, although they are very thin, is their peculiar structure. The entire axon is covered with a thin

sheath of a fatlike substance, *myelin*, which is interrupted at short intervals to expose the nerve membrane. The exposed sites are known as *nodes*, and the distance between the nodes is from a fraction of a millimeter up to a few millimeters (Figure 11.11). The myelin sheath is formed from glial (or supporting) cells, which grow into a many-layered wrapping from which the protoplasm disappears so that multiple layers of the glial cell membrane remain.

Let us examine the conduction of an impulse in a myelinated axon. At a node the action potential

FIGURE 11.11 Diagram of a myelineated nerve fiber (top). The myelin sheath is developed from a supporting cell, which grows around the nerve fiber until it is covered by a multilayered wrapping of the membrane of the supporting cell (bottom). [Aidley 1989]

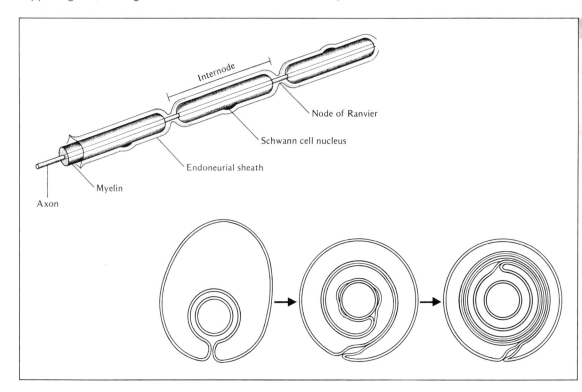

is exactly like any other action potential (i.e., there is a local depolarization of the membrane). This means that this node, relative to the neighboring node, appears negative (the usual positive membrane charge at our node has disappeared). This instantaneously sets up an electric current between this and the neighboring node, sufficient to initiate depolarization and trigger an action potential at the neighbor. This is a virtually instantaneous process, so that the action potential appears to jump from one node to the next. The depolarization at the node itself is less rapid, but as it develops, it in turn causes depolarization at the next node, and so on. The rapid transmission from node to node is known as *saltatory conduction* (from Latin *saltare* = to

dance or jump), and because it takes place almost without delay, the result is a rapid transmission of action potentials in fibers whose small diameter would otherwise give a very low conduction velocity.

The conduction velocity in a myelinated frog axon has been accounted for as follows. A myelinated axon of 10 μm diameter has a conduction velocity of 20 m s^{-1}, and the delay at each node is approximately 0.06 ms. Because the nodes in a frog axon of this size are spaced about 1.6 mm apart, it can be seen that most of the conduction time is accounted for by the delay at the nodes; transmission from node to node occupies only a small fraction of the time (Tasaki 1959).

FIGURE 11.12 The speed of conduction in myelinated fibers of the cat, plotted against fiber diameter. [Hursh 1939]

FIGURE 11.13 The conduction velocities in myelinated and nonmyelinated fibers are different functions of fiber diameter. Below a certain diameter the myelinated fiber therefore conducts more slowly than a nonmyelinated fiber of the same diameter. The broken line is an extrapolation from data in Figure 11.12; the solid line is a theoretical curve based on the fastest observed nonmyelinated C-fibers from the cat. [Rushton 1951]

This conclusion can be further supported by a rather elegant method that involves cooling the nerve. This slows the rate of propagation of an impulse. However, if the cooling is restricted to the area between two nodes, the duration of the transmission between the nodes is nearly unchanged. This result is again consistent with the hypothesis that the conduction between the nodes is an electric phenomenon and that the potential change at the node itself is of the same nature as the action potential in a nonmyelinated axon (Hodler et al. 1951).

The conduction velocity in myelinated axons still depends on the fiber diameter, but in these fibers the conduction velocity is directly proportional to fiber diameter (Figure 11.12). The reason the conduction velocity is higher in the larger-diameter myelinated axon seems to be that the distance between the nodes increases with the diameter of the axon. Because virtually all delay is at the nodes, a smaller number of nodes permits a faster overall propagation of the action potential (Rushton 1951).

The fact that, in myelinated fibers, velocity is linearly related to the fiber diameter, but in nonmyelinated fibers is proportional to the square root of the diameter, has an interesting consequence. For very small diameters, less than about μm,

myelinated axons conduct more slowly than nonmyelinated axons (Figure 11.13). This coincides with the lower limit for the actual size of myelinated fibers in the organism, which seem never to be smaller than 1 μm in diameter. The so-called C-fibers of the sympathetic nervous system are nonmyelinated, and the fastest and largest C-fibers have a diameter of 1.1 μm and conduct at 2.3 m s^{-1}. This point is marked with a circle on Figure 11.13.

The greatest advantage of myelinated axons comes from their small size, which allows a highly complex nervous system with high conduction velocities without undue space occupied by the conduits. Let us say that we wish to increase the conduction velocity 10-fold in a given nonmyelinated fiber. This would require a 100-fold increase in its diameter, and the volume of nerve per unit length would in turn be increased 10 000-fold.

Obviously, if an increased diameter were used for increasing conduction velocity, a nerve trunk that should contain hundreds or thousands of nonmyelinated axons would be unreasonably voluminous. Imagine the size of the optic nerve in humans if a high conduction velocity were to be achieved without myelination. This nerve has a diameter of 3 mm; if it were to contain the same numbers of fibers without myelination and conduct at the same speed, it would require a diameter of 300 mm (or 12 inches).

Invertebrate myelinated nerves

We have seen that vertebrates achieve rapid nerve conduction with the aid of a myelin sheath and saltatory conduction. Many invertebrates achieve rapid conduction by increasing the fiber diameter, (i.e., they employ giant axons when rapid conduction is essential, such as in escape reactions). This, however, is not the full story.

Invertebrate axons show two kinds of structural modification related to fast conduction. First, as we have already discussed, there are giant axons whose rapid conduction is directly related to their cable properties (Hodgkin 1954). Second, some invertebrate nerve fibers are covered with multiple layers of sheaths that are strikingly similar to the myelination of vertebrate nerves. Such nerves covered by ''myelin'' have been found in insects, earthworms, crabs, and prawns.

The speed of conduction in these fibers is much greater than in other invertebrate nerves of similar diameter. In fibers from prawns (*Palaemonetes vulgaris*) with a diameter of 26 μm the speed is as high as 18 to 23 m s^{-1} (Holmes et al. 1941), far faster than the speed recorded in nerve fibers of such small diameter in other invertebrates. However, it is not as great as the conduction velocity in myelinated axons of most vertebrates.

The sheaths of the prawn nerve are somewhat

FIGURE 11.14 Cross section of a shrimp axon, covered with multiple layers of myelinlike sheaths. [Heuser and Doggenweiler 1966. Courtesy of J. E. Heuser and C. F. Doggenweiler, and The Rockefeller University Press]

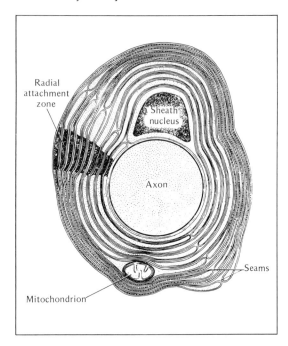

different from vertebrate myelination (Figure 11.14). The layers are arranged more loosely, and there is an extracellular space between the inner layer of the sheath and the axon. Because of the loose arrangement, fewer membranes are present in a prawn sheath of a given thickness than in a vertebrate myelin sheath of the same thickness (Heuser and Doggenweiler 1966).

Another difference is that the spaces form a reservoir for ions that increases the capacitance of the nerve sheath. This influences the cable properties and offers a reasonable explanation for the slower rate of conduction in prawn fibers than in vertebrate fibers. Also, the spacing of nodes in prawn fibers seems to be shorter than is usually found in vertebrates, and because the greatest delay in saltatory conduction is at the node, the greater number of nodes explains the lower conduction velocity. Nevertheless, the structure is remarkably similar to vertebrate myelinated axons, and for a given fiber diameter the conduction velocity is strikingly high.

THE SYNAPSE: EXCITATION, INHIBITION, AND COMPUTATION

We now come to the most important aspect of the function of a nervous system: How is information transferred from one neuron to another? We have seen that signals travel in axons, and the point where an axon makes contact with another neuron is the synapse.

The transmission of signals at synapses is of two distinct kinds: *electrical* or *chemical*. Electrical transmission is not as widespread as chemical transmission, but it has characteristics that in certain situations confer considerable biological advantage.

Electrical synapses

At electrical synapses the terminal end of an axon comes so close to the membrane of the next neuron that there is electrical continuity between the interior of the axon and the next neuron (Figure 11.15). Such connections are called *gap junctions* because there are gaps in the two adjoining membranes where the space separating them is no more than 2 nm (0.002 μm). This space forms a low-resistance pathway that provides a shunt for current to flow from the axon terminal into the next cell. In addition to a low-resistance pathway between the interior of one neuron and the next, electrical transmission also requires a high resistance that prevents sideways spread of the current at the site of contact.

Electrical transmission takes place without measurable delay and therefore is not subject to the characteristic time lag that occurs at chemical synapses. This is important in certain synapses in the

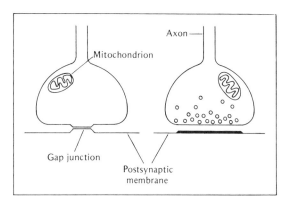

FIGURE 11.15 Diagram of the difference between synapses that effect electrical (left) and chemical transmission (right).

nervous systems of both vertebrates and invertebrates.

Electrical synapses were first demonstrated in the abdominal nerve cord of crayfish (Furshpan and Potter 1959) and have since been found in several arthropods, annelids, coelenterates, and molluscs. In fish, electrical transmission occurs at certain large nerve cells, known as Mauthner cells, where rapid transmission is important in escape reactions. These will be discussed later in this chapter. Electrical transmission is probably common in certain parts of the vertebrate central nervous system where it may have major functional significance (Krnjevic 1986).

Many electrical synapses conduct equally well in both directions, but in others the contact area permits current to flow only from the presynaptic to the postsynaptic site, and not in the reverse direction. Thus, some but not all electrical synapses show rectifying properties; they allow impulses to pass in only one direction, like chemical synapses, which also permit conduction in one direction only.

Chemical synapses

The structure of chemical synapses is important for the understanding of their function (Figure

FIGURE 11.16 At the chemical synapse (the point where the axon from one neuron makes contact with another neuron) the end of the axon is swollen into a synaptic knob. Impulses are transmitted to the receiving neuron by release of a chemical transmitter substance contained in the synaptic vesicles.

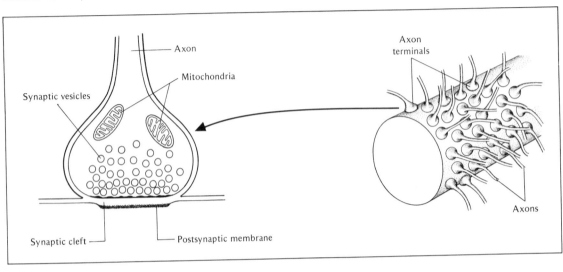

11.16). The terminal end of the axon spreads out and forms the *axon knob,* which in turn makes contact with a dendrite or a cell body of another neuron. The axon and the other cell do not fuse; there remains a narrow space or gap, the *synaptic cleft,* which has a width of about 20 nm. The knob is referred to as *presynaptic;* the dendrite or neuron is the *postsynaptic* structure.

Both the appearance of the synaptic cleft and its width are surprisingly similar throughout the animal kingdom, but the cleft is so small that the nature of its contents has remained virtually unknown. It has been suggested that the relatively constant width of the cleft can best be explained by assuming that it is not merely a fluid-filled space, but contains a structure of oriented molecules arranged between the two nerve membranes.

The synaptic knob has an important structural characteristic: It contains a large number of small vesicles that are usually about 20 to 100 nm in diameter (Figure 11.16). The occurrence of vesicles of this nature is so typical of all chemical synapses that their presence is used by electron microscopists as a criterion for the presence of a synapse.

The transmission of an impulse from the presynaptic knob to the postsynaptic neuron is achieved through the release from the synaptic vesicles of a *chemical transmitter substance* that diffuses across the synaptic cleft and affects the postsynaptic membrane. Most of what we know about the mechanism of synaptic transfer comes from studies of the neuromuscular junction because this junction is more amenable to experimental study than most other synapses.

When a nerve impulse arrives at the presynaptic knob, the change in membrane potential briefly allows *calcium ions* to flow into the terminal. The ions pass through calcium channels that normally are closed, but open in response to the change in membrane voltage; we therefore say that these channels are *voltage-gated.* The rise in calcium concentration is extremely brief because the knob contains apparatus that rapidly sequesters calcium ions. The momentary rise in calcium ions affects

SYNAPSE IN THE AUDITORY SYSTEM OF A CAT The synaptic terminal is the structure that is filled with numerous small, oblong vesicles and a cluster of mitochondria (M). The flattened shape of the vesicles suggests that this is an inhibitory synapse. The active part of the synaptic membrane stretches horizontally across the lower part of the picture. The diameter of the terminal is about 3 μm, and the synaptic cleft (C) is about 0.02 μm wide. [Courtesy of Nell Cant, Duke University]

the vesicles, which fuse with the presynaptic membrane and release their contents of the chemical transmitter substance.

The calcium dependency of the transmitter release has been studied by varying the calcium ion concentration in the surrounding medium. In a low-calcium medium little or no transmitter is released in response to a nerve impulse, and at higher calcium concentrations more transmitter is released. These studies have permitted an estimate of how many single synaptic vesicles are released in the normal transmission process (del Castillo and Katz 1954). At the neuromuscular junction, about 300 vesicles are released for each nerve impulse.

At the neuromuscular junction the transmitter is *acetylcholine*. Each synaptic vesicle contains about 10 000 molecules of acetylcholine, which rapidly diffuse across the synaptic cleft to the postsynaptic membrane. Because of the short distance the diffusion takes place in a small fraction of a millisecond (although noticeably slower than the transmission at electric synapses). At the postsynaptic membrane the acetylcholine molecules bind to specific *acetylcholine receptors*. The receptor molecules, which are a constituent protein of the post-synaptic membrane, form channels that normally are closed, but open in response to acetylcholine and then permit sodium and potassium ions to flow through. These channels are therefore said to be *chemically gated* or *ligand gated,* as opposed to the voltage-gated calcium channels in the presynaptic membrane.

Each channel molecule requires two acetylcholine molecules to open; it remains open for a brief period during which an average of perhaps 20 000 sodium and potassium ions flow through it before it again closes. As a result of the ion movements, the potential difference across the membrane is reduced and may approach zero. How great the decrease in potential is depends on the number of open channels and how long they remain open.

We can now see why chemical synapses can transmit impulses only in one direction. Because transmission depends on the release of a transmitter substance that is present only in the presynaptic knob, there is no means for transferring impulses in the opposite direction.

Our discussion of the information transfer at a single synapse does not properly reflect the immense complexity of the function of the nervous

FIGURE 11.17 Diagram showing a motor neuron cell body and dendrites branching from it. The synaptic knobs are the terminals of axons, or impulse- carrying fibers, from other nerve cells. The motor neuron axon leads to a muscle fiber. [Eccles 1965]

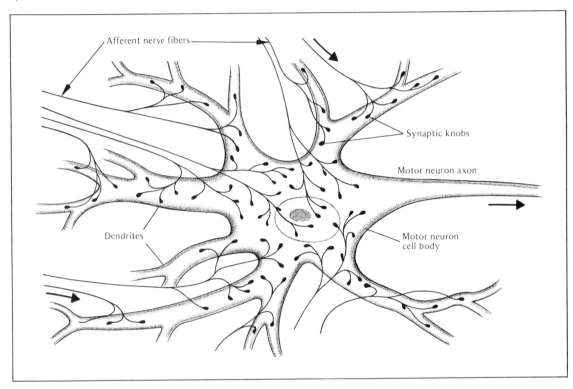

system. Most axons are highly branched and connect to a large number of other neurons. Although the single neuron gives off only one axon, this axon may branch widely and connect to a large number of other neurons. As a consequence, each single neuron must also receive a large number of axon branches that terminate on the neuron or its dendrites (Figure 11.17).

There are often hundreds of synapses on a single neuron in the central nervous system, and other neurons in the vertebrate spinal cord may have thousands of synapses. Some specialized neurons are so densely covered with synaptic knobs that it has been estimated that a single neuron may have about 10 000 synaptic connections. This, viewed in combination with the enormous number of neurons (estimated at more than 10 000 000 000 in the central nervous system of humans), should give an idea of the complexity of the central nervous system. It must be emphasized, however, that the connections are by no means random; they are highly specific and form precisely functioning tracts within the central nervous system. These have been studied and described in great detail by neuroanatomists as well as neurophysiologists.

In the discussion of neuromuscular transmission we found that the transmitter substance is acetylcholine. However, several other substances are

SYNAPTIC KNOBS This scanning electron micrograph shows the multitude of synaptic knobs that impinge on the surface of a single neuron in the nervous system of *Aplysia,* a marine snail. The diameter of each disclike terminal knob is about 1 μm. [Courtesy of Edwin R. Lewis, University of California, Berkeley]

known to be transmitter substances in other parts of the nervous system, and we shall return to some of these in the next chapter. It is worth mentioning, however, that the action of transmitter substances depends on specific *receptor proteins* in the postsynaptic membrane and the function of chemically gated channels. These channels in turn may have different selectivities. Some pass sodium and potassium about equally well, but others may be highly selective, and the resulting potential changes therefore depend on whether potassium ions move out or sodium ions move in through the appropriate channels. Synapses in the brain also differ in other ways so that the action of the same substance may be excitatory at some synapses and inhibitory at others. We should therefore examine in greater detail the postsynaptic potentials that arise in response to transmitter substances.

Postsynaptic potentials

Some of the most interesting features of synaptic transmission pertain to the events at the postsynaptic membrane and, in fact, the entire postsynaptic neuron. It was suggested many years ago that some of the incoming axons to a neuron carry inhibiting impulses, while others are exciting. This suggestion has been amply confirmed, and the mechanisms responsible for inhibition and excitation have been clarified. There is no difference in the impulses carried in excitatory and inhibitory axons. Both axons carry the same kind of action potentials, and the opposite effects are achieved by the release of different chemical transmitter substances from the presynaptic endings.

The understanding of synaptic transmission became possible when methods were developed to make electric recordings directly from single neurons. Briefly, the technique is as follows. Extremely fine glass pipettes with tip diameters of less than 1 μm are prepared. These are filled with a salt solution (usually concentrated potassium chloride), making microelectrodes that can be used to record any potential change between the tip and a common "ground." As such a microelectrode is slowly moved toward and inserted into a neuron, it first shows the same potential as the ground, but the moment it penetrates the nerve membrane, it shows a negative potential relative to the ground. That is, the outside of the membrane is positively charged relative to the inside, exactly like the axon membrane. The membrane seems to seal around the glass, and it is possible to record from a neuron for several hours while it continues to function apparently normally. By using pipettes with two or more channels, instead of a single one, it is even possible to apply minute amounts of chemicals at the site from which the recording is being made.

With this technique we can study the events at the synapse. As an impulse arrives at the presynaptic membrane, there is a delay of a fraction of a millisecond before a change occurs in the potential at the postsynaptic membrane. This latter potential is the *postsynaptic potential* or *PSP* (Figure 11.18).

FIGURE 11.18 The arrival of an action potential at the terminal knob (a) gives rise to a potential at the postsynaptic membrane, the postsynaptic potential (b), which is smaller in magnitude but has a much longer duration. [Stevens 1966]

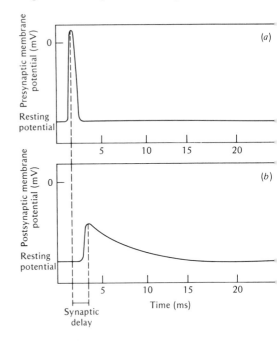

The PSP initially rises rapidly, and then decays but at a much slower rate.

PSPs differ from action potentials in two important respects: They are usually much smaller in amplitude, and they have a much longer duration, sometimes as much as 10 or 100 times as long. These characteristics have two important consequences: A single PSP is rarely (if ever) sufficient to cause an action potential in the postsynaptic neuron, and the long duration permits a great deal of interaction with other PSPs in the same neuron, both at the same synapse and at neighboring synapses on the same neuron. This interaction is very important and requires a more detailed discussion of the postsynaptic events.

What causes the PSP? We will assume that the PSP is caused by the transmitter substance released from the presynaptic membrane and that the delay corresponds to the release of the transmitter and the diffusion time across the synaptic cleft.

The best known and understood transmitter substance is *acetylcholine,* whose function in synapses and at neuromuscular junctions has been studied in great detail. In the synapse, the magnitude of the PSP seems to be directly related to the amount of acetylcholine released. The system could not function, however, unless the acetycholine were rapidly removed again, for otherwise it would gradually accumulate and maintain a continuous PSP. The enzyme *acetylcholinesterase* is always present at the synapse and serves to hydrolyze and thus remove the transmitter. As the transmitter substance disappears, the PSP gradually decays.

Summation

If a second impulse arrives at the same axon terminal before the preceding PSP has fully decayed, the amount of transmitter substance is increased, and the PSP is greater. We therefore observe a summation of the two impulses, and because this is a summation in time, it is referred to as *temporal summation.*

Assume that a long train of impulses arrives at an axon terminal at a constant rate. The PSPs are summed, giving an overall PSP that is directly related to the frequency of impulse arrival. If the frequency is increased, the PSP increases, and a new steady-state PSP is established at a level where the decay rate of the transmitter equals the release rate from the axon terminal. The magnitude of the PSP, therefore, is a direct expression of the impulse frequency in the axon; in other words, the PSP is a *frequency modulated* potential.

This clarifies how a series of action potentials, each an all-or-none event of constant magnitude, can be used to transmit information about a changing signal. For example, an increase in the magnitude of a signal from a sensory neuron is encoded and transmitted in its axon as an increase in the frequency of action potentials; on arrival at the synapse it is then decoded as a PSP of a magnitude

directly related to the magnitude of the original signal.

Let us return to the postsynaptic neuron and remember that there are numerous synapses on its surface. The PSP at any one synapse is not completely restricted to the area under the axon knob; it spreads to the immediate neighborhood with a decreasing magnitude as the distance increases. This spatial spread means that there is a slight change in membrane potential in the neighborhood of the synapse. Should an impulse now arrive at another synapse within this area, the new PSP is added to the existing potential. Such summation in space is known as *spatial summation.*

We have now seen that various inputs to a neuron, arriving at different synapses, influence each other and can be summated through both temporal and spatial summation. These two types of summation form the basis for computations in each neuron and thus in the entire nervous system.

Before we discuss these integration processes further, we must know what effect postsynaptic potentials have on the neuron. We said that the PSPs are usually insufficient to cause an action potential in the postsynaptic neuron. There is, however, an area of the neuron, the *"hillock,"* near the origin of its axon, which fires action potentials if it is sufficiently stimulated by depolarization of the membrane.

If many stimulatory impulses arrive at various synapses on a neuron, a combination of temporal and spatial summation may lead to a PSP sufficiently great to cause the passive spread to reach and depolarize the *axon hillock,* thus setting up an action potential in the axon.

Excitation and inhibition

Until now we have not considered the polarity of the PSP. A PSP can, in fact, either decrease or increase the normal membrane potential. If the postsynaptic membrane potential is slightly decreased, or depolarized, by the PSP, the change is in the direction that normally leads to an action potential. This is referred to as an *excitatory post-*

FIGURE 11.19 When an EPSP reaches a certain threshold value (left), it induces the discharge of a complete action potential. If an IPSP occurs at the same time, the combined effect is to prevent the depolarization from reaching the threshold value (right), and no action potential is generated. [Keynes and Aidley 1981]

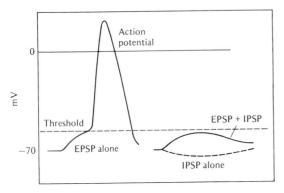

synaptic potential (*EPSP*). If the PSP causes an increase in the normal membrane potential, or a hyperpolarization, its effect is the opposite of that which would lead to an action potential. This is known as an *inhibitory postsynaptic potential* (*IPSP*) because it tends to inhibit the generation of an action potential.

Although there are several different transmitter substances, both excitatory and inhibitory, it was earlier believed that any given presynaptic terminal always would release one and only one transmitter substance, characteristic of that kind of synapse. This concept has been known as the "one neuron–one messenger" dogma. At the vertebrate neuromuscular junction, where acetylcholine is the transmitter, this is undoubtedly true.

However, evidence is now overwhelming for the coexistence of multiple chemical messengers in single nerve cells, in both the peripheral and the central nervous systems (Millhorn and Hökfelt, 1988).

What makes this concept important is that transmitter substances, and therefore the individual synaptic knob, can be both excitatory and inhibitory.

We shall shortly discuss the importance of excitatory and inhibitory signals; at this point we shall only mention that the coexistence of known excitatory and inhibitory transmitters has been demonstrated in pinched-off nerve terminals (known as synaptosomes) from rat brain (Docherty et al. 1987).

The interaction of excitatory and inhibitory impulses is indicated in Figure 11.19. An EPSP alone causes a partial depolarization of the postsynaptic membrane, and if the depolarization reaches the threshold, a full action potential is initiated. If an IPSP also arrives, the summation of the two opposing charges keeps the membrane from reaching the threshold for firing an action potential.

Whether or not an action potential will be fired depends on the impulses a neuron receives from its many different synapses and their timing. EPSPs transform an arriving train of impulses into a sustained depolarization; IPSPs convert a train of impulses into a sustained hyperpolarization. A given neuron responds differently to arriving information, depending on what information it has recently received or is receiving from other sources. Under circumstances when a neuron is already partly depolarized, an additional excitatory potential may be sufficient to fire an action potential from the hillock, although at other times the same EPSP would be far from sufficient.

These characteristics explain how a single neuron can carry out extensive integration of information received from various sources. Although the synapse is a one-way valve, the amount of integration and computation that results from excitatory and inhibitory synapses, in combination with spatial and temporal summation, makes the single neuron a formidable device in the computation processes in the nervous system.

Presynaptic inhibition

We must consider one further characteristic of neurons: an important process known as *presynaptic inhibition*. This can occur when an inhibitory fiber connects directly to the terminal knob of an excitatory fiber (Figure 11.20). An impulse in the

FIGURE 11.20 Impulses in an inhibitory axon can exert their effect, either postsynaptically directly on a neuron, or presynaptically on an excitatory terminal knob.

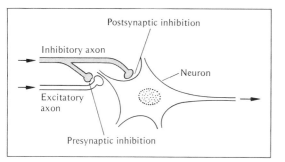

inhibitory fiber then causes hyperpolarization of the excitatory knob. As a result, an impulse that now arrives in the excitatory axon causes less than the usual amount of transmitter to be released, and therefore a reduced PSP. The reason for the reduced PSP is located in the excitatory terminal knob, as indicated by the term *presynaptic inhibition.*

At first glance it may seem trivial whether inhibition is pre- or postsynaptic, but there are some important differences. Presynaptic inhibition is highly selective, for it affects only signals arriving at that particular synapse. Postsynaptic inhibition, on the other hand, works by subtracting from all excitatory PSPs that arrive at the neuron; it therefore is a nonselective inhibition. Presynaptic inhibition has one further characteristic: It is not a simple additive inhibition (as is postsynaptic summation), for it influences the transmission of each impulse in a larger train of an arriving signal. Presynaptic inhibition therefore increases both the specificity and the complexity of the integration that takes place at the neuronal level.

Escape: a flip of the tail

The normal escape reaction of a fish is a sudden flip of the tail, followed by ordinary undulatory swimming. The reaction is directed from two remarkable large nerve cells, the *Mauthner cells,* lo-

cated in the brain of teleost fish. Their role is to integrate and relay information from sense organs and the brain to the motor nerves that drive the body muscles.

The Mauthner cells receive a rich supply of nervous inputs, many forming electric rather than chemical synapses. The main output from the Mauthner cell is through a large axon that from each Mauthner cell crosses to the opposite side and runs down along the spinal cord, where it forms synapses with motor neurons that innervate the main swimming muscles (Diamond 1971).

The importance of the Mauthner cell is in its reaction to sensory information, particularly information produced by mechanical disturbances in the water. In response, the Mauthner cell discharges a single impulse to the large axon, which crosses to the opposite side and causes a rapid, vigorous muscle contraction (i.e., a fast flip of the tail, followed by normal swimming movements). This escape reaction can readily be observed in aquarium fish by tapping on the glass side.

What would happen if both Mauthner cells fired impulses simultaneously? If the muscles on both sides contract at the same time, there will be no movement. This is where the importance of an inhibitory input comes in.

It is virtually impossible for a sensory input to be completely bilaterally symmetrical, both in amplitude and timing, and one Mauthner cell is therefore usually activated an instant ahead of the other. Each Mauthner cell sends inhibitory connections to the Mauthner cell on the opposite side, and as one fires an impulse, the other cell is inhibited. This is where the speed of electrical synaptic transmission is important, for it makes the transmission nearly instantaneous. A chemical synapse would involve a delay, and the inhibition of the second Mauthner cell might not occur soon enough to prevent it from firing an impulse.

The superiority of electric transmission at the synapse is its speed; the delay characteristic of chemical synapses is eliminated. This saving of time explains the importance of the mechanism for fast processing of signals in the nervous system. When a Mauthner cell fires, it almost instantaneously inhibits the other Mauthner cell, thus preventing a paralyzing contraction of both sides simultaneously.

Except where there is a great need for speed, chemical transmission appears to be superior. The greatest advantage is probably in regard to the duration of a change, whether it is excitatory or inhibitory, for the presence of a chemical transmitter permits integration over much longer periods. The preponderance of chemical transmission implies some functional superiority, and we can assume that the greater refinement in temporal interactions is a rational explanation.

So far, our discussion of nerve transmission and integration has centered on the role of action potentials. We have noted the important roles of the release of chemical transmitter substances in the communication between neurons. The role of acetylcholine in synaptic transmission and in neuromuscular transmission is only an example of a multitude of physiologically highly active substances. We shall soon see that the neurons of the central nervous system produce and release a wide variety of substances that have profound effects as hormones. In fact, the entire endocrine system is controlled and directed by the nervous system, both in vertebrates and in many invertebrates.

REFERENCES

Aidley, D. J. (1971) *The Physiology of Excitable Cells.* Cambridge: Cambridge University Press. 468 pp.

Baker, P. F., Hodgkin, A. L., and Ridgway, E. B. (1971) Depolarization and calcium entry in squid giant axons. *J. Physiol. 218:*709–755.

Baker, P. F., Hodgkin, A. L., and Shaw, T. I. (1961) Replacement of the protoplasm of a giant nerve fibre with artificial solutions. *Nature, Lond. 190:*885–887.

Baker, P. F., Hodgkin, A. L., and Shaw, T. I. (1962) Replacement of the axoplasm of giant nerve fibres with artificial solutions. *J. Physiol. 164:*330–354.

Bullock, T. H., and Horridge, G. A. (1965) *Structure and Function in the Nervous Systems of Invertebrates,* vol. I and II. San Francisco: Freeman. 1719 pp.

Caldwell, P. C., Hodgkin, A. L., Keynes, R. D., and Shaw, T. I. (1960) The effects of injecting 'energy-rich' phosphate compounds on the active transport of ions in the giant axons of *Loligo. J. Physiol., Lond. 152:*561–590.

de Castillo, J., and Katz, B. (1954) Quantal components of the end-plate potential. *J. Physiol. 124:*560–573.

Diamond, J. (1971) The Mauthner cell. In *Fish Physiology,* vol. 5, *Sensory Systems and Electric Organs* (W. S. Hoar and D. J. Randall, eds.), pp. 265–346. New York: Academic Press.

Docherty, J., Bradford, H. F., and Wu, J.-Y. (1987) Co-release of glutamate and aspartate from cholinergic and GABAergic synaptosomes. *Nature, Lond. 330:*64–66.

Eccles, J. (1965) The synapse. *Sci. Am. 212:*56–66.

Furshpan, E. J., and Potter, D. D. (1959) Transmission at the giant motor synapses of the crayfish. *J. Physiol. 145:*289–325.

Goldman, D. E. (1943) Potential, impedance, and rectification in membranes. *J. Gen. Physiol. 27:*37–60.

Heuser, J. E., and Doggenweiler, C. F. (1966) The fine structural organization of nerve fibers, sheaths, and glial cells in the prawn, *Palaemonetes vulgaris. J. Cell. Biol. 30:*381–403.

Hodgkin, A. L. (1954) A note on conduction velocity. *J. Physiol. 125:*221–224.

Hodgkin, A. L., and Horowicz, P. (1959) The influence of potassium and chloride ions on the membrane potential of single muscle fibres. *J. Physiol., Lond. 148:*127–160.

Hodgkin, A. L., and Huxley, A. F. (1952) The dual effect of membrane potential on sodium conductance in the giant axon of *Loligo. J. Physiol. 116:*497–506.

Hodgkin, A. L., and Keynes, R. D. (1955) Active transport of cations in giant axons from *Sepia* and *Loligo. J. Physiol. 128:*28–60.

Hodler, J., Stämpfli, R., and Tasaki, I. (1951) Ueber die Wirkung internodaler Abkühlung auf die Erregungsleitung in der isolierten markhaltigen Nervenfaser des Frosches. *Pflügers Arch. 253:*380–385.

Holmes, W., Pumphrey, R. J., and Young, J. Z.

(1941) The structure and conductance velocity of the medullated nerve fibres of prawns. *J. Exp. Biol. 18:*50–54.

Hursh, J. B. (1939) Conduction velocity and diameter of nerve fibers. *Am. J. Physiol. 127:*131–139.

Keynes, R. D. (1958) The nerve impulse and the squid. *Sci. Am. 199:*83–90.

Keynes, R. D., and Aidley, D. J. (1981) *Nerve and Muscle.* Cambridge: Cambridge University Press. 163 pp.

Krnjevic, K. (1986) Ephaptic interactions: A significant mode of communications in the brain. *News Physiol. Sci. 1:*28–29.

Millhorn, D. E., Hökfelt, T. (1988) Chemical messengers and their coexistence in individual neurons. *News Physiol. Sci. 3:*1–5.

Rushton, W. A. H. (1951) A theory of the effects of fibre size in the medullated nerve. *J. Physiol. 115:*101–122.

Stevens, C. F. (1966) *Neurophysiology: A Primer.* New York: Wiley.

Tasaki, I. (1959) Conduction of the nerve impulse. In *Handbook of Physiology,* sect. 1, *Neurophysiology,* vol. 1 (J. Field, H. W. Magoun, and V. C. Hall, eds.), pp. 75–121, Washington, D.C.: American Physiological Society.

Treherne, J. E., and Pichon, Y. (1972) The insect blood–brain barrier. In *Advances in Insect Physiology,* vol. 9. (J. E. Treherne, M. J. Berridge, and V. B. Wigglesworth, eds.), pp. 257–313. London: Academic Press.

ADDITIONAL READING

Aidley, D. J. (1978) *The Physiology of Excitable Cells,* 2d ed. Cambridge: Cambridge University Press. 530 pp.

Bullock, T. H., and Horridge, G. A. (1965) *Structure and Function in the Nervous Systems of Invertebrates,* vols. 1 and 2. San Francisco: Freeman. 1719 pp.

Bullock, T. H., Orkand, R., and Grinnell, A. (1977)

Introduction to Nervous Systems. San Francisco: Freeman. 559 pp.

Changeux, J.-P. (1986) *Neuronal Man: The Biology of Mind.* New York: Oxford University Press. 348 pp.

Cooke, I., and Lipkin M., Jr. (1972) *Cellular Neurophysiology: A Source Book.* New York: Holt. 1039 pp.

Eccles, J. C. (1973) *The Understanding of the Brain.* New York: McGraw-Hill, 238 pp.

Edelman, G. M., Gall, W. E., and Cowan, W. M. (eds.) (1987) *Synaptic Function.* New York: Wiley. 789 pp.

Field, J., Magoun, H. W., and Hall, V. C. (eds.) (1959–1960) *Handbook of Physiology,* sect. 1, *Neurophysiology:* vol. 1, pp. 1–780 (1959); vol. 2, pp. 781–1440 (1960); vol. 3, pp. 1441–1966 (1960). Washington, D.C.: American Physiological Society.

Gershenfeld, H. M. (1973) Chemical transmission in invertebrate central nervous systems and neuromuscular junctions. *Physiol. Rev. 53:*1–119.

Hodgkin, A. L. (1964) The ionic basis of nervous conduction. *Science 145:*1148–1154.

Huxley, A. F. (1964) Excitation and conduction in nerve: Quantitative analysis. *Science 145:*1154–1159.

Kandel, E. R., and Schwartz, J. H. (1981) *Principles of Neural Science.* New York: Elsevier/North-Holland. 731 pp.

Kandel, E. R., and Schwartz, J. H. (1985) *Principles of Neural Science,* 2d ed. Amsterdam: Elsevier. 979 pp.

Krieger, D. T., Brownstein, M. J., and Martin, J. B. (eds.) (1983) *Brain Peptides.* New York: Wiley–Interscience. 1032 pp.

Kuffler, S. W., and Nicholls, J. G. (1976) *From Neuron to Brain.* Sunderland, Mass.: Sinauer. 486 pp.

Kuffler, S. W., Nicholls, J. G., and Martin, A. R. (1984) *From Neuron to Brain: A Cellular Approach to the Function of the Nervous System,* 2d ed. Sunderland, Massachusetts: Sinauer Associates Inc. 651 pp.

Shepherd, G. M. (1988) *Neurobiology,* 2d ed. New York: Oxford University Press. 640 pp.

Young, D. (1989) *Nerve Cells and Animal Behaviour.* Cambridge: Cambridge University Press. 236 pp.

CHAPTER TWELVE

Hormonal control

Many physiological functions are under hormonal control. This is often referred to as chemical control, as opposed to nervous control. The term is unfortunate and actually misleading. We have already seen that the transmission of nerve signals at the synapse typically is of a chemical nature, and we shall soon see that the nervous system not only is an essential factor in the control of hormone production, but itself is a major producer of hormones. Furthermore, some chemical substances that normally are formed in the body are not hormones. Carbon dioxide, for example, has effects that are important in the regulation of respiration, but it is not a hormone and has no role in endocrine function.

What, then, do we mean by the term hormone? A *hormone* is best defined as a substance that is released from a well-defined organ or structure and has a specific effect on some other discrete structure or function.

For a hormone to affect a specific target or organ, the target must be able to receive the signal, the cells must have receptors that respond to the signal. Equally important, other organs that are exposed to the same concentration of the hormone must be unresponsive; they must lack the receptors that are essential in starting the cellular machinery that causes the cells to respond.

Many physiological functions are under both endocrine and nervous control. The greatest difference is that the nervous system is like a telephone network where the signals follow specific wires and reach only the intended receiver. The endocrine system is more like radio; the signals are broadcast indiscriminately via the bloodstream and specific receptors are needed to receive each kind of signal.

HOW IS ENDOCRINE FUNCTION STUDIED?

Until recently, endocrinology and hormone research were empirical sciences centered around a fairly uniform procedure. A presumed endocrine organ is removed, depriving the organism of the normal source of its hormone, and the effects on the organism are observed. This is in effect the procedure that has been in common use for thousands of years in the castration of the males of domestic animals. Removing the testes at an early age not only makes the animal incapable of reproduction, it also reduces or eliminates many male characteristics, both in anatomy and in behavior. In this way castrated horses and bulls (oxen) become much more tractable as work animals than uncastrated males.

The changes caused by castration can at least in part be reversed by transplantation of the appropriate glands (in this case the testes) or by injection of extracts of the glands that contain the characteristic male hormones.

The technique of removing an organ suspected of producing a hormone has long been the common approach in endocrine research. The results are observed and described, and if the symptoms can be relieved by injections of extracts of the organ, further work consists in purification of the extract, chemical isolation of the active component, and eventually its chemical synthesis.

This general approach is quite successful in research on organs that can be removed without hazard to life, such as the gonads. However, many hormone-producing organs have functions that are essential in other respects, and their removal leads to serious difficulties. The liver and kidneys, for example, produce hormones, but their removal rapidly leads to fatal effects unconnected with their role as sources of hormones.

Radioimmunoassay

The introduction of powerful new techniques has totally changed the approach to research in endocrinology. The use of radioisotopes as tracers, and especially the use of what is known as radioimmunoassay, have expanded endocrinological research into areas that a few years ago were not even imagined.

The method of *radioimmunoassay* was introduced by Rosalyn Yalow in the 1970s and is now widely used for the detection and assay of the extremely low concentrations of hormones (and other substances) that are found in the circulating blood. For example, the hormone gastrin can be determined in a concentration of 0.1 pg in 1 cm^3 of blood plasma (Yalow 1978).

In principle the radioimmunoassay technique is quite simple. The initial step is to produce an antibody to the substance under study. This is achieved by injecting the substance to be studied into a suitable animal, for example, a rabbit. (The rabbit serves only as a factory to produce an antibody needed for the assay.) The injected substance should be an antigen – for example, a protein or polypeptide – that induces the formation of an antibody in the rabbit. If the substance is not itself an antigen, it is first linked to an antigenic substance such as a foreign protein, and the antibody is formed to the complex. The specific antibody produced by the rabbit can now readily be obtained from its blood plasma.

When the antibody is mixed with a sample that contains the substance under study (the original antigen), an insoluble complex is immediately formed between the antibody and the antigen. A known amount of the antigen, labeled with a radioisotope, should also be present. The two chemically identical antigens, one labeled and one unlabeled, compete for the binding sites on the antibody. The amount of the unlabeled antigen is then calculated from the ratio between the labeled and the unlabeled antigen bound to the antibody.

Radioisotope tracers

Another important approach in endocrine research pertains to the use of *radioisotope tracers* to study the cellular receptor sites where hormones exert their initial action. By using hormones that

TABLE 12.1 Important vertebrate hormones.

Source	Hormone[a]	Major functions
Adenohypophysis	Adrenocorticotropic hormone (ACTH)	Stimulates adrenal cortex
	Thyrotropic hormone (TSH)	Stimulates thyroid
	Follicle-stimulating hormone (FSH)	Stimulates ovarian follicle development; seminiferous tubule develpment in testes
	Luteinizing hormone (LH)	Stimulates conversion of ovarian follicle to corpus luteum; stimulates progesterone and testosterone production
	Prolactin	Stimulates milk production; osmoregulation in fish
	Melanocyte-stimulating hormone (MSH)	Stimulates dispersion of melanin in amphibian skin pigment cells
	Growth-stimulating hormone (GSH)	Stimulates growth (acts via liver, see below)
Hypothalamus	Releasing and release-inhibiting hormones acting on adenohypophysis	Hormones delivered via portal circulation to adenohypophysis; for functions, see Table 12.4
Hypothalamus (via neurohypophysis)	Antidiuretic hormone (ADH)[b]	Stimulates water reabsorption in kidney
	Oxytocin	Stimulates contraction of uterine muscle; releases milk
Liver	Somatomedin	Stimulates growth (is mediator for hypophyseal GSH)
Kidney	Renin	Increases blood pressure; stimulates secretion from adrenal cortex
	Dihydroxycholecalciferol	Affects calcium absorption and bone calcification
Heart	Atrial natriuretic factor (ANF)	Increases renal sodium excretion
Adrenal cortex	Glucocorticoids (corticosterone, cortisone, hydrocortisone, etc.)	Regulate carbohydrate metabolism
	Mineralocorticoids (aldosterone, deoxycorticosterone, etc.)	Regulate sodium metabolism and excretion
	Cortical androgens, progesterone	Stimulate secondary sexual characteristics, predominantly male
Ovary	Estrogens	Initiate and maintain female secondary sexual characteristics; initiate periodic thickening of uterine mucosa; inhibit release of FSH
	Progesterone	Cooperates with estrogens in stimulating female secondary sexual characteristics; supports and glandularizes uterine mucosa, inhibits release of LH and FSH
	Relaxin	Causes relaxation of pelvic ligaments before parturition
Testis	Testosterone	Initiates and maintains male secondary sexual characteristics

TABLE 12.1 Important vertebrate hormones.

Source	Hormone[a]	Major functions
Thyroid	Thyroxine, triiodothyronine	Stimulate oxidative metabolism; stimulate amphibian metamorphosis; inhibit release of TSH
	Calcitonin[c]	Inhibits excessive rise in blood calcium
Parathyroid	Parathormone	Increases blood calcium
Stomach	Gastrin	Stimulates secretion of gastric juice
Duodenum	Secretin	Stimulates secretion of pancreatic juice
	Cholecystokinin (= pancreozymin)	Stimulates release of bile by gallbladder; stimulates secretion of pancreatic enzymes
	Enterogastrone	Inhibits gastric secretion
Pancreas	Insulin	Reduces blood glucose; stimulates formation and storage of carbohydrates
	Glucagon	Increases blood glucose by mobilization of glycogen from liver
Adrenal medulla	Adrenaline, noradrenaline	Augment sympathetic function (vasodilation in muscle, liver, lungs; vasoconstriction in many visceral organs); increase blood sugar

[a] The terms *hormone* and *tropin* are equivalent (e.g., thyrotropic hormone is also known as thyrotropin). In analogy with other uses of the suffix *-tropin* (from *tropein* = to turn or change), the preferred spelling is *tropin,* rather than *trophin* (from *trophos* = feeder) (cf. *phototropic* and *hypertrophy*) (Stewart and Li 1962).
[b] ADH is also known as vasopressin, a name it received because in high concentration it stimulates contraction of vascular smooth muscle and causes increased blood pressure.
[c] In mammals, from the C-cells of the thyroid; in lower vertebrates, from the ultimobranchial bodies.

have been radioactively labeled, it is possible to trace their fate in the organism and pinpoint the exact site where they are bound to the cells where they act.

We shall later return to this and see that the receptor sites for some hormones are located on cell membranes, while others are located within the cell structure or within the nucleus of the cell. A given hormone acts only on those cells that possess receptor sites for this particular hormone, and other cells that lack the proper receptors remain unaffected. This is the mechanism by which hormones are so very specific in their action: Only those cells that have the appropriate receptors can respond.

These two modern methods have revolutionized endocrine research and form the basis for most of what we know today about the details of hormone action.

ENDOCRINE SYSTEMS IN VERTEBRATES

A wide variety of physiological functions are under endocrine control or are affected by hormones. In addition, many hormones modify or influence other endocrine functions, and new and often unexpected connections are continuously being discovered. Some of the most important vertebrate hormones and their major functions are listed in Table 12.1.

TABLE 12.2 Major functional areas under endocrine control with some well-known hormones listed as examples.

Digestion and related metabolic functions
Secretin
Gastrin
Insulin
Glucagon
Noradrenaline
Thyroxine
Adrenal corticoids
Osmoregulation, excretion, water and salt
metabolism
Vasopressin
Prolactin
Aldosterone
Calcium metabolism
Parathormone
Calcitonin
Growth and morphological changes
Growth hormone
Adrenocortical androgens
Thyroxine (amphibian metamorphosis)
Melanocyte-stimulating hormone (amphibian
color change)
Reproductive organs and reproduction
Follicle-stimulating hormone
Luteinizing hormone
Estrogen
Progesterone
Prolactin
Testosterone

TABLE 12.3 Classification of vertebrate hormones on the basis of structure and chemical nature. These groups are important for the cellular mechanism of hormone action.

FAT SOLUBLE	
Steroid hormones	
Testosterone	
Estrogen	
Progesterone	
Corticosteroids	
Vitamin D_3	
WATER SOLUBLE	
Peptide and protein hormones	
Peptides	
Hypothalamic hormones	(3 to 14 amino acids)
Angiotensin	(8 amino acids)
Somatostatin	(14 amino acids)
Gastrin	(17 amino acids)
Secretin	(27 amino acids)
Glucagon	(29 amino acids)
Calcitonin	(32 amino acids)
Insulin	(51 amino acids)
Parathormone	(84 amino acids)
Larger proteins	
Growth hormone	
Prolactin	
Luteinizing hormone	
Follicle-stimulating hormone	
Thyrotropic hormone	
Tyrosine-derived hormones	
Catecholamines	
Noradrenaline	
Adrenaline	
Thyroid hormones	
Thyroxine	
Triiodothyronine	

Areas of endocrine control

The physiological functions that are under mainly endocrine control can conveniently be divided into five broad areas (Table 12.2). The first three categories have to do with metabolic functions and maintenance of steady states; the remaining two with growth, development, and reproduction.

Many of these hormones are similar or identical throughout all vertebrates classes; others have specific functions that differ characteristically from group to group. One example is prolactin, which in mammals stimulates milk secretion, in pigeons stimulates the formation of the crop ''milk,'' and

in fish affects renal function and osmotic permeability of the gills.

Chemical nature of vertebrate hormones

In spite of the large number of vertebrate hormones and the great variety in their actions, in regard to structure and chemical nature they fall into three distinct groups: (1) steroids, (2) peptides and

Cholesterol   Estradiol, a major estrogen   Testosterone

proteins, and (3) tyrosine-derived hormones (Table 12.3). These groupings are important, for the structure of a hormone is closely related to the process through which its action is communicated to the cells of its target organ (see later in this chapter).

Steroid hormones. All the steroid hormones are derived from cholesterol, which has a characteristic basic structure consisting of three six-membered and one five-membered carbon rings. Modifications in this basic structure give rise to a whole series of important hormones in which a slight change can lead to radical changes in physiological effects. Compare, for example, the small structural differences between estradiol and testosterone, which are sex hormones with diametrically differing effects. Hormones of this group include adrenocortical androgens, estrogens, progesterones, and corticosteroids such as cortisone and aldosterone.

Peptide and protein hormones. Several hormones from the hypothalamus exert their action by causing the release of other hormones from their respective endocrine glands. These releasing hormones are all peptides that contain from 3 to 14 amino acids. The thyrotropin releasing hormone (T-RH) contains only 3 amino acids, growth hormone releasing hormone (GH-RH) contains 10, and growth hormone release-inhibiting hormone (GH-RIH) contains 14. Several other peptide and protein hormones are also listed in Table 12.3.

The major hormones originating from the adenohypophysis are protein in nature and may contain several hundred amino acids. Human growth hormone, for example, contains 191 amino acids

and has a molecular weight of about 22 000. Some are glycoproteins that in addition to characteristic peptide chains also contain carbohydrate components. As proteins go, they are not very large; some have molecular weights around 30 000, but it is often difficult to say whether active components that have been isolated from the gland are in fact identical with the functional hormone in the living organism.

Tyrosine-derived hormones. The structures of the two common catecholamines, noradrenaline and adrenaline, are shown in Figure 12.5. These two hormones differ only in a methyl group ($-CH_3$) that is absent in noradrenaline and present in adrenaline.

The formation of the catecholamines begins with the amino acid tyrosine and in a few steps leads to the formation of noradrenaline and adrenaline.

Tyrosine is also the raw material for the syntheses of the thyroid hormones triiodothyronine and thyroxine. These hormones are not catecholamines, but form a group by themselves; they are formed from tyrosine by the condensation of two six-membered carbon rings, which after iodination become the active hormones.

Brain control of endocrine function

There is a constant interaction between the endocrine organs and the central nervous system. In the following discussion we shall see how the brain, directly or indirectly, influences and controls endocrine function.

We should not forget, however, that the reverse is also true: Hormones have profound effects on the function of the central nervous system. Just consider that a female dog in heat accepts the mating behavior of a male dog, although at other times the same signals from the male leave her indifferent or antagonistic. The simple fact that the same signal can lead to completely different behaviors depends on hormonal effects that can be mimicked by the injection of appropriate hormones.

The central nervous system in turn has an essential role in the control of endocrine function. Present evidence shows that the region of the brain known as the *hypothalamus* plays a completely dominant role in this control. Adjacent parts of the brain, the supraoptic and paraventricular nuclei, are part of this system.

The hypothalamic control system

The hypothalamus is located at the base of the brain, immediately above the hypophysis (pituitary gland),* just posterior to the optic chiasm, where it forms the floor of the third ventricle (Figure 12.1).

The hypothalamus is the seat of several nervous control functions, notably temperature regulation and the regulation of intake of water and food. Control of body temperature has already been discussed as a feedback system. The role in regulation of food intake can be demonstrated by destruction of certain parts of the hypothalamus and by electric

---

*The terms *hypophysis* and *pituitary gland* are equivalent. The organ was known to the ancient Greeks as the hypophysis, which refers to its location under the brain. The name pituitary came into use later because it was mistakenly believed that this gland secretes the mucus of the nose (Latin *pituita* = mucus or slime). The use of the terms *anterior* and *posterior* in connection with the hypophysis is unfortunate, for the relative position of these two major parts of the gland differs in different vertebrates. The *adenohypophysis,* also called the anterior lobe, is that part which is of glandular origin, formed by an invagination from the digestive tract (Greek *aden* = gland). The *neurohypophysis,* also called the posterior lobe, is that part of the hypophysis which is of neural origin. Further confusion is caused by the term *intermediate lobe,* a part that arises with the adenohypophysis but is included with the neurohypophysis in the posterior lobe.

stimulation, which, if properly located, make animals ingest huge quantities of food and grow abnormally obese.

Regulation of water intake is shown similarly; electric stimulation or the injection of small amounts of hypertonic salt solution into certain areas of the hypothalamus causes animals to drink excessively. Goats, for example, have in this way been induced to drink, within minutes, 40% of their body weight in water.

The hypothalamus is of major importance in the endocrine system, for it controls the function of the hypophysis, which has been called the master gland of the endocrine system. This control is mediated to the *neurohypophysis via neural connections* and to the *adenohypophysis via special blood vessels,* known as the *portal circulation.*

The *neurohypophysis* contains two well-known hormones, vasopressin and oxytocin. It has been known for more than half a century that the neurohypophysis contains substances that affect the reabsorption of water in the kidney and are necessary for the formation of a concentrated urine. It is now known that the mammalian *antidiuretic hormone,* ADH, is identical with *vasopressin.* It received this name because, if injected in large amounts, it causes a marked rise in blood pressure owing to the constriction of arterioles. Another substance in the neurohypophysis, *oxytocin,* causes contraction of the smooth muscle of the uterus in the pregnant female at term.

Vasopressin and oxytocin are octapeptides (i.e., they consist of eight amino acids). Both are formed in nerve cells near the hypothalamus and are transported along axons to the nerve endings in the neurohypophysis, from where they are released into the blood. The neurohypophysis therefore serves as a storage and release organ (*a. neurohemal organ*) for the hormones that are produced in the adjacent parts of the brain.

The *adenohypophysis,* in contrast, produces hormones, and their release into the bloodstream is regulated by the hypothalamus. This control is

FIGURE 12.1 The location of the hypophysis and the hypothalamus in the human brain. The adenohypophysis is controlled by neurohormones originating in the hypothalamus; the neurohypophysis is a neurohemal organ that releases hormones produced by neurons in the adjoining parts of the brain.

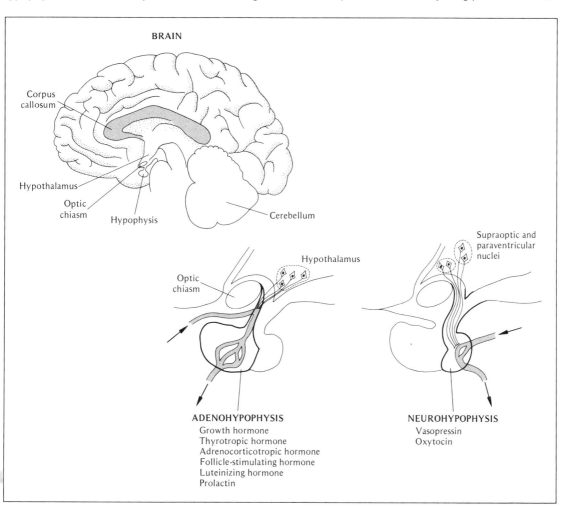

achieved by hormones that are produced in the hypothalamus and reach the adenohypophysis via the portal circulation. It is now well established that there are 10 hypothalamic regulating hormones involved in the system that controls the hypophysis (Table 12.4).

Three of the hormones from the adenohypo-

physis – *growth hormone* (GH), *prolactin* (P), and *melanocyte-stimulating hormone* (MSH) – are under dual hypothalamic control, one inhibitory and one stimulatory. The release of these three hormones thus is not regulated by simple feedback systems, although feedback signals undoubtedly are involved in their control.

TABLE 12.4 Hypothalamic hormones that control the release of hormones from the adenohypophysis. RH, releasing hormone; RIH, release-inhibiting hormone. [Schally et al. 1973]

Hormone	Abbreviation
Growth hormone releasing hormone	GH-RH
Growth hormone release-inhibiting hormone	GH-RIH
Prolactin releasing hormone	P-RH
Prolactin release-inhibiting hormone	P-RIH
Melanocyte-stimulating hormone releasing hormone	MSH-RH
Melanocyte-stimulating hormone release-inhibiting hormone	MSH-RIH
Corticotropin (ACTH) releasing hormone	C-RH
Thyrotropin releasing hormone	T-RH
Luteinizing hormone releasing hormone	LH-RH
Follicle-stimulating hormone releasing hormone	FSH-RH

The release of the four other hormones seems to depend on a regular negative feedback system. *Corticotropin* (ACTH), *thyrotropin* (TSH), *luteinizing hormone* (LH), and *follicle-stimulating hormone* (FSH) have as their targets the adrenal cortex, the thyroid, and the gonads, respectively. These glands when stimulated release the appropriate hormones into the blood. These hormones (corticosteroids, thyroxine, and sex steroids) in turn inhibit, by negative feedback, the secretion from the adenohypophysis of the tropic hormones. There is now strong evidence that this inhibition acts via the hypothalamus (except that thyroxine may have a shorter feedback loop and act via the adenohypophysis).

The central role of the hypothalamus in endocrine regulation prompts the question of how this important organ in turn is controlled. We know that there are extensive nerve connections to the hypothalamus from other parts of the brain and that these are influenced by many environmental as well as emotional factors, by light cycles, seasons, and so on. We can therefore conclude that the major portion of the entire endocrine system is under nervous control, acting through the central role of the hypothalamus in the control system.

Endocrine glands

We should now turn to a discussion of some of the endocrine glands whose roles have long been known and appear relatively uncomplicated.

The reproductive glands

The reproductive glands and the endocrine activities associated with them are concerned with a broad range of functions where coordination is necessary for successful reproduction of the species. The following steps are essential:

1. Development and maturation of eggs and sperm
2. Development of reproductive organs
3. Preparation of male and female for mating
4. Synchronization of events that lead to fertilization of mature egg
5. In mammals, preparation of the female tract for pregnancy, ending in birth
6. Caring for young: in mammals, secretion of milk; in other vertebrates, nest building, development of brood pouches, etc.

All these steps involve complex endocrine functions that affect the development of each step, the proper synchronization of events, and the necessary associated behavior.

The timing of reproduction is, in most animals, adjusted to the most favorable time of the year. In many vertebrates this is the spring and summer because that is the most favorable time of the year in regard to both available food and favorable temperature.

An important environmental clue to the seasonal change is the length of the day-and-night light cycle. Many mammals and birds come into breeding condition in response to a lengthening daylight period and the associated shortening of the dark period. The effect of a prolonged daylight period has long been known and utilized in, for example, the egg production of domestic hens. When the light period is artificially lengthened by a few hours, egg laying continues through most of the winter, a season when, in the absence of artificial light, egg production would decrease or cease. It is interesting that the duration of the light period, not the light intensity, is the important factor. Even a moderate amount of light for a few extra hours has the desired effect.

There is good evidence that the effect of light is transmitted to the hypothalamus from the *pineal gland*. The hypothalamus in turn controls the adenohypophysis and the entire reproductive process. The hypothalamus, through the appropriate releasing hormones, stimulates the adenohypophysis to release two gonadotropins: follicle-stimulating hormone (FSH) and luteinizing hormone (LH). FSH stimulates growth of the ovarian follicles and their maturation; LH stimulates the persistence of the follicle. Estrogens produced by the ovarian follicle prepare the uterus for reception and implantation of the fertilized egg and trigger other secondary changes in the female reproductive organs.

In some mammals the egg is released from the follicle as a consequence of the proper endocrine stimulus, but in others (e.g., rabbit and cat) ovulation occurs in response to mating and the associated neural stimulation of the hypothalamus.

In most fish the eggs are released into the water in response to the proper mating behavior of the male, which accomplishes the external fertilization. In some, however, there is internal fertilization and development of the young within the female. There is a tremendous amount of variation in the details, and in some fish and amphibians the young are cared for by the males.

In birds the eggs are normally placed in a nest and incubated by one or both parents. The number of eggs is carefully adjusted to a specific clutch size, which may be a single egg (the emperor penguin) or as many as a dozen or twenty.

Toward the end of the pregnancy in mammals the neurohypophysis releases the hormone *oxytocin,* which causes contraction of the uterus and birth of the young. Preceding this, the hormone *relaxin,* a polypeptide originating in the corpus luteum, has caused relaxation of the fused bones of the female pelvis. Also, relaxin inhibits the contractility of the uterine muscle so that birth does not occur prematurely.

The production of milk in mammals is stimulated by the hormone *prolactin,* from the adenohypophysis. This hormone received its name because of its effect in mammals. However, prolactin is produced in nearly all vertebrates and serves various functions. Cyclostomes seem to be the only vertebrates that lack prolactin. In pigeons and in penguins prolactin promotes a secretion from the crop that is fed to the young by regurgitation, the secretion being known as "pigeon's milk," In teleost fish, prolactin affects the permeability of the gills and is important in osmotic regulation.

Adrenal cortex

The adrenal gland consists of two parts that are distinctly different in origin. The inner part, the adrenal medulla, is of neural origin; it will be discussed later. At the moment we will be concerned with the outer part, the adrenal cortex. This is derived from mesodermal cells, and in some vertebrates it remains separate from the adrenal medulla. The cells that embryologically form the adrenal cortex are derived in close relation to cells that form the ovaries and testes. All these cells synthesize steroids, and there is a great deal of similarity in the steroid hormones produced by the ovaries, the testes, and the adrenal cortex.

The adrenal cortex is essential to life. Its major hormones – cortisol, corticosterone, and aldosterone – affect the metabolism of carbohydrates, protein, fat, electrolytes, and water. The adrenocortical hormones are often divided into *glucocorticoids* and *mineralocorticoids. Cortisol* and *corti-*

costerone are glucocorticoids and have striking effects on glucose metabolism. Cortisol deficiency leads to hypoglycemia; excess cortisol produces hyperglycemia and increased resistance to insulin. Cortisol also promotes the metabolism of proteins and affects nitrogen balance and growth. Cortisol has some effect on electrolyte and water excretion, but is much less potent than aldosterone.

The main mineralocorticoid is *aldosterone,* which promotes the retention of sodium and the excretion of potassium by the kidney. Aldosterone has marked effects on the sodium and potassium distribution in the body, but little action on carbohydrate or protein metabolism.

Aldosterone has no antiinflammatory effect, which is an important characteristic of cortisol. The antiinflammatory effect is widely used in suppressing inflammatory responses, but it does not eliminate the cause of inflammation, and in large doses cortisol delays healing while apparently alleviating the symptoms.

The release of cortisol and corticosterone from the adrenal cortex is controlled by the adenohypophysis through its secretion of the adrenocorticotropic hormone (ACTH), a polypeptide composed of 39 amino acids. ACTH is often released in response to stress and alleviates many symptoms through the release of adrenal glucocorticoids. The release of ACTH from the adenohypophysis is, in turn, controlled through a negative feedback loop because the released glucocorticoid hormones act as inhibitors of the secretion of ACTH.

The thyroid gland

The thyroid gland produces the hormones *thyroxine* and the closely related *triiodothyronine.* The gland releases the hormones in response to the thyrotropin releasing hormone (T-RH) from the adenohypophysis, and an increase in the concentration of thyroid hormones in the blood plasma acts as a negative feedback loop that suppresses the release of T-RH.

The thyroid hormones are characterized by their contents of iodine. Thyroxine contains four and triiodothyronine three iodine atoms. Their main effect in higher vertebrates is to stimulate oxidative metabolism; a high concentration of the hormones increases the metabolic rate, and a lowering below the normal concentration reduces the metabolic rate and is accompanied by other deficiency symptoms.

The thyroid gland avidly takes up iodine and stores it for use in the production of the hormones. An often discussed consequence of nuclear accidents is a release into the atmosphere of radioactive iodine, which, when taken up by the thyroid gland of exposed individuals, may cause serious radiation damage. A suggested countermeasure to this potential danger is that everybody who might become exposed to radioactive iodine should take a sufficient dose of nonradioactive iodide (in the form of potassium iodide) to saturate the gland. The gland would then in case of exposure sequester less radioactive iodine. One objection to this measure is that some persons are hypersensitive to iodine in any form and may display serious reactions to the ''protective'' measure. Avoidance of nuclear accidents seems a better policy.

In amphibians the thyroid gland has an important role that has no counterpart in mammals: It is essential for the metamorphosis of the larvae to the adult form. In frogs and toads, metamorphosis begins with the appearance of hindlimb buds, disappearance of the external gills, followed by growth of forelimbs with gradual resorption of the tail. When this process is completed, the tadpole has become a small frog, which breathes with lungs and is a normal terrestrial frog.

If the thyroid is removed from a tadpole, the animal does not metamorphose into an adult frog, but remains in the larval form. However, if the tadpole is fed small pieces of thyroid gland or is injected with thyroid hormones, metamorphosis takes place as normal. If such treatment is given to a tadpole before the time it should normally change, it develops into a midget adult.

The thyroid hormones are important also in the development of tailed salamanders. The axolotl (*Ambystoma mexicanum*) is a salamander that normally remains as an aquatic animal with external gills even when a mature adult. If it is fed thyroid hormone, however, metamorphosis results in loss of the gills and development of air breathing.

The pancreas

The pancreas is one of the major digestive glands and produces important digestive enzymes. It also produces two important hormones that have no direct role in digestion but are extremely important in carbohydrate metabolism. These two hormones, *insulin* and *glucagon,* are produced by small groups of endocrine cells, known as the *Islets of Langerhans,* which are found dispersed throughout the secretory glandular tissue of the pancreas.

Insulin is produced by and released from the *beta cells* of the islets. The most obvious effect of insulin is that it stimulates the formation and deposition of glycogen in muscle and liver from the blood glucose. If the glucose concentration in the blood rises above the normal level (e.g., after ingestion of carbohydrates), insulin is released, glucose uptake in the muscles and the formation of glycogen is stimulated, and blood sugar decreases again. The release of insulin is therefore a negative feedback system.

Insulin or the lack of it is the essential agent in the disease diabetes mellitus, which is characterized by an abnormally high glucose level in the blood. In the most serious type of diabetes there is no insulin produced by islets. As a result the blood glucose remains grossly elevated and glucose is excreted in the urine, often in vast quantities. The lack of insulin has additional serious effects in the intermediary metabolism, and if not treated with insulin injections, this type of diabetes, known as type one or juvenile diabetes, is lethal within months or a few years.

In the second kind of diabetes, known as type two or mature-onset diabetes, the insulin production is adequate, but there is a deficiency of insulin receptors on the cells that normally should respond to insulin. This type has normal or elevated insulin production – it is the response to insulin that is deficient. In many cases this more benign type can be treated conservatively by suitable adjustments of the diet.

If, in the normal individual, the glucose level falls too low, the hormone *glucagon* is released from the *alpha cells* of the islets. Glucagon causes mobilization of glucose from the liver and an increase in the blood glucose concentration. The glucagon effect is the opposite of the effect of insulin; the normal blood glucose level thus is regulated by two antagonistic hormones, one inhibitory and one stimulating, each acting through its own negative feedback loop.

A third type of islet cells, the *delta cells,* produce *somatostatin,* the hormone we already know from the hypothalamic regulation of endocrine function. It received its name because it inhibits the secretion of hypophyseal growth hormone. However, it also is an inhibitor of both insulin and glucagon secretion, and furthermore, it decreases the rate of absorption of nutrients from the gastrointestinal tract.

Parathyroids and thyroid

The parathyroid glands, which are located near or embedded in the thyroid gland, produce a hormone, *parathormone,* which causes the calcium ion concentration in the blood plasma to increase. An antagonistic hormone, *calcitonin,* which lowers the plasma calcium ion concentration, is released from the C-cells of the thyroid in mammals.

An excess of parathormone causes the plasma calcium to increase to an abnormal level, the calcium being taken primarily from the bones. This in turn results in an increased renal excretion of calcium, and because of the excessive loss of calcium the bones become decalcified. Removal of the parathyroids has the opposite effect, leading to a fall in blood calcium concentration, followed by tetanic muscle cramps caused by the low blood calcium level.

The normal secretion from the parathyroids is apparently controlled by a simple feedback mechanism, because a high plasma level of calcium inhibits release of the hormone from the parathyroids, and a low level stimulates release. The effect of the calcium level on the release of calcitonin is the opposite; a high calcium level stimulates hormone release and a low level inhibits release. The calcium ion concentration in the blood plasma is thus controlled by the balance between two negative feedback systems.

The kidney

A further understanding of calcium metabolism came about when the role of vitamin D and how it acts in the organism became understood, primarily through the work of De Luca and his collaborators (De Luca 1981). Vitamin D_3, also known as calciferol or cholecalciferol, whether taken in the diet or generated in the skin by ultraviolet radiation, must be hydroxylated in the liver in preparation for function. The hydroxylated vitamin is then transported in the blood by a specific transport protein to the kidney, where it is further hydroxylated to the active form of the hormone, *dihydroxycholecalciferol*. The final hormone is then transported to the intestine, where it stimulates calcium absorption, and to bone, where it is necessary for the normal mineralization of bone.

In combination with the parathyroid hormones dihydroxycalciferol is responsible for maintenance of plasma calcium and phosphorus levels and for the normal function of the neuromuscular system.

Another endocrine function of the kidney was first revealed through the discovery that an injection of extracts of the renal cortex causes a prolonged rise in arterial pressure. It was later shown that restriction of the blood flow to the kidney by a clamp placed on the renal artery releases a substance that causes persistent hypertension. This substance, called *renin*, is an enzyme that acts on a plasma protein to release an octapeptide called *angiotensin*. Angiotensin has a number of effects. It stimulates the secretion of aldosterone and therefore influences sodium and water metabolism; it causes vasoconstriction and therefore raises the blood pressure; and it causes the release of catecholamines from the adrenal medulla.

Angiotensin apparently has an additional effect on salt and water metabolism because infusion of angiotensin causes water-replete rats to drink. It therefore seems that angiotensin is a renal factor in the induction of thirst.

Atrial natriuretic factor

It has long remained an unsolved problem how the blood volume is regulated. The problem is easily explained. How can a highly compliant system, such as the vascular system, sense its own volume, which must be done in order for regulation to be achieved? We could imagine that a volume expansion could be sensed through an increase in the pressure within the system, but it has been difficult to demonstrate this experimentally.

There was a suspicion that the atrium of the heart was involved and could respond to an increased return of blood from the venous system, a hypothesis that is now confirmed. When there is a volume expansion, a peptide is released from the wall of the atrium, causing increased excretion of sodium and diuresis, thus tending to return blood volume to the normal.

The peptide became known as the *atrial natriuretic factor*. There are in fact several peptides, ranging from 21 to 33 amino acids. These have a common sequence of 21 amino acids, a peptide that is also known as *atriopeptin I* (Currie et al. 1984). As with so many other peptide hormones, a natriuretic peptide has been found also in the brain, where its function remains uncertain (Sudoh et al. 1988).

Adrenal medulla

The cells of the adrenal medulla are of neural origin and belong to the sympathetic nervous system. The adrenal medulla secretes two similar hormones, *noradrenaline* and *adrenaline*, which are also known as norepinephrine and epinephrine (see Figure 12.5). Adrenaline causes acceleration of the heartbeat, increased blood pressure, increase in blood sugar through conversion of glycogen to glucose, vasodilation and increased blood flow in heart muscle, lungs, and skeletal muscle. However, it causes vasoconstriction and decreased blood flow in smooth muscle, digestive tract, and skin. These effects of adrenaline are well known as the fight-or-flight syndrome, which occurs in response to fear, pain, and anger. These reactions help mobilize the physical resources of the body in response to an emergency situation.

Noradrenaline has very similar effects, and the differences between the two hormones are mostly quantitative. Adrenaline, for example, has a more potent effect on the heart rate, and in some organs noradrenaline causes more vasoconstriction. The two hormones act through receptor sites known as alpha receptors (most affected by noradrenaline) and beta receptors (most affected by adrenaline); many organs have both alpha and beta receptors, and in addition, each receptor type is to some degree sensitive to both hormones. This explains why it is difficult to give a definite listing of similarities and differences between the two hormones; there is too much overlap in their effects.

Noradrenaline has been found in the nervous systems of insects and annelids and may be widely distributed among invertebrates. In the protozoan *Tetrahymena* adrenaline acts by stimulation of adenylate cyclase, which is the initial biochemical step of adrenaline action in vertebrate cells (see below). This effect in *Tetrahymena* is blocked by a drug, propranolol, which is a specific adrenaline antagonist because it acts as a blocking agent for the adrenaline receptor site (Rozensweig and Kindler 1972).

The cascade effect

The control of metabolic functions by the endocrine system can lead to a cascade, or step-by-step amplification, that makes it possible to control a final process through minute amounts of the initial hormone.

An example is shown in Figure 12.2, where the approximate amounts are given for a series of hormones that lead to deposition of glycogen. A minute amount, 0.1 µg, of the hypothalamic corticotropin-releasing hormone (C-RH) causes a series of subsequent events in which the final step is the formation of 5600 µg of glycogen in the liver.

FIGURE 12.2 An example of the biological amplification possible in the endocrine system. The release of a minute amount of hypothalamic hormone leads to the deposition of glycogen in the liver, with a total amplification of 56 000 times. [Data from Bentley 1976]

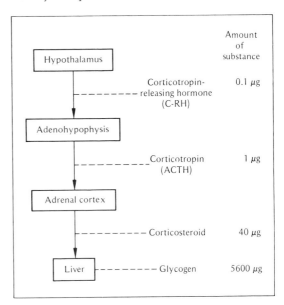

The initial amount of C-RH required is so small that the final response represents an amplification of 56 000 times.

Hormonal interactions with target cells

We have discussed the origin, release, and transport of hormones, but we have not discussed details of how they affect a target organ. For a hormone to have an effect on a specific target, and not on other organs exposed to the same concentration of the hormone in the blood, the responding cells must possess specific receptor sites that interact with the hormone, and other cells must lack these receptors.

In regard to the effect on target cells, hormones can be classified in two categories: (1) those, the catecholamines and peptide hormones, that act via appropriate receptors on the cell surface, and (2) those, including the steroid and thyroid hormones,

FIGURE 12.3 The adrenaline-induced release of glucose from the cells of the liver. Adrenaline interacts with a receptor site in the cell membrane, which sets off the release of cyclic AMP as a "second messenger." The last step shown, the rate-limiting phosphorylase a, includes two additional enzymes, phosphoglucomutase and glucose-6-phosphatase. [Sutherland 1972]

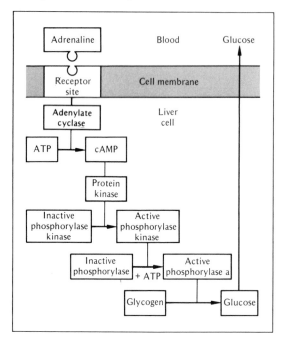

that penetrate into the cell and exert their effect directly on the cell nucleus and the mechanism for cellular protein synthesis.

The discovery of the steps through which adrenaline acts was made during studies of the action of adrenaline on the liver, where it causes the conversion of glycogen to glucose (Sutherland 1972). The hormone acts by controlling the formation of the cyclic nucleotide *3′, 5′-adenosine monophosphate (cyclic AMP* or *cAMP).*

The process depends on a series of enzymes, of which one is phosphorylase, the rate-limiting step in the process (Figure 12.3). The active enzyme, *phosphorylase a,* is formed from a precursor,

phosphorylase b, which through the action of a phosphorylase kinase reacts with adenosine triphosphate (ATP) to form the active enzyme. The whole process is initiated by the binding of adrenaline to the receptor site in the cell membrane. This causes the release of the enzyme *adenylate cyclase,* which in turn catalyzes the formation of cyclic AMP from ATP.

Cyclic AMP has been called a "second messenger" in the process of hormone action, the hormone itself being the first messenger. cAMP and adenylate cyclase have been found in several different tissues from vertebrates as well as invertebrates from many different phyla and in bacteria (Sutherland et al. 1962). The initiating event in the action of those hormones that act via cAMP is always the release of adenylate cyclase from the hormone receptor site in the cell membrane.

Cell function can also be modulated by other receptor-activated mechanisms that do not involve cAMP. One such modulating process depends on the generation of *inositol triphosphate* and the mobilization of calcium ions from an intracellular calcium pool. In this system the calcium ions and *phosphoinositol* serve as two second messengers (Berridge and Irvine, 1984).

The *steroid hormones,* which include the male and female sex hormones and hormones secreted by the adrenal cortex, act through a different mechanism. As we saw before, the hormones affect those cells that contain the appropriate receptor sites and leave other cells unaffected. Estradiol binds to receptors in the uterus, testosterone to receptors in the prostate, progesterone to the oviducts of birds, and so on. However, at the cell surface these hormones form a complex with a receptor protein, and this complex is rapidly moved from the cell membrane to the cell nucleus, where its presence stimulates or induces the expression of genetic information (Jensen et al. 1971).

Recent studies of the receptor sites for steroid hormones have revealed some fascinating information. Steroid hormones secreted by the gonads of a newborn rat can be traced to target cells in specific areas of the brain. The hormones interact with these receptor areas and induce the development of nervous circuits that later control whether the adult animal shows male or female behavior.

This sexual differentiation of nerve circuits in the brain lays down a morphological basis for the activation of certain types of behavior and the suppression of others (McEwen 1976). During development the functioning of the central nervous system is already modulated by hormones, and as we have seen, the central nervous system is in turn a major regulator of endocrine function throughout the body.

TRANSMITTER SUBSTANCES

We saw earlier that the transmission of signals from the terminals of motor nerves to the muscles is achieved through the release of a transmitter substance, acetylcholine. Likewise, synaptic transmission between nerve cells often is accomplished by acetylcholine acting as a transmitter substance. However, the nervous system uses several additional transmitter substances; one of particular interest is noradrenaline.

The effects of noradrenaline released into the blood from the adrenals resemble the effects of a general stimulation of the sympathetic nervous system.* This similarity is easy to understand when we realize that the cells of the adrenal medulla are of neural origin and analogous to the large sympathetic ganglia located along the spinal column. The similarity is even more clearly understood when we learn that noradrenaline is the terminal trans-

mitter substance of the sympathetic nervous system as a whole.

While the sympathetic nerves release noradrenaline from their nerve terminals, the parasympathetic nerves release acetylcholine. These two substances have antagonistic effects; when one stimulates, the other inhibits. One of the best known examples of such antagonistic action is the control of the heartbeat. The heart is innervated by two nerves. One is a branch of the splanchnic nerve (sympathetic), which when stimulated releases noradrenaline and speeds the heartbeat; the other is a branch of the vagus nerve, which is parasympathetic and when stimulated releases acetylcholine from its nerve endings and causes the heart to slow.

The effect of acetylcholine in slowing the heart is a striking example of the fact that the response to a given substance depends on the target organ rather than on the nature of the transmitter substance. Ordinary striated muscle is stimulated to contract through the release of acetylcholine; the effect of acetylcholine on the heart is to reduce the frequency of contraction.

The major aspects of transmitter action in the peripheral nervous system of vertebrates are summarized in Figure 12.4. We can see that acetylcholine is the common transmitter substance for the following: (A) spinal nerves that stimulate the adrenal medulla to secretion of adrenaline and noradrenaline, (B) sympathetic nerves that stimulate ganglia in the sympathetic chain along the vertebral column, (C) parasympathetic spinal nerves that stimulate peripheral parasympathetic ganglia, and (D) motor nerves at the neuromuscular junction. Thus, in nerve-to-nerve transmission outside the brain and spinal cord, acetylcholine seems to be the ubiquitous transmitter substance in vertebrates. Within the central nervous system the situation is different.

Although acetylcholine is so widespread in peripheral synaptic transmission, it is only one of several transmitters within the vertebrate central nervous system. Also in invertebrates several other

---

*The major components of the peripheral vertebrate nervous system are sensory nerves, motor nerves, and the autonomic nervous system. The autonomic nervous system in turn is divided into the *sympathetic* and the *parasympathetic* nervous systems, which jointly control the function of internal organs such as heart, stomach, glands, intestine, kidneys, and so on (i.e., functions outside voluntary control). These organs are innervated by both sympathetic and parasympathetic nerves, which act antagonistically, one being stimulatory and the other inhibitory.

FIGURE 12.4 Diagram of the actions of peripheral transmitter subtances in vertebrates. For details, see text. ACh, acetylcholine; Adr, adrenaline; Noradr, noradrenaline.

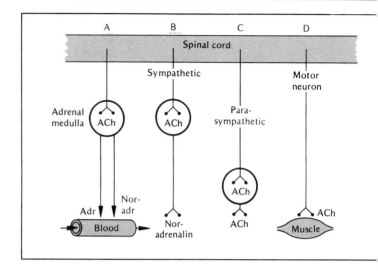

substances, most of them amines, are important transmitters. The structures of a few of these substances are shown in Figure 12.5. The reason the information about transmitter substances is uncertain is the extreme difficulty of establishing with certainty that a given substance has a role as a transmitter.

The greatest problem is that many substances that occur in the organism in small amounts have rather profound physiological effects, particularly on the nervous system. The fact that a substance causes a reaction that is similar to some normally occurring nervous phenomenon does not establish this substance as a transmitter. A number of criteria should be fulfilled before this can be established; among these criteria are:

1. The substance or a precursor should be present in the neuron from which the suspected transmitter is released.
2. The substance should be present in the extracellular fluid in the region of the activated synapse.
3. When applied to the postsynaptic structure, the substance should mimic the action of the transmitter.
4. A mechanism for removal or inactivation of

the substance should be present, either through enzymatic inactivation or through specific uptake or reabsorption.

In addition to these classical criteria, modern techniques make it possible to study the receptor sites in greater detail. If a receptor for a specific compound is identified, it implies the existence of a transmitter system, but much additional information is needed before it can be considered verified.

Another tool is the use of pharmacological agents that help clarify mechanisms by interfering with synthesis, release, removal, inactivation, or target. However, they do not provide direct evidence for transmitter function.

The crucial demonstration of a transmitter substance consists in proving that, on the arrival of an action potential at the presynaptic ending, the substance is released in sufficient quantity to produce the observed physiological effect on the postsynaptic structure.

It is indeed difficult to meet all these criteria, and very often the technical obstacles have been insurmountable. As a result much work on suspected transmitter substances, especially in invertebrate animals, has been unsatisfactory because it

FIGURE 12.5 Structural formulas of some compounds known or believed to be transmitter substances.

has not progressed much beyond the demonstration of an effect of some suspected agent. Therefore, present information about transmitter substances is insufficient for generalizations. Nevertheless, two substances, *5-hydroxytryptamine* and *gamma-aminobutyric acid,* should be mentioned because they are invariably recognized as being highly active in many invertebrates.

5-Hydroxytryptamine, often written as 5HT, is found in the neurons of many invertebrates, notably molluscs and arthropods. 5HT is not restricted to nervous structures, but the concentration in other tissues is usually much lower.

The heart of bivalve molluscs is quite sensitive to 5HT, and there is good evidence that this substance is a normal transmitter for acceleration of the mollusc heart. The catch muscle of bivalves, which can remain contracted for extended periods

of time with minimal energy consumption, relaxes rapidly in response to 5HT.

Large amounts of 5HT are found in certain animal toxins, such as those from octopus and the snail *Murex,* and in the venom of wasps and other insects. The very painful effects of some insect stings is in accord with 5HT as a potent stimulant of pain nerve endings in human skin.

The effect of 5HT on vertebrate blood pressure (which it increases) was known before the substance was chemically identified. It was then called *serotonin,* a name that is still in use. 5HT is also believed to be a transmitter within the vertebrate central nervous system, but its normal role in vertebrates remains uncertain.

Gamma-aminobutyric acid, often written GABA, may be a transmitter substance in the central nervous system of vertebrates, and it almost certainly is a transmitter at inhibitory synapses of crustaceans. If the inhibitory nerve to crustacean muscle is stimulated, its terminals release GABA, and the effect can be mimicked by direct application of GABA (Otsuka et al. 1966; Takeuchi and Takeuchi 1967).

GABA is found in rather high concentration in certain regions of the mammalian brain, but this does not prove that it has a normal function as a transmitter substance. A great deal of research is currently underway to clarify the role of GABA and other putative neurotransmitters in the central nervous system and especially the brain. There is good evidence that GABA is synthesized in cells of the cerebellum and released from their terminals, where the effect is inhibitory. It is probable that GAGA is also the transmitter substance in presynaptic inhibition in general.

Several amino acids have been suggested as transmitter substances, both for vertebrate and invertebrate nervous tissues, but substantial evidence for their normal role is sometimes lacking. The amino acids that are recognized as transmitters fall into two general groups: (1) *excitatory amino acids,* including *glutamic* and *aspartic acids,* which depolarize the membranes of most neurons in the vertebrate central nervous systems; and (2) *inhibitory amino acids,* including *GABA* and *glycine,* which cause hyperpolarization of neuron membranes and thus inhibition. One major difficulty in establishing amino acids as transmitter substances is that they also occur normally in the animals, and their mere presence therefore carries no particular significance. Another problem is the extreme difficulty of establishing an inactivation mechanism specifically for amino acids as related to their suspected role as transmitter substances, for all the amino acids are regular participants in normal metabolic pathways.

Other neuroactive substances

It has become increasingly evident that the central nervous system produces more chemical substances with specific physiological actions than

anybody imagined a few years ago. A series of remarkable discoveries has come from studies of the action of such well-known opiate drugs as *morphine* and *heroin*. These are used to suppress pain, but they also cause euphoria and, as everybody knows, they are addictive.

The most important basic discovery grew out of a search for specific receptors in the brain that would react with opiate drugs and thus provide a basis for further study of their characteristic effects. If such specific receptor sites could be found in the central nervous system, one would know where to begin the further search for the mechanism of action of the opiates (Pert and Snyder 1973).

The search was successful and set the stage for the next step. It was reasonable to assume that the receptors must have some normal physiological role and that there might be endogenous materials that normally act at these sites. This hypothesis led to the tracking down of a whole series of endogenous substances that participate in the normal functioning of the nervous system – substances that may be involved not only in the suppression of pain, but also in changes in mood and emotions, and possibly in certain types of mental disease.

The idea that there must be a normal function behind the existence of specific opiate receptors has paid off well. A number of substances that have been isolated from mammalian brain have an affinity for these receptors and have morphinelike effects. The first substances to be isolated were two very similar pentapeptides, called *enkephalins*. The only difference between them is that the terminal amino acid in one is methionine and in the other leucine. Accordingly, their names are methionine enkephalin *met-enkephalin*, and leucine enkephalin, *leu-enkephalin*.

Several other naturally occurring polypeptides that interact with opiate receptors have been isolated. As a group they are called *endorphins,* to indicate that they are endogenous substances that have morphinelike effects. These various endorphins seem to be derived from common larger precursor molecules, one being the 91-amino-acid polypeptide

FIGURE 12.6 The amino acid sequence of the polypeptide beta-lipotropin. This chain of 91 amino acids contains the sequences of several endogenous opiates. The 31-unit end, from site 61 to 91, is beta-endorphin. The shorter chain, from 61 to 76, is alpha-endorphin, and the pentapeptide from 61 to 65 is met-enkephalin. The lipotropin chain also contains the entire sequence of beta-melanotropin (41 to 57), a hormone named for its darkening effect on the skin of fish and amphibians (by causing expansion of the black pigment cells, the melanophores). Melanotropin is produced in the mammalian hypophysis, but its function in mammals is obscure.

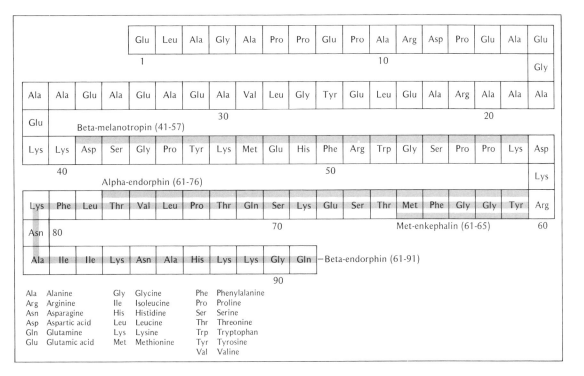

Ala	Alanine	Gly	Glycine	Phe	Phenylalanine
Arg	Arginine	Ile	Isoleucine	Pro	Proline
Asn	Asparagine	His	Histidine	Ser	Serine
Asp	Aspartic acid	Leu	Leucine	Thr	Threonine
Gln	Glutamine	Lys	Lysine	Trp	Tryptophan
Glu	Glutamic acid	Met	Methionine	Tyr	Tyrosine
				Val	Valine

beta-lipotropin, which serves to mobilize fat from adipose tissue (Figure 12.6). Beta-lipotropin is probably itself part of a larger precursor protein, and the smaller, active peptides are snipped out of the long chains of amino acids.

The structure of morphine is entirely different from the structure of the neuroactive endorphins, which all are polypeptides. There is one similarity, however: Morphine contains a benzene ring that probably corresponds to the ring structure of the terminal amino acid tyrosine, which all endorphins and enkephalins have in common (at site 61 of beta-lipotropin).

The enkephalins, the five-amino-acid peptides, are unstable in brain tissue and have half lives of seconds or a fraction of a minute. Beta-endorphin, which contains 30 amino acid residues, is much more stable, a difference that undoubtedly is of functional significance but not well understood.

The effects of endorphins as well as of opiates* can be counteracted by a synthetic compound,

---

*Opiates are drugs and other compounds that are derived from the opium poppy and chemically are related to morphine. Chemicals that have physiological effects similar to the opiates but chemically are unrelated to them are called opioids. Thus, the endorphins and enkephalins are opioids.

naloxone, a molecule that structurally is very similar to morphine but lacks its characteristic euphoric effects. Naloxone acts by binding to the opiate receptors and thus inhibits the effects of morphine.

A method for relief of pain, *acupuncture,* used for thousands of years in China, may be related to the normal action of endorphins. Acupuncture consists in inserting slender needles at various peripheral parts of the body, such as the earlobe or between the fingers. Slow rotation of these needles has been used as a method of relieving more serious pain. One explanation for the effectiveness of acupuncture is that the slight peripheral pain causes a central release of endorphins, serving to relieve other pain. The effects of acupuncture can be antagonized by naloxone, which suggests that endorphins are indeed involved. It is therefore particularly interesting to compare this effect with pain suppression induced by hypnosis. In hypnosis the suppression of pain is not antagonized by naloxone, suggesting that the blockage in this case occurs at a different and probably higher level of the central nervous system.

In general, painful sensations from the periphery are transmitted through thin, unmyelinated nerve fibers, which in the spinal cord cause the release of *substance P,* a polypeptide of 11 amino acid residues. Substance P has been implicated as a transmitter of pain. Its release from the dorsal root of the spinal cord is brought about by noxious stimuli, and the terminals that contain substance P also possess opiate receptors. When morphine binds to these receptors, it presumably reduces the release of substance P and thus explains the analgesic properties of morphine and other opium derivatives.

When small amounts of true morphine were isolated from the brain of rats and cattle, the immediate question was whether it originated internally or from some external source. Presumably the opiate receptors would eagerly sequester even the minutest amount of opiates from the feed or any other source of contamination. However, careful studies of the metabolism of chemical precursors

has shown that morphine and codeine can be synthesized by rat liver, although not by the brain (Weitz et al. 1987). Thus, true opiates can indeed be of endogenous origin, adding further puzzlement to the questions surrounding the role of opiates.

Brain–digestive tract interactions

It may seem incongruous that endorphins and other neuroactive polypeptides occur also outside the central nervous system. Many are found elsewhere in the body, especially in the digestive tract. However, some very interesting connections are gradually being revealed.

It has long been known that several of the hormones involved in the digestive processes are polypeptides. One of these, *secretin,* was discovered in 1902 by the English investigators Bayliss and Starling. It has become a classic because it was the first known instance where a chemical substance produced in one organ was carried in the blood and had a clearly demonstrated effect on a different organ. This discovery led to the use of the term *hormone,* which in Greek designates a messenger. The message carried by secretin, which is released by the intestine, is to stimulate secretion from the pancreas.

Another hormone released from the intestine is *cholecystokinin* (CCK), a 33-amino-acid polypeptide, often written CCK-33. CCK causes contraction of the gallbladder so that bile is released into the intestine. It also stimulates the secretion of enzymes from the pancreas, an action at one time ascribed to a separate enzyme, pancreozymin. Once the amino acid sequence of these polypeptides was known, however, it was realized that they are identical.

The surprise came when it was discovered that the brain contains receptor sites for CCK, a hormone known only for its role in digestion. The brain contains a trypsinlike enzyme that separates off from CCK-33 its terminal amino acids, forming CCK-8. This fragment seems to be of importance in regulation of food intake and satiety. Infusion of CCK-

8 inhibits eating in hungry rats and sheep. This finding appeared even more challenging when it became known that genetically obese mice have less CCK-8 in the cerebral cortex than nonobese littermates (Straus and Yalow 1979).

CCK-8 also inhibits stress-induced eating in rats. This is especially interesting because stress-induced eating is mediated through opiates and, as perhaps could be expected, naloxone blocks this effect of stress.

Another intestinal hormone, *vasoactive intestinal peptide* (VIP, a 28-amino-acid chain), is also found in high concentration in the cerebral cortex. This again points to a vast area of close connection between brain and digestive functions.

Most research on active peptides has been concerned with mammals. However, there is a growing body of knowledge about active peptides in other vertebrates and in various invertebrates. There are strong indications that peptide molecules occur widely as transmitting and modulating signals throughout the animal kingdom.

One example is the effect of opiates on the process known as pinocytosis in amebas. *Pinocytosis* is the process through which an ameba takes in small portions of the surrounding medium by an invagination of the cell membrane, which is then closed off, leaving a droplet that is incorporated in the cell. The process of pinocytosis can be induced by sodium ions and then suppressed by beta-endorphin, met-enkephalin, and morphine. These agents must act through typical opiate receptors because their effect is abolished by naloxone, the opiate antagonist discussed earlier. This suggests a normal role of opiates in amebas, but by no means can it be considered proof.

An example of a widely occurring polypeptide is insulin, which has been found in insects, annelids, molluscs, and several unicellular organisms. What the role of insulin is in these organisms remains to be clarified. Other peptides, well known from higher vertebrates, have been found in a wide variety of invertebrates, including unicellular organisms and even yeasts and molds. The function of these various peptides is, for the most part, still quite obscure, and we can expect that further studies will reveal a great deal of interesting information.

More than anything else, the wide distribution of active peptides suggests that communication by means of biochemical agents arose very early in evolution and that these mechanisms have been conservatively maintained in higher organisms. What have emerged in complex organisms are anatomical systems where specialization has taken place, with differentiation of specialized secretory cells and target cells. This in turn has permitted the development of specialized communication systems.

The wide occurrence in the nervous system of what are known as hormones emphasizes the overlap of the nervous and endocrine systems and shows clearly, not only that the two systems are interdependent, but also that many of the means of communication between the cells are similar or even identical.

We will now turn to a few examples of endocrine function among invertebrates. Among these the role of the nervous system in the production of hormones and control of endocrine function is even more prominent than among vertebrates.

INVERTEBRATES: CONTROL AND INTEGRATION

In the more highly organized invertebrates physiological function is under endocrine as well as nervous control. As in vertebrates, the nervous system serves for rapid communication, which is essential for actions related to escape, feeding, mating, and so on. As in vertebrates, an endocrine system produces hormones that control slower processes, such as growth, maturation, and many other metabolic functions. The nervous system has a direct and primary role in hormone production, and the nervous and endocrine systems are closely integrated.

Function and role of nervous systems

We have seen that the central nervous system of vertebrates coordinates a large number of separately controlled functions, both nervous and endocrine. The apparent contrast between nervous control (usually rapid) and hormonal control (usually slow to very slow) breaks down under closer scrutiny. Not only is the transmission from neuron to neuron, and from neuron to effector, usually of a chemical nature, but the central nervous system itself is a major producer of hormones and exerts much of its overall regulatory function in this way.

When we examine the role of nervous systems in the less highly organized invertebrates, we will need a more precise definition of nervous system than we used for vertebrates, where the meaning of the term is self-evident.

A nervous system can be defined as an assembly of neurons specialized for repeated and organized transmission of information from sensory receptor sites to neurons, or between neurons, or from neurons to effectors (e.g., muscles, glands, etc.). This definition precludes the existence of a nervous system in unicellular animals (protozoans), although functions within such cells may be coordinated with the aid of conducting organelles.

It is, on the whole, reasonably easy to recognize a neuron or nerve cell, based on its membrane potential and the ability to produce repeated action potentials. However, it may be difficult to recognize whether a ''neurosecretory'' cell is truly a modified nerve cell. Its location in association with nerve cells is helpful, and further information can be gained from recognition of action potentials, structural characteristics, and the function of similar structures in related species.

Invertebrate neurons have the same characteristics as vertebrate neurons. They are able to produce *action potentials* that consist of changes in membrane permeability and potential. Such action potentials are of an *all-or-none* nature and can be propagated in a nerve fiber, or axon. Typical neurons are able to produce repeated brief action potentials that last from a fraction of a millisecond to

several milliseconds, and such neurons occur in all major phyla except protozoans and sponges.

The transfer of information from one neuron to another takes place at *synapses,* which usually show evidence of *chemical transmission.* As in vertebrates, the nerve membranes of the two neurons most often remain spatially separated, and transmission is achieved through the release of an intermediary *chemical transmitter substance.* It appears, however, that there is direct electric transmission at certain synapses where the nerve membranes of the two cells become contiguous, a situation that also prevails at some vertebrate synapses.

Typical *postsynaptic potentials* seem to be universally present in at least four major phyla: molluscs, annelids, arthropods, and chordates. These may be *excitatory* or *inhibitory postsynaptic potentials.* In general, the excitatory postsynaptic potentials consist in a decrease in membrane polarization (in the direction of complete depolarization), and inhibitory potentials consist in an increase in membrane potential (hyperpolarization). These general characteristics are in principle similar in invertebrate and vertebrate neurons and nervous systems.

Neurosecretion

The importance of specialized *neurosecretory cells* and organs was recognized in invertebrates long before their importance was generally established in vertebrates. Neurosecretory cells are difficult to define and characterize so that they are easily and unequivocally recognized; the major approach for locating and pinpointing such cells still depends on the use of certain stains and other substances that have a special affinity for neurosecretory cells.

Neurohemal organs

Neurosecretory organs are groups of neurons that are the source of secretion. These cells have nerve fibers in which the secreted agent is transported and which usually terminate in close association with a vascular structure. Here they form a *neurohemal organ*, where the secreted product is stored and released. Neurohemal organs are very widespread, but it is not certain that they are present in connection with all neurosecretory organs.

In all the more highly organized animal phyla neurosecretory systems are central to the control and operation of endocrine mechanisms. We have already seen the importance of neurosecretion in the control of vertebrate endocrine organs; among invertebrates neurosecretion seems to be even more important.

We shall discuss later in this chapter examples of endocrine function in some insects, where neurosecretion and neurohemal organs have a central role in the control systems. Neurosecretory systems are also important in annelids and crustaceans, in which they are involved in control of reproduction, metabolism, molting, pigmentation, and so on.

Although some of the endocrine glands of vertebrates as well as invertebrates are outside direct neuroendocrine control, the normal function of most or all glands ultimately depends on regulation from the nervous system, either directly through nerve pathways or through hormonal mechanisms. It seems that control by neurosecretory mechanisms is the general rule in all animals in which endocrine mechanisms play a major role.

Hormones and endocrine function

Endocrine function is best understood and has been most extensively studied in those invertebrates that are morphologically highly organized. There are two reasons for this. One is that a highly organized animal such as an insect needs more detailed control and integration than a less organized animal such as a sea anemone. The other reason is that in the more highly organized animals many functions are delegated to specialized organs. This permits experiments such as removal of organs, reimplantation, extraction of active compounds, and other procedures that are difficult or impossible in less highly organized animals.

The groups in which organs of internal secretion

have been clearly demonstrated and are reasonably well understood are molluscs (notably cephalopods), annelids, crustaceans, insects, and tunicates. All these are highly organized animals where the complexity of function and the specialization of organs facilitate endocrine research. In particular, insects have been studied in great detail for reasons that include their easy accessibility, the ease of collecting and breeding them, their tolerance to drastic surgical procedures, and their economic importance.

Most invertebrate hormones are different from vertebrate hormones, both in chemical constitution and in effects. Many vertebrate hormones that profoundly affect vertebrate growth, development of gonads, metabolic processes, and so on, have no effect whatever if introduced into invertebrate animals.

There is, however, a rapidly increasing body of information that suggests many similarities in certain areas. Messenger molecules as well as appropriate receptors known from vertebrates have been identified in widely different invertebrates and even in unicellular organisms and plants. Adrenaline stimulates the adenylate cyclase system in protozoans, and the effect is blocked by a specific receptor antagonist, propranolol, exactly as in vertebrates.

This indicates a fundamental similarity in the biochemical mechanism, although its exact function in protozoans remains unknown. Likewise, endorphins alter the feeding behavior of amebas, an effect that is blocked by naloxone, an inhibitor of the specific receptor site for endorphins in vertebrates (Josefsson and Johansson 1979). It is unfortunate that we remain virtually ignorant of the normal role, if any, of these substances in protozoans.

Insulin, or materials that in radioimmunoassays react identically to insulin, has been found in insects, annelids, molluscs, and unicellular eukaryotes (Roth et al. 1982). Another vertebrate polypeptide hormone, cholecystokinin, has been identified in blowflies and snails. Again we can only assume that these hormones have a normal but so far unknown function in invertebrates (Osborne et al. 1982).

The unexpected fact that several messenger substances known from vertebrates occur widely among invertebrates as well as in unicellular organisms suggests that these substances are evolutionarily much more ancient than was previously thought. The hormones and endocrine systems as we know them in vertebrates appear to have forerunners in some of the simplest organisms known – organisms without glands, nerves, or circulatory systems. It now appears that vertebrate endocrine systems may be highly specialized forms of general biological mechanisms of very ancient origin.

Insect endocrinology

Molting, pupation, metamorphosis

Hormones play a major role in the physiology of insects, especially in growth, molting, pupation, and metamorphosis into the mature adult form. These phenomena have been studied in many different species, but considering that there may be around 1 million different insect species, the fraction is very small indeed.

Nevertheless, there is a striking similarity in the endocrine function of different insects. For example, the *hemimetabolous insects,* which go through a number of molts as they change from the newly hatched form and gradually develop into adults, employ the same hormones as the *holometabolous insects,* which remain in the completely larval form through a number of molts, and then transform into pupae from which the fully formed adults emerge. We shall discuss both types.

As an example of a hemimetabolous insect we shall use the South American bloodsucking bug, *Rhodnius,* a relative of the ordinary bedbug. *Rhodnius* hatches from the egg as a tiny blood-sucking bug, a nymph, which through five stages (*instars*) gradually develops into the adult form. In each stage the nymph must obtain a meal of blood before it can develop into the next stage by shedding its old cuticle. After it has sucked blood, about 4 weeks pass by; the nymph then sheds the old cuticle and

increases in size by filling the tracheal system with air before the new cuticle hardens. It is then ready to suck blood again, but if the opportunity does not occur, it can survive for many months. However, if it obtains blood, it molts again about 4 weeks later.

The molt of *Rhodnius* (see Figure 12.7) is stimulated by a hormone, *ecdysone* (also known as molting hormone), which is secreted after the blood meal has been ingested. This hormone is secreted by the two *prothoracic glands,* located in the thorax of the insect. The prothoracic glands, in turn, are stimulated by a brain hormone (PTTH) secreted by specialized neurosecretory cells in the brain.

The adult *Rhodnius* has wings and mature gonads, and it differs from the larval forms in other respects as well. What directs this development? The lack of adult characters in the earlier nymphal stages is attributable to another hormone, the *juvenile hormone,* which is secreted by the *corpus allatum,* a tiny cluster of cells just behind the brain.

The juvenile hormone determines that the new cuticle will have nymphal characteristics, and its presence prevents the formation of adult characters. Thus, in all the early nymphal stages, when juvenile hormone is present, ecdysone causes molting that results in a larger nymph. In the fourth nymphal stage the amount of juvenile hormone declines, and the fifth nymphal stage shows the beginning of the development of wings. In the very last molt the juvenile hormone is absent, and an adult emerges (Figure 12.8).

Let us examine the evidence for endocrine control of these events. If a fifth-stage *Rhodnius* is decapitated shortly after a blood meal, the final molt does not take place, although the headless animal may live for more than a year. This makes it appear that the molting hormone, ecdysone, is produced in the head. However, if the decapitation takes place after a certain ''critical period,'' which occurs around the seventh day after the meal, the animal will molt and develop into a headless adult. Thus, the head is needed only early in the period and not for the molt itself.

The difference can be shown to result from en-

FIGURE 12.7 The endocrine system of insects involved in metamorphosis. The hormone PTTH (prothoracicotropic hormone) is secreted by the corpora cardiaca. It stimulates secretion of alpha-ecdysone, which in the fat body is hydroxylated to form beta-ecdysone. The epidermis digests the old cuticle and secretes new cuticle under the influence of beta-ecdysone. The absence or presence of juvenile hormone (JH) determines the morphology of the new cuticle.

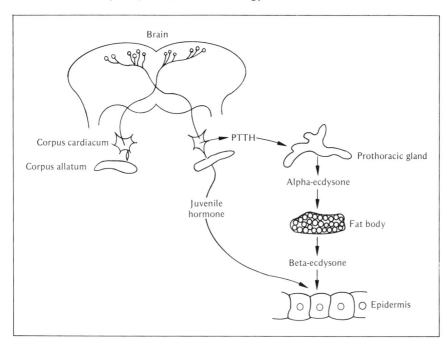

docrine factors, for if blood from an insect decapitated after the critical period is transferred to an insect decapitated before the critical period, the latter is induced to molt. The explanation is that during the critical period the brain releases a hormone, the *brain hormone,** which stimulates the prothoracic gland. If the decapitation occurs early, the prothoracic gland is never stimulated; if decapitation is later, the brain hormone has taken effect and is not needed, and ecdysone is released and induces molting in due time.

---

*The brain is the source of several hormones, but it is conventional among insect endocrinologists to use the term *brain hormone* for the agent that specifically activates the prothoracic gland and leads to release of ecdysone. It is also known as the *prothoracicotropic hormone (PTTH)*.

The simplest way of performing a blood transfusion in these insects is to connect them with a short glass capillary that can be fixed into the neck of the decapitated insect with a droplet of wax (Figure 12.9). If we join together a fifth-stage nymph that has had a blood meal and has passed through the critical period, and a small first-stage nymph recently emerged from the egg and not yet fed, the small nymph will develop adult characteristics such as wings and genital organs and appear as a midget adult. If the experiment is repeated with an older first-stage nymph, this animal is again induced to molt, but does not acquire adult characteristics because by this time it has produced and released enough juvenile hormone to molt into a juvenile nymph.

Ecdysone is obviously needed for each molt, and

FIGURE 12.8 The bloodsucking bug *Rhodnius* develops into an adult through five molts. Each molt is caused by release of the hormone ecdysone from the prothoracic glands. In the first four molts the juvenile hormone (from the corpora allata) causes the new cuticle to be of the larval type, thus preventing the formation of an adult. In the last molt the absence of juvenile hormone results in a fully developed adult.

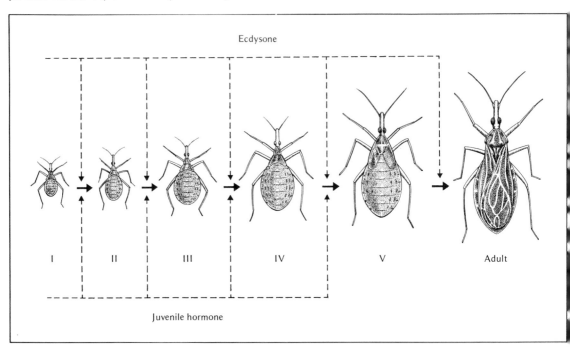

FIGURE 12.9 The transfer of hormones via the blood can be demonstrated by joining together two animals. In this case a fourth-stage *Rhodnius* nymph (left) with only the tip of the head removed (leaving the brain intact) has been joined to a headless fifth-stage nymph (right). [Wigglesworth 1959]

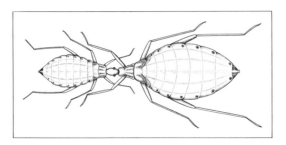

its effects are merely modified by the juvenile hormone, which is secreted later. Therefore, if decapitation is carried out after the brain hormone has been released and has stimulated the prothoracic gland, molting is induced; the type of molt depends on whether the decapitation occurred before or after juvenile hormone was secreted. By accurately timing the decapitation in relation to the blood meal and the critical period, it has been possible to obtain small, headless *Rhodnius* with adult morphological characteristics.

In summary, we can see that a molt is caused by the release of ecdysone from the prothoracic gland, this release being initiated by the brain hormone, which in turn was released when *Rhodnius* became

FIGURE 12.10 The tobacco hornworm (*Manduca sexta*) molts in response to the molting hormone, ecdysone. Whether the molt results in another larva, a pupa, or an adult is determined during a relatively short, critical period of a day or two (shaded area). If juvenile hormone is present above the threshold level, the larva molts into a larger, but otherwise similar larva (top panel). If the juvenile hormone has fallen below threshold before the critical period, the next molt leads to a pupa (center panel). In the continued absence of juvenile hormone, development progresses to the next stage and an adult moth emerges from the pupa (lower panel). [Diagram courtesy of H. F. Nijhout, based on Nijhout and Wheeler 1982]

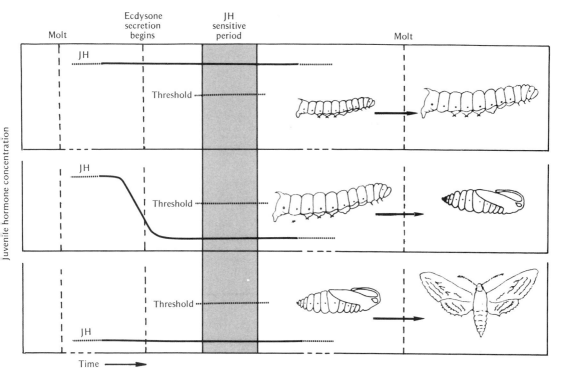

engorged with blood. The hypothesis that the release of brain hormone is caused by the stretching of the insect as it becomes engorged has been shown experimentally in another bug, the milkweed bug (*Oncopeltus*). If saline is injected into the abdomen of this bug, the stretching of the abdominal wall induces molting, even in the absence of feeding (Nijhout 1979).

In holometabolous insects the problem is not only when to molt, but also how to develop into the correct next stage, be it larva, pupa, or adult. The endocrine function that controls these stages has been extensively studied in several moths, including the silk moth (*Bombyx mori*), the cecropia moth (*Hyalophora cecropia*), and the tobacco hornworm (*Manduca sexta*). The larvae of these moths usually go through four molts (five stages), all initiated by the secretion of the brain hormone, which in turn causes the prothoracic gland to release ecdysone.

Whether a molt leads to a larva, a pupa, or an adult depends on the presence or absence of juvenile hormone (Figure 12.10). The secretion of ecdysone, and thus the events leading to a molt, is

SILK WORM The larva of the silk moth, *Hyalophora cecropia,* in the first, third, and fifth instars. After the fifth instar the larva pupates and goes through a complete metamorphosis, whereupon the adult moth emerges. [Courtesy of Charles Walcott, State University of New York, Stony Brook]

initiated during a relatively short critical period when the brain hormone is secreted (horizontal bars). In the earlier larval stages juvenile hormone is present, and the molt leads to another and larger larva. In the last instar the level of juvenile hormone has fallen sharply before the brain hormone appears, and a pupa is formed (center panel). The final molt, the pupa developing into an adult, depends on the absence of juvenile hormone throughout (lower panel).

One question remains: What causes the decrease in juvenile hormone in the last larval stage? The answer appears to be that secretion of juvenile hormone stops when the animal reaches a critical size.

In the younger larval stages the *corpora allata* release juvenile hormone, so that the larva molts into another larva, which because of its loose and wrinkled skin can grow and increase in size as it feeds. At the end of the fifth larval stage, the corpora allata cease to secrete juvenile hormone, and at the next molt, a harder cuticle is formed (i.e., a pupa). In the pupa the larval tissues break down and are transformed into pupal structures, and after a final molt, when no juvenile hormone remains, a fully developed adult moth emerges (Figure 12.10).

The simplest way to confirm the function of the corpus allatum is to remove this organ in the young silkworm during one of the early instars. If this is done, molting takes place as usual (the brain–prothoracic system is untouched), but instead of developing another larva, the molt results in a diminutive pupa, from which, in turn, a tiny adult emerges (Figure 12.11). There is no other effect of the removal of the corpus allatum (and thus the juvenile hormone) so that, except for size, a normal-looking adult develops from the pupa.

The role of the corpus allatum can be confirmed by transplantation. If several corpora allata from early larvae are implanted into the last larval instar, the next molt does not produce a pupa, but instead an oversized larva. This larva may continue to grow, and by introducing additional corpora allata it is possible to produce a giant larva that can molt into a giant adult. There is a limit to this development, and it is unlikely that an animal will survive beyond an artificial seventh instar.

The chemical structure of the hormones we have discussed should be mentioned for comparison with vertebrate hormones. Ecdysone is a steroid (Karlson and Sekeris 1966) that was originally isolated and crystallized by extraction of 500 kg silkworm

FIGURE 12.11 If the corpora allata are removed from a young moth larva, the larva is deprived of juvenile hormone, and instead of going through a larval molt, the animal pupates and a small adult emerges. [Highnam and Hill 1969]

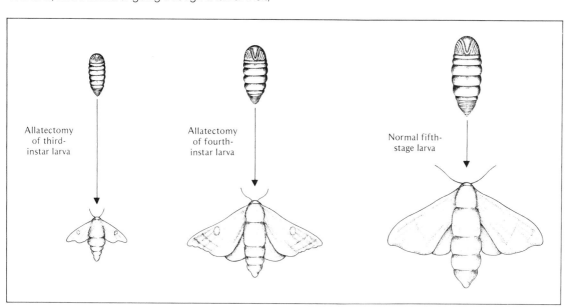

Allatectomy of third-instar larva

Allatectomy of fourth-instar larva

Normal fifth-stage larva

pupae, which yielded 25 mg of the crystalline hormone. The full chemical structure of the hormone was finally established on material extracted from a batch of 1000 kg dried silkworm pupae (corresponding to about 4 ton fresh weight). It is interesting that ecdysone is a steroid because of the widespread occurrence of hormones of steroid nature in vertebrates.

Ecdysone

The juvenile hormone, on the other hand, is entirely different and is similar in structure to terpenes. A large number of compounds with juvenile hormone activity have been isolated, and many have been synthesized, some apparently far more potent than the naturally occurring ones. The major natural hormone from cecropia has the following structure (Röller et al. 1967):

The synthesis of compounds with juvenile hormone activity is very challenging because they may be potential insecticides. These compounds are effective in extremely small amounts, and if they are applied at suitable periods during the life history of insects, they can prevent normal adult development and thus reproduction. The advantage of juvenile hormones as insecticides is that insects are less likely to develop immunity to substances on which they normally depend than to the various poisonous substances now in use.

Because some of the artificial juvenile hormones act more specifically on some insects than on others, it may be feasible to develop insecticides that are more specific for one insect species than for another – a far more attractive prospect than the use of highly toxic substances such as DDT, which kill desirable and undesirable insects alike. Also, as juvenile hormone has little or no effect on many other animals, at least not on vertebrate predators, such substances probably will have less disastrous ecological effects.

The paper factor

An accidental discovery of a naturally occurring juvenile hormone substance suggests that plants in fact already use such compounds for their own protection. The discovery was made when a Czech investigator, Dr. Karel Sláma, spent a year at Harvard University and brought with him his stock of the bug *Pyrrhocoris*. In the laboratory these bugs are reared on linden seeds with paper towels as a climbing surface. At Harvard they would not develop into adults after the fifth instar, but developed into oversized sixth or even seventh instar immature forms.

It appeared as if the bugs had unintentionally been exposed to contamination with juvenile hormone, and a systematic comparison of all possible differences between conditions in the United States and in Europe revealed that the paper towels used at Harvard contained a substance with juvenile hormone activity (Sláma and Williams 1966).

It turned out that all American-made paper, including newsprint, contained the substance. It was not added during manufacture, but originated in the wood from fir trees used for paper manufacture in the United States. Paper of Japanese or European origin had no such effect. The substance, called the *paper factor*, has been isolated and has some chemical resemblance to the juvenile hormone. This

suggests that many terpenes, which occur commonly in plant material, especially in evergreen trees, may be natural protective substances that have evolved as a countermeasure to insect pests.

The insect hormones discussed so far are all concerned with the control of growth and development, but many other physiological processes are also under endocrine control. As an example, we shall briefly describe a hormone concerned with the regulation of water excretion in a bloodsucking insect, which, because of its feeding habits, is periodically subjected to a heavy water load.

Diuretic hormone

We remember that, in mammals, the antidiuretic hormone is important for water conservation. In contrast, elimination of excess water in insects is often controlled by a *diuretic hormone.*

The bloodsucking bug *Rhodnius* is faced with the problem of too much water each time it takes a blood meal. Immediately after feeding it excretes a large volume of dilute urine, eliminating most of the excess water. This sudden surge in the excretion of water is regulated by a hormone that stimulates water excretion.

The action of this hormone has been studied on isolated Malpighian tubules of *Rhodnius*. These organs can continue to secrete urine if they are placed in blood from *Rhodnius* or in a suitable saline solution. If they are placed in the blood from a recently fed *Rhodnius,* the tubules secrete a copious urine at a much higher rate than if they are placed in blood from an unfed animal. This suggests that the blood contains a hormone that is directly responsible for stimulating the secretion of urine (i.e., a *diuretic hormone*).

Extracts of various known hormone-producing organs of *Rhodnius* have no effect on urine secretion, but extracts of the ganglia from the first abdominal segment are effective. These ganglia contain several groups of large neurosecretory cells, and 97% of the diuretic activity is contained in the most posterior group of these.

After a *Rhodnius* has sucked blood, urine production increases within less than a minute, showing that the hormone must be released very rapidly, undoubtedly stimulated by nervous reflexes from the feeding process or from the abdominal expansion caused by the feeding (Maddrell 1963).

The rapid excretion of surplus water is especially important for a flying insect, such as a mosquito. A mosquito can take in a blood meal that is twice its body weight, and flying and maneuvering would be a problem unless it rapidly got rid of the excess water.

Blood-feeding mosquitoes begin to urinate at high rates even before the blood meal has been completed. The diuresis is mediated by a low-molecular peptide from the head that stimulates sodium excretion. The hemolymph of mosquitoes taken during a blood meal stimulates fluid secretion in isolated Malpighian tubules. However, the hemolymph from a mosquito that drinks a sucrose solution lacks the diuretic effect, indicating that it is the natriuretic effect that is the primary factor in elimination of excess fluid from a blood meal (Beyenbach and Petzel 1987).

These few examples of insect hormones give only an indication of the complexity of endocrine regulation. The importance of the nervous system in hormone production is remarkable, not only in insects, but in many other invertebrates. An increasing number of hormones have been isolated, chemically identified, and finally synthesized. However, the field of invertebrate endocrinology is so vast that this area of physiological research will continue to expand and remain an area of fruitful research for many years.

REFERENCES

Bentley, P. J. (1976) *Comparative Vertebrate Endocrinology.* Cambridge: Cambridge University Press. 415 pp.

Berridge, M. J., and Irvine, R. F. (1984) Inositol triphosphate, a novel second messenger in cellular signal transduction. *Nature, Lond. 312:*315–321.

Beyenbach, K. W., and Petzel, D. H. (1987) Diuresis in mosquitoes: Role of a natriuretic factor. *News Physiol. Sci. 2:*171–176.

Currie, M. G., Geller, D. M., and Cole, B. R. (1984) Purification and sequence analysis of bioactive atrial peptides (Atriopeptins). *Science 223:*67–69.

De Luca, H. F. (1981) Recent advances in the metabolism of vitamin D. *Annu. Rev. Physiol. 43:*199–209.

Highnam, K. C., and Hill, L. (1969) *The Comparative Endocrinology of the Invertebrates.* London: Edward Arnold. 270 pp. Second ed. (1977), 368 pp.

Jensen, E. V., Numata, M., Brecher, P. I., and Desombre, E. R. (1971) Hormone-receptor interaction as a guide to biochemical mechanism. In *The Biochemistry of Steroid Hormone Action,* Biochemical Society Symposium no. 32 (R. M. S. Smellie, ed.), pp. 133–159. London: Academic Press.

Josefsson, J.-O., and Johansson, P. (1979) Naloxone-reversible effect of opioids on pinocytosis in *Amoeba proteus. Nature, Lond. 282:*78–80.

Karlson, P., and Sekeris, C. E. (1966) Ecdysone, an insect steroid hormone, and its mode of action. *Recent Prog. Horm. Res. 22:*473–502.

Maddrell, S. H. P. (1963) Excretion in the blood-sucking bug, *Rhodnius prolixus* Stål. *J. Exp. Biol. 40:*247–256.

McEwen, B. S. (1976) Interactions between hormones and nerve tissue. *Sci. Am. 235:*48–58.

Nijhout, H. F. (1979) Stretch-induced moulting in *Oncopeltus fasciatus. J. Insect Physiol. 25:*277–281.

Nijhout, H. F., and Wheeler, D. E. (1982) Juvenile hormone and the physiological basis of insect polymorphisms. *Quart. Rev. Biol. 57:*109–133.

Osborne, N. N., Cuello, A. C., and Dockray, G. J. (1982) Substance P and cholecystokinin-like peptides in *Helix* neurons and cholecystokinin and serotonin in a giant neuron. *Science 216:*409–411.

Otsuka, M., Iversen, L. L., Hall, Z. W., and Kravitz, E. A. (1966) Release of gamma-aminobutyric acid from inhibitory nerves of lobster. *Proc. Natl. Acad. Sci. U.S.A. 56:*1110–1115.

Pert, C. B., and Snyder, S. H. (1973) Opiate receptor: Demonstration in nervous tissue. *Science 179:*1011–1014.

Röller, H., Dahm, K. H., Sweely, C. C., and Trost, B. M. (1967) The structure of the juvenile hormone. *Angew. Chem: (Engl.) 6:*179–180.

Roth, J., LeRoith, D., Shiloach, J., Rosenzweig,

J. L., Lesniak, M. A., and Havrankova, J. (1982) The evolutionary origins of hormones, neurotransmitters, and other extracellular chemical messengers. *N. Engl. J. Med. 306:*523–527.

Rozensweig, Z., and Kindler, S. H. (1972) Epinephrine and serotonin activation of adenyl cyclase from *Tetrahymena pyriformis. FEBS Lett. 25:*221–223.

Schally, A., Arimura, A., and Kastin, A. J. (1973) Hypothalamic regulatory hormones. *Science 179:*341–349.

Sláma, K., and Williams, C. M. (1966) The juvenile hormone. 5. The sensitivity of the bug, *Pyrrhocoris apterus,* to a hormonally active factor in American paperpulp. *Biol. Bull. 130:*235–246.

Stewart, J., and Li, C. H. (1962) On the use of -tropin or -trophin in connection with anterior pituitary hormones. *Science 137:* 336–337.

Straus, E., and Yalow, R. S. (1979) Cholecystokinin in the brains of obese and nonobese mice. *Science 203:*68–69.

Sudoh, T., Kangawa, K., Minamino, N., and Matsuo, H. (1988) A new natriuretic peptide in porcine brain. *Nature, Lond. 332:*78–81.

Sutherland, E. W. (1972) Studies on the mechanism of hormone action. *Science 177:*401–408.

Sutherland, E. W., Rall, T. W., and Menon, T. (1962) Adenyl cyclase. 1. Distribution, preparation, and properties. *J. Biol. Chem. 237:*1220–1227.

Takeuchi, A., and Takeuchi, N. (1967) Anion permeability of the inhibitory post-synaptic membrane of the crayfish neuromuscular junction. *J. Physiol. 191:*575–590.

Weitz, C. J., Faull, K. F., and Goldstein, A. (1987) Synthesis of the skeleton of the morphine molecule by mammalian liver. *Nature, Lond. 330:*674–677.

Wigglesworth, V. B. (1959) *The Control of Growth and Form: A Study of the Epidermal Cell in an Insect.* Ithaca, N.Y.: Cornell University Press. 140 pp.

Yalow, R. S. (1978) Radioimmunoassay: A probe for the fine structure of biologic systems. *Science 200:*1236–1245.

ADDITIONAL READING

Adler, N. T. (1981) *Neuroendocrinology of Reproduction: Physiology and Behavior.* New York: Plenum Press. 555 pp.

American Society of Zoologists (1983) Evolution of endocrine systems in lower vertebrates, a symposium honoring professor Aubrey Gorbman. *Am. Zool. 23:*593–748.

American Society of Zoologists (1985) Advances in crustacean endocrinology. *Am. Zool. 25:*153–284.

American Society of Zoologists (1986) Evolution in hormone diversity. *Am. Zool. 26:*923–1054.

Barrington, E. J. W. (1975) *An Introduction to General and Comparative Endocrinology,* 2d ed. Oxford: Clarendon Press.

Bentley, P. J. (1982) *Comparative Vertebrate Endocrinology,* 2d ed. Cambridge: Cambridge University Press. 485 pp.

Bloom, F. E. (ed.) (1980) *Peptides: Integrators of Cell and Tissue Function.* New York: Raven Press. 257 pp.

Chung, S.-H., and Dickenson, A. (1980) Pain, enkephalin and acupuncture. *Nature, Lond. 283:*243–244.

Ganong, W. F., and Martini, L. (eds.) (1978) *Frontiers in Neuroendocrinology,* vol. 5. New York: Raven Press. 416 pp.

Gilbert, L. I. (ed.) (1975) *The Juvenile Hormones.* New York: Plenum Press.

Gorbman, A., Dickhoff, W. W., Vigna, S. R., Clark, N. B., and Ralph, C. L. (1983) *Comparative Endocrinology.* New York: Wiley. 572 pp.

Greep, R. O., and Astwood, E. B., (eds.) (1972–1976). *Handbook of Physiology,* sect. 7, *Endocrinology,* vols. 1–7. Washington, D.C.: American Physiological Society.

Highnam, K. C., and Hill, L. (1977) *The Comparative Endocrinology of the Invertebrates,* 2d ed. Baltimore: University Park Press. 368 pp.

Höllt, V. (1986) Opioid peptide processing and receptor selectivity. *Annu. Rev. Pharmacol. Toxicol. 26:*59–77.

Imura, H., and Nakai, Y. (1981) ''Endorphins'' in pituitary and other tissues. *Annu. Rev. Physiol. 43:*265–278.

Koob, G. F., and Bloom, F. E. (1982) Behavioral effects of neuropeptides. Endorphins and vasopressin. *Annu. Rev. Physiol. 44:*571–582.

Krieger, D. T. (1983) Brain peptides: What, where and why? *Science 222:*975–985.

Krieger, D. T., and Martin, J. B. (1981) Brain peptides. 1 and 2. *N. Engl. J. Med. 304:*876–885; 944–951.

Snyder, S. H. (1980) Brain peptides as neurotransmitters. *Science 209:*976–983.

Sutherland, E. W. (1972) Studies on the mechanism of hormone action. *Science 177:*401–408.

Treherne, J. E. (ed.) (1980) Neurotransmission, neurotransmitters, and neuromodulators: A discussion meeting organized by E. A. Kravitz and J. E. Treherne, Woods Hole, Mass., April 1980. *J. Exp. Biol. 89:*1–286.

Wigglesworth, V. B. (1970) *Insect Hormones*. Edinburgh: Oliver & Boyd. 159 pp.

CHAPTER THIRTEEN

Information and Senses

In the first parts of this book we discussed aspects of the environment that are important to animals: oxygen, food, temperature, water. We then discussed physiological processes and how they are controlled and integrated by nerves and hormones.

We shall now ask the question of how animals obtain information about their environment and how this information is used.

Virtually all animals depend on information about their surroundings. They need to find food and mates and escape from predators. They must find their way about and also assess important qualities of the environment – temperature, light, oxygen, and so on.

We shall first be concerned with the kind of information that is available to animals. Then we shall consider how such information is received, processed, and passed on to the central nervous system.

Most information about the environment is obtained through specialized sensory organs. Traditionally, sense organs are separated into *exteroceptors* that respond to stimuli coming from the outside, such as light and sound, and *proprioceptors* that refer to internal information, such as the position of the limbs. This separation does not have much inherent meaning and, at best, is a matter of convenience.

Another traditional classification of senses is based on the five most obvious senses of humans: vision, hearing, taste, smell, and touch. In reality our sensory equipment is not nearly so limited.

In this context the question of whether information can reach the central nervous system via other avenues, outside the sensory organs (extrasensory perception, or ESP), is irrelevant. In this chapter we are dealing with measurable physical quantities that can be recognized, described, and manipulated in controlled ways. Although some sensory mechanisms may still be unknown (this is true of some responses to magnetic fields), the word *extrasensory* by definition means that no sensory structure is involved.

SENSORY QUALITIES

A list of external stimuli to which at least some animals respond is quite extensive (Table 13.1). The listed categories are not discrete, and the separation is somewhat arbitrary. However, for convenience we shall follow this sequence in our discussion of the possibilities and limitations that apply to the use of the various kinds of available information. The information naturally falls into three major categories: electromagnetic and thermal energy, mechanical energy and mechanical force, and chemical agents.

The wide variety of stimuli and the striking structural differences in the sense organs seem confusing until we realize that the general sequence of events is much the same for all of them (Figure 13.1). The external stimulus impinges on an accessory structure, which may be highly complex such as the eye or the ear, or much simpler such as the touch receptors of the skin.

Through the accessory structures the external stimulus reaches one or more sensory neurons. These have many properties in common, but they differ from organ to organ in that each type is particularly sensitive to one specific kind of stimulus. The sensory neurons of the retina are highly sensitive to light, those of the ear to vibrations at frequencies in the range we identify as "sound," and so on. However, all sensory neurons respond in the same way; they translate the stimulus into nerve impulses that, via the appropriate sensory nerves, are transmitted to the central nervous system.

We now come to the intriguing fact that the nerve impulses carried in the different sensory nerves are all of the same fundamental nature. For example, the optic nerve carries the same kind of nerve impulses as the auditory nerve. Both carry similar action potentials; it is the central nervous system that sorts out what the original stimulus was.

If the auditory nerve is stimulated artificially, the induced nerve impulses are perceived by the central nervous system as "sound," and artificial stimulation of the optic nerve is perceived as "light." Most of us have experienced this kind of

TABLE 13.1 Environmental stimuli significant in sensory perception.

Electromagnetic and thermal energy
Light
Infrared radiation
Thermal: "heat," "cold"
Electric
Magnetic
Mechanical energy and force
Sound and sonar
Touch and vibration
Pressure
Gravity
Inertia
Chemical agents
Taste
Smell
Humidity

interpretation; we know that mechanical pressure on the eyeball is perceived as light and that a sharp blow on the eye makes us "see stars," although no light is involved.

Returning to the sequence in Figure 13.1, we say that the sensory neurons are the *transducers* that receive the external information and *encode* it as impulses in the sensory nerves.* In the central nervous system the nerve signals are again *decoded* and the pertinent information is integrated and utilized. The most interesting events are the transducing and encoding of information on the one hand, and the decoding and processing on the other. The sensory nerves are transmission lines and in this context the least interesting part of the system.

In this chapter we shall first consider what kinds of information can be obtained and the possibilities

---

* A *transducer,* in a general sense, is a device that transforms or "translates" energy supplied in one form into another form. For example, a microphone is a transducer that translates sound energy into electric energy. A transducer often requires an auxiliary energy supply and thus, strictly, is an energy controller. Furthermore, the output is often at a higher energy level than the input, i.e., the transducer is also an amplifier.

FIGURE 13.1 Simplified diagram of the sequence of events in sensory perception. Light perception is used as an example, although in higher animals vision is far more complex than indicated here.

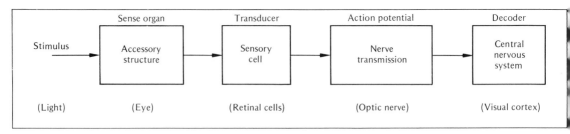

and limitations inherent in the nature of the information. The sensory information is encoded and transmitted as nerve signals and again decoded in the central nervous system. The astounding fact is that these various operations are all carried out by nerve cells that function in the same fundamental way, based on the electric potential across the cell membrane and the generation of action potentials.

GENERAL PRINCIPLES

Determination of direction and distance

The direction from which a stimulus comes is of utmost importance, for this permits an animal to move toward the source of the stimulus or to evade it, as need may be. Without knowledge of direction, this is impossible. Distance is also important, although usually not by itself but in combination with information about direction. Based on the nature of the stimulus, there are strict physical limitations to what is possible and what the accessory structure can help to achieve in this regard.

The most directional sense is vision, to a majority of animals probably the most important single sense. The reason for the directionality of vision is, of course, that light rays travel in straight lines. In addition to information about the location, the sense of vision gives much information about the nature of various objects. Although vision is the most highly directional sense, information about direction can also be obtained with other senses, but there are certain inevitable limitations.

In principle, there are only three ways in which directional information can be obtained:

1. The sense organ itself can be directional, which means that it must give different signals if turned toward or away from the source of the stimulus, as the eye does.
2. Signals obtained from a pair of similar sense organs can be compared. Directional information about sound is obtained in this way by the use of two ears, although the hearing organ itself is not very directional.
3. Signals can be compared in space and time; this is achieved by successively assessing the condition at a series of points in space.

Information about distance is more difficult to obtain. Here again vision is superior, and with the aid of two eyes and the differences in the images they receive, distance can be evaluated. Most other sense organs can achieve only a rough estimate of distance, based on the attenuation of the signals. Certain auditory mechanisms can, however, be used for extremely precise information about distance as well as direction, as we shall see later.

CHEMICAL SENSES: TASTE AND SMELL

We find some kind of chemical sense throughout the animal phyla; even very simple organisms and protozoans respond to chemical stimuli. Obviously, it is of interest to an ameba whether or not a particle is digestible and thus should be ingested. For this it needs chemical cues. However, the simplest animals do not have discrete chemical sensory organs; these are most highly developed in vertebrates and arthropods.

The chemical senses cannot be used for scanning the environment, the way light and sound are used, for the chemical compounds to be sensed must first arrive at the animal's sensory organ, either by diffusion or by mass movement of the medium. Light has, for physiological purposes, the quality of instantaneous arrival, and sound arrives very rapidly, but with such a delay that the delay itself carries valuable information (as in directional hearing and echo location). The limitations on the use of the chemical senses are much more severe, and detailed information is usually obtained only over extended periods of time, often in combination with the animal's moving about.

The most useful information obtained with the chemical senses concerns finding food or mates and locating enemies. The chemical senses are useful primarily as alerting or warning devices, allowing the efforts of other senses to be concentrated.

The presence of a chemical sense in less highly organized animals is easily demonstrated by observation of their behavior. However, virtually all our information about the physiological mechanisms in the chemoreceptor organs has been obtained from arthropods and vertebrates. The reason is that the chemosensitive organs in these animals form discrete structures that, especially in insects, are easily accessible and eminently well suited for electrophysiological studies.

To a human the sense of taste refers to materials in contact with the mouth; smell refers to gaseous substances that reach the nose via the air. The separation is highly subjective, for when we speak about the "taste" of food, much of the information is obtained through the nose.* When we extend the use of the terms taste and smell to other air-breathing vertebrates, the analogy is justified, for their sense organs have a great deal of structural similarity to ours.

For aquatic animals a distinction between taste and smell rapidly becomes meaningless. For example, some fish have chemical senses of unbelievable acuity, but should we say that they "smell" or "taste" the water? In common language it is often convenient to use the term smell for objects at a distance and taste for materials in direct contact with the animal, in particular with the mouth. However, a catfish has sensitive chemoreceptors located all over the body, and we find it natural to call these receptors tastebuds (Figure 13.2).

The chemoreceptors of fish are highly sensitive to amino acids, which undoubtedly is of importance in finding food. Electrophysiological tests of the responses of the chemoreceptors on the barbels of the channel catfish (*Ictalurus punctatus*) show different sensitivities for different amino acids with estimated threshold concentrations between 10^{-9} and 10^{-11} mol per liter (1 to 100 μg per liter). These are the lowest electrophysiological thresholds reported for taste in any vertebrate (Caprio 1977). Humans can barely imagine what such dilutions mean. To reach a concentration of 10^{-10} mol per liter, we would dissolve in an olympic-size swimming pool no more than 23 mg of alanine, or less than one-hundredth of a teaspoonful.

Such high sensitivity is not unique to fish. The spiny lobster, *Panulirus,* is especially sensitive to the amino acid taurine, which is common in many marine animals (Fuzessery et al. 1978). Its sensitivity threshold for taurine is about 10^{-10} molar, requiring that we dissolve 33 mg of this amino acid

---

*There are more taste qualities than the simple sweet, sour, salty, and bitter that have been popularized as the only taste qualities perceived in the mouth. Furthermore, the texture of the food is also an important quality in "taste"; if familiar foods have been homogenized, they often are very difficult to recognize.

FIGURE 13.2 The catfish has tastebuds distributed all over the body. Each black dot represents 100 tastebuds. Higher concentrations on lips and bar- bels are indicated by small dots and solid black, respectively. Top, side view; bottom, dorsal (left) and ventral (right) views. [Atema 1971]

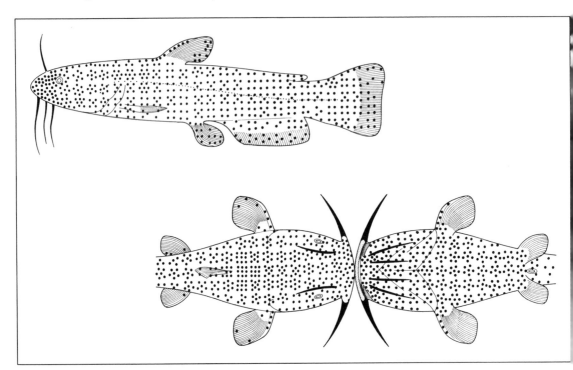

in the olympic pool. Note that an olympic pool holds 2600 m^3, and this amount of sea water contains about 92 tons of sodium chloride and other salts. It is difficult to imagine that the few milligrams of taurine can still be discovered by the sensory equipment of the spiny lobster in the presence of this immense amount of salt.

Air-breathing animals

Let us separate taste and olfaction according to the medium that carries the chemical stimulus and say that the transfer medium for taste is water and for olfaction is air. We can then say that taste and olfactory receptors in the insect are distinctly separate structures. They can be located almost anywhere – on the mouth parts, the antennae, the feet.

The taste organs are usually bristles or hairs, a fraction of a millimeter long, open at the tip, and with one or more sensory neurons at the base. The olfactory organs may be hair- or bristlelike, or may consist of thin-walled pegs or pits in a wide variety of forms and shapes.

The taste organs usually require the stimulating molecules to be in much higher concentrations than the olfactory receptors. The latter in some cases are unbelievably sensitive to specific odoriferous molecules that are important for locating food, prey and mates.

Substances used to find other individuals of the same species, as well as for a variety of social communication activities in colonial insects, are called *pheromones* (Greek *pherein* = to carry).

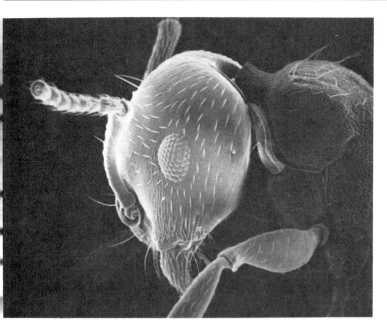

ANT HEAD The head of a small harvester ant, *Pheidole bicarinata*. The many hairs on the head, thorax, and legs presumably are sensory organs, but it is not known which are chemoreceptors and which are mechanoreceptors. The beaded structure in the center of the head is the compound eye. The total body length of this ant is 1.3 mm [Courtesy of Diana Wheeler, Duke University]

Pheromones constitute a chemical language that is of great importance in laying trails, in recognizing individuals from the same nest, in marking the location of a food source, as alarm substances, and so on.

Insect taste and olfaction

The sensory hairs on the proboscis of the ordinary blowfly (*Phormia regina*) make an excellent experimental model. Severing the head from the fly and attaching it to a fine micropipette filled with saline provides a way of mounting it, as well as an electrical connection to the inside. By penetrating the wall of a sensory hair, at some distance from its tip, with a microelectrode, the electrical activity in the hair can be recorded. A minute glass tube, filled with a solution to be tested, can then with the aid of a micromanipulator be brought into contact with the open tip of the hair, and if the solution contains a suitable stimulant, it causes electric activity, which the microelectrode records (Morita and Yamashita 1959).

A typical *taste hair* (Figure 13.3) may have five

sensory neurons at its base. One of these is always a mechanoreceptor, sensitive to deflection of the hair; of the others, one is usually sensitive to sugar, one to water, and one or two to salts and various other compounds (Hodgson 1961). By recording directly from the taste hair, it is possible to determine which substances stimulate the receptor and what concentration is necessary.

The olfactory receptors are often, but not always, located on the antennae. The ability of certain male moths to locate females of the same species, even at distances of several kilometers, is well known. The female gives off a sex attractant, which spreads downwind, forming a plume like the smoke from a smoke stack. The male is equipped with giant antennae that have a primary role in sensing the sex attractant in immensely dilute concentrations (Figure 13.4).

In the male polyphemus moth (*Telea polyphemus*) each antenna carries about 70 000 sensory organs with about 150 000 sensory cells. The female, in contrast, has much smaller antennae, each with only about 14 000 sensory organs and 35 000

FIGURE 13.3 The proboscis of the blowfly carries a large number of sensory hairs; these hairs can sense various chemicals and thus are taste receptors. At the base of the hollow hair are several sensory neurons, each more or les specialized in the range of compounds it responds to. One of them is always a touch receptor.

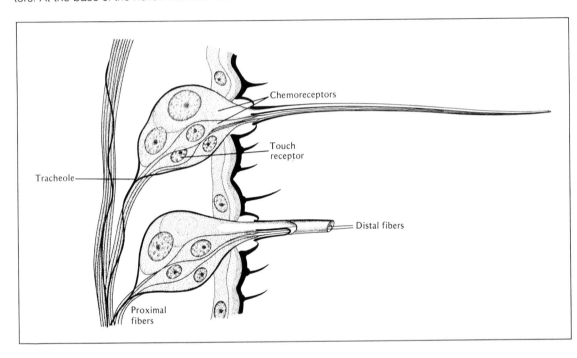

sensory cells (Boeckh et al. 1960). Approximately two-thirds of the receptor cells of the male are specialized for sensing the female sex attractant. The shape of the antennae makes them serve as sieves, virtually combing through the air for the appropriate molecules. With the aid of radioactively labeled sex attactant, it has been demonstrated that more than 25% of the sex attractant molecules from the air streaming through the antennae are absorbed (Schneider et al. 1968).

Experiments with male silk moths (*Bombyx*) have provided an estimate of how many molecules are necessary to cause a reaction in the moth. If the air contains 1 molecule of the sex attractant per 10^{15} air molecules, and the air flows at a speed of 60 cm per second for a total time of 2 seconds, the male responds. The number of molecules that reaches the antennae corresponds approximately to a level where each of the 10 000 receptors has received a maximum of one molecule of the attractant during the 2 seconds (Schneider 1969).

A male moth that senses a female from a distance of several kilometers cannot possibly use concentration gradients in the air to discover the direction in which he can find the female. The typical behavior of the male is to take off upwind, and this gets him started in the right direction. He then follows the aerial trail of the attractant in a characteristic zigzag flight across it.

Animals that follow an odoriferous trail commonly zigzag; when they pass out of the trail and lose it, they tend to turn in the reverse direction so that they enter the trail again. Leaving the trail thus automatically brings them back in again, so that

FIGURE 13.4 The large antennae of the male polyphemus moth (*Telea polyphemus*) are sensory organs that are highly sensitive to the sex attractant of the female moth. Their shape serves to attain contact with the largest possible volume of air.

ANTHERAEA MOTH The large antennae of the male moth (*Antheraea pernyi*) are used in locating the female of the same species by means of the sex attractant she gives off. [Courtesy of Muriel V. Williams, Harvard University]

MALE

FEMALE

their path becomes a sinuous crossing in and out of the trail. When they have reached the immediate vicinity of their goal, the final localization can take place with the aid of concentration gradients.

The localization of the source of a chemical signal (such as the sex attractant) at a long distance requires a long-lasting, relatively stable molecule. For alarm substances, which should act briefly and in a localized area, the molecule should be relatively unstable to avoid long-lasting residual effects that would be confusing. When the odoriferous substance is unstable and decays rapidly, it results in steep gradients in the vicinity of the source, which therefore is more easily pinpointed.

VIBRATION AND SOUND

In trying to define sound, we easily end up with a circular definition. We know that sound consists of regular compression waves that we perceive with

the ears; in turn, an ear can be defined as an organ sensitive to sound. However, compression waves can be transmitted in air, in water, and in solids, and we cannot restrict the word "hearing" to perception in air only.

To humans sound is very important and constitutes our primary channel of communication, but for some animals other kinds of mechanical disturbances are primary channels of sensory information.

A few examples will suffice. Snakes possess two sensory systems that respond to both airborne sound and substrate vibration. Their auditory system is not very sensitive to sound, but the sensitivity to head vibration is remarkable; Hartline has shown that at the best frequency a peak-to-peak amplitude of 0.1 nm is above threshold (Hartline 1971).

Another highly sensitive system is the lateral line of fishes. In a school of fish the movements of the individuals are extremely well coordinated and, to a human observer, they appear nearly perfectly synchronized. Fish in a school often swim at a constant pace and maintain characteristic individual distances, and the school as a whole executes complicated maneuvers that require individuals to respond exceedingly fast to velocity and direction changes of their neighbors. It is the mechanoreceptors of the lateral line that are used for monitoring the swimming speed and direction of travel of the neighboring fish. If a fish is blinded, it is nevertheless able to match the velocity changes of its neighbors; it shows as high a correlation as do controls. However, section of the nerves to the lateral line makes a fish unable to respond normally to its neighbors (Partridge and Pitcher 1980).

An elegant technique has demonstrated the importance of mechanical signals for sex discrimination in a water strider (*Gerris remigis*). Males of this species can produce surface-wave signals at about 90 Hz (hertz = cycles per second), and these signals attract receptive females. The male's surface waves can be mimicked by gluing a tiny magnet to the leg of a female water strider. Through an oscillating magnetic field the foreleg can be made to oscillate vertically, producing surface waves with the characteristic male frequency and amplitude.

If a masked male approaches a female with an inactive magnet, he will grasp her and readily mate with her. If the magnetic field is activated, he treats her as a male. If the magnetic field is activated only after he has grasped her, but before copulation has occurred, he immediately lets go, apparently believing he has made a homosexual mistake (Wilcox 1979).

Surface waves are used also by fish to establish direction and distance to prey, such as a struggling insect that has accidentally landed on the water surface (Bleckman and Schwarz 1982). The clawed frog, *Xenopus*, can readily learn to respond to surface waves between 5 and 30 Hz and to discriminate between frequencies if given appropriate rewards or light punishments. The *Xenopus* can learn, not only to discriminate between frequencies, but also to remember the exact frequency used during reinforcement training. This feat is an aquatic equivalent of absolute pitch in human hearing (Elephandt 1986).

In describing human hearing, we meet no great conceptual difficulties. Our ears are sensitive to regular compression waves in air in a range from approximately 40 to 20 000 Hz. However, a dog can perceive higher frequencies that are completely inaudible to humans, up to 30 000 or 40 000 Hz. Bats can perceive frequencies as high as 100 000 Hz, and we still consider it hearing. There are two good reasons. One is that dogs and bats have ears very similar to ours. The other is that the frequency range perceived is continuous; our own ear is just designed so that it is insensitive to the very high frequencies many other animals hear.

However, we know that compression waves at frequencies below those we think of as sound can often be perceived as a vibration (i.e., we still sense frequencies lower than those that we consider hearing).

When a fish senses footsteps at the edge of the water, is this hearing? The vibration from the steps is transmitted through ground and water, and these low-frequency vibrations are probably perceived by sensory organs in the lateral line rather than in the ear. Is it then correct to say that the fish "hears" or "feels" the approaching steps? In general, we tend to speak about hearing when animals have specially developed organs that are sensitive to what we consider to be sound, but this is obviously not a good way to define hearing.

Defining hearing becomes much more difficult in relation to invertebrate animals, in which the organs responsible for perception of vibration are very different from the vertebrate ear and in many cases have not been identified at all.

What kinds of information can be obtained from sound waves? Humans can perceive sound waves and their temporal pattern, determine intensity, distinguish different frequencies (pitch), and carry out very complex frequency analysis. We are also able to determine the direction from which sound comes. To some extent we can also obtain a rough idea of distance from the strength and the quality of familiar sounds.

We must keep in mind, however, that the sense organs of animals differ from our own and that there are possibilities for discrimination that are not intuitively obvious to us. For example, codfish can determine the direction of a sound source without ambiguity. Direction is obtained by the otoliths of the ear, which are strongly directional in their response to oscillatory movement. In addition, codfish apparently can also obtain information about the distance to a sound source, provided by the phase difference between particle acceleration and sound pressure, an achievement that our own ears are incapable of (Schuijf and Hawkins 1983).

Some animals use self-generated sound for obtaining detailed information about the physical structure of the environment, as we shall soon see.

The complexity of frequency analysis of which humans are capable is astonishing if we consider that what we hear basically is nothing more than a minute in-and-out motion of the eardrum. In spite of this fundamental physical simplicity, the average musical ear can distinguish individual instruments in an orchestra with a large number of simultaneously performing instruments.

Insects can produce and perceive a wide variety of sounds, which are used extensively for communication, often with the opposite sex. The hearing organs of insects vary greatly in structure and may be located in various parts of the body. Apparently, insect hearing organs are not sensitive to pitch the way our ear is; instead, information is transmitted mainly as changes in intensity, duration, and pattern of the sound.

Records of sound produced by male insects to attract females can be played back with so much frequency distortion that they are unrecognizable to the human ear, and still attract the female insects. The meaningful information seems to be carried in the pattern of pulses, not in pitch and tone quality. The pattern of insect sound is relatively fixed and is highly species-specific. It wouldn't be very useful for a male to attract females of a different species.

The human ear is inherently unsuited for perceiving the important features of insect sound. However, as electronic equipment for recording and analysis has become available, the pattern of insect song can be transcribed to visual patterns that we understand better and can evaluate more intelligently.

Directional information

Determining the direction of sound depends mainly on having two ears, separated in space. Some information is derived from the fact that the signals arriving at the two ears often differ in intensity, but more important is the difference in time of arrival of the sound from a given source.

For a complex signal the interpretation of direction may be brought about by a simultaneous comparison of intensity, frequency spectrum, phase, and time of arrival of the signals at the two ears. The comparisons must be made by the central nervous system, which must also be capable of separating the important signal from other noise in the environment (Erulkar 1972).

Humans cannot locate sounds as precisely as, for example, predators that normally work in the dark, such as cats or owls. The barn owl, for example, can locate prey in total darkness, using only the sense of hearing, with an error of less than 1° in both the vertical and the horizontal planes. For this accurate location of the source of a sound, the owl depends on frequencies about 5000 Hz, which are more directional than lower frequencies (Payne 1971).

Sonar

Owls use sounds coming from the prey to determine its location, but some vertebrates obtain information from the faint reflections, or echoes, of sound they themselves produce. This *echo location* or *animal sonar* is particularly well developed in bats, but exists also in whales, dolphins, shrews, and a few birds.

It has long been known that bats can fly unhindered around in a completely darkened room, and it makes no difference whether or not their poorly developed eyes are covered. This puzzle was re-solved in the 1940s by a graduate student at Harvard University, Donald Griffin.

One of the best studied bats is the big brown bat (*Eptesicus fuscus*). As this bat cruises in the dark, it emits pulses of sound that are inaudible to man, for the frequency is between some 25 000 and 50 000 Hz. Each pulse or click last about 10 or 15 ms and may be given off at a rate of some five single clicks per second. The bat obtains information about the environment from the echoes and is able to avoid all sorts of obstacles. In captivity it can even fly between wires strung across the room.

If the bat approaches a particularly "interesting" object, such as a flying insect, the number of clicks increases and the duration of the single click is shortened, so that the rate may be as high as 200 clicks per second, each click lasting less than 1ms.

One advantage of using a high-frequency sound (short wavelength) is that the directional precision is much better than for low-frequency sound, which spreads too widely and gives diffuse reflections unsuited for pinpointing accurately the location of objects. Furthermore, in order to reflect sound waves an object must be above a certain size, relative to the wavelength. The shorter the wavelength, the smaller is the size of the object that gives a distinct reflection, and the highest frequencies therefore permit the detection of the smallest objects.

The bat's impressive skill at avoiding fine wires can be tested in a long room where the animal can fly back and forth. If the center of the room is divided by vertical wires, spaced 30 cm apart, a bat flying without information about the wires has a 35% random chance of missing a wire. Wires several millimeters in diameter are successfully avoided, but if the diameter of the wires is reduced to less than 0.3 mm, the flying bat hits or touches the wires more often. If the wires are very thin, 0.07 mm, the number of hits and touches increases and the number of misses decreases to the chance value of 35%, the same as in experiments with bats made temporarily deaf by plugging their ears (Figure 13.5).

FIGURE 13.5 Obstacle avoidance of the little brown bat (*Myotis lucifugus*) when flying in darkness through a row of vertical wires spaced 30 cm apart. The statistical chance of missing the wires was 35%. The ability to avoid the wires deteriorated when the diameter of the wires was decreased to less than 0.3 mm. [Griffin 1958]

The shortest wavelength used by the little brown bat, which was used in these experiments, is about 3 mm (100 000 Hz). This means that it detects objects no more than about 0.1 wavelength in diameter.

Acoustic orientation is, of course, particularly useful for animals that are active in the dark, and it is also of importance in deep or murky waters. Both dolphins and whales use acoustic echoes to avoid colliding with objects and with the ocean bottom, as well as for finding food. When they swim in open water, there are few obstacles, but at night and at depths where little or no light penetrates, their sonar helps them avoid collision with the bottom. It is, of course, very valuable to be able to locate food both in the dark and when the visibility is low.

How dolphins use acoustic information has been tested in captive animals. A trained dolphin can find a dead fish thrown into the water and reach it within seconds from the opposite end of the tank, even if the water is so murky that a man cannot see the fish at all. If a panel of clear plastic is used to block its path, the dolphin can immediately find an opening, which it swims through without hesitation (Kellogg 1958; Norris et al. 1961).

One particular species, the blind river dolphin of the Ganges and Indus river systems, is particularly interesting. It normally swims on its side, probably a useful adaptation because the normal up-and-down movement of the tail fluke of dolphins might be impractical for an animal that spends much of its time very close to the river bottom. In the murky river water vision is of little or no use, and acoustic cues seem to give all the necessary information about the surrounding world. This dolphin's eyes are vestigial and lack a lens; they could at best serve as light detectors.

Other mammals that use acoustic orientation or echo location include the shrews, which employ sound pulses of short duration and high frequency (to over 50 000 Hz) (Gould et al. 1964).

At least two species of birds use echo location: the oilbirds (*Steatornis*) of South America and the cave swiftlets (*Collocalia*) of Southeast Asia. They are not closely related, but both live and nest in deep caves. The best known are the oilbirds or guacharos, which can fly around freely in dark caves without hitting the walls or other obstacles. They use a sonar system, much like bats, except that their sounds have a frequency of only about 7000 Hz. Therefore, they are fully audible to man, and their sound has been described as something like the ticking of a typewriter. If the birds' ears are plugged, they lose their sense of orientation in the dark, but they can still fly about in a lighted room and evidently use their eyes for orientation (Griffin 1953, 1954).

Infrasound

Echo location depends on frequencies in the range we can hear or much higher. What about frequencies below what we perceive as sound, so-called infrasound?

It has recently been discovered that homing pigeons can detect extremely low sound frequencies,

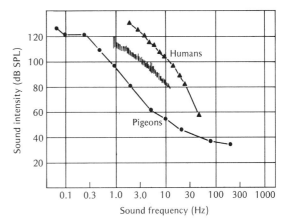

FIGURE 13.6 Thresholds for low-frequency sound detection by homing pigeons and by humans. The intensity of loud natural infrasound events is indicated by hatching. These are well within the hearing level of pigeons but too low for human perception. [Kreithen and Quine 1979]

as low as 0.05 Hz. As 1 Hz is 1 cycle per second, 0.05 Hz corresponds to 1 cycle per 20 seconds. The sensory response may be related to the structure of the ear because surgical removal of middle- or inner-ear structures reduces or eliminates the response to infrasound (Kreithen and Quine 1979).

The sensitivity to infrasound may be very important to the birds, for natural infrasounds originate from many sources, including thunderstorms, earthquakes, jetstreams, and wind over mountain ranges. The attenuation of sound is inversely related to the square of the wavelength, and infrasounds can therefore travel over long distances and can be detected at hundreds or even thousands of kilometers from their source.

The threshold for low-frequency sound detection by homing pigeons is shown in Figure 13.6, which also shows the much higher threshold for humans. The hatched region indicates the intensity of many naturally occurring infrasounds, well above the threshold for detection by pigeons, but well below what humans can perceive. The difference between the two curves is some 30 or 40 dB (deci-bels), or a 1000- to 10 000-fold difference in sound intensity.

We can now ask whether the unexpected sensitivity to infrasound is a component of the highly complex navigational system of homing pigeons. Could infrasounds that originate, for example, from strong winds blowing over a mountain range give useful geographical information? Direction of the sound cannot be obtained by comparing the phase shift or time of arrival for signals received by the two ears because we are dealing with wavelengths of several hundred meters.

However, a flying bird could use the differences in the Doppler shift of the infrasound frequency caused by its own flight velocity if it flew toward and then away from the source of the infrasound. A pigeon flying at 20 m per second can generate as much as a 12% shift in frequency. As its threshold for detection is 7% at 1 Hz, such a shift is fully adequate to allow the pigeon to determine directions (Quine and Kreithen 1981).

Infrasound, meaning sound at lower frequencies that we can hear, may be of greater importance than we have imagined. Records from captive Asian elephants show that they make calls with fundamental frequencies of 14 to 24 Hz, which is below the range of human hearing.

Many observers of wild African elephants have reported that these animals communicate with signals that humans can neither see nor hear; use of infrasound is a probable explanation. Records from African elephants in nature show that they make sounds at fundamental frequencies of 14 to 35 Hz (Poole et al. 1988). Because these low frequencies undergo relatively little attenuation in air, they may be audible to other elephants several kilometers away. In a forest habitat such signals would be well suited for communication between individuals in a herd of these highly social animals.

The discovery that some animals are sensitive to infrasound may lead to a reevaluation of the many anecdotal reports of abnormal animal behavior prior to earthquakes. Current information is at best suggestive, but because sensitivity to infrasound has just barely been discovered, it would be unwise to predict future developments.

Other mechanical senses

We have seen that hearing can provide a great deal of detailed information about the environment – in fact, far more than we initially would think. There is no doubt that those animals that use echo location obtain a three-dimensional impression of their immediate environment, somewhat analogous to what humans obtain through vision, except that the range is more limited.

Other senses provide additional information, but mostly of a less comprehensive nature. These include a variety of mechanical senses about which we have only rudimentary information.

Sensitivity to *touch,* which actually detects a deformation of the skin surface, is widespread and gives information about objects in immediate contact with the body. The sensation of vibrations is similar in nature. The sensations picked up by the lateral line of fish belong in the same category. The lateral line is, in·fact, a highly complex organ that is used for a multiplicity of purposes, among others to sense the rate of flow of water over the fish, thus giving information about speed (Cahn 1967).

Sensitivity to *hydrostatic pressure* has been demonstrated in a variety of aquatic invertebrates. Some are quite sensitive to small pressure changes, but the mechanism involved remains obscure. It is easy to understand how an animal that carries a volume of gas, such as the swimbladder, can be sensitive to pressure changes, for the resulting distortion can easily be sensed with the aid of mechanoreceptors. However, the majority of plankton organisms that are pressure-sensitive carry no bodies of air that can explain their high sensitivity to small pressure changes (Enright 1963).

Sensitivity to the force of *gravity* is extremely widespread and gives information about what is up and down. More complex animals have gravity receptors, known as *statocysts,* which are uniformly built on the same principle (Figure 13.7). A heavy

CILIARY BUNDLES IN FISH EAR Ciliary bundles of hair cells in the ear of a squirrel fish (*Adioryx xantherythrus*). The bundles, all oriented in the same direction, are from the saccule, one of the inner ear vesicles that in fish are associated with hearing. Each bundle contains 40 to 60 short, graded stereocilia. In these bundles the kinocilium (to the far left of each bundle) is only slightly longer than the stereocilia. [Courtesy of A. N. Popper, Georgetown University]

FIGURE 13.7 The statocyst of a pelagic snail. Within the statocyst, a heavy body of calcium carbonate, the statolith, rests on sensitive sensory cells. These respond to changes in the position of the statolith as the animal changes its position in the field of gravity. [Tschachotin 1908]

Statocyst nerve

Ciliary bundle

Statolith

body, the statolith, often of calcium carbonate, rests on sensory hairs. Any displacement relative to the force of gravity causes the statolith to stimulate the ciliary bundles of other sensory cells, which then send information to the central nervous system.

Angular acceleration is sensed in a similar way; fluid within a suitable structure is set in motion and deflects the ciliary cells in a direction that depends on the direction of acceleration. In vertebrates the organ responsible for sensation of angular acceleration is located in the inner ear and consists of the three semicircular canals, oriented in three planes at right angles to each other. They are filled with fluid, and any angular movement causes the inertia of the fluid to stimulate the sensory cells in the plane of the movement. With the canals oriented in three different planes, any angular acceleration can be perceived.

The receptor cells involved in hearing, in perception of position and equilibrium, and in the de-

FIGURE 13.8 Structure of the lateral line organs of fish. Top: several sensory cells, each carrying a sensory hair, are grouped together with the protruding sensory hairs covered by a jellylike cupula. Bottom: longitudinal section of lateral line. Movement of water causes deformation of the cupulae (dotted). [Dijkgraaf 1963]

FIGURE 13.9 The common sensory cells of mechanoreceptors, known as hair cells, have a longer kinocilium placed at one side of a group of smaller stereocilia. Deformation of the protruding cilia gives rise to action potentials in the sensory cell. [Flock 1971]

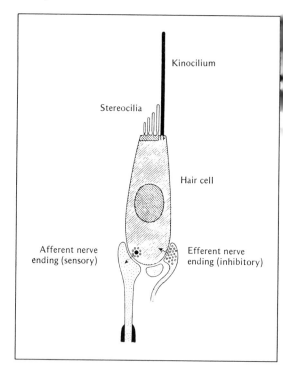

tection of water movements by the lateral line of a fish are all of the same basic type. These receptor cells possess ciliumlike processes, and movements of these processes lead to appropriate sensory outputs. Let us, as an example, examine the typical receptor cell in the lateral line of fish, a system used in the perception of movements relative to the surrounding water.

Each sensory receptor cell has two types of processes: a group of relatively short *stereocilia* and one longer *kinocilium*. The kinocilium is a true cil-

ium and contains two central and nine peripheral filaments (i.e., the characteristic structure of a cilium). The stereocilia, in contrast, lack this characteristic structure of true cilia.

These sensory cells are usually placed together in a group, and the protruding cilia are covered by a gelatinous projection, the *cupula* (Figure 13.8). When the cupula is deformed by movements in the adjacent water, the cilia are bent, and the movement excites the sensory cell.

The kinocilium is always placed to one side of the group of stereocilia (Figure 13.9), and because of this orientation, the response of the receptor cell is polarized. The sensory cell is excited when the cilia are bent toward the side where the kinocilium is located. In the lateral line adjacent hair cells are oriented back-to-back: Some are therefore stimulated by the movement of water in one direction; other hair cells are stimulated by movement in the opposite direction.

LIGHT AND VISION

Vision undoubtedly can give more detailed information than any other sense. With the aid of suitable accessory structures (eyes), light can form a detailed image of the environment, both near and distant.

Information obtained with the aid of light depends on differences in *intensity,* for a uniformly luminous environment conveys no information of interest. Much additional information is contained in differences in *wavelength* or *color.* Finally, for animals that have appropriate accessory organs the *plane of polarization* of the light can also be highly informative.

As far as animals are concerned, the transmission of light is so fast (300 000 km per second) that it involves no delay. This means that changes in the environment can be perceived instantaneously, except for the time used in transducing the signal and transmitting it to the central nervous system. Compare this, for example, with the slowness with which a change in an olfactory signal is perceived.

Image-forming eyes

Sensitivity to light is extremely widespread; even many unicellular organisms are sensitive to light. Multicellular animals usually have this sensitivity concentrated in certain spots (eye-spots). If an eye-spot is shielded from light on one side, it becomes direction-sensitive as well.

More highly organized animals have increasingly complex light-sensitive structures that, in their ultimate form, are excellent image-forming eyes. Well-developed image-forming eyes are found in five different animal phyla: coelenterates, annelids, molluscs, arthropods, and vertebrates. The various eyes differ in design and development, and it must be assumed that they have evolved independently in these animals.

An image-forming eye can be based on one of two different principles: either a multifaceted eye as in insects, or a single-lens, camera-like structure as in vertebrates (Figure 13.10). The multifaceted eye has the same angle of resolution whether an object is distant or close by, but the lens-type eye must have some device for focusing if it is to perform equally well at different distances. It is interesting that among the few worms that have well-developed, image-forming eyes, both single-lens and multifaceted eyes occur.

Wavelength sensitivity

What humans perceive as light lies within a very narrow wavelength band, 380 to 760 nm, out of the wide spectrum of electromagnetic radiation that ranges from the extremely short gamma rays to long-wave radio waves (Figure 13.11).

All other animals' visual sensitivity lies within or very close to the same range of wavelengths as humans'. The remarkable fact is that not only animals, but also plants, respond to light within this same range. This includes both photosynthesis and phototropic growth of plants.

The reason for this universal importance of a very narrow band of electromagnetic radiation is simple. The energy carried by each quantum of radiation is inversely related to the wavelength. Therefore, the longer wavelengths do not carry sufficient energy in each quantum to have any appreciable photochemical effect, and the shorter wavelengths (ultraviolet and shorter) carry so much energy that they are destructive to organic materials. The universal biological use of what we know as ''light'' is a result of the unique suitability of these particular wavelengths.

The range is not exactly the same for all animals. The vision of insects, for example, extends into the near ultraviolet range, to slightly shorter wavelengths than the vertebrate eye. This is evident from the ability of honeybees to distinguish

FIGURE 13.10 The eyes of vertebrates and some invertebrates have a lens and function on the same principle as a camera. Insects have compound eyes in which the single elements combine to form the image (a single ommatidium is shown in detail).

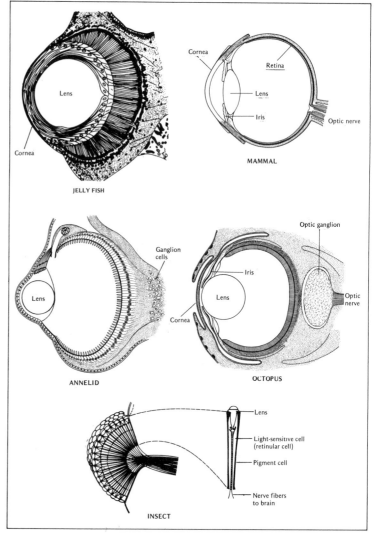

any spectral color between 313 and 650 nm from white light, an ability that is unrelated to light intensity and therefore must depend on wavelength discrimination (i.e., ''color'' vision) (Kühn 1927).

The mammalian retina is sensitive to ultraviolet light, but these wavelengths do not penetrate to the retina, primarily because of a slight yellowness of the lens, which acts as a filter. Persons whose lens has been surgically removed because of cataract can perceive light in the near ultraviolet that previously was invisible to them.

This narrowing of the band of wavelengths that can be perceived is probably advantageous. A lens, if made of a uniform material, refracts short wavelength radiation more strongly than longer wavelengths. This means that the various wavelengths

FIGURE 13.11 The spectrum of electromagnetic radiation ranges from the shortest cosmic rays to long-wave radio waves. Only a narrow band between about 380 and 760 nm is perceived as light. The energy carried by each quantum of electromagnetic radiation increases 10-fold for each 10-fold decrese in wavelength. V, violet; B, blue; G, green; Y, yellow; O, orange; R, red.

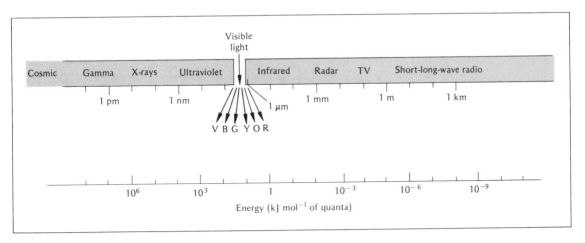

Light absorption: retinal pigments

For the sensory cells of the retina to be stimulated by light and respond, a sufficient number of light quanta must be absorbed. This is achieved with the aid of a light-absorbing pigment. The best known such pigment is *rhodopsin,* which is found in the rods of the vertebrate eye. (The rods are the retinal elements responsible for vision in dim light; the cones, which will be discussed below, are involved in vision in bright light and the perception of color.)

Rhodopsin can be isolated from retinas of animals whose eyes are fully dark-adapted. It is light-sensitive, and when exposed to light breaks down to *retinene,* a molecule closely related to vitamin A, and a protein called *opsin.* This process is the initial step in the perception of light by the rods of the vertebrate eye.

The strongest argument for the role of rhodopsin in the visual process is that, if the rhodopsin in the retina is depleted by exposing the eye to strong light, the eye is insensitive to dim light until the rhodopsin has regenerated. Another strong argument for the role of rhodopsin comes from its absorption spectrum, which has its maximum at 497 nm. This maximum and the detailed absorption spectrum coincide with the maximum sensitivity and the sensitivity spectrum of the human eye in dim light (Figure 13.12).

It has been more difficult to establish the role of pigments in the perception of light by the cones, primarily because of difficulties in extracting appropriate pigments. It has long been understood that

cannot be brought into focus simultaneously. In man-made lenses this difficulty is known as chromatic aberration. It can be corrected for by the use of composite lenses consisting of several elements with different refractive indexes. For an eye that does not have color correction, the simplest way of reducing the difficulty of simultaneously focusing different wavelengths is to narrow the wavelength band that is permitted to enter. It is then best to eliminate the shortest wavelengths, for which the refraction error is the greatest. In insects, which have nonfocusing composite eyes, this is irrelevant, for the resolution is determined by the angular distance between the single elements of the eye.

FIGURE 13.12 The relative sensitivity of the dark-adapted human eye at different wavelengths (open circles) compared with the absorption spectrum of the visual pigment. [Crescitelli and Dartnall 1953]

INSECT VISION What a bee can perceive of the color pattern of a flower is remarkably different from what the human eye sees. To the human eye the marsh marigold (*Caltha palustris*) appears as an even, bright yellow flower (top photograph); an ultraviolet photograph of the same flowers (bottom) reveals a strong pattern that serves as a nectar guide for bees in search of food. [Courtesy of Thomas Eisner, Cornell University]

the perception of color requires more than one pigment, and that three pigments with absorption peaks in blue, green, and red are sufficient to explain the perception of all colors we see (the trichromatic vision theory).

The difficulties have been overcome by making measurements of the absorption spectra directly in the human eye. By using an ingenious instrumentation, measurements on single cones have revealed that there are three kinds of cones: blue-sensitive, green-sensitive, and red-sensitive cones. Presumably, these three types of cones are responsible for human color vision in bright light (Brown and Wald 1964). The maximum absorbance for the three types has been determined as: "blue" cones 420 nm, "green" cones 534 nm, and "red" cones 563 nm (Bowmaker and Dartnall 1980).

The spectral sensitivity of the human eye is in accord with the three-pigment theory and the measured absorption spectra. Furthermore, what is known about deficiencies in the color vision of hu-

mans (often called color blindness) is satisfactorily explained on the basis of the existence of these same three pigments (Wald 1964; Rushton 1972).

FIGURE 13.13 The spectral sensitivity of the eye of the sphingid moth *Deilephila* indicates the presence of three visual pigments, one of them suitable for ultraviolet reception and none for the red part of the spectrum. [Höglund et al. 1973]

It is well established that many insects can distinguish ultraviolet light. The moth *Deilephila* has ultraviolet receptors with maximum absorption sensitivity at 350 nm, well into the ultraviolet part of the spectrum (Figure 13.13). This moth lacks a red-sensitive pigment, in accord with the lack of sensitivity to red among many insects.

It has long been thought that birds have color vision, and it is well established that hummingbirds and pigeons perceive ultraviolet light. In addition to the usual red-, green-, and blue-sensitive pigments, they have a fourth pigment that is sensitive to shorter wavelengths (Fager and Fager 1981). A more recent study of a wide range of bird species from 10 different families has revealed that they have cones that are maximally sensitive at 370 nm, well within the ultraviolet region (Chen et al. 1984).

Polarized light

One quality of light that humans are unaware of is its plane of polarization.* For this reason, the

*In polarized light the vibrations of the propagated wave (ray) are all in one plane, perpendicular to the direction of the ray. In ordinary unpolarized light the vibrations are in all directions

information carried by the polarization of light and its importance for animals remained unnoticed until the German zoologist Karl von Frisch discovered that honeybees use the plane of polarization as a directional cue.

Von Frisch studied communication among bees and how information is relayed from a bee that has found a good food source to other bees. Usually the bees indicate the direction of the food source relative to the direction of the sun (i.e., they use the sun as a compass), and they have an internal clock that compensates for the movements of the sun across the sky during the day.

Von Frisch found, however, that the bees could orient correctly, even in the absence of a visible sun, provided a small piece of the blue sky could be seen. The light from the blue sky is polarized to a degree that gives information about the actual position of the sun, and this is used by the bees. One way to show the dependence on the plane of polarization of the light is to reflect the light by a mirror. This makes the bees orient in the opposite of the correct direction, as expected from the reversal of the plane of polarization caused by the mirror. Also, by imposing filters that change the direction of polarization, any degree of deviation from the correct orientation can be imposed on the bees (von Frisch 1948).

After the importance of polarized light had been demonstrated, it was found that insects in general are sensitive to polarized light. Surprisingly, several aquatic animals are also sensitive to the polarization of light. This was first observed in the horseshoe crab, *Limulus* (Waterman 1950). Many other aquatic animals, including octopus and fish,

(yet perpendicular to the ray). Unpolarized light may become polarized by reflection from water and other nonmetallic surfaces and by transmission through certain materials. The plane of polarization can carry information not contained in unpolarized light. It appears that, in nature, the most important information animals can obtain in this way is attributable to the fact that the light from the blue sky is polarized, the polarization at any point in the sky depending on its position relative to the sun. Indirectly, this tells the location of the sun, information that is most important in animal orientation and navigation.

are also sensitive to polarized light (Waterman and Forward 1970).

The observation that homing pigeons can be trained to discriminate the plane of polarization of light has added another possible cue to the several natural sources of information that they can use in their amazing ability to orient and home correctly from unfamiliar sites of release (Kreithen and Eisner 1978). There is now a substantial amount of evidence that, for a large number of both terrestrial and aquatic animals, the ability to perceive the plane of polarization is important in orientation and navigation.

Infrared and thermal energy

When walking up to a hot stove, we clearly perceive a sensation of warmth that we interpret as radiating from the stove. This is indeed what the warmth is: radiation in the infrared part of the spectrum. How is it we perceive this radiation that our eyes cannot see? What is it our senses tell us when we evaluate the hot or cold condition of our surroundings?

We think we are able to sense the thermal condition of the environment, and our behavior often seems to indicate that this is so. What is sensed, however, is not the environmental temperature, but the temperature of the skin at the depth of the appropriate receptors. A good example of our inability to judge air temperature is provided by the kind of electric hand drier that is installed in some public washrooms. The drier blows a stream of hot air (as high as 100°C) over the hands, and as long as the skin is wet, the air feels moderate, but as soon as the skin is dry, the airstream becomes painfully hot.

Heat and cold are perceived by different sense organs in the skin. This can be shown by an examination of the activity in the individual nerve fibers from the temperature-sensitive skin receptors. The fibers from a heat-sensitive receptor show no activity until the temperature increases above a certain point. From then on the activity increases, roughly in proportion to the temperature rise, until

FIGURE 13.14 Response curves obtained from single warm-receptors and cold-receptors in the scrotal skin of rats. [Iggo 1969]

a certain upper limit is reached, above which the nerve activity again decreases.

Records obtained from the scrotal skin of the rat show exactly this (Figure 13.14). The warm-receptors are "silent" below approximately 37°C. Above this temperature their output increases rapidly with increasing temperature, but above 43°C their output falls off. Below 37°C the warm-receptors remain silent, and cold-receptors show an increasing response, again roughly proportional to the decrease in temperature. Below a certain level, the cold-receptors show a decreased activity that eventually falls to zero.

Within the narrow neutral range, which corresponds to a normal scrotal skin temperature for a rat in heat balance, there is a low level of activity from both warm- and cold-receptors. A minor change in either direction can therefore be perceived with rapidity and precision. This provides the makings of a sensitive control system based on obtaining accurate information about slight changes in skin temperature and hence on changes in heat transfer between the skin and the environment.

Infrared radiation

Cutaneous temperature receptors of the kind we have just discussed are sensitive to the local skin temperature. They give information about the environment only indirectly, for skin temperature is a result of heat flow from the core of the animal as well as the conduction and radiation to the environment. They discover radiation only through its effect on skin temperature, as we know when we can feel the radiation from a hot object. To some extent we can also sense a loss of heat by radiation. For example, a cold brick wall in a house may give a definite feeling of ''cold''; we can feel that the chill comes from the wall and we often interpret it as a ''cold draft,'' although obviously no air flows through the wall.

Infrared radiation can be perceived directly by a few animals that have specialized sense organs that respond to this type of radiation. The so-called *facial pits* or *pit organs* on the head of some snakes are such specialized infrared receptors.

When a rattlesnake strikes, the direction of the strike seems to be guided by the infrared radiation from its prey. A rattlesnake strikes only at warm-blooded prey, and when the prey is dead and at room temperature, the snake will not strike. However, a blindfolded snake strikes correctly at a dead rat that is pulled across the cage, provided the rat is warmer than the surroundings. Blindfolded, the snake cannot be guided by vision. Nor is it guided by the sense of smell, for it will strike correctly even at a moving, cloth-wrapped electric bulb. The pit organ is evidently involved in sensing the location of warm objects.

All snakes that have pit organs feed preferentially on warm-blooded prey, and this further supports the view that these organs are infrared sensors. In the rattlesnake the pit organ is located, one on each side, between the nostril and the eye; it is richly innervated, and this in itself suggests a sensory role for the organ.

The sensitivity of the facial pit has been examined by recording the activity in the nerve leading from the organ. A variety of stimuli, such as sound, vibration, or light of moderate intensity (with the infrared part of the spectrum filtered out), has no detectable effect on the activity in the nerve. How-

FIGURE 13.15 The pit organs on the head of a rattlesnake are specialized receptors for infrared radiation used to locate warm-blooded prey. The two pits are located, facing forward, one on each side of the head, between the eye and the nostril. The pit has an outer and an inner chamber, separated by a thin membrane. [Gamow and Harris 1973. Copyright © 1973 by Scientific American, Inc. All rights reserved]

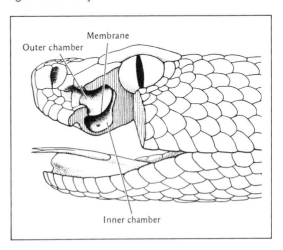

ever, if objects of a temperature different from the surroundings are brought into the receptive field around the head, there is a striking change in nerve activity, regardless of the temperature of the intervening air (Bullock and Cowles 1952; Barrett et al. 1970).

How is the infrared radiation sensed? The pit is covered by a thin transparent membrane (Figure 13.15), and it has been suggested that a rise in temperature in the chamber behind the membrane could cause an expansion of the gas with a consequent deformation of the membrane. This in turn could be sensed by a suitable receptor. This hypothesis is highly improbable, for a cut in the membrane that opens the chamber to the outside air causes no loss in responsiveness, a result that is incompatible with the hypothesis that a pressure change is sensed.

We are left with two other possibilities to consider: Either the effect is photochemical, which means that the infrared radiation is absorbed by a specific compound, analogous to the light-sensitive pigments in the eye, or the pit organ is sensitive to the slight temperature rise caused when infrared radiation reaches it.

The infrared radiation emitted from a mammalian body has its peak around 10 000 nm, and the low quantum energy in this wavelength makes any photochemical effect on a pigment extremely unlikely. Pure infrared radiation can be produced by a laser, and experiments with such radiation of known wavelength provide strong evidence that the mode of reception in the facial pit organ is entirely thermal (Harris and Gamow 1971).

Can the pit organs be used for stereoscopic perception of an infrared source, the way our two eyes are used for steroscopic vision? This seems highly likely, not only from observations of the precision with which a snake can strike, but also from studies of its brain activity. When infrared radiation falls on the facial pit organ, electric potentials can be recorded from the optic tectum, the part of the brain with which the optic nerve is connected. This in itself is interesting; although the nerve from the pit organ (a branch of the trigeminal nerve) is completely separate from the optic nerve, the same part of the brain seems to handle visual and infrared information.

Many of the neurons in the tectum respond to stimulation of the pit organ on the opposite side of the head. This is reminiscent of the way information from the eyes is handled; the crossover of the optic nerve in the optic chiasma is essential for stereoscopic vision and interpretation of distance.

Information from the two pit organs is apparently coordinated and interpreted in a similar way, a conclusion in agreement with recorded changes in the neural activity in the tectum when the position of the infrared source is changed. The response is particularly noticeable when the source is in a position to irradiate both pits at once. It therefore appears that the facial pits indeed provide stereoscopic perception and substantially aid in the precision of estimating the location of prey (Goris and Terashima 1973).

Some of the neurons in the optic tectum respond only to stimulation of the pit organ and others only to sensory input from the retina. However, some of the neurons respond only to simultaneous input of the two different modes of stimulation. Such bimodal neurons are probably important in the central processing of sensory information that is useful in giving attention to and recognizing objects that emit both thermal and visual signals (Newman and Hartline 1981).

It is unlikely that specialized infrared-sensitive organs are of widespread importance among animals. In particular, it is almost unthinkable that any similar organ could function in aquatic animals, partly because of the very low penetration of infrared radiation in water, and also because the direct contact of the body surface with a medium of high thermal conductivity and thermal capacity would make it impossible to perceive the small amounts of heat involved.

ANIMAL ELECTRICITY

It is well known that some fish can produce strong electric shocks. Ancient peoples such as the Greeks and Egyptians knew and wrote about the powerful shocks delivered by the electric ray (*Torpedo*) and the electric catfish (*Malapterurus*), but the phenomenon must have appeared utterly mysterious until the nature of electricity became known.

Powerful electric discharges serve obvious offensive and defensive purposes. The strongest shocks are produced by the South American electric eel (*Electrophorus*). It can deliver discharges of between 500 and 600 V, which are powerful enough to kill other fish and possibly animals the size of a person. Such strong electric discharges can be delivered only by a few species of fish. These are not particularly closely related: Some are teleosts; others are elasmobranchs; and they occur both in fresh water and in the sea.

As electric fish have been studied more closely, it has become apparent that even weak discharges – too weak to have any direct effect on other fish – serve a variety of useful purposes. Obviously,

FIGURE 13.16 Feeding responses of a small shark (*Scyliorhinus canicula*). (*a*) When a hungry shark passes in the vicinity of a flounder completely buried in the sand, it detects the flounder and immediately attacks it. (*b*) To exclude olfactory cues, the flounder is covered with an agar chamber perfused with water that exits at some distance (broken arrow). The shark nevertheless attacks the correct location of the prey. (*c*) When pieces of fish are placed in the agar chamber, the shark searches for food where the perfusing water exits. (*d*) If the agar chamber is covered with a plastic film, the signal is attentuated and the shark passes by without noticing. (*e*) An artificial electric field of the same magnitude as generated by the breathing movements of the flounder excites the shark, which immediately attacks this "prey." [Kalmijn 1971]

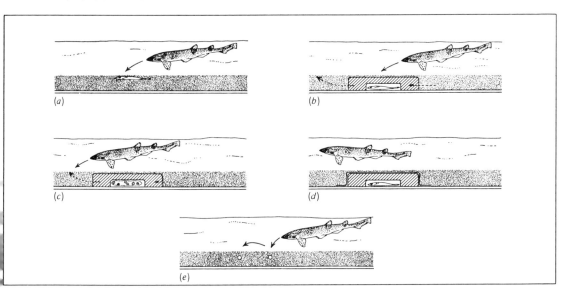

very weak discharges cannot stun prey and are useless as weapons of offense or defense. However, it is now clear that they can be used to obtain information about the environment in general.

Electric discharges are particularly suitable for distinguishing between nonconducting objects and good conductors, such as another animal. A freshwater animal, say, a fish, has a much higher electric conductivity (because of the salt content of its body fluids) than the poorly conducting water. Weak electric discharges can be used for navigation in murky water, as well as for the location of predators and prey. Furthermore, electric discharges are used in communication between individuals, and in some cases a fish will respond with sexual behavior only to the electric signals from the opposite sex (Hopkins 1972).

Finally, it has recently become known that many fish that cannot produce electric signals themselves are quite sensitive to the weak electric activity that results from the ordinary muscle function of other organisms. In this way sharks and rays are able to locate another fish, even if it is at rest. A hungry dogfish (the shark *Scyliorhinus*) will even respond to a flounder that is resting at the bottom of an aquarium, completely covered by sand. If the shark passes within less than 15 cm of the flounder, it makes a sudden turn toward the hidden prey, removes the sand by sucking it up and expelling it, and seizes the flounder (Figure 13.16).

Careful experiments have shown that this ability to locate the flounder is not associated with the sense of smell. Apparently the dogfish responds to the minute electric potentials produced by the breathing movements of the hidden flounder, for it will react to an artificial discharge of 4 μA (the same order of magnitude as that produced by a living flounder) and try to dig out this imaginary prey (Kalmijn 1971).

To summarize, electric fish can use their discharges (1) for defense and to stun prey, (2) to scan their environment and locate enemies or prey (much more common), and (3) for communication. Furthermore, nonelectric fish can use their extremely sensitive electroreceptors to locate prey by means of the action potentials caused by normal muscle activity.

Production of the electric discharges

In nearly all electric fish the discharges are produced by discrete electric organs, which consist of modified muscle. These organs have been carefully studied in the electric eel, which can produce discharges of over 500 V. The large electric organs of the electric eel run along most of the body, one large mass on each side, and make up about 40% of the animal's volume.

These organs consist of thin, waferlike cells, known as *electroplaques* or *electroplates*. The electroplaques are stacked in columns that may contain between 5000 and 10 000 plaques, and there are some 70 such columns on each side of the body. The two faces of each electroplaque are markedly different; one face is innervated by a dense network of nerve terminals, and the other is deeply folded and convoluted.

When the organ is at rest, the two surfaces of the electroplaque are positively charged on the outside, each being +84 mV relative to the inside (Figure 13.17). The overall potential from the outside of one membrane to the outside of the other therefore is zero. During a discharge, however, the potential on the innervated surface is reversed, and the total voltage across the single electroplaque becomes about 150 mV.

With the serial arrangement of the electroplaques, the voltage adds up as when we connect a number of batteries in series. With several thou-

FIGURE 13.17 The electric organs of fish consist of thin, waferlike cells (electroplaques) stacked in columns of several thousand. When a plaque is inactive and at rest (top), both faces are positively charged, the outsides being at +84 mV. Therefore, no potential difference exists between the two outside faces. During discharge (bottom) the potential on the posterior face of the plaque is reversed and becomes −67 mV on the outside. The potential difference between the two outside faces therefore is 84 + 67 mV = 151 mV. [Keynes and Martins-Ferreira 1953]

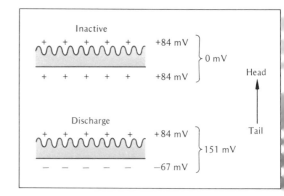

sand electroplaques arranged in this way, the electric eel can reach several hundred volts. (Because of internal losses the voltage does not reach the full magnitude of an ideal serial connection of the electroplaques.)

The arrangement of electroplaques in parallel columns increases the current flow. In the electric eel the current may be about 1 A. The giant electric ray (*Torpedo*) lives in sea water, which has a much lower resistance than fresh water. Its voltage is lower (some 50 V), but because it has some 2000 columns in parallel in each electric organ, the total current is extraordinarily high, up to 50 A. A current of 50 A at 50 V corresponds to a power of 2500 W during the discharge (Keynes and Martins-Ferreira 1953). A much larger number of fish produce very weak discharges that could not possibly serve any defensive action or disable prey. They

ELECTRIC FISH A pair of the South American electric fish *Sternarchoramphus*. The background shows the simultaneous record of the discharge from the electric organs. The top trace begins at a rate of about 5 Hz. [Courtesy of Gerald Langner, Technische Hochschule, Darmstadt]

can, however, be used both for communication and to obtain information about the environment.

Electroreceptors

Most electric fish live in dark and murky water where the visibility is poor, and they are often nocturnal and have poorly developed eyes. The value of an electric sense is therefore obvious. It permits scanning of the environment when vision is inadequate, and it is independent of the day-and-night light cycle. The greatest disadvantage of an electric sense is its very limited range – usually no more than a few meters.

One of the first fish in which electric reception was studied in detail is the African electric fish, *Gymnarchus niloticus* (Machin and Lissman 1960). This fish, which lacks a common name, discharges a continuous stream of pulses at a frequency of about 300 to 400 per second. During each discharge the tip of the tail is momentarily negative with reference to the head, so that an electric current flows into the surrounding water. The configuration of the electric field depends on the conductivity of the

surroundings and is distorted if an object with higher or lower conductivity than the water is introduced into the field. The lines of current flow converge on an object of higher conductivity than water and diverge around a poorer conductor.

What are the electroceptive organs responsible for this performance? The receptors are located in the skin and are of two kinds: tuberous and ampullary. *Tuberous receptors* are found only in electric fish. They respond specifically to the high-frequency discharge rates (several hundred hertz) characteristic of electric fish, and in each species the receptors' frequency is in the range of the characteristic discharge rate of the same species (Figure 13.18).

The other type, *ampullary electroreceptors,* are found in both electric and nonelectric fish. They respond to much lower frequencies and to changes in direct-current fields. They open externally through minute pores in the skin that through jelly-filled canals lead to ampullae that contain the sensory cells. In marine sharks and rays the canals and receptors are very conspicuous and were known as

FIGURE 13.18 The electroreceptor units of electric fish have their best receptive frequencies in the same range as the normal electric discharge frequency of the same species. [Hopkins 1976]

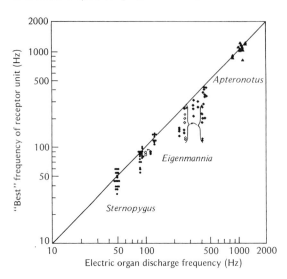

FIGURE 13.19 Location of sense organs on the head of the dogfish *Scyliorhinus*. Openings of the ampullae of Lorenzini are shown by black dots. The open circles indicate the pores of the lateral line system and the black lines its location. [Dijkgraff and Kalmijn 1963]

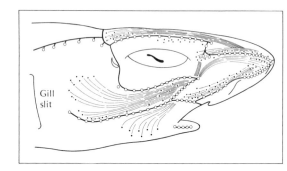

ampullae of Lorenzini long before their function was understood (Figure 13.19). In fresh-water electrosensitive fish, including fresh-water rays, the canals are much shorter, and the less conspicuous organs are referred to as microampullae.

The difference in electroreceptor structure between marine and fresh-water fish is easily understood. In sea water, where the fish has a higher resistivity than the surrounding water, the current lines diverge around the fish. A long canal filled with jelly of high conductance helps to maximize the voltage drop across the sensory unit.

In fresh water, the current lines converge on the fish because of its lower resistivity. A long canal could function if there was a high skin resistance and a low canal resistance, but this is difficult to achieve. It would require either a high salt concentration in the gel to keep its resistance low, or an inordinately high wall insulation. Accordingly, all fresh-water electrosensitive fish have receptors with canals just long enough to penetrate the skin (Figure 13.20).

The electroreceptors can be remarkably sensitive and respond to voltage gradients as low as 10^{-8} V per centimeter. The meaning of negative powers of 10, such as 10^{-8}, is not always intuitively obvious to us; the following illustration may help. Imagine an ordinary 1.5-V flashlight battery placed with one end in New York and the other end 1500 km away in Florida; the voltage gradient will be 1 V per 1000 km, or 10^{-8} V per centimeter. This is the magnitude of the minute fields sensed by the fish electroreceptors!

What about interference caused by electric signals from other fish? If many individual electric fish live in the same area, how can they avoid confusing their own signals with those of others? If an electric fish is subjected to an artificial electric pulsing of the same frequency as its own, it responds by shifting its frequency away from the artificial signal. This presumably can prevent the fish from confusing its own signals with those originating from other fish. This conclusion is supported by the tendency of an electric fish in nature to shift and find its own individual private frequency when approached by another fish with the same frequency (Bullock et al. 1972).

The ability to use self-generated impulses to scan

FIGURE 13.20 Diagram of three different types of cutaneous ampullary receptors of an electric fish (*Eigenmannia virescens*). The receptor cells (shown stippled, with their nerves in black) are located at the bottom of a canal connecting to the surface of the skin. [Lissmann and Mullinger 1968]

the environment has a superficial similarity to the use of sound pulses by bats. The electric system is different, however, because there is no reflection or echo involved; it depends on the distortion of the electric field.

It has been assumed that only fish possess an electric sense, but there is now excellent evidence that the lateral line system of salamanders, for example the axolotl (*Ambystoma*), also provides electroreception. The lateral line of the axolotl has two types of sensory units, electrosensitive and mechanosensitive. The mechanosensitive units are extremely sensitive to movements in the water. The electrosensitive units, which react to minute voltage gradients, do not respond to any but the roughest mechanical stimulation (Münz et al. 1984).

The function of the electrosensitive units in salamanders resembles those in fish, in both frequency range and sensitivity. The axolotl feeds on small invertebrates, tadpoles, and fish, and the currents generated by these animals are of the same order of magnitude as the sensitivity of the electroreceptive system of the axolotl.

Another vertebrate that feeds on live aquatic organisms is the egg-laying mammal, the duck-billed platypus. The platypus seeks its food in muddy streams, and when it dives it usually has eyes, ears, and nose shut. It has long been known that the bill of the platypus contains densely packed receptor organs, and it has been assumed that these are extremely sensitive mechanoreceptors. This is undoubtedly true, but recently the presence of electrosensitivity has been demonstrated in the platypus. It can home in on weak electric fields, whether emitted by prey animals or by an artificial source (Scheich et al. 1986).

It is most likely that the electroreception in the platypus has evolved independently from electroreception in fish. This appears so because the innervation of the electrosensitive units in the platypus is carried in the trigeminal, the fifth cranial nerve, while in fish the innervation of the electrosensitive organs is in the acoustic, the eighth cranial nerve (Gregory et al. 1987).

Electromagnetic sensitivity

It was long ago suggested by the English zoologist Hans Lissmann that the electrosensitivity of fish is high enough to serve another purpose: the detection of the magnetic field of the earth.

Whenever a conductor moves through a magnetic field, an electric current is induced. A fish is a conductor, and when it swims through the earth's magnetic field, the induced current is well within the sensitivity of electroreception. If a fish moves passively with an ocean current, the polarity of the electric field may inform the drifting animal of the upstream or downstream direction. If a fish is swimming actively, the induced current may indicate the actual compass directions.

An experimental test of the ability to use electromagnetic information has been carried out by Kalmijn (1978). He placed stingrays in a circular fiberglass pool and trained them to receive a food reward in a small enclosure in magnetic east; they were punished for entering a similar enclosure in the west. The magnetic nature of a trained animal's response became evident when the magnetic field was reversed. The ray would now enter the enclosure in new magnetic east, although this was the enclosure where it had previously been punished. The field sensed by the ray is induced by its own swimming movements and therefore shows what is called active electromagnetic orientation. Later studies have demonstrated that the electromagnetic sensitivity, as might be expected, is located in the ampullae of Lorenzini.

The electromagnetic sensing of marine fish depends on the high conductivity of sea water. Because of the much lower conductivity of fresh water, electromagnetic sensing is less promising for freshwater fish, and the principle should not apply at all to land animals because the insulation provided by air precludes any channel for the return current. What is interesting is that entirely different principles may apply to the ability of land animals to perceive the magnetic compass direction.

Perception of magnetic fields

There is a great deal of anecdotal and semianecdotal information about animals able to sense the magnetic compass direction. Homing pigeons most definitely obtain compass direction from the sun. However, in heavy overcast weather, many pigeons can still home successfully. Possibly they could use the magnetic field of the earth as a backup compass. This suggestion is supported by experiments in which artificial magnetic fields have been imposed on the pigeons.

In laboratory experiments, where single environmental cues can be better isolated and controlled, the effect of magnetic fields can be studied more accurately. The European robin (*Erithacus rubecula*) displays a period of migration restlessness in the spring and again in the fall. If it is placed in an indoor circular cage, it tends to move toward the direction that would be its natural migratory direction for that time of year.

The preferred flight direction is maintained in a completely closed room without any optical reference points or other known cues that would make the bird orient correctly. However, if the test cage is moved to an all-steel chamber, which provides

a complete shield to magnetic fields, the birds can no longer find the natural migratory direction (Fromme 1961). Furthermore, artificially generated magnetic fields influence the directional choice made by the bird. Caged indigo buntings orient during their migration period according to the natural geomagnetic field, orienting toward the north. If the horizontal component of the magnetic field is artificially deflected 120° clockwise, the birds orient accordingly toward east-southeast (Emlen et al. 1976). There are also reports that free-flying nocturnal migrants are affected by natural disturbances in the geomagnetic field (Moore 1977).

It appears that these birds are not sensitive to the polarity of the magnetic field; that is, they do not sense north the same way a compass needle points north. Instead, they seem to use the inclination of the axial direction of the field lines for deriving the information about north–south polarity (Wiltschko and Wiltschko 1972).

The magnetic sensitivity of honeybees has been studied in connection with their foraging activities, and a recent discovery may open the way for new experimental approaches. Honeybees carry in their abdomen discrete bodies of ferromagnetic material, presumably magnetite (Gould et al. 1978). This suggests that their magnetic sense is distinctly different from the electromagnetic, or induced field sensitivity, described above. Similar magnetic material, presumably magnetite (Fe_3O_4), has also been identified in pigeons (Walcott et al. 1979) and in a variety of other animals, including birds, mammals, amphibians, and fish. The presence of particles of magnetic material by no means proves that they play any role in magnetic orientation; it will be necessary to establish how they work and what role they play, if any.

The main impetus to the search for magnetic material came from the striking discovery of magnetic bacteria that are common in marine muds. When observed in a water droplet under the microscope, these bacteria consistently swim toward magnetic north. Artificial changes in the field prompt corresponding changes in the swimming direction of the bacteria. It is beyond any doubt that their orientation is magnetic (Blakemore 1975).

A biologist will immediately ask: What good is it for a bacterium that swims distances of micrometers and millimeters to know magnetic north? The answer is simple: These bacteria are anaerobic, and as they orient along the axis of the magnetic field, which at our latitude is at about 60° from the horizontal, swimming toward north means swimming into the mud and away from the dangerous oxygen.

These bacteria are roughly spherical and about 1 μm in diameter. They contain chains of small magnetic particles, presumably consisting of magnetite. The northward orientation is entirely passive, for killed bacteria remain aligned with the magnetic field and follow imposed changes in its direction.

Other magnetic microbes have been isolated from fresh water, and in these too, chains of magnetite crystals are the cause of a passive magnetic orientation, clearly different from the electromagnetic phenomenon exhibited by elasmobranch fishes (Blakemore et al. 1979).

As these bacteria swim to the north to avoid oxygen, what happens to similar microbes in the southern hemisphere? They behave like bacteria in the northern hemisphere, but with the striking difference that their polarity is reversed. The bacteria are south-oriented and migrate downward by swimming along the inclined magnetic field lines in the southern hemisphere (Blakemore et al. 1980).

Several of the sensory modalities discussed here – polarized light, infrared radiation, electric and magnetic fields – cannot be perceived by humans and were unknown until recently. Several had been proposed and tested without success, and had therefore been rejected as impossible. Theoretical arguments were also used to demonstrate the impossibility of some of those sensory channels that we now realize do exist.

An obvious question therefore is: Can any new sensory modalities be discovered in the future? It

may seen unlikely, but it would be risky to suggest that it is impossible. What is necessary is to find discrete, measureable signals that can be experimentally manipulated and to identify the corresponding sensory receptor system.

TRANSMISSION AND SORTING OF SENSORY INFORMATION

We have now seen that a wide variety of external information can be received by the sensory organs. What happens to all this information? How is it processed?

In each case the sensory organ serves as a transducer that converts the external signal into a change in membrane potential of the receptor cell, a change we refer to as the *receptor potential*.

Information about stimulus strength

We saw in Chapter 11 that an action potential is always of the same magnitude, irrespective of the strength of the stimulus that produces the action potential, once the threshold is exceeded. If the action potentials are always of the same magnitude, how is it possible to convey information about variations in the strength of the stimulus – information that obviously is of the greatest interest?

The answer is that a change in frequency of the action potentials in the axon can be used as an indicator of stimulus strength. This is illustrated in Figure 13.21, which shows that a pressure-sensitive skin receptor cell in the finger of a person responds to an increased stimulus strength with increasing frequency of the action potentials in its axon. This particular receptor had a threshold of about 0.5 g, and therefore showed no response to a force of 0.2 g. A force of 0.6 g, however, gave a clear response, which increased in frequency with the application of increasing forces. Thus, we can say that the magnitude of a sensory stimulus is coded and transmitted as a *frequency-modulated signal*.

Some receptor cells show *spontaneous activity;* they produce action potentials and nerve impulses

FIGURE 13.21 Action potentials recorded from a single pressure-sensitive receptor on the human hand. A plastic rod 1 mm in diameter was pressed against the skin with the force indicated on each record. This receptor did not respond to temperature changes (i.e., it was a specific mechanoreceptor). [Hensel and Boman 1960]

in the absence of stimulation. At first glance this may seem meaningless and counterproductive; in reality, it has important advantages.

The most important consequence is that it increases the *sensitivity* of the receptor. A very small stimulus – too small to cause a depolarization by itself – increases the rate of impulse discharge. This means that there is no such thing as a subthreshold stimulus; any slight increase in stimulus changes the discharge frequency.

The other advantage of spontaneous receptor activity is that a change in either direction can be encoded. Consider, for example, the infrared receptor of the rattlesnake. In a completely uniform environment the receptor shows spontaneous activity, and a small change in the environmental radiation in either direction can be encoded as a frequency change. The sensitivity of the system is high enough so that it will respond every time the snake suddenly faces a suitable surface with a temperature of 0.1°C above or below the previous one (Bullock and Diecke 1956). Thus spontaneous activity not only brings about a highly sensitive system but also permits determination of the polarity of a change.

We can now summarize some basic principles:

1. Sensory organs serve as transducers where external signals generate receptor potentials.
2. Information is conveyed in the sensory nerves as action potentials.
3. Action potentials in all sensory nerves are of the same nature and magnitude.
4. The magnitude of the action potentials is unaffected by stimulus intensity.
5. Stimulus intensity is coded by frequency modulation of the action potentials.

Let us proceed to the question of how sensory information is further processed.

Sorting and processing of information

If all information available to the sensory organs were transmitted to the central nervous system, the mass of signals would be overwhelming and probably utterly unmanageable. However, a great deal of screening, filtering, and processing takes place before the signals are passed on, beginning at the sensory neuron and continuing at several levels on the way to the brain. The filtering networks pass on only selected portions of the information they receive. Furthermore, they carry out certain steps of processing that improve on the information that is transmitted to higher levels. We shall discuss these principles using the processing of visual information as an example.

In spite of the complexity of the visual system, the processing of visual signals is better understood than that of other complex senses. There are several reasons. One is that artificial stimuli (light) can be directed, timed, and quantified with great accuracy. Another reason is the use of the arthropod eye as experimental material, because the structure of the compound eye permits ready access to single sensory receptor units. However, even the complex signal processing in the vertebrate eye is amazingly well understood.

Lateral inhibition

The horseshoe crab, *Limulus,* has a compound eye in which individual receptor units can readily

FIGURE 13.22 Action potentials recorded simultaneously from two adjacent ommatidia, A and B, in the compound eye of the horseshoe crab *Limulus*. The discharge rate is indicated at the right of each record. Top: illumination of ommatidium A alone. Center: illumination of A and B, resulting in lateral inhibition (see text). Bottom: illumination of B alone. [Hartline and Ratliff 1957]

FIGURE 13.23 Diagram showing that lateral inhibition causes an enhanced edge effect. For details, see text.

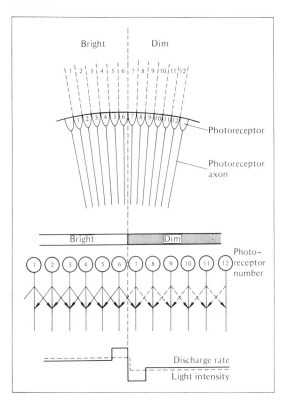

be stimulated by a fine beam of light. It is also possible to record the impulses from that fiber of the optic nerve which connects to the particular receptor in question. It turns out that the signals in the axon do not completely represent the stimulus; the pattern also depends on the amount of light falling on other receptor units. This is because each visual receptor is connected to its neighbors and inhibits their activity. This characteristic, known as *lateral inhibition,* has the effect of enhancing the contrast between the amounts of light falling on two adjacent receptors.

This may need some further explanation. Figure 13.22 shows the records from two adjacent receptor units of the *Limulus* eye. The top record shows the regularly spaced discharges when a single receptor (A) was stimulated. The rate of discharge was 35 impulses per second. When the adjacent receptor (B) was also stimulated, the discharge rate in unit A decreased to 29. Unit B was likewise influenced by its neighbor A, for when the stimu-

lus was removed from A, the discharge rate from receptor B increased from 23 to 31 impulses per second. Thus, there was a mutual lateral inhibition between these two units.

The effect of lateral inhibition can be understood from the diagram in Figure 13.23. Consider 12 units that are stimulated by uniform light of two intensities, there being a sharp transition between units 6 and 7. In this system, units 2, 3, 4, and 5 are all bordered by units receiving bright illumination and are therefore subject to lateral inhibition. Unit 6, however, is not subjected to the same degree of lateral inhibition from unit 7, which is within the dimly stimulated zone. As a result, its discharge frequency is higher.

Next, consider the dimly lit units, of which receptors 8, 9, 10, and 11 receive the same small amount of lateral inhibition. Unit 7, however, is inhibited by its brightly lit neighbor, receptor 6, and therefore discharges at a lower frequency than the other dimly lit receptors. The overall effect is that the transition between brightly and dimly lit receptors is emphasized. As a result, the messages in the optic nerve give a correct picture of the edge, but with an emphasized contrast between the two zones.

Another aspect of visual reception should be emphasized before we move on to the vertebrate eye. It has been observed in the scallop (*Pecten*), which has well-developed image-forming eyes along the edge of the mantle, that the retina has two layers. One layer, as expected, responds to light as a normal stimulus. The other layer of the retina, however, behaves in a very different way; it does not respond to increased light, but is sensitive only to a decrease in the intensity of illumination. The importance of this phenomenon can easily be imagined. When a shadow suddenly falls on an animal, it often means the approach of a predator, and in this case, a decrease in stimulus intensity is far more important to the animal than any other information.

We are now familiar with two relatively simple but important aspects of visual systems: the amplification of edge effects and the importance of dark areas in the visual fields.

Information processing

The frog's eye. The frog's retina has no central fovea,* and rods and cones are uniformly distributed so that the structure of the retina is much the same from place to place. The photoreceptors connect to several kinds of neurons located within the retina at various levels.

---

*In the mammalian eye a small, central area of the retina, the *fovea centralis,* contains only cones. It is the area that provides the most acute vision in strong light, but because of the absence of rods the fovea is insensitive to dim light.

We shall be concerned only with those neurons that are known as *ganglion cells* and whose axons make up the optic nerve. There are about half a million of these ganglion cells, a number that corresponds approximately to the number of fibers in the optic nerve (Maturana 1959). However, there are more than 1 million receptor cells (rods and cones) in the retina. Clearly, the optic nerve cannot carry a complete point-by-point picture of the image that falls on the retina; the analogy between the retina and the photographic film in a camera is not fully valid, for there must be some discrimination or processing before the signals are sent on to the brain.

The fibers in the optic nerve can be divided into five different classes on the basis of the responses of the ganglion cells. They have been clearly identified because they respond differently to specific kinds of stimulation of the retina. Some fibers respond only to the onset of illumination; these are called *on fibers*. Other fibers respond only to the termination of a light stimulus and are called *off fibers*. Still other fibers respond to either onset of a light stimulus or its termination; these are known as *on–off fibers*. Their response to the movement over the retina of a linear shape is marked, and they could therefore also be called *moving-edge detectors*.

There are also fibers that respond to the presence of a sharp edge in the visual field, whether it is stationary or moving. These *edge receptors* differ distinctly from the on–off fibers. Finally, there are fibers that could be called *bug detectors* because they respond to small, dark, moving objects, but not to large, dark objects or to stationary objects. The two last categories do not react to changes in the general intensity of the light; even switching a light on or off does not affect them (Figure 13.24).

We can now see that the frog's retina can carry out a great deal of analysis of the visual signals before information is transmitted to the optic nerve, and if we examine this information in view of what may be important to the frog, the system makes sense.

FIGURE 13.24 In the eye of a frog a high degree of processing and filtering of information takes place in the retina. The various fibers in the optic nerve carry different kinds of information. Certain fibers do not respond to movement of large objects (*a*), to movement of the whole visual field (*b*), or to on and off light. They do respond to small moving objects if these are darker than the background (*c*), but not if they are lighter (*d*). The response is independent of the general level of illumination (*e*), and there is no response to movement of objects with fuzzy edges (*f*). [Bullock and Horridge 1965. Courtesy of T. H. Bullock and G. A. Horridge and W. H. Freeman and Company. Copyright © 1965]

From the frog's viewpoint the insects on which it feeds are some of the most relevant objects in its life. Therefore, a small dark object is important, particularly if it moves, and the frog is equipped to respond specifically to this stimulus through the bug-detector fibers. Small stationary objects, such as spots, shadows, or pebbles, are of no special interest, and information about these does not reach that part of the system designed for a rapid response to live prey. The most meaningful processing of signals related to food has therefore already taken place in the retina and is transmitted as specialized information to the central nervous system (Lettvin et al. 1961). We shall not discuss the details of the other types of retinal detection systems but instead move on to the mammalian visual system.

Processing in the retina. The eye of a cat, and of many other mammals, has about 100 million receptor cells in the retina. The optic nerve carries about 1 million axons. This immediately tells us that the brain cannot receive separate information from each individual receptor cell; a great deal of sorting and processing takes place before the information is sent to the central nervous system.

In spite of the complexity, what occurs in the mammalian optical system is reasonably well known. In a simplified way we can regard the transmission as taking place over six levels: three in the retina of the eye, one in the lateral geniculate body of the brain, and two in the visual cortex of the brain (Figure 13.25).

Let us follow these levels and examine the processing that takes place before information reaches the visual cortex, which in turn is connected with other parts of the central nervous system.

The three levels of processing in the vertebrate retina are (1) absorption of light quanta in the receptor cells and a change in their membrane potential (always a hyperpolarization), (2) transmission of the potential change to a complex layer of neurons that carries out a great deal of processing, and (3) transmission of the processed information to the ganglion cells, the neurons whose axons make up the optic nerve.

FIGURE 13.25 Diagram of human brain, seen from below, indicating the main pathway between the eye and the visual cortex.

FIGURE 13.26 Action potentials from a retinal ganglion cell. When a small spot of light falls on an "on" receptor cell in the center of a receptive field, the ganglion cell fires a rapid train of impulses (upper tracing). If a light spot falls on the surrounding "off" area, the ganglion cell shows no response while the light is on but fires after the light eases (lower tracing). [Kuffler 1953]

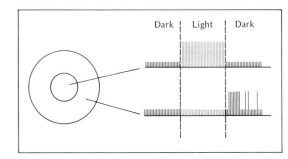

How the retina organizes the information received by the receptor cells, the rods and cones, has been most successfully studied in an aquatic salamander, the mudpuppy (*Necturus maculosus*). This is because the neurons of the retina in this animal are much larger than in other vertebrates, and microelectrodes can therefore be inserted into the individual neurons of the retina. After a recording has been made, a small amount of dye is injected through the hollow microelectrode so that the particular cell from which the recording was made can later be identified microscopically (Werblin and Dowling 1969).

We remember that there are a hundred times fewer ganglion cells and fibers in the optic nerve than there are receptor cells. Because there are more than 100 receptor cells for each ganglion cell, each ganglion cell must receive information from a large number of receptor cells. Together, such a group of receptor cells makes up a *receptive field*, which

designates that area of the retina with which one particular ganglion cell is connected.

One of the most striking characteristics of the ganglion cells is that they are spontaneously active. In the dark they send a steady stream of impulses into the optic nerve at a rate of about 20 to 30 per second. There are two main types of ganglion cells: In some, a light on the retina causes an increased rate of action potentials; in others, light causes a decreased rate.

Let us consider a ganglion cell that responds with an increased rate firing, an *on-center cell*. The response is best when a small spot of light falls in the center of the receptive field of that particular ganglion cell (Figure 13.26). If the light spot falls within a concentric ring around the "on" spot, there is no response from the ganglion cell until the light is terminated. Then a train of impulses is fired (Kuffler 1953).

If two light spots are used, one falling at the "on" spot and one in the area around the "on" spot, discharges from both areas are reduced, and the output from the ganglion cell is less than from a single light spot. Because of the inhibition from

FIGURE 13.27 Simplified diagram of the connections between the receptor cells (R), the bipolar cells (B), and the ganglion cells (G) in the retina of the eye. Also shown are horizontal cells (H), amacrine cells (A), and a glial or supporting cell (GL). [Courtesy of John E. Dowling, Harvard University]

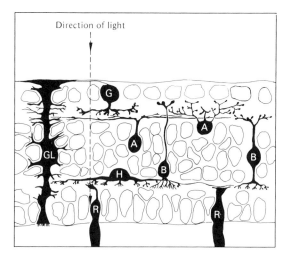

the area surrounding the "on" receptor cell, this area is called the *inhibitory surround*. The result of the arrangement is that illumination of an entire receptive field gives a weaker response than is obtained from the "on" spot alone, and the inhibition may even lead to no response at all. This example of lateral inhibition appears similar to what we discussed before in relation to the compound eye of *Limulus*.

The other type of ganglion cell behaves in the reverse fashion. In response to a small spot of light in the center of its receptive field, its activity decreases or ceases. It is called an *off-center cell*, and its activity ceases in response to illumination of the center of the receptive field and is unaffected or increases when the surround is illuminated.

We can now see the basis for some of the interesting and most useful responses of the eye. Uniform illumination of the entire retina does not have any striking effect on the number of impulses from the ganglion cells. However, a small spot of light

falling on one receptive field has a very marked effect on the corresponding ganglion cell and on the message passed on to the brain.

So far we have not discussed the layers of neurons that form the connections between the receptor cells and the ganglion cells. These neurons can be divided into three major groups: *horizontal cells, bipolar cells,* and *amacrine cells* (Figure 13.27).

It is a curious fact that the receptors cells (rods and cones) of the vertebrate retina never fire action potentials. They always respond to illumination by becoming hyperpolarized (i.e., their negative membrane potential becomes more negative). The receptor cells connect to both bipolar and horizontal cells. The bipolar cells may respond to the hyperpolarization of a receptor cell either by an increased or by a decreased membrane potential, depending on which receptor cell was stimulated. The bipolar cells receive information also from the horizontal cells, which have extensive lateral processes and form a network that carries information from the surround area. The bipolar cells, like the receptor cells, produce no action potentials, but affect the ganglion cells both directly and through the amacrine cells. The response of the ganglion cells is the inverse of that in the corresponding bipolar cells: They depolarize and produce action potentials when the bipolar cells are hyperpolarized, and they become hyperpolarized when the bipolar cells are depolarized (Figure 13.28).

The meaning of this highly complex signal evaluation or processing is obvious. It is of no particular interest to record and inform the brain about the general intensity of uniform light; the details of light and dark contrast are of much greater importance. What we have is a case of *selective destruction of information*, carried out in the eye. The eye selects what kind of information is to be transmitted in the optic nerve and thus reduces the information load on the brain.

Central processing

The fourth level in the processing of visual stimuli is in the brain, in the *geniculate body*. The cells

FIGURE 13.28 The response of retinal cells of the mudpuppy to illumination of a small spot of the retina (left) and to added illumination of the surrounding area (right). The receptor cell responds to a spot of light by hyperpolarization (more negative). Light on the surround inhibits the response to light on center. The bipolar cell (in this experiment) also hyperpolarizes in response to the spot, but slightly depolarizes in response to light on the surround. The ganglion cell responds to the spot with increased rate of firing action potentials and to light on the surround by inhibition; however, after cessation of illumination it resumes firing. Duration of light signal (0.6 to 0.8 second) marked by horizontal bars. [Werblin and Dowling 1969]

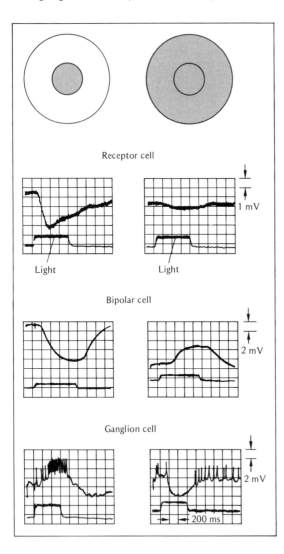

of the geniculate body correspond to the receptive fields of the retina. They have some characteristics in common with the retinal ganglion cells, and they function to increase the contrast between small spots of light and changes in diffuse lighting. For our purposes, however, we shall regard the geniculate body merely as a way station, consisting of synaptic connections between the optic nerve and the visual cortex of the brain.

Of the many different kinds of cells in the *visual cortex,* we shall discuss only two: simple cells and complex cells. *Simple cells* respond primarily to contrasting lines, such as light bars on a dark background or dark bars on a light background, and to sharp light–dark edges.

Whether or not a given simple cell responds to such a bar depends on the orientation and position of the bar within the receptive field of the retina. For example, if a vertical bar gives a response in a given simple cell of the visual cortex, the cell will not respond if the bar is rotated some 10 or 15° or if the bar is displaced slightly to one side or the other. A careful testing of the retina with minute light spots has shown that the simple cells respond to arrangements such as are shown in Figure 13.29.

Because the response of a given simple cell depends on the orientation as well as the position on the retina of a straight line, it constitutes a built-in filter that removes or destroys a great deal of information and transmits only specialized information about certain kinds of contrasting lines. Each simple cell thus has a specific function. It corresponds to one restricted part of the retina and responds only to one particular kind of information falling on this spot. There is no evidence that in the cat's retina there is any preferred orientation; the receptive fields may occur in all possible orientations with no pref-

FIGURE 13.29 Various arrangements of receptive fields in the visual cortex. Areas giving an "on" response are indicated by crosses; those giving an "off" response by triangles. The orientation of the receptive fields, whether of the "on" or "off" type, is indicated by straight lines through the center of the field. [Hubel and Wiesel 1962]

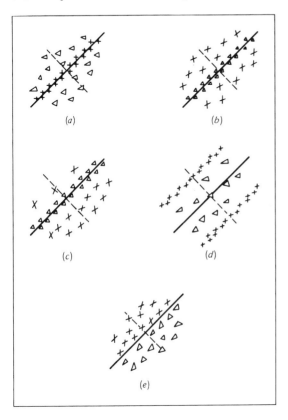

erence for vertical, horizontal, or oblique orientations.

The *complex cells* of the visual cortex respond, like simple cells, to the orientation of a given stimulus. If a complex cell responds to a certain orientation of a light bar on the retina, rotation of the bar causes the cell to cease its activity. If the bar is moved over the retina with unchanged orientation, however, the complex cell continues to be active.

This can most readily be explained by assuming that the complex cell receives information from a large number of simple cells, all of which are of the same general type and have the same orientation of their fields. This requires an enormous complexity of the connections in the visual cortex, but all the evidence that has been obtained shows that the cells are indeed located in such anatomically well-defined groups.

We can now summarize the main steps in the processing of visual information. there are "on"-type and "off"-type receptors, which do not individually send information to the central nervous system, but are arranged in fields that readily process information about contrast between light and dark, rather than about the general level of illumination. The several levels of transmitting neurons within the retina select the most pertinent information and reject less meaningful information.

The selected information is transmitted to the geniculate body, which accentuates the information obtained from the "on-center" and "off-center" receptive fields. The visual cortex, finally, rearranges the information from the geniculate body so that lines and contours are selected for further processing. Here the first step (in the simple cells) is concerned with the orientation of lines; then the information converges on the complex cells, from which the information is transmitted to other parts of the brain for integration with the general functioning of the body.

The processing and sorting of sensory signals carried out before the signals are passed on to the central nervous system remove much irrelevant information and greatly reduce the load on the control systems responsible for adequate and proper responses. We shall now pass on to one simple control system that can serve as an example of more complex coordination.

Sensory coordination of vertebrate muscle

We saw earlier that crustacean muscles are controlled by a small number of axons (often three to five). Vertebrate muscles, in contrast, are innervated by a large number of axons, each branching to a small number of muscle fibers (*a motor unit*). This difference in innervation causes much of the control of crustacean muscles to take place peripherally, whereas vertebrate muscle is essentially under central control, directed by a highly complex central nervous system.

The difference is readily explained. Myelinated axons permit the fast transmission of impulses without undue increase in axon diameter; the necessary large number of connections can be carried in a nerve of moderate size. For the crustacean muscle, in contrast, rapid conduction requires a substantial increase in fiber size. A finely graded central control that depends on a large number of fibers in the motor nerve is therefore not feasible. We have already described how crustacean muscles nevertheless are under precise control, although they may receive no more than two or three stimulatory and one or two inhibitory axons. We shall now examine the control of vertebrate muscle.

The contraction of vertebrate muscle is under central control, and there is no need for inhibitory axons to reach the muscle and modify the degree of contraction.* The vertebrate muscle system acts under the direction of a central computer that receives a great deal of information from various sources, including a very important part that comes as feedback directly from the muscles themselves. We shall use the admirable performance of this system as an example of what can be achieved through central control and integration.

---

*Although vertebrate skeletal muscle apparently has no inhibitory innervation, this is not true for other kinds of vertebrate muscle. The heart, which is a striated muscle, has double innervation, both stimulatory and inhibitory. All smooth muscle receives both adrenergic and cholinergic fibers from the autonomic nervous system. For some smooth muscle (e.g., bronchial muscle) the adrenergic fibers are inhibitory; for others (vascular smooth muscle) they are stimulatory. Because cholinergic fibers are always antagonistic to the adrenergic fibers, they may likewise be either inhibitory or stimulatory.

The nerve to a vertebrate muscle has fibers that carry the impulses for stimulation of contraction; in addition, it contains other fibers that carry sensory information from the muscle to the central nervous system. There are two major kinds of such sensory fibers (Figure 13.30). Some come from small sensory units located in the tendons, the *tendon organs;* other come from a specialized type of muscle fiber, known as *muscle spindles.*

The tendon organs seem to be used for sensing the deformation produced by tension in the tendon and give information about the force of muscle contraction. The muscle spindles, which we shall discuss in greater detail, are used for obtaining information about length, but as we shall see, they are even more useful because of the way their information is used in the feedback control of muscle contraction. In fact, it has been stated that the muscle spindles, next to the eye and the ear, are the most complex sensory organs in the body.

Let us use a highly simplified diagram (Figure 13.31). A person can easily, in response to a constant load, maintain a constant force in a muscle. If an additional load is unexpectedly added, it causes a stretching of the muscle that almost instantly is compensated for by an increased force of contraction in the muscle. This is caused by a reflex that originates in the muscle spindles. When stretched, the spindles send impulses to the spinal cord, where the sensory fibers make synaptic contact with motor neurons, which in turn cause sufficient contraction in the muscle to compensate for the initial stretching.

In reality, more than one synapse is usually involved, and there are connections from the spinal cord that carry information to different levels of the central nervous system, including the brain. However, we need not be concerned with this complexity. The important matter is that the muscle spindles can serve to maintain a constant length of the muscle in the face of variations in the load – an obvious advantage in coordinating movements of all sorts.

Another important characteristic of the muscle

FIGURE 13.30 A vertebrate muscle is equipped with sensory organs that transmit information to the central nervous system. The tendon organs provide information about the force of contraction; the muscle spindles primarily about the length of the muscle. This information is used as important feedback signals in the precise control of the muscle and its contraction. [Merton 1972]

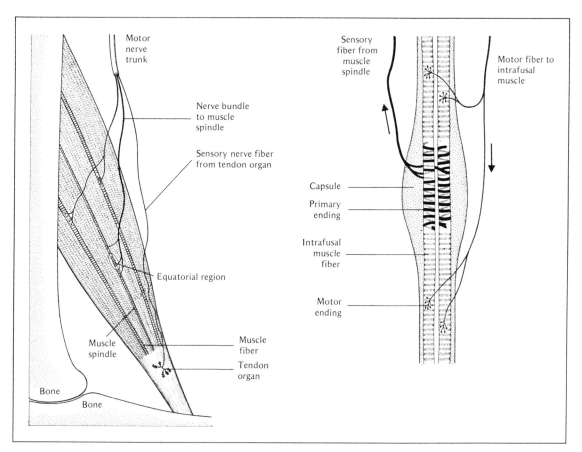

spindles is that these are themselves contractile and innervated by motor axons that are separate from the axons that cause contraction of the regular muscle fibers. The detailed structure of the muscle spindles is important (Figure 13.30). They consist of a bundle of a few modified muscle fibers, called the *intrafusal muscle fibers* (Latin *fusus* = spindle). These fibers are contractile, but in the middle of the fiber the contractile apparatus is absent and the fiber has sensory nerve endings wrapped around it.

Let us say that an intrafusal fiber is stimulated to contract; this stretches its central region, which sends to the spinal cord a volley of impulses. These in turn stimulate the motor neurons, which respond by sending action potentials in the motor nerve, causing contraction in the muscle. When the contraction in the muscle has reached the same level as the contraction in the intrafusal fiber, the equatorial region is no more stretched and is therefore silenced.

FIGURE 13.31 Feedback control of muscle contraction. A constant load on a muscle is balanced by a constant force (a). A sudden increase in the load (b) stretches the muscle; the muscle spindles send signals back to the spinal cord, where additional motor neurons are excited and send impulses back in the motor nerve, causing increased contraction to balance the load (c). [Merton 1972]

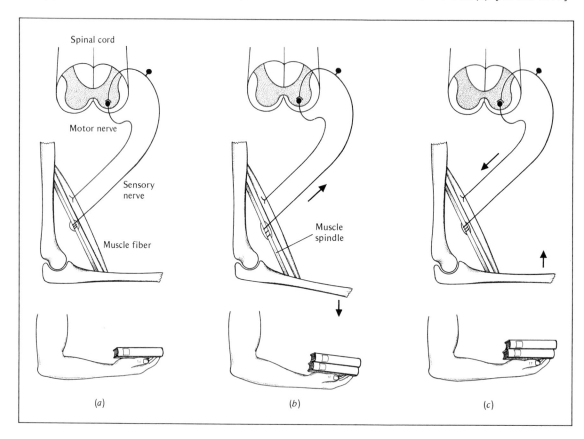

This elegant system in a way is analogous to a mechanical servosystem. For example, in the power steering of a heavy automobile a small movement of the steering wheel is used to direct a motor that performs the heavy work of turning the wheels on the road. A servomechanism is no more than an automatic feedback control mechanism in which the controlled variable is the position of a mechanical device. Such systems are used widely in the design of all sorts of mechanical and industrial control processes.

The characteristics of intrafusal fibers provide a clever mechanism for maintaining a graded speed of contraction against a variable load. If the contraction of a given muscle is directed by a certain slow contraction of the intrafusal fibers, an increase in the load on the muscle only momentarily slows its rate of contraction. The intrafusal fibers continue to contract at a set rate, but if the muscle follows too slowly because of the increased load, the equatorial region is stretched and sends additional impulses to the spinal cord, which in turn increases the contraction of the muscle. The result is an automatic compensation for increases or de-

FIGURE 13.32 When two antagonistic muscles (A and B) participate in the movement of a limb, the contraction of muscle A causes impulses to be sent to the spinal cord. Here an inhibitory interneuron, by inhibiting the appropriate motor neuron, causes relaxation of the antagonistic muscle (B).

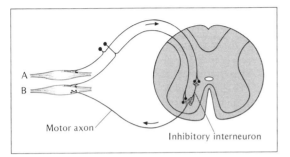

Motor axon

Inhibitory interneuron

creases in the load that occur during contraction of the muscle.

When we turn to the question of how the muscles are used in the movements of the limbs, we find a need for additional control. The muscles that serve in maintaining posture normally maintain some degree of contraction; if they did not, the body would collapse, as happens when a person faints. Therefore, to move a limb, it is not sufficient that a given set of muscles contract; it is also necessary that the antagonistic muscles relax. This is achieved through a relatively simple reflex mechanism shown in schematic form in Figure 13.32.

Suppose that stimulation of the muscle spindles in muscle A evokes contraction of this muscle. Information about this contraction is received by certain neurons in the spinal cord, which are called *interneurons** and act by inhibiting other neurons. The pathways are arranged in such a way that these *inhibitory interneurons* cause the inhibition of the motor neurons of the antagonistic muscle (B). The reciprocal action of antagonistic muscles, with one relaxing as the other contracts, is functionally very

---

*Neurons whose axons do not extend outside the central nervous system are known as *interneurons*. They may be excitatory or inhibitory.

useful and could not take place without the aid of inhibitory neurons or other inhibition.

In addition to the control of antagonistic muscles, inhibition serves other important purposes. For example, if a contracted muscle is stretched very forcefully, the tendon organs are activated and initiate impulses that reach inhibitory interneurons in the spinal cord. This causes inhibition of contraction in the muscle that is being stretched; the active resistance to stretching is thereby decreased. This reflex contributes to the protection of the muscles from rupture when they are suddenly loaded. Without this protection there would be far more ''pulled muscles'' than we now see.

The integration of muscular movement in locomotion is more complex than described in the examples mentioned here. It depends on various commands from the brain and on information received from sensory organs such as eyes, ears, skin, and so on. The central nervous system is a highly intricate control system that acts on information and feedback signals from many sources.

REFERENCES

Atema, J. (1971) Structures and functions of the sense of taste in the catfish (*Ictalurus natalis*). *Brain Behav. Evol. 4*:273–274.

Barrett, R., Maderson, P. F. A., and Meszler, R. M. (1970) The pit organs of snakes. In *Biology of the Reptilia*, vol. II (C. Gans and T. S. Parsons, eds.), pp. 277–300. London: Academic Press.

Blakemore, R. (1975) Magnetotactic bacteria. *Science 190*:377–379.

Blakemore, R. P., Franekl, R. B., and Kalmijin, A. J. (1980) South-seeking magnetotactic bacteria in the southern hemisphere. *Nature, Lond. 286*:384–385.

Blakemore, R. P., Maratea, D., and Wolfe, R. S. (1979) Isolation and pure culture of a freshwater magnetic spirillum in chemically defined medium. *J. Bacteriol. 140*:720–729.

Bleckmann, H., and Schwarts, E. (1982) The functional significance of frequency modulation within a wave train for prey localization in the surface-feeding fish *Aplocheilus lineatus* (Cyprinodontidae). *J. Comp. Physiol.* 145:331–339.

Boeckh, J., Kaissling, K.-E., and Schneider, D. (1960) Sensillen und Bau der Antennengeissel von *Telea polyphemus* (Vergleiche mit weiteren Saturniden: *Antheraea, Platysamia* und *Philosamia*). *Zool. Jahrb. Abt. Anat.* 78:559–584.

Bowmaker, J. C., and Dartnall, H. J. A. (1980) Visual pigments of rods and cones in a human retina. *J. Physiol.* 298:501–511.

Brown, P. K., and Wald, G. (1964) Visual pigments in single rods and cones of the human retina. *Science* 144:45–52.

Bullock, T. H., and Cowles, R. B. (1952) Physiology of an infrared receptor: The facial pit of pit vipers. *Science* 115:541–543.

Bullock, T. H., and Diecke, F. P. J. (1956) Properties of an infra-red receptor. *J. Physiol.* 134:47–87.

Bullock. T. H., Hamstra, R. H., Jr., and Scheich, H. (1972) The jamming avoidance response of high frequency electric fish. 1. General features. *J. Comp. Physiol.* 77:1–22.

Bullock, T. H., and Horridge, G. A. (1965) *Structure and Function in the Nervous System of Invertebrates*, vols. 1 and 2. San Francisco: Freeman. 1719 pp.

Cahn, P. H. (ed.) (1967) *Lateral Line Detectors.* Bloomington: Indiana University Press. 496 pp.

Caprio, J. (1977) Electrophysiological distinctions between the taste and smell of amino acids in catfish. *Nature, Lond.* 266:850–851.

Chen, D.-M., Collins, J. S., and Goldsmith, T. H. (1984) The ultraviolet receptor of bird retinas. *Science* 225:337–340.

Crescitelli, P., and Dartnall, H. J. A. (1953) Human visual purple. *Nature, Lond.* 172:195–197.

Dijkgraaf, S. (1963) The functioning and significance of the lateral-line organs. *Biol. Rev.* 38:51–105.

Dijkgraaf, S., and Kalmijn, A. J. (1963) Untersuchungen über die Funktion der Lorenzinischen Ampullen an Haifischen. *Z. Physiol.* 47:438–456.

Elepfandt, A. (1986) Wave frequency recognition and absolute pitch for water waves in the clawed frog, *Xenopus laevis. J. Comp. Physiol.* 158:235–238.

Emlen, S. T., Wiltschko, W., Demong, N. J., Wiltschko, R., and Bergman, S. (1976) Magnetic direction finding: Evidence for its use in migratory indigo buntings. *Science* 193:505–508.

Enright, J. T. (1963) Estimates of the compressibility of some marine crustaceans. *Limnol. Oceangr.* 8:382–387.

Erulkar, S. D. (1972) Comparative aspects of spatial localization of sound. *Physiol. Rev.* 52:237–360.

Fager, L. Y., and Fager, R. S. (1981) Chicken blue and chicken violet, short wavelength sensitive visual pigments. *Vision Res.* 21:581–586.

Flock, A. (1971) Sensory transduction in hair cells. In *Handbook of Sensory Physiology,* vol. 1, *Principles of Receptor Physiology* (W. R. Lowenstein, ed.), pp. 396–441. Berlin: Springer-Verlag.

von Frisch, K. (1948) Gelöste and ungelöste Rätsel der Bienensprache. *Naturwissenschaften* 35:38–43.

Fromme, H. G. (1961) Untersuchungen über das Orientierungsvermögen nächtlich ziehender Kleinvögel (*Erithacus rubecula, Sylvia communis*). *Z. Tierpsychol.* 18:205–220.

Fuzessery, Z. M., Carr, W. E. S., and Ache, B. W. (1978) Antennular chemosensitivity in the spiny lobster, *Panulirus argus:* Studies of taurine sensitive receptors. *Biol. Bull.* 154:226–240.

Gamow, R. I., and Harris, J. F. (1973) The infrared receptors of snakes. *Sci. Am.* 228:94–101.

Goris, R. C., and Terashima, S. (1973) Central response to infra-red stimulation of the pit receptors in a crotaline snake, *Trimeresurus flavoviridis. J. Exp. Biol.* 58:59–76.

Gould, E., Negus, N. C., and Novick, A. (1964) Evidence for echo location in shrews. *J. Exp. Zool.* 156:19–39.

Gould, J. L., Kirschvink, J. L., and Deffeyes, K. S. (1978) Bees have magnetic remanence. *Science* 201:1026–1028.

Gregory, J. E., Iggo, A., McIntyre, A. K., and Proske, U. (1987) Electroreceptors in the platypus. *Nature, Lond.* 326:386–387.

Griffin, D. R. (1953) Acoustic orientation in the oil bird, *Steatornis. Proc. Natl. Acad. Sci. U.S.A.* 39:884–893.

Griffin, D. R. (1954) Bird sonar. *Sci. Am.* 190:78–83.

Griffin, D. R. (1958) *Listening in the Dark: The Acoustic Orientation of Bats and Men.* New Haven, Conn.: Yale University Press. 413 pp.

Harris, J. F., and Gamow, R. I. (1971) Snake infrared receptors: Thermal or photochemical mechanism? *Science* 172:1252–1253.

Hartline, H. K., and Ratliff, F. (1957) Inhibitory interaction of receptor units in the eye of *Limulus*. *J. Gen. Physiol. 40:*357–376.

Hartline, P. H. (1971) Physiological basis for detection of sound and vibration in snakes. *J. Exp. Biol. 54:*349–371.

Hensel, H., and Boman, K. K. A. (1960) Afferent impulses in cutaneous sensory nerves in human subjects. *J. Neurophysiol. 23:*564–578.

Hodgson, E. S. (1961) Taste receptors. *Sci. Am. 204:*125–144.

Höglund, G., Hamdorf, K., and Rosner, G. (1973) Trichromatic visual system in an insect and its sensitivity control by blue light. *J. Comp. Physiol. 86:*265–279.

Hopkins, C. D. (1972) Sex differences in electric signaling in an electric fish. *Science 176:*1035–1037.

Hopkins, C. D. (1976) Stimulus filtering and electroreception: Tuberous electroreceptors in three species of gymnotoid fish. *J. Comp. Physiol. 111:*171–207.

Hubel, D. H., and Wiesel, T. N. (1962) Receptive fields, binocular interaction and functional architecture in the cat's visual cortex. *J. Physiol. 160:*106–154.

Iggo, A. (1969) Cutaneous thermoreceptors in primates and sub-primates. *J. Physiol. 200:*403–430.

Kalmijn, A. J. (1971) The electric sense of sharks and rays. *J. Exp. Biol. 55:*371–383.

Kalmijn, A. J. (1978) Electric and magnetic sensory world of sharks, skates, and rays. In *Sensory Biology of Sharks, Skates, and Rays* (E. S. Hodgson and R. F. Mathewson, eds.), pp. 507–528. Washington, D.C.: Office of Naval Research, Department of the Navy.

Kellogg, W. N. (1958) Echo ranging in the porpoise. *Science 128:*982–988.

Keynes, R. D., and Martins-Ferreira, H. (1953) Membrane potentials in the electroplates of the electric eel. *J. Physiol. 119:*315–351.

Kreithen, M. L., and Eisner, T. (1978) Ultraviolet light detection by the homing pigeon. *Nature, Lond. 272:*347–348.

Kreithen, M. L., and Quine, D. B. (1979) Infrasound detection by the homing pigeon: A behavioral audiogram. *J. Comp. Physiol. 129:*1–4.

Kuffler, S. W. (1953) Discharge patterns and functional organization of mammalian retina. *J. Neurophysiol. 16:*37–68.

Kühn, A. (1927) Ueber den Farbensinn der Bienen. *Z. Vergl. Physiol. 5:*762–800.

Lettvin, J. Y., Maturana, H. R., Pitts, W. H., and McCulloch, W. S. (1961) Two remarks on the visual system of the frog. In *Sensory Communication* (W. A. Rosenblith, ed.), pp. 757–776. New York: M.I.T. Press and Wiley.

Lissmann, H. W., and Mullinger, A. M. (1968) Organization of ampullary electric receptors in Gymnotidae (Pisces). *Proc. R. Soc. Lond. B. 169:*345–378.

Machin, K. E., and Lissmann, H. W. (1960) The mode of operation of the electric receptors in *Gymnarchus niloticus*. *J. Exp. Biol. 37:*801–811.

Maturana, H. R. (1959) Number of fibers in the optic nerve and the number of ganglion cells in the retina of anurans. *Nature, Lond. 183:*1406–1407.

Merton, P. A. (1972) How we control the contraction of our muscles. *Sci. Am. 226:*30–37.

Moore, F. R. (1977) Geomagnetic disturbance of the orientation of nocturnally migrating birds. *Science 196:*682–684.

Morita, H., and Yamashita, S. (1959) Generator potential of insect chemoreceptors. *Science 130:*922.

Münz, H. Claas, B., and Fritzsch, B. (1984) Electroreceptive and mechanoreceptive units in the lateral line of the axolotl *Ambystoma mexicanum*. *J. Comp. Physiol. 154:*33–44.

Newman, E. A., and Hartline, P. H. (1981) Integration of visual and infrared information in bimodal neurons of the rattlesnake optic tectum. *Science 213:*789–791.

Norris, K. S., Prescott, J. H., Asa-Dorian, P. V., and Perkins, P. (1961) An experimental demonstration of echo-locating behavior in the porpoise, *Tursiops truncatus* (Montagu). *Biol. Bull. 120:*163–176.

Partridge, B. L., and Pitcher, T. J. (1980) The sensory basis of fish schools: Relative roles of lateral line and vision. *J. Comp. Physiol. 135:*315–325.

Payne, R. S. (1971) Acoustic location of prey by barn owls. *J. Exp. Biol. 54:*535–573.

Poole, J. H., Payne, K., Langbauer, W. R., and Moss, C. J. (1988) The social contexts of some very low frequency calls of African elephants. *Behav. Ecol. Sociobiol. 22:*385–392.

Quine, D. B., and Kreithen, M. L. (1981) Frequency shift discrimination: Can homing pigeons locate infrasounds by Doppler shifts? *J. Comp. Physiol. 141:*153–155.

Rushton, W. A. H. (1972) Pigments and signals in colour vision. *J. Physiol. 220:*1–31P.

Scheich, H., Langner, G., Tidemann, C., Coles, R. B., and Guppy, A. (1986) Electroreception and electrolocation in platypus. *Nature, Lond.* 319:401–402.

Schneider, D. (1969) Insect olfaction: Deciphering system for chemical messages. *Science 163:*1031–1037.

Schneider, D., Kasang, G., and Kaissling, K.-E. (1968) Bestimmung der Riechschwelle von *Bombyx mori* mit Tritium-markiertem Bombykol. *Naturwissenschaften 55:*395.

Schuijf, A., and Hawkins, A. D. (1983) Acoustic distance discrimination by the cod. *Nature, Lond. 302:*143–144.

Tschachotin, S. (1908) Die Statocyste der Heteropoden. *Z. Wissensch. Zool. 90:*343–422.

Walcott, C., Gould, J. L., and Kirschvink, J. L. (1979) Pigeons have magnets. *Science 205:*1027–1029.

Wald, G. (1964) The receptors of human color vision. *Science 145:*1007–1016.

Waterman, T. H. (1950) A light polarization analyzer in the compound eye of *Limulus. Science 111:*252–254.

Waterman, T. H., and Forward, R. B., Jr. (1970) Field evidence for polarized light sensitivity in the fish *Zenarchopterus. Nature, Lond.* 228:85–87.

Werblin, F. S., and Dowling, J. E. (1969) Organization of the retina of the mudpuppy, *Necturus maculosus.* 2. Intracellular recording. *J. Neurophysiol.* 32:339–355.

Wilcox, R. S. (1979) Sex discrimination in *Gerris remigis:* Role of a surface wave signal. *Science 206:*1325–1327.

Wiltschko, W., and Wiltschko, R. (1972) Magnetic compass of European robins. *Science 176:*62–64.

ADDITIONAL READING

Autrum, H., et al (editorial board) (1971–1978) *Handbook of Sensory Physiology:* vol. 1, *Principles of Receptor Physiology;* vol. 2, *Somatosensory System;* vol. 3, *Enteroceptors;* vol. 4, *Chemical Senses;* vol. 5, *Auditory System;* vol. 6, *Vestibular System;* vol. 7, *Photochemistry of Vision;* vol. 8, *Perception;* vol. 9, *Development of Sensory Systems.* New York: Springer-Verlag.

Barlow, H. G. (1981) Critical limiting factors in the design of the eye and visual cortex. *Proc. R. Soc. Lond. B 212:*1–34.

Barlow, H. B., and Mollon, J. D. (eds.) (1982) *The Senses.* Cambridge: Cambridge University Press. 490 pp.

Bench, R. J., Pye, A., and Pye, J. D. (eds.) (1975) *Sound Reception in Mammals.* London: Academic Press. 340 pp.

Bennett, M. V. L. (1970) Comparative physiology: Electric organs. *Annu. Rev. Physiol. 32:*471–528.

Bullock, T. H., (1982) Electroreception. *Annu. Rev. Neurosci. 5:*121–170.

Bullock, T. H. and Heiligenberg, W. (1986) *Electroreception.* New York: Wiley and Sons. 722 pp.

Busnel, R.-G., and Fish, J. F. (eds.) (1979) *Animal Sonar Systems.* New York: Plenum Press. 1135 pp.

Catton, W. T. (1970) Mechanoreceptor function. *Physiol. Rev. 50:*297–318.

Cold Spring Harbor Laboratory (1965) Sensory receptors. *Cold Spring Harbor Symp. Quant. Biol. 30:*1–649.

Darian-Smith, I. (ed.) (1984) *The Nervous System:* vol. 3, *Sensory Processes. Handbook of Physiology,* Sect. 1. Bethesda, Md.: American Physiological Society. 1190 pp.

Davson, H. (1972) *The Physiology of the Eye,* 3d ed. New York: Academic Press. 643 pp.

Daw, N. W. (1973) Neurophysiology of color vision. *Physiol. Rev. 53:*571–611.

Dowling, J. E. (1987) *The Retina: An Approachable Part of the Brain.* Cambridge: Harvard University Press. 282 pp.

Erulkar, S. D. (1972) Comparative aspects of spatial localization of sound. *Physiol. Rev. 52:*237–360.

Griffin, D. R. (1986) *Listening in the Dark: Acoustic Orientation of Bats and Men.* Ithaca, N.Y.: Cornell University Press. 464 pp.

Horridge, G. A. (ed.) (1975) *The Compound Eye and Vision of Insects.* Oxford: Clarendon Press. 595 pp.

Lythgoe, J. N. (1979) *The Ecology of Vision.* New York: Clarendon Press. 244 pp.

Roeder, K. D. (1967) *Nerve Cells and Insect Behavior,* rev. ed. Cambridge, Mass.: Harvard University Press. 238 pp.

Sales, G., and Pye, D. (1974) *Ultrasonic Communication by Animals.* London: Chapman & Hall. 281 pp.

Stebbins, W. C. (1983) *The Acoustic Sense of Animals.* Cambridge, Harvard University Press. 192 pp.

Stoddart, D. M. (ed.) (1980) *Olfaction in Mammals.* Symposia of the Zoological Society of London, no. 45. London: Academic Press. 363 pp.

APPENDIXES

APPENDIX A

Measurements and units

TABLE A.1 Base units

Physical quantity	SI base unit	Symbol
length	meter	m
mass	kilogram	kg
time	second	s
thermodynamic temperature	kelvin	K
amount of substance	mole	mol
electric current	ampere	A
luminous intensity	candela	cd

The purpose of physiological measurement is to determine the magnitude of certain physical quantities. To express such measurements, we need generally accepted units. Such units, their symbols, and the mathematical signs connected with their use, are standarized and internationally adopted through a long-standing permanent international body, *The General Conference on Weights and Measures* (Conférence Générale des Poids et Mésures). The system of units is known as *The International System of Units,* or commonly the *SI* system (= Système Internationale).

This appendix provides (1) names of units and their symbols, (2) a brief list of conversion factors, and (3) a list of some physical quantities and the recommended symbols for their designation.

UNITS

The SI system uses seven *base units* for seven dimensionally independent basic physical quantities; each base unit is represented by a *symbol.* Use of these symbols is *mandatory* (Table A.1).

The SI System recognizes several *derived units* (as well as some supplementary units). Although some of these have special symbols, they are derived from and defined in terms of the base units. A few of particular interest in physiology are listed in Table A.2.

The following should be noted about the use of these units and symbols:

TABLE A.2 Derived units

Physical quantity	SI derived unit	Symbol for unit	Definition in terms of base units
force	newton	N	$m\ kg\ s^{-2}$
energy	joule	J	$m^2\ kg\ s^{-2}$
power	watt	W	$m^2\ kg\ s^{-3}$
pressure	pascal	Pa	$m^{-1}\ kg\ s^{-2}$
viscosity	poise	P	$m^{-1}\ kg\ s^{-1} \times 10^{-1}$
electric potential difference	volt	V	$m^2\ kg\ s^{-3}\ A^{-1}$
electric resistance	ohm	Ω	$m^2\ kg\ s^{-3}\ A^{-2}$
electric charge	coulomb	C	$s\ A$
frequency	hertz	Hz	s^{-1}

1. The symbols are printed in roman (upright) letters. The symbols are not abbreviations and are not followed by a period, except at the end of a sentence.
2. The plural form of a symbol is unchanged, and the ending $-s$ should not be used.
3. Certain units are derived from proper names (e.g., Volta); the symbols for these units are capital roman (upright) letters (V), but the unit names are not capitalized (volt).

There is frequent need for smaller and larger decimal multiples of the units. These are obtained by the use of the prefixes listed in Table A.3.

These prefixes may be attached to any SI base or derived unit. (Large multiples of seconds, however, are rarely expressed this way.) The combination of prefix and symbol is regarded as a single symbol, and compound prefixes are not allowed.

The decimal sign is usually printed as a point (.) in English, but as a comma (,) in French and German. In the SI system a point or comma should be used only for the decimal sign. To facilitate the reading of long numbers the digits may be divided into groups of three, but not separated by commas.

Example: 212 489.143 *not* 212,489.143

TABLE A.3 Prefixes

Prefix	Symbol	Equivalent
tera	T	10^{12}
giga	G	10^{9}
mega	M	10^{6}
kilo	k	10^{3}
deci	d	10^{-1}
centi	c	10^{-2}
milli	m	10^{-3}
micro	μ	10^{-6}
nano	n	10^{-9}
pico	p	10^{-12}
femto	f	10^{-15}
atto	a	10^{-18}

When the decimal sign is placed before the first digit of a number, a zero should always be placed before the decimal sign.

Example: 0.143 872 *not* .143 872

Products or quotients of two or more units are treated according to the rules of mathematics. Note that the use of more than one solidus(/) is ambiguous. Thus, a specific rate of oxygen consumption can be written:

ml/kg s, or ml $(kg\ s)^{-1}$, or ml $kg^{-1}s^{-1}$
not ml/kg/s

Several circumstances make it impractical, at the present time, to express all physiological measurements in SI base and derived units. One reason is the large body of existing information, including widely known and used tables of physical and chemical constants, which use other units (e.g., calorie). Another reason is that methods in common use make it simpler to use certain traditional units (e.g., gram-force, rather than the SI unit, the newton). This practice should progressively be abandoned in favor of SI units.

USEFUL CONVERSION FACTORS

Traditional units are commonly used in those fields of physiology that concern energy (e.g., in the area of metabolic rates) and pressure (e.g., in the area of gas exchange).

Force
    1 kilogram-force = 9.807 N
    1 newton = 0.102 kgf

Energy, work (force × distance)
    1 cal = 4.184 J
    1 joule = 0.239 cal

    1 liter O_2 (at 0°C, 760 mm Hg), when used in oxidative metabolism, is often equated with 4.8 kcal; thus 1 liter $O_2 \simeq 20.083$ kJ

Power (work per unit time)
    1 cal s^{-1} = 4.184 J s^{-1} = 4.184 W
    1 kcal h^{-1} = 1.163 W
    1 watt = 0.239 cal s^{-1} = 14.34 cal min^{-1}
        = 860.4 cal h^{-1}

Pressure (force per unit area)
    1 atm = 1.013×10^5 N m^{-2} = 101.3 kPa
    1 mm Hg at 0°C (1 torr) = 1.333×10^2
        N m^{-2} = 133.3 Pa
    1 kN m^{-2} = 7.50 mm Hg = 9.869×10^{-3} atm

Volume
    1 liter is (since 1964) defined as exactly equal to 1 dm^3 (0.001 m^3); thus, ml = cm^3 and μl = mm^3
    1 U.S. gallon = 3.785 liter
    1 U.S. quart = 0.946 liter

Length
    1 inch = 25.4 mm
    1 foot = 304.8 mm = 0.3048 m
    1 yard = 0.914 m
    1 mile = 1609.3 m
    1 ångström (Å) = 10^{-10} m = 0.1 nm

Mass
    1 pound (avoirdupois) = 0.4536 kg
    1 ounce = 0.0283 kg

Temperature
    The temperature unit of 1 degree Celsius (°C)* equals the temperature unit of 1 kelvin (K).
    The Celsius temperature is defined as the excess of the thermodynamic temperature over 273.15 K.
    For practical purposes, the melting point of ice at 1 atm pressure is 0°C; the triple point of ice, water, and water vapor is 0.01°C; and the boiling point of water at 1 atm pressure is 100°C.

PHYSICAL QUANTITIES

We often need symbols to represent physical quantities. This is especially true when relationships are expressed in the form of equations. An example is Poiseuille's law, which states that the rate of flow ($\dot{Q}$) of a fluid through a tube is proportional to the pressure (p) and inversely proportional to the resistance (R), or $\dot{Q} = p/R$.

The names of some physical quantities and *recommended* symbols are given in Table A.4.

The symbol for a physical quantity should be a single letter of the Latin or Greek alphabet and should be printed in sloping (italic) type. When

*Also incorrectly known as "centigrade."

TABLE A.4 Physical quantities and recommended symbols. Alternative symbols are separated by commas.

Physical quantity	Symbol
Length	l
Height	h
Radius	r
Area	A
Volume	V
Mass	m
Density (m/V)	ρ
Pressure	p
Time	t
Velocity	u, v
Frequency	f
Force	F
Work	W
Energy	E, W
Power	P
Temperature	T

necessary, subscripts (and/or superscripts) are used to indicate a specific meaning of a symbol. For example: body temperature, T_b; blood volume, V_{bl}; rate of oxygen consumption, $\dot{V}_{O_2}$.

Equations that represent physical or physiological relationships use these and similar symbols for physical quantities. Symbols for units (m, kg, s, etc.) should *never* be used in such equations.

REFERENCES

Goldman, D. T., and Bell, R. J. (eds.) (1986) *The International System of Units (SI)*. NBS Special Publ. 330. Washington, D.C.: GPO. 51 pp.

Mechtly, E. A. (1973) *The International System of Units. Physical Constants and Conversion Factors*, 2d revision. NASA SP-7012. Washington, D.C.: National Aeronautics and Space Administration. 21 pp.

Symbols Committee of the Royal Society (1975) *Quantities, Units, and Symbols*, 2d. ed. London: The Royal Society. 55 pp.

APPENDIX B

Diffusion

Diffusion is a fundamental process in the movement of materials. Those diffusion processes that are of physiological importance take place over short distances, a fraction of a millimeter, a few micrometers, or less. Examples are the diffusion of respiratory gases across the respiratory membranes or the diffusion of ions across the nerve membrane during the action potential. Over longer distances, diffusion is physiologically unimportant, and transport takes place by mass movement (convection) (e.g., protoplasmic streaming or circulation of blood). Why is diffusion limited to short-distance transport? The reason is that diffusion time increases with the square of the diffusion distance.

Let us consider the time it takes for particles to travel a given distance by diffusion.

The average distance (R) covered by a diffusing substance depends on the average length (l) of each track in the random thermal motion of the particles (the mean free path) and on the number of individual tracks (n). The relationship is $R = l\sqrt{n}$. Because the number of individual tracks is proportional to time (t), we can write $R \propto l\sqrt{t}$, or $t \propto R^2 l^{-2}$. This means that diffusion time (t) increases with the square of the average distance (R) covered by the diffusing substance.

What does this mean for diffusion in the tissues of an organism? Suppose we impose a stepwise increase in oxygen concentration at a given point, A. Oxygen will diffuse into the surrounding tissues and will attain an average diffusion distance of 1 μm in about 10^{-4} s. An average diffusion distance 10 times as long will require 100 times the time, and so on. Thus, for an average diffusion distance of 10 μm the time required will be 10^{-2} s; for 1 mm it will be 100 s; for 10 mm, 10^4 s (nearly 3 h); and for 1 m, the time will be 10 000 times as long again, or more than 3 years. The distance from the lungs of a human to his extremities is about 1 m; clearly, transport by diffusion alone would be impossibly slow and mass movement is necessary.

It is important to realize that we have discussed the time needed for diffusion over a certain distance when we impose a change in a concentration

571

FIGURE B.1 Diagram of steady-state diffusion through a homogeneous membrane.

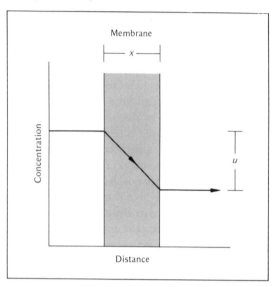

of the diffusing substance at one point and observe the effect of this change at another point at a certain distance. Thus, the diffusion of the neurotransmitter acetylcholine across the synaptic cleft (about 0.02 μm) is very fast; in contrast, if the hormone adrenaline were to reach its target somewhere in the body by diffusion from its source, the adrenal gland, the process would be so slow as to be useless.

Let us now consider a steady-state diffusion over an unchanged gradient (as we find in many physiological processes).

The diffusion over a constant concentration gradient (Figure B.1) is expressed by Fick's equation:

$$dQ = -D A \frac{du}{dx} dt$$

in which Q is the quantity of a substance diffused, A is the area through which diffusion takes place, u is the concentration at point x (du/dx thus being the concentration gradient), t is time, and D is the diffusion coefficient. The minus sign is a conven-

tion that makes the flux positive in the direction in which the gradient is negative. Diffusion coefficients for various substances can be found in standard tabular works.*

From this equation we can see that the amount of a substance diffusing is proportional to elapsed time. This also makes sense, intuitively. Consider a physiological example. Alveolar air has a partial pressure of oxygen of 100 mm Hg, and mixed venous blood arriving at the lung may have an oxygen partial pressure (tension) of 40 mm Hg. Normally, these concentrations remain very constant, and oxygen therefore diffuses from the air through the alveolar wall at a constant rate. It is now obvious that in 2 minutes twice as much oxygen diffuses as in 1 minute, in 10 minutes 10 times as much, and so on. Thus, if a constant gradient is maintained, the amount of the substance diffusing is directly proportional to time.

Another consequence of Fick's equation is that if the diffusion distance (x) is increased, the amount of the substance diffusing (Q) decreases. If x is doubled, Q is reduced to half; if x is 10 times as large, Q is one-tenth, and so on. The amount diffusing is thus inversely proportional to the diffusing distance, but again, this applies only when a steady state is maintained. This requires that we must have, on one side of the membrane, a constant supply or an infinite pool of the diffusing substance, and on the other side a constant lower concentration maintained by the continuous removal of, or an infinite sink for, the diffusing substance. This, for practical purposes, applies to our example of the lung.

The diffusion constant. The diffusion constant (K), which was used for the calculation on page 17 differs from the diffusion coefficient (D) used in Fick's equation. The need for the use of the diffusion constant in physiology arose in connection with the study of diffusion of gases between air and tis-

---

*At 20°C the diffusion coefficient D, for oxygen in air is $D = 0.196$ cm^2 s^{-1}, and for oxygen in water $D = 0.183 - 10^{-4}$ cm^2 s^{-1}.

TABLE B.1 The diffusion constant (K) for oxygen at 20 °C in various materials. Note that these figures are not diffusion coefficients. [Krogh 1919]

Material	K (cm^2 atm^{-1} min^{-1})
Air	11
Water	34×10^{-6}
Gelatin	28×10^{-6}
Muscle	14×10^{-6}
Connective tissue	11.5×10^{-6}
Chitin[a]	1.3×10^{-6}
Rubber	7.7×10^{-6}

[a]The diffusion constant for chitin often is given incorrectly as 10 times the value listed here. This is the result of a misprint in a book by Krogh (1941) which has been copied by many later authors.

sues, notably in the mammalian lung. For this purpose it is practical to refer the diffusion rate of a gas, say oxygen, to the partial pressure of the gas in the alveolar air. The diffusion coefficient (D) cannot be used in this case, for it requires knowledge of the concentration gradient (du/dx) in the tissue. The concentration in the tissue is difficult to determine directly and cannot be calculated unless the solubility for the particular gas in the particular tissue is known. The diffusion constant circumvents this problem by relating the rate of diffusion directly to the partial pressure of the gas in the gas phase.

The diffusion constant (K) is defined as the number of cubic centimeters of a gas that diffuses in 1 minute through an area of 1 cm^2 when the pressure gradient is 1 atm per centimeter. These units can be written out and reduced to a simpler form as follows:

$$\frac{cm^3}{min \times cm^2 \times atm \times cm^{-1}} = cm^2 \ atm^{-1} \ min^{-1}$$

Diffusion constants have been determined for only a small number of animal tissues; some examples are listed in Table B.1 These constants are empirically determined and include the solubility of gas in the tissues in question. The figures show that the diffusion in those tissues that have relatively high water content (e.g., muscle and connective tissue) is somewhat lower than for water, but of the same order of magnitude. The diffusion through chitin (the hard cuticle of insects), on the other hand, is lower by an order of magnitude.

It is worth noting that the rate of diffusion of oxygen through rubber is considerable, about one-half the rate for muscle. This is of importance in physiological experimentation, for if rubber tubing is used for connections, oxygen can diffuse through the walls of the tubing, and carbon dioxide can diffuse even faster because of its high solubility. The actual rate of diffusion depends on the kind of rubber or plastic used; some kinds are more permeable and others much less permeable.

REFERENCES

Krogh, A. (1919) The rate of diffusion of gases through animal tissues, with some remarks on the coefficient of invasion. *J. Physiol. 52*:391–608.
Krogh, A. (1968) *The Comparative Physiology of Respiratory Mechanisms*, reprint ed. New York: Dover. 172 pp.

Logarithmic and exponential equations

LOGARITHMIC EQUATIONS

Logarithmic equations are of the general form:

$$y = a \cdot x^b \tag{1}$$

The logarithmic form of this equation is:

$$\log y = \log a + b \log x \tag{2}$$

Equation (2) shows that $\log y$ is a linear function of $\log x$; that is, by plotting $\log y$ against $\log x$ we obtain a straight line with the slope b.

Example: Rate of oxygen consumption ($\dot{V}_{O_2}$) plotted against body mass (M_b) (Figure 5.12, page 198).

$$\dot{V}_{O_2} = k \cdot M_b{}^b \tag{3}$$

The independent variable, body mass (M_b), is customarily plotted on the abscissa. The rate of oxygen consumption per unit mass is obtained by dividing both sides of equation (3) by M_b:

$$\frac{\dot{V}_{O_2}}{M_b} = k \frac{M_b{}^b}{M_b} = k\, M_b{}^{(b-1)} \tag{4}$$

Example: Figure 5.10, page 194.

EXPONENTIAL EQUATIONS

Exponential equations are of the general form:

$$y = b \cdot a^x \tag{5}$$

The logarithmic form of this equation is:

$$\log y = \log b + x \log a \tag{6}$$

Equation (6) shows that $\log y$ is a linear function of x, and plotting $\log y$ against x gives a straight line. Use of semilog graph paper (linear abscissa and logarithmic ordinate) for plotting y against x gives the same result.

Example: Rate of oxygen consumption plotted against temperature (Figure 6.4, page 221).

APPENDIX D

Thermodynamic expression of temperature effects

TABLE D.1 The change in Q_{10} for temperature intervals between 0 °C and 50 °C, calculated for a constant activation energy (μ) of 12 286.

T_1 (°C)	T_2 (°C)	Q_{10}
0	10	2.22
10	20	2.10
20	30	2.00
30	40	1.91
40	50	1.84

The term Q_{10} (defined on page 219) is widely used in biology, for it gives a simple and easily understood expression of the effect of temperature on rates (e.g., rate of oxygen consumption). In physical chemistry, however, temperature effects are expressed differently, based on thermodynamic considerations of the frequency of molecular collision as a function of temperature. The equation, as developed by Arrhenius, is usually given in the following form:

$$K_{T_2} = K_{T_1} \cdot e^{\mu/R[(T_2 - T_1)/T_2 T_1]} \tag{1}$$

in which K stands for rate constant, T for thermodynamic temperature (absolute temperature, in kelvin), μ for activation energy, R for the universal gas constant, and e is the base of the natural logarithm.

We can obtain a thermodynamically correct expression for Q_{10} (the ratio of two rates 10 °C apart) by solving equation (1) for K_{T_2}/K_{T_1} when $T_2 - T_1 = 10$. We then find:

$$Q_{10} = \frac{K_{T + 10}}{K_T} = e^{\mu/R\{10/[T(T + 10)]\}} \tag{2}$$

When we examine equation (2), we find that Q_{10} is a function of and changes with temperature (T) and thus cannot be a constant. R, e, and (for a given process) μ are constants, and Q_{10} therefore decreases with increasing temperature. An example of this change in Q_{10} with temperature is given in Table D.1 for $\mu = 12\ 286$ (a value of μ chosen to correspond to $Q_{10} = 2.00$ for the temperature interval between 20 and 30°C). If $\mu = 19\ 506$, Q_{10} is 3.0 for the interval between 20 and 30°C, but will be higher at low temperatures and lower at high temperatures. The use of Q_{10} to describe temperature effects therefore lacks a theoretical foundation in thermodynamics; however, Q_{10} is a highly descriptive term that is convenient and much used in biology.

575

Solutions and Osmosis

Water is the universal solvent for virtually all biological reactions and related phenomena. The following brief summary therefore is restricted to aqueous solutions. For a more extensive treatment, the reader is referred to an introductory chemistry or physical chemistry text.

SOLUTIONS AND CONCENTRATIONS

A 1 *molar* solution contains 1 mol of solute per liter of solution. Thus a 1 molar solution of sodium chloride contains 58.45 g NaCl per liter solution. In biological work it is common to refer to a one-thousandth of a 1 molar solution, a millimolar solution. This helps to eliminate small decimal fractions.

A 1 *molal* solution contains 1 mol of solute per kilogram water. For example, a solution of 58.45 g NaCl in 1 kg of water is a 1 molal solution. Because the volume of this solution is more than 1 liter, its concentration is less than 1.0 molar.

The molarity designation is often used in biological work and is convenient because it is easy to determine the volume of a given sample of a biological fluid, whereas it is cumbersome or impossible to determine its exact water content. For example, the volume of a blood sample is easily measured, and its content of sodium can readily be determined. Referring the sodium concentration to the amount of water in the sample is more difficult, for both plasma and red cells contain many other solutes, including substantial amounts of proteins. In physical chemistry, the molality concept is more useful; this includes the consideration of osmotic phenomena.

Concentrations are sometimes given as *percent* (%). Unfortunately, this expression is ambiguous and there is no accepted convention for how it should be used. This often leads to confusion. For example, does a 1% solution of sodium chloride contain 1 g NaCl and 99 g water? Or does it contain 1 g NaCl in 100 ml solution? Or, as the term *percent* means *per hundred*, does it mean 1 g sodium chloride per 100 g water? Even more confusing is the expression *milligram percent* (mg %), which logically should mean "milligrams per 100 milligrams." Usually it means the number of milligrams of a substance per 100 ml solution. Ambiguities can be avoided by using the appropriate units (e.g., 120 mg glucose per 100 ml blood) or often better, the molar concentration.

Addition of modifiers such as "weight percent" helps little. It commonly, but not always, refers to parts of solute by weight per 100 parts of solution by weight. Accordingly, a 1% solution of sucrose contains 1 g sucrose in 99 g water. However, if a salt that contains water of crystallization is weighed out, both the amount of the salt and the total amount of water may be uncertain.

Strictly, the term *percent* should express a nondimensional fraction. This means that the ratio of the two quantities should be expressed in the same units that cancel out in the division. It should be noted that gas concentrations can properly be expressed in percent. For example, dry atmospheric air contains 20.95% oxygen, whether the quantity of oxygen relative to the total is given in number of molecules, in milliliters, or in pressure units.

OSMOSIS

Osmosis is the movement of water between two solutions with different solute concentrations when the solutions are separated by a semipermeable membrane (i.e., a membrane permeable to water but impermeable to the solute).

Consider the system in Figure E.1. Water will tend to move from compartment B into compartment A, causing a pressure, the osmotic pressure, to develop in A. The movement of water can be counteracted by pressure exerted on the piston, and when this pressure exactly balances the osmotic pressure there is no net movement of water.

The force (π) that tends to move water is defined by van't Hoff's equation:

$$\pi V = n R T$$

FIGURE E.1 When two solutions are separated by a semipermeable membrane (M), water tends to move from the more dilute solution to the more concentrated solution. If the membrane is rigid, the water movement can be prevented by exerting a pressure on the piston (P) equal to the difference in osmotic pressure between the two solutions.

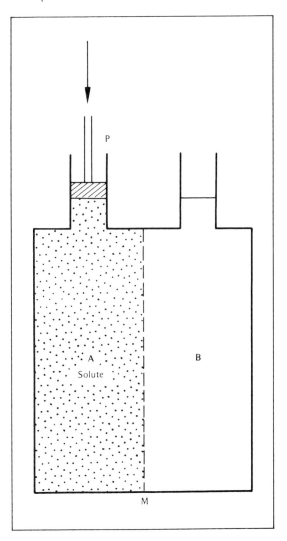

in which V is volume, n is number of moles of solute particles, R is the universal gas constant, and T is absolute temperature. This equation is similar to the gas equation $PV = nRT$ (in which P is pressure).

A 1 molal solution of an ideal nonelectrolyte has an osmotic pressure of 22.4 atm at 0°C (273 K). This equals the pressure exerted by 1 mol of an ideal gas when the volume is 1 liter.

Solutes depress the freezing point of water, and a 1 molal solution of an ideal nonelectrolyte has a freezing point of -1.86°C. Solutes also reduce the vapor pressure and increase the boiling point of a solution.

The osmotic pressure of a solution ideally depends only on the number of particles present, and not on their size or nature. An electrolyte such as sodium chloride, which in water ionizes into Na^+ and Cl^-, therefore has an osmotic concentration higher than expected from its molality.

Unfortunately, observations of actual osmotic pressures in solutions of electrolytes do not completely conform to the expected results as estimated from the expected number of ions. This is because of interaction between the positive and negative ions and interaction between the ions and the water molecules. The osmotic pressure of a solution of an electrolyte therefore cannot be calculated directly from the molal concentration of the electrolyte, and it is necessary to introduce an empirical *osmotic coefficient* (ϕ).

If sodium chloride were completely ionized in solution, we would expect a 1.00 molal solution to have a freezing point depression of 2×1.86°C, or 3.72°C. In fact, a 1 molal solution of NaCl freezes at -3.38°C. Its osmotic coefficient therefore is $3.38/3.72 = 0.91$. The osmotic coefficient varies with the concentration of the electrolyte and differs from one electrolyte to another.

THE DONNAN EQUILIBRIUM

Assume a system similar to that in Figure E.1, but with the compartments separated by a rigid membrane permeable to water and electrolytes, but not to a large ionized molecule, say a negatively

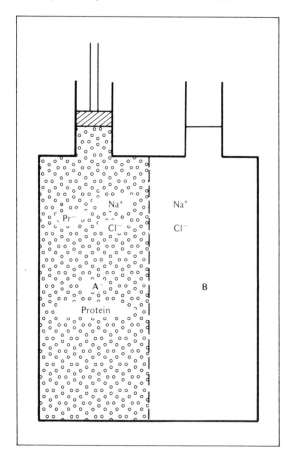

FIGURE E.2 An osmotic system consisting of two chambers, A and B, separated by a rigid membrane that is freely permeable to water and small ions but impermeable to large ions such as a negatively charged protein (Pr^-). The distribution of ions and the osmotic pressure at equilibrium (the Donnan equilibrium) are described in the text.

charged protein, Pr^- (Figure E.2). Let the electrolyte be sodium chloride.

If the compartments have rigid walls, pressure on the piston P prevents any net movement of water through the membrane. It can now be shown that at equilibrium:

$$[Na^+]_A \times [Cl^-]_A = [Na^+]_B \times [Cl^-]_B$$

Rearranging the terms, we find that the ratio of the diffusible cations on one side to those on the other is inversely proportional to the ratio of the anions on the two sides:

$$\frac{[Na^+]_A}{[Na^+]_B} = \frac{[Cl^-]_B}{[Cl^-]_A} = r$$

These ratios are known as the *Donnan constant (r)*.

Assume that the initial concentration of sodium chloride is the same on the two sides of the membrane, and that compartment A also contains sodium proteinate, NaPr, dissociated as Na^+ and Pr^-. Thus, the total concentration of Na^+ in compartment A is higher than in B, and sodium ions will diffuse into B. This will give an excess of Na^+ ions relative to Cl^- ions in compartment B, and an equal number of chloride ions will therefore cross the membrane into compartment B. However, this tendency is counteracted by the resulting higher chloride concentration in B than in A. At equilibrium there will be unequal concentrations of ions across the membrane and a net negative charge in chamber A relative to chamber B. This equilibrium situation with the unequal distribution of diffusible ions and the resulting electric charge is known as the *Donnan equilibrium*.

We should note that the potential difference is attributable to the presence of the nondiffusible anion in chamber A. Because the system is in equilibrium, the potential is maintained without expenditure of energy and remains indefinitely.

The potential difference (E, in millivolts) can be calculated from Nernst's equation:

$$E = \frac{RT}{zF} \ln \frac{[Na^+]_A}{[Na^+]_B}$$

in which R is the universal gas constant (8.3144 J K^{-1} mol^{-1}), T is absolute temperature, z is valency of the ion (in this case unity), and F is Faraday's constant (96 500 coulomb per mole).

The pressure applied to the piston in Figure E.2 necessary to prevent osmotic flow of water into compartment A is known as the *colloidal osmotic pressure*. In the event that the pressure on the plunger exceeds the colloidal osmotic pressure, water and electrolytes are forced out through the membrane while the nonpermeable protein is withheld in A. This process is called *ultrafiltration*. The osmotic pressure of the ultrafiltrate will equal the osmotic pressure of the solution in compartment B.

ADDITIONAL READING

Overbeek, J. Th. G. (1956) The Donnan equilibrium. *Progr. Biophys.* 6:57–84.

Hammel, H. T., and Scholander, P. F. (1976) *Osmosis and Tensile Solvent*. New York: Springer-Verlag. 133 pp.

The Animal Kingdom

An abbreviated classification of living animals

PHYLUM PROTOZOA
  CLASS FLAGELLATA
   Flagellates
  CLASS SARCODINA
   Protozoans with pseudopods. *Amoeba*
  CLASS CILIATA
   Ciliates. *Paramecium*

PHYLUM PORIFERA. Sponges

PHYLUM COELENTERATA (or Cnidaria)
  CLASS HYDROZOA
   Hydrozoans. *Hydra*
  CLASS SCYPHOZOA
   Jellyfishes
  CLASS ANTHOZOA
   Sea anemones and corals

PHYLUM CTENOPHORA. Comb jellies

PHYLUM PLATYHELMINTHES. Flat-
  worms
  CLASS TURBELLARIA
   Free-living flatworms
  CLASS TREMATODA
   Flukes
  CLASS CESTODA
   Tapeworms

PHYLUM NEMATODA. Roundworms.
  Ascaris

PHYLUM MOLLUSCA. Molluscs
  CLASS AMPHINEURA
   Chitons
  CLASS GASTROPODA
   Snails
  CLASS PELECYPODA
   Bivalve molluscs, clams, mussels, etc.
  CLASS CEPHALOPODA
   Squids, octopuses, and *Nautilus*

PHYLUM ANNELIDA. Segmented worms
CLASS POLYCHAETA
  Marine worms, tubeworms, etc.
CLASS OLIGOCHAETA
  Earthworms and many fresh-water annelids
CLASS HIRUDINEA
  Leeches

PHYLUM ARTHROPODA. Arthropods
CLASS XIPHOSURA
  Horseshoe crabs. *Limulus*
CLASS ARACHNIDA
  Spiders, ticks, mites, scorpions, etc.
CLASS CRUSTACEA
  Crustaceans. Lobster, *Daphnia, Artemia*
CLASS CHILOPODA
  Centipedes
CLASS DIPLOPODA
  Millipedes
CLASS INSECTA
  Insects

PHYLUM ECHINODERMATA. Echinoderms
CLASS ASTEROIDEA
  Sea stars
CLASS OPHIUROIDEA
  Brittle stars
CLASS ECHINOIDEA
  Sea urchins and sand dollars
CLASS HOLOTHUROIDEA
  Sea cucumbers

PHYLUM CHORDATA. Chordates

SUBPHYLUM UROCHORDATA (or TUNI-CATA). Tunicates

SUBPHYLUM CEPHALOCHORDATA. Amphioxus

SUBPHYLUM VERTEBRATA. Vertebrates
CLASS AGNATHA
  Jawless fishes. Lampreys and hagfish
CLASS CHONDRICHTHYES
  Elasmobranchs. Sharks, skates, rays
CLASS OSTEICHTHYES
  Bony fishes
Subclass Crossopterygii. *Latimeria*
Subclass Dipnoi. Lungfishes
Subclass Actinopterygii. Higher bony fishes
CLASS AMPHIBIA
  Amphibians. Frogs, toads, salamanders
CLASS REPTILIA
  Reptiles
Order Chelonia. Turtles
Order Crocodylia. Crocodiles and alligators
Order Squamata. Lizards, snakes
CLASS AVES
  Birds
CLASS MAMMALIA
  Mammals
Subclass Prototheria
  Order Monotremata. Egg-laying mammals (echidna and platypus)
Subclass Metatheria
  Order Marsupialia (kangaroos, opossum, etc.)
Subclass Eutheria
  Orders: Insectivores (moles, shrews, etc.), bats, primates (lemurs, monkeys, apes, humans), sloths and armadillos, rabbits and hares, rodents, whales and dolphins, carnivores and seals, elephants, manatees and sea cows, ungulates

REFERENCE

Rothschild, L. (1965) *A Classification of Living Animals,* 2d ed. London: Longmans, Green. 134 pp.

Index